T0180637

Lecture Notes in Computer Science

Lecture Notes in Artificial Intelligence 13936

Founding Editor

Jörg Siekmann

The series Lecture Notes in Artificial Intelligence (LNAI) was established in 1988 as a topical subseries of LNCS devoted to artificial intelligence.

The series publishes state-of-the-art research results at a high level. As with the LNCS mother series, the mission of the series is to serve the international R & D community by providing an invaluable service, mainly focused on the publication of conference and workshop proceedings and postproceedings.

Hisashi Kashima · Tsuyoshi Ide · Wen-Chih Peng
Editors

Advances in Knowledge Discovery and Data Mining

27th Pacific-Asia Conference
on Knowledge Discovery and Data Mining, PAKDD 2023
Osaka, Japan, May 25–28, 2023
Proceedings, Part II

Editors
Hisashi Kashima
Kyoto University
Kyoto, Japan

Wen-Chih Peng
National Chiao Tung University
Hsinchu, Taiwan

Tsuyoshi Ide
IBM Research, Thomas J. Watson Research Center
Yorktown Heights, NY, USA

ISSN 0302-9743 ISSN 1611-3349 (electronic)
Lecture Notes in Artificial Intelligence
ISBN 978-3-031-33376-7 ISBN 978-3-031-33377-4 (eBook)
https://doi.org/10.1007/978-3-031-33377-4

LNCS Sublibrary: SL7 – Artificial Intelligence

This Springer imprint is published by the registered company Springer Nature Switzerland AG
The registered company address is: Gewerbestrasse 11, 6330 Cham, Switzerland

General Chairs' Preface

On behalf of the Organizing Committee, we were delighted to welcome attendees to the 27th Pacific-Asia Conference on Knowledge Discovery and Data Mining (PAKDD 2023), held in Osaka, Japan, on May 25–28, 2023. Since its inception in 1997, PAKDD has long established itself as one of the leading international conferences on data mining and knowledge discovery. PAKDD provides an international forum for researchers and industry practitioners to share their new ideas, original research results, and practical development experiences across all areas of Knowledge Discovery and Data Mining (KDD). PAKDD 2023 was held as a hybrid conference for both online and on-site attendees.

We extend our sincere gratitude to the researchers who submitted their work to the PAKDD 2023 main conference, high-quality tutorials, and workshops on cutting-edge topics. We would like to deliver our sincere thanks for their efforts in research, as well as in preparing high-quality presentations. We also express our appreciation to all the collaborators and sponsors for their trust and cooperation.

We were honored to have three distinguished keynote speakers joining the conference: Edward Y. Chang (Ailly Corp), Takashi Washio (Osaka University), and Wei Wang (University of California, Los Angeles, USA), each with high reputations in their respective areas. We enjoyed their participation and talks, which made the conference one of the best academic platforms for knowledge discovery and data mining. We would like to express our sincere gratitude for the contributions of the Steering Committee members, Organizing Committee members, Program Committee members, and anonymous reviewers, led by Program Committee Co-chairs: Hisashi Kashima (Kyoto University), Wen-Chih Peng (National Chiao Tung University), and Tsuyoshi Ide (IBM Thomas J. Watson Research Center, USA). We feel beholden to the PAKDD Steering Committees for their constant guidance and sponsorship of manuscripts.

Finally, our sincere thanks go to all the participants and volunteers. We hope all of you enjoyed PAKDD 2023 and your time in Osaka, Japan.

April 2023

Naonori Ueda
Yasushi Sakurai

PC Chairs' Preface

It is our great pleasure to present the 27th Pacific-Asia Conference on Knowledge Discovery and Data Mining (PAKDD 2023) as the Program Committee Chairs. PAKDD is one of the longest-established and leading international conferences in the areas of data mining and knowledge discovery. It provides an international forum for researchers and industry practitioners to share their new ideas, original research results, and practical development experiences from all KDD-related areas, including data mining, data warehousing, machine learning, artificial intelligence, databases, statistics, knowledge engineering, big data technologies, and foundations.

This year, PAKDD received a record number of 869 submissions, among which 56 submissions were rejected at a preliminary stage due to policy violations. There were 318 Program Committee members and 42 Senior Program Committee members involved in the reviewing process. More than 90% of the submissions were reviewed by at least three different reviewers. As a result of the highly competitive selection process, 143 submissions were accepted and recommended to be published, resulting in an acceptance rate of 16.5%. Out of these, 85 papers were primarily about methods and algorithms and 58 were about applications. We would like to thank all PC members and reviewers, whose diligence produced a high-quality program for PAKDD 2023. The conference program featured keynote speeches from distinguished researchers in the community, most influential paper talks, cutting-edge workshops, and comprehensive tutorials.

We wish to sincerely thank all PC members and reviewers for their invaluable efforts in ensuring a timely, fair, and highly effective PAKDD 2023 program.

April 2023

Hisashi Kashima
Wen-Chih Peng
Tsuyoshi Ide

Organization

General Co-chairs

Naonori Ueda	NTT and RIKEN Center for AIP, Japan
Yasushi Sakurai	Osaka University, Japan

Program Committee Co-chairs

Hisashi Kashima	Kyoto University, Japan
Wen-Chih Peng	National Chiao Tung University, Taiwan
Tsuyoshi Ide	IBM Thomas J. Watson Research Center, USA

Workshop Co-chairs

Yukino Baba	University of Tokyo, Japan
Jill-Jênn Vie	Inria, France

Tutorial Co-chairs

Koji Maruhashi	Fujitsu, Japan
Bin Cui	Peking University, China

Local Arrangement Co-chairs

Yasue Kishino	NTT, Japan
Koh Takeuchi	Kyoto University, Japan
Tasuku Kimura	Osaka University, Japan

Publicity Co-chairs

Hiromi Arai	RIKEN Center for AIP, Japan
Miao Xu	University of Queensland, Australia
Ulrich Aivodji	ÉTS Montréal, Canada

Proceedings Co-chairs

Yasuo Tabei	RIKEN Center for AIP, Japan
Rossano Venturini	University of Pisa, Italy

Web and Content Chair

Marie Katsurai	Doshisha University, Japan

Registration Co-chairs

Machiko Toyoda	NTT, Japan
Yasutoshi Ida	NTT, Japan

Treasury Committee

Akihiro Tanabe	Osaka University, Japan
Aya Imura	Osaka University, Japan

Steering Committee

Vincent S. Tseng	National Yang Ming Chiao Tung University, Taiwan
Longbing Cao	University of Technology Sydney, Australia
Ramesh Agrawal	Jawaharlal Nehru University, India
Ming-Syan Chen	National Taiwan University, Taiwan
David Cheung	University of Hong Kong, China
Gill Dobbie	University of Auckland, New Zealand
Joao Gama	University of Porto, Portugal
Zhiguo Gong	University of Macau, Macau
Tu Bao Ho	Japan Advanced Institute of Science and Technology, Japan
Joshua Z. Huang	Shenzhen Institutes of Advanced Technology, Chinese Academy of Sciences, China
Masaru Kitsuregawa	University of Tokyo, Japan
Rao Kotagiri	University of Melbourne, Australia
Jae-Gil Lee	Korea Advanced Institute of Science & Technology, Korea

Tianrui Li	Southwest Jiaotong University, China
Ee-Peng Lim	Singapore Management University, Singapore
Huan Liu	Arizona State University, USA
Hady W. Lauw	Singapore Management University, Singapore
Hiroshi Motoda	AFOSR/AOARD and Osaka University, Japan
Jian Pei	Duke University, USA
Dinh Phung	Monash University, Australia
P. Krishna Reddy	International Institute of Information Technology, Hyderabad (IIIT-H), India
Kyuseok Shim	Seoul National University, Korea
Jaideep Srivastava	University of Minnesota, USA
Thanaruk Theeramunkong	Thammasat University, Thailand
Takashi Washio	Osaka University, Japan
Geoff Webb	Monash University, Australia
Kyu-Young Whang	Korea Advanced Institute of Science & Technology, Korea
Graham Williams	Australian National University, Australia
Raymond Chi-Wing Wong	Hong Kong University of Science and Technology, Hong Kong
Min-Ling Zhang	Southeast University, China
Chengqi Zhang	University of Technology Sydney, Australia
Ning Zhong	Maebashi Institute of Technology, Japan
Zhi-Hua Zhou	Nanjing University, China

Contents – Part II

Graphs and Networks

Semi-supervised and Unsupervised Learning

Theoretical Foundations

Transfer Learning and Meta Learning

Graphs and Networks

Improving Knowledge Graph Entity Alignment with Graph Augmentation

Feng Xie, Xiang Zeng, Bin Zhou(✉), and Yusong Tan

College of Computer, National University of Defense Technology, Changsha, China
{xiefeng,zengxiang,binzhou,ystan}@nudt.edu.cn

Abstract. Entity alignment (EA) which links equivalent entities across different knowledge graphs (KGs) plays a crucial role in knowledge fusion. In recent years, graph neural networks (GNNs) have been successfully applied in many embedding-based EA methods. However, existing GNN-based methods either suffer from the structural heterogeneity issue that especially appears in the real KG distributions or ignore the heterogeneous representation learning for unseen (unlabeled) entities, which would lead the model to overfit on few alignment seeds (i.e., training data) and thus cause unsatisfactory alignment performance. To enhance the EA ability, we propose GAEA, a novel EA approach based on graph augmentation. In this model, we design a simple Entity-Relation (ER) Encoder to generate latent representations for entities via jointly modeling comprehensive structural information and rich relation semantics. Moreover, we use graph augmentation to create two graph views for margin-based alignment learning and contrastive entity representation learning, thus mitigating the negative influence caused by structural heterogeneity and sparse seeds. Extensive experiments conducted on benchmark datasets demonstrate the effectiveness of our method. Our codes are available at https://github.com/Xiefeng69/GAEA.

Keywords: Knowledge Graph · Entity Alignment · Graph Neural Networks · Graph Augmentation · Knowledge Representation

1 Introduction

Knowledge graphs (KGs) can effectively organize and represent facts about the world in a structured fashion. More and more KGs have been constructed based on different data sources or for different purposes. Therefore, the knowledge contained in different KGs is far from complete yet complementary [22]. Entity alignment (EA) which aims to link semantically equivalent entities located on different KGs has attracted increasing attention since it could facilitate knowledge integration and thus promote knowledge-driven applications, such as question answering, recommender systems, and semantic search.

In recent years, embedding-based EA methods [2,3,11,14,20,22,26,30] have achieved decent results. The general pipeline can be summarized into two steps: (I) generating low-dimensional embeddings (latent representations) for entities

H. Kashima et al. (Eds.): PAKDD 2023, LNAI 13936, pp. 3–14, 2023.
https://doi.org/10.1007/978-3-031-33377-4_1

via KG encoder (e.g., TransE [1]), and then (II) pulling two KGs into a unified embedding space through prior alignment seeds and pairing each entity by distance metrics (e.g., Euclidean distance). Moreover, some works further improve the EA performance by introducing extra information, such as entity names [29], attributes [5,12], and literal descriptions [24], while these discriminative features are usually privacy sensitive, noise polluted, and hard to collect [8].

Due to the powerful structure learning capability, Graph Neural Networks (GNNs) like GCN [4] and GAT [17] have been employed as the encoder with Siamese architecture (i.e., shared-parameter) for many embedding-based models [6,14,20,26]. KGs are heterogeneous, especially in real KG distributions, which means entities that have the same referent in different KGs usually have dissimilar relational neighborhood. To address this problem, existing GNN-based models modify and improve GNN variants to better capture structural information in KGs, e.g., AliNet [14] adopts multi-hop aggregation with gating mechanism to expand neighborhood ranges and RDGCN [21] incorporates relation features via attention interactions for embedding learning. However, these models introduce a large number of neural network operations and ignore representation learning for unseen entities, which will tend to make the models overfit on few alignment seeds and thus undermine their generalization and performance.

In this paper, we propose GAEA, a novel knowledge graph entity alignment model based on graph augmentation. Firstly, we design an Entity-Relation (ER) Encoder to generate entity representations via jointly leveraging neighborhood structures and relation semantics in KGs. Then, we apply graph augmentation to increase the structural diversity of input KG in the alignment learning process, which encourages the model to capture the semantic importance of different neighbors and enforces the model to obtain stable representations against structure perturbation, thus mitigating overfitting issue to some extent. Moreover, since graph augmentation can inherently generate two distinct graph views without extra parameters, we can let the model perceive structural differences and further improve the feature learning for (unseen) entities by applying contrastive entity representation learning to maximize the consistency between the original KG and augmented KG [19,25]. Our experiments on benchmark datasets OpenEA [15] show that GAEA outperforms the existing state-of-the-art embedding-based EA methods. We also conduct thorough auxiliary analyses to demonstrate the effectiveness of incorporating graph augmentation techniques.

2 Related Works

Entity alignment is a fundamental task to identify the same entities across different KGs, which has attracted increasing attention in recent years. The existing embedding-based methods can be roughly divided into two categories:

1. **Structure-based models.** These models solely rely on the original structure information of KGs (i.e., triples) to align entities. Previous methods mainly use knowledge representation learning to generate low-dimensional embeddings for entities [2,7,30]. For example, MTransE [2] applies TransE [1] to

embed different KGs into independent vector spaces and constructs transitions via proposed alignment modules. Inspired by the powerful structure learning ability of Graph Neural Networks (GNNs), a large body of works begin to focus on employing GNNs as the encoder. GCN-Align [20] incorporates GCN [4] to capture entities' neighborhood structures for the first time and achieves promising results. Subsequent works not only apply various GNN variants, like GAT [17], but also improve the structure awareness by overcoming heterogeneity of different KGs [3,14,21], capturing multi-context structural features [22], and infusing relation semantics [6,11].
2. **Enhancement-based models.** These models aim to build a high-accuracy alignment system using designed alignment strategies or extra information. BootEA [13] applies iterative learning to find potential alignments and adds them to the training set for data augmentation. CEA [28] formulates alignment inference as a stable matching problem to model collective signals, successfully guaranteeing 1-to-1 alignment. Other effective models introduce extra information to enhance the alignment performance, including entity names [29], attributes [5,12], and literal descriptions [24].

In this work, we aim to improve the performance and efficiency of entity alignment only utilizing structural contexts which are abundant and always available without privacy issues in the real-world KGs.

3 Preliminaries

Knowledge Graph. A knowledge graph (KG) is formalized as $G = (E, R, T)$, where E and R refer to the set of entities and the set of relations, respectively. $T = E \times R \times E = \{(h, r, t)|h, t \in E \wedge r \in R\}$ is the set of triples, where h, r, and t denote the head entity, connected relation, tail entity, respectively.

Entity Alignment. Given two KGs: $G_s = (E_s, R_s, T_s)$ as the source KG and $G_t = (E_t, R_t, T_t)$ as the target KG, and few alignment seeds (aka pre-aligned entity pairs) $S = \{(e_i, e_j)|e_i \in E_s \wedge e_j \in E_t \wedge e_i \equiv e_j\}$, where $\equiv$ means equivalence relationship, entity alignment (EA) aims to seek remaining equivalent entities located on different KGs via entity representations.

Augmented Graph. Graph augmentation techniques will generate a perturbed version of the original graph, i.e., augmented graph, by augmentation strategies (e.g., node dropping, edge perturbation). In order not to introduce wrong facts, we only choose edge dropping in this work. At each training iteration, we randomly drop out some triples based on the deletion ratio $r \sim uniform(0, pr)$, where pr is a preset upper bound of the deletion ratio. The augmented graphs for G_s and G_t are denoted as G_s^{aug} and G_t^{aug}, respectively. Note that we do not consider deleting the triples associated with entities whose degree is less than 2, because these long-tail entities have sparse neighborhood structures inherently.

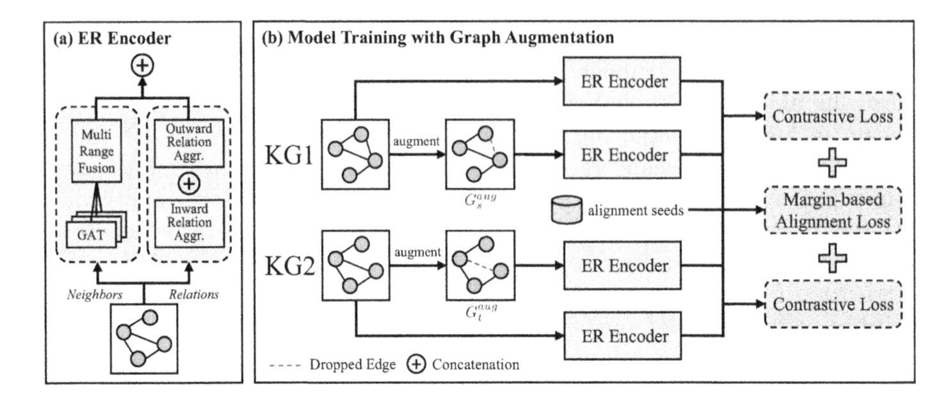

Fig. 1. The framework of our proposed GAEA.

4 Methodology

This section details our proposed method, termed as GAEA, which is drawn in Fig. 1: (a) Entity-Relation (ER) Encoder which generates latent representations for entities by capturing neighborhood structures and relation semantics jointly; (b) the training process of GAEA can be decomposed into multiple epochs, and in each epoch, we incorporate graph augmentation to conduct margin-based alignment learning and contrastive entity representation learning.

Initialization. At the beginning, we randomly initialize entity embeddings $\mathbf{H}^{ent} \in \mathbb{R}^{(|E_s|+|E_t|)\times d_{ent}}$ and relation embeddings $\mathbf{H}^{rel} \in \mathbb{R}^{|R_s \cup R_t|\times d_{rel}}$, where d_{ent} and d_{rel} are the embedding dimension of entities and relations, respectively.

4.1 Entity-Relation Encoder

Here, we present the Entity-Relation Encoder (ER Encoder for short), which aims to fully capture the contextual information of entities using two aspects jointly: (I) neighborhood structures and (II) relation semantics.

Neighborhood Aggregator. First, we aggregate neighbor entities' information to the central entity. The rationality of neighborhood aggregator lies in the structure assumption that, equivalent entities tend to have similar neighbor structures [20]. Moreover, leveraging multi-range neighborhood structures is capable of providing more alignment evidence and mitigating the structural heterogeneity issue. In this work, we apply Graph Attention Network (GAT) [17] to allow the central entity to learn the importance of different neighbors and thus selectively aggregate surrounding information, and we then recursively capture multi-range neighbor information by stacking multiple layers:

$$\mathbf{h}_{e_i}^{(l)} = \sum_{e_j \in N_{e_i}} \alpha_{ij} \mathbf{h}_{e_j}^{(l-1)}, \tag{1}$$

$$\alpha_{ij} = \frac{\exp(\mathrm{LeakyReLU}(\mathbf{a}^\top [\mathbf{W}_g \mathbf{h}_{e_i} \oplus \mathbf{W}_g \mathbf{h}_{e_j}]))}{\sum_{e_k \in N_{e_i}} \exp(\mathrm{LeakyReLU}(\mathbf{a}^\top [\mathbf{W}_g \mathbf{h}_{e_i} \oplus \mathbf{W}_g \mathbf{h}_{e_k}]))}, \tag{2}$$

where $\top$ represents transposition, $\oplus$ is the concatenation operation, $\mathbf{W}_g$ and $\mathbf{a}$ are the transformation parameter and attention transformation vector, respectively. N_{e_i} means the neighbor set of entity e_i in KG, and α_{ij} indicates the learned importance of entity e_j to entity e_i. $\mathbf{h}_{e_i}^{(l)}$ denotes the embedding of e_i at l-th layer (total L layers) with $\mathbf{H}^{(0)} = \mathbf{H}^{ent}$. Note that here we remove the feature transformation and nonlinear activation that act on input embeddings in vanilla GAT since we mainly focus on information aggregation. We only use $\mathbf{W}_g$ and $\mathbf{a}$ to make each entity aware of its neighborhood contexts.

After multi-layer GAT, we obtain the multi-range neighborhood structural representation matrix for each entity, i.e., $\mathbf{H}_{e_i}^m = [\mathbf{h}_{e_i}^{(1)}, ..., \mathbf{h}_{e_i}^{(L)}] \in \mathbb{R}^{L \times d_{ent}}$ for e_i. Since different neighborhood ranges have different contributions to characterize the central entity, it is necessary to employ a mechanism to adaptively control the flow of each range and thus reduce noise. Inspired by the skipping connections in neural networks [10,14,23], we firstly utilize a Scaled Dot-Product Attention mechanism [16] to learn the importance of each range, and then fuse small-range and wide-range representations by weighted average:

$$[\hat{\mathbf{h}}_{e_i}^{(1)}, ..., \hat{\mathbf{h}}_{e_i}^{(L)}] = \text{softmax}(\frac{(\mathbf{H}_{e_i}^m \mathbf{W}_q)(\mathbf{H}_{e_i}^m \mathbf{W}_k)^\top}{\sqrt{d_{ent}}})\mathbf{H}_{e_i}^m \tag{3}$$

$$\mathbf{h}_{e_i}^n = \frac{1}{L} \sum_{l=1}^{L} \hat{\mathbf{h}}_{e_i}^{(l)}, \tag{4}$$

where $1/\sqrt{d_{ent}}$ is the scaling factor, $\mathbf{W}_q$ and $\mathbf{W}_k$ are the learnable parameter matrices, and $\mathbf{h}_{e_i}^n$ is the output of neighborhood aggregator.

Relation Aggregator. Relation-level information which carries rich semantics is vital to align entities in KGs [24,29] because two equivalent entities may share overlapping relations. MRAEA [6] pointed out that relation directions impose extra but delicate constraints on the head and tail entity individually. Therefore, in this work, we directly use two mean aggregators to gather outward relation semantics and inward relation semantics separately to provide supplementary alignment signals for heterogeneous KGs:

$$\mathbf{h}_{e_i}^r = \frac{1}{|N_{e_i}^{r+}|} \sum_{r \in N_{e_i}^{r+}} \mathbf{h}_r^{rel} \oplus \frac{1}{|N_{e_i}^{r-}|} \sum_{r \in N_{e_i}^{r-}} \mathbf{h}_r^{rel}, \tag{5}$$

where $N_{e_i}^{r+}$ and $N_{e_i}^{r-}$ are the outward and inward relation set of e_i, respectively.

Feature Fusion. Finally, we concatenate two aspects of information:

$$\tilde{\mathbf{h}}_{e_i} = \mathbf{h}_{e_i}^n \oplus \mathbf{h}_{e_i}^r, \tag{6}$$

where $\tilde{\mathbf{h}}_{e_i} \in \mathbb{R}^{d_{ent}+2\times d_{rel}}$ is the final output representation of ER Encoder for e_i. In the following training process, the ER Encoder is shared for G_s, G_t, and their augmented graphs, and given an entity e_i, we denote by $\tilde{\mathbf{h}}_{e_i}$ its representation generated by ER Encoder with the original graph as input, and $\tilde{\mathbf{h}}_{e_i}^{aug}$ its representation generated with the augmented graph as input.

4.2 Model Training with Graph Augmentation

Graph augmentation learning has been demonstrated to promote the performance of graph learning, such as overcoming overfitting and oversmoothing issues [9], and being used for graph contrastive learning [25]. We apply graph augmentation for EA and highlight two main enhancements contributed by it: (I) injecting perturbations into the original KG can increase the diversity of the structural differences, thus preventing the model from overfitting to the training data during alignment process to some extent as well as enforcing the model to produce robust entity representations against structural changes; (II) graph augmentation inherently generates two graph views without extra parameters, which facilitates conducting contrastive learning to promote heterogeneous representation learning for (unseen) entities by contrasting different views.

Margin-Based Alignment Loss. In order to make equivalent entities close to each other and unmatched entities pull away from each other in a unified embedding space. Following previous works [5,6,20], we apply the margin-based alignment loss supervised by pre-aligned entity pairs S. Notably, here, we use the output of ER Encoder based on augmented graphs to make the model avoid overfitting and behave durable against edge changes:

$$\mathcal{L}_a = \sum_{(e_i,e_j)\in S} \sum_{(\bar{e}_i,\bar{e}_j)\in \bar{S}_{(e_i,e_j)}} \left[||\tilde{\mathbf{h}}_{e_i}^{aug} - \tilde{\mathbf{h}}_{e_j}^{aug}||_{L2} + \rho - ||\tilde{\mathbf{h}}_{\bar{e}_i}^{aug} - \tilde{\mathbf{h}}_{\bar{e}_j}^{aug}||_{L2} \right]_+, \quad (7)$$

where ρ is a hyper-parameter of margin, $[x]_+ = \max\{0, x\}$ is to ensure non-negative output, and $\bar{S}_{(e_i,e_j)}$ denotes the set of negative entity alignments constructed by corrupting the ground-truth alignment (e_i, e_j), i.e., replacing e_i or e_j with another entity in G_s or G_t via negative sampling strategy.

Contrastive Loss. Contrastive learning is a good means to explore supervision signals from the vast unlabeled data. Many graph learning works [18,19,25] apply it to learn representations by contrasting different views and then maximizing feature consistency between them. RAC [27] is an effective EA model which incorporates contrastive learning to ameliorate the alignment performance. However, RAC needs to employ two separate graph encoders with the same architecture to model different views of the structural features of entities, which will bring twice the parameters and damage the diversity of graph views. Graph augmentation inherently provides two different views (i.e., original graph view and augmented graph view) without extra parameters. Therefore, we define the contrastive loss to improve entity representation learning by maximizing the feature consistency between the original structure and augmented structure:

$$\mathcal{L}_c = \sum_{z=\{s,t\}} \frac{1}{2|E_z|} \sum_{e_i\in E_z} (\mathcal{L}_{c,e_i}^{(G_z,G_z^{aug})} + \mathcal{L}_{c,e_i}^{(G_z^{aug},G_z)}), \quad (8)$$

$$\mathcal{L}_{c,e_i}^{(G_z,G_z^{aug})} = -\log \frac{\exp(\langle \mathrm{proj}(\tilde{\mathbf{h}}_{e_i}), \mathrm{proj}(\tilde{\mathbf{h}}_{e_i}^{aug}) \rangle)}{\sum_{e_k\in E_z} \exp(\langle \mathrm{proj}(\tilde{\mathbf{h}}_{e_i}), \mathrm{proj}(\tilde{\mathbf{h}}_{e_k}^{aug}) \rangle)}, \quad (9)$$

where $\langle\cdot\rangle$ means inner product, and proj$(\cdot)$ is a shared projection head consisting of a linear layer and a ReLU activation function to map entity representations to low-dimensional vector space [25]. The definition of the symmetric contrastive loss term $\mathcal{L}_{c,e_i}^{(G_z^{aug},G_z)}$ is similar with Eq. (9).

Model training. We combine the margin-based alignment loss and the contrastive loss, arriving at the final objective of our model:

$$\mathcal{L} = \mathcal{L}_a + \lambda \mathcal{L}_c, \tag{10}$$

where $\lambda \geq 0$ is a tunable parameter weighting the two objectives. The training process of GAEA is outlined in Algorithm 1, where negative sample set and augmented graphs will be updated every iteration (10 epochs as an iteration).

Algorithm 1: Training Procedure of GAEA

Input: Knowledge graph G_s and G_t, pre-aligned entity pairs S.

1 Initialize entity embeddings and relation embeddings;
2 **while** *Not Converge* **do**
3   **for** *each Epoch* **do**
4   **if** *Epoch % 10 == 0* **then** `// 10 epochs as an iteration`
5   Generate augmented graphs G_s^{aug} and G_t^{aug} for G_s and G_t;
6   Generate negative sample set $\bar{S}$ based on S;
7   Generate entity representations using ER Encoder;
8   Calculate $\mathcal{L}_a$ using S and $\bar{S}$ via Eq.(7);
9   Calculate $\mathcal{L}_c$ using Eq.(8) and Eq.(9);
10     $\Theta \leftarrow$ BackProp$(\mathcal{L}_a + \lambda\mathcal{L}_c)$; ▷ Adam step
11 **return** Model parameters Θ;

4.3 Alignment Inference

After pulling embeddings from two KGs into a unified vector space and making them comparable, alignment relationships can be inferred by measuring the distance between two entities. In this work, we use Euclidean Distance to be the distance metric, i.e., for $e_i \in E_s$ and $e_j \in E_t$, the distance between entity pair (e_i,e_j) is calculated by $||\tilde{\mathbf{h}}_{e_i} - \tilde{\mathbf{h}}_{e_j}||_{L2}$. In order to find e_i' alignment relationship, we calculate its distance to all entities belonging to G_t and perform the nearest neighbor (NN) search to identify e_i' counterpart entity in G_t:

$$e_j = \arg \min_{e_j' \in E_t} ||\tilde{\mathbf{h}}_{e_i} - \tilde{\mathbf{h}}_{e_j'}||_{L2}. \tag{11}$$

Notably, we use the original KG structures in the inference phase instead of augmented versions to generate final entity representation $\tilde{\mathbf{h}}$. We apply Faiss[1] to accelerate the alignment inference process.

[1] https://github.com/facebookresearch/faiss.

5 Experimental Setup

5.1 Experimental Setup

Datasets. We use the 15K benchmark dataset (V1) in OpenEA [15] for evaluation since the entities thereof follow the degree distribution in real-world KGs. It contains two cross-lingual settings, i.e., EN-FR-15K (English-to-French) and EN-DE-15K (English-to-German), and two monolingual settings, i.e., D-W-15K (DBPedia-to-Wikidata) and D-Y-15K (DBPedia-to-YAGO). Following the data splits in OpenEA, we use the same split setting where 20%, 10%, and 70% alignments are harnessed for training, validation, and testing, respectively.

Metrics. We adopt *Hits@k* ($k = 1{,}5$) and *Mean Reciprocal Rank* (*MRR*) as the evaluation metrics. *Hits@k* is to measure the alignment accuracy, while *MRR* measures the average performance of ranking over all test samples. The higher the *Hits@k* and *MRR*, the better the alignment performance.

Baselines. We choose some GNN variants and several existing state-of-the-art embedding-based EA models as baselines: GCN [4] and GAT [17] are the classic variants of GNNs; MTransE [2] and SEA [7] are triple-based methods that capture the local semantics information of relation triples via knowledge representation learning; GCN-Align [20], AliNet [14], HyperKA [11], and KE-GCN [26] are the neighborhood-based methods which apply GNNs to explore neighborhood structure information; IPTransE [30] and RSNs [3] both are path-based methods that extract the long-term dependencies across relation paths; IMEA [22] is the recent strong baseline which uses Transformer-like architecture to capture multiple structural contexts in an end-to-end manner.

We should note here that our model and the above baselines all mainly focus on the structural information of KGs. Therefore, for a fair comparison, we do not consider the models which utilize extra information (e.g., attributes, literals) for enhancement, such as AttrGNN [5], HMAN [24], MultiKE [29].

Implementation Details. All programs are implemented using *Python* 3.6.13 and *PyTorch* 1.10.2 with *CUDA* 11.3 on an *NVIDIA GeForce RTX 3090* GPU. Following OpenEA [15], we report the average results of five-fold cross-validation. We initialize trainable parameters with the Xavier initializer, and we train the model using Adam optimizer with weight decay 1e$-$5 and perform early stopping to terminate training based on the *MRR* score tested every 10 epochs on the validation data. As for hyper-parameters, the learning rate is set to 0.001, the dropout rate is 0.2, the layer number of GAT L is 2, the number of negative samples for each entity is 5, the negative sampling strategy is ϵ-Truncated Uniform Negative Sampling [13] with $\epsilon = 0.9$, the margin ρ is 1, the balance parameter λ is 100, and the embedding dimension of entities d_{ent} and relations d_{rel} are set to 256 and 128, respectively. The pr is searched in $\{0.05, 0.1, 0.15\}$. Following the convention, the default alignment direction is from left to right. Taking D-W-15K as an example, we regard DBpedia as the source KG and seek to find the counterparts of source entities in the target KG Wikidata.

Table 1. Entity alignment results in cross-lingual and monolingual settings. The results with $\dagger$ are retrieved from [15], and $\ddagger$ from [22]. Results labeled by $*$ are reproduced using the released source codes. The **boldface** indicates the best result of each column and underlined the second-best.

Models	EN-FR-15K			EN-DE-15K			D-W-15K			D-Y-15K		
	Hit@1	Hit@5	MRR	Hit@1	Hit@5	MRR	Hit@1	Hit@5	MRR	Hit@1	Hit@5	MRR
GCN*	.210	.414	.304	.304	.497	.394	.208	.367	.284	.343	.503	.416
GAT*	.297	.585	.426	.542	.737	.630	.383	.622	.489	.468	.707	.573
MTrasnE†	.247	.467	.351	.307	.518	.407	.259	.461	.354	.463	.675	.559
SEA†	.280	.530	.397	.530	.718	.617	.360	.572	.458	.500	.706	.591
IPTransE†	.169	.320	.243	.350	.515	.430	.232	.380	.303	.313	.456	.378
RSNs†	.393	.595	.487	.587	.752	.662	.441	.615	.521	.514	.655	.580
GCN-Align†	.338	.589	.451	.481	.679	.571	.364	.580	.461	.465	.626	.536
AliNet‡	.364	.597	.467	.604	.759	.673	.440	.628	.522	.559	.690	.617
HyperKA‡	.353	.630	.477	.560	.780	.656	.440	.686	.548	.568	.777	.659
KE-GCN‡	.408	.670	.524	.658	.822	.730	.519	.727	.608	.560	.750	.644
IMEA‡	.458	.720	.574	.639	.827	.724	.527	.753	.626	**.639**	**.804**	**.712**
GAEA	**.486**	**.746**	**.602**	**.684**	**.854**	**.760**	**.562**	**.768**	**.654**	.608	.791	.688
w/o *rel.*	.324	.626	.458	.593	.785	.678	.409	.666	.521	.502	.743	.605

5.2 Experimental Results

Performance Comparison. Table 1 reports the comparison results on the OpenEA 15K datasets. Experimental results show that our proposed GAEA outperforms other models in most tasks, especially in cross-lingual settings. There is a phenomenon that the performance of models utilizing knowledge representation learning as the encoder, e.g., MTransE, SEA, and IPTransE, are inferior compared with the models applying GNNs as the encoder like AliNet and KE-GCN, and have on-par or even worse performance than vanilla GCN and GAT, which demonstrates the GNNs' powerful representation ability in EA. We also notice that, compared with some methods applying GCN as the encoder (e.g., GCN-Align, AliNet), the vanilla GCN fails to surpass them, which shows the significance of designing a more effective encoder for representing entities in KGs. IMEA is a strong baseline that captures abundant structure contexts and it obtains excellent results on D-Y-15K task. However, IMEA introduces carefully designed data processing (e.g., entity paths encoding) and becomes a complicated network due to the Transformer-like architecture, which will inevitably increase the training difficulty and overfitting risk. Additionally, we compare the model size (denoted as #Params) in Table 2. GAEA greatly reduces the number of parameters compared to IMEA while acquiring decent alignment performance. This is because GAEA designs a simple Entity-Relation Encoder to capture multi-range neighborhood structures to mitigate heterogeneity and infuse relation semantics to provide more comprehensive signals for alignment. Moreover, GAEA further facilitates producing expressive and robust entity representations by integrating graph augmentation to achieve alignment learning supervised by alignment seeds and contrastive representation learning for unseen entities. In summary, our proposed GAEA is a light and powerful solution for EA.

Table 2. #Params comparison.

Models	#Params (M)
GCN	~7.81M
AliNet	~16.18M
IMEA	~20.44M
GAEA (ours)	~8.10M

Table 3. Ablation study results.

Models	EN-DE-15K			D-W-15K		
	Hit@1	Hit@5	MRR	Hit@1	Hit@5	MRR
GAEA	**.684**	**.854**	**.760**	**.562**	**.768**	**.654**
−*gaal.*	.674	.848	.751	.557	.764	.650
$-\mathcal{L}_c$	.665	.841	.744	.544	.755	.639

Ablation Study. In the above experiments, the overall effectiveness of GAEA is proved. In this section, we conduct ablation analyses to demonstrate the validity of each component of GAEA. First, Table 1 also gives the results of a variant of GAEA (denoted as w/o *rel.*), which means the original GAEA eliminates relation injection. The ablation results clearly show the effectiveness of relation embedding learning, which identifies the relation semantics can help in enriching the expressiveness of entity representations. Next, Table 3 gives the ablation results about graph augmentation. −*gaal.* and $-\mathcal{L}_c$ represent the variants by removing graph augmentation in alignment learning (i.e., Eq. (7)) or removing contrastive objective (i.e., Eq. (8)), respectively (the results of removing graph augmentation are illustrated in the next section). The results show that utilizing graph augmentation can have positive impacts on EA and consistently get better performance. By introducing graph augmentation into EA training process, the model not only is encouraged to learn useful and robust entity representations but also lets the scarce yet valuable alignment seeds and vast unlabeled entities in KGs jointly provide abundant supervision for model learning.

Parameter Analysis. Considering that our model employs edge dropping to generate augmented graphs for margin-based alignment learning and contrastive entity representation learning. We investigate how the alignment performance varies with the upper bound of the deletion ratio. We evaluate upper bound *pr* in {0, 0.05, 0.1, 0.15}, and the results measured by *Hit@1* and *MRR* are drawn in Fig. 2. The performance is worst on all three tasks when $pr = 0$, i.e., without any graph augmentation enhancement, indicating that graph augmentation can

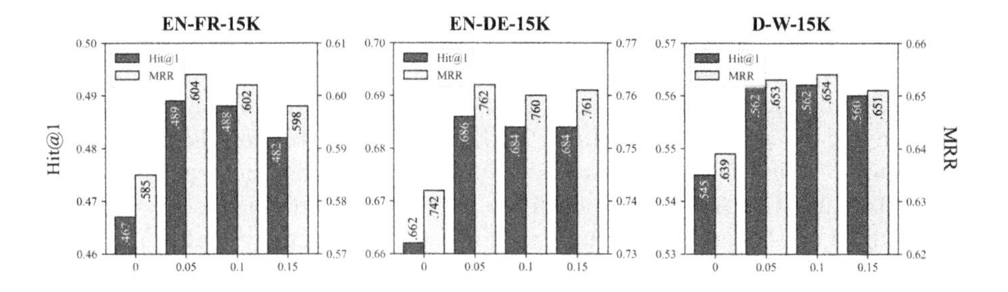

Fig. 2. Parameter analysis results of *pr* measured by Hit@1 (green bar with left axis) and MRR (yellow bar with right axis). (Color figure online)

do benefit for alignment learning. We can see that the alignment effect is best when pr equals 0.05 or 0.1, increasing pr to 0.15 will not further improve the performance, and even bring performance drops. One potential reason is that when pr becomes large, edge dropping will lead to losing more semantic knowledge and structural information, thus bringing an adverse impact on neighborhood aggregation and model training. Therefore, we need to set pr as a suitably small value to ensure information retention as well as performance improvement.

6 Discussion and Conclusion

In this paper, we propose GAEA, a novel entity alignment method based on graph augmentation. Specifically, we design an Entity-Relation (ER) Encoder to generate latent representations for entities via jointly capturing neighborhood structures and relation semantics. Meanwhile, we apply graph augmentation to create two graph views for margin-based alignment learning and contrastive entity representation learning, thus improving the model's alignment performance. Finally, experimental results verified the effectiveness of our method.

Although GAEA achieves promising results, it still has limitations that need further investigation. First, our experimental results show that graph augmentation learning can bring some performance gains, but the supervision signals provide key performance bases in the alignment learning process. Thus, it is worth further studying how to amplify the improvement brought by graph augmentation when there no alignment seeds are given. Besides, we currently apply edge dropping as the only graph augmentation strategy, which exposes a new problem, that is, how to conduct graph augmentation learning in a highly structured KG to improve performance without introducing logic errors.

Acknowledgments. We thank reviewers for their helpful feedback. This work is supported by the National Natural Science Foundation of China No. 62172428.

References

1. Bordes, A., Usunier, N., Garcia-Duran, A., Weston, J., Yakhnenko, O.: Translating embeddings for modeling multi-relational data. In: Proceedings of NIPS (2013)
2. Chen, M., Tian, Y., Yang, M., Zaniolo, C.: Multilingual knowledge graph embeddings for cross-lingual knowledge alignment. In: Proceedings of IJCAI (2016)
3. Guo, L., Sun, Z., Hu, W.: Learning to exploit long-term relational dependencies in knowledge graphs. In: Proceedings of ICML (2019)
4. Kipf, T.N., Welling, M.: Semi-supervised classification with graph convolutional networks. arXiv preprint arXiv:1609.02907 (2016)
5. Liu, Z., Cao, Y., Pan, L., Li, J., Chua, T.S.: Exploring and evaluating attributes, values, and structures for entity alignment. In: Proceedings of EMNLP (2020)
6. Mao, X., Wang, W., Xu, H., Lan, M., Wu, Y.: Mraea: an efficient and robust entity alignment approach for cross-lingual knowledge graph. In: Proceedings of WSDM (2020)

7. Pei, S., Yu, L., Hoehndorf, R., Zhang, X.: Semi-supervised entity alignment via knowledge graph embedding with awareness of degree difference. In: Proceedings of WWW (2019)
8. Pei, S., Yu, L., Yu, G., Zhang, X.: Graph alignment with noisy supervision. In: Proceedings of WWW (2022)
9. Rong, Y., Huang, W., Xu, T., Huang, J.: DropEdge: towards deep graph convolutional networks on node classification. arXiv preprint arXiv:1907.10903 (2019)
10. Srivastava, R.K., Greff, K., Schmidhuber, J.: Highway networks. arXiv preprint arXiv:1505.00387 (2015)
11. Sun, Z., Chen, M., Hu, W., Wang, C., Dai, J., Zhang, W.: Knowledge association with hyperbolic knowledge graph embeddings. In: Proceedings of EMNLP (2020)
12. Sun, Z., Hu, W., Wang, C., Wang, Y., Qu, Y.: Revisiting embedding-based entity alignment: a robust and adaptive method. IEEE TKDE (2022)
13. Sun, Z., Hu, W., Zhang, Q., Qu, Y.: Bootstrapping entity alignment with knowledge graph embedding. In: Proceedings of IJCAI (2018)
14. Sun, Z., et al.: Knowledge graph alignment network with gated multi-hop neighborhood aggregation. In: Proceedings of AAAI (2020)
15. Sun, Z., et al.: A benchmarking study of embedding-based entity alignment for knowledge graphs. arXiv preprint arXiv:2003.07743 (2020)
16. Vaswani, A., et al.: Attention is all you need. In: Proceedings of NIPS (2017)
17. Veličković, P., Cucurull, G., Casanova, A., Romero, A., Lio, P., Bengio, Y.: Graph attention networks. arXiv preprint arXiv:1710.10903 (2017)
18. Veličković, P., Fedus, W., Hamilton, W.L., Liò, P., Bengio, Y., Hjelm, R.D.: Deep graph infomax. In: Proceedings of ICLR (2018)
19. Wan, S., Pan, S., Yang, J., Gong, C.: Contrastive and generative graph convolutional networks for graph-based semi-supervised learning. In: Proceedings of AAAI (2021)
20. Wang, Z., Lv, Q., Lan, X., Zhang, Y.: Cross-lingual knowledge graph alignment via graph convolutional networks. In: Proceedings of EMNLP (2018)
21. Wu, Y., Liu, X., Feng, Y., Wang, Z., Yan, R., Zhao, D.: Relation-aware entity alignment for heterogeneous knowledge graphs. In: Proceedings of IJCAI (2019)
22. Xin, K., Sun, Z., Hua, W., Hu, W., Zhou, X.: Informed multi-context entity alignment. In: Proceedings of WSDM (2022)
23. Xu, K., Li, C., Tian, Y., Sonobe, T., Kawarabayashi, K.I., Jegelka, S.: Representation learning on graphs with jumping knowledge networks. In: ICML (2018)
24. Yang, H.W., Zou, Y., Shi, P., Lu, W., Lin, J., Sun, X.: Aligning cross-lingual entities with multi-aspect information. In: Proceedings of EMNLP (2019)
25. You, Y., Chen, T., Sui, Y., Chen, T., Wang, Z., Shen, Y.: Graph contrastive learning with augmentations. In: Proceedings of NIPS (2020)
26. Yu, D., Yang, Y., Zhang, R., Wu, Y.: Knowledge embedding based graph convolutional network. In: Proceedings of WWW (2021)
27. Zeng, W., Zhao, X., Tang, J., Fan, C.: Reinforced active entity alignment. In: Proceedings of CIKM (2021)
28. Zeng, W., Zhao, X., Tang, J., Lin, X.: Collective entity alignment via adaptive features. In: Proceedings of ICDE (2020)
29. Zhang, Q., Sun, Z., Hu, W., Chen, M., Guo, L., Qu, Y.: Multi-view knowledge graph embedding for entity alignment. In: Proceedings of IJCAI (2019)
30. Zhu, H., Xie, R., Liu, Z., Sun, M.: Iterative entity alignment via knowledge embeddings. In: Proceedings of IJCAI (2017)

MixER: MLP-Mixer Knowledge Graph Embedding for Capturing Rich Entity-Relation Interactions in Link Prediction

Thanh Le[1,2](✉), An Pham[1,2], Tho Chung[1,2], Truong Nguyen[1,2], Tuan Nguyen[1,2], and Bac Le[1,2]

[1] Faculty of Information Technology, University of Science, Ho Chi Minh City, Vietnam
{lnthanh,lhbac}@fit.hcmus.edu.vn,
{pntan19,cttho19,nttruong19,nhtuan19}@clc.fitus.edu.vn
[2] Vietnam National University, Ho Chi Minh City, Vietnam

Abstract. Knowledge graphs are often incomplete in practice, so link prediction becomes an important problem in developing many downstream applications. Therefore, many knowledge graph embedding models have been proposed to predict the missing links based on known facts. Convolutional neural networks (CNNs) play an essential role due to their excellent performance and parameter efficiency. Previous CNN-based models such as ConvE and KMAE use kernels to capture interactions between embeddings, yet they are limited in quantity. In this paper, we propose a novel neural network-based model named MixER to exploit more additional interactions effectively. Our model incorporates two types of multi-layer perceptions (i.e., channel-mixing and token-mixing), which extract spatial information and channel features. Hence, MixER can seize richer interactions and boost the link prediction performance. Furthermore, we investigate the characteristics of two core components that benefit in capturing additional interactions in diverse regions. Experimental results reveal that MixER outperforms state-of-the-art models in the branch of CNNs on three benchmark datasets.

Keywords: Knowledge graph · Link prediction · Convolutional neural network · Multi-layer perception

1 Introduction

A knowledge graph (KG) is a special kind of graph structure that includes entities and relations in the form of triples. Examples of KGs include WordNet [1], YAGO [2], and Freebase [3]. Even though real-world KGs contain a large quantity of data, they are still incomplete. As a result, it is crucial to predict missing triples to complete the KGs automatically. Knowledge graph embedding (KGE) techniques have been utilized as prevalent methods to address this problem by embedding entities and relations as low-dimensional vectors, allowing efficient

H. Kashima et al. (Eds.): PAKDD 2023, LNAI 13936, pp. 15–27, 2023.
https://doi.org/10.1007/978-3-031-33377-4_2

computation and analysis of the graph. Subsequently, a scoring function is used to give valid triples higher scores than invalid triples. Finding these missing links is significant in many practical applications, including recommender systems, question answering, and gene-protein interaction prediction [23,24].

KGs store data as a collection of triples (s, r, o), in which $s, o \in \mathcal{E}$ (where $\mathcal{E}$ denotes the entity set) are subject, object entities in turn and $r \in \mathcal{R}$ (where $\mathcal{R}$ denotes the relation set) is relation connecting s with o. Figure 1 illustrates an example of a KG about the film industry. Intuitively, we can straightforwardly predict that Marvel Studios is an organization based on two existing triples—*(Disney, Is_a, Organization)* and *(Disney, Is_parent_of, Marvel Studios)*—by human knowledge. However, a model can not understand triples at the semantic level as humans. Therefore, it relies on mathematical theories or neural networks to predict missing entities or relations, e.g., *(Marvel_Studios, Is_a, ?)* or *(Marvel_Studios, ?, Organization)*.

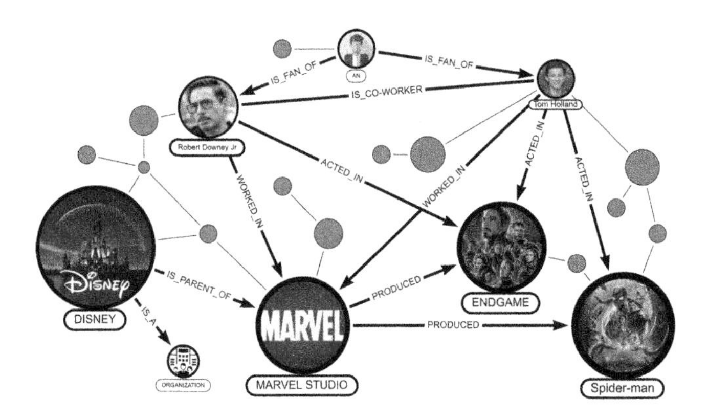

Fig. 1. An example of a knowledge graph in real life.

In recent years, many proposals have attempted to enhance interactions between embeddings but do not efficiently exploit them. To mitigate this drawback, we suggest a novel model that increases interaction efficiency based on the recently proposed computer vision architecture named MLP-Mixer [4], described in Sect. 3. To the best of our knowledge, we are the first to adapt MLP-Mixer to graph-related tasks, especially link prediction (LP). We also examine its compatibility with graphs in terms of grasping interactions and information integration ability.

Contribution. Our main contributions are as follows:

- To deal with the LP task, we propose a new convolution-free model in the branch of the neural network.
- We provide a detailed description of the vital properties of two primary components in MixER, which extract spatial information and channel features to boost interconnection between entities and relations.

- Our model can be considered a potential synergy between computer vision and graphs, a foundation that could accelerate the appearance of more well-designed networks.
- MixER is evaluated on three benchmark datasets and experiments show that our model outperforms state-of-the-art neural network-based baselines.

Organization. We discuss the relevant research of three taxonomy, including translation-based models, matrix factorization-based models, and neural network-based models in Sect. 2. Section 3 provides a detailed illustration of our proposed model and explores its aspects. We evaluate and compare our model against baselines and analyze the results in Sect. 4. Finally, Sect. 5 summarises our work and potential future work directions.

2 Related Work

2.1 Translation-Based Approaches

In the translation group, KGE models project entities and relations into the mathematical space and then consider relations as distance-related transformations used to determine the plausibility of facts. TransE [7] is the first model of this kind that embeds entities and relations as vectors in Euclidean space. The idea of TransE is simple, but it can not model multi-fold relations (i.e., 1-to-N, N-to-1, N-to-N). Many extension models based on TransE have been proposed to tackle this problem, namely TransH [8] and TransD [9]. RotatE [10] is another model that overcomes the shortcoming of TransE further by treating the relations as the rotation of entities in the complex vector space. To take the strengths of both TransH and RotatE, RotatPRH [11] first projects embedded entities onto the relational hyperplane and then performs rotation.

2.2 Matrix Factorization-Based Approaches

Matrix factorization is also a method that considers KG as a 3-way binary tensor with the value 1 indicating an existing triple and otherwise 0. Then to handle the LP task, the KG tensor could be decomposed into a composition of embeddings of both entities and relations that are lower dimensional tensors/vectors. RESCAL [22], an early work in this group, takes one matrix to represent all entities and another to describe the hidden interactions between them. However, the problems with RESCAL model are parameter explosion and overfitting. DistMult [15] overcomes this drawback by embedding relations in terms of a diagonal matrix instead of a regular matrix. A further improvement is ComplEx [16] as it extends DistMult to complex spaces and thus can model other types of relations (e.g., symmetric, inverse relations).

2.3 Neural Network-Based Approaches

The neural network has been applied in many models because of its superior performance. Such models are able to extract features automatically and utilize that information to predict entities. ConvE [6] uses convolutional kernels on the concatenation of embeddings to extract feature maps before feeding them into a fully-connected layer. In contrast, ConvKB [12] and CapsE [13] capture interactions in a very different manner. They convolve on the concatenation of the subject entity, relation, and object entity embeddings. In recent years, KMAE [20] extends the Gaussian kernel separately in entity and relation attributes to achieve better representations fed into a multi-attention neural network. To develop ConvKB, DMACM [14] is a proposed method to extract implicit fine-grained characteristics in a triple and exploit directional information.

Due to the potential performance, the neural network is an excellent method for almost all tasks, including LP. In this paper, our work extends the standard framework named MLP-Mixer to further enrich interactions of embeddings for LP in KGs. In particular, the proposal uses mixing operators to capture beneficial interactions inherent in the subject entity and the corresponding relation.

Table 1. Different KGE models with the scoring function $\psi(s,r,o)$. $\mathbf{e}^s, \mathbf{e}^r, \mathbf{e}^o$ are the embedding of the subject entity, relation, and object entity, and $\overline{\mathbf{e}^s}, \overline{\mathbf{e}^r}, \overline{\mathbf{e}^o}$ denote the $2D$ reshaped matrices of $\mathbf{e}^s, \mathbf{e}^r$, and $\mathbf{e}^o$ in turn; $\star$ is a convolution operator; $||$ denotes left component or right component.

Model	Scoring Function $\psi(s,r,o)$	Relation parameter
TransE	$\lVert\mathbf{e}^s + \mathbf{e}^r - \mathbf{e}^o\rVert_p$	$\mathbf{e}^r \in \mathbb{R}^{d_r}$
TransH	$\lVert(\mathbf{e}^s - \mathbf{w}_r{}^T\mathbf{e}^s\mathbf{w}_r) + \mathbf{e}^r - (\mathbf{e}^o - \mathbf{w}_r{}^T\mathbf{e}^o\mathbf{w}_r)\rVert$	$\mathbf{w}_r, \mathbf{e}^r \in \mathbb{R}^{d_r}$
TransD	$\mathbf{e}^{sT} M_r \mathbf{e}^o$	$\mathbf{e}^r \in \mathbb{R}^{d_r}$
RESCAL	$\mathbf{e}^{sT} M_r \mathbf{e}^o$	$\mathbf{e}^r \in \mathbb{R}^{d_e^2}$
DistMult	$\mathbf{e}^{sT} diag(\mathbf{e}^r)\mathbf{e}^o$	$\mathbf{e}^r \in \mathbb{R}^{d_e}$
ComplEx	$Re(\mathbf{e}^{sT} diag(\mathbf{e}^r)\overline{\mathbf{e}^o})$	$\mathbf{e}^r \in \mathbb{C}^{d_r}$
ConvE	$f(vec(f([\overline{\mathbf{e}^s}, \overline{\mathbf{e}^r}] \star \mathbf{w}))\mathbf{W})\mathbf{e}^o$	$\mathbf{e}^r \in \mathbb{R}^{d_r}$
KMAE	$g(Mul([\phi(\overline{\mathbf{e}^s}); \overline{\mathbf{e}^r}] \Vert [\overline{\mathbf{e}^s}; \phi(\overline{\mathbf{e}^r})]))\mathbf{e}^o$	$\mathbf{e}^r \in \mathbb{R}^{d_r}$
MixER (ours)	$f(Mixer(\phi_s[\varphi(\mathbf{e}^s), \varphi(\mathbf{e}^r)])\mathbf{W})\mathbf{e}^o$	$\mathbf{e}^r \in \mathbb{R}^{d_r}$

3 Methodology

3.1 Problem Formulation and Notations

A KG contains multi-relational data formalizing as a set of triples $\mathcal{G} = \{(s,r,o)|(s,o) \in \mathcal{E}, r \in \mathcal{R}\}$. In that, $\mathcal{E}$ is the set of entities, and $\mathcal{R}$ is the set of relations. n_e and n_r are the number of entities and relations while d_e and d_r are the dimensionalities of entity embedding and relation embedding respectively. The LP problem, which aims to predict the missing entities or relations based on the existing triples, can be regarded as the rank problem. A triple (s,r,o) is evaluated by a scoring function $\psi(s,r,o)$ to show the plausibility of that triple. Due to the difference in model categories, the corresponding scoring function can be designed to tailor the model's properties. Table 1 summarizes the scoring function of the aforementioned models in Sect. 2.

3.2 Overall Architecture Design

In this section, we propose a model to capture interactions more effectively by taking advantage of the ideas behind MLP-Mixer (Multi-Layer Perceptron - Mixer) [4]. The main components contain the channel-mixing layer, the token-mixing layers, and the latent attention mechanism. In the context of our scenario, a token refers to a segment of an embedding, while a channel is simply defined as an embedding. The channel-mixing layer is a simple MLP layer, which can be considered as 1×1 convolutional kernels, leveraging the communications between embeddings and thus improving interactions between them. In contrast, the token-mixing layer allows the communication of different receptive fields on 2D embeddings by single-channel depth-wise convolutions. Besides, the latent attention mechanism is that the information in the channel-mixing and token-mixing layers are weighted by MLPs, showing the different importances within embeddings or between entity-relation embeddings.

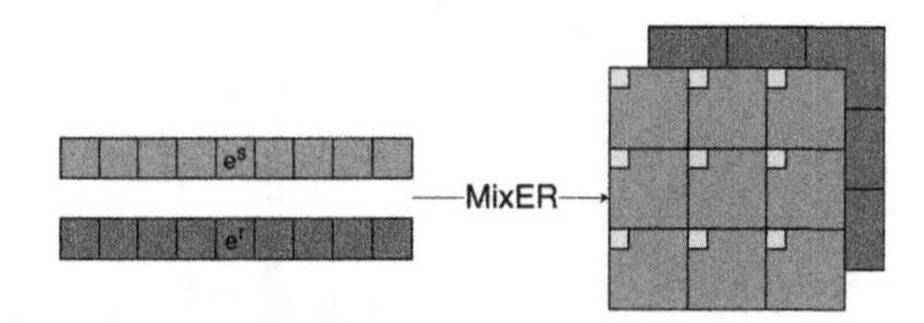

Fig. 2. MixER operation on the entity-relation matrix. The embeddings of relation and entity are respectively represented by the blue and red parts, while the advanced shapes of convolutional kernels are depicted by the green grid. (Color figure online)

The main merit of MixER over CNN-based models is the ability to model more entity-relation interactions. Convolutional filters can only model interactions between the boundary regions of entity embedding and relation embedding in the case of concatenation. Otherwise, for given stacking, even though filters can grab more interactions, the captured information is localized. However, the size of the convolutional kernel also heavily impacts the result. MixER, on the other hand, is able to automatically capture interactions in multiple regions through token- and channel-mixing MLPs. These MLPs encourage within- and between-patch communications. Therefore, the extracted features by MixER are more generalized than those of CNN-based models. The overview of our model's operation is depicted in Fig. 2.

Moreover, our proposal also regards a simple superposition of MLP-layers as a hidden attention mechanism, which prevents it from focusing on areas that are not effective for predictions and are updated during the learning phase. Since the linear transformation and the reshaping process rapidly grow the number of parameters, certain superfluous elements do not make valuable contributions to the final output. With this observation, it can be inferred that MixER implicitly distills more meaningful interactions, enhancing its performance subsequently.

3.3 Model Architecture

In this section, we describe our proposed model and its components in detail. The overall framework is depicted in Fig. 3. First, MixER learns low-dimensional continuous vectors (i.e., $\mathbf{e}^s \in \mathbb{R}^{d_e}$, $\mathbf{e}^r \in \mathbb{R}^{d_r}$) to represent entities and relations in the KG through the embedding layer. With the expectation of more expressive representations of original data, we utilize a linear transformation matrix $\mathbf{W} \in \mathbb{R}^{d\times(H\times W)}$ ($d = \{d_e, d_r\}$) and a translational vector $\mathbf{b} \in \mathbb{R}^{(H\times W)}$ to project embeddings onto a new higher-dimensional continuous space. In that, H and W are the height and width of a reshaped matrix. The new embeddings of the subject entity and relation are subsequently reshaped and stacked before being divided into patches. These patches are transposed twice in every Mixer layer to capture spatial and channel information in embeddings. The final linear transformation layer receives the feature output vector from the previous layer to predict the object and then performs entity matrix multiplication on the predicted object to give a score for the triple (s, r, o).

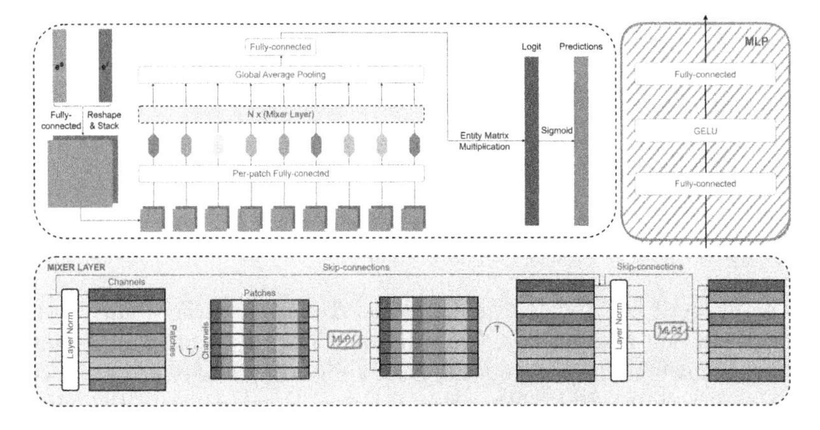

Fig. 3. An overview of the MixER framework. MixER consists of a transformation module, a series of Mixer layers, a global average pooling, and a finally densely-connected layer. The top-left part is the main architecture. Components in each Mixer layer are represented on the bottom part. Each MLP block includes two fully-connected layers and nonlinearity, as the top-right frame shows. Given an entity embedding $\mathbf{e}^s$ and a relation embedding $\mathbf{e}^r$, MixER performs a linear transformation, and a reshape operation $\varphi(\cdot)$ over these embeddings. MLP-Mixer is employed to divide patches and extract information through N Mixer layers. Then the generated output is input into a fully-connected layer which predicts the object embedding $\mathbf{e}^o$.

Embedding Layer. We represent both entities s, o, and relations r as embedding vectors $\mathbf{e}^s, \mathbf{e}^o \in \left\{\mathbf{e}^e_1, \mathbf{e}^e_2, ..., \mathbf{e}^e_{n_e} | \mathbf{e}^e_i \in \mathbb{R}^{d_e}\right\}$ and $\mathbf{e}^r \in \left\{\mathbf{e}^r_1, \mathbf{e}^r_2, ..., \mathbf{e}^r_{n_r} | \mathbf{e}^r_i \in \mathbb{R}^{d_r}\right\}$ respectively. Then, the embeddings of the subject and relation $\mathbf{e}^s, \mathbf{e}^r$ are projected onto a higher dimensional embedding space in the next layer.

Linear Transformation. This layer aims to create higher dimensional vectors for entities and relations, i.e., $\mathbf{e}' \in \mathbb{R}^{(H \times W)}$, $(H \times W) > d_e$. A reshaping function $\varphi : \mathbb{R}^{(H \times W)} \to \mathbb{R}^{H \times W}$ transforms these vectors into 2D square matrices (i.e., $H = W$) before stacking. This process can be formalized as following equations:

$$\phi_s \left[\varphi(\mathbf{e}^s), \varphi(\mathbf{e}^r)\right] \tag{1}$$

$$\varphi(\mathbf{e}^s) = \mathbf{e}^s \mathbf{W}_s + \mathbf{b}_s \tag{2}$$

$$\varphi(\mathbf{e}^r) = \mathbf{e}^r \mathbf{W}_r + \mathbf{b}_r \tag{3}$$

where $\phi_s[\cdot]$ represents a stack operator, $\mathbf{W}_s, \mathbf{W}_r \in \mathbb{R}^{d \times (H \times W)}$ and $\mathbf{b}_s, \mathbf{b}_r \in \mathbb{R}^{(H \times W)}$ are the linear transformation parameters for the subject entity and relation, respectively. To overcome the CNN-based models' limitation and encourage more interactions between channels, we stack two 2D matrices instead of concatenating them. The impact of a linear transformation on a vector can increase the importance of contributive elements. In the training process, the parameters in linear matrices $\mathbf{W}_*$ and translation vectors $\mathbf{b}_*$ are learned so that they can better encapsulate the original data points. As a result, the MixER can utilize the data more effectively. The entity-relation matrices are then processed using per- and cross-location operations in Mixer layers with the purpose of extracting different features for predictions.

Patch Division. A patch is a small resolution region $(P \times P)$ of an image or a 2D matrix in our case. The 2D stacked matrix is divided into S non-overlapping patches, and S must be a positive integer. The i-th patch is then linearly projected to become a C-dimensional vector $\mathbf{p}_i$ by the i-th per-patch transformation layer. As a result, we obtain a two-dimensional table $\mathbf{X} \in \mathbb{R}^{S \times C}$. The number of patches is calculated as follows:

$$S = \frac{H \times W}{P^2} \tag{4}$$

All patch vectors $\{\mathbf{p}_1, \mathbf{p}_2, ..., \mathbf{p}_S\}$ are fed into N Mixer layers including the token-mixing and channel-mixing MLPs. The number of patches can be changed with respect to the hyperparameter P.

Mixer Layer. The original MLP-Mixer model comprises two main layers:

- Token-mixing layer: receptive fields on the per-channel 2D matrix are weighted by cross-location operations. This layer operates on columns of $\mathbf{X}$ by an MLP, which shares parameters across all columns. It can be formulated as a mapping function: $\mathbb{R}^S \to \mathbb{R}^S$.
- Channel-mixing layer: this layer purports to improve interactions between two embeddings, which neural network-based approaches such as ConvE, and ConvKB do. Contrary to the token-mixing MLP, the channel-mixing MLP acts on rows of $\mathbf{X}$, and it also shares parameters across all rows. It can be formulated as a mapping function: $\mathbb{R}^C \to \mathbb{R}^C$.

Each Mixer layer includes two MLP blocks. In each block, there are two fully-connected layers and a non-linear activation layer between them. To summarize, Mixer layers can be formalized as follows:

$$\mathbf{U}_{*,i} = \mathbf{X}_{*,i} + \mathbf{W}_2 g(\mathbf{W}_1 LayerNorm(\mathbf{X})_{*,i}),\ \forall i = 1, 2, ..., C \tag{5}$$

$$\mathbf{Y}_{j,*} = \mathbf{U}_{j,*} + \mathbf{W}_4 g(\mathbf{W}_3 LayerNorm(\mathbf{U})_{j,*}),\ \forall j = 1, 2, ..., S \tag{6}$$

where $g(\cdot)$ is the GELU activation function, $\mathbf{W}_1, \mathbf{W}_2 \in \mathbb{R}^{C \times D_S}$ and $\mathbf{W}_3, \mathbf{W}_4 \in \mathbb{R}^{S \times D_C}$ are the parameters of token-mixing and channel-mixing layers respectively. D_S, D_C are the number of hidden perceptions. In order to avoid the overfitting problem, the last fully-connected layer in the MLP is sandwiched between two added dropout layers.

Scoring Function. The output of the n-th Mixer layer is mapped into the entity space by the final fully-connected layer and then multiplied with the candidate object matrix $\mathbb{R}^{n_e \times d_e}$. In other words, for a given triple, we use the scoring function to calculate candidate triples' scores. Formally, our scoring function is defined as follows:

$$\psi(s, r, o) = f(Mixer(\phi_s[\varphi(\mathbf{e}^s), \varphi(\mathbf{e}^r)])\mathbf{W})\mathbf{e}^o \tag{7}$$

where $Mixer(\cdot)$ denotes the modified MLP-Mixer network, $\phi_s[\cdot]$ represents vector stacking operator, $\mathbf{e}^o$ is the object entity embedding vector and $\mathbf{W}$ is a linear transformation matrix used to project the result of $Mixer(\cdot)$ onto entity embedding space $\mathbb{R}^{d_e}$. The function $f(\cdot)$ is a non-linear activation function.

Loss Function. We use the standard binary cross entropy to train our model in combination with the Adam optimizer and label smoothing, as suggested by ConvE.

$$\mathcal{L} = -\frac{1}{n_e} \sum_{o \in \mathcal{E}} (y_o \log(p_o) + (1 - y_o) \log(1 - p_o)) \tag{8}$$

where y_o denotes a binary label $\{0, 1\}$ with 0 for an invalid triple and 1 for a valid triple, and p_o denotes the predicted score by Eq. (7) after applying a sigmoid function, meaning $p_o = \sigma(\psi(s, r, o))$.

4 Experiments

4.1 Datasets

To verify our model, we use three datasets namely WN18RR [6], FB15k-237 [5], and YAGO3-10 [2]. The statistics of these datasets are represented in Table 2.

Table 2. Statistics of three benchmarks.

Dataset	$\lvert\mathcal{E}\rvert$	$\lvert\mathcal{R}\rvert$	Train	Valid	Test
WN18RR	40943	11	86835	3034	3134
FB15k-237	14541	237	272115	17535	20466
YAGO3-10	123182	37	1079040	5000	5000

4.2 Evaluation Protocol and Metric

We employ an uniform sampling method to generate negative triples in the test set $\mathcal{G}' = \{(s', r, o)|(s' \in \mathcal{E} \setminus s) \wedge (r \in \mathcal{R})\} \cup \{(s, r, o')|(o' \in \mathcal{E} \setminus o) \wedge (r \in \mathcal{R})\}$. Next, the scores for these corrupted triples are arranged in descending order. Our evaluation protocol follows the filter setting where the sampled triples already existing in the original KG dataset are filtered out. Then, the ranks of the correct triples are recorded to calculate the evaluation measures in our experiments. We use Mean Ranking (MR), Mean Reciprocal Ranking (MRR), and Hits@K with $K \in \{1, 3, 10\}$ as the main evaluation metrics [7].

Table 3. Results on FB15k-237 and WN18RR. The best results are highlighted in bold, while the second best results are underlined.

Model	FB15k-237					WN18RR				
	MRR	MR	H@1	H@3	H@10	MRR	MR	H@1	H@3	H@10
TransE	**.334**	323	<u>.238</u>	**.371**	.465	.226	<u>2300</u>	.042	.406	.501
DistMult	.241	254	.155	.263	.419	.241	5100	.390	.440	.490
ComplEx	.247	339	.151	.275	.428	.440	5261	.410	.460	.510
R-GCN	.249	–	.151	.264	.417	–	–	–	–	–
ConvE	.317	<u>244</u>	.237	.356	.501	.430	4187	.400	.440	.520
KBGAN	.277	–	–	–	.458	.215	–	–	–	.469
A2N	.317	–	.232	.348	.486	.450	–	.400	.440	.520
KMAE	<u>.326</u>	235	.240	.358	<u>.502</u>	<u>.448</u>	4441	<u>.415</u>	<u>.465</u>	<u>.524</u>
DMACM	.270	<u>244</u>	–	–	.440	.230	**552**	-	-	**.540**
MixER	**.334**	**214**	**.244**	<u>.366</u>	**.510**	**.457**	5842	**.426**	**.473**	.518

Table 4. Results on YAGO3-10. Due to the absence of reported scores in this dataset, MixER is only evaluated against three baselines.

Model	YAGO3-10				
	MRR	MR	H@1	H@3	H@10
DistMult	.340	5926	.240	.380	.540
ComplEx	.360	6351	.260	–	.550
ConvE	<u>.440</u>	**1671**	<u>.350</u>	<u>.490</u>	<u>.620</u>
MixER	**.527**	<u>3617</u>	**.442**	**.577**	**.676**

4.3 Hyperparameters and Baselines

Hyperparameters are set on the three training, validation, and test datasets as follows: embedding space $d \in \{200, 400\}$, number of negative samples $n \in \{500, 1000, 4000\}$, batch size $b \in \{128, 256, 512\}$, number of epochs $e = 500$, learning rate $\eta \in \{0.01, 0.001, 0.0001\}$, hidden dropout $h \in \{0.2, 0.5\}$, MLP dropout $m \in \{0.2, 0.25, 0.4\}$, and label smoothing $l \in \{0, 0.1\}$. For better optimization, we use the Adam optimizer with L_2 norm for regularization. In the case of MLP-Mixer, we set hidden space $C \in \{32, 64, 128\}$, token-mixing MLP dimension $D_s = 4C$ and channel-mixing MLP dimension $D_C = 0.5C$, the number of Mixer layers $N \in \{4, 6, 8, 12\}$, the 2D reshaped matrix size $H = W \in \{32, 64, 128\}$, the patch resolution $P \in \{8, 16, 32, 64\}$.

In our experiment, we compare MixER against various baselines, including TransE [7], DistMult [15], ComplEx [16], R-GCN [17], ConvE [6], KBGAN [18], A2N [19], KMAE [20], and DMACM [14].

4.4 Results and Discussion

Performance Comparison. Tables 3 and 4 show the result of our method on the LP task compared with the others. All results of baselines are from original papers. In general, our model achieves the best scores across all three datasets except Hits@3 on FB15k-237, MR and Hits@10 on WN18RR, and MR on YAGO3-10. One noticeable disadvantage of the MR metric is that almost NN-based models are significantly worse than the translational-based models, showing the inefficiency of this metric. In contrast, MixER obtains the highest MR score on FB15k-237.

Because MixER is an NN-based model, we initially emphasize the result with the recent proposal KMAE [20] to verify the model's effectiveness. It reveals that MixER outperforms KMAE on the FB15k-237 dataset, three out of five metrics for WN18RR. To summarize, MixER offers increases of 2%, 2% in MRR, 1.6%, 2.6% in Hits@1, and 2.2%, 1.7% in Hits@10 on FB15k-237, WN18RR respectively. The results explain the hypothesis that the more interactions between head and embeddings, the better the result returns. In addition, our model embraces the capacity to enrich interactions by using token-mixing and channel-mixing, which are MixER's cores.

4.5 Analysis

Hyperparameter Analysis. To further understand the effect of the patch sizes $\{16, 32, 64\}$ on LP. Figure 4 shows that with the patch size of 16 on FB15k-237, 32 on WN18RR, and 32 on YAGO3-10, MRR is slightly higher than other sizes. It can also be deduced that the optimal patch size for capturing various entity-relation interactions may differ based on the dataset used. Notably, in the final stages of the iterative process, the points of convergence for the three distinct patch sizes on the WN18RR dataset are almost equivalent.

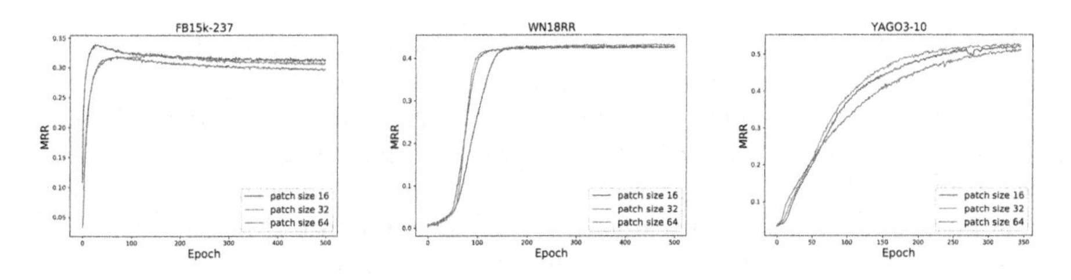

Fig. 4. MRR across epochs in FB15k-237, WN18RR, and YAGO3-10.

Interaction Analysis. In order to verify its performance, we examine our model on each relation type on the WN18RR dataset. Table 5 reports the values of MRR and Hits@10 metrics used to evaluate MixER, ConvE, and ComplEx across each relation on WN18RR. MixER performs prominently with the best MRR and Hits@10 scores on all relations of type S and type C. For instance, *hypernym* and *has_part* are transitive relations that MixER performs well on. Compared to ConvE and ComplEx on relation type R, despite our model obtaining three per four best MRR scores, Hits@10 is slightly lower. According to the hierarchical order of the relation type R > S > C [21], it can infer that our model outperforms both complex and simple relations.

Table 5. MRR and H@10 metrics were evaluated on each relation on WN18RR. The complexity of relation types is ordered as R > S > C, where R, S, and C are short for highly related, generalized specialization, and generalized context-shift respectively.

Relation	Type	#Triple	ConvE		ComplEx		MixER	
			MRR	H@10	MRR	H@10	MRR	H@10
verb_group	R	39	.956	.974	.647	**.987**	**.974**	.974
derivationally_related_form	R	1074	.947	.965	.900	**.969**	**.955**	.967
also_see	R	56	**.667**	.705	.499	**.741**	.656	.679
similar_to	R	3	**1**	**1**	.496	**1**	**1**	**1**
instance_hypernym	S	122	.342	.512	.194	.328	**.363**	**.516**
hypernym	S	1251	.085	**.181**	.113	.177	**.119**	**.181**
synset_domain_topic_of	C	114	.301	.443	.200	.316	**.333**	**.452**
member_of_domain_usage	C	24	.293	.417	.281	.396	**.329**	**.521**
member_of_domain_region	C	26	.340	.423	.247	.327	**.371**	**.500**
member_meronym	C	253	.170	.322	.099	.188	**.204**	**.370**
has_part	C	172	.134	.247	.105	.212	**.167**	**.291**

5 Conclusion and Future Work

In this work, we introduce a novel KGE model named MixER, which can extract more meaningful interactions between embeddings. MixER is a scalable model regarding information integration ability since it can combine with much additional information in the form of a 2D matrix without significantly altering the model's framework. Experiments show that MixER achieves consistent improvements compared to a range of baselines on benchmark datasets.

Nevertheless, despite better capturing interactions, the parameter efficiency is not considered, which can demand substantial computational resources. Therefore, in the future, we plan to research MLP-Mixer's variants further, reduce the number of parameters, investigate the reciprocal effect of head entities and relations, and integrate more meaningful information, such as neighbor-related information.

Acknowledgement. This research is supported by the research funding from the Faculty of Information Technology, University of Science, Ho Chi Minh city, Vietnam.

References

1. Miller, G.A.: WordNet: a lexical database for English. Commun. ACM (1995)
2. Suchanek, F.M., Kasneci, G., Weikum, G.: Yago: a core of semantic knowledge. In: ACM (2007)
3. Bollacker, K., Evans, C., Paritosh, P., Sturge, T., Taylor, J.: Freebase: a collaboratively created graph database for structuring human knowledge. In: ACM (2008)
4. Tolstikhin, I.O., Houlsby, N., et al.: MLP-Mixer: an all-MLP architecture for vision. In: NeurIPS (2021)
5. Toutanova, K., Chen, D.: Observed versus latent features for knowledge base and text inference. In: ACL (2015)
6. Dettmers, T., Minervini, P., Stenetorp, P., Riedel, S.: Convolutional 2D knowledge graph embeddings. In: AAAI (2018)
7. Bordes, A., Usunier, N., Garcia-Duran, A., Weston, J., Yakhnenko, O.: Translating embeddings for modeling multi-relational data. In: NIPS (2013)
8. Wang, Z., Zhang, J., Feng, J., Chen, Z.: Knowledge graph embedding by translating on hyperplanes. In: AAAI (2014)
9. Ji, G., He, S., Xu, L., Liu, K., Zhao, J.: Knowledge graph embedding via dynamic mapping matrix. In: ACL (2015)
10. Sun, Z., Deng, Z.H., Nie, J.Y., Tang, J.: RotatE: knowledge graph embedding by relational rotation in complex space. In: ICLR (2019)
11. Le, T., Huynh, N., Le, B.: Link prediction on knowledge graph by rotation embedding on the hyperplane in the complex vector space. In: Farkaš, I., Masulli, P., Otte, S., Wermter, S. (eds.) ICANN 2021. LNCS, vol. 12893, pp. 164–175. Springer, Cham (2021). https://doi.org/10.1007/978-3-030-86365-4_14
12. Nguyen, D.Q., Nguyen, T.D., Nguyen, D.Q., Phung, D.: A novel embedding model for knowledge base completion based on convolutional neural network. In: NAACL-HLT (2018)
13. Vu, T., Nguyen, T.D., Nguyen, D.Q., Phung, D.: A capsule network-based embedding model for knowledge graph completion and search personalization. In: ACL (2019)
14. Huang, J., Zhang, T., Zhu, J., Yu, W., Tang, Y., He, Y.: A deep embedding model for knowledge graph completion based on attention mechanism. In: NCA* (2021)
15. Yang, B., Yih, W.T., He, X., Gao, J., Deng, L.: Embedding entities and relations for learning and inference in knowledge bases. In: ICLR (2015)
16. Trouillon, T., Welbl, J., Riedel, S., Gaussier, É, Bouchard, G.: Complex embeddings for simple link prediction. In: PMLR (2016)

17. Schlichtkrull, M., Kipf, T.N., Bloem, P., van den Berg, R., Titov, I., Welling, M.: Modeling relational data with graph convolutional networks. In: Gangemi, A., et al. (eds.) ESWC 2018. LNCS, vol. 10843, pp. 593–607. Springer, Cham (2018). https://doi.org/10.1007/978-3-319-93417-4_38
18. Cai, L., Wang, W.Y.: KBGAN: adversarial learning for knowledge graph embeddings. In: ACL (2018)
19. Bansal, T., Juan, D.C., Ravi, S., McCallum, A.: A2N: attending to neighbors for knowledge graph inference. In: ACL (2019)
20. Jiang, D., Wang, R., Yang, J., Xue, L.: Kernel multi-attention neural network for knowledge graph embedding. In: KBS (2021)
21. Allen, C., Balažević, I., Hospedales, T.: Interpreting knowledge graph relation representation from word embeddings. In: ICLR (2021)
22. Nickel, M., Tresp, V., Kriegel, H.P.: A three-way model for collective learning on multi-relational data. In: ICML (2011)
23. Nickel, M., Tresp, V., Kriegel, H.P.: A comprehensive overview of knowledge graph completion. In: KBS (2022)
24. Hogan, A., et al.: Knowledge graphs. In: ACM Computing Surveys (2021)

GTEA: Inductive Representation Learning on Temporal Interaction Graphs via Temporal Edge Aggregation

Siyue Xie[1(✉)], Yiming Li[1], Da Sun Handason Tam[1], Xiaxin Liu[2], Qiufang Ying[2], Wing Cheong Lau[1], Dah Ming Chiu[1], and Shouzhi Chen[2]

[1] The Chinese University of Hong Kong, Hong Kong, China
xs019@ie.cuhk.edu.hk
[2] Tencent Technology Co. Ltd., Shenzhen, China

Abstract. In this paper, we propose the Graph Temporal Edge Aggregation (GTEA) framework for inductive learning on Temporal Interaction Graphs (TIGs). Different from previous works, GTEA models the temporal dynamics of interaction sequences in the continuous-time space and simultaneously takes advantage of both rich node and edge/ interaction attributes in the graph. Concretely, we integrate a sequence model with a time encoder to learn pairwise interactional dynamics between two adjacent nodes. This helps capture complex temporal interactional patterns of a node pair along the history, which generates edge embeddings that can be fed into a GNN backbone. By aggregating features of neighboring nodes and the corresponding edge embeddings, GTEA jointly learns both topological and temporal dependencies of a TIG. In addition, a sparsity-inducing self-attention scheme is incorporated for neighbor aggregation, which highlights more important neighbors and suppresses trivial noises for GTEA. By jointly optimizing the sequence model and the GNN backbone, GTEA learns more comprehensive node representations capturing both temporal and graph structural characteristics. Extensive experiments on five large-scale real-world datasets demonstrate the superiority of GTEA over other inductive models.

Keywords: Edge Embedding · Graph Neural Networks · Self-attention · Temporal Dynamics Modeling · Temporal Interaction Graphs

1 Introduction

Representation learning on temporal graphs is a hot topic in the community of graph learning, where researchers have devoted to mining temporal correlations from graphs and achieve great successes across different domains [6,10,15,31]. However, many methods [14,30] only work for a fixed topology (transductive settings), while in product scenarios, a temporal graph usually evolves as new

S. Xie and Y. Li—Equal contributions.

H. Kashima et al. (Eds.): PAKDD 2023, LNAI 13936, pp. 28–39, 2023.
https://doi.org/10.1007/978-3-031-33377-4_3

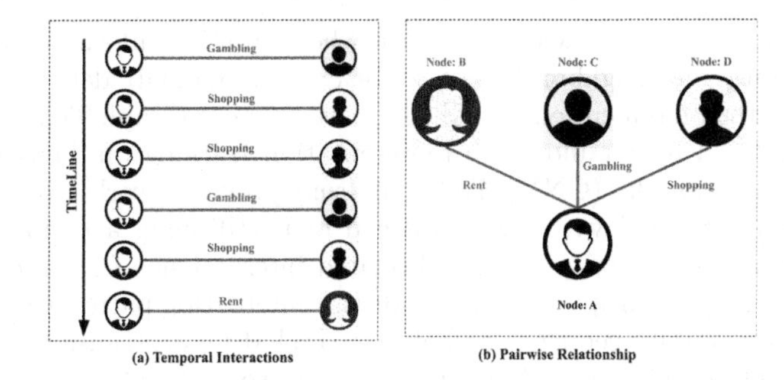

Fig. 1. Motivations of capturing pairwise relationships: there are different kind of interaction between nodes in TIG, e.g., node A behaves normally with node B and D, while conducting (illicit) gambling activities (what we are interested in) with node C.

nodes/ edges added, which requires a model to work inductively. Therefore, in this work, we focus on the inductive learning on Temporal Interaction Graphs (TIGs), where each edge includes all interaction records between two nodes over the history. Applications on TIGs are common in real-world environments, such as recommendation systems [11], social network analytics [9,18], etc.

Although researchers have made substantial advances in processing temporal graphs, it is still challenging to learn discriminative and fine-grained embeddings for TIGs. Previous works commonly preprocess a temporal graph by compressing time-related records within a regular time interval, which yields a spatial-temporal graph with multiple snapshots. However, interactions/ events in TIGs usually occur irregularly along time. Such snapshots are coarse approximations of temporal interactions and resulting in great losses of fine-grained temporal patterns, which prevents spatial-temporal methods [10,21,28] from being generalized to TIGs. Some works [14,27] circumvent such a drawback by grouping all interactions associated with a node to form a consecutive time series for temporal dynamics analyses. Although it preserves the time granularity of interactions, it mixes the temporal behaviors of different neighbors of a node and sometimes obfuscates some explicit temporal relationships between two nodes. Instead, modeling the interaction dynamics between a pair of adjacent nodes can be more helpful to capture temporal relation patterns from TIGs. As the example in Fig. 1, a gambler involves in many interactions, but abnormal behaviors can be readily captured by the pairwise interaction dynamics between node A and C, which in turn, helps identify the roles or illicit activities of these nodes.

Another drawback of previous works is that the edge information is usually underestimated or even ignored for graph learning. However, one should expect that edges carry rich interactional information of TIGs, which can be instrumental in the learning process. Following this intuition, some works [2,20] take edge information into account by concatenating both node and edge features for neighborhood aggregations. Others try to distinguish important connections by introducing dense attention mechanism [19,25,29]. However, naive feature concatenations can be inferior for learning. Dense attention inevitably introduces noises during aggregation, which may overwhelm critical information in some tasks where only few neighbors are of interests (e.g., anomaly detection).

To handle the aforementioned challenges, we propose **G**raph **T**emporal **E**dge **A**ggregation (GTEA) for inductive representation learning on TIGs based on Graph Neural Networks (GNN). Different from previous works, we present a new perspective to deal with TIGs. Instead of partitioning a temporal graph into multiple snapshots or grouping all related interactions of a target node to form a time series, we propose to mine pairwise interaction patterns from edges. Specifically, we adapt a sequence model (e.g., LSTM [5] or Transformer [24]) to mine the temporal interaction dynamics between two adjacent nodes. This helps capture complex interactional patterns of a node pair over the history. In addition, we integrate a time-encoding scheme [13] with the sequence model, enabling GTEA to learn continuous and irregular time patterns for interaction events. To jointly learn topological dependencies and temporal dynamics, we utilize a GNN to capture the relationships among nodes, where embeddings outputted by the sequence model are taken as edge features and incorporated into the neighborhood aggregation process. Furthermore, we adapt a sparsity-inducing attention mechanism to augment the aggregation, which refines neighborhood information by filtering out noises raised by unimportant neighbors. By training GTEA in an end-to-end manner, all modules can be jointly optimized, which yields discriminative node representations for downstream tasks. Extensive experiments are conducted on five real-world datasets across different tasks, where results demonstrate the superiority of GTEA over other inductive models. The contributions of our work are summarized as follows:

- We present a novel perspective for modeling TIGs, which helps capture fine-grained interaction patterns from node pairs.
- We propose a general framework, GTEA, for inductive learning on TIGs, which yields discriminative node representations for downstream tasks.
- We conduct extensive experiments on five large-scale datasets. Experimental results demonstrate the great effectiveness of GTEA.

2 Related Works

2.1 Temporal Dynamics Modeling on Graph-Structured Data

Temporal graphs are ubiquitous in real-world scenarios, which motivates researchers to extend the target from learning from static graphs to the temporal domain [14–16]. A common way is to form multiple static graph snapshots to approximate the time-continuous temporal graph [10,21,28]. However, critical details, e.g., the time granularity of interactions in TIGs, may lose in the simplification process. Instead of learning from snapshots, [22,23,31] learn node representations using temporal point process. [9,11,18,30] progressively update node embeddings as a new event/interaction occurs. TGAT [27] adopts a time-encoder with self-attention to aggregate interactions of neighbors. However, many of above models only work for transductive tasks, which restricts their generalization ability. Node embeddings updated event-by-event may be biased by noises and lose focus on the information of interest, e.g., the rare gambling activities as shown in Fig. 1. Although TGAT adopts attention on different interactions, it lumps all and may fail to distinguish the interactions of the same instance, e.g., in Fig. 1, there are two different gambling interactions but both are with node C. Different from previous methods, GTEA attempts to inductively model the temporal dynamics by looking at the complete interaction history between each pair of adjacent nodes, which enables it to capture specific and fine-grained mutual interaction patterns. We make more discussions and comparisons in Appendix 1 [1]

2.2 Representation Learning on Graphs with Edge Features

Edges are natural reflections of relationships among instances but the information they carried is usually underestimated. Therefore, some pioneers propose to mine from edges to enhance a model, e.g., ECConv [20] attempts to generate a set of edge-specific weights for Graph Convolutional Networks (GCN), while EGNN [2] constructs a weighted graph for each edge features' dimension. EdgeConv [26] and AttE2Vec [1] learns edge features to describe relationships between adjacent nodes. Instead of learning from an edge directly, CensNet [7] converts a original graph into a line graph, where edges are mapped as nodes in the new graph. Motivated by previous works, GTEA is extended to learn from edges in the temporal domain by modeling the mutual interaction dynamics, which improves the model performance and node representation power.

3 Proposed Methods

3.1 Problem Formulation

A **Temporal Interaction Graph** (TIG) is an attributed graph $\mathcal{G} = (\mathcal{V}, \mathcal{E})$ where $\mathcal{V}$ is a set with N nodes and $\mathcal{E}$ a set with M edges. A node u is associated with features $\mathbf{x}_u \in \mathbb{R}^{D_N}$, while an edge between u and v corresponds to a sequence of interaction events, denoted as $\{\mathbf{e}^k_{uv} = (t^k_{uv}, \mathbf{f}^k_{uv}); k = 1, 2, ..., S_{uv}\}$, where t^k_{uv} is the time stamp of the k-th event, $\mathbf{f}^k_{uv} \in \mathbb{R}^{D_E}$ the interaction features and S_{uv} the length of the sequence. Given the interaction history, the goal of GTEA is to learn discriminative node representations, which can be used in downstream tasks such as node classifications or future link predictions.

3.2 Overview of GTEA

The architecture of GTEA is shown in Fig. 2. In TIGs, interaction events occur between two nodes from time to time, which motivates us to mine fine-grained interaction patterns from pairwise events. Targeting on this goal, GTEA utilizes a sequence model to learn the dynamics of pairwise interactions to represent

[1] Appendix can be found in: https://github.com/xslangley/GTEA.

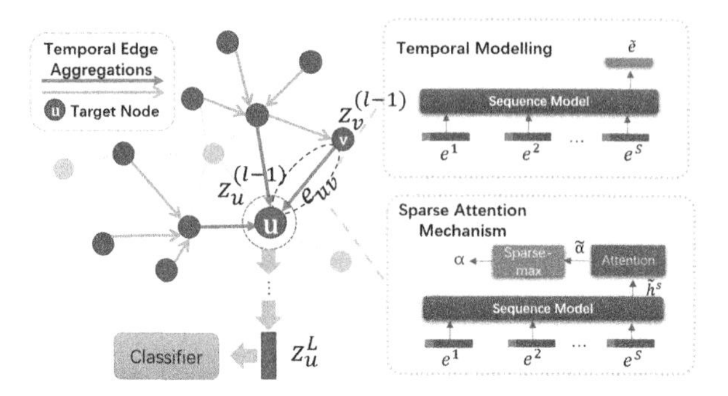

Fig. 2. The framework of GTEA, where a sequence model enhanced by a time-encoder is proposed to learn embeddings for edges. The learned edge embeddings will be aggregated together with node attributes by the GNN backbone with a sparse attention mechanism, which helps yields discriminative node embeddings.

edges. An additional time-encoder is further introduced to capture irregular temporal patterns. Learned edge embeddings are fed into a GNN, along with a sparse-inducing attention mechanism for neighbor aggregations, which jointly captures both topological and time-related dependencies among nodes. With these designs, GTEA is able to yield discriminative representations for TIGs.

3.3 Learning Edge Embeddings for Interaction Sequences

Interaction Dynamics Modeling with Sequence Models. In TIGs, the types of interactions of a node involved can vary greatly with different neighbors. However, interaction patterns of two specific nodes are usually consistent even as time goes. Therefore, it is more reasonable to model the interaction behaviours edge-by-edge instead of mixing all interactions from different neighbors. Given the interaction history $[\mathbf{e}_{uv}^1, ..., \mathbf{e}_{uv}^{S_{uv}}]$ of edge (u, v), we adopt a sequence model $Enc_i(\cdot)$ to learn the interaction dynamics as follows:

$$\tilde{\mathbf{e}}_{uv} = Enc_i([\mathbf{e}_{uv}^1, ..., \mathbf{e}_{uv}^{S_{uv}}]), \tag{1}$$

where $Enc_i(\cdot)$ indicates the interaction encoder and $\tilde{\mathbf{e}}_{uv}$ is the edge embedding to represent the interaction sequence. In our experiments, we implement $Enc_i(\cdot)$ by LSTM [5] and Transformer [24]. In LSTM, we represent $\tilde{\mathbf{e}}_{uv}$ by the hidden output of the last time unit. As for Transformer, an interaction is correlated with all other interactions following self-attention. Therefore, it is sufficient for us to represent $\tilde{\mathbf{e}}_{uv}$ by the embedding with respect to the last interaction. In this way, interactions of the same node pair can be completely reviewed by the sequence model, which helps capture specific interactional patterns for any two connected nodes. More technical details refer to Appendix 2.1.

Edge Feature Enhancement with Time Encoding. Sequence models implicitly assume the time gap between consecutive inputs is regular along the timeline, while interactions happened in TIGs do not follow. Therefore, to capture more complex time-related interaction patterns, we enhance GTEA by integrating $Enc_i(\cdot)$ with a time-encoder, which is adapted from Time2Vec (**T2V**) [13]. Specifically, for any given time t, a time embedding $\tau(t) \in \mathbb{R}^{l+1}$ can be generated through:

$$\tau(t)[i] = \begin{cases} \omega_i t + \varphi_i, & \text{if } i = 0. \\ \cos(\omega_i t + \varphi_i), & \text{if } 1 \leq i \leq l, \end{cases} \tag{2}$$

where ω_i and φ_i are trainable parameters. We append the time embedding to the raw features of each interaction, denoted by $\hat{\mathbf{e}}_{uv}^k = (\tau(t_{uv}^k), \mathbf{f}_{uv}^k)$. The enhanced edge embedding can therefore be formulated as:

$$\tilde{\mathbf{e}}_{uv} = Enc_i([\hat{\mathbf{e}}_{uv}^1, ..., \hat{\mathbf{e}}_{uv}^{S_{uv}}]). \tag{3}$$

The time embedding inherits some good properties of Random Fourier Feartures (RFF) [17], which enables GTEA to capture more fine-grained temporal behaviours, e.g., time periodicity. Detailed analyses refer to Appendix 2.2.

3.4 Representation Learning with Temporal Edge Aggregation

Sparsity-Inducing Attention for Neighbors Filtering. Common GNN models learn mutual relationships by iteratively aggregating neighborhood information of a target node. To distinguish important nodes in aggregations, dense attention [25] is usually applied to calculate an attentive weight for each neighbor. However, in real-world tasks for TIGs, e.g., anomaly detection in mobile payment networks, the target node can interact with a large number of neighbors but only few of them are of interest. In this case, considerable noisy neighbors, who are assigned small but non-zero attentive weights, can overwhelm the few important, which degrades the representative power of the learned node embeddings.

With this concern, we propose a sparse attention strategy, motived by [12], for GTEA to enhance neighbor aggregations. The operations are formulated as:

$$\tilde{\alpha}_{uv} = \mathbf{a}^\intercal \mathbf{h}_{uv}, \quad \mathbf{h}_{uv} = Enc_a([\hat{\mathbf{e}}_{uv}^1, ..., \hat{\mathbf{e}}_{uv}^{S_{uv}}]), \tag{4}$$

$$\boldsymbol{\alpha}_{u:} = \text{Sparse}(\tilde{\boldsymbol{\alpha}}_{u:}), \tag{5}$$

where $Enc_a(\cdot)$ maps the interaction sequence of (u, v) to the hidden space, $\mathbf{a}$ is a trainable weight vector, $\boldsymbol{\alpha}_{u:}$ is the attentive weight vector for all neighbors of node u. In experiments, we align $Enc_a(\cdot)$ with $Enc_i(\cdot)$ but keep independent parameters. Sparse$(\cdot)$ is a sparsification operator (details refer to Appendix 2.3). The key idea of Sparse$(\cdot)$ is to truncate the input by a dynamic threshold. It induces GTEA to learn sparse but normalized attentive weights $\boldsymbol{\alpha}_{u:}$ for neighbors. This forces the model to distinguish the few important neighbors and discard the trivial mass, which refines the information for neighbor aggregations.

Neighbors Aggregation with Temporal Edge Embeddings. With the sparse attention mechanism, neighbors can be selectively aggregated. In addition, we incorporate learned edge embeddings into aggregation. But instead of simply concatenating node and edge embeddings as the input to the aggregator [27,30], we propose a new method to correlate node and edge information:

$$\mathbf{z}_{\mathcal{N}(u)}^{(l)} = \sum_{v \in \mathcal{N}(u)} \alpha_{uv} \mathrm{MLP}_1([\mathbf{z}_v^{(l-1)} || \tilde{\mathbf{e}}_{uv}]), \tag{6}$$

$$\mathbf{z}_u^{(l)} = \mathrm{MLP}_2([\mathbf{z}_u^{(l-1)} || \mathbf{z}_{\mathcal{N}(u)}^{(l)}]), \tag{7}$$

where $\mathbf{z}_u^{(l-1)}$ is the node embedding of u in the $(l-1)$-th layer, $\mathcal{N}(u)$ is the neighbors set of u and $\mathrm{MLP}_1(\cdot)$ and $\mathrm{MLP}_2(\cdot)$ are multi-layer perceptrons (MLP), which enables GTEA to fuse features from different latent spaces. The introduction of edge embeddings forces GTEA to correlate both temporal and topological dependencies in a TIG and therefore learn more comprehensive and discriminative node representations. We prove that Eq. 6 has more powerful representation ability than the naive concatenation operator. Details refer to Appendix 2.4.

3.5 Model Training for Different Graph-Related Tasks

By iteratively stacking L GNN layers, GTEA can generate node embeddings with high-level semantics for downstream tasks. In this work, we focus on two tasks: node classifications and future link predictions. For node classifications, the category probability vector is computed based on:

$$\mathbf{y}_u = \mathrm{Softmax}(\mathrm{MLP}_3(\mathbf{z}_u^{(L)})). \tag{8}$$

For future link predictions, we predict the probability of a future link between u and v by:

$$y_{uv} = \mathrm{Sigmoid}(\mathbf{z}_u^{(L)\intercal} \mathbf{z}_v^{(L)}). \tag{9}$$

GTEA can then be trained in an end-to-end manner by cross-entropy loss. The steps to train GTEA are summarized in Algorithm 2 in the appendix. Different from some previous works [3,30], GTEA does not need to maintain a memory to update embeddings, which enables it to work for real-world inductive tasks.

4 Experiments

4.1 Experimental Setup

Datasets In our experiments, we formulate node classification as a task to identify illicit users/ nodes. We evaluate GTEA on a payment dataset, denoted as Mobile-Pay, which is provided by a major mobile payment provider. We additionally assess GTEA on two Ethereum phishing datasets [2] with different scales, denoted as Phish-L(arge) and Phish-S(mall). For future link predictions, we use Wikipedia and Reddit datasets [9] evaluation. *Note that we align all other*

[2] Raw data: https://www.kaggle.com/xblock/ethereum-phishing-transaction-network.

Table 1. Experimental Results of Node Classifications and Future Link Predictions

Tasks	Node Classifications						Future Link Predictions			
Datasets	Mobile-Pay		Phish-S		Phish-L		Wikipedia		Reddit	
Model	Acc	F1	Acc	F1	Acc	F1	Acc	F1	Acc	F1
GCN	0.7481	0.7480	0.9077	0.9077	0.9298	0.8683	0.6472	0.6259	0.5369	0.4285
GraphSAGE	0.7474	0.7472	0.9405	0.9405	0.9753	0.9569	0.5986	0.5953	0.6424	0.6334
GAT	0.7265	0.7264	0.9405	0.9405	0.9631	0.9375	0.6167	0.5975	0.6396	0.6252
ECConv	0.7399	0.7399	0.9554	0.9559	0.9700	0.9480	0.6426	0.6424	0.6232	0.6219
EGNN	0.7549	0.7538	0.9479	0.9477	0.9659	0.9393	0.6401	0.6259	0.5484	0.4406
$GTEA_{HE}$	0.7519	0.7516	0.9673	0.9673	0.9777	0.9615	0.6169	0.6123	0.6515	0.6495
TGAT	0.7212	0.7212	0.9673	0.9673	0.9740	0.9559	0.7253	0.7256	0.8418	0.8414
$GTEA_{L}$	0.7848	0.7847	0.9836	0.9836	**0.9805**	**0.9668**	0.7988	0.7981	0.8809	0.8807
$GTEA_{L}$+T	**0.7990**	**0.7990**	0.9777	0.9777	0.9789	0.9640	**0.8149**	**0.8145**	0.885	0.8849
$GTEA_{TX}$	0.7676	0.7670	**0.9851**	**0.9851**	0.9801	0.9658	0.7841	0.7832	**0.8865**	**0.8864**
$GTEA_{TX}$+T	0.7758	0.7758	0.9792	0.9792	0.9769	0.9603	0.7869	0.7864	0.8747	0.8746

four datasets' setting with Mobile-Pay to follow the product scenarios. Thus, our results may not be directly compared with the numbers reported in other works. **More details and statistics of dataset splits refer to Appendix 3.1.**

Compared Methods. We compare GTEA with different methods, including GNN baselines (GCN [8], GraphSAGE [4] and GAT [25]), edge-feature-involved methods (ECConv [20] and EGNN [2]) and a state-of-the-art temporal graph learning model (TGAT [27]). We also implement a GTEA variant, denoted as $GTEA_{HE}$, by replacing the learned edge embedding with handcrafted edge features. *Note that we mainly focus on the methods applicable to Mobile-Pay, i.e., can learn the overall status of all past interactions. Therefore, models cannot fit the product environments, e.g., transductive* [14,30] *or event-by-event learning methods* [9,18]*, are not included.* More details refer to Appendix 3.2.

Implementations. We implement two variants of GTEA (regarding the sequence model) by LSTM and Transformer, which are denoted as $GTEA_{L}$ and $GTEA_{TX}$, respectively. We use '+T' to mark variants enhanced by the time encoder. All hyperparameters are tuned through grid-search on the validation set and we report the best **accuracy** and **Macro-F1** on the test set. **More implementation details can be found in Appendix 3.3.**

4.2 Experimental Results of Overall Performance

Node Classifications. Table 1 shows the results for node classifications. We can clearly observe that variants of GTEA consistently outperform all other models. We owe such superiority to the edge embedding module and the joint integration of both topological and temporal information. With these designs, GTEA can model the interaction dynamics for node pairs, and therefore be more effective to capture discriminative behavior patterns of a node. An evidence is that GTEA performs much better in the Mobile-Pay than that of others. This

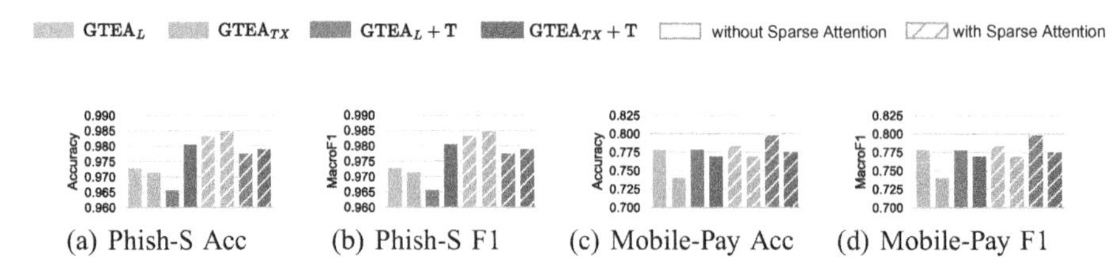

(a) Phish-S Acc (b) Phish-S F1 (c) Mobile-Pay Acc (d) Mobile-Pay F1

Fig. 3. Effects of the sparse attention mechanism on Phish-S and Mobile-Pay datasets.

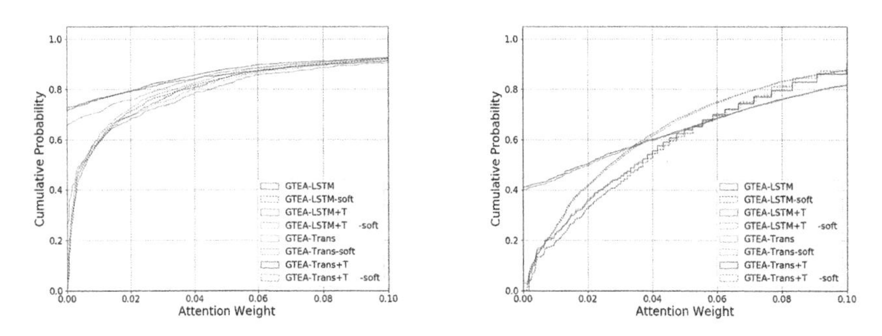

Fig. 4. Distributions of attention weights (with/without Sparse(·)) on Phish-S (left) and Mobile-Pay (right). The variant with dense attention is denote by "soft".

may due to that actions of phishing is mostly naive and instantaneous, while illicit payment interactions are associated with more complex temporal patterns, which can be captured more effectively by temporal modeling. Even though, GTEA still dominates other competitors over all datasets, which demonstrates its effectiveness.

Future Link Predictions Results are shown in the right of Table 1. It can be observed that GTEA achieves the best performance in this task. Note that TGAT and GTEA perform much better than other competitors that do not incorporate temporal information, which shows the importance of temporal modeling. Even though, GTEA still outperforms TGAT by a large margin. This is because TGAT mixes all interactions from different neighbors of a target node in temporal modeling, which is hard to distinguish the interactions from the same instance. In contrast, we adopt a pairwise interaction modeling scheme, which elaborately exploits the relationship patterns for each neighbor. This endows GTEA the power to learn connection features for node embeddings and therefore is more effective for the future link prediction task.

4.3 Experiments Analyses

Effects of Edge Features/Embeddings From Table 1, we observe that models incorporating edge features mostly perform better than those who learn only from node attributes. It is not surprised as edge features carry rich semantics about connections and interactions. Information such as user behaviors and potential relations can be mined from edges. We also notice that GTEA performs much better than other edge-feature-involved models. This is because high-level semantics is encoded into the embedding, which extracts critical information for other modules. The additional MLP module introduced in the aggregation process (Eq. 6) also helps GTEA to align inputs from different feature domains, which enables it to learn discriminative embeddings more efficiently.

Effects of the Temporal Dynamics Modeling An advantage of GTEA is to model the temporal dynamics for TIGs, where Table 1 shows the benefit. Compared with GTEA_{HE}, which substitutes the learned temporal edge embedding by handcrafted edge features, all GTEA variants perform much better. It implies that modeling the temporal dynamics of an interaction sequence is critical for analyzing TIGs. The great performance improvement (over GTEA_{HE}) also demonstrates the effectiveness of the temporal dynamics modeling scheme of GTEA. In addition, we observe that the time encoder works better in GTEA's LSTM variants than transformer variants. We speculate that this is because the function of the time encoder partially overlaps with the position encoder in transformer. Instead, vanilla LSTM doesn't encode the position information and therefore benefit much by introducing the time encoder. However, when it comes to an environment with more complex and diverse temporal behaviors, e.g., the Mobile-Pay dataset, the time-encoder can be more powerful to capture irregular but discriminative patterns, such as different periodicities, for each interaction sequence, which enhances the representation ability of all GTEA variants.

Effects of the Sparse Attention Aggregation. We conduct additional experiments on Phish-S and Mobile-Pay datasets to validate the effectiveness of the sparse attention mechanism. Specifically, we replace the Sparse$(\cdot)$ operator in Eq. 5 by the Softmax function, which generates dense attentive weights. Quantitative results are shown in Fig. 3. In most cases, models with Sparse$(\cdot)$ achieve a better performance. This is reasonable as redundant and noisy signals of irrelevant neighbors are discarded in aggregation, which encourages the model to yield discriminative node embeddings. We additionally visualize the attention weights' distributions of the sparse and dense attention mechanisms, as shown in Fig. 4. With dense attention, neighbors with small attentive weights, e.g., 0.1 in our cases, account for over 80% of all. In contrast, for Sparse$(\cdot)$, around 70% of neighbors are truncated (attentive weights are zeroed) in Phish-S, while 40% in Mobile-Pay. Such quantitative and qualitative results demonstrates that noises are substantially suppressed, which explains the effectiveness of GTEA.

5 Conclusions

In this paper, we propose GTEA for inductive representation learning on Temporal Interaction Graphs (TIGs). Different from previous works, GTEA learns an edge embedding for temporal interactions between each pair of adjacent nodes by adopting an enhanced sequence model. By incorporating the learned edge embeddings into the aggregation of a GNN, which is driven by a sparse attention mechanism, GTEA is encouraged to exploit both temporal and topological dependencies in TIGs. As a general framework, GTEA is evaluated on different graph-related tasks and extensive experimental results show its effectiveness.

Acknowledgements. This research is supported in part by the Innovation and Technology Committee of HKSAR under the project#ITS/244/16, the CUHK MobiTeC R&D Fund and a gift from Tencent.

References

1. Bielak, P., Kajdanowicz, T., Chawla, N.V.: Attre2vec: unsupervised attributed edge representation learning. arXiv preprint arXiv:2012.14727 (2020)
2. Gong, L., Cheng, Q.: Exploiting edge features for graph neural networks. In: Proceedings of the IEEE Conference on Computer Vision and Pattern Recognition, pp. 9211–9219 (2019)
3. Grover, A., Leskovec, J.: node2vec: Scalable feature learning for networks. In: Proceedings of the 22nd ACM SIGKDD International Conference on Knowledge Discovery and Data Mining, pp. 855–864 (2016)
4. Hamilton, W., Ying, Z., Leskovec, J.: Inductive representation learning on large graphs. In: Advances in Neural Information Processing Systems, pp. 1024–1034 (2017)
5. Hochreiter, S., Schmidhuber, J.: Long short-term memory. Neural Comput. **9**(8), 1735–1780 (1997)
6. Huang, S., Bao, Z., Culpepper, J.S., Zhang, B.: Finding temporal influential users over evolving social networks. In: 2019 IEEE 35th International Conference on Data Engineering (ICDE), pp. 398–409. IEEE (2019)
7. Jiang, X., Zhu, R., Li, S., Ji, P.: Co-embedding of nodes and edges with graph neural networks. IEEE Trans. Pattern Anal. Mach. Intell. (2020)
8. Kipf, T.N., Welling, M.: Semi-supervised classification with graph convolutional networks. arXiv preprint arXiv:1609.02907 (2016)
9. Kumar, S., Zhang, X., Leskovec, J.: Predicting dynamic embedding trajectory in temporal interaction networks. In: Proceedings of the 25th ACM SIGKDD International Conference on Knowledge Discovery and Data Mining, pp. 1269–1278 (2019)
10. Li, Y., Yu, R., Shahabi, C., Liu, Y.: Diffusion convolutional recurrent neural network: data-driven traffic forecasting. In: International Conference on Learning Representations (ICLR) (2018)
11. Ma, Y., Guo, Z., Ren, Z., Tang, J., Yin, D.: Streaming graph neural networks. In: Proceedings of the 43rd International ACM SIGIR Conference on Research and Development in Information Retrieval, pp. 719–728 (2020)
12. Martins, A., Astudillo, R.: From softmax to sparsemax: a sparse model of attention and multi-label classification. In: International Conference on Machine Learning, pp. 1614–1623 (2016)
13. Mehran Kazemi, S., et al.: Time2vec: learning a vector representation of time. arXiv preprint arXiv:1907.05321 (2019)
14. Nguyen, G.H., Lee, J.B., Rossi, R.A., Ahmed, N.K., Koh, E., Kim, S.: Continuous-time dynamic network embeddings. In: Companion Proceedings of the Web Conference 2018, pp. 969–976 (2018)
15. Qiu, Z., Hu, W., Wu, J., Liu, W., Du, B., Jia, X.: Temporal network embedding with high-order nonlinear information. In: Proceedings of the AAAI Conference on Artificial Intelligence, vol. 34, pp. 5436–5443 (2020)

16. Qu, L., Zhu, H., Duan, Q., Shi, Y.: Continuous-time link prediction via temporal dependent graph neural network. In: Proceedings of the Web Conference 2020, pp. 3026–3032 (2020)
17. Rahimi, A., Recht, B.: Random features for large-scale kernel machines. In: Advances in Neural Information Processing Systems, pp. 1177–1184 (2008)
18. Rossi, E., Chamberlain, B., Frasca, F., Eynard, D., Monti, F., Bronstein, M.: Temporal graph networks for deep learning on dynamic graphs. arXiv preprint arXiv:2006.10637 (2020)
19. Shi, Y., Huang, Z., Wang, W., Zhong, H., Feng, S., Sun, Y.: Masked label prediction: unified message passing model for semi-supervised classification. arXiv preprint arXiv:2009.03509 (2020)
20. Simonovsky, M., Komodakis, N.: Dynamic edge-conditioned filters in convolutional neural networks on graphs. In: Proceedings of the IEEE Conference on Computer Vision and Pattern Recognition, pp. 3693–3702 (2017)
21. Singer, U., Guy, I., Radinsky, K.: Node embedding over temporal graphs. In: Proceedings of the Twenty-Eighth International Joint Conference on Artificial Intelligence, IJCAI-19, pp. 4605–4612. International Joint Conferences on Artificial Intelligence Organization, July 2019. https://doi.org/10.24963/ijcai.2019/640
22. Trivedi, R., Dai, H., Wang, Y., Song, L.: Know-evolve: Deep temporal reasoning for dynamic knowledge graphs. arXiv preprint arXiv:1705.05742 (2017)
23. Trivedi, R., Farajtabar, M., Biswal, P., Zha, H.: Dyrep: learning representations over dynamic graphs (2018)
24. Vaswani, A., et al.: Attention is all you need. In: Advances in Neural Information Processing Systems, pp. 5998–6008 (2017)
25. Veličković, P., Cucurull, G., Casanova, A., Romero, A., Lio, P., Bengio, Y.: Graph attention networks. arXiv preprint arXiv:1710.10903 (2017)
26. Wang, Y., Sun, Y., Liu, Z., Sarma, S.E., Bronstein, M.M., Solomon, J.M.: Dynamic graph CNN for learning on point clouds. ACM Trans. Graph. (TOG) **38**(5), 1–12 (2019)
27. Xu, D., Ruan, C., Korpeoglu, E., Kumar, S., Achan, K.: Inductive representation learning on temporal graphs. arXiv preprint arXiv:2002.07962 (2020)
28. Yu, B., Yin, H., Zhu, Z.: Spatio-temporal graph convolutional networks: a deep learning framework for traffic forecasting. In: Proceedings of the 27th International Joint Conference on Artificial Intelligence (IJCAI) (2018)
29. Zhang, J., Shi, X., Xie, J., Ma, H., King, I., Yeung, D.Y.: GAAN: gated attention networks for learning on large and spatiotemporal graphs. arXiv preprint arXiv:1803.07294 (2018)
30. Zhang, Z., et al.: Learning temporal interaction graph embedding via coupled memory networks. In: Proceedings of The Web Conference 2020, pp. 3049–3055 (2020)
31. Zuo, Y., Liu, G., Lin, H., Guo, J., Hu, X., Wu, J.: Embedding temporal network via neighborhood formation. In: Proceedings of the 24th ACM SIGKDD International Conference on Knowledge Discovery and Data Mining, pp. 2857–2866 (2018)

You Need to Look Globally: Discovering Representative Topology Structures to Enhance Graph Neural Network

Huaisheng Zhu[1], Xianfeng Tang[2], TianXiang Zhao[1], and Suhang Wang[1](✉)

[1] Pennsylvania State University, University Park, PA 16802, USA
{hvz5312,tkz5084,szw494}@psu.edu
[2] Amazon, Palto Alto, CA, USA
xianft@amazon.com

Abstract. Graph Neural Networks (GNNs) have shown great ability in modeling graph-structured data. However, most current models aggregate information from the local neighborhoods of a node. They may fail to explicitly encode global structure distribution patterns or efficiently model long-range dependencies in the graphs; while global information is very helpful for learning better representations. In particular, local information propagation would become less useful when low-degree nodes have limited neighborhoods, or unlabeled nodes are far away from labeled nodes, which cannot propagate label information to them. Therefore, we propose a new framework GSM-GNN to adaptively combine local and global information to enhance the performance of GNNs. Concretely, it automatically learns representative global topology structures from the graph and stores them in the memory cells, which can be plugged into all existing GNN models to help propagate global information and augment representation learning of GNNs. In addition, these topology structures are expected to contain both feature and graph structure information, and they can represent important and different characteristics of graphs. We conduct experiments on 7 real-world datasets, and the results demonstrate the effectiveness of the proposed framework for node classification.

Keywords: Graph Neural Network · Global · Node Classification

1 Introduction

Over the past few years, Graph Neural Networks (GNNs) [9,11,19] have shown great success in modeling graph data for a wide range of applications such as social networks [17]. GNNs typically follow the message passing mechanism, which aggregates the neighborhood representation of a node to enrich the node's representation. Hence, the learned representations capture both local topology information and node attributes, which benefits various tasks [11].

Supplementary Information The online version contains supplementary material available at https://doi.org/10.1007/978-3-031-33377-4_4.

H. Kashima et al. (Eds.): PAKDD 2023, LNAI 13936, pp. 40–52, 2023.
https://doi.org/10.1007/978-3-031-33377-4_4

Despite the success of GNNs in modeling graphs, most of them can only help nodes aggregate local neighbors' information. *First*, long-range or global information can be used to learn better representations. For example, two structurally similar nodes can offer strong predictive power to each other but might be very distant from each other [6]. *Second*, in node classification tasks, we only have partially labeled nodes on graphs. Nodes are often sparsely labeled as it is time-consuming, expensive and sometimes requires domain knowledge to label. In this case, labeled nodes may only propagate their label information to their local neighbors based on local aggregation, which may result in the misclassification of nodes distant from labeled nodes [5]. Therefore, it is important to design GNNs to capture global information and long-range dependencies. Several works [1,12] about aggregating node information from a wider range has been proposed to improve the expressive power of GNNs. However, methods of aggregating information from a wider range cannot explicitly distinguish relevant nodes from lots of distant neighborhoods, which will also result in over-smoothing issues. Thus, how to capture global information needs further investigation.

In real-world graphs, for each class, there are usually some representative ego-graphs. For each ego-graph, it contains one central node and its neighbors from the original graph together with their edge relations, which would be helpful to provide global information about each class in the graph. For example, for malicious account detection, one representative pattern for malicious accounts is that they tend to connect to each other and also try to connect to benign accounts to pretend that they are benign accounts; similarly, benign accounts also have several representative graph patterns. Therefore, it's important to extract and use global representative ego-graphs to improve the performance of GNN models. Though promising, the work on extracting global patterns to facilitate GNN representation learning is rather limited [20]. However, MemGCN [20] only learns global feature information but loses graph structure information.

Therefore, in this paper, we study a novel problem of learning global representative patterns to improve the performance of GNNs. There are several critical challenges: (i) how can we efficiently extract both different structures and features as global information automatically? (ii) how can we make all nodes or even nodes with low-degree to find and utilize highly relevant global information? (iii) how can we use these extracted ego-graphs to improve current GNN models? To fill this gap, we propose a novel framework Graph Structure Memory Enhanced Graph Neural Network, GSM-GNN. It utilizes a clustering algorithm to select representative ego-graphs from the original graph and they are stored in memories. Then, query vectors are generated by preserving topology and node feature information, and they can be used to find relevant global information. Finally, based on query vectors, relevant global information from stored ego-graphs is obtained, and neighborhood patterns about the representative graph structure are also used to augment current GNNs. The main contributions are as follows:

- We study a problem with using global patterns in the graph to improve the performance of GNNs based on local aggregation.
- We develop a novel framework that extends current GNNs with global information. The adoption of memories learns and propagates both global feature and structure information to enrich the representation of nodes.

– Experimental results on seven real-world datasets demonstrate the effectiveness of the proposed approach.

2 Related Works

Graph Neural Networks. Graph Neural Networks have shown great success for various applications such as social networks [11,17]. Generally, existing GNNs can be categorized into two categories, i.e., spectral-based [3,11] and spatial-based [1,9,19]. Spectral-based approaches are defined according to graph signal processing [3]. A first-order approximation is utilized to simplify the graph convolution via GCN [11]. Spatial-based GNN models aggregate information of the neighbor nodes [9]. Despite differences between spectral-based and spatial-based approaches, most GNN variants can be summarized with the message-passing framework [7]. The high-level idea of the message passing mechanism is to propagate the information of nodes through the graph based on pattern extraction and interaction modeling within each layer. However, most of these works only utilize local neighbors' information. Thus, many works about utilizing global information or high-order neighbors for GNNs [1,12] were proposed. There is still little work on using global ego-graph patterns for node classification tasks.

Memory Augmented Neural Networks. Memory Augmented Neural Networks use the memory mechanism with differentiable reading operations to store past experiences and show advantages in many applications [20]. Their implementations of memory on different tasks are inspired by key-value memory [16] and array-structured memory [8]. Also, there have been several works on GNNs that utilized memory-based design for different tasks recently [4,20]. For node classification, memory nodes are introduced in [20] to store global characteristics of nodes, which can learn high-order neighbors' information in the message passing process. In summary, their memory mechanisms try to store important node feature information to improve models' performance. However, there are also important global graph structure patterns like ego-graphs with node feature and their edge relationships. Unlike the aforementioned approaches, our proposed GSM-GNN can learn global ego-graph patterns.

3 Problem Formulation

We use $\mathcal{G} = (\mathcal{V}, \mathcal{E}, \mathbf{X})$ to denote an attributed graph, where $\mathcal{V} = \{v_1, \ldots, v_N\}$ is the set of N nodes, $\mathcal{E}$ is the set of edges and $\mathbf{X}$ is the attribute matrix for nodes in $\mathcal{G}$. The i-th row of $\mathbf{X}$, i.e., $\mathbf{x}_i \in \mathbb{R}^{1\times d_0}$, is the d_0 dimensional features of node v_i. $\mathbf{A} \in \mathbb{R}^{N\times N}$ is the adjacency matrix. $A_{ij} = 1$ if node v_i and node v_j are connected; otherwise $A_{ij} = 0$. We denote a k-hop ego-graph centered at node v_i as $g_i(\mathcal{V}_{g_i}, \mathbf{A}_{g_i})$, where $\mathcal{V}_{g_i}$ include v_i and the set of nodes within k-hop distance with v_i, $\mathbf{A}_{g_i}$ is the corresponding adjacency matrix of the ego-graph.

In real-world graphs, for each class, there are usually some representative ego-graphs, which would represent global information of the graph. Thus, in this paper, we utilize memory mechanisms to learn and store representative ego-graphs and then propagate this information through the whole graph. Memory

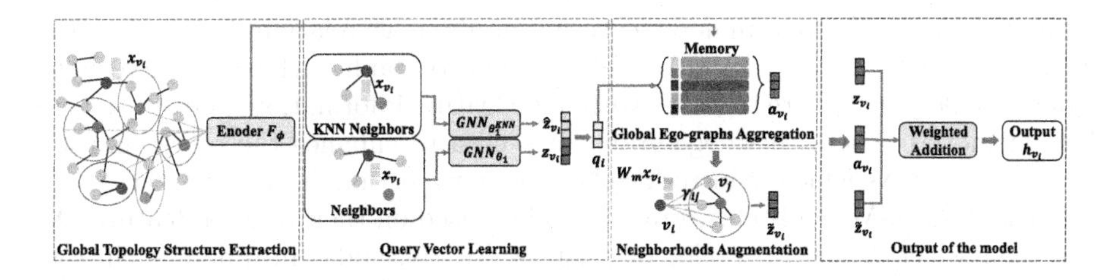

Fig. 1. An illustration of the proposed GSM-GNN.

can be seen as an array of objects, where each object in our paper represents a ego-graph. Each ego-graph $g^i(\mathcal{V}_{m_i}, \mathbf{A}_{m_i})$ is centered at node v'_{m_i} with its k-hop neighbors drawn from the original graph $\mathcal{G}$, where $\mathcal{V}_{m_i} = \left\{v^1_{m_i}, \ldots, v^{B'_i}_{m_i}\right\}$ and each node in this set is from $\mathcal{V}$ with their edges in the adjacency matrix $\mathbf{A}_{m_i}$. B'_i is the number of nodes in memory i and $\mathbf{A}_{m_i} \in [0,1]^{B'_i \times B'_i}$ to represent the adjacency matrix for ego-graph g_i. Note that memories are updated during the training process. Our memory module in the last epoch is defined as $\mathcal{M}_T = \{g^1_T(\mathcal{V}_{m_1}, \mathbf{A}_{m_1}), \ldots, g^B_T(\mathcal{V}_{m_B}, \mathbf{A}_{m_B})\}$, where B is the number of memories, T is the number of training epochs. In semi-supervised node classification, only a subset of nodes are labeled. We denote the labeled set as $\mathcal{V}_L \in \mathcal{V}$ with $\mathcal{Y}_L$ being the corresponding label set of the labeled nodes. The remaining nodes $\mathcal{V}_U = \mathcal{V} \backslash \mathcal{V}_L$ are the unlabeled set. The problem is formally defined as:

Given an attributed graph $\mathcal{G} = (\mathcal{V}, \mathbf{A}, \mathbf{X})$ and the partial labels $\mathcal{Y}_L$, we aim to learn a node classifier Q_θ via plugging our proposed memories into current GNNs. Q_θ should model edges together with node features accordingly via learning global ego-graphs during the label prediction process $Q_\theta(\mathcal{V}, \mathbf{A}, \mathbf{X}) \rightarrow \mathcal{Y}$.

4 Methodology

In this section, we introduce the details of the proposed GSM-GNN, which stores representative ego-graph structures and propagates their information to enhance the representation learning of GNNs. An overview is shown in Fig. 1. Our model can be split into two parts: Global Topology Structure Extraction, and Graph Structure Memory Augmented Representation Learning. Firstly, we store global topology structure information in memories by nodes with their neighborhoods as ego-graphs from the center of clusters, which are obtained via the clustering algorithm on both original and learnt nodes features. Then, we use the query vectors based on feature and structure information to obtain relevant global information from memories, which can enhance the expressive power of GNNs based on local aggregation. The details of them will be introduced below.

4.1 Global Topology Structure Extraction

To mitigate limits of GNNs based on local aggregation, we propose to extract global information from the original graph to enhance GNN models. For graph data, it doesn't only contain node features but also edge relations between nodes. Thus, our model is to extract representative ego-graphs, which preserve both global topology structure together with node feature information, and store them into memories. Then, these ego-graphs are used to enhance the representation learning of GNNs which is introduced in the next section. To extract these ego-graphs, we do k-Medoids clustering on nodes based on their original features $\mathbf{X}$. We then select B central nodes of clusters. Firstly, nodes in different clusters have different information about the graph so a set of clusters are related to different global patterns, and central nodes represent important characteristics of clusters. Furthermore, information in one single node is limited, and neighbors of central nodes and their relationship patterns are also important. Thus, these central nodes with their k-hop neighbors, which form ego-graphs, are treated as global topology graph structure information and stored in a memory set $\mathcal{M}_0$.

However, knowledge of extracting representative ego-graphs only on the original feature is limited so it is necessary to update stored ego-graph based on more informative representation vectors during the training process. Therefore, we use k-Medoids clustering algorithms on the hidden representation $\mathbf{Z}$ which is the output representation of GNN models. Also, central nodes may represent the characteristics of some global patterns in the whole graph. Central nodes and their one-hop or two-hop neighbors as ego-graphs are stored in $\mathcal{M}_t$ at the training epoch t. In our experiment, memories are updated every 100 epochs. An update of memories is utilized to automatically extract representative ego-graph patterns on the whole graph. Then, ego-graphs stored in the memory are used to enhance the representation learning of GNNs.

4.2 Graph Structure Memory Augmented Representation Learning

By extracting and storing representative ego-graphs, we propose to propagate their global information to enhance representation learning of GNNs. The high level idea is to use stored ego-graphs in the memory to improve the expressive power of GNNs. To achieve this purpose, we need to query and aggregate relevant global information from memories and use it to augment representation learning of GNNs. And it can be split into three parts: (1) **Query Vector Learning**, which learns the query vector of each node, and it will be used to obtain relevant global information; (2) **Global Ego-graphs Aggregation**, which encodes ego-graphs and aggregate their information by the similarity between query vectors and encoded feature vectors of ego-graphs in memories; and (3) **Neighborhoods Augmentation**, which utilizes the neighborhood patterns to augment the representation, and can add long-range interactions for distant nodes through this way. Specifically, we firstly obtain representation vectors $\mathbf{z}_{v_i}$ for node v_i via one-layer GNN with local aggregation by:

$$\mathbf{z}_{v_i} = GNN_{\theta_1}(\mathbf{X}, \mathbf{A})_{v_i}, \tag{1}$$

where GNN can be flexible to different GNNs like GCN, GraphSage and GAT.

Query Vector Learning. The purpose of our model is to use ego-graphs stored in memories to enhance the representation learning of GNNs. But for a node v_i, not all memory elements are relevant to v_i. Thus, we first need to learn query vectors of nodes to get relevant information in memories. To guarantee that nodes can query relevant memories, query vector learning should preserve nodes' features and structural information. Firstly, local neighborhood patterns are represented as a vector z_{v_i} with the message passing process of GNNs and can be immediately treated as the query vector for node v_i. However, nodes in the graph only have a small number of neighbors and their local neighborhood patterns may contain bias. Thus, we further augment the query vector with feature information, which can help low-degree nodes be more representative [10]. Concretely, a KNN graph is constructed based on the cosine similarity. For a given node pair (v_i, v_j), their cosine similarity is calculated as $\mathbf{O}_{ij} = \frac{\mathbf{x}_i^\top \mathbf{x}_j}{\|\mathbf{x}_i \cdot\|_2 \|\mathbf{x}_j\|_2}$. We choose $k \in \{20, 30\}$ nearest neighbors via cosine similarity for each node and get a KNN graph. The adjacency matrix of the KNN graph is denoted as $\hat{A}$. Then, node similarity information is aggregated via GNN on KNN graph as:

$$\hat{\mathbf{z}}_{v_i} = GNN_{\theta_1^{KNN}}(\mathbf{X}, \hat{\mathbf{A}})_{v_i}, \tag{2}$$

where θ_1^{KNN} is the learnable parameter for learning KNN graph information, which is denoted as $\hat{\mathbf{z}}_{v_i}$. To learn query vectors from feature similarity and structural information from $\mathbf{z}_{v_i}$ and $\hat{\mathbf{z}}_{v_i}$, these two vectors are concatenated together and used as an input for an MLP layer to obtain the query vector as:

$$\mathbf{q}_i = [\mathbf{z}_{v_i} \| \hat{\mathbf{z}}_{v_i}]\mathbf{W}_q + \mathbf{b}_q \tag{3}$$

where $\mathbf{W}_q \in R^{2d \times d}$ and $\mathbf{b}_q \in R^d$ are learnable parameters, d is the dimension of vectors $\mathbf{z}_{v_i}, \hat{\mathbf{z}}_{v_i}$. Then, $\mathbf{q}_i$ will be used to query relevant global information.

Global Ego-Graphs Aggregation. Furthermore, to query relevant ego-graphs in memories via $\mathbf{q}_i$, we also need to encode the ego-graph information into vectors. To achieve this goal, we utilize one-layer GCN to obtain representation vectors for all ego-graphs in memories:

$$\mathbf{m}_i = F_\phi(\mathbf{X}_{m_i}, \mathbf{A}_{m_i}), \tag{4}$$

where $F_\phi(*)$ represents one-layer GCN with graph pooling to get the representation vectors of ego-graphs in memories, $\mathbf{m}_i \in \mathbb{R}^{1 \times d}$. $\mathbf{X}_{m_i}$ is the representation matrix for nodes in memory i, where $\mathbf{X}_{m_i}[j,:] \in \mathbb{R}^{1 \times d_0}$ is the representation vectors of node $v_{m_i}^j$ which is obtained from the original feature matrix $\mathbf{X}$. Note that the pooling method here is the mean pooling method. Then we calculate the similarity between $\mathbf{q}_i$ and $\mathcal{M}_t$ in the layer l of the training epoch t as:

$$\mathbf{s}_i = Softmax(\mathbf{q}_i(\mathbf{M})^T), \tag{5}$$

where $\mathbf{M} \in \mathbb{R}^{B_i' \times d}$ is the representation matrix for B_i' memories. The similarity scores measure the importance of each global pattern in the memory. Any pattern with a higher attention score is more similar to the local structural patterns of nodes. The representation vector of global information for node v_i is then constructed from the weighted sum of all global patterns in the memory $\mathcal{M}_t$ as

$$\mathbf{a}_{v_i} = \sum_{j=1}^{B} s_{i,j}\mathbf{m}_j, \tag{6}$$

where $\mathbf{a}_{v_i}$ represents global ego-graph information of node v_i. $s_{i,j}$ is the similarity score between node i and memory j.

Neighborhoods Augmentation. Each memory module contains one representative central node with its one or two-hop neighbors and their edge relations. Even though the central node of each memory node has aggregated information from its neighbors, we will also lose some information about their neighbors. Thus, it's important to explore neighborhood distributions from relevant ego-graphs, which can further capture ego-graphs' information. However, it will be time-consuming and introduce more noisy neighbors' information if we add all neighbors' of nodes in memories to one node's augmented neighbors. Thus, we select the most relevant memory modules as $r_i = \arg\max_j s_{i,j}$, where r_i is the index of the most relevant memory module. Then, we obtain the most relevant ego-graph $g_t^{r_i}(\mathcal{V}_{m_{r_i}}, \mathbf{A}_{m_{r_i}})$ of the training epoch t based on the similarity of structural information between local ego-graph patterns and global ego-graph patterns. Central nodes have aggregated their neighbors' information and are treated as representation vectors in memories which are added as global information in Eq. (6). Instead of aggregating central nodes' information again, neighborhood nodes of them in $\mathcal{V}_{m_{r_i}}$ are treated as the augmented neighbors for the node i and their information is aggregated for node v_i. Enhanced neighbors may contain noisy information so an attention mechanism is utilized to assign different weights to augmented neighbors:

$$\gamma_{ij} = \frac{\exp\left(\text{LeakyReLU}\left(\mathbf{u}^T\left[\mathbf{x}_{v_i}\mathbf{W}_m \| \mathbf{x}_{v_j}\mathbf{W}_m\right]\right)\right)}{\sum_{k\in\mathcal{V}_{m_{r_i}}}\exp\left(\text{LeakyReLU}\left(\mathbf{u}^T\left[\mathbf{x}_{v_i}\mathbf{W}_m \| \mathbf{x}_{v_k}\mathbf{W}_m\right]\right)\right)}, \tag{7}$$

where $\mathbf{W}_m \in \mathbb{R}^{d_0\times d}$ is the learnable parameters, $\mathbf{u} \in \mathbb{R}^{1\times 2d}$ is the learnable weight vector, γ_{ij} represents the weight for nodes j in $\mathcal{V}_{m_{r_i}}$ when node i aggregates information from node j. To mitigate noise from augmented neighbors, we aggregate these neighbors via the weight γ_{ij}. The aggregation process of augmented neighbors is:

$$\tilde{\mathbf{z}}_{v_i} = \sum_{j\in\mathcal{V}_{m_{r_i}}\setminus v'_{m_{r_i}}} \gamma_{ij}\mathbf{x}_{v_j}\mathbf{W}_m, \tag{8}$$

where $v'_{m_{r_i}}$ is the central node of the ego-graph in the memory r_i.

Finally, we get the local representation $\mathbf{z}_{v_i}$, global representation $\mathbf{a}_{v_i}$ and augmented neighbors representation $\tilde{\mathbf{z}}_{v_i}$. Different nodes might rely on different information. For example, low degree nodes may need more global information; while high degree nodes with enough similar neighborhoods information may only need more local information. Thus, we propose to assign different weights to get the final node representation. Specifically, the weight is calculated as:

$$\boldsymbol{\beta}_i = Softmax(\mathbf{W}_o[\tilde{\mathbf{z}}_{v_i}||\mathbf{a}_{v_i}||\mathbf{z}_{v_i}] + \mathbf{b}_o), \tag{9}$$

where $\mathbf{W}_o$ and $\mathbf{b}_o$ are learnable parameters, $\boldsymbol{\beta}_i \in \mathbb{R}^3$ is the weight for different representation vectors. Then, different representation vectors are added together based on their weights to get $\mathbf{h}_{v_i}$ as:

$$\mathbf{h}_{v_i} = \beta_{i,0}\mathbf{z}_{v_i} + \beta_{i,1}\mathbf{a}_{v_i} + \beta_{i,2}\tilde{\mathbf{z}}_{v_i}. \tag{10}$$

With the above operation, GNN models can help nodes aggregate both local neighborhood information and global information adaptively from memories. Therefore, our memory modules can store representative global information and help propagate this information on the whole graph. Note that our memory module can be also added in more layers, to reduce the computational cost, we only use it to augment the representation of one layer.

4.3 Objective Function of GSM-GNN

With the representation $\mathbf{H}$ capturing both local and global information, we add another GNN layer together with Softmax function to predict the class probability vector of each node v_i as:

$$\hat{\mathbf{y}}_{v_i} = Softmax(GNN_{\theta_p}(\mathbf{H}, \mathbf{A})_{v_i}) \tag{11}$$

where $\hat{\mathbf{y}}_{v_i}$ is the predicted label's probability by passing the output from the final GNN layer to a softmax function. θ_p represents the parameters of our model's final prediction layer. The cross-entropy loss function for node classification is:

$$\min_{\theta} \mathcal{L}_c(\theta) = - \sum_{v_i \in \mathcal{V}_L} \sum_{c=1}^{C} \mathbf{y}_{v_i}^c \log \hat{\mathbf{y}}_{v_i}^c, \tag{12}$$

where C is the number of classes, $\mathbf{y}_{v_i}$ is the one hot encoding of v_i's label and $\mathbf{y}_{v_i}^c$ is the c-th element of $\mathbf{y}_{v_i}$. θ denotes the set of parameters.

5 Experiments

In this section, we conduct experiments on real-world datasets to demonstrate the effectiveness of GSM-GNN. In particular, we aim to answer the following research questions: (**RQ1**) Can the proposed memory mechanism improve node classification performance? (**RQ2**) Is the designed approach flexible to be applied in various GNN variants? (**RQ3**) What are the contributions of the proposed module in this paper for GSM-GNN?

Table 1. Node classification performance (Accuracy (%) ± Std.) on all graphs.

Method	Cora	Citeseer	Computer	Photo	Physics	Chameleon	Squirrel
MLP	45.44 ±1.55	52.61 ±0.80	67.35 ±0.71	79.10 ±0.74	92.11 ±0.11	48.00 ±1.5	34.02 ±2.13
GCN	74.65 ±1.91	65.20 ±0.74	80.80 ±0.29	87.90 ±0.58	94.24 ±0.10	63.50 ±1.93	47.48 ±2.00
GraphSage	75.43 ±2.08	65.63 ±0.35	73.47 ±0.34	86.31 ±0.15	94.56 ±0.08	48.36 ±2.08	35.88 ±1.20
GAT	71.30 ±0.92	64.55 ±2.32	76.47 ±1.49	85.74 ±1.32	94.21 ±0.07	64.12 ±1.82	48.18 ±4.14
Mixhop	70.40 ±2.83	62.44 ±1.14	75.88 ±1.00	87.92 ±0.53	94.41 ±0.23	60.71 ±1.55	44.11 ±1.10
ADA-UGNN	74.29 ±1.95	65.30 ±1.28	79.88 ±0.92	88.08 ±0.78	94.70 ±0.11	52.19 ±1.85	34.84 ±1.36
H2GCN	72.45 ±0.46	66.10 ±0.48	78.22 ±0.75	86.94 ±0.47	94.59 ±0.09	59.13 ±2.00	36.91 ±1.10
FAGCN	69.53 ±0.12	61.07 ±0.32	81.09 ±0.14	85.33 ±0.12	94.67 ±0.07	65.57 ±4.80	48.73 ±2.50
Simp-GCN	74.24 ±1.32	66.24 ±1.05	74.23 ±0.12	82.41 ±1.36	94.43 ±0.07	64.71 ±2.30	42.81 ±1.20
GCN-MMP	72.23 ±2.19	64.95 ±1.69	79.55 ±1.48	87.56 ±1.03	94.42 ±0.12	66.17 ±1.68	50.93 ±1.45
GSM-GCN	**75.93 ± 0.65**	**66.33 ± 0.49**	**81.32 ± 0.34**	**88.87 ± 0.35**	**94.72 ± 0.07**	**67.20 ± 1.85**	**54.14 ± 1.60**

5.1 Datasets

We conduct experiments on seven publicly available benchmark datasets. **Cora and Citeseer** [11] are two datasets for citation networks **Computers and Photo** [18] are two datasets for the Amazon co-purchase graph [14]. **Physics** [18] is a larger co-authorship graph. **Chameleon and Squirrel** [15] are two datasets for the web pages in Wikipedia. They are used for heterophilous graphs. For Chameleon and Squirrel, we follow the 10 standard splits from [21]. For other datasets, we randomly split the dataset into train/val/test as 2.5%/2.5%/95%. The random split is conducted 5 times and average performance will be reported.

5.2 Experimental Setup

Baselines. We compare GSM-GNN with representative methods for node classification, which includes MLP, GCN [11], GraphSage [9], GAT [19], Mixhop [1], ADA-UGNN [13]. We also compare GSM-GNN with the following representative and state-of-the-arts GNN models on heterophilous graphs, which includes H2GCN [21], FAGCN [2], Simp-GCN [10] and GCN-MMP [4].

Configurations. All experiments are conducted on a 64-bit machine with Nvidia GPU (Tesla V100, 1246 MHz, 16 GB memory). For a fair comparison, we utilize a two-layer neural network for all methods, and the hidden dimension is set as 64. The learning rate is initialized to 0.001. Besides, all models are trained until converging, with the maximum training epoch being 1000. The implementations of all baselines are based on Pytorch Geometric or their original code. For our method, the update epoch of memories is fixed at 100 and B is set by grid search from 5 to 30 for all datasets. The hyperparameters of all methods are tuned on the validation set. We adopt accuracy (ACC) as the metric.

5.3 Performance on Node Classification

In this subsection, we compare the performance of the proposed method with baselines for node classification on the heterophilous and homophilous graphs introduced in Sect. 5.1, which aims to answer **RQ1**. For Cora, Citeseer, Computers, Photo and Physics, each experiment is conducted 5 times and for Chameleon

Table 2. Node classification accuracy with different GNNs.

	Cora	Citeseer	Chameleon	Squirrel
GCN	74.65 ± 1.91	65.20 ± 0.74	64.80 ± 1.93	47.48 ± 0.20
GSM-GCN	75.93 ± 0.65	66.33 ± 0.49	67.20 ± 1.85	54.14 ± 1.60
GraphSage	75.43 ± 2.08	65.63 ± 0.35	48.36 ± 2.08	35.88 ± 1.20
GSM-Sage	76.77 ± 0.62	67.01 ± 0.29	50.59 ± 2.81	37.86 ± 1.47
GAT	71.30 ± 0.92	64.55 ± 2.32	64.12 ± 1.82	48.18 ± 4.14
GSM-GAT	73.38 ± 1.82	65.55 ± 1.20	64.67 ± 1.62	53.51 ± 2.78

and Squirrel, each experiment is conducted 10 times. The average results with standard deviation are reported in Table 1. Note that GSM-GCN uses the GCN as the backbone of our proposed memory modules. From the table, we make the following observations: (1) Compared with GCN and other GNN models, GSM-GCN can consistently improve the performance of GCN on all datasets, which demonstrates the effectiveness of the proposed memory module. Furthermore, GSM-GCN outperforms all baselines on all datasets, especially for the Squirrel dataset. This is because the proposed memory module can capture representative ego-graphs to facilitate GNNs to capture global information. (2) Both Simp-GNN and GSM-GCN utilize the information of node features' similarity. GSM-GCN significantly outperforms Simp-GNN on all datasets. This is because GSM-GCN adopts similarity information as a query vector to query global information, which shows the effectiveness of our query mechanism based on feature similarity. (3) GCN-MMP also designs a memory mechanism on nodes to improve GNNs' performance. The proposed GSM-GCN consistently outperforms GCN-MMP on all datasets. This is because GCN-MMP only stores feature vectors in memories while our method utilizes both feature and structure information.

5.4 Flexibility of GSM-GNN for Various GNNs

To answer **RQ2**, we conduct experiments with different architectures of GSM-GNN by inserting our memory module into different GNN variants. Specifically, we test our memory modules on GCN, GraphSage, and GAT layers. For GCN, GraphSage, and GAT, we utilize a two-layer graph network with 64 hidden dimensions. For a fair comparison, all models use the same settings. For all the methods, hyperparameters are tuned via the performance of the validation set. Each experiment is conducted 5 times on Cora and Citeseer datasets, and 10 times on Chameleon and Squirrel. The average performance with standard deviation is reported in Table 2. From the table, we have the following observations: (i) GSM-GNN can consistently improve the performance on these four datasets with all backbones which indicates that our proposed memory mechanism is effective when incorporated into other GNN variants and demonstrates the flexibility and advantage of the proposed method; and (ii) in particular, the proposed memory module can significantly improve the GCN, GraphSage and

Table 3. Compared with different information from memory modules.

Method	Cora	Citeseer	Computer	Photo	Physics	Chameleon	Squirrel
GCN	74.65 ± 1.91	65.20 ± 0.74	80.80 ± 0.29	87.90 ± 0.58	94.24 ± 0.10	63.50 ± 1.93	47.48 ± 2.00
Without $\mathbf{a}_{v_i}$	75.27 ± 1.26	65.82 ± 0.70	80.93 ± 0.21	86.67 ± 0.29	94.63 ± 0.07	67.09 ± 1.63	54.04 ± 1.20
Without $\tilde{\mathbf{z}}_{v_i}$	73.63 ± 1.59	66.01 ± 0.49	80.87 ± 0.17	88.04 ± 0.34	94.47± 0.40	64.56 ± 2.16	53.87 ± 0.87
GSM-GCN	**75.93 ± 0.65**	**66.33 ± 0.49**	**81.32 ± 0.34**	**88.87± 0.35**	**94.72 ± 0.07**	**67.20 ± 1.85**	**54.14 ± 1.60**

GAT on the heterophilous graphs. This shows that our memory module can use global information to improve current GNNs.

5.5 Ablation Study

To answer **RQ3**, in this section, we conduct an ablation study to evaluate the influence of each queried information from memories including $\mathbf{a}_{v_i}$ and $\hat{\mathbf{z}}_{v_i}$ in GSM-GNN. First, to investigate how the global ego-graph information ($\mathbf{a}_{v_i}$) influences the performance of node classification, we only encode ego-graph information and add it with local information via the attention mechanism. Then, we also query augmented neighbors and aggregate this information as global information $\hat{\mathbf{z}}_{v_i}$. We use GCN as the backbone for ablation studies. The experiments are conducted on all graphs. The average performance in terms of Accuracy is shown in Table 3. From the table, we observe that: (**i**) Comparing GCN with "Without $\mathbf{a}_{v_i}$", augmented neighbors from ego-graphs in memories can lead to a little improvement on almost all graphs. It means that augmented neighbors from memories can provide more similar nodes information during the aggregation process; (**ii**) "Without $\tilde{\mathbf{z}}_{v_i}$" only utilizes ego-graph information in memories and it also performs better than the original GCN. It is because global patterns based on ego-graph may contain important label information which can be used to improve the performance of the original GCN; (iii) Finally, our proposed GSM-GCN has the best performance on all datasets because GSM-GCN can aggregate local information, global information from ego-graphs, and augmented neighbors adaptively. This ablation study further proves the effectiveness of our proposed method to capture global information from the whole graph.

6 Conclusion

In this paper, we study a novel problem of learning global patterns and propagating global information to improve the performance of GNNs. We propose a novel framework Graph Structure Memory Enhanced Graph Neural Network (GSM-GNN) which stores representative global patterns with nodes and graph structure, and can be used to augment the representation learning of GNNs. Through extensive experiments, we validate the advantage of the proposed GSM-GNN, which can utilize a memory network to store and propagate global information.

Acknowledgements. This material is based upon work supported by, or in part by, the National Science Foundation (NSF) under grant number IIS-1909702, the Army Research Office (ONR) under grant number W911NF21-1-0198, and Department of Homeland Security (DNS) CINA under grant number E205949D. The findings in this paper do not necessarily reflect the view of the funding agencies.

References

1. Abu-El-Haija, S., et al.: Mixhop: higher-order graph convolutional architectures via sparsified neighborhood mixing. In: International Conference on Machine Learning, pp. 21–29. PMLR (2019)
2. Bo, D., Wang, X., Shi, C., Shen, H.: Beyond low-frequency information in graph convolutional networks. arXiv preprint arXiv:2101.00797 (2021)
3. Bruna, J., Zaremba, W., Szlam, A., LeCun, Y.: Spectral networks and locally connected networks on graphs. arXiv preprint arXiv:1312.6203 (2013)
4. Chen, J., Liu, W., Pu, J.: Memory-based message passing: Decoupling the message for propogation from discrimination. arXiv preprint arXiv:2202.00423 (2022)
5. Dai, E., Jin, W., Liu, H., Wang, S.: Towards robust graph neural networks for noisy graphs with sparse labels. arXiv preprint arXiv:2201.00232 (2022)
6. Donnat, C., Zitnik, M., Hallac, D., Leskovec, J.: Learning structural node embeddings via diffusion wavelets. In: Proceedings of the 24th ACM SIGKDD International Conference on Knowledge Discovery and Data Mining, pp. 1320–1329 (2018)
7. Gilmer, J., Schoenholz, S.S., Riley, P.F., Vinyals, O., Dahl, G.E.: Neural message passing for quantum chemistry. In: International Conference on Machine Learning, pp. 1263–1272. PMLR (2017)
8. Graves, A., Wayne, G., Danihelka, I.: Neural turing machines. arXiv preprint arXiv:1410.5401 (2014)
9. Hamilton, W., Ying, Z., Leskovec, J.: Inductive representation learning on large graphs. In: Advances in Neural Information Processing Systems, vol. 30 (2017)
10. Jin, W., Derr, T., Wang, Y., Ma, Y., Liu, Z., Tang, J.: Node similarity preserving graph convolutional networks. In: Proceedings of the 14th ACM International Conference on Web Search and Data Mining, pp. 148–156 (2021)
11. Kipf, T.N., Welling, M.: Semi-supervised classification with graph convolutional networks. arXiv preprint arXiv:1609.02907 (2016)
12. Liu, M., Wang, Z., Ji, S.: Non-local graph neural networks. IEEE Trans. Pattern Anal. Mach. Intell. (2021)
13. Ma, Y., Liu, X., Zhao, T., Liu, Y., Tang, J., Shah, N.: A unified view on graph neural networks as graph signal denoising. In: Proceedings of the 30th ACM International Conference on Information and Knowledge Management (2021)
14. McAuley, J., Targett, C., Shi, Q., Van Den Hengel, A.: Image-based recommendations on styles and substitutes. In: Proceedings of the 38th International ACM SIGIR Conference on Research and Development in Information Retrieval (2015)
15. Pei, H., Wei, B., Chang, K.C.C., Lei, Y., Yang, B.: GEOM-GCN: geometric graph convolutional networks. arXiv preprint arXiv:2002.05287 (2020)
16. Pritzel, A., et al.: Neural episodic control. In: International Conference on Machine Learning, pp. 2827–2836. PMLR (2017)
17. Qu, L., Zhu, H., Zheng, R., Shi, Y., Yin, H.: ImGAGN: imbalanced network embedding via generative adversarial graph networks. In: Proceedings of the 27th ACM SIGKDD Conference on Knowledge Discovery and Data Mining (2021)

18. Shchur, O., Mumme, M., Bojchevski, A., Günnemann, S.: Pitfalls of graph neural network evaluation. arXiv preprint arXiv:1811.05868 (2018)
19. Veličković, P., Cucurull, G., Casanova, A., Romero, A., Lio, P., Bengio, Y.: Graph attention networks. arXiv preprint arXiv:1710.10903 (2017)
20. Xiong, T., Zhu, L., Wu, R., Qi, Y.: Memory augmented design of graph neural networks (2020)
21. Zhu, J., Yan, Y., Zhao, L., Heimann, M., Akoglu, L., Koutra, D.: Beyond homophily in graph neural networks: Current limitations and effective designs. Adv. Neural. Inf. Process. Syst. **33**, 7793–7804 (2020)

UPGAT: Uncertainty-Aware Pseudo-neighbor Augmented Knowledge Graph Attention Network

Yen-Ching Tseng[1,2(✉)], Zu-Mu Chen[1], Mi-Yen Yeh[1], and Shou-De Lin[2]

[1] Institute of Information Science, Academia Sinica, Taipei, Taiwan
franklyn.chen@gmail.com, miyen@iis.sinica.edu.tw

[2] Department of Computer Science and Information Engineering, National Taiwan University, Taipei, Taiwan
{r0822a10,sdlin}@csie.ntu.edu.tw

Abstract. The uncertain knowledge graph (UKG) generalizes the representation of entity-relation facts with a certain confidence score. Existing methods for UKG embedding view it as a regression problem and model different relation facts independently. We aim to generalize the graph attention network and use it to capture the local structural information. Yet, the uncertainty brings in excessive irrelevant neighbor relations and complicates the modeling of multi-hop relations. In response, we propose UPGAT, an uncertainty-aware graph attention mechanism to capture the probabilistic subgraph features while alleviating the irrelevant neighbor problem; introduce the pseudo-neighbor augmentation to extend the attention range to multi-hop. Experiments show that UPGAT outperforms the existing methods. Specifically, it has more than 50% Weighted Mean Rank improvement over the existing approaches on the NL27K dataset.

Keywords: uncertain knowledge graph · embedding · graph attention

1 Introduction

Knowledge Graph (KG) embedding has been widely studied in recent years, allowing machine learning models to leverage structural knowledge. As a generalized form, *Uncertain Knowledge Graphs* (UKG) no longer simply consider the existence of relational facts. Instead, they also express the corresponding plausibility. For example, the occurrence of protein interactions is probabilistic; if two words are synonymous is also probabilistic due to lexical ambiguity.

All existing uncertain knowledge graph embedding methods, including [2–4,9,13], model each relation fact independently. To be more concrete, the model samples one single relation fact, termed *triplet*, at a time, and predicts its confidence score. The simplicity of these methods helps avoid the caveat of overfitting. However, the probabilistic nature of UKG complicates the structural

This work was supported in part by National Science and Technology Council, Taiwan, R.O.C., under grants 110-2628-E-001-001 and 111-2628-E-001-001-MY2.

H. Kashima et al. (Eds.): PAKDD 2023, LNAI 13936, pp. 53–65, 2023.
https://doi.org/10.1007/978-3-031-33377-4_5

information and also makes the graph denser since there can be many extra uncertain relations between each entity pair. Therefore, we claim modeling subgraph information is indispensable for advancing UKG embedding quality.

Among all methods that incorporate subgraph information on embedding deterministic KG, graph attention network (GAT) [12] can aggregate neighboring triplets and use the attention mechanism to weigh each triplet based on their importance. This feature is highly desirable for uncertain and dense graphs, as it can filter out implausible or irrelevant neighbors on an as-needed basis.

Nonetheless, present GAT methods such as [14] are designed for the deterministic KG only. They cannot be directly applied to UKG due to the following challenges we aim to tackle in this study. First, it is challenging to retain the uncertainty information of the subgraph while filtering out unrelated ones. We presume that many neighboring triplets surrounding a given triplet can be uninformative for embedding such a triplet due to the uncertainty and high density of the uncertain KG. We show in the experiment section that, in reality, the plausibility does not imply relevance - it is groundless to focus on the more plausible neighbors trivially. Second, unlike deterministic KGs where the existence of a path is a binary question, it is difficult to determine the confidence score of a multi-hop relation without predefined inference rules.

To address these challenges, the conventional practice will devise a dedicated model structure that embeds confidence scores. In contrast, we take an innovative perspective to encode the uncertainty information implicitly and reach an even better performance proven in experiments. Our contributions are: (i) our proposed model, *UPGAT*, is the first to incorporate subgraph features and generalize GAT for the uncertain KG, (ii) rather than formulating the confidence score for neighbor triplets, the proposed graph attention with *the attention baseline mechanism* can catch the uncertainty while distinguishing unrelated information, and (iii) without using rule-based inferences, *the pseudo-neighbor augmented graph attention* overcomes the difficulty of identifying and leveraging multi-hop relations, where we add predicted n-hop neighbors, termed pseudo-neighbors, into the neighbor aggregation mechanism.

Experiments on three public datasets depict that our proposed method is superior to the existing technology on both the link prediction and confidence score prediction tasks. Specifically, UPGAT shows more than 50% Weighted Mean Rank (WMR) improvement on the NL27K dataset.

2 Preliminaries

2.1 Problem Statement

Definition 1 (Uncertain knowledge graph). *Let $\mathcal{G}$ be an uncertain knowledge graph, such that $\mathcal{G} = \{(l, s_l) | h, t \in \mathcal{E}, r \in \mathcal{R}, s_l \in [0, 1]\}$, where $l = (h, r, t)$ is a triplet containing head and tail entities h and t with relation r; $\mathcal{E}$ is the entity set, $\mathcal{R}$ is the relation set, and s_l is the confidence score of triplet l.*

Definition 2 (Negative and positive samples). *An uncertain relation fact sample,* (l, s_l)*, is a negative sample if* $s_l = 0$ *and is a positive sample otherwise.*

Note that as existing datasets only include positive samples, most studies randomly draw out-of-dataset triplets as negative samples. If the true confidence score of such a sample is non-zero, it is termed as a *false-negative triplet* [4].

Given an uncertain knowledge graph $\mathcal{G}$, the *Uncertain Knowledge Graph Embedding* problem aims to encode each entity and relation in a low-dimensional space while preserving the probabilistic graph structure. Note that the deterministic KG is a special case of UKG, where the confidence scores are either one or zero. Hence, applying deterministic KG embedding methods to UKG can be incompatible and/or have degraded performance. Moreover, the inherent higher density of uncertain KG could lead to more false-negative triplets.

Most existing embedding methods for uncertain knowledge graphs [3,4,13] follow the paradigm of traditional KG embedding methods to estimate the confidence score of a single triplet, $l = (h, r, t)$, using a parameterized score function $S(\hat{h}, \hat{r}, \hat{t})$, where $\hat{h}$, $\hat{r}$, and $\hat{t}$ are the respective embedding vectors. As an example, UKGE [3] applies the DistMult [15] scoring function for S and uses Mean Square Error (MSE) loss to learn the confidence score s_l as a regression model: $\mathcal{J}^+ = ||S(\bar{h}, \bar{r}, \bar{t}) - s_l||^2$.

Definition 3 (Subgraph Features). *Given an uncertain relation fact,* (l, s_l)*, where* $l = (h, r, t)$*, let the subgraph feature be any connected subgraph* $G' \subset G$*, where* $(l, s_l) \in G'$ *and* $|G'| > |\{(l, s_l)\}|$.

The above mentioned methods for embedding uncertain KGs model different relation facts independently. Nonetheless, studies on deterministic KGs, such as [6,7,11,12,14], have shown the benefit for node classification and link prediction to aggregate the neighboring triplets around an entity. We believe the complex and continuous graph structure of uncertain KGs makes it even more essential to model subgraph features explicitly.

2.2 Motivations and Challenges

To exploit subgraph features, we found graph attention network (GAT)-based method is well-suitable yet has not been studied for UKGs. Concerning the probabilistic nature of UKGs, where a triplet can be surrounded by implausible neighboring relation facts, GAT can learn the importance of each neighboring relation fact and assigns different attention scores accordingly. Other ways to model multi-hop relationships are subject to various limitations, which we have more discussions about in Sect. 2.3.

KBGAT [14] is a variant of graph attention network (GAT). Different from GAT that only considers the node features in the graph, KBGAT can encode the edge (relation) features in the knowledge graph. Given an entity h_i and one of its neighbor entities h_j connected with relation r_k, let their embeddings be $\bar{h}_i$, $\bar{h}_j$, and $\bar{g}_k$ respectively. The corresponding neighbor feature, $\bar{c}_{ijk}$, attention value, b_{ijk}, and the attention score, α_{ijk}, are as follows: $\bar{c}_{ijk} = \mathbf{W_1}[\bar{h}_i||\bar{h}_j||\bar{g}_k]$; $b_{ijk} = \text{LeakyReLU}(\mathbf{W_2}\bar{c}_{ijk})$; $\alpha_{ijk} = \text{softmax}_{jk}(b_{ijk})$. Finally, the new presentation of entity h_j is $\bar{h}'_i = \sigma(\sum_{j \in \mathcal{N}_i} \sum_{k \in \mathcal{R}_{ij}} \alpha_{ijk}\bar{c}_{ijk})$, where $\mathcal{N}_i$ denotes the neighbor entity set of entity h_i, $\mathcal{R}_{ij}$ denotes the set of relations connecting entities h_i and h_j, and σ represents any non-linear function.

However, components of KBGAT are not well-generalizable to UKG, resulting in worse performance, for which we have identified two major issues.

First, KBGAT does not take confidence score information into account in its aggregation mechanism. It is presumable that due to the uncertainty and high density, irrelevant triplets are prevailing in the uncertain KG. The intuitive assumption is that there exist positive correlations between the confidence score and attention score. Therefore, a naive solution is to explicitly formulate a confidence score in the attention mechanism. For instance, let s_{ijk} be the confidence score of triplet (h_i, r_k, h_j), we have explored the following three setups: (1) concatenate the confidence score into the feature vector: redefine $\bar{c}_{ijk}$ as $\bar{c}_{ijk} = \mathbf{W_1}[\bar{h}_i||\bar{h}_j||\bar{g}_k||s_{ijk}]$, (2) linearly weight confidence score with the learned attention value: Redefine $\bar{h}'_i$ as $\bar{h}'_i = \sigma(\sum_{j \in \mathcal{N}_i} \sum_{k \in \mathcal{R}_{ij}} s_{ijk}\alpha_{ijk}\bar{c}_{ijk})$, and (3) let the attention value be the confidence score: $\bar{h}'_i = \sigma(\sum_{j \in \mathcal{N}_i} \sum_{k \in \mathcal{R}_{ij}} s_{ijk}\bar{c}_{ijk})$. Analyses in Sect. 4.3 show none of them is perceivably better over KBGAT [14].

In fact, these naive assumptions are questionable. First, the embeddings of neighbor triplets should have contained their confidence score information and obviated the need to model it explicitly. We only formulated confidence scores in the loss function. Second, there is no clear evidence that weighting neighbors according to their plausibility helps the attention process. Low-confidence triplets are not necessarily irrelevant. Hence, we avoid adding constraints between the attention and the confidence scores. Rather, we aim to propose an uncertainty-aware attention mechanism to properly handles irrelevant neighbors.

Second, finding multi-hop relations is infeasible using path-searching algorithms due to uncertainty. To broaden the scope of attention beyond 1-hop neighbors, Xie et al. [14] propose the n-hops auxiliary neighbor mechanism. However, given uncertain relations, such n-hops operation is difficult to be realized without extra domain knowledge. For instance, Chen et al. [3] apply human-defined first-order logic to imply the plausibility of multi-hop relations. Please note this problem is not unique to attention-based models but to most existing methods.

With regard to the above issues, therefore, we propose a robust attention-based model to reduce the impact of irrelevant neighbors and extend its attention range beyond one hop, even if relation links are uncertain.

2.3 Related Work

Traditional embedding methods for deterministic KGs, such as [1,15], focus primarily on modeling single triplets. Subsequent studies explore more about subgraph modeling. For example, [7] exploits paths as subgraph features. There are also logic rule-based methods, e.g. [10,16], which can model more subgraph

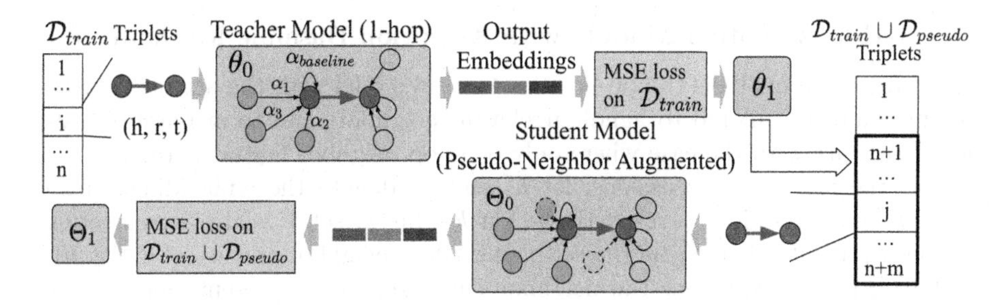

Fig. 1. The overview of UPGAT. α is the attention score. Dashed circles and arrows represent pseudo-neighbors.

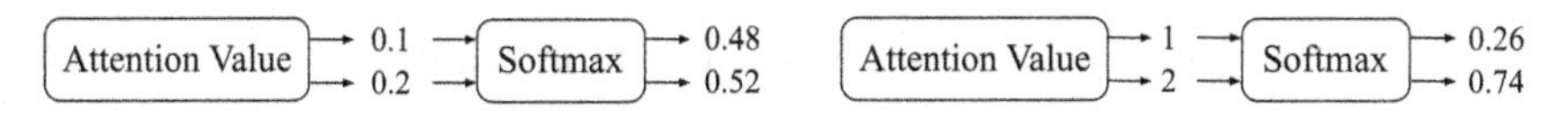

Fig. 2. The attention value on the left is lower than that on the right (0.1 vs 1.0), but, after softmax, the attention score on the left becomes significantly higher (0.48 vs 0.26)

patterns other than paths with well-designed rule templates. Yet, such templates rely on prior domain knowledge. Recently, neural-network-based methods with deeper and more complex artificial neural networks are proposed, e.g., [8,11,12,14]. Some of them can model larger subgraphs without extra human knowledge.

Existing works for uncertain knowledge graph embedding, to our knowledge, only focuses on modeling single triplets; none of them explore how to utilize subgraph features. UKGE [3], the first method for embedding uncertain knowledge graphs, applies probabilistic logic rules (PSL) to infer unseen triplets. PASSLEAF [4] aims to alleviate the false-negative problem. SUKE [13] proposes an evaluator-generator architecture. BEUrRE [2] models each entity as a box and relations as affine transforms to capture the uncertainty. FocusE [9] adds an extra layer after the scoring layer, balancing the high-confidence and uncertain graph structure.

3 Approach

3.1 Overview

The uncertainty-aware pseudo-neighbor augmented knowledge graph attention network (UPGAT) is a generalization of the graph attention network featuring the ability to incorporate uncertain subgraph features and exploit multi-hop relations. Particularly, for the issues discussed in Sect. 2.2, the attention baseline mechanism alleviates the negative impact of irrelevant neighbors; pseudo-neighbor augmentation overcomes the difficulty of identifying multi-hop paths.

UPGAT consists of two pipelined training stages depicted in Fig. 1. First, train a 1-hop graph attention model, θ, to generate the pseudo-neighbor triplets, indexed $n + 1$ to $n + m$. Second, train a multi-hop model, Θ, based on the augmented knowledge graph to explore both one- and multi-hop relations.

3.2 1-Hop Attention Module with Attention Baseline Mechanism

This subsection explains the proposed attention model with the attention baseline mechanism, tailored to handle irrelevant neighbors and to increase robustness. We start with the case where only one-hop neighbor features are used.

For the following discussions, let $\bar{h}_i$ and $\bar{g}_k$ denote the embeddings of an entity h_i and a relation r_k, respectively. Let $\mathcal{R}_{ij}$ be the set of relations connecting entities h_i and h_j. Let $\mathcal{N}_i$ be the set of the 1-hop neighbor entities of entity h_i.

Given an entity h_i, let the neighbor entity-relation representation $\bar{c}_{ijk}$ be $\mathbf{W_1}[\bar{h}_j \circ \bar{g}_k]$, for a neighbor entity h_j and the connecting relation g_k, where $\mathbf{W_1} \in \mathbb{R}^{d \times d}$ and d is the embedding size. The corresponding attention value, i.e., the importance of the neighbor feature, is defined in Eq. (1), where $\mathbf{W_a} \in \mathbb{R}^{1 \times d}$.

$$a_{ijk} = \text{LeakyReLU}(\mathbf{W_a}[\bar{h}_i \circ \tanh(\bar{c}_{ijk})]). \tag{1}$$

We aggregate neighbor features with an element-wise product to preserve both the semantic interaction and confidence score information between the entity and the relation in the embedding vectors. We pass this information to the attention mechanism and let it decides what is important instead of explicitly modeling the confidence score. Compared to concatenating embedding vectors as in GAT [12] and KBGAT [14], it also reduces the feature dimension, enhancing the robustness against noises from uncertainty.

To mitigate the impact of many irrelevant uncertain neighbors, we introduce the attention baseline mechanism, before normalizing the attention values by softmax to get the attention score. The normalized attention score is essentially a convex combination among all neighboring features. Consequently, when all neighbors are irrelevant, the attention mechanism still has to "choose" from one of them, as shown in Fig. 2. Thus, the attention baseline mechanism serves as the "none of them" option. Each neighbor can get a high normalized attention score only when its attention value is higher than the baseline value.

Formally, for each entity h_i, add a baseline value a_i^{baseline} as follows where g_0 is the embedding for the special self-loop relation of the attention baseline:

$$\begin{aligned} \bar{c}_i^{\text{baseline}} &= \mathbf{W_1}[\bar{h}_i \circ \bar{g}_0], \\ a_i^{\text{baseline}} &= \text{LeakyReLU}(\mathbf{W_a}[\bar{h}_i \circ \tanh(\bar{c}_i^{\text{baseline}})]). \end{aligned} \tag{2}$$

For the attention score, normalize the baseline value and the attention value of other neighbors together as below in Eq. 3. Then, get the new embedding, $\bar{h}_i{}'$, of entity h_i as shown in Eq. (4), composed of the weighted neighbor and self representation.

$$\begin{aligned} \alpha_{ijk} &= \frac{\exp(a_{ijk})}{\exp(a_i^{\text{baseline}}) + \sum_{m \in \mathcal{N}_i} \sum_{n \in \mathcal{R}_{im}} \exp(a_{imn})}, \\ \alpha_i^{\text{baseline}} &= \frac{\exp(a_i^{\text{baseline}})}{\exp(a_i^{\text{baseline}}) + \sum_{m \in \mathcal{N}_i} \sum_{n \in \mathcal{R}_{im}} \exp(a_{imn})}. \end{aligned} \tag{3}$$

$$\bar{h}_i{}' = \sigma(\alpha_i^{\text{baseline}} \bar{c}_i^{\text{baseline}} + \sum_{j \in \mathcal{N}_i} \sum_{k \in \mathcal{R}_{ij}} \alpha_{ijk} \bar{c}_{ijk}). \tag{4}$$

The non-linear function σ we choose is ELU. [5].

$\alpha_i^{\text{baseline}}$ in Eq. (4) can be regarded as a gate that controls the amount of neighbor information. As the neighbor's attention values are lower, the new embedding $\bar{h_i}'$ contains less neighbor information, and vice versa. A similar concept is the self-attention mechanism proposed by [12]. However, it cannot be applied to multi-relational graphs, so this mechanism is removed in KBGAT and replaced by residual connections. Nevertheless, the residual connection cannot be used to control the intensity of neighbor information. Our method can be viewed as a generalization of self-attention for multi-relational graphs.

Finally, let the output entity embedding be $\bar{h_i}^o = \mathbf{W_E}\bar{h}_i + \bar{h}_i'$ where $\mathbf{W_E} \in \mathbb{R}^{d\times d}$; $\bar{h_i}$ and $\bar{h_i}'$ are the initial and new entity embedding respectively. Similarly, let the output relation embedding be $\bar{g_k}^o = \mathbf{W_R}\bar{g}_k$ where $\mathbf{W_R} \in \mathbb{R}^{d\times d}$.

This is the proposed attention module for 1-hop neighbors. The attention module beyond 1-hop will be described in Sect. 3.4.

3.3 Confidence Score Prediction and Training Objective

Confidence Score Prediction. To estimate the confidence score of a given triplet $l = (h_i, r_k, h_j)$, firstly, use DistMult [15] as the scoring function to get the score $p(l) = \bar{h_i}^o \cdot (\bar{g}_k^o \circ \bar{h}_j^o)$. Then, map $p(l)$ to the [0,1] range to get the confidence score prediction: $S(l) = S(p(l)) = \text{Sigmoid}\Big(w \cdot p(l) + b\Big)$, where w and b are the weight and bias for the linear transformation.

Training Objective . Given a triplet $l = (h, r, t)$ and its confidence score s_l, where $h, t \in \mathcal{E}, r \in \mathcal{N}, s_l \in [0, 1]$, use the mean-square-error loss function to make $S(l)$ approximate s_l.

For a triplet l from the training triplet set $\mathcal{L}^+$ and its confidence score s_l, $\mathcal{J}^+ = \frac{1}{|\mathcal{L}^+|}\sum_{l\in\mathcal{L}^+} ||S(l) - s_l||^2$. For a triplet l from the randomly drawn negative triplet set $\mathcal{L}^-$, assume $s_l = 0$ such that $\mathcal{J}^- = \frac{1}{|\mathcal{L}^-|}\sum_{l\in\mathcal{L}^-} ||S(l)||^2$. Finally, sum the two terms up to get the total loss for the 1-hop model: $\mathcal{J} = \mathcal{J}^+ + \lambda_1\mathcal{J}^-$, where λ_1 is a hyper-parameter set to 1 by default.

3.4 Pseudo-neighbor Augmented Graph Attention Network

This subsection covers how to extend the base 1-hop attention module to n-hop and concludes the proposed UPGAT model. In deterministic graphs, multi-hop features can be represented as auxiliary relations, which are the summation of the relation embeddings in the path between any n-hop entity pair. However, in an uncertain graph, the confidence score of these relations is unknown without data-dependent inference rules. We use pseudo-neighbors to overcome the limitation.

After training the 1-hop model, build the *pseudo-neighbor augmented uncertain knowledge graph*, consisting of the original KG and the predicted false-negative triplets. Precisely, use the 1-hop model as the teacher model to predict false-negative tail entities for all $(head, relation)$ pairs in the training data. Then, select those with top-k predicted confidence scores to form the new augmented dataset. These pseudo-labeled triplets are the missing neighbors that may span across one or multiple hops in the original graph. The rationale for using the top-k filtering is that it causes more damage to the model to mispredict triplets with higher confidence scores as negative.

Finally, train a student model from scratch on the pseudo-neighbor augmented uncertain knowledge graph as the final model of the UPGAT. Pseudo-neighbors enable the graph attention to extend its attention range beyond 1-hop, no longer limited by the uncertainty of the paths. Note that the pseudo-labeling process follows a two-staged teacher-student schema to improve training stability. However, other semi-supervised learning strategies can be applied as well.

Formally, for the attention module beyond 1-hop, modify Eqs. (3, 4) by replacing $\mathcal{N}_i$ and $\mathcal{R}_{im}$ in the summation with $\mathcal{N}_i' = \mathcal{N}_i \cup \mathcal{N}_i^{pseudo}$ and $\mathcal{R}_{im}' = \mathcal{R}_{im} \cup \mathcal{R}_{im}^{pseudo}$ where $\mathcal{N}_i^{pseudo}$ is the predicted false-negative neighbor set of entity h_i; $\mathcal{R}_{ij}^{pseudo}$ is the set of predicted relations bridging h_i and h_j, which can be both 1-hop and multi-hop in the original graph.

$\mathcal{J}^{pseudo}$ and $\mathcal{J}^{semi}$ shown below are the loss function of the predicted samples $\mathcal{L}^{pseudo}$ and the total loss for the semi-supervised learning of the student model respectively, where λ_2 is the hyper-parameter.

$$\begin{aligned} \mathcal{J}^{pseudo} &= \frac{1}{|\mathcal{L}^{pseudo}|} \sum_{l \in \mathcal{L}^{pseudo}} ||f(l) - s_l^{pseudo}||^2. \\ \mathcal{J}^{semi} &= \mathcal{J}^{+} + \lambda_1 \mathcal{J}^{-} + \lambda_2 \mathcal{J}^{pseudo}. \end{aligned} \tag{5}$$

4 Experiment

We evaluate our model with two tasks: confidence score prediction (CSP) and tail entity prediction (TEP). Sect. 4.2 compares UPGAT with existing works. The ablation studies in Sect. 4.3 intend to answer the following questions: (i) How well native solutions perform; (ii) If the attention baseline mechanism boosts the performance; (iii) If the pseudo-neighbor augmented graph attention successfully models multi-hop relations. Lastly, in Sect. 4.4, we verify if the proposed method works well on the deterministic KG.

4.1 Settings

We used the same training and testing datasets, CN15K/NL27K/PPI5K, and the same evaluation metrics as used by [4]. The CSP task is a regression problem to predict the confidence score given a triplet, so we take MSE as the metric. The TEP task is a ranking problem to predict the tail entity given a head entity and a relation so we choose weighted mean rank, weighted Hit@k, and NDCG as the metrics. Chen et al. [4] proposed new evaluation metrics such as WMR (Weighted Mean Rank) and WH@K, which are MR (mean rank) and Hit@K linearly weighted by confidence scores. The new metrics are claimed to be more suitable for the uncertain knowledge graph. We keep this setting for compatibility as these metrics show identical trends as the original ones, e.g., MR and H@K.

Table 1. Tail Entity Prediction & Confidence Score Prediction. (MSE in 0.01). Boldface indicates the best value for a metric.

		TEP				CSP	
datasets	models	WMR	WH@20	WH@40	NDCG	MSE+	MSE-
CN15K	UKGE_{logi} [3]	1676.0	32.2%	38.5%	29.7%	28.2	**0.17**
	PASSLEAF [4]	1326.3	34.2%	41.3%	**30.4%**	23.8	0.36
	SUKE [13]	1849.5	32.3%	38.3%	29.8%	30.4	**0.05**
	UPGAT (ours)	**1098.7**	**36.0%**	**44.4%**	28.7%	**18.3**	0.27
NL27K	UKGE_{logi} [3]	288.6	70.4%	76.8%	71.7%	7.9	0.32
	PASSLEAF [4]	242.3	71.8%	77.9%	**74.5%**	5.5	0.38
	SUKE [13]	268.7	71.5%	78.2%	73.8%	4.1	**0.03**
	UPGAT (ours)	**109.2**	**72.0%**	**78.4%**	73.3%	**2.6**	0.10
PPI5K	UKGE_{logi} [3]	38.6	42.6%	68.8%	43.9%	0.76	0.28
	PASSLEAF [4]	34.9	45.1%	70.6%	44.5%	0.51	0.30
	SUKE [13]	37.0	45.9%	71.3%	**45.3%**	0.52	**0.17**
	UPGAT (ours)	**32.4**	**46.1%**	**72.2%**	44.6%	**0.34**	0.22

Our deterministic KGs settings follow those in [3] to binarize the uncertain KGs of CN15K/NL27K/PPI5K with the thresholds of 0.8/0.8/0.85. For evaluation metrics, we use the MR (Mean Rank) and Hit@K. The comparison of UKG embedding methods with DKG methods is unfair and beyond the scope of this paper, as they are optimized for the unique characteristics of DKG.

The UKGE_{logi} model is UKGE [3] trained without PSL-enhanced data to get the initial entity and relation embeddings. For other hyper-parameters, we choose Adam optimizer with learning rate = (5e−4), batch size = 512, embeddings size = 512, $k = 20$ pseudo-labeled triplets, and negative sampling ratio = 10.

4.2 Results and Analysis

The experimental results of the TEP and CSP task are presented in Table 1. MSE+ and MSE− are the MSE on in-dataset positive samples and randomly drawn negative samples, respectively. We choose models with similar score function as the baselines, which, therefore, shares similar mathematical structures in the embedding space. Values for UKGE_{logi} and PASSLEAF (with Distmult) are from [4]; values of SUKE are our reproduced results without PSL augmented data. FocusE [9] is not listed as it can be an add-on layer to most models.

Experimental results show that our method outperforms the existing methods in most of the evaluation metrics on the TEP task; there is over 50% WMR improvement on the NL27K dataset. Note that the NDCG score is dominated by a very small number of candidates with top confidence scores. The discounted factor of NDCG is a logarithmic function. Therefore, as the ranking goes up, the penalties for the candidates with lower confidence scores become significantly less. For example, the ratio of the discounted factor of the 3rd to 1st candidate is the same as the 99th to 9th candidate.

Table 2. Ablation Study - TEP & CSP (MSE in 0.01). Boldface indicate better value than existing methods on a metric.

		TEP				CSP	
dataset	models	WMR	WH@20	WH@40	NDCG	MSE+	MSE-
CN15K	1-hop w/o AB	**1215.6**	32.8%	**42.0%**	27.4%	**19.7**	0.29
	1-hop w/o AB + Naive	**1210.3**	32.9%	**42.1%**	27.7%	**19.7**	0.28
	1-hop	**1160.8**	**35.5%**	**43.3%**	28.3%	**19.1**	0.27
	1-hop+TS	**1111.9**	**35.8%**	**44.3%**	28.6%	**19.0**	0.26
	UPGAT (pseudo-neighbor+TS)	**1098.7**	**36.0%**	**44.4%**	28.7%	**18.3**	0.27
NL27K	1-hop w/o AB	**160.6**	70.9%	78.0%	71.1%	**2.9**	0.10
	1-hop w/o AB + Naive	**160.7**	70.8%	78.0%	71.1%	**2.9**	0.10
	1-hop	**151.5**	71.3%	78.2%	73.2%	**2.9**	0.10
	1-hop+TS	**119.3**	**71.8%**	**78.4%**	73.2%	**2.7**	0.10
	UPGAT (pseudo-neighbor+TS)	**109.2**	**72.1%**	**78.4%**	73.3%	**2.6**	0.10
PPI5K	1-hop w/o AB	35.1	44.2%	70.4%	43.1%	**0.42**	0.30
	1-hop w/o AB + Naive	**34.8**	45.2%	71.4%	43.0%	**0.41**	0.31
	1-hop	**33.2**	45.9%	70.9%	43.1%	**0.35**	0.24
	1-hop+TS	**32.9**	**46.0%**	**71.7%**	44.1%	**0.36**	0.27
	UPGAT (pseudo-neighbor+TS)	**32.4**	**46.1%**	**72.2%**	44.6%	**0.34**	0.22

For the CSP task, experimental results show that our method outstrips the existing methods in MSE+, reducing by up to 36% relative to the best model on NL27K. Contrarily, we incline not to emphasize the MSE of negative samples (MSE-) since there are no ground-truth negative labels in these three datasets. SUKE performs better on the negative MSE of the three datasets because it uses the evaluator to assign all low confidence scores to zero values, which we consider to be a post-processing method and does not imply better CSP accuracy.

4.3 Ablation Study

Table 2 is the result of the ablation study for the questions at the start of Sect. 4.

For the model name abbreviations, [1-hop] indicates the 1-hop attention model, depicted in Sect. 3.2; [1-hop w/o AB] refers to the 1-hop model without the attention baseline mechanism. Note that even the [1-hop w/o AB] model outperforms existing methods on most metrics, verifying our claimed advantage of incorporating subgraph features for UKG.

The naive solutions to model uncertainty: [1-hop w/o AB] vs [1-hop w/o AB +Naive]. [1-hop w/o AB + Naive] represents the three naive solutions discussed in Sect. 2.2 that explicitly model confidence score in the attention mechanism. As they have similar performance, only the one with the best WMR is shown. These three solutions achieve limited or no improvement, supporting the founding assumption that the plausibility of a neighbor cannot imply how relevant it is to the centered triplet.

The attention baseline: [1-hop] vs [1-hop w/o AB]. From the two models, it is evident that adding the attention baseline mechanism greatly improves

Table 3. TEP with the deterministic settings depicted in Sect. 4.1. (H@20 in %)

	CN15K		NL27K		PPI5K	
models	MR	H@20	MR	H@20	MR	H@20
UKGE$_{logi}$	3586.4	27.6	335.5	70.6	57.9	52.9
SUKE	3033.1	27.3	330.1	70.7	57.9	53.0
PASSLEAF	2402.0	27.7	312.6	71.5	55.2	54.8
UPGAT (ours)	**2368.5**	**28.0**	**288.1**	**72.1**	**53.3**	**55.6**

Table 4. Ablation Study for the Attention Baseline Mechanism in TPE. The (.) indicate the relative change w.r.t. (1-hop w/o AB).

		Deterministic		Uncertain	
	models	MR	H@20	WMR	WH@20
CN15K	1-hop w/o AB	2389.0	27.6%	1215.6	32.8%
	1-hop	2386.6 (−0.1%)	27.7% (+0.4%)	1160.8 (−4.5%)	35.5% (+8.2%)
NL27K	1-hop w/o AB	312.7	71.4%	160.6	70.9%
	1-hop	307.9 (−1.5%)	71.6% (+0.2%)	151.5 (−5.6%)	71. 3%(+0.6%)
PPI5K	1-hop w/o AB	55.3	54.7%	35.1	44.2%
	1-hop	55.3 (−0.1%)	54.8% (+0.2%)	33.2 (−5.2%)	45.9% (+3.8%)

its performance in most metrics. We attribute this to the mitigation of the irrelevant neighbor problem. The attention baseline mechanism strengthens the ability to extract the relevant information, outperforming the baseline methods that explicitly model the confidence score.

Pseudo-neighbor augmented graph attention: [UPGAT] vs [1-hop + TS] vs [1-hop]. We break down the contributions of the teacher-student semi-supervised learning (TS) and pseudo-neighbors. [1-hop + TS] generates pseudo-neighbors as UPGAT does, but the data are not used in the neighbor aggregation. Results show that applying graph attention over pseudo-neighbor augmented graph can further advance in all metrics, given that TS has already brought notable improvement over the [1-hop] model. Such results suggest that it is viable and effective to model multi-hop relations with the pseudo-neighbors in a data-driven manner.

4.4 Deterministic Settings

To verify if our proposed method is compatible with the deterministic KG (DKG), we compare UPGAT with existing UKG methods on DKGs in the TEP task. The result is presented in Table 3. UPGAT outperforms other methods on all metrics. As the number of ground truth labels is only 20% of the original uncertain settings, only H@20 is shown.

The ablation study shown in Table 4 further compares the attention baseline mechanism on the deterministic and uncertain KG. The attention baseline improves UKGs but has limited benefits for DKGs. For example, this mechanism brings 3.8% WH@20 improvement for the uncertain setting and only 0.2% H@20 improvement for the deterministic setting in the PPI5k dataset. This agrees with our argument that the attention baseline is effective for extending the graph attention to UKGs where uncertainty complicates the neighboring features.

5 Conclusion and Future Work

The proposed Uncertainty-Aware Pseudo-neighbor Augmented Knowledge Graph Attention Network (UPGAT) is the first work to model the subgraph feature on uncertain KGs. The attention baseline mechanism generalizes the self-attention model for multi-relational graphs with uncertainty; The pseudo-neighbors successfully model multi-hop relations. Our model gets promising improvements over existing works on both uncertain and deterministic KGs. While this paper focuses on a mechanism of fixed attention weight for each entity, we believe the weightings of different relation queries are also important. Namely, the attention mechanism must be "relation query-aware", which we leave for future studies.

References

1. Bordes, A., Usunier, N., Garcia-Durán, A., Weston, J., Yakhnenko, O.: Translating embeddings for modeling multi-relational data. In: NIPS, pp. 2787–2795 (2013)
2. Chen, X., Boratko, M., Chen, M., Dasgupta, S.S., Li, X.L., McCallum, A.: Probabilistic box embeddings for uncertain knowledge graph reasoning. In: NAACL'21 (2021)
3. Chen, X., Chen, M., Shi, W., Sun, Y., Zaniolo, C.: Embedding uncertain knowledge graphs. In: AAAI (2019)
4. Chen, Z.M., Yeh, M.Y., Kuo, T.W.: Passleaf: a pool-based semi-supervised learning framework for uncertain knowledge graph embedding. AAAI **35**(5) (2021)
5. Clevert, D.A., Unterthiner, T., Hochreiter, S.: Fast and accurate deep network learning by exponential linear units (elus) (2016)
6. Kipf, T.N., Welling, M.: Semi-supervised classification with graph convolutional networks. In: ICLR (2017)
7. Lin, Y., Liu, Z., Luan, H., Sun, M., Rao, S., Liu, S.: Modeling relation paths for representation learning of knowledge bases. In: EMNLP (2015)
8. Nguyen, D.Q., Nguyen, T.D., Nguyen, D.Q., Phung, D.: A novel embedding model for knowledge base completion based on convolutional neural network. In: NAACL (Short Papers), vol. 2, pp. 327–333 (2018)
9. Pai, S., Costabello, L.: Learning embeddings from knowledge graphs with numeric edge attributes. In: IJCAI, pp. 2869–2875 (2021)
10. Qu, M., Tang, J.: Probabilistic Logic Neural Networks for Reasoning. Curran Associates Inc. (2019)
11. Schlichtkrull, M., Kipf, T.N., Bloem, P., van den Berg, R., Titov, I., Welling, M.: Modeling relational data with graph convolutional networks. In: ESWC (2018)
12. Veličković, P., Cucurull, G., Casanova, A., Romero, A., Liò, P., Bengio, Y.: Graph attention networks. In: ICLR (2018)

13. Wang, J., Nie, K., Chen, X., Lei, J.: Suke: embedding model for prediction in uncertain knowledge graph. IEEE Access **9**, 3871–3879 (2021)
14. Xie, Z., Zhu, R., Zhao, K., Liu, J., Zhou, G., Huang, J.X.: Dual gated graph attention networks with dynamic iterative training for cross-lingual entity alignment. ACM Trans. Inf. Syst. **40**(3) (2021)
15. Yang, B., tau Yih, W., He, X., Gao, J., Deng, L.: Embedding entities and relations for learning and inference in knowledge bases (2015)
16. Zhang, Y., et al.: Efficient probabilistic logic reasoning with graph neural networks. In: ICLR (2020)

Mining Frequent Sequential Subgraph Evolutions in Dynamic Attributed Graphs

Zhi Cheng[1](✉), Landy Andriamampianina[1,2], Franck Ravat[2], Jiefu Song[2], Nathalie Vallès-Parlangeau[3], Philippe Fournier-Viger[4], and Nazha Selmaoui-Folcher[5]

[1] Activus Group, 1 Chemin du Pigeonnier de la Cépière, 31100 Toulouse, France
{zhi.cheng,landy.andriamampianina}@activus-group.fr

[2] IRIT-CNRS (UMR 5505) - Université Toulouse 1 Capitole (UT1), 2 Rue du Doyen Gabriel Marty F-31042 Toulouse Cedex 09, Toulouse, France
{franck.ravat,jiefu.song}@irit.fr

[3] Université de Pau et des Pays de l'Adour, Pau, France
nathalie.valles-parlangeau@univ-pau.fr

[4] Shenzhen University, Shenzhen, China
philfv@szu.edu.cn

[5] University of New Caledonia, ISEA, BP R4, 98851 Nouméa, New Caledonia
nazha.selmaoui@unc.nc

Abstract. Mining patterns in a dynamic attributed graph has received more and more attention recently. However, it is a complex task because both graph topology and attributes values of each vertex can change over time. In this work, we focus on the discovery of frequent sequential subgraph evolutions (FSSE) in such a graph. These FSSE patterns occur both spatially and temporally, representing frequent evolutions of attribute values for general sets of connected vertices. A novel algorithm, named FSSEMiner, is proposed to mine FSSE patterns. This algorithm is based on a new strategy (graph addition) to guarantee mining efficiency. Experiments performed on benchmark and real-world datasets show the interest of our approach and its scalability.

Keywords: dynamic attributed graph · frequent sequential pattern · graph mining

1 Introduction

Dynamic attributed graphs have recently received a lot of attention [6]. The reason is that this graph type provides a rich representation of real-world phenomena. It has been widely used to describe many complex datasets (e.g. spatio-temporal data, health data, biological data or social data) [2,3,8]. A dynamic attributed graph depicts a time-ordered sequence of graphs to capture the evolution of a real-world phenomena. Specifically, vertices and edges between vertices of each graph of the sequence represent respectively objects and spatial relations or other types of interactions between objects that are valid at the graph timestamp. Attributes are used to complete the semantics of vertices. Objects and their relations may

H. Kashima et al. (Eds.): PAKDD 2023, LNAI 13936, pp. 66–78, 2023.
https://doi.org/10.1007/978-3-031-33377-4_6

evolve over time. Indeed, changes may happen in two levels: at the topological level, there may be addition and removal of objects and relations; at the object level, changes may also happen in attribute values. Mining patterns in a dynamic attributed graph allows analysing how objects, relations and attribute values of objects evolve over time.

Existing pattern mining approaches in dynamic attributed graphs allow following sequential evolutions within an individual vertex [8] or a set of vertices [1,5,7] that occur frequently over time. None of these approaches allows finding frequent sequential evolutions for general sets of connected vertices (i.e., frequent subgraphs). For instance, in the case of monitoring the spread of a virus, existing patterns may reveal sequential changes of individuals' health status within a specific group. However, if this specific group is special (for example, they have innate resistance to this virus, or they were vaccinated), existing patterns lose general representativeness and cannot provide meaningful analysis for virus transmission. Indeed, the objective of doctors is to understand how virus spread among general groups instead of specific ones. Therefore, our objective is to find frequent sequential evolutions for general sets of connected vertices. To do so, we propose a novel pattern type denoted as *frequent sequential subgraph evolutions (FSSE)*. After describing a formal representation of FSSE (Sect. 3), we present the *FSSEMiner* algorithm to find FSSE patterns in a dynamic attributed graph (Sect. 4). The scalability of the algorithm is studied through some experimental assessments based on both real-world and benchmark datasets (Sect. 5).

2 Related Work

Several research works are proposed to analyse dynamic attributed graphs. Desmier et al. [2] defined a *cohesive co-evolution pattern*. This pattern represents a set of vertices with the same attribute variations and a similar neighbourhood during a time interval. The authors extended their work in [3] by integrating constraints on topology and on attribute values to extract maximal dynamic attributed subgraphs. Yet, these patterns do not represent sequential evolutions.

Kaytoue et al. [8] defined the *triggering pattern* problem, which allows finding temporal relationships between attribute values and topological properties of vertices. The TRIGAT algorithm allows mining triggering patterns by using a projection strategy. However, triggering patterns do not consider neighbouring vertices, neither their evolutions since they focus on a single vertex.

Fournier-Viger et al. [5] proposed *significant trend sequences* patterns. Such patterns allow discovering the influence of attribute variations of a single vertex on its neighbours. The authors extended these patterns in [7] by defining *attribute evolution rules* (AER) to discover the influence of multiple vertices on other vertices. The AER-Miner algorithm allows mining AER patterns by using breadth first search (BFS) strategy. Yet, AER patterns represent sequential evolutions of attributes, but not sequential evolutions of connected vertices.

Cheng et al. [1] proposed a *recurrent pattern*, which is a frequent sequence of attribute variations for a set of connected vertices. The RPMiner algorithm, based on successive graph intersections strategy, allows mining recurrent patterns. However, recurrent patterns focus only on the evolutions of a specific set of vertices, as they depend on their vertices' temporal occurrences instead of considering the spatio-temporal occurrences.

In response to the previous limitations, we propose a novel pattern denoted as *frequent sequential subgraph evolutions (FSSE)*. Compared to existing work, the main advantage of this pattern is to consider evolutions independently of subgraphs in which they occur. In a spatio-temporal context, it means that such pattern would highlight phenomena independently of their locations. However, none of the previous algorithms mines FSSE patterns. Indeed, mining such patterns in a dynamic attributed graph is so complex that all existing strategies cannot guarantee both the mining efficiency and the completeness of patterns. For this purpose, we propose a novel algorithm called *FSSEMiner* to mine FSSE. The algorithm is based on a new strategy called *graph addition* to guarantee mining efficiency. It requires traversing only once each graph, instead of an exponential graph traversing operation.

3 Notations

3.1 Dynamic Attributed Graph

A dynamic attributed graph, denoted as $\mathcal{G} = \langle G_{t_1}, G_{t_2}, ..., G_{t_{max}} \rangle$, represents the evolution of a graph over a set of ordered and consecutive timestamps $T = \{t_1, t_2, ..., t_{max}\}$. It is composed of the set of vertices denoted as $\mathcal{V}$. The set of attributes $\mathcal{A}$ is used to describe all the vertices. Each attribute $a \in \mathcal{A}$ is associated with a domain value. A domain value $\mathbb{D}_a$ (numerical or categorical) is associated to each vertex and attribute $a \in \mathcal{A}$. For each time $t \in T$, $G_t = (V_t, E_t, \lambda_t)$ is an attributed undirected graph where: $V_t \subseteq \mathcal{V}$ is the set of vertices, $E_t \subseteq V_t \times V_t$ is the set of edges, and $\lambda_t : V_t \rightarrow 2^{\mathcal{A}\mathbb{D}}$ is a function that associates each vertex of V_t with a set of attribute values $\mathcal{A}\mathbb{D} = \cup_{a \in \mathcal{A}}(a \times \mathbb{D}_a)$.

When attribute values are numerical, a dynamic attributed graph is usually preprocessed to derive trend values from these attributes, since we are not interested in their absolute variations [6]. A trend is an increase (+), decrease (−) or stability (=) which means the value of a vertex's attribute increase, decrease or does not change between two consecutive timestamps.

3.2 A New Pattern Domain

Definition 1 (frequent subgraph). Let $(\lambda, Occurrence(\lambda)\, in\, T')$ be an attributed subgraph of $\mathcal{G}$, where λ is a set of attribute values (trends or categorical values) representing a pattern and $Occurrence(\lambda)\, in\, T'$ represents the occurrences of λ in the set of times $T' \subseteq T$. More precisely, $Occurrence(\lambda)\, in\, T'$ is a set of subgraphs such that $Occurrence(\lambda) \subseteq V_t$ where $t \in T', V_t \subseteq \mathcal{V}$. As shown in Fig. 1, $\langle \{(a_1 - a_2+, a_1 - a_2+, a_1 + a_2 =)\}, \{t_1 : (1, 2, 3)|t_1 : (7, 8, 10)|t_1 : (13, 14, 15)|t_2 : (1, 2, 3)|t_2 : (7, 8, 10)\}\rangle$ is an attributed subgraph in t_1 and t_2.

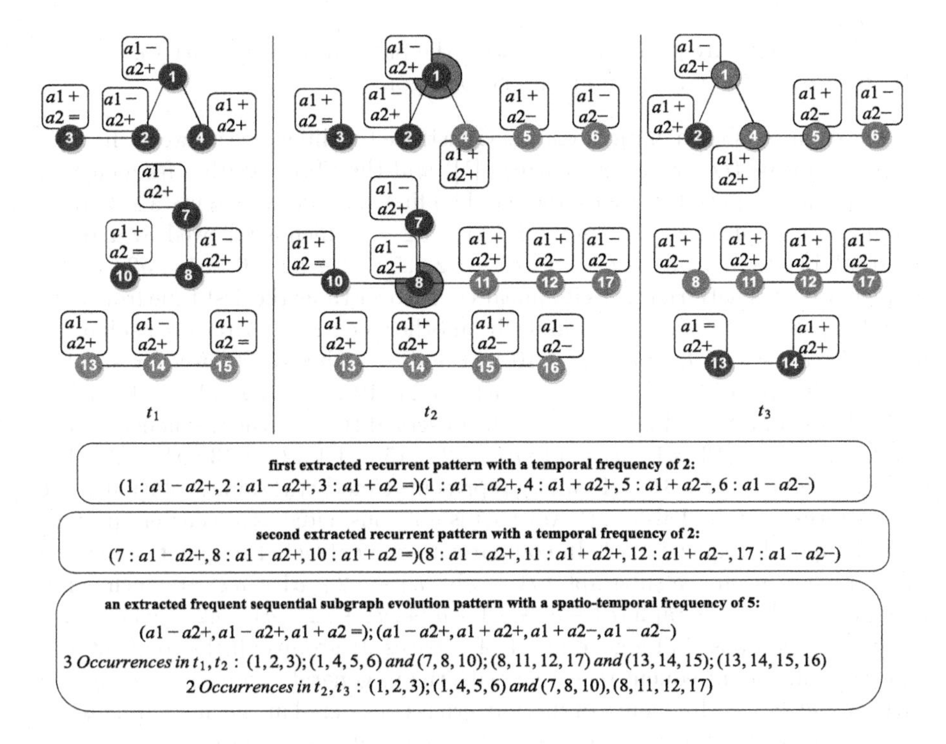

t_1
t_2
t_3
first extracted recurrent pattern with a temporal frequency of 2:
(1 : a1 − a2+, 2 : a1 − a2+, 3 : a1 + a2 =)(1 : a1 − a2+, 4 : a1 + a2+, 5 : a1 + a2−, 6 : a1 − a2−)
second extracted recurrent pattern with a temporal frequency of 2:
(7 : a1 − a2+, 8 : a1 − a2+, 10 : a1 + a2 =)(8 : a1 − a2+, 11 : a1 + a2+, 12 : a1 + a2−, 17 : a1 − a2−)
an extracted frequent sequential subgraph evolution pattern with a spatio-temporal frequency of 5:
(a1 − a2+, a1 − a2+, a1 + a2 =); (a1 − a2+, a1 + a2+, a1 + a2−, a1 − a2−)
3 Occurrences in t1, t2 : (1, 2, 3); (1, 4, 5, 6) and (7, 8, 10); (8, 11, 12, 17) and (13, 14, 15); (13, 14, 15, 16)
2 Occurrences in t2, t3 : (1, 2, 3); (1, 4, 5, 6) and (7, 8, 10), (8, 11, 12, 17)

Fig. 1. Dynamic attributed graph.

The set of attribute values $\lambda = (a_1 - a_2+, a_1 - a_2+, a_1 + a_2 =)$ represents a size-1 pattern, i.e., a pattern composed of only one set of attribute values. $Occurrence(\lambda)\, in\, t_1 = \{t_1 : (1,2,3)|t_1 : (7,8,10)|t_1 : (13,14,15)\}$ represents the set of occurrences of λ in t_1 (in red and orange). $Occurrence(\lambda)\, in\, t_2 = \{t_2 : (1,2,3)|t_2 : (7,8,10)\}$ represents the set of occurrences of λ in t_2 (in red). So the frequency of the subgraph is 5 because it occurs 5 times in time and space.

Definition 2 (frequent sequential subgraph evolution). A frequent sequential subgraph evolution (FSSE) of $\mathcal{G}$ appearing in the time intervals $\{T_1, \ldots, T_k\}$, $T_i \subseteq T$, is a sequence $P = \langle\{\lambda_1; ...; \lambda_n\}, \{T_1 : Occurrence_1(\lambda_1); \ldots; Occurrence_1(\lambda_n), | \ldots |T_k : Occurrence_k(\lambda_1); \ldots; Occurrence_k(\lambda_n)\}\rangle$. The first set represents the frequent sequential subgraphs where each subgraph is separated by semicolons. The second set is composed of all patterns' occurrences, where each occurrence is separated by vertical bars. For each $T_i = \{t_i, ..., t_j\}, 1 \leq i \leq j \leq |T|$, t_i represents the starting time of this occurrence and t_j represents the end time of this occurrence. As shown in Fig. 1, $\langle\{(a1 - a2+, a1 - a2+, a1 + a2 =); (a1 - a2+, a1 + a2+, a1 + a2-, a1- a2-)\}$, $\{t_1, t_2 : (1,2,3); (1,4,5,6)|t_1, t_2 : (7,8,10); (8,11,12,17)|t_1, t_2 : (13, 14, 15); (13, 14, 15, 16)|t_2, t_3 : (1,2,3); (1,4,5,6)|t_2, t_3 : (7,8,10); (8,11,12,17)\}\rangle$ is a FSSE starting at t_1 and t_2. It is a sequence of size 2 composed of a frequent subgraph with the pattern $(a1 - a2+, a1 - a2+, a1 + a2 =)$ in $\{t_1, t_2\}$ (in red and orange) and a frequent subgraph with the pattern $(a1-a2+, a1+a2+, a1+a2-, a1-a2-)$ in $\{t_2, t_3\}$ (in green and orange).

The example shown in Fig. 1 can also illustrate the difference between the closest pattern type (i.e., recurrent pattern [1]) and the FSSE. With a frequency of 2, two recurrent patterns are extracted. The first pattern represents a subgraph in red $(1 : a1-a2+, 2 : a1-a2+, 3 : a1+a2 =)$ which is followed by another subgraph in green $(1 : a1-a2+, 4 : a1+a2+, 5 : a1+a2-, 6 : a1-a2-)$. The frequency of this pattern is 2, as it appears twice over time: the first time from t_1 to t_2 and the second from t_2 to t_3. Similarly, another recurrent pattern is a subgraph in red $(7 : a1-a2+, 8 : a1-a2+, 10 : a1+a2 =)$ which is followed by another subgraph in green $(8 : a1-a2+, 11 : a1+a2+, 12 : a1+a2-, 17 : a1-a2-)$ with a frequency of 2. Moreover, it can be observed that another sequence of subgraphs in orange, $(13 : a1-a2+, 14 : a1-a2+, 15 : a1+a2 =)(13 : a1-a2+, 14 : a1+a2+, 15 : a1+a2-, 16 : a1-a2-)$, represents exactly the same evolution as the two patterns extracted above. However, it is not considered as a recurrent pattern as its temporal frequency is 1. In comparison, one extracted frequent sequential subgraph evolution is a subgraph $(a1-a2+, a1-a2+, a1+a2 =)$ which is followed by another subgraph $(a1-a2+, a1+a2+, a1+a2-, a1-a2-)$ where each subgraph is composed of a general set of vertices. It groups all the frequent and especially all the infrequent recurrent patterns to generate a much more general pattern. Indeed, the frequency of this pattern is considered in one more dimension, the spatial one. So, the spatio-temporal frequency of this pattern is 5.

3.3 Interesting Measures and Constraints

Let $P = \langle\{\lambda_1; ...; \lambda_n\}, \{T_1 : Occurrence_1(\lambda_1); \ldots; Occurrence_1(\lambda_n) | \ldots | T_k : Occurrence_k(\lambda_1); \ldots; Occurrence_k(\lambda_n)\}\rangle$ be a pattern. Several measures and constraints are defined for two purposes: (i) to let the user express his preferences to select patterns via a set of constraints, and (ii) to reduce the search space and improve the efficiency of the algorithm.

Spatio-temporal Frequency. The frequency constraint, denoted as ***minsup***, is a user-defined threshold to filter patterns which occur more than a minimum number in time and in space. The frequency of P is the number of occurrences of pattern P, $sup(P) = k$. Consequently, P is a frequent evolution iff $sup(P) \geq minsup$. For example, in Fig. 1, the frequency of the sequence $(a1-a2+, a1-a2+, a1+a2 =); (a1-a2+, a1+a2+, a1+a2-, a1-a2-)$ is 5, as the pattern appears 5 times.

Connectivity. During pattern extraction, vertices should be connected by edges to extract potentially correlated evolutions among a set of objects. In Fig. 1, the pattern $(a1-a2+, a1-a2+, a1+a2 =); (a1-a2+, a1+a2+, a1+a2-, a1-a2-)$ occurs in $\{t_1, t_2\}$ on a sequence of sets of connected vertices such as $(1, 2, 3); (1, 4, 5, 6)$.

Volume. The volume measure defines the number of vertices of a subgraph. Let $vol(P) = \min_{\forall i \in [1,n]} |\lambda_i|$ be the volume of a pattern P. A pattern P is sufficiently voluminous iff $vol(P) \geq minvol$, where $\boldsymbol{minvol}$ is a minimum number of vertices of a subgraph defined by the user. The user can also define the maximum number of vertices of a subgraph, denoted as $\boldsymbol{maxvol}$, such as $vol(P) \leq maxvol$. For example, the pattern $(a1-a2+, a1-a2+, a1+a2=); (a1-a2+, a1+a2+, a1+a2-, a1-a2-)$ has a volume of 3.

Temporal Continuity. An evolution may include different vertices at each timestamp. However, it is difficult for end users to interpret the evolution of vertices without a direct relation between them at each step. Hence, it is desirable to study evolution around a common core of vertices. To do so, a constraint, denoted as $\boldsymbol{mincom}$ and set by the user, is defined to follow a minimum number of common vertices over time. Let denote $Occurrence_j(P)$ is a jth instance of pattern P and $com(Occurrence_j(P)) = |\cap_{\forall i \in 1,...,n} Occurrence_j(\lambda_i)|$ be the common number of vertices occurring in the instance sequence j. P is a continuous pattern iff $\forall j \in \{1,..,k\}$ $com(Occurrence_j(P)) \geq mincom$. Consider P the pattern $(a1-a2+, a1-a2+, a1+a2=); (a1-a2+, a1+a2+, a1+a2-, a1-a2-)$ in Fig 1. All instances of the pattern P have at least one common vertex. For instance, the subgraphs of the occurrence $(7, 8, 10); (8, 11, 12, 17)$, at t_1 and t_2, have one common vertex, which is 8.

4 Mining Frequent Sequential Subgraph Evolutions

In this section, we propose an algorithm, called FSSEMiner, to mine FSSE patterns in a dynamic attributed graph. This algorithm allows dealing with the following mining problem: Given a dynamic attributed graph $\mathcal{G}$, the problem is to extract the complete set of frequent sequential subgraph evolutions in $\mathcal{G}$, denoted as Sol, such that $\forall P \in Sol$, (i) P is frequent (i.e., $sup(P) \geq minsup$); (ii) the occurrences of P are connected at each time; (iii) P is sufficiently voluminous (i.e., $minvol \leq vol(P) \leq maxvol$); (iv) P is centered around a core of vertices sufficiently large (i.e., $com(P) \geq mincom$), where $minvol, maxvol, minsup$ and $mincom$ are user-defined thresholds.

This algorithm solves the above-mentioned problem in three steps: (i) identify subgraphs (Sect. 4.1); (ii) count the spatio-temporal frequency of subgraphs (Sect. 4.2); (iii) construct sequences of subgraphs using frequent subgraphs (Sect. 4.3). The sequence of the three steps is illustrated via the FSSEMiner algorithm (Algorithm 1).

4.1 Extraction of Subgraph Candidates

The first step of the algorithm is to construct all possible candidate subgraphs (frequent and infrequent) based on a dynamic attributed graph $\mathcal{G}$ (Lines 1-2). More precisely, the algorithm constructs all possible sets of patterns λ_i whose volume is between $minvol$ and $maxvol$. For each generated pattern λ_i, a depth-first search (DFS) strategy is used to compute its occurrences $Occurrence(\lambda_i)$ in each $G_t \in \mathcal{G}$. The anti-monotonicity property is respected to find anti-monotonic subgraphs [4]. The result is the set of subgraphs satisfying the *volume* and

Algorithm 1: *FSSEMiner* : Mining frequent sequential subgraph evolutions

```
Input: 𝒢 : a dynamic attributed graph, minsup, minvol, maxvol, mincom
Output: Sol: set of frequent sequential subgraph evolutions satisfying the constraints
/* Step 1: Extraction of subgraph candidates                                    */
1  $ = {$_i set of subgraphs of G_t, t ∈ T | ∀s_i ∈ $_i, s_i = (λ_i, Occurrence(λ_i) in t), minvol ≤
     |Occurrence(λ_i)| ≤ maxvol}
2  Cand_i = ∅, ∀i ∈ {1, 2, ..., |T|}
/* Step 2: Generation size-1 patterns by graph addition                          */
3  for k = 1 to |T| do
4      for each T_1^k ⊆ T such as t_1 ∈ T_1^k do
5          P_union = {$_union set of frequent subgraphs in T_1^k, | ∀s_union ∈ $_union, s_union =
             (λ, Occurrence_union), Occurrence_union = Union(Occurrence_t) where (t ∈
             T_1^k) and |Occurrence_union| ≥ minsup}
6          Cand_1 = Cand_1 ∪ P_union
7      end
8  end
/* Step 3: Extension of patterns                                                 */
9  Sol_i = ∅, ∀i ∈ {1, 2, ..., |T|}
10 for i = 2 to |T| do
11     for each T_i^k ⊆ T such as t_i ∈ T_i^k do
12         for each P_i ∈ P_union = {$_union set of frequent subgraphs in T_i^k | ∀s_union ∈
             $_union, s_union = (λ, Occurrence_union), Occurrence_union =
             Union(Occurrence_t), where (t ∈ T_i^k) and |Occurrence_union| ≥ minsup} do
13             for each P such as P ∈ Cand_{i-1} do
14                 P' = ExtendWith(P, P_i)
15                 if com(P') ≥ mincom and |P'| ≥ minsup then
16                     Cand_i = Cand_i ∪ {P'}
17                 end
18                 else
19                     Sol_{i-1} = Sol_{i-1} ∪ {P}
20                     Cand_i = Cand_i ∪ {P_i}
21                 end
22             end
23         end
24     end
25 end
26 Sol = MergeUpdate(⋃_{∀i∈T} Sol_i)
```

connectivity constraints and denoted as $\$ = \{\$_i \text{ set of subgraphs of } G_t, t \in T \mid \forall s_i \in \$_i, s_i = (\lambda_i, Occurrence(\lambda_i)\, in\, t), minvol \leq |Occurrence(\lambda_i) \leq maxvol\}$. For example, occurrences of $(a1 - a2+, a1 - a2+, a1 + a2 =)$ are extracted at each time. They are represented by the two red connected subgraphs in t_1 and t_2 in Fig. 1.

4.2 Generation of Size-1 Patterns by Graph Addition

The second step of the algorithm is to combine the candidate subgraphs generated in the previous step (Sect. 4.1) to create size-1 patterns (i.e., sequences composed of one subgraph) (Line 3-8). The construction of size-1 patterns is the fundamental building block for constructing the final patterns. To do so, a new strategy, called graph addition, is proposed. It consists in adding the occurrences (of different times) of candidate subgraphs having the same pattern. Then, the algorithm verifies if the spatio-temporal frequency of this subgraph union respects the *minsup* constraint.

Let be n times $t_i, ..., t_j \in T$, where $1 \leq n \leq |T|$ and $1 \leq i < j \leq |T|$. The addition of n subgraphs $s_i = (\lambda_i, Occurrence(\lambda_i)\, in\, t_i)$,..., $s_j = (\lambda_j, Occurrence(\lambda_j)\, in\, t_j)$ is denoted as $s_{union} = (\lambda, Occurrence_{union}(\lambda))$ where $\lambda = \lambda_i = ... = \lambda_j$ and $Occurrence_{union}(\lambda) = Occurrence(\lambda_i)\, in\, t_i \cup ... \cup Occurrence(\lambda_j)\, in\, t_j$. s_{union} is a subgraph composed of the union of occurrences of the n initial subgraphs having the same pattern (i.e., same attribute values). If $|Occurrence_{union}(\lambda)| \geq minsup$, the algorithm keeps s_{union} in the mining process. For the special case where $n = 1$ and $i = j$ the result of the addition of a subgraph is itself. However, this case is necessary because a size-1 pattern (one subgraph) could also be spatially frequent in one timestamp.

Graph addition is applied to all sets of time combinations, denoted as T^k, $1 \leq k \leq |T|$. The number of linear additions is $|T^k| = 2^{|T|} - 1$ and depends only on the number of timestamps in $\mathcal{G}$. The advantage of the graph addition strategy is to avoid performing a huge amount of subgraph traversals for the generation of patterns of size 1.

Let us suppose that $minsup = 4$. In Fig. 1, there is the subgraph $s_1 = \langle\{(a1 - a2+, a1 - a2+, a1 + a2 =)\}, \{t_1 : (1, 2, 3)|t_1 : (7, 8, 10)|t_1 : (13, 14, 15)\}$ in t_1 and $s_2 = \langle\{(a1 - a2+, a1 - a2+, a1 + a2 =)\}, \{t_2 : (1, 2, 3)|t_2 : (7, 8, 10)\}$ in t_2 having the same attribute values. By adding s_1 and s_2, the pattern $s_{union} = \langle\{(a1 - a2+, a1 - a2+, a1 + a2 =)\}, \{t_1 : (1, 2, 3)|t_1 : (7, 8, 10)|t_1 : (13, 14, 15)|t_2 : (1, 2, 3)|t_2 : (7, 8, 10)\}$ is obtained. It can be observed that s_1 and s_2 are infrequent. However, s_{union} is frequent after the addition of the subgraphs.

4.3 Extension of Patterns

The final step of the algorithm is to construct the complete sequential patterns by extending each size-1 pattern of each successive set of times generated in the previous step (Sect. 4.2) (Line 9-26). To do this, size-1 patterns are iteratively extended by checking the *mincom* and *minsup* constraints to connect other consecutive patterns to build sequences of frequent subgraphs. This extension can be achieved by processing the times incrementally. Figure 2 shows an incremental construction of a pattern beginning from $\{t_1, t_2\}$. This figure displays the parallel extensions of a pattern which occurs at t_1 and t_2. Let s, s' and s^* be frequent subgraphs extracted in graph additions. Additions between S_1 and S_2 result in a set of frequent subgraphs, such that $s = (\lambda, Occurrence(\lambda)\, in\, t_1, t_2) \in S_1 \cup S_2$. Candidate extensions for these subgraphs can only be at t_2 and t_3 respectively (since gaps are not allowed). Now consider times t_2, t_3 and suppose that $s' = (\lambda', Occurrence(\lambda)'\, in\, t_2, t_3)$ is a frequent subgraph of $S_2 \cup S_3$. If s and s' have at least *minsup* occurrences verifying the temporal continuity constraint, then we can extend s with s' to obtain $P = \langle\{\lambda; \lambda'\}, \{t_1, t_2 : Occurrence(\lambda)\, in\, t_1, t_2; t_2, t_3 : Occurrence(\lambda')\, in\, t_2, t_3\}\rangle$. The process continues until no more extensions can be performed. At each iteration, subgraphs can be used to extend patterns from the previous iteration, but they can also be "starting points" for new patterns. For a sequence of 3 subgraphs, the pattern will be constructed and extended seven times (from $\{t_1\}$, from $\{t_2\}$, from $\{t_3\}$, from $\{t_1, t_2\}$, from $\{t_2, t_3\}$, from $\{t_1, t_3\}$, from $\{t_1, t_2, t_3\}$).

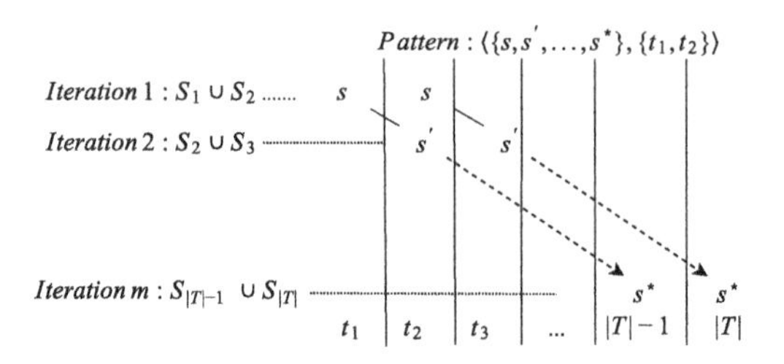

Fig. 2. Additions and extensions of patterns from $\{t_1, t_2\}$

Although the study of the combination $\{t_1, t_2\}$ does not bring more information compared to $\{t_1, t_2, t_3\}$, it allows discovering other patterns to be extended. All these time combinations are therefore necessary. This highlights the importance of the proposed graph addition strategy described above, which requires only $|T|$ times instead of $2^{|T|} - 1$ graph traversals in a dynamic attributed graph.

5 Experiments

In this section, the performance of the FSSEMiner algorithm has been evaluated. The algorithm was implemented in C ++. Experiments were conducted on a PC (CPU: Intel(R) Core(T:) 3.5 GHz) with 16 GB of main memory.

Datasets. Benchmark data were generated by varying different parameters of a dynamic attributed graph: the number of vertices/attributes/edges and the number of graphs of the sequence (or the number of timestamps) by setting the other parameters $(minsup, minvol, maxvol, mincom)$. Edges (pairs of vertices) and attribute values follow a uniform distribution. The first real-world dataset is the Domestic US Flights traffic dataset during the Katrina hurricane period (from 01/08/2005 to 25/09/2005) [3]. It is composed of 280 vertices (airports) and 1206 edges in average (flight connections) per timestamp, 8 timestamps (data are aggregated by weeks) and 8 attributes (e.g. number of departures/arrivals/cancelled flights). The second real-world dataset is composed by a travel flows in China dataset[1] and a COVID-19 daily cases dataset[2] during two periods (from 25/01/2020 to 20/03/2020 and from 15/04/2022 to 15/05/2022). It is composed of 232 vertices (cities) and 13260 edges (travel flows between cities) in average, 6 timestamps (data are aggregated every 3 days) and 4 attributes: the size of the city (small, medium-sized, big and megacity according to the population), the total number of new COVID cases, of deaths, of recoveries since 25/01/2020, with 4 values (=:no new cases, +:]0,5],++:]5,15] and +++:]15,]).

Quantitative Results. We have conducted a quantitative analysis of patterns extracted from benchmark datasets to evaluate the scalability of the proposed

[1] https://dataverse.harvard.edu/dataset.xhtml?persistentId=doi:10.7910/DVN/FAEZIO.

[2] https://dataverse.harvard.edu/dataset.xhtml?persistentId=doi:10.7910/DVN/MR5IJN.

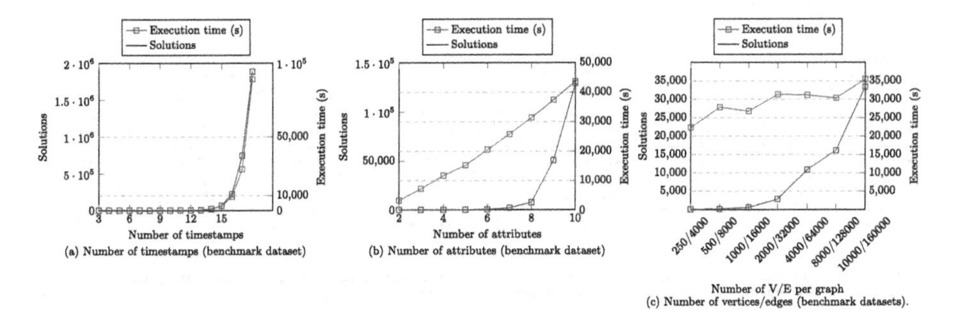

Fig. 3. Impacts of parameters on execution times of FSSEMiner

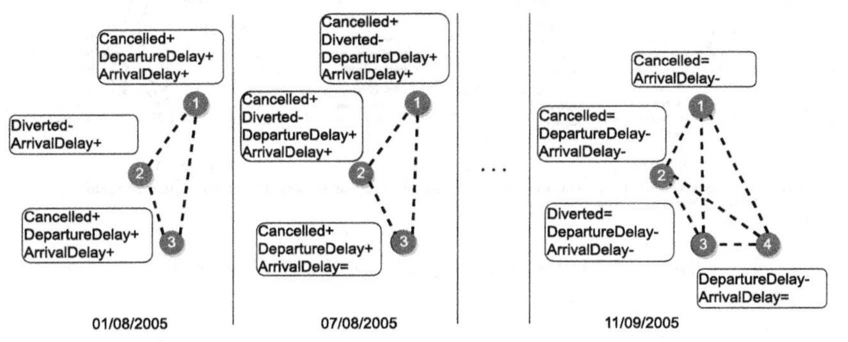

Fig. 4. A pattern in US Flights dataset

algorithm. Figure 3 (a) shows the impact of the number of timestamps on the algorithm's runtimes for 2000 vertices and 8000 edges per graph, 2 attributes and $minvol = maxvol = 2, mincom = 1, minsup = 60\%$ of the average number of vertices. This impact is high, but performance of the algorithm is comparable with the one in [1] since it generates more general and more complex patterns. Figure 3 (b) shows the impact of the number of attributes on the algorithm's runtimes for 8 timestamps and the other parameters are fixed as before. Execution times remain low for less than 8 attributes, so for most of the real-world datasets. Figure 3 (c) shows the impact of the number of vertices and edges at each timestamp on the algorithm's runtimes (the other parameters are fixed as before). It can be noticed that the algorithm remains efficient for large dense graphs (10,000 vertices and 160,000 edges at each timestamp).

Qualitative Results. We have conducted a qualitative analysis of patterns extracted on the real-world datasets. Figure 4 shows an example of a pattern extracted from the US Flights traffic dataset ($minvol = 2, maxvol = 4, minsup = 25, mincom = 1$). This pattern appears 28 times in the dataset. It shows the impact of hurricanes on US airport traffic for 6 weeks. First, it is

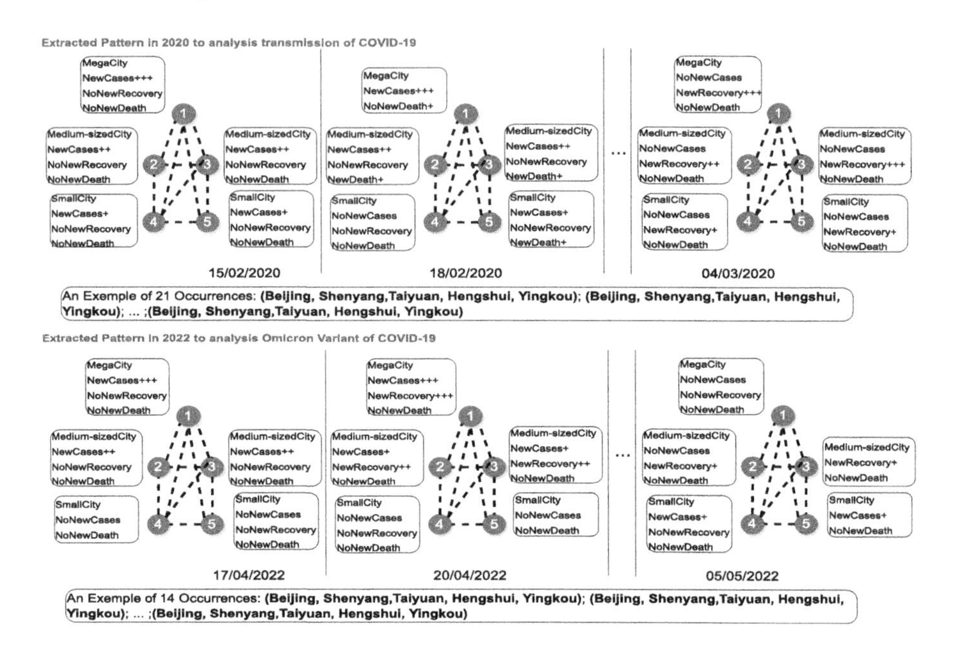

Fig. 5. Two patterns extracted from COVID dataset

observed that delays and cancellations increased at destination and arrival airports, while diverted flights always remained the same when the hurricane came. It shows that hurricanes have strong impact on cancellations but have hardly any impact on flight diversion. Second, it is noticed that cancellations and diverted flights became stable while delays decreased when the hurricane became weaker at the end. Third, this pattern shows the evolution in terms of network (the set of three airports became four). It is observed that new flight routes via a new airport were added by airlines from 11/09/2005. Moreover, referring to map, we notice that the new added airport (for example, Chicago) in the airline network is usually located at the centre position of the previous airport network (for example, Kalamazoo, Detroit and Minneapolis), which ensures that airline connections are more convenient as it is close to all other airports.

In Fig. 5, we compare two extracted patterns to analyse the transmission of COVID and its variant Omicron ($minvol = 5, maxvol = 8, minsup = 15, mincom = 5$). This pattern appears 21 times in the dataset. First, the two chosen patterns highlight the transmission of COVID in a mixed city network which is composed of mega, medium and small cities. We note that the city size and new case numbers are strongly correlated. In 2020 and 2022, COVID spreads very quickly in medium and mega cities, while new cases in small cities shown almost zero growth. It is probably because in small cities, the transport connections are much easier to control. For example, a small city has in general only two train stations and one airport while in medium-sized and big cities, it could have up to 109 train stations and 12 airports. Moreover, the flow is in general 30 times higher than small cities, which makes it much more difficult to miss any positive case. Second, it is observed that the COVID caused many severe consequences in 2020, as the death began to increase in three days after the emergence of new COVID cases, and it took in average more than 10 days for recovery. While in 2022, the virulence of variant Omicron became much weaker, as there are almost no deaths, and the recoveries began to emerge only three days after new detected cases. This analysis is very useful for anti-epidemic measures. Indeed, many countries began to cancel isolation policies in 2022. To conclude, these patterns allow studying virus transmission in different scale of cities (among big cities, small cities, medium-sized or mixed city network).

6 Conclusion

This paper has proposed a novel type of patterns called *frequent sequential subgraph evolutions* (FSSE) in a dynamic attributed graph. Its main advantage is to represent evolutions of general groups of objects. To mine FSSE, we have proposed the *FSSEMiner* algorithm. The latter is based on a novel mining strategy, called *graph addition*, to save computation time. Experiments on both benchmark and real-world datasets have shown the scalability of the algorithm and the interest of these patterns. In the short-term, we plan to make new applications by using new datasets and study the explainability of the found patterns. In the long-term, a distributed version of the algorithm could be developed to further improve the performance.

References

1. Cheng, Z., Flouvat, F., Selmaoui-Folcher, N.: Mining recurrent patterns in a dynamic attributed graph. In: Kim, J., Shim, K., Cao, L., Lee, J.-G., Lin, X., Moon, Y.-S. (eds.) PAKDD 2017. LNCS (LNAI), vol. 10235, pp. 631–643. Springer, Cham (2017). https://doi.org/10.1007/978-3-319-57529-2_49
2. Desmier, E., Plantevit, M., Robardet, C., Boulicaut, J.-F.: Cohesive co-evolution patterns in dynamic attributed graphs. In: Ganascia, J.-G., Lenca, P., Petit, J.-M. (eds.) DS 2012. LNCS (LNAI), vol. 7569, pp. 110–124. Springer, Heidelberg (2012). https://doi.org/10.1007/978-3-642-33492-4_11
3. Desmier, E., Plantevit, M., Robardet, C., Boulicaut, J.-F.: Trend mining in dynamic attributed graphs. In: Blockeel, H., Kersting, K., Nijssen, S., Železný, F. (eds.) ECML PKDD 2013. LNCS (LNAI), vol. 8188, pp. 654–669. Springer, Heidelberg (2013). https://doi.org/10.1007/978-3-642-40988-2_42
4. Fiedler, M., Borgelt, C.: Support computation for mining frequent subgraphs in a single graph. In: MLG (2007)
5. Fournier-Viger, P., Cheng, C., Cheng, Z., Lin, J.C.W., Selmaoui-Folcher, N.: Mining significant trend sequences in dynamic attributed graphs. Knowl.-Based Syst. **182**, 104797 (2019)
6. Fournier-Viger, P., et al.: A survey of pattern mining in dynamic graphs. WIREs Data Min. Knowl. Discovery **10**(6), e1372 (2020)

7. Fournier-Viger, P., He, G., Lin, J.C.-W., Gomes, H.M.: Mining attribute evolution rules in dynamic attributed graphs. In: Song, M., Song, I.-Y., Kotsis, G., Tjoa, A.M., Khalil, I. (eds.) DaWaK 2020. LNCS, vol. 12393, pp. 167–182. Springer, Cham (2020). https://doi.org/10.1007/978-3-030-59065-9_14
8. Kaytoue, M., Pitarch, Y., Plantevit, M., Robardet, C.: Triggering patterns of topology changes in dynamic graphs. In: International Conference on Advances in Social Networks Analysis and Mining, pp. 158–165. IEEE (2014)

CondTraj-GAN: Conditional Sequential GAN for Generating Synthetic Vehicle Trajectories

Nils Henke[1(✉)], Shimon Wonsak[1], Prasenjit Mitra[2], Michael Nolting[1], and Nicolas Tempelmeier[1]

[1] Volkswagen AG, Lister Straße 17, 30163 Hannover, Germany
{nils.henke1,shimon.wonsak,michael.nolting,nicolas.tempelmeier}@volkswagen.de
[2] L3S Research Center, Appelstraße 9a, 30167 Hannover, Germany
mitra@l3s.de

Abstract. While the ever-increasing amount of available data has enabled complex machine learning algorithms in various application areas, maintaining data privacy has become more and more critical. This is especially true for mobility data. In nearly all cases, mobility data is personal and therefore the drivers' privacy needs to be protected. However, mobility data is particularly hard to anonymize, hindering its use in machine learning algorithms to its full potential. In this paper, we address these challenges by generating synthetic vehicle trajectories that are not subject to personal data protection but have the same statistical characteristics as the originals. We present CondTraj-GAN– **Cond**itional **Traj**ectory **G**enerative **A**dversarial **N**etwork. – a novel end-to-end framework to generate entirely synthetic vehicle trajectories. We introduce a specialized training and inference procedure that enables the application of GANs to discrete trajectory data conditioned on their sequence length. We demonstrate the data utility of the synthetic trajectories by comparing their spatial characteristics with the original dataset. Finally, our evaluation shows that CondTraj-GAN reliably outperforms state-of-the-art trajectory generation baselines.

Keywords: Generative models · road networks · vehicle trajectories

1 Introduction

Over the past years, the unprecedented amount of available data has enabled numerous breakthroughs for machine learning models in many domains. Prominent examples include image generation models such as DALL-E 2 [15]. In the mobility domain, numerous applications rely on large mobility datasets to efficiently train deep learning models. *Trajectories*, i.e., sequences of locations coordinates, are widely adopted to capture mobility behavior. Examples of trajectory-based models include transportation demand prediction [6], traffic flow prediction [10], and next location prediction [11]. However, trajectory data is hard to obtain for two reasons. First, collecting vehicle trajectory data usually requires

H. Kashima et al. (Eds.): PAKDD 2023, LNAI 13936, pp. 79–91, 2023.
https://doi.org/10.1007/978-3-031-33377-4_7

large vehicle fleets. Access to such fleets is often limited to a few companies such as car manufacturers or mobility service providers. Secondly, trajectories often reveal personal data, e.g., the driver's home or work address. Trajectory data, hence, is protected by regulations like the EU GDPR which requires consent from the drivers for all data processing activities. Strong data protection currently hinders the development of trajectory-based models in practice. Current approaches anonymize existing trajectories or generate synthetic trajectories to lift the data protection from trajectory data. Anonymization approaches include Differential Privacy [21] or k-anonymity [17] techniques. Considering synthetic generation, established approaches build on Markov models to infer trajectories [9,14]. More recently, generative models like *Generative Adversarial Networks* (GAN) [7] have shown impressive results. We observe high interest in adopting GANs to trajectory data [2,3,16,19]. Unlike model-based approaches, GANs do not suffer from simplified assumptions often used to model our mobility patterns. However, these approaches are not able to learn trajectory patterns in an end-to-end manner, require tedious preprocessing steps, e.g., mapping trajectories to artificial images, or need real trajectories during generation.

This paper addresses the need for anonymization and publication of trajectory data by introducing the novel CondTraj-GAN end-to-end framework for synthetic vehicle trajectory generation. We train CondTraj-GAN on real-world trajectory data and generate realistic synthetic trajectories. Our generated trajectories are fully synthetic and not subject to personal data protection regulations. We propose a novel 2-step pre-training procedure to capture the spatial-sequential patterns provided by trajectories in the training dataset. CondTraj-GAN introduces a *topology learning* step to encapsulate the spatial characteristics of the road network and, further, we capture sequential patterns by leveraging *sequential GAN* architecture using a *policy gradient loss* [20]. Finally, during trajectory generation, we introduce the *transition mask* procedure to adapt the GAN inference step to a road network setting. In summary, the contributions of this work are as follows:

- We propose a novel end-to-end trajectory generation framework for the generation of realistic synthetic trajectories. To the best of our knowledge, this is the first end-to-end approach for road network vehicle trajectory generation that does not require real trajectories during inference.
- We introduce a novel GAN training procedure explicitly accounting for the road network and trajectory patterns and propose the transition mask procedure.
- We conduct a thorough evaluation comparing synthetic to real-world vehicle trajectory data and propose a set of evaluation metrics quantifying the synthetic data quality. Our CondTraj-GAN framework reliably outperforms state-of-the-art baselines for vehicle trajectory generation.

2 Problem Definition

In this section, we formally define the task of *trajectory generation*. Given a source real-world dataset, trajectory generation aims at creating a synthetic dataset that follows the same spatial distributions as the source dataset, e.g., concerning the trajectories' length, origins, and destinations. A *road network* is an undirected graph $\mathcal{G} = (\mathcal{E}, \mathcal{V})$ with a set of edges $\mathcal{E}$ which can be interpreted as street segments and a set of nodes $\nu \in \mathcal{V} \cup \{\nu_b\}$ which can be interpreted as junctions between street segments. ν_b serves as a start node with an id but no location and has a connection to all other nodes, i.e. $(\nu_b, \nu_j) \in \mathcal{E}, \forall \nu_j \in \mathcal{V}$. All other nodes can be identified using a unique id and provide location information via geo-coordinates with shape $(latitude, longitude)$. Furthermore, we define the binary adjacency matrix $\mathcal{A}$ that encodes the connection of two nodes with 1 if $(\nu_i, \nu_j) \in \mathcal{E}$ and 0 otherwise. Next, we define a *trajectory* $\tau_{0:T} = (\nu_0, \nu_1, \ldots, \nu_T)$ as a sequence of nodes where ν_t is the node in $\mathcal{V}$ passed at time step t, with $t = 0 \leq t \leq T$ and T is the total number of discrete steps. In this paper, we focus on trajectories that are locatable on road networks only.

Finally, we define *trajectory generation* as follows: Given a source real-world trajectory dataset $\mathcal{T}^r = \{\tau_0, \tau_1, .., \tau_n\}$ with an underlying spatial distribution p, learn a function $\hat{p}$ that approximates p such that a synthetic trajectory τ^s drawn from $\hat{p}$ meets the following conditions: (1) τ^s contains only sequences of adjacent nodes, $\forall(\nu_t, \nu_{t+1}) \in \tau^s : (\nu_t, \nu_{t+1}) \in \mathcal{E}$. (2) τ^s represents spatial characteristics provided by the source dataset, e.g., patterns of commuting from home to work. We denote the set of synthetic trajectories as $\mathcal{T}^s$. Although the trajectories do not provide time information, there is still value for spatial applications.

3 The CondTraj-GAN Framework

This section introduces our proposed CondTraj-GAN framework which is presented in Fig. 1. We propose a specialized training procedure to capture the spatial-sequential patterns of vehicle trajectories within road networks. Then, we describe the inference process to generate synthetic trajectories. CondTraj-GAN builds on the established GAN setup consisting of a stochastic generator function G_θ and a discriminator function D_ϕ with θ and ϕ denoting the respective model parameters. To adjust the generated trajectory length and thus match the source data characteristics better, we apply the Conditional GAN [12] architecture. Therefore, we condition the generation process on the sequence length l and the current node index t. Formally, we define the generator function as $G_\theta(\nu_t|\tau_{0:t-1}, l, t)$ where ν_t is the next node of the sequence conditioned on the historic trajectory $\tau_{0:t-1}$. Considering the *discriminator* D_ϕ, we aim to distinguish synthetic from real trajectories. Formally, we define the discriminator function as $D : \mathcal{T}^r \cup \mathcal{T}^s \mapsto \{True, False\}$. After training, we utilize the generator G_θ to iteratively generate a synthetic dataset $\mathcal{T}^s$.

3.1 Training

This section describes the training process of CondTraj-GAN consisting of *topology learning*, *trajectory learning* and *adversarial training*.

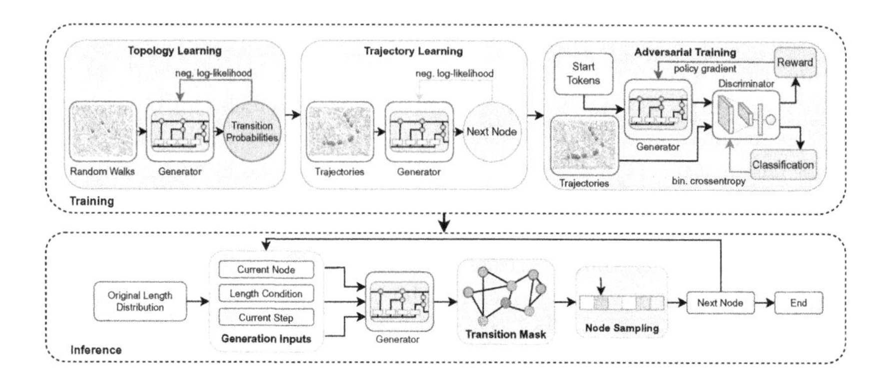

Fig. 1. The training and inference diagram. During *topology learning* the model is trained on 3-hop random walks. Afterwards, real trajectories are used for *trajectory learning* and *adversarial training*. The transition mask is applied only during inference.

Topology Learning. Topology learning is a self-supervised generator pre-training step with two goals. First, the topology learning step constitutes a simple but effective way to capture adjacency information in the generator function. Secondly, we particularly counteract data sparseness and mitigate the exclusion of nodes that are not covered in the source trajectory set $\mathcal{T}^r$. The model is trained on 3-hop random walks generated by the road network adjacency matrix without repetitions. The goal is to help the generator find the transition probabilities only for adjacent nodes as a multi-label classification task. Therefore, we calculate the transition probability vector from ν_{t-1} to all possible others: $\bar{p}_{\nu_t} = G_\theta(\nu_t|\tau_{0:t-1}, l = 3, t)$, where $\bar{p}_{\nu_t}$ is the probability of going from node ν_{t-1} to any other node in the node set $\mathcal{V}$. We then take the probability vector as input to calculate the *negative log-likelihood* loss $\mathcal{L}_{top} = -\frac{1}{N \cdot T}\sum_{n=0}^{N}\sum_{t=0}^{T}\mathcal{A}_{\nu_{t-1}} \cdot log(\bar{p}_{\nu_t})$, where $\mathcal{A}_{\nu_{t-1}}$ is the row in the adjacency matrix corresponding to the neighborhood of the source node ν_{t-1}. We aim to obtain high probabilities for nodes adjacent to node ν_{t-1}.

Trajectory Learning. The second pre-training phase aims to learn the sequential patterns of real trajectories. In particular, we aim to capture realistic travel patterns like the way from home to work. To learn the semantic long-term relationships between non-adjacent nodes, we shift the original trajectory by one time step $\tau_{1:T+1}$ and ask the model to predict the next node of the original trajectory. We roll out a complete trajectory to obtain all transition probabilities $\bar{p}^g_{\tau_{1:T+1}}$: $\bar{p}^g_{\nu_t} = G_\theta(\nu_t|\tau_{0:t-1}, l, t)$, where $\bar{p}^g_{\nu_t}$ is the probability vector for the next node given the historic trajectory containing information of the historic travel pattern. Since we want to draw trajectories at the end from a probability distribution $\hat{p}$, the problem can also be solved with the maximum likelihood estimation. We obtain the transition probabilities $\bar{p}^o_\tau$ of the origin trajectory τ^r by assigning a fixed position to each node ν in the node-set $\mathcal{V}$

and then one-hot encoding it. Finally, we apply the *negative log-likelihood* loss: $\mathcal{L}_{traj} = -\frac{1}{N \cdot T} \sum_{n=0}^{N} \sum_{t=0}^{T} \bar{p}_{\nu_t}^{o} \cdot log(\bar{p}_{\nu_t}^{g})$

Adversarial Training. Adversarial training aims at generating novel trajectories that are not part of the source dataset $\mathcal{T}^r$. We follow the established GAN training procedure of jointly training the generator G_θ and the discriminator D_ϕ. Standard GAN loss functions that rely on real-valued data do not apply to discrete trajectories. Instead, we adapt the policy gradient loss proposed in [20] for sequential categorical data with our before-mentioned conditions l and t:

$$\mathbb{E}[R_T|\nu_b, \theta] = \sum_{\nu_1 \in \mathcal{V}} G_\theta(\nu_1|\nu_b, l, t) \cdot Q_{D_\phi}^{G_\theta}(\nu_b, \nu_1, l, t), \tag{1}$$

Since an intermediate reward is not helpful because short-term goals can be abandoned to achieve long-term ones, we also adopt the idea of *Monte Carlo search*. Starting from a starting state ν_b we wish to maximize the expected end reward. M-time Monte-Carlo search MC is used to find the next action value a coming from state s, with the generator G_β determining the roll-out policy:

$$Q_{D_\phi}^{G_\theta}(s = \tau_{0:t-1}, a = \nu_t, l, t) = \\ \begin{cases} \frac{1}{M} \sum_{m=1}^{M} D_\phi(\tau_{0:T}^m|l), \tau_{0:T}^m \in MC^{G_\beta}(\tau_{0:t}, M), & \text{for } t < T, \\ D_\phi(\tau_{0:T}|l), & \text{for } t = T \end{cases} \tag{2}$$

For the discriminator, we rely on the well-known GAN training objective. After each generator training update, the discriminator can be updated as follows:

$$\min_\phi -\mathbb{E}_{\tau \sim p}[log(D_\phi(\tau|l))] + \mathbb{E}_{\tau \sim G_\theta}[log(1 - D_\phi(\tau|l)), \tag{3}$$

Analogous to the generator, we feed the conditional sequence length l into the discriminator.

3.2 Trajectory Inference

This section presents the inference phase for synthetic trajectory generation.

Generation Inputs. After training, we decouple the generator from the discriminator to create trajectories auto-regressively. First, we sample trajectory lengths from the original sequence length distribution. Based on the start node, length condition, and generation step information a new trajectory is generated.

Transition Mask. During inference, we apply a probability mask after each node generation step to guarantee that constructed trajectories have solely transitions to adjacent nodes: $\bar{p}_{\nu_t} = \mathcal{A}_{\nu_{t-1}} \cdot G_\theta(\nu_t|\tau_{0:t-1}, l, t)$, where $\bar{p}_{\nu_t}$ is the masked probability vector for sampling the next node ν_t in the sequence. We obtain $\bar{p}_{\nu_t}$ by calculating the Hadamard product of the adjacency matrix $\mathcal{A}_{\nu_{t-1}}$ and the output probability distribution of the generator.

Node Sampling. Since we condition the generation process on node ids and no prior distribution, we obtain stochastic properties in our model by sampling subsequent nodes from a multinomial distribution $\mathcal{MN}$: $\nu_t \sim \mathcal{MN}(\bar{p}_{\nu_t})$.

4 Evaluation Setup

This section presents the datasets, the model setup for CondTraj-GAN training, and the metrics used to assess the synthetic dataset generation performance.

4.1 Dataset

The CondTraj-GAN framework aims to train the topology of the road network and the spatial-sequential patterns of real trajectories. Therefore, we use two different datasets: a road network dataset based on Open Street Map (OSM) and a trajectory dataset. The road network training data is obtained from OSM and used for topology learning. We calculate an adjacency matrix and construct 50k 3-hop random walks with no repetitions allowed. We employ a proprietary real-world trajectory dataset provided by a ride-hailing service in Hamburg, Germany, which is used for trajectory learning, adversarial training, and evaluation. The drivers and passengers consented to the data collection and the data is anonymized. The dataset consists of more than 17.5k GPS traces (sample 1 Hz) from several vehicles with multiple drivers. The trajectory lengths are between 8 and 50 nodes. In the service area, about 65% of all nodes in the road network are covered by the trajectories. The data covers the time between October 2019 and July 2021. All traces are map-matched to road network nodes. The dataset is split into 90% training and 10% test data.

4.2 Model Setups

Here, we describe the architectures and parameters of CondTraj-GAN's generator and discriminator functions.

Architectures. The generator is an LSTM network. We sample the length condition input from the original trajectory data length distribution obtained from the training set. The length condition and the current generation step information are separately embedded and added. The result is concatenated with the node embedding and fed into the LSTM layer followed by a linear layer with softmax activation. At each generation step, the model outputs a vector of probabilities for all node transitions from which the next node is sampled. The transition mask is only applied during the inference phase, not while training. The discriminator model is a CNN. Input to the discriminator is the trajectory and the sequence length, both separately embedded, then added and fed into a convolutional layer with pooling and a linear classification output.

Model Parameters. The generator and discriminator embedding and hidden dimensions, and batch size are all set to 64. We run both pre-training phases with 200 epochs each. The discriminator is pre-trained with the real trajectory dataset and synthetic samples for 3 steps with 2 epochs each. Adversarial training runs for 200 epochs with 9,600 synthetic samples. We use the Adam optimizer with a learning rate (LR) of 1e-4 for generator training and an LR of 1e-5 for discriminator training, as a higher discriminator LR leads to convergence failure.

4.3 Evaluation Metrics

This section presents the evaluation metrics to assess the synthetic data utility.

Average Trip Length. The trip length d_{tl} is the sum of all distances d between consecutive nodes in a trajectory $(\nu_t, \nu_{t+1}) \in \tau$. The *average trip length* (av. TL) of a dataset is the mean of all trip lengths. We calculate the delta $\Delta \overline{d_{tl,set}} = \frac{1}{|\mathcal{T}^s|}\sum_j d^s_{tl,j} - \frac{1}{|\mathcal{T}^r|}\sum_i d^r_{tl,i}$, between the original and synthetic dataset. $|\mathcal{T}^r|$ is the number of real trips, $|\mathcal{T}^s|$ is the number of synthetic trips, $d^r_{tl,i}$ is the trip length of a real trip i and $d^s_{tl,j}$ is the trip length of a synthetic trip j.

Average Start-to-End Node Distance. The *average start-to-end node distance* (av. SE) is calculated as the mean of all distances d_{se} between the start and end node of a trip in a dataset. We compare with the test data by calculating the delta $\Delta \overline{d_{se,set}} = \frac{1}{|\mathcal{T}^s|}\sum_j d^s_{se,j} - \frac{1}{|\mathcal{T}^r|}\sum_i d^r_{se,i}$, where $|\mathcal{T}^r|$ is the number of real trips, $|\mathcal{T}^s|$ is the number of synthetic trips, $d^r_{se,i}$ is the start-end distance of a real trip i and $d^s_{se,j}$ is the start-end distance of a synthetic trip j. The start-to-end node distance indicates how targeted a trip is.

Trip Length Distribution. We calculate the Jensen-Shannon distance J of trip lengths (JSD TL) to analyze the variety of trajectory lengths in the dataset: $J = \sqrt{\frac{D(p^r_{tl}\|m)+D(p^s_{tl}\|m)}{2}}$, where p^r_{tl} and p^s_{tl} are the trip length distributions of the real and synthetic trajectory datasets, m is their point-wise mean and D is the Kullback-Leibler divergence.

Start Node Distribution. We calculate the Wasserstein distance W between the start point probability distributions (WD SP)to evaluate the similarity of the real and synthetic origins, p^r_{sp} and p^s_{sp}: $W(p^r_{sp}, p^s_{sp}) = \min_\gamma \sum \gamma_{i,j} M_{i,j}$, which can be interpreted as transport costs from one distribution to the other [18]. γ are the transportation costs and M is the distance matrix between samples i and j in p^r_{sp} and p^s_{sp}, respectively.

Normalized Harmonic Mean. We calculate the normalized harmonic mean (NHM) to assess the overall model performance along all metrics: $nhm = \frac{n}{\sum_i \frac{x_{i,\min}}{x_i}}$. Since the metrics operate in a different range, we normalize each value x_i regarding the minimum value $x_{i,\min}$ scored in a metric i. n is the number of metrics.

4.4 Baselines

In this section, we want to describe the baselines compared with CondTraj-GAN. For all models, we generate a dataset of 2,500 synthetic trajectories.

Markov Chain. The *Markov Chain* baseline randomly samples a node from OSM as a beginner state. Like in [22], the transition matrix is based on the road network and the real trajectories inferring the next node from a current state. The sequence length is sampled from the original sequence length distribution.

Shortest Path. The *Shortest Path* baseline randomly samples a start and end node from OSM and calculates the shortest length path between them.

TrajGen. The *TrajGen* [2] model consists of a GAN that generates images of trajectories and a Sequence-to-Sequence model that sorts the node ids extracted from the images into the correct order. We implemented TrajGen with the public code and the paper's description.

LSTM-TrajGAN. The generator of the *LSTM-TrajGAN* model [16] takes embedded real trajectories and noise to generate a synthetic trajectory via LSTM modeling. Since the model was originally proposed for point-of-interest trajectories, we adapt it to only spatial features and map-match the synthetic lat/lon pairs to obtain a sequence of node ids. During inference, the baseline requires real trajectories as input which we randomly sample from the training set.

5 Evaluation

The evaluation aims to assess CondTraj-GAN's capability of generating realistic trajectories. First, we discuss the performance regarding the evaluation metrics compared with the baselines. Then, we conduct an ablation study to investigate the contribution of CondTraj-GAN's components. We run each experiment 5 times with random initialization and report the average and standard deviation.

5.1 Trajectory Generation Performance

Table 1 presents the overall trajectory generation performance of CondTraj-GAN and all baselines. In the av. TL metric, CondTraj-GAN performs best, indicating the usefulness of the *length condition* and *step information* in constructing realistic trajectory lengths. Since the Shortest Path baseline randomly samples start and end nodes, it creates the longest trips. LSTM-TrajGAN performs worse than CondTraj-GAN although it generates the same number of nodes as the input trajectory. Since the model is not bound to the road network it may create distance errors that propagate through each trajectory. Regarding av. SE LSTM-TrajGAN performs exceptionally well. Here, the absence of the road network doesn't affect the result significantly. The Markov Chain only achieves a poor av. SE score. Since node transitions are only based on the current node, its trips are unfocused and result in very short start-end distances. In JSD TL, all models with close av. TL show good performance, except for the TrajGen model, which fails to offer much variety in the generated trajectories. Regarding the WD SP, Markov Chain and Shortest Path perform poorly since they randomly sample start nodes from an equal distribution over the whole road network. Figure 2 compares the start node probability distributions of the original trajectory dataset and all GAN-based models, visualizing the scores in WD SP from Table 1. The CondTraj-GAN distribution can capture the original start point distribution best. This indicates the effectiveness of the trajectory learning step in the training phase for estimating the first node transition. The distribution of LSTM-TrajGAN looks like a blurred version of the original due to mixing real origin locations with noise as

Table 1. Overall trajectory generation performance with respect to average trip length, average start-end distance, Jensen-Shannon distance of trip lengths, Wasserstein distance of start points, and NHM of all metrics (values close to 0 are best).

Model	av. TL [km]	av. SE [km]	JSD TL	WD SP	NHM
Markov Chain	0.55 (±0.03)	–0.95 (±0.01)	0.16 (±5.6e–3)	3.56e–4 (±2.4e–5)	2.74
Shortest Path	4.49 (±0.08)	4.08 (±0.06)	0.49 (±3.1e–3)	4.62e–4 (±1.5e–5)	6.49
TrajGen	–0.59 (±0.85)	–0.71 (±0.41)	0.63 (±5.8e–2)	266.10e–4 (±1.1e–2)	8.85
LSTM-TrajGAN	0.38 (±0.19)	**0.001** (±0.1)	0.14 (±2.8e–2)	2.87e–4 (±6.6e–5)	1.43
CondTraj-GAN	**0.15** (±0.04)	–0.16 (±0.03)	**0.12** (±3.8e–3)	**1.56e–4** (±9.4e–6)	**1.33**

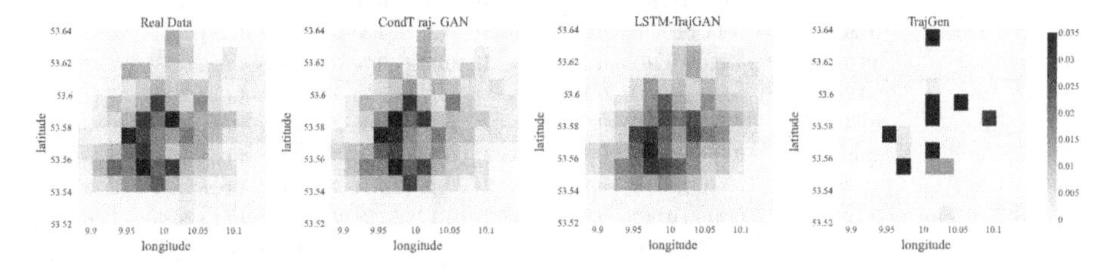

Fig. 2. Start point probability distribution of real trajectory data and the synthetic datasets generated by CondTraj-GAN, LSTM-TrajGAN and TrajGen.

model input. TrajGen achieves only a poor WD SP score. We find that TrajGen generates many duplicate trajectories, i.e. there are less than 1% unique trajectories in the 2,500 trajectory test set, which indicates mode collapse.

In summary, CondTraj-GAN offers the best overall performance at NHM and achieves the best scores in three out of four metrics and the second-best performance in one metric. It shows high similarity with the real dataset indicating potential usage of CondTraj-GAN's trajectories for downstream applications. In contrast to LSTM-TrajGAN, CondTraj-GAN does not require any trajectory information during the inference phase and can capture complex spatial relationships along non-linear road networks.

5.2 Ablation Study

In this section we investigate the importance of *transition mask* (*tm*), *step information* (*si*), *length condition* (*lc*) and *topology learning* (*tl*) by training different model setups. Table 2 shows the results of the ablation study. To mitigate discriminator inferiority and thus assuring an effective GAN-training generator pre-training is necessary. Therefore, our minimum model consists of a *vanilla GAN + trajectory learning.* We aim to particularly investigate the impact of *length condition* and *step information* on the node sequence length. Similar to JSD TL, we introduce the Jensen-Shannon distance of node sequence lengths (JSD NL). Overall, the full CondTraj-GAN model shows the best performance, leading in 3 out of 5 metrics and the overall performance at NHM of all metrics. The *Vanilla + tr + to + tm* model generally builds longer trips

Table 2. Ablation study of different model components. Each row adds different components to the vanilla GAN model: trajectory learning (tr), topology learning (to), length condition (lc), step information (si), and transition mask (tm). The CondTraj-GAN model consists of all components. Values close to 0 are best.

Model	av. TL [km]	av. SE [km]	JSD TL	JSD NL	WD SP	NHM
Vanilla + tr	11.49 (±0.23)	1.60 (±0.06)	0.40 (±1.4e–2)	0.29 (±2.6e–2)	1.60e–4 (±2.0e–5)	3.07
+ to	5.54 (±0.23)	1.23 (±0.12)	0.34 (±8.8e–3)	0.24 (±3.9e–3)	1.84e–4 (±2.8e–5)	2.95
+ lc	20.65 (±4.07)	1.80 (±0.14)	0.40 (±6.1e–2)	0.19 (±6.4e–2)	1.52e–4 (±2.3e–5)	2.67
+ si	13.13 (±0.77)	1.63 (±0.09)	0.40 (±5.6e–3)	0.21 (±9.3e–3)	1.56e–4 (±1.2e–5)	2.77
+ tm	–2.84 (±0.07)	–2.24 (±0.03)	0.69 (±1.9e–2)	0.73 (±2.3e–3)	2.38e–4 (±1.8e–5)	5.00
+ to + lc	1.63 (±1.50)	0.39 (±0.69)	0.24 (±4.7e–2)	0.21 (±2.7e–2)	11.7e–4 (±6.2e–4)	3.21
+ to + si	2.81 (±0.32)	0.96 (±0.20)	0.29 (±8.1e–3)	0.16 (±9.7e–3)	5.04e–4 (±3.2e–4)	3.13
+ to + tm	0.39 (±0.03)	**0.11** (±0.01)	0.21 (±1.2e–3)	0.21 (±4.3e–3)	1.85e–4 (±1.2e–5)	1.57
+ lc + si	19.01 (±1.62)	1.72 (±0.10)	0.37 (±7.8e–3)	0.15 (±9.7e–3)	1.46e–4 (±2.6e–5)	2.37
+ lc + tm	–0.24 (±0.26)	–0.57 (±0.12)	0.14 (±1.8e–2)	0.17 (±3.7e–2)	1.46e–4 (±2.1e–5)	1.52
+ si + tm	0.34 (±0.06)	–0.15 (±0.02)	0.19 (±1.4e–2)	0.19 (±1.4e–2)	1.42e–4 (±2.2e–5)	1.48
+ to + lc + si	2.38 (±0.38)	0.76 (±0.15)	0.24 (±1.6e–2)	0.13 (±6.9e–3)	2.67e–4 (±1.3e–4)	2.35
+ to + lc + tm	–0.28 (±0.57)	–0.46 (±0.31)	0.18 (±4.8e–2)	0.19 (±5.1e–2)	7.79e–4 (±5.2e–4)	2.23
+ to + si + tm	0.72 (±0.19)	0.20 (±0.08)	0.20 (±2.6e–2)	0.17 (±1.7e–2)	2.74e–4 (±8.8e–5)	1.96
+ lc + si + tm	0.23 (±0.17)	–0.37 (±0.08)	0.14 (±8.8e–3)	0.15 (±1.8e–2)	**1.33e–4** (±1.6e–5)	1.39
CondTraj-GAN	**0.15** (±0.04)	–0.16 (±0.03)	**0.12** (±3.8e–3)	**0.12** (±1.5e–3)	1.56e–4 (±9.4e–6)	**1.10**

since it has no *length condition* or *step information* applied, leading to higher values in av. TL, JSD TL, and JSD NL, but resulting in a better av. SE score. Model setups that are missing *topology learning* are showing better results in WD SP (e.g. the *Vanilla + tr + lc + si + tm* model), indicating that learning an equal distribution of start points degrades this score. With our model, two factors are mainly contributing to the trip length: the node sequence length and the distance of two consecutive nodes. Realistic trajectories contain only node transitions to adjacent nodes and also have a similar node sequence length distribution to the original dataset, leading to similar trip lengths and trip length distributions. Adding *topology learning* improves the trip length substantially because it enables the handling of unseen nodes which are not contained in the training data and, thus, can lead to gaps in the generated trajectories. Without *transition mask* the distance of subsequent nodes is larger which results in a higher av. TL, av. SE and worse JSD TL. Models containing *length condition* or *step information* show increasing quality in the JSD NL, therefore leading to realistic trip lengths as long as *topology learning* or *transition mask* are also applied. In summary, the *length condition* and *step information* help construct the right node sequence length, while the *topology learning* and *transition mask* handle the transition to adjacent nodes. All components of CondTraj-GAN contribute to generating realistic synthetic vehicle trajectories.

6 Related Work

This section discusses the related work in the areas of synthetic trajectory generation and data anonymization.

Synthetic Trajectory Generation. With TimeGeo [9] and DITRAS [14] two mechanistic models show promising results in trajectory generation. The existing literature also addresses synthetic trajectory generation using GAN and Variational Autoencoder-based models. [2] use a GAN and a seq2seq model to generate trajectories with spatio-temporal information. A DCGAN creates images of travel patterns which are sequenced into trajectories using map information and a seq2seq network. In [16] a combination of an LSTM and GAN is used to generate sequences of visited POIs from original trajectories with an end-to-end model that uses original trajectories during generation. [19] developed a two-stage GAN model to generate vehicle trajectories that match the street network. Multiple papers propose Variational Autoencoders (VAEs) to generate trajectories from real starting points [3] or in combination with seq2seq models [8]. [13] leverages a *location-major* representation to encode trajectories onto a 2D map. These trajectories are fed into a GAN model to generate a synthetic dataset. In contrast, we propose an end-to-end GAN framework for trajectory generation that does not rely on multiple stages or intermediate images, and no original trajectories are used during generation.

Data Anonymization. The existing literature investigated methods for anonymizing mobility or trajectory data. This data is typically critical for privacy attacks as it can contain personal information like home or work address which can be inferred by frequently visited places. Differential privacy (DP) [5] offers the opportunity to protect users' locations while still preserving the characteristics of the dataset. [1] protects the privacy of the user's location by achieving geo-indistinguishability by adding random noise to the user's location. [21] investigate trajectory data publishing, especially with multiple repetitive trajectories of the same users in a dataset, and propose of the use DP. However, privacy preservation usually comes with the cost of diminished data utility. Therefore, we do not focus on anonymizing existing trajectories but propose a framework to generate synthetic trajectories to maintain both privacy and data utility.

7 Conclusion and Future Work

In this paper, we have presented CondTraj-GAN- a novel end-to-end framework for synthetic trajectory generation. At the core of CondTraj-GAN, we proposed a novel training process including *topology learning* and *trajectory learning* to capture the complex spatial and sequential patterns of vehicle trajectories. CondTraj-GAN introduces a trajectory *length condition* supported by *step information* to guide the trajectory generation process into realistic trajectory lengths. Our evaluation shows that CondTraj-GAN reliably outperforms state-of-the-art models for trajectory generation. This work is focused on the spatial dimension of trajectories only which is relevant for, e.g., popular route discovery [4]. In future work, we plan to expand CondTraj-GAN to complement the generated trajectories with time information. Further, we would like to investigate the benefit of using synthetic trajectories in downstream applications.

Acknowledgements. This work was partially funded by the German Federal Ministry for Economic Affairs and Climate Action (BMWK) under the project "CampaNeo" (grant ID 01MD 19007A).

References

1. Andrés, M.E., Bordenabe, N.E., Chatzikokolakis, K., Palamidessi, C.: Geo-indistinguishability: differential privacy for location-based systems. In: Proceedings of the ACM SIGSAC (2013)
2. Cao, C., Li, M.: Generating Mobility Trajectories with Retained Data Utility. In: Proceedings of the ACM SIGKDD (2021)
3. Chen, X., Xu, J., Zhou, R., Chen, W., Fang, J., Liu, C.: TrajVAE: a variational AutoEncoder model for trajectory generation. Neurocomputing **428**, 332–339 (2021)
4. Chen, Z., Shen, H.T., Zhou, X.: Discovering popular routes from trajectories. In: 2011 IEEE 27th International Conference on Data Engineering, pp. 900–911 (2011)
5. Dwork, C.: Differential privacy. In: Bugliesi, M., Preneel, B., Sassone, V., Wegener, I. (eds.) Automata, Languages and Programming (2006)
6. Fu, H., Wang, Z., Yu, Y., Meng, X., Liu, G.: Traffic flow driven spatio-temporal graph convolutional network for ride-hailing demand forecasting. In: Advances in Knowledge Discovery and Data Mining (2021)
7. Goodfellow, I., et al.: Generative adversarial nets. In: Proceedings of NIPS (2014)
8. Huang, D., et al.: A variational autoencoder based generative model of urban human mobility. In: 2019 IEEE Conference on Multimedia Information Processing and Retrieval (MIPR), pp. 425–430 (2019)
9. Jiang, S., Yang, Y., Gupta, S., Veneziano, D., Athavale, S., González, M.C.: The TimeGeo modeling framework for urban mobility without travel surveys. In: Proceedings of the National Academy of Sciences (2016)
10. Li, M., Tong, P., Li, M., Jin, Z., Huang, J., Hua, X.S.: Traffic flow prediction with vehicle trajectories. In: Proceedings of the AAAI (2021)
11. Lin, Y., Wan, H., Guo, S., Lin, Y.: Pre-training context and time aware location embeddings from spatial-temporal trajectories for user next location prediction. In: Proceedings of the AAAI (2021)
12. Mirza, M., Osindero, S.: Conditional generative adversarial nets. CoRR (2014)
13. Ouyang, K., Shokri, R., Rosenblum, D.S., Yang, W.: A non-parametric generative model for human trajectories. In: IJCAI International Joint Conference on Artificial Intelligence 2018-July, pp. 3812–3817 (2018)
14. Pappalardo, L., Simini, F.: Data-driven generation of spatio-temporal routines in human mobility. Data Min. Knowl. Disc. **32**(3), 787–829 (2017). https://doi.org/10.1007/s10618-017-0548-4
15. Ramesh, A., Dhariwal, P., Nichol, A., Chu, C., Chen, M.: Hierarchical text-conditional image generation with CLIP latents. CoRR (2022)
16. Rao, J., Gao, S., Kang, Y., Huang, Q.: LSTM-TrajGAN: a deep learning approach to trajectory privacy protection. In: LIPIcs (2020)
17. Terrovitis, M., Mamoulis, N.: Privacy preservation in the publication of trajectories. In: Proceedings of the IEEE International Conference on Mobile Data Management (2008)
18. Villani, C.: Optimal transport: old and new, vol. 338. Springer (2009). https://doi.org/10.1007/978-3-540-71050-9

19. Wang, X., Liu, X., Lu, Z., Yang, H.: Large scale GPS trajectory generation using map based on two stage GAN. J. Data Sci. **19**, 126–141 (2021)
20. Yu, L., Zhang, W., Wang, J., Yu, Y.: SeqGAN: sequence generative adversarial nets with policy gradient. In: Proceedings of the AAAI 2017 (2017)
21. Zhao, J., Mei, J., Matwin, S., Su, Y., Yang, Y.: Risk-aware individual trajectory data publishing with differential privacy. IEEE Access (2021)
22. Zhao, X., Spall, J.C.: A markovian framework for modeling dynamic network traffic. In: 2018 Annual American Control Conference (ACC) (2018)

A Graph Contrastive Learning Framework with Adaptive Augmentation and Encoding for Unaligned Views

Yifu Guo and Yong Liu(✉)

Heilongjiang University, Harbin, China
2211907@s.hlju.edu.cn, 2010023@hlju.edu.cn

Abstract. Recently, graph contrastive learning has emerged as a successful method for graph representation learning, but it still faces three challenging problems. First, existing contrastive methods cannot preserve the semantics of the graph well after view augmentation. Second, most models use the same encoding method to encode homophilic and heterophilic graphs, failing to obtain better-quality representations. Finally, most models require that the two augmented views have the same set of nodes, which limits flexible augmentation methods. To address the above problems, we propose a novel graph contrastive learning framework with adaptive augmentation and encoding for unaligned views, called GCAUV in this paper. First, we propose multiple node centrality metrics to compute edge centrality for view augmentation, adaptively removing edges with low centrality to preserve the semantics of the graph well. Second, we use a multi-headed graph attention network to encode homophilic graphs, and use MLP to encode heterophilic graphs. Finally, we propose g-EMD distance instead of cosine similarity to measure the distance between positive and negative samples. We also perform adversarial training by adding perturbation to node features to improve the accuracy of GCAUV. Experimental results show that our method outperforms the state-of-the-art graph contrastive methods on node classification tasks.

Keywords: Contrastive Learning · Graph Representation Learning · Homophilic Graph · Heterophilic Graph

1 Introduction

In recent years, graph representation learning has emerged as an effective method for analyzing graph-structured data. Graph representation learning aims to convert high-dimensional node features into low-dimensional embeddings while preserving the graphs' topological structures for downstream tasks. Due to scarce labeled data in graphs, many graph contrastive learning methods [4,13,14,18–20] have been proposed to extract semantic information-rich knowledge from graphs.

Although graph contrastive learning have achieved remarkble performance in downstream tasks, they still have three important problems to be resolved. First, due to the non-Euclidean structure of the graph, the augmentation methods in

H. Kashima et al. (Eds.): PAKDD 2023, LNAI 13936, pp. 92–104, 2023.
https://doi.org/10.1007/978-3-031-33377-4_8

the CV and NLP domains cannot be directly applied to graph data. We have to design the augmentation methods for graph data. GraphCL [17] designed four general augmentation methods (node deletion, edge perturbation, feature masking, and subgraph sampling) for graph contrastive learning. However, we found that the above augmentation methods only has good results for some datasets. This indicates that these random augmentation methods cannot preserve the semantics of the graph well; in contrast, they may erase important intra-graph information during the augmentation process, which will affect the ability of the model to learn knowledge from the graph. Second, most contrastive learning framework use the same encoding method(GCN) for both homophilic and heterophilic graphs. However, we find that encoding method has a great impact on the performance in downstream tasks and different types of graphs should use different coding methods. Finally, in the contrastive learning model where the contrast level is node-to-node, the same nodes in two views are generally treated as a pair of positive samples, and the other nodes in two views are treated as negative samples, which requires that the augmented views have the same set of nodes. This rigid requirement will hinder the flexibility and diversity of view sampling and augmentation and limit the expressive power of graph contrastive learning.

To address the above three problems, we propose a novel graph contrastive learning framework with adaptive augmentation and encoding for unaligned views, called GCAUV. First, we generate subgraphs by performing a random walk with restart for each node in the graph, where subgraphs sampled by the same central node are considered as a pair of positive samples, and subgraphs sampled by different central nodes are negative samples. We then augment the subgraph with edge dropping and feature masking. For edge dropping, we calculate the edge centrality of each edge in the subgraph based on node centrality measures, so that the edges can be dropped adaptively in terms of their centrality. To improve the accuracy of the node classification task, we train the model adversarially by adding perturbation to the node features. Then we use a multi-head graph attention network to encode the nodes in the homophilic graph and use MLP to encode the nodes in the heterophilic graph. After obtaining node representations, we propose g-EMD distance as a contrastive metric to measure the distance between positive and negative samples, where the g-EMD distance can be modeled as the minimum generation cost required to convert the node attribute distribution of one view to the node attribute distribution of another view. We make all the code publicly available at https://github.com/GCAUV-PyTorch/GCAUV. Our main contributions are as follows.

- We propose a novel graph contrastive learning framework based on adaptive augmentation called GCAUV.
- We design different encoding methods for homophilic and heterophilic graphs respectively and illustrate their effectiveness. We propose g-EMD distance to measure the distance between views, allowing to have different set of nodes between views.
- Extensive experimental results on homophilic and heterophilic datasets demonstrate the superior performance of our method.

2 Related Work

2.1 Graph Contrastive Learning

With the great success of contrastive learning in CV and NLP domains, experts have worked on applying contrastive learning to graph structures to obtain better-quality representations. Initially, DGI [14] obtained node or graph representations by maximizing the Mutual Information between graph-level and node-level representations. BGRL [13] measures the Mutual Information by parameterizing the Mutual Information estimator to obtain representations. Subsequently, MVGRL [4] proposed learning node-level and graph-level representations by performing node diffusion and comparing node representations with augmented graph representations. MNCI [9] and ConMNCI [10] proposed a new kind of inductive network representation learning method by mining neighborhood and community influences in temporal networks. GRACE [18] applied node dropping and feature masking to propose a node-to-node contrastive learning framework. The above methods do not consider how to preserve the semantic and important information of the graph in the view augmentation part, resulting in poor performance. GCA [19] first used node centrality for adaptive augmentation of views, but GCA [19] requires both views to have the same set of nodes that loses the flexibility of view sampling and augmentation. RoSA [20] first proposed to use g-EMD distance instead of cosine similarity to ensure the flexibility and diversity of view sampling. However, it does not consider how to preserve the semantics of the graph better when augmenting the views, which may affect the ability of the model to learn important knowledge in the graph.

This paper performs adaptive augmentation based on the node centrality measure. In the contrastive process, we use g-EMD distance instead of cosine similarity to measure the distance between positive and negative samples, which ensures the flexibility and diversity of view augmentation and also improves the ability of the model to learn important knowledge in the graph.

2.2 Adversarial Training

Adversarial training is generally used to improve the model's resistance to interference. The main idea of adversarial training is to add noise to the original samples to generate adversarial samples during model training, so that the training samples include the original samples and the adversarial samples. In the beginning, our neural network may misclassify these adversarial samples, and the purpose of adversarial training is to adapt the neural network to these changes and classify the adversarial samples correctly, and as the training proceeds, the model's resistance to interference is improved. [8] Kong et al. have used a supervised approach to apply adversarial training to graph structures. SimGrace [16] proposed adding perturbation to the graph encoder parameters to improve the robustness of the model. In this paper, we propose to add perturbation to the node features for adversarial training to improve the accuracy of the model on node classification tasks.

3 Method

In this section, we introduce GCAUV in detail. The framework of our model is shown in Fig. 1.

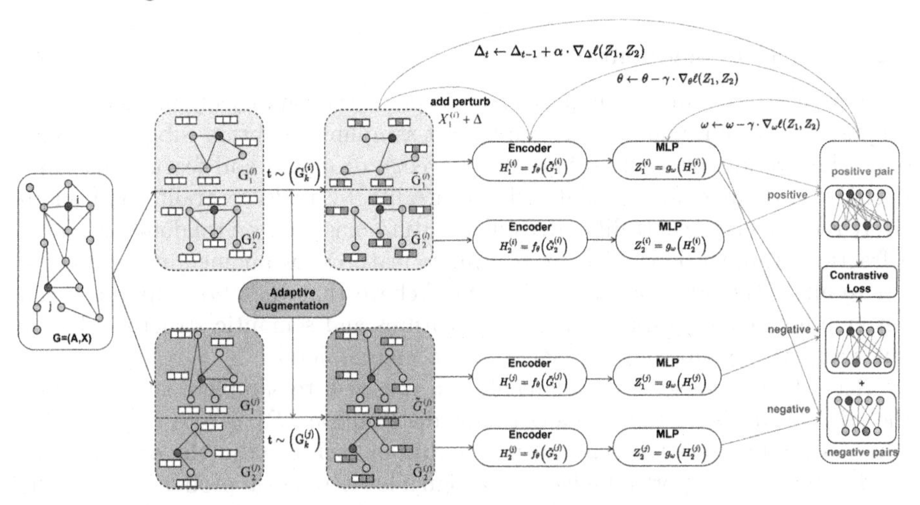

Fig. 1. Overview of GCAUV model: we first generate subgraphs by performing random walk with restart sampling for each node, where the subgraphs obtained by sampling the same central node are considered as a pair of positive samples, and the subgraphs obtained by sampling different central nodes are negative samples. Subsequently, we perform adaptive augmentation of the subgraph based on node centrality. Then we put the homophilic and heterophilic graphs into different graph encoders and the same projection head to obtain the node representations, respectively. We update the graph encoder parameters by contrastive loss based on g-EMD distance. We also introduce adversarial training for the model to improve the accuracy of node classification.

3.1 Preliminaries

Notation. Let $\mathcal{G} = (\mathcal{V}, \mathcal{E})$ denote a graph, where $V = \{v_1, v_2, \cdots, v_N\}$ denotes the set of nodes and $\mathcal{E} \subseteq \mathcal{V} \times \mathcal{V}$ denotes the set of edges. We use $\mathbf{X} = \{\mathbf{x}_1, \mathbf{x}_2, \ldots, \mathbf{x}_N\} \in R^{N\times d}$ denotes the node feature matrix of the graph, and the feature size of each node is d, $\mathbf{A} \in \mathbb{R}^{N\times N}$ denotes the adjacency matrix of the graph, where $\mathbf{A}_{i,j} = 1$ if an edge exists between node i and node j, else $\mathbf{A}_{i,j} = 0$. Each node in the graph will be treated as a central node for sampling, and we denote the sampled subgraph as $\mathcal{G}_k^{(i)}$, where i denotes the central node and k denotes the k-th subgraph obtained by sampling. Our objective is to train a GNN graph encoder $f_\theta(\boldsymbol{X}, \boldsymbol{A}) \in \mathbb{R}^{N\times F'}$ $(F' \ll d)$. The node feature matrix and adjacency matrix are input to obtain a low-dimensional node representation, and the obtained node representations are applied to downstream tasks.

Homophilic and Heterophilic Graph. We define homophilic and heterophilic graphs by edge homophily [15]. Edge homophily is the proportion of edges connecting two nodes of the same class in a graph. Edge homophily is in the range [0, 1], with values close to 1 indicating strong homophily, and values close to 0 indicating strong heterophily. We refer the graph with high homophily as homophilic graph and the graph with low homophily as heterophilic graph.

3.2 Adaptive Augmentation

Since we need to obtain unaligned views to prove the validity of our method, we propose to utilize random walk with restart sampling to obtain subgraphs. Subsequently, we augment the subgraph with edge dropping and feature masking. For edge dropping, we calculate edge centrality by node centrality and adaptively remove lower centrality edges. For feature masking, we randomly masking a fraction of dimensions with zeros in node features for augmentation in order not to affect the adversarial training. We believe that the above approach can better preserve the graph's topological features and semantic structure. In the following, we describe our augmentation scheme in detail.

Regarding topology-level augmentation, we wish to sample a subset $\tilde{\varepsilon}$ from the original edge set ε through a Bernoulli distribution : $P\{(u,v) \in \tilde{\mathcal{E}}\} = 1 - p^e_{uv}$, where p^e_{uv} represents the removal probability of each edge in the original edge set. For the more important edges, we assign a lower removal probability, and the less important edges are assigned a higher removal probability so that p^e_{uv} can reflect the importance of each edge. With this augmentation, the more important topologies in the graph can be preserved.

In networks, node centrality is a common way to measure the importance of nodes. We define $\varphi_c(\cdot) : \mathcal{V} \rightarrow \mathbb{R}^+$ as the node centrality metric function and calculate edge centrality by the centrality of two connected nodes. In directed graphs, we use the centrality of the trailing node as the edge centrality because the importance of edges is usually expressed by nodes they are pointing to, i.e. $w^e_{uv} = \varphi_c(v)$. In undirected graphs, we use the average of two adjacent nodes' centrality scores, i.e. $w^e_{uv} = (\varphi_c(u) + \varphi_c(v))/2$.

After obtaining the edge centrality, we can calculate the dropping probability of each edge. Before that, we normalize the calculated edge centrality to prevent the effect of different centrality metrics with different orders of magnitude, i.e. $s^e_{uv} = \log w^e_{uv}$. Finally, we calculate the dropping probability of edges in the following way.

$$p^e_{uv} = \min\left(\frac{s^e_{\max} - s^e_{uv}}{s^e_{\max} - \mu^e_s} \cdot p_e, p_\tau\right), \tag{1}$$

where $s^e_{\max}$ and u^e_s are the maximum and average values of s^e_{uv}, p_e is the hyperparameter that controls the overall probability of removing edges, and $p_\tau < 1$ is used to control the cutoff probability of edge removal because if too many edges are removed from the graph, the semantic structure of the graph may be severely damaged.

For the node centrality measure, we use the three measures of GCA [19], which are degree centrality, PageRank centrality, and eigenvector centrality. In addition, we design two novel node centrality measures, closeness centrality and betweenness centrality. We will introduce them in the following.

Closeness Centrality. The closeness centrality reflects the proximity between a node and other nodes in the network. Suppose the shortest path distance from node u to all other nodes in the graph is small. In that case, node u dosen't need to rely excessively on other nodes when transmitting information to other nodes, indicating that node u is important and has a high degree of closeness centrality. We calculate the closeness centrality of a node by Eq. (2).

$$C(u) = \frac{n-1}{\sum_{v=1}^{n-1} d(u,v)}, \tag{2}$$

we express the closeness centrality of node u as the reciprocal of the average shortest path distance between node u and $n-1$ reachable nodes in the graph, where $d(u,v)$ denotes the shortest distance from node u to node v, and n is the number of nodes that node u can reach in the graph.

Betweenness Centrality. The interactions between two non-adjacent nodes of the network depend on other nodes, especially those on the path between nodes. They have a controlling and constraining effect on the interactions between two non-adjacent nodes. Therefore, the idea of betweenness centrality is that if node v is located on multiple shortest paths between other nodes, then node v is a core node, indicating that node v has a large betweenness centrality. We calculate the betweenness centrality of a node by Eq. (3).

$$g(v) = \sum_{s=1}^{n-1} \sum_{t=1}^{n-1} \frac{\sigma_{st}(v)}{\sigma_{st}}, \tag{3}$$

where $\sigma_{st}(v)$ denotes the number of shortest paths from node s to node t through node v, σ_{st} is the number of shortest path from node s to node t.

3.3 Encoding Methods for Homophilic and Heterophilic Graphs

After adaptive augmentation of the subgraphs, we design different encoding methods for the nodes in homophilic and heterophilic graphs. For homophilic graphs, we use a multi-headed graph attention network [12] to encode them, and for heterophilic graphs, we use MLP to encode them.

Previous studies found that when GCN [7] aggregates the information of neighbor nodes, the weights of neighbor nodes are calculated only related to the topology structure but not to the node features. Since the topology structure of each graph is not the same, the generalization of GCN on the graph structure is poor. In contrast, GAT [12] uses more feature information of the nodes in calculating attention coefficients rather than all graph structure information. Therefore, GAT is an aggregation method that partially depends on the graph structure. The introduction of multi-headed attention enhances its expressiveness, solving the problem that GCN completely depends on the degree matrix and adjacency matrix, and the coefficients are not learnable. The encoding process of the homophilic graph is as follows.

First, we calculate the attention coefficients of all neighbor nodes of the target node by Eq. (4), where *LeakyReLU* is a nonlinear activation function, $\tilde{\mathbf{a}}$

is a weight vector of size $2F'$, W is a weight matrix, h_i denotes the node features of the target node, h_j denotes the node features of the neighbor node j connected to the target node, and N_i denotes all the nodes connected to the target node.

$$\alpha_{ij} = \frac{\exp\left(LeakyReLU\left(\tilde{\mathbf{a}}^T\left[\mathbf{W}\vec{h}_i \parallel \mathbf{W}\vec{h}_j\right]\right)\right)}{\sum_{k\in\mathcal{N}_i}\exp\left(LeakyReLU\left(\tilde{\mathbf{a}}^T\left[\mathbf{W}\vec{h}_i \parallel \mathbf{W}\vec{h}_k\right]\right)\right)} \tag{4}$$

We introduce multi-headed attention [12] to improve the expression of GAT [12]. After calculating the multiple attention coefficients, we obtain the node representations by weighted average, and calculate them by Eq. (5), where σ is the Sigmoid nonlinear activation function, W^k is the weight matrix, and α_{ij} is the attention coefficient.

$$\vec{h}'_i = \sigma\left(\frac{1}{K}\sum_{k=1}^{K}\sum_{j\in\mathcal{N}_i}\alpha_{ij}^k\mathbf{W}^k\vec{h}_j\right) \tag{5}$$

Unlike the homophilic graph, the heterophilic graph emphasizes that the nodes are less similar to their neighbors, and the node representations and labels are more different. If we use GCN to encode the nodes in the heterophilic graph, GCN will aggregate the information of all neighbor nodes during the encoding process. The difference between target and neighbor nodes in the heterophilic graph is large. If the multi-hop neighbor nodes of the target node have the same type or similar features as the target node, then once we use multiple layers of GCN to encode the nodes in the heterophilic graph, the homophilic and heterophilic information in the heterophilic graph will be mixed. This mixed information cannot help our model extract the important information in the graph. In contrast, it generates interference signals that affect the ability of GNN to learn knowledge in the graph and degrade the performance of downstream tasks. Therefore, the node information aggregation method of GCN is unsuitable for heterophilic graphs. Since the features of the nodes themselves in the heterophilic graph are sufficient for the neural network to classify the nodes in the graph, we use two-layer MLP to encode the heterophilic graph, as shown in Eq. (6).

$$\mathbf{H} = \left(\sigma(BatchNorm(\mathbf{XW}))\right)\mathbf{W}', \tag{6}$$

where X is the initial node feature, W and W' are the weight matrix of the first and second layers, respectively, σ is the Relu nonlinear activation function.

3.4 G-EMD-based Contrastive Loss

After obtaining the representations of the two unaligned views, we use the g-EMD distance in RoSA [20] as the contrastive metric to measure the distance between positive and negative samples. Most previous contrastive learning studies have used cosine similarity as the contrastive metric. However, cosine similarity is limited to the contrast between aligned views, which inevitably limits the diversity and flexibility of view sampling and augmentation. Therefore, we use g-EMD distance to solve this problem.

EMD is the measure of the distance between two discrete distributions, it can be interpreted as the minimum cost to move one pile of dirt to the other. Since g-EMD distance can directly calculate the distance between representations of views, it can solve the problem that nodes between views must be aligned. The calculation of g-EMD can be formulated as a linear optimization problem. In the contrastive of positive and negative samples, the two augmented views have feature mappings $\mathbf{X} \in \mathbb{R}^{M \times d}$ and $\mathbf{Y} \in \mathbb{R}^{N \times d}$, respectively, and the goal is to measure the distance from converting X to Y. Assume that for each node $\boldsymbol{x}_i \in \mathbb{R}^d$, it has $\boldsymbol{t}_i$ units to transport and that node $\boldsymbol{y}_j \in \mathbb{R}^d$ has $\boldsymbol{r}_j$ units to receive. For a given pair of nodes $\boldsymbol{x}_i$ and $\boldsymbol{y}_j$, the unit transport cost is $\mathbf{D}_{ij}$ and the transport volume is $\boldsymbol{\Gamma}_{ij}$. We define the problem as follows:

$$\min_{\Gamma} \sum_{i=1}^{M} \sum_{j=1}^{N} \mathbf{D}_{ij} \boldsymbol{\Gamma}_{ij}. \tag{7}$$

We design the cost matrix $\mathbf{D}_{ij}$ incorporating the topological distance and find the optimal $\tilde{\Gamma}$ by the *Sinkhorn Algorithm* [20] with entropy regularizer. After obtaining the cost matrix $\mathbf{D}_{ij}$ and the optimal $\tilde{\Gamma}$, we can compute the g-EMD distance that converts X to Y. Due to space limitation, the method of calculating the g-EMD distance is described in Supplementary Material[1].

In order to map the node representations of different views into the same contrastive space, we send the obtained node representations into a projection head (i.e. a two-layer MLP) to obtain $\mathrm{Z}_1^{(n)}$, $\mathrm{Z}_2^{(n)}$. The contrastive loss of node $\boldsymbol{v}_i$ is shown in Eq. (8), where s(x,y)=g-EMD(x,y) is used to calculate the similarity between x and y, $\mathbb{I}$ is an indicator function that returns 1 if $i = k$; otherwise returns 0, τ is the temperature parameter. The overall contrastive loss is shown in Eq. (9).

$$\ell\left(\mathbf{Z}_1^{(i)}, \mathbf{Z}_2^{(i)}\right) = -\log\left(\frac{e^{s\left(\mathbf{Z}_1^{(i)}, \mathbf{Z}_2^{(i)}\right))/\tau}}{\sum_{k=1}^{N} e^{s\left(\mathbf{Z}_1^{(i)}, \mathbf{Z}_2^{(k)}\right))/\tau} + \sum_{k=1}^{N} \mathbb{I}_{[k \neq i]} e^{s\left(\mathbf{Z}_1^{(i)}, \mathbf{Z}_1^{(k)}\right))/\tau}}\right) \tag{8}$$

$$\mathcal{J} = \frac{1}{2N} \sum_{i=1}^{N} \left[\ell\left(\mathbf{Z}_1^{(i)}, \mathbf{Z}_2^{(i)}\right) + \ell\left(\mathbf{Z}_2^{(i)}, \mathbf{Z}_1^{(i)}\right)\right] \tag{9}$$

3.5 Adversarial Training on GCAUV

We introduce adversarial training to the model by adding perturbation to the node features. When adding perturbation, we need to find the perturbation that makes the maximum loss. The purpose is to make the added perturbation have as much interference effect on the neural network as possible so that it can have the effect of adversarial training. Initially, the neural network under perturbation may misclassify the adversarial samples. Still, as the training proceeds, our model

[1] https://github.com/GCAUV-PyTorch/GCAUV/blob/main/Supplementary%20Material%20.pdf.

adapts to this perturbation, and the model's performance is improved in the process. We can formulate it as the following optimization problem.

$$\min_{\theta,\omega} \mathcal{L}\left(Z_1^{(i)}, Z_2^{(i)}\right) = \frac{1}{M}\sum_{i=1}^{M} \max_{\Delta_t} \ell_i \left(f\left(\tilde{A}_1^{(i)}, \tilde{X}_1^{(i)} + \Delta_t\right), f\left(\tilde{A}_2^{(i)}, \tilde{X}_2^{(i)}\right)\right), \quad (10)$$

where ℓ_i is the contrastive loss, f is the graph encoder and Δ_t is the added perturbation. We update the perturbation by iterating internally through the gradient ascent algorithm M times to find the perturbation that maximizes the contrastive loss. As the perturbation is determined, the outer updates the weights of the graph encoder and MLP by gradient descent.

4 Experiments

4.1 Experimental Setup

Datasets. We conduct experiments on seven public benchmark datasets, including four homophilic datasets, Cora, Citseer, Amazon-Photo, and Amazon-Computers, and three heterophilic datasets, Cornell, Texas, Wisconsin. All the datasets we used are from the Pytorch Geometry Library (PyG) [2]. For more information about the above datasets in Table 1.

Table 1. Statistics of datasets used in experiments.

Dataset	#Nodes	#Edges	#Features	#Classes
Cora	2,708	5,429	1,433	7
Citeseer	3,327	4,732	3703	6
Amazon-Photo	7,650	119,081	745	8
Amazon-Computers	13,752	245,861	767	10
Cornell	183	280	1,703	5
Texas	183	295	1,703	5
Wisconsin	251	466	1,703	5

Evaluation Protocol. We measure the the models' performance by node classification accuracy. To evaluate the trained graph encoders, we use the linear evaluation protocol [14], first train the model in an unsupervised manner and then train a separate classifier on the learned node representation. The homophilic graph dataset is trained using an l2-regularization LogisticRegression classifier, and the heterophilic graph dataset is trained using a layer of MLP through 100 Epochs. We randomly split the nodes in the homophilic dataset (10%/10%/80%) for training/validation/testing, and for the heterophilic dataset, we use the standard data splits processed by Geom-GCN [11]. We experiment with hyperparameters for optimal performance, and the detailed hyperparameter settings are published in github. We perform 10 experiments on the model for different dataset splits, report each dataset's average performance for evaluation, and report the average accuracy with standard deviation.

Baselines. On the homophilic dataset, we compare the GCAUV model with traditional baseline methods, including node2vec [3] and DeepWalk [1] and DeepWalk with embeddings concatenated with input node features. Comparisons also were made with existing deep learning methods such as GAE [6], DGI [14], GRACE [18], GCA [19], BGRL [13], MVGRL [4], and RoSA [20]. To reflect the effectiveness of the GCAUV model, we also compare it with two supervised representative models, GCN [7] and GAT [12]. For the heterophilic dataset, we compare DGI [14], SUBG-CON [5], and ROSA [20] as baselines for our model. The hyperparameters of each baseline are set according to the original paper.

4.2 Performance on Node Classification

Results for Homophilic Datasets. Table 2 summarizes the node classification performance of the GCAUV model on the four homophilic datasets. The experimental results show that the GCAUV model outperforms baseline methods in node classification accuracy. This result is mainly attributed to the four components of our framework: (1) Adaptive augmentation of subgraphs by node centrality metrics can better preserve the semantics within the graph and keep the important knowledge in the graph. (2) The information aggregation approach of GAT in encoding nodes in the homophilic graph is beneficial. (3) Using g-EMD distance instead of cosine similarity can ensure the flexibility and diversity of view sampling. (4) Adversarial training can effectively improve the model's performance on the node classification task. We also observe that variants using

Table 2. Performance summary on homophilic graphs, where A, X, and Y correspond to the node features, adjacency matrix, and labels, respectively. GCAUV-DE/PR/EV/CL/BT denote the five variants of GCAUV with different node centrality of degree/PageRank/eigenvector/closeness/betweenness for adaptive augmentation.

Method	Training Data	Cora	Citeseer	Amazon-photo	Amazon-Computers
node2vec	A	74.8	52.3	89.67±0.12	84.39±0.08
DeepWalk	A	75.7	50.5	89.44±0.11	85.68±0.06
DW+fea	X,A	73.1	47.6	90.05±0.08	86.28±0.07
GCN	X,A,Y	82.8	72.0	92.42±0.22	86.51±0.54
GAT	X,A,Y	83.00±0.70	72.50±0.70	92.56±0.35	86.93±0.29
GAE	X,A	71.50±0.40	65.80±0.40	91.62±0.13	85.27±0.19
DGI	X,A	82.60±0.40	71.80±0.70	91.61±0.22	83.95±0.47
Grace	X,A	83.30±0.40	72.10±0.50	92.15±0.24	87.46±0.22
GCA	X,A	83.80±0.80	72.20±0.70	92.53±0.16	87.85±0.31
MVGRL	X,A	83.50±0.40	72.60±0.70	91.74±0.07	87.52±0.11
BGRL	X,A	83.83±1.61	72.32±0.89	92.95±0.07	87.89±0.10
RoSA	X,A	83.34±0.81	72.80±0.63	92.92±0.13	88.90±0.19
GCAUV-DE	X,A	84.61±0.77	73.34±0.35	93.01±0.11	89.08±0.05
GCAUV-PR	X,A	84.59±0.68	72.88±0.45	**93.27±0.09**	89.23±0.03
GCAUV-EV	X,A	**85.03±0.35**	**73.37±0.40**	93.16±0.01	**89.45±0.05**
GCAUV-CL	X,A	84.56±0.49	73.29±0.51	93.13±0.15	89.27±0.11
GCAUV-BT	X,A	84.28±0.74	72.84±0.70	93.05±0.06	89.22±0.22

Table 3. Performance of the node classification task using GCN (left) and MLP (right) on heterophilic graphs.In this case, we show the highest performance in bold.

Method	Cornell (GCN)	Texas (GCN)	Wisconsin (GCN)	Cornell (MLP)	Texas (MLP)	Wisconsin (MLP)
DGI	56.3±4.7	56.9±6.3	50.9±5.5	58.1±4.1	57.8±5.2	52.1±6.3
SUBG-CON	54.1±6.7	56.9±6.9	48.3±4.8	58.7±6.8	61.1±7.3	59.0±7.8
RoSA	58.92±4.56	59.19±4.90	52.35±3.40	70.0±3.91	69.19±5.16	**71.18±5.41**
GCAUV-DE	59.46±3.63	**60.27±4.69**	55.29±4.37	**71.08±8.11**	70.54±5.73	70.78±4.51
GCAUV-PR	59.46±4.19	59.73±3.51	52.35±4.39	70.54±7.88	70.81±5.64	70.20±4.09
GCAUV-EV	59.46±2.96	**60.27±4.02**	52.55±5.10	**71.08±8.11**	70.54±5.73	70.98±4.28
GCAUV-CL	**60.54±3.86**	59.73±4.75	**55.49±4.12**	70.81±4.65	70.54±5.73	70.98±4.62
GCAUV-BT	**60.54±3.24**	59.46±4.83	52.94±4.72	70.81±8.09	**71.08±5.68**	70.98±4.37

different node centrality measures outperform baseline methods on all datasets, indicating that our model is not limited to a specific choice of node centrality measures, illustrating the generalizability of our model.

Results for Heterophilic Datasets. Table 3 summarizes the node classification performance of the GCAUV model on the three heterophilic datasets. Among them, to illustrate that MLP is a more suitable encoding method for heterophilic graphs, we use GCN and MLP as encoders for the GCAUV model and other baselines, respectively. The experimental results show that the performance of the model encoded by MLP is always higher than that of the model encoded by GCN on the heterophilic graph, both for the the GCAUV model and for baseline methods. It fully illustrates the effectiveness of MLP. The GCAUV model outperforms the other baselines in the case of GCN coding, and when coded by MLP, it outperforms other baselines on the Cornell and Texas datasets. In comparison, the performance on the Wisconsin dataset is slightly lower than RoSA but equally competitive.

4.3 Ablation Studies

To illustrate the effectiveness of each component in the model, we perform ablation experiments at the same hyperparameter settings, remove or replace key

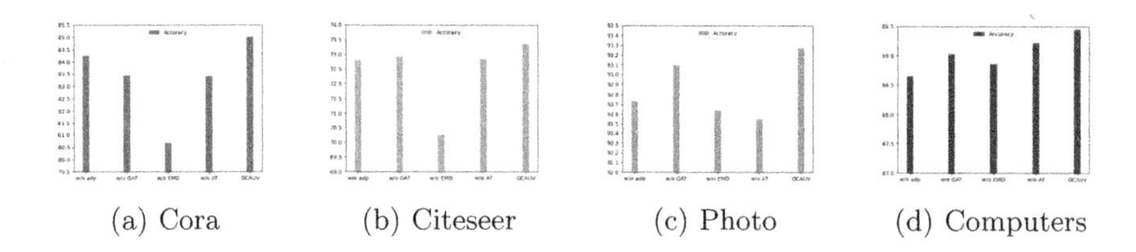

(a) Cora (b) Citeseer (c) Photo (d) Computers

Fig. 2. Ablation Study on GCAUV.

components for each of these variants. First, we replace the adaptive augmentation with random edge dropping and feature masking in the graph, named it (w/o adp). Second, we replace the GAT with GCN to encode the nodes and named it (w/o GAT). Subsequently, we replace the g-EMD distance with cosine similarity to measure the distance between positive and negative samples and named it (w/o EMD). Finally, we remove the adversarial training to observe the model's performance and named it (w/o AT). The results are shown in Fig. 2. The experimental results show that the node classification accuracy decreases when we remove a key component. This fully illustrates the effectiveness of each component in the GCAUV model. We also perform sensitivity analysis for the hyperparameters, which we show in Supplementary Material.

5 Conclusion

In this paper, we propose a novel graph contrastive learning framework based on adaptive augmentation and node self-alignment. We design multiple node centrality measures to calculate edge centrality, and obtain adaptive augmentation views by assigning different dropping probability to edges in terms of their centrality. We propose different encoding methods for homophilic (GAT) and heterophilic (MLP) graph respectively to obtain high-quality representations. We also use g-EMD distance instead of cosine similarity to measure the distance between positive and negative samples, thus allowing to have the different set of nodes between different views. Extensive experiments on homophilic and heterophilic graph datasets demonstrate the excellent performance of our model.

Acknowledgements. This work was supported by the National Natural Science Foundation of China (No. 61972135), the Natural Science Foundation of Heilongjiang Province in China (No. LH2020F043).

References

1. Bryan Perozzi, R.A.R., Skiena, S.: DeepWalk: Online learning of social representations. In: Proceedings of KDD, pp. 701–710 (2014)
2. Fey, M., Lenssen, J.E.: Fast graph representation learning with PyTorch geometric. arXiv e-prints arXiv:1903.02428 (2019)
3. Grover, A., Leskovec, J.: node2vec: Scalable feature learning for networks. In: Proceedings of KDD, pp. 855–864 (2016)
4. Hassani, K., Ahmadi, A.H.K.: Contrastive multi-view representation learning on graphs. In: Proceedings of ICML, pp. 4116–4126 (2020)
5. Jiao, Y., Xiong, Y., Zhang, J., Zhang, Y., Zhang, T.: Sub-graph contrast for scalable self-supervised graph representation learning. In: Proceedings of ICDM (2020)
6. Kipf, T.N., Welling, M.: Variational graph auto-encoders. arXiv e-prints arXiv:1611.07308 (2016)
7. Kipf, T.N., Welling, M.: Semi-supervised classification with graph convolutional networks. In: Proceedings of ICLR (2017)
8. Kong, K., Li, G., Ding, M., Wu, Z., Zhu, C.: Robust optimization as data augmentation for large-scale graphs. arXiv e-prints arXiv:2010.09891 (2020)

9. Liu, M., Liu, Y.: Inductive representation learning in temporal networks via mining neighborhood and community influences. In: Proceedings of SIGIR (2022)
10. Liu, M., Quan, Z.W., Wu, J.M., Liu, Y., Han, M.: Embedding temporal networks inductively via mining neighborhood and community influences. In: Applied Intelligence 52 (2022)
11. Pei, H., Wei, B., Chang, K.C.-C., Lei, Y., Yang, B.: Geom-GCN: geometric graph convolutional networks. arXiv e-prints arXiv:2002.05287 (2020)
12. Veličković, P., Cucurull, G., Casanova, A., Romero, A., Liò, P., Bengio, Y.: Graph attention networks. In: Proceedings of ICLR (2018)
13. Thakoor, S., Tallec, C., Azar, M.G., Munos, R., Veličković, P., Valko, M.: Bootstrapped representation learning on graphs. In: Proceedings of ICLR (2021)
14. Velickovic, P., Fedus, W., Hamilton, W.L., Liò, P., Bengio, Y., Hjelm, R.D.: Deep graph infomax. In: Proceedings of ICLR (2019)
15. Wang, H., Zhang, J., Zhu, Q., Huang, W.: Augmentation-free graph contrastive learning with performance guarantee. arXiv e-prints arXiv:2204.04874 (2022)
16. Xia, J., Wu, L., Chen, J., Hu, B., Li, S.Z.: SimGRACE: a simple framework for graph contrastive learning without data augmentation. In: Proceedings of WWW (2022)
17. You, Y., Chen, T., Wang, Z., Shen, Y.: Graph contrastive learning with augmentations. In: Proceedings of NIPS, pp. 5812–5823 (2020)
18. Zhu, Y., Xu, Y., Yu, F., Liu, Q., Wu, S., Wang, L.: Deep graph contrastive representation learning. In: Proceedings of ICML (2020)
19. Zhu, Y., Xu, Y., Yu, F., Liu, Q., Wu, S., Wang, L.: Graph contrastive learning with adaptive augmentation. In: Proceedings of WWW, pp. 2069–2080 (2021)
20. Zhu, Y., Guo, J., Wu, F.: Rosa: a robust self-aligned framework for node-node graph contrastive learning. In: Proceedings of IJCAI, pp. 3795–3801 (2022)

MPool: Motif-Based Graph Pooling

Muhammad Ifte Khairul Islam[1(✉)], Max Khanov[2], and Esra Akbas[1]

[1] Department of Computer Science, Georgia State University, Atlanta, GA, USA
mislam29@student.gsu.edu, eakbas1@gsu.edu
[2] University of Wisconsin-Madison, Madison, WI, USA

Abstract. Recently, Graph Neural Networks (GNNs) have emerged as a powerful technique for various graph-related tasks. Current GNN models apply different graph pooling methods that reduce the number of nodes and edges to learn the higher-order structure of the graph in a hierarchical way. However, these methods primarily rely on the one-hop neighborhood and do not consider the higher-order structure of the graph. To address this issue, in this work, we propose a multi-channel Motif-based Graph Pooling method named (MPool) that captures the higher-order graph structure with motif and also considers the local and global graph structure through a combination of selection and clustering-based pooling operations. In the first channel, we develop node selection-based graph pooling by designing a node ranking model considering the motif adjacency of nodes. In the second channel, we develop cluster-based graph pooling by designing a spectral clustering model using motif adjacency. Finally, the result of each channel is aggregated into the final graph representation. We perform extensive experiments and demonstrate that our proposed method outperforms the baseline methods for graph classification tasks on eight benchmark datasets.

Keywords: Graph Neural Network · Graph Classification · Pooling · Motif

1 Introduction

Recently, Graph Neural Networks (GNNs) have emerged as a powerful technique for various graph-related tasks. With message propagation along the edges, while some GNN [11] models learn the node-level representation for node classification [9,11,25], some others learn graph-level representation for graph classification [4,7,14]. Graph classification is the task of predicting graph labels by considering node features and graph structure. Motivated from the pooling layer in Convolutional Neural Networks [12], graph pooling methods have been used to reduce the number of nodes and edges to capture the local and global structural information of the graph in the graph representation in a hierarchical way.

There are mainly two types of hierarchical pooling methods for the graph in the literature: clustering-based and selection-based methods. While clustering-based

Supplementary Information The online version contains supplementary material available at https://doi.org/10.1007/978-3-031-33377-4_9.

H. Kashima et al. (Eds.): PAKDD 2023, LNAI 13936, pp. 105–117, 2023.
https://doi.org/10.1007/978-3-031-33377-4_9

methods merge similar nodes into super nodes using a cluster assignment matrix, selection-based methods calculate a score for each node, which represents their importance, and select the top k nodes based on the score by discarding other nodes from the graph. All these methods primarily rely on Graph Convolution Networks (GCNs) with layer-wise propagation based on the one-hop neighbors to calculate the assignment matrix in the clustering-based method and score in the selection-based method.

Despite the success of these models, there are some limitations. The selection-based model mainly focuses on preserving the local structure of the node while the clustering-based method basically focuses on the global structure of the graph. Moreover, while selection-based models may lose information by selecting only some portion of the nodes, clustering-based models may include some redundant information including noise and over-smoothing. Further, the current methods fail to incorporate the higher-order structure of the graph in pooling. There are different ways to model higher-order graph structures [3] such as hypergraphs, simplicial complexes [1], and motifs [16]. Among them, motifs (graphlets) are small, frequent, and connected subgraphs that are mainly used to measure the connectivity patterns of nodes [6] (see Fig. 1 for a preview). They capture the local topology around the vertices, and their frequency can be used as the global fingerprints of graphs. Although motifs have been used for different graph mining tasks, including classification [13], and community detection [15], to the best of our knowledge, they have not been used in graph pooling operations. On the other hand, utilizing these structures for pooling provides crucial information about the structure and the function of many complex systems that are represented as graphs [20,22].

In this paper, to address these problems, we propose a multi-channel Motif-based Graph Pooling method named (`MPool`) that captures the higher-order graph structure with motif and also local and global graph structure with a combination of selection and clustering-based pooling operation. We utilize motifs to model the relation between nodes and use this model for message passing and pooling in GNN. We develop three motif-based graph pooling models ($\texttt{MPool}_S$ $\texttt{MPool}_C$, and $\texttt{MPool}_{cmb}$): selection and clustering based and combined model ($\texttt{MPool}_{cmb}$). For the selection-based graph pooling model, we design a node ranking model considering motif-based relations of nodes. Based on the ranks, we select the top k nodes to create the pooled graph for the next layer. For clustering-based graph pooling, we design a motif-based clustering model that learns a differentiable soft assignment based on learned embedding from the convolution layer.

After learning the assignment matrix, we group the nodes in the same cluster to create a coarsened graph. By combining these selecting and clustering-based methods into one model, we learn both local and global graph structures. All models incorporate higher-order graph structure in graph representation with taking motifs into consideration while pooling. We further demonstrate detailed experiments on eight benchmark datasets. Our results show that the proposed pooling methods show better accuracy than the current baseline pooling methods for graph classification tasks.

2 Related Work

Graph Pooling: Recent GNN with pooling methods learn graph representation hierarchically and capture the local substructures of graphs. There are two different hierarchical pooling methods in the literature: clustering-based and selection-based pooling. Clustering-based pooling methods [2,4,23,27] do the pooling operation by calculating the cluster assignment matrix using node features and graph topology. After calculating the cluster assignment matrix, they build the coarse graph by grouping the nodes on the same cluster. For example, while DiffPool [27] calculates the cluster assignment matrix using a graph neural network, MinCutPool [4] calculates the cluster assignment matrix using a multi-label perception.

Selection-based pooling methods [7,8,14,24,28] compute the importance scores of nodes and select top k nodes based on their scores and drop other nodes from the graph to create the pooled graph. For example, while gPool [7] calculates the score using node feature and a learnable vector, SAGPool [14] uses an attention mechanism to calculate the scores. SUGAR [24] uses a subgraph neural network to calculate the score and select top-K subgraphs for pooling operation. All these methods use the classical graph adjacency matrix to propagate information and calculate the score.

Motifs in Graph Neural Network. Motifs are the most common higher-order graph structure used in various graph mining problems. A few works have used motif structure in GNNs as well [13,17,21,26]. In these works, they use motifs to learn the representation of nodes or subgraphs and use this representation for node classification. But in our method, we use motif structure while defining pooling operation on graph for graph classification problem.

3 Methodology

In this section, first, we discuss the problem formulation of graph classification and preliminaries. Then we present our motif-based pooling models.

3.1 Preliminaries and Problem Formulation

We denote a graph as $G(V, A, X)$ where V is the node-set, $A \in \mathbb{R}^{N \times N}$ is the adjacency matrix, and $X \in \mathbb{R}^{N \times d}$ is the feature matrix with d dimensional node feature and N is the number of nodes in the graph. We denote a graph collection as $(\mathcal{G}, Y)$ where $\mathcal{G} = \{G_0, G_1, ..., G_n\}$ with G_i's are graphs and Y is the set of the graph labels. In this paper, we work on the graph classification problem, whose goal is to learn a function $f : \mathcal{G} \to Y$ to predict the graph labels with a graph neural network in an end-to-end way.

Graph Neural Network for Graph Classification: GNN for graph classification has two modules: message-passing and pooling. For message-passing operations, Graph convolution network (GCN) [11] is the most widely used model where it combines the features of each node from its neighbors as follows:

$$H^{(l+1)} = \sigma(\tilde{D}^{-\frac{1}{2}} \tilde{A} \tilde{D}^{-\frac{1}{2}} H^{(l)} \theta^{(l)}) \tag{1}$$

where $H^{(l+1)}$ is the node representation matrix for layer $(l+1)$, σ is an activation function, $\tilde{A} = A + I$ is the adjacency matrix with self-loop, $\tilde{D} \in \mathbb{R}^{\mathbb{N}\times\mathbb{N}}$ is the normalized degree matrix of $\tilde{A}$, $\theta^{(l)}$ is trainable weight for $l^{(th)}$ layer and $H^{(l)}$ is the input node representation matrix for $l + 1^{th}$ layer obtained from previous layer. $H_0 = X$ is the initial input node feature matrix of the input graph. We utilize GCN for message-passing operations in our model.

The second module of GNNs for graph classification is the pooling operation that helps to learn the graph features. The main idea behind graph pooling is to coarsen the graph by reducing the number of nodes and edges to encode the information of the whole graph. In the literature, there are two types of hierarchical graph pooling methods: selection-based and clustering-based methods. Selection-based methods calculate a score (attention) using a scoring function for every node that represents their importance. Based on the calculated scores, the top k nodes are selected to construct a pooled graph. They use a classical graph adjacency matrix to propagate information and calculate the score.

Clustering-based pooling methods learn a cluster assignment matrix $S \in R^{N\times K}$ using graph structure and/or node features. Then, they reduce the number of nodes by grouping them into super nodes by $S \in R^{N\times K}$ to construct the pooled graph at $(l+1)^{th}$ layer as follows

$$A^{(l+1)} = S^{(l)^T} A^{(l)} S^{(l)}, \qquad H^{(l+1)} = S^{(l)^T} H^{(l)}. \tag{2}$$

Motifs and Motif-based Adjacency Matrix: Motifs (graphlets) are small, frequent, and connected subgraphs that are mainly used to measure the connectivity patterns of nodes [6]. Motifs of sizes 2-4 are shown in Fig. 1. To include higher-order structural information between nodes, we create the motif adjacency matrix M_t for a motif t where $(M_t)_{i,j}$ represents the # of the motif containing nodes i and j.

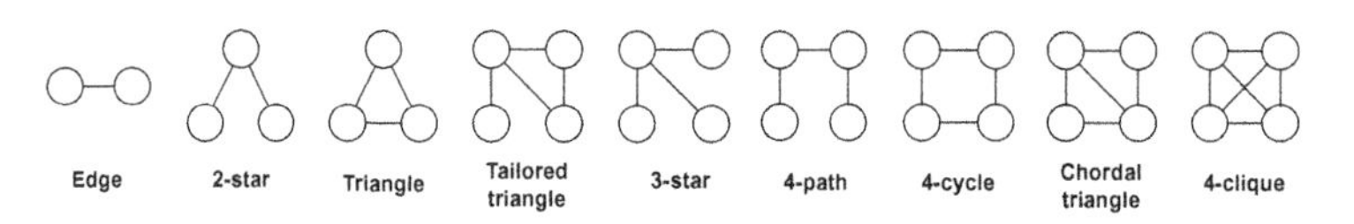

Fig. 1. Motif Networks with size 2-4.

3.2 Motif Based Graph Pooling Models

We propose a hierarchical pooling method based on motif structure. As the first layer, graph convolution (GCN) takes the adjacency matrix A and feature matrix X of the graph as input and then updates the feature matrix by propagating the features through the neighbors and aggregating features coming from adjacent nodes. After getting the updated feature matrix from the convolution layer, our proposed graph pooling layer, `MPool`, operates coarsen on the graph. These steps are repeated l steps, and outputs of each pooling layer are aggregated with readout function [5] to obtain a fixed-sized graph representation. After

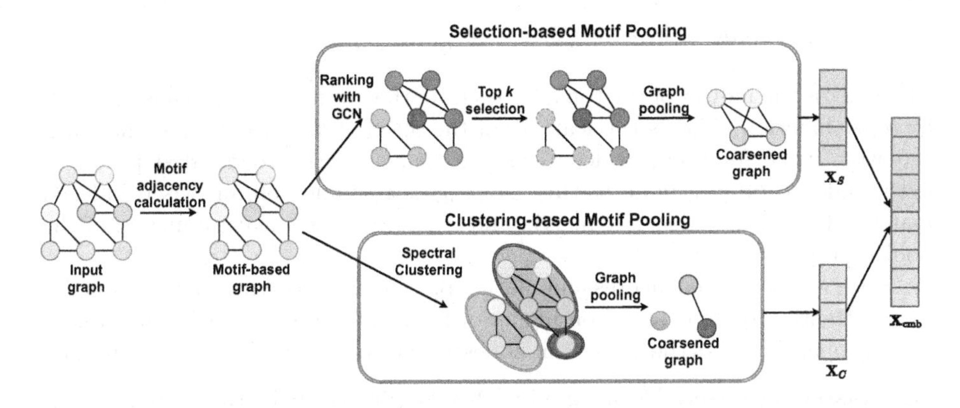

Fig. 2. An illustration of our motif-based pooling methods.

concatenating the results of readouts, it is fed to the multi-layer perceptron (MLP) layer for the graph classification task. We develop three types of motif-based graph pooling methods: (1) $\texttt{MPool}_S$ is the selection-based method, (2) $\texttt{MPool}_C$ is the clustering-based method, and (3) $\texttt{MPool}_{cmb}$ is the combined model. These are illustrated in Fig. 2. In this paper, we adopt the model architectures from SAGPool [14] as the selection-based and MinCutPool [4] as the clustering-based model. On the other hand, our method is compatible with any graph neural network that we show later in our experiment section.

A. Selection-based Pooling via Motifs ($\texttt{MPool}_S$): Previous selection-based methods [7,14] do the pooling operation using a classical adjacency matrix. However, higher-order structures like motifs show great performance on graph convolution network [13] and are also important structures for graph classification. Therefore, in our selection method, we first calculate the motif adjacency matrix for a particular motif type, e.g., triangle, from the original graph as we discuss in Sect. 3.1. Then, we calculate the motif attention score for each node by considering the motif adjacent. Based on these scores, we select the top k nodes for pooling and construct the coarsened graph using the pooling function. Figure 2 presents the overview of our selection-based graph pooling method. We use a graph convolution network to calculate the motif attention score for each node where we use node attributes and also motif-based graph topological information instead of pair-wise edge information. *Motif attention* score is defined as follows

$$Z = \sigma(D^{-\frac{1}{2}}\tilde{M}D'^{-\frac{1}{2}}X\theta_{att}) \tag{3}$$

where σ is an activation function, $\tilde{M} \in \mathbb{R}^{N\times N}$ is the motif adjacency matrix with self loop where $\tilde{M} = M + I_N$, $D' \in \mathbb{R}^{N\times N}$ is the degree matrix of M, and $\theta_{att} \in R^{d\times 1}$ is the learnable parameter matrix.

Based on the motif attention score, we select the top k nodes from the graph following the node selection method in [7]. The top $k = \alpha \times N$ nodes are selected based on the Z value where α is the pooling ratio between 0 and 1. Thus, we obtain the pooling graph as follows

$$idx = topK(Z, [\alpha \times N])X_{out} = X_{idx,:} \odot Z_{idx}, \quad A_{out} = A_{idx,idx} \tag{4}$$

where idx is the indices of the top k nodes from the input graph which is returned by $topK$ function, X_{idx} is the features of the selected k nodes, Z_{idx} is the motif attention value for those nodes. $\odot$ is the element-wise broadcasted product,: is the indexed of each node feature. $A_{idx,idx}$ is row and column wised indexed matrix, A_{out} is the adjacency matrix and X_{out} is the new feature matrix of the pooled graph.

Since we use graph features and motif adjacency matrix with convolution, the motif attention score is based on higher-order graph structures and features. So pooling operation gets the important nodes with respect to higher-order structure.

B. Clustering-based Pooling via Motifs ($\texttt{MPool}_C$**):** In this paper, as the base for our clustering-based pooling methods, we use MinCutPool [4] that is defined based on Spectral clustering (SC) by minimizing the overall intra-cluster edge weights. MinCUTpool proposes to use GNN with a custom loss function to compute cluster assignment with relaxing the normalized minCut problem. However, they consider only the regular edge-based adjacency matrix to find clusters. On the other hand, considering edge-type relations between nodes may result in ignoring the higher-order relations. Including higher-order relations like motifs for clustering may produce better groups for pooling.

In our clustering-based method, we calculate the cluster assignment matrix S utilizing motif adjacency information. We adopt spectral clustering method [4] where we use multi-layer perceptron (MLP) by inputting node feature matrix X. We use the softmax function on the output layer of MLP. This function maps each node feature X_i into the ith row of a soft cluster assignment matrix S

$$S = MLP(X; \theta_{MLP}) = softmax(ReLU(XW_1)W_2) \tag{5}$$

However, as it is seen in Eq. 5, we do not use adjacency but use attributes of nodes obtained from the convolution part. Therefore, to include motif information in the pooling layer, we use the motif adjacency matrix in the convolution layer while passing the message to neighbors as $X = GNN(X, \tilde{M}; \theta_{GNN})$ where $\tilde{M}$ is the normalized motif adjacency matrix, and θ_{GNN} and θ_{MLP} are learnable parameter.

We also incorporate motif information in the optimization. Parameters of the convolution layer and pooling layer are optimized by minimizing a loss function $\mathcal{L}$ including the usual supervised loss function $\mathcal{L}_s$ and also an unsupervised loss function [4] $\mathcal{L}_u$ as $\mathcal{L}_u = \mathcal{L}_c + \mathcal{L}_o$ where

$$\mathcal{L}_c = -\frac{Tr(S^T MS)}{Tr(S^T DS)} \quad \text{and} \quad \mathcal{L}_o = \left\| \frac{S^T S}{||S^T S||_F} - \frac{I_K}{\sqrt{K}} \right\|_F \tag{6}$$

$\mathcal{L}_c$ is the cut loss that encourages strongly connected nodes in motif adjacency to be clustered together. Here, $Tr(S^T MS) = \frac{1}{k}\sum_{k=1}^{K} S_k^T MS_k$ and $Tr(S^T DS) = \frac{1}{k}\sum_{k=1}^{K} S_k^T DS_k$ where K is the number of clusters and D is the degree matrix and M is the motif adjacency matrix. L_o is the orthogonality loss, which helps the clusters to become similar in size. I_K is a (rescaled) clustering matrix $I_K = \widehat{S}^T\widehat{S}$, where $\widehat{S}$ assigns exactly N/K points to each cluster. After calculating the cluster assignment matrix, we compute the coarsened graph adjacency matrix and attribute matrix using Eq. 2.

C. Combined model ($\texttt{MPool}_{cmb}$): Selection-based models mainly focus on preserving the local structure of the node by selecting top-K representative nodes while cluster-based methods basically focus on the global structure of the graph by assigning nodes into K-clusters. To utilize the benefits of the selection-based and cluster-based models at the same time, we combine our selection-based and cluster-based motif pooling model into one model. As a result graph representation from the combined model encoded local structure information from the selection-based model and the global structure model from the cluster-based model. In this model we concatenate the graph-level representation from the selection-based motif pooling method and cluster-based motif pooling method into one final representation as follows:

$$X_{cmb} = X_S \oplus X_C \tag{7}$$

where X_S is the graph-level representation from $\texttt{MPool}_S$ model and X_C from $\texttt{MPool}_C$ method and, $\oplus$ is the concatenation operation.

3.3 Readout Function and Output Layer

To get a fixed-sized representation from different layers' pooled graph, we apply a readout function [14] that aggregates the node features as follows: $Z = \frac{1}{N} \sum_{i=1}^{N} x_i || \max_{i=1}^{N} x_i$ where N is the number of nodes, x_i is the ith node feature and $||$ denotes concatenation. After concatenating the results of all readout functions as a representation of the graph, it is given as an input to a multilayer perceptron with the softmax function to get the predicted label of the graph as $\hat{Y} = softmax(MLP(Z))$ where Z is the graph representation. For graph classification, parameters of GNNs and pooling layers are optimized by a supervised loss as $\mathcal{L}_s = -\sum_{i=1}^{L}\sum_{j=1}^{C} Y_{i,j} log\hat{Y}_{i,j}$ where Y is the actual label of the graph.

4 Experiment

We evaluate the performance of our models in graph classification problems and compare our results with the baseline methods for selection-based and clustering-based on different datasets. We also give the results for the variation of our model by utilizing different message-passing models. Further, we analyze the effect of the motif types on the results of the pooling. More experiments can be found on supplements.

Datasets: We use eight benchmark graph datasets in our experiments commonly used for graph classification [18]. Among these, three datasets are social networks (SN); IMDB-BINARY, REDDIT-BINARY, and COLLAB, and five other datasets are biological and chemical networks (BN) ;D&D, PROTEINS (PROT), NCI1, NCI109, and Mutagenicity (MUTAG) .

Baseline: We use five graph pooling methods as baseline methods. Among them, gPool [7] and SAGPool [14] are selection-based method and MinCutPool (MCPool) [4], DiffPool [27] and ASAP [23] are clustering-based method. We also combined SAGPool and MCPool models in a single model and use as a baseline model.

Experimental Setup: To evaluate our models for the graph classification task, we randomly split the data for each dataset into three parts. We use 80% data for the training set, 10% data for the validation set, and 10% data for the test set. We do the splitting process 10 times using 10 random seed values. We implement our model using PyTorch and PyTorch Geometric library. For optimizing the model, we use Adam optimizer [10]. In our experiments, we take node representation size as 128 for all datasets. Our hyperparameters are as follows: learning rate in {1e–2, 5e–2, 1e–3, 5e–3, 1e–4, 5e–4}, weight decay in {1e–2, 1e–3, 1e–4, 1e–5}, and pooling ratio in {1/2, 1/4}. We find the optimal hyperparameters using grid search. We run the model for a maximum of 100K epochs, and there is an early stopping condition if the validation loss does not improve for 50 epochs. Our model architecture consists of three blocks, and each block contains one graph convolution layer and one graph pooling layer like [14]. We use the same model architecture and hyperparameters with MinCuT and SAGPool models.

4.1 Overall Evaluation

Performance on Graph Classification: In this part, we evaluate our proposed graph pooling methods for the graph classification task on the given eight datasets. Each dataset contains a certain number of input graphs and their corresponding label. In the graph classification task, we classify the input graph by predicting the label of the graph. We use node features of the graph as the initial features of the model. If a dataset does not contain any node feature, we use node degrees as initial features using one-hot encoding. Table 1 and Table 2 show the average graph classification accuracy, standard deviation, and ranking of our models and other baseline models for all datasets. We can observe from the tables that our motif-based pooling methods consistently outperform other state-of-art models, and our models get the first rank for almost all datasets.

Table 1 shows the results for our motif-based models and other graph pooling models on biochemical datasets. We obtain the reported results for gPool and DiffPool from the SAGPool paper since our model architecture and hyperparameters are the same as SAGPool. Also, for the ASAP method, we obtain the results from the initial publication ("-") means that results are not available for that dataset. As we see from the table, $\texttt{MPool}_{cmb}$ gives the highest result for all biochemical networks. In particular, $\texttt{MPool}_{cmb}$ achieves an average accuracy of 81.2% on D&D and 77.4% on NCI1 datasets which are around 4% improvements over the $\texttt{MPool}_{C}$ method as the second-best model. We can also see $\texttt{MPool}_{cmb}$ gives very good accuracy compared to baseline models for all biochemical datasets. Especially for D&D, NCI1, and NCI109 datasets $\texttt{MPool}_{cmb}$ gives 5.8%, 5.8%, and 3.9% improvements over the best model of baseline models for these datasets. From this result, we can say that incorporating global

Table 1. Comparison of our models with baseline pooling methods for biochemical datasets.

Model	D&D	NCI1	NCI109	PROT	MUTAG	Rank
gPool	75.0±0.9/7	67.0±2.3/8	66.1 ±1.6/8	71.1 ±0.9/8	71.9 ±3.7/8	7.8
SAGPool	75.7±3.7/6	68.7±3.0/7	71.0±3.4/5	72.5 ±4.0/7	74.9±3.9/7	6.4
MCPool	76.7±3.0/5	73.1 ±1.4/4	71.5 ±2.7/4	76.3 ±3.6/3	75.9 ±2.7/6	4.4
DiffPool	66.9 ±2.4/9	62.2 ±1.9/9	62.0±2.0/9	68.2 ±2.0/9	77.6 ±2.6/3	7.8
ASAP	76.9 ±0.7/4	71.5±0.4/5	70.1 ±0.6/7	74.2 ±0.8/5	-	4.2
Combined	74.5±9.8/8	74.1 ±1.2/3	72.0 ±2.1/3	75.6±2.1/4	76.5±3.2/4	4.4
$\texttt{MPool}_{cmb}$	**81.2** ±2.1/1	**77.4**± 1.9/1	**73.5**±2.5/1	**79.3** ±3.3/1	**79.6** ±3.7/1	1
$\texttt{MPool}_{S}$	77.2 ±4.6/3	71.0±3.4/6	70.8±2.1/6	72.7 ±4.2/6	76.4 ±3.1/5	5.2
$\texttt{MPool}_{C}$	78.5 ±3.3/2	74.4±1.8/2	73.1±2.5/2	78.1 ±3.3/2	78.8 ±2.1/2	2

Table 2. Comparison of our models with baseline pooling methods for social network datasets.

Model	IMDB-B	REDDIT-B	COLLAB	Avg. Rank
gPool	73.40±3.7 (3)	74.70±4.5 (7)	77.58 ±1.6 (3)	4.3
SAGPool	73.00±4.06 (4)	84.66±5.4 (2)	70.10±2.5 (7)	4.3
MinCutPool	70.78±4.7 (8)	75.67 ±2.7 (6)	69.91 ±2.3 (8)	7.3
DiffPool	68.40 ±6.1 (9)	66.65 ±7.7(8)	74.83 ±2.0 (4)	7
ASAP	72.74 ±0.9 (5)	-	78.95 ±0.7 (2)	3.5
Combined	71.20±4.50(7)	**88.40**±0.22(1)	71.85±3.73(6)	4.7
$\texttt{MPool}_{cmb}$	**74.20** ±2.8 (1)	84.10± 5.0 (3)	74.13±2.3 (5)	2.6
$\texttt{MPool}_{S}$	73.44 ±3.9 (2)	83.89±4.3 (4)	68.95±2.7 (9)	5
$\texttt{MPool}_{C}$	71.44 ±4.0 (6)	78.77±5.0 (5)	**83.62**±5.2 (1)	4

and local structures of the graph in the combined model gives better results for graph classification on biochemical data. We further calculate the average rank for all models, where our model $\texttt{MPool}_{cmb}$ average rank is the lowest at 1 and our model $\texttt{MPool}_{C}$ is the second lowest.

Table 2 shows the performance comparison with our models and other baseline models on social network datasets. As we see from the table, our proposed methods outperform all the baseline methods for all datasets except ReDDIT-BINARY, where our model is the third best with giving very close to the second one, SAGPool. For IMDB-BINARY and REDDIT-BINARY $\texttt{MPool}_{cmb}$ model gives better accuracy than the $\texttt{MPool}_{S}$ and $\texttt{MPool}_{C}$ model while for COLLAB dataset $\texttt{MPool}_{C}$ give much higher accuracy than our other two models. For both types of datasets, our selection-based method $\texttt{MPool}_{S}$ gives better accuracy than the selection-based baseline methods SAGPool and gPool for most of the datasets. In particular, $\texttt{MPool}_{S}$ achieves an average accuracy of 77.21% on D&D and 76.42% on MUTAG datasets which is around 2% improvement over the SAGPool method which is our base model. Similarly, our cluster-based model outperforms the baseline methods of cluster-based methods for most of the datasets. Especially, $\texttt{MPool}_{C}$ achieves an average accuracy of 83.62% on COLLAB datasets, which is around 5%

Table 3. $\texttt{MPool}_S$ $\texttt{MPool}_C$ and $\texttt{MPool}_{cmb}$ performance with different GNN models.

	$\texttt{MPool}_S$		$\texttt{MPool}_C$		$\texttt{MPool}_{cmb}$	
GNN Model	**NCI1**	**IMDB-B**	**NCI1**	**IMDB-B**	**NCI1**	**IMDB-B**
$MPool_{GCN}$	70.98	73.44	74.44	71.44	76.09	73.90
$MPool_{GraphConv}$	74.20	73.50	**75.93**	71.90	74.7	73.00
$MPool_{SAGE}$	**74.69**	73.00	74.13	**72.22**	**78.80**	**74.00**
$MPool_{GAT}$	67.15	**74.00**	–	–	–	–

Table 4. $\texttt{MPool}_S$ $\texttt{MPool}_C$ and $\texttt{MPool}_{cmb}$ performance with different motifs.

Model	Motif	DD	NCI1	Mutag	IMDB-B
$\texttt{MPool}_S$	2-star	**77.21**	69.48	70.11	73.00
	Triangle	75.63	**70.98**	**76.42**	**73.44**
	2-star+triangle	75.63	69.82	72.39	69.64
$\texttt{MPool}_C$	2-star	**78.48**	73.56	73.56	71.20
	Triangle	75.80	**74.44**	**78.77**	**71.44**
	2-star+triangle	74.21	74.20	76.00	70.96
$\texttt{MPool}_{cmb}$	2-star	**81.20**	**77.36**	**79.60**	**74.20**
	Triangle	80.50	76.09	77.90	73.90
	2-star+triangle	79.95	76.75	78.42	73.40

improvement over the ASAP method as the second-best model and around 14% improvement over the MinCutPool, which is our base model.

Furthermore, when we compare our selection-based model $\texttt{MPool}_S$ and clustering-based model $\texttt{MPool}_C$ results from Tables, we can see that $\texttt{MPool}_C$ outperforms $\texttt{MPool}_S$ for all biochemical datasets. While $\texttt{MPool}_S$ gives better accuracy for two social networks, IMDB-BINARY and REDDIT-BINARY, $\texttt{MPool}_C$ have 15% better accuracy than $\texttt{MPool}_S$ on the COLLAB dataset.

Ablation Study: While we use GCN as the base model for message passing, our pooling model can integrate other GNN architectures. In order to see the effects of different GNN models in our methods, we utilize the other four most widely used convolutional graph models: Graph convolution network (GCN) [11], Graph-SAGE [9], GAT [25], and GraphConv [19]. Table 3 shows average accuracy results for these GNN models using $\texttt{MPool}_S$ $\texttt{MPool}_C$ and $\texttt{MPool}_{cmb}$ on NCI1 and IMDB-BINARY datasets. As there is no dense version of Graph attention network(GAT), we use it only for selection-based model $\texttt{MPool}_S$. For this experiment, we use triangle motifs for the motif adjacency matrix calculation. As we see in the table, the effects of GNN models and which model gives the best result depend on the dataset. For the NCI1 dataset, Graph-SAGE gives the highest accuracy on $\texttt{MPool}_S$ and $\texttt{MPool}_{cmb}$ model while GraphConv gives the highest accuracy on $\texttt{MPool}_C$model. For IMDB-BINARY, all the graph convolutional models give very close results for all of our pooling models. For $\texttt{MPool}_C$

and $\texttt{MPool}_{cmb}$ Graph-SAGE gives better accuracy than the other GNN models while GAT gives the highest accuracy for $\texttt{MPool}_S$ model.

We further study the effect of the motif type for pooling. In this experiment, we use 2-star, triangle, and a combination of 2-star and triangle motifs, as these motifs are observed the most in real-world networks. We present the graph classification accuracy for different motifs using $\texttt{MPool}_S$ $\texttt{MPool}_C$ and $\texttt{MPool}_{cmb}$ in Table 4. As we see in the table, we get the highest accuracy for $\texttt{MPool}_S$ and $\texttt{MPool}_C$ with the triangle motif for three datasets NCI1, MUTAG, and IMDB-BINARY. For D&D, we get the highest accuracy with 2-star motif adjacency on $\texttt{MPool}_S$ and $\texttt{MPool}_C$. We also observe that for D&D, the accuracy of the selection-based model does not vary much compared to the clustering-based model. For MUTAG, different motifs have a large effect on the accuracy, where triangle motif adjacency gives around 4% and 3% higher accuracy than the 2-star motif adjacency for the selection-based method and for the clustering-based model, respectively. For IMDB-BINARY, 2-star and triangle motifs give similar accuracy for both methods, and 2-star+triangle motif adjacency gives less accuracy for the clustering-based method. For our combined model $\texttt{MPool}_{cmb}$ the 2-star motif gives the highest accuracy for all datasets whereas other motifs give very close results to the 2-star motif.

5 Conclusion

In this work, we introduce a novel motif-based graph pooling method, MPool, that captures the higher-order graph structures for graph-level representation. Our proposed method includes hierarchical graph pooling models for both selection-based and clustering-based methods. Additionally, we combine these methods to develop a hybrid model. The selection-based pooling method employs a motif attention mechanism, while the clustering-based method uses motif-based spectral clustering with the mincut loss function. Our experiments demonstrate that our proposed methods outperform the baseline models on a majority of the datasets.

Acknowledgment. This work is funded partially by National Science Foundation (NSF) under Grant No 2104720.

References

1. Aktas, M.E., Nguyen, T., Jawaid, S., Riza, R., Akbas, E.: Identifying critical higher-order interactions in complex networks. Sci. Rep. **11**(1), 1–11 (2021)
2. Bacciu, D., Conte, A., Grossi, R., Landolfi, F., Marino, A.: K-plex cover pooling for graph neural networks. Data Min. Knowl. Disc. **35**(5), 2200–2220 (2021). https://doi.org/10.1007/s10618-021-00779-z
3. Benson, A.R., Gleich, D.F., Higham, D.J.: Higher-order network analysis takes off, fueled by classical ideas and new data. arXiv preprint arXiv:2103.05031 (2021)

4. Bianchi, F.M., Grattarola, D., Alippi, C.: Spectral clustering with graph neural networks for graph pooling. In: International Conference on Machine Learning, pp. 874–883. PMLR (2020)
5. Cangea, C., Veličković, P., Jovanović, N., Kipf, T., Liò, P.: Towards sparse hierarchical graph classifiers. arXiv preprint arXiv:1811.01287 (2018)
6. Elhesha, R., Kahveci, T.: Identification of large disjoint motifs in biological networks. BMC Bioinform. **17**(1), 1–18 (2016)
7. Gao, H., Ji, S.: Graph U-nets. In: International Conference on Machine Learning, pp. 2083–2092. PMLR (2019)
8. Gao, X., Dai, W., Li, C., Xiong, H., Frossard, P.: iPool-information-based pooling in hierarchical graph neural networks. IEEE Trans. Neural Netw. Learn. Syst. **PP**, 1–13 (2021)
9. Hamilton, W., Ying, Z., Leskovec, J.: Inductive representation learning on large graphs. Advances in Neural Information Processing Systems 30 (2017)
10. Kingma, D.P., Ba, J.: Adam: a method for stochastic optimization. In: 3rd International Conference on Learning Representations (2015)
11. Kipf, T.N., Welling, M.: Semi-supervised classification with graph convolutional networks. In: Proceedings of the ICLR (2017)
12. Krizhevsky, A., Sutskever, I., Hinton, G.: ImageNet classification with deep convolutional neural networks. NeurIPS (2012)
13. Lee, J.B., Rossi, R.A., Kong, X., Kim, S., Koh, E., Rao, A.: Graph convolutional networks with motif-based attention. In: Proceedings of the 28th ACM International Conference on Information and Knowledge Management, pp. 499–508 (2019)
14. Lee, J., Lee, I., Kang, J.: Self-attention graph pooling. In: International Conference on Machine Learning, vol. 97 (2019)
15. Li, P.Z., Huang, L., Wang, C.D., Lai, J.H.: EdMot: an edge enhancement approach for motif-aware community detection. In: Proceedings of the 25th ACM SIGKDD International Conference on Knowledge Discovery & Data Mining (2019)
16. Milo, R., Shen-Orr, S., Itzkovitz, S., Kashtan, N., Chklovskii, D., Alon, U.: Network motifs: simple building blocks of complex networks. Science **298**(5594), 824–827 (2002)
17. Monti, F., Otness, K., Bronstein, M.M.: MotifNet: a motif-based graph convolutional network for directed graphs. In: IEEE Data Science Workshop (DSW) (2018)
18. Morris, C., Kriege, N.M., Bause, F., Kersting, K., Mutzel, P., Neumann, M.: TUDataset: a collection of benchmark datasets for learning with graphs. arXiv preprint arXiv:2007.08663 (2020)
19. Morris, C., et al.: Weisfeiler and leman go neural: Higher-order graph neural networks. In: Proceedings of the AAAI Conference on Artificial Intelligence, vol. 33 (2019)
20. Paranjape, A., Benson, A.R., Leskovec, J.: Motifs in temporal networks. In: ACM International Conference on Web Search and Data Mining, pp. 601–610 (2017)
21. Peng, H., Li, J., Gong, Q., Ning, Y., Wang, S., He, L.: Motif-matching based subgraph-level attentional convolutional network for graph classification. In: Proceedings of the AAAI Conference on Artificial Intelligence, vol. 34 (2020)
22. Prill, R.J., Iglesias, P.A., Levchenko, A.: Dynamic properties of network motifs contribute to biological network organization. PLoS Biol. **3**(11), e343 (2005)
23. Ranjan, E., Sanyal, S., Talukdar, P.: ASAP: adaptive structure aware pooling for learning hierarchical graph representations. In: Proceedings of the AAAI Conference on Artificial Intelligence, vol. 34, pp. 5470–5477 (2020)

24. Sun, Q., et al.: Sugar: subgraph neural network with reinforcement pooling and self-supervised mutual information mechanism. In: Proceedings of the Web Conference 2021, pp. 2081–2091 (2021)
25. Veličković, P., Cucurull, G., Casanova, A., Romero, A., Liò, P., Bengio, Y.: Graph attention networks. In: International Conference on Learning Representations (2018)
26. Yang, C., Liu, M., Zheng, V.W., Han, J.: Node, motif and subgraph: leveraging network functional blocks through structural convolution. In: IEEE/ACM International Conference on Advances in Social Networks Analysis and Mining (ASONAM) (2018)
27. Ying, Z., You, J., Morris, C., Ren, X., Hamilton, W., Leskovec, J.: Hierarchical graph representation learning with differentiable pooling. NeurIPS 31 (2018)
28. Zhang, Z., et al.: Hierarchical graph pooling with structure learning. arXiv preprint arXiv:1911.05954 (2019)

Anti-Money Laundering in Cryptocurrency via Multi-Relational Graph Neural Network

Woochang Hyun(✉), Jaehong Lee, and Bongwon Suh

Seoul National University, Seoul, Republic of Korea
{woochang,jhlee0105,bongwon}@snu.ac.kr

Abstract. The cryptocurrency market has been growing exponentially. However, due to their anonymity, cryptocurrencies are frequently abused for laundering money obtained from illegal activities. Although recent approaches based on Graph Neural Networks (GNNs) have shown remarkable achievements in fraud detection, investigating cryptocurrency transaction networks is subject to the following challenges: 1) There is a lack of useful node features as cryptocurrencies block access to user information in principle. 2) The classification tasks are extremely disproportionate since fraudsters are very few compared to benign addresses. 3) Lastly, the computational cost must be considered for large-scale analysis in real-world scenarios. This study presents a novel approach for examining financial transactions to detect anomalies in cryptocurrency networks. We design a multi-relational GNN that incorporates the orientation and characteristics of edges, such as the amount or frequency of transactions. In addition, an adaptive neighbor sampler is designed to improve spotting performance while effectively containing computational costs. Experiments on real-world datasets demonstrate that our proposed method outperforms state-of-the-art GNN-based fraud detectors.

Keywords: Graph Neural Network · Cryptocurrency · Fraud Detection

1 Introduction

Since the introduction of Bitcoin in 2008 [9], the cryptocurrency ecosystem has been proliferating. The recent adoption of cryptocurrencies has been growing much faster than ever before. Global transaction volume through cryptocurrencies in 2021 is estimated to be $15.8 trillion [1], an increase of more than 550% compared to the previous year.

A key feature of cryptocurrency is supporting monetary transactions via decentralized peer-to-peer networks. However, due to their anonymity, cryptocurrencies have been widely abused for various crimes, such as scams, phishing, or ransomware. Cybercriminals laundered $8.6 billion worth of cryptocurrency in 2021, which is a 30% increase in money laundering over 2020 [1]. Unreported

The original version of this chapter was revised: the typesetting errors were corrected in Table 2, Table 3, and in Figure 3. The correction to this chapter is available at https://doi.org/10.1007/978-3-031-33377-4_43

H. Kashima et al. (Eds.): PAKDD 2023, LNAI 13936, pp. 118–130, 2023.
https://doi.org/10.1007/978-3-031-33377-4_10

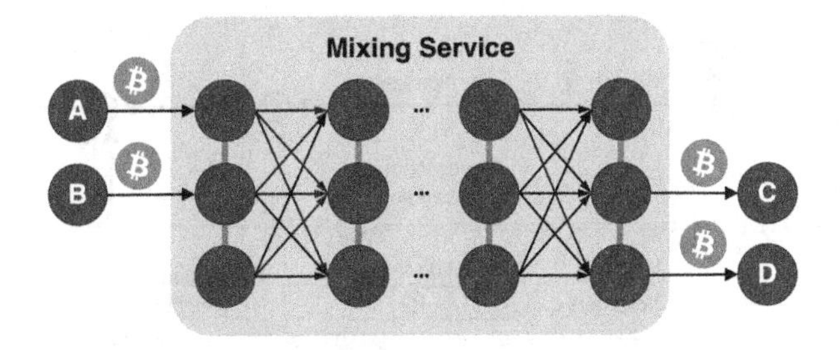

Fig. 1. Example of Bitcoin mixing process. Solid arrows indicate actual transactions. In this study, we draw cooperative edges (marked by orange lines) between nodes with a common sender or receiver. With the cooperative edges, we better understand mixing behaviors and construct a GNN mixing detector. (Color figure online)

issues could be larger. According to [3], by 2019, 25% of all Bitcoin users and 44% of existing Bitcoins may have been associated with illicit activities.

The most representative way to laundry cryptocurrency is to use a mixing service. As shown in Fig. 1, mixing services split and blend transaction flows to avoid traceability. Mixing services are initially intended for the privacy of ordinary users, but they have been easily involved in cryptocurrency crimes. Therefore, detecting abnormally used accounts is very important to prevent innocent damage.

One promising solution is the application of Graph Neural Networks (GNNs). Along with remarkable achievements of the recent GNN techniques in various fields, GNN-based fraud detectors have also been proposed to detect fraudsters in online reviews [2,7,12], fake news [10], and defaulters in credit payment platforms [8,14,15,17]. However, their solutions are limited to be employed in this study because investigating cryptocurrency transactions is more challenging for the following reasons: 1) There is a lack of node features as cryptocurrency systems inherently do not provide users' information. Sometimes addresses are disposable only for one-time use. On the other hand, most existing solutions mainly select neighbors with the highest similarity utilizing rich node information. 2) The classification tasks are extremely disproportionate. Since there is very little data about known fraudsters, it is subject to severe class imbalance problems in which positive rates are lower than 0.5%. Finally, scalability and computational cost must be considered for large-scale analysis in real-world scenarios. Methods that are too complex cannot be employed in practice.

This study presents a multi-relational GNN framework (BitcoNN) for examining financial transactions to detect anomalies in cryptocurrency networks. Figure 2 illustrates the pipeline of the proposed BitcoNN. 1) In the graph construction stage, we propose cooperative edges that can better reflect the proximity of nodes, and separate existing transactive edges according to the direction. 2) In the representation embedding stage, BitcoNN learns neighbors' representation and the dissimilarity between center nodes and incorporates edge features such as the amount or frequency of transactions. 3) In the inter-relation aggregation stage, BitcoNN evaluates the contribution score of each relation and then feeds it back to the neighbor sampling for the next epoch. The adaptive neighbor sampler contributes to improving model performance while suppressing the computation costs efficiently. Experiments on the dominant cryptocurrency system,

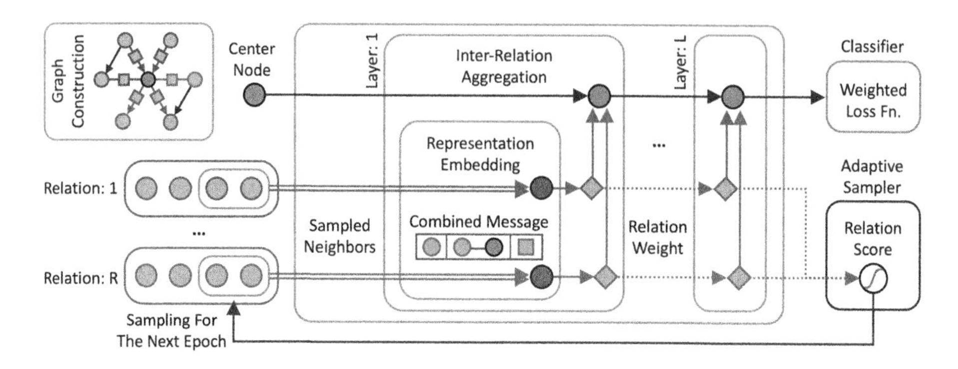

Fig. 2. Framework of the proposed BitcoNN. We first construct a multi-relational graph as input of GNN. Sampled neighbors under each relation send messages which are the combination of their representation, the difference from the center node, and the edge feature. In the inter-relation aggregation, the contribution of each relation is evaluated, and the scores are fed-back to the adaptive neighbor sampler for the next epoch.

i.e., Bitcoin, demonstrate that our proposed method outperforms state-of-the-art GNN-based fraud detection baselines.

2 Methodology

2.1 Graph Construction

Financial data generally consists of a list of transaction records. Many graph-based studies turn this into a graph topology by converting senders and receivers to nodes and transactions to non-directional edges. In this study, we leverage a multi-relational structure by suggesting transactive and cooperative relations.

Transactive Relations. At first, we make edges between the nodes of the forward transactions. Edge $e^1_{ij} \in \mathcal{E}_1$ means that node v_j has an incoming transaction from v_i. In directed GNNs, however, the edge conveys a message unilaterally only from v_i to v_j. Thus to enable information transfer in the opposite direction, we also need to specify outgoing edge set $\mathcal{E}_2$ as follows:

$$\mathcal{E}_2 = \{(v_i, v_j) \mid e^1_{ji} \in \mathcal{E}_1\} \tag{1}$$

Cooperative Relations. To take advantage of richer structural information, we propose cooperative relations by connecting nodes that have common transaction partners in a short time. We consider a time window between the transactive edges to prevent an explosion of additional edges and to capture more closely related two-hop neighbors. That is, if nodes v_i and v_j commonly trade with another node v_k and their time interval does not exceed a time window δ, we define that the pair (v_i, v_j) is in a cooperative relation. The cooperative relations are divided into two categories, namely, co-input edges $\mathcal{E}_3$ and co-output edges $\mathcal{E}_4$

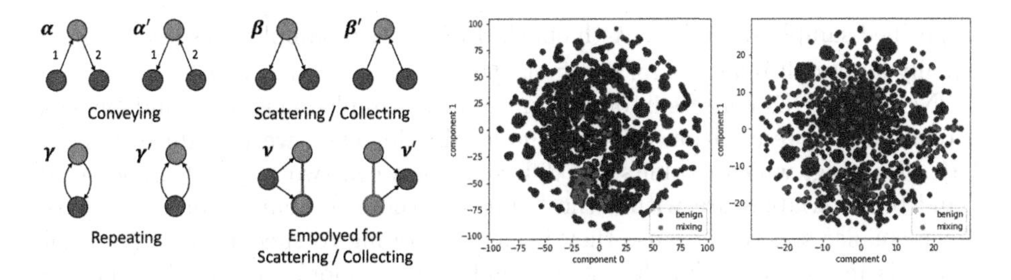

Fig. 3. Types of two-edge temporal motifs for a centered (orange) node in a financial transaction network (Color figure online).

Fig. 4. T-SNE visualization of the crafted node features (left) and node2vec embeddings (right) from randomly sampled 20k data (containing 137 mixing nodes) of the Bitcoin 2014.

as defined in (2) and (3), respectively. t_{ij} is the timestamp of the corresponding edge e^1_{ij}. Note that the cooperative relation is non-directional.

$$\mathcal{E}_3 = \{(v_i, v_j) \mid \exists (e^1_{ik}, e^1_{jk}), |t_{ik} - t_{jk}| \leq \delta\} \tag{2}$$

$$\mathcal{E}_4 = \{(v_i, v_j) \mid \exists (e^1_{ki}, e^1_{kj}), |t_{ki} - t_{kj}| \leq \delta\} \tag{3}$$

We introduce cooperative relations for two reasons: First, we assume that nodes that have joint transaction partners would have more similarities. The second reason is that by including the cooperative edges, we present novel and meaningful temporal motifs. Temporal motifs are the smallest subgraphs extracted considering the sequence and time interval to capture minute information in dynamic networks [11]. The previous work [16] derived the proportion of the simplest two-edge temporal motifs centered on each node (see motif $\alpha \cdot \alpha'$, $\beta \cdot \beta'$ and $\gamma \cdot \gamma'$ in Fig. 3). Each motif can represent the role of the corresponding node. We unveil important but unexposed roles of nodes through the temporal motifs created by incorporating cooperative relations. In Fig. 3, ν and ν' are newly proposed two-edge temporal motifs centering a node formulated by one transactive edge and one cooperative edge. They articulate that the node engages as a branch in transactions of money diffusion or collection, which are closely associated with money laundering. As such, nodes in these types of relations are not directly connected but work together. That's why we refer to them as cooperative relations.

2.2 Representation Embedding

After the graph construction, BitcoNN trains the representation of the nodes through the GNN methodology. To deal with large and complex networks, we only use information from a fixed number of neighbors in our calculation. S is the total number of samples that BitcoNN takes from all relations. R is the number of relations defined in the graph. $\mathcal{N}_r(v)$ is a neighbor set of node v in relation r. $\mathcal{S}^{(k)}_r(v) \subset \mathcal{N}_r(v)$ is a set of sub-sampled neighbors from relation r according to a specified sample size $s_r^{(k)}$ at k-th epoch. Initially, the sample sizes are assigned equally across all relations for the first epoch (i.e., $s_r^{(1)}|_{r=1}^{R} = S/R$).

Next, BitcoNN collects information from the sampled neighbors and embeds a relational representation of center nodes. In this case, we introduce dissimilarity learning, which exploits the difference between two nodes. Camouflaged fraudsters' variables are within the bounds of those of benign nodes but have much smaller spectra because they behave according to specific rules (we analyze this in Sect. 3.2). Furthermore, fraudsters are more connected to benign nodes. Therefore, by leveraging the difference in representation with neighboring nodes, fraudsters can be detected more effectively. In addition, since edge features in this graph contain transaction information between two nodes, they need to be embedded in the representation. In summary, the message transmitted from a neighbor node consists of three elements: its representation, the difference between itself and the center node, and edge features, as (4) shows.

$$m_{uv}^{r(l)} = concat\big(h_u^{(l-1)}, |h_u^{(l-1)} - h_v^{(l-1)}|, y_{uv}^r\big) \tag{4}$$

where $m_u^{r(l)}v$ is the message from node u to a center node v under relation r at l-th layer. $h_u^{(l-1)}$ and $h_v^{(l-1)}$ are the node representations at the previous layer. y_{uv}^r is the edge feature of the corresponding edge e_{uv}^r.

We aggregate the messages from sampled neighbors to obtain a relational representation. We use the max-pooling aggregator as (5). W_r^P is a learnable pooling weight for each relation.

$$h_{v,r}^{(l)} = max\big(ReLU(W_r^P m_{uv}^{r(l)}), \forall u \in \mathcal{S}_r(v)\big) \tag{5}$$

2.3 Inter-relation Aggregation

Considering the diverse contribution of relations, we define the inter-relation aggregation as follows:

$$h_v^{(l)} = concat\big(W^S h_v^{(l-1)}, W_r^A h_{v,r}^{(l)}|_{r=1}^{R}\big) \tag{6}$$

where $h_v^{(l)}$is the updated node representation at l-th layer, W^S and W_r^A are learnable weights in the aggregation for self-node and each relation r, respectively. Since W_r^A can be regarded as the contribution of relation r we devise an adaptive neighbor sampler that automatically adjusts the sampling size from each relation by assessing relation scores.

2.4 Adaptive Neighbor Sampler

How do we determine the sampling size for each relation when using a subsampling approach in multi-relational graphs? It is an important but rarely addressed question. Should we gather more samples depending on the number of relations? It will inevitably increase the computational costs in proportion to the increased sample neighbors. We need an adaptive sampling module that properly distributes a fixed number of total samples according to the contribution of each relation in multi-relational conditions.

With a simple assumption, we evaluate the contribution of each relation through the size of relation weight obtained at the end of inter-relation aggregation at k-th epoch, $W_r^{A(k)}$. First, the L2-distance of each weight is calculated to turn it into a scalar. The softmax function then normalizes the distances to the range $[0, 1]$ and gives them attentive differences.

$$D_r^{(k)} = \left\| W_r^{A(k)} \right\|_2 \tag{7}$$

$$a_r^{(k)} = softmax(D_r^{(k)}) = \frac{exp(D_r^{(k)})}{\sum_{r=1}^{R} exp(D_r^{(k)})} \tag{8}$$

$a_r^{(k)}$ is the relation score that reflects the contribution of each relation proportionately. The scores are applied to determine the sample size for the next epoch per relation. A minimum sample size p is assigned to prevent completely excluding relations with low importance, as shown in (9).

$$s_r^{(k+1)} = max\big(p, round(a_r^{(k)} S)\big) \tag{9}$$

The intuition underlying the adaptive neighbor sampler is to increase the chance of capturing relevant neighbors by exploring the broadened area in significant relations. At the same time, by reducing the number of samples from less important relations, the computational costs are restrained efficiently. We discuss the effectiveness of the adaptive neighbor sampler in Sect. 3.5.

After the sampling and aggregation, an MLP classifier yields predicted labels, and GNN is optimized to minimize loss. Here, most GNNs apply the standard cross-entropy loss as a loss function. We adopt a weighted cross-entropy loss to deal with severe class imbalance.

3 Experiment

3.1 Experimental Setup

Dataset. The objective of this study is to spot financial anomalies, such as money laundering accounts from monetary transaction networks. We select the Bitcoin transaction dataset published in [16] to validate the proposed framework. The dataset is publicly available at http://xblock.pro/bitcoin/. It consists of a considerable number of addresses that were part of mixing services (e.g., *BitcoinFog*, *BitLaunder*, *HelixMixer*, and *BitMixer*) and snapshots of extensive Bitcoin transaction records crawled at the time those mixing addresses were active (e.g., 2014, 2015, and 2016). Descriptive statistics of the dataset are in Table 1.

For node features, we extract 32 variables, including basic graph statistics and the temporal motifs in Fig. 3. For edge features, we assign 10 variables regarding the amount, frequency, and weight of transactions between nodes. Although the feature extraction process is rather hand-crafted, but highly effective. We find that it better characterizes mixing behavior than other techniques that automatically embed graph structure information (e.g., node2vec and eigenvalue decomposition), as shown in Fig. 4. In addition, current implementations for graph embedding are barely applicable to such giant data.

Baselines. We compare BitcoNN with various state-of-the-art GNNs to demonstrate the ability for financial fraud detection. We choose GCN [6], GAT [13], and GraphSAGE [4] as primary GNN baselines, and CARE-GNN [2] and PC-GNN [7] as the latest applications for fraud detection.

Experiment Settings. We conduct experiments in two conditions: the transductive and the inductive setting. The transductive setting is a laboratory condition where train, valid, and test sets are randomly split from the given dataset with a ratio of 40:20:40, respectively. The train-test split is based on stratified sampling to maintain the positive rate constantly. The inductive setting is closer to a real-world usage scenario. We split the dataset into 10 equal-sized buckets over time so that each bucket contains the same number of nodes. The first four buckets are used to train the model, and then the following two and four buckets are used for validation and test.

Table 1. Descriptive statistics

#Nodes (Mix%)	Relation	#Edges	Feat. Sim.	Label Sim.
Bitcoin 2014	*Trans.*	14,167k	.340	.115
	incoming	16,192k	.348	.096
	outgoing	16,192k	.328	.142
2,513k (0.24%)	*Coop.*	85,599k	.673	.696
	co-input	52,220k	.732	.905
	co-output	35,146k	.491	.037
Bitcoin 2015	*Trans.*	14,653k	.433	.112
	incoming	14,631k	.438	.084
	outgoing	14,631k	.423	.165
2,528k (0.15%)	*Coop.*	93,884k	.593	.418
	co-input	32,116k	.710	.969
	co-output	62,478k	.513	.034
Bitcoin 2016	*Trans.*	10,182k	.416	.378
	incoming	10,218k	.383	.463
	outgoing	10,218k	.438	.319
2,506k (0.16%)	*Coop.*	70,962k	.811	.756
	co-input	50,813k	.821	.994
	co-output	20,304k	.546	.059

Hyperparameters are as follows: embedding size = 64, number of layers = 2, learning rate = 0.001, max epoch = 50, Adam optimizer, and batch size = 4,096. Since positive labels are too sparse, a smaller batch size can hardly be adopted to distribute positive labels in all batches. The minimum sample size $p = 2$ and time window $\delta = 3$ h since it is reported that mixing processes are usually completed within 3 h [16]. For BitcoNN and GraphSAGE, the total sample size is set to 20. For CARE-GNN and PC-GNN, learning rate is set to 0.01. For all the experiments, we present the average value of 10 runs.

BitcoNN is implemented in Pytorch 1.6.0 with Python 3.7 and DGL library. All the experiments are run with a single GeForce Titan RTX GPU with CUDA 11.3 and 128 GB RAM. GCN, GAT, and GraphSAGE are implemented based on DGL. CARE-GNN and PC-GNN are implemented using the source code provided by the authors.

3.2 Demystifying Mixing Behavior

For graph construction, we suggest a multi-relational approach to supplement the structural information of transaction graphs. In Table 1, we compute the feature similarity of mixing nodes and their neighbors under each relation based on the feature vectors' Euclidean distance. Although cooperative nodes are not directly connected, their overall feature similarity is higher than transactive neighbors. This supports our assumption that cooperative relations lead to nodes with more similar states.

Looking at the label similarity column, mixing nodes rarely connect with each other through actual transactions. However, under the co-input relation, they have outstanding connectivity. Almost all of the mixing nodes are connected indirectly through a common counterpart. On the other hand, the co-output relation shows the lowest label similarity. Taken together, we summarize that mixing nodes act according to the following rules: 1) In remittance transactions, they collect money together with other mixing nodes in a short time. 2) In receiving transactions, they hide among benign nodes.

Figure 4 shows the distribution of node features of mixing nodes. The features extracted from their behavior fall into the utterly normal range. Note that there are significantly fewer mixing nodes than benign nodes. As a result, conventional anomaly detection methods are rarely effective for this task since mixing nodes do not deviate from the normal feature distribution. A clue is that the spectrum of mixing nodes is very small since they act according to the specific rules mentioned above. Therefore, by aggregating the differences in the representation of a central node and its neighbors, we could distinguish the characteristics of mixing nodes and benign nodes more efficiently, especially through the highly connected relations of mixing nodes. Taking advantage of this in the GNN architecture, we successfully detect mixing nodes.

3.3 Performance Comparison

Table 2 shows the performance of the proposed BitcoNN and various GNN baselines in the financial anomaly detection task on three datasets of Bitcoin transactions. Best results are in bold, and results within the 95% confidence intervals of the best results are underlined. We observe that BitcoNN outperforms the baselines under most metrics and different settings. Further analysis of the results follows:

Primary GNN Models. We present the results from GCN, GAT, and GraphSAGE (GS) in the first block of each experimental setting in Table 2. Being basic models, they use only the single topology of transactive edges. The subscript *WL* indicates that the weighted loss function is applied to the models, as with the proposed BitcoNN. Due to the severe class imbalance of the tasks, vanilla GNNs with standard cross-entropy loss cannot find a mixing node at all. However, just applying the weighted cross entropy loss significantly improves the performance of these models. The weighted loss function brings competitive results even in inductive settings where the positive rate in the test set is unknown. It supports the effectiveness of our design choice for the class imbalance problem.

Latest GNN-based Fraud Detectors. We test the latest GNN-based fraud detectors, CARE-GNN and PC-GNN. Since they are multi-relational models, all relations in the graph are equally applied, as in BitcoNN. They achieve high AUC and recall. However, they show limited performance in terms of precision. One possible explanation is again the severe class imbalance. Since both models have their own modified loss function, the weighted loss is not applied. Those

Table 2. Performance (%) comparison for Bitcoin mixing detection

	Dataset	Bitcoin 2014			Bitcoin 2015			Bitcoin 2016		
	Metric	AUC	Recall	Prec.	AUC	Recall	Prec.	AUC	Recall	Prec.
Transductive	GCN_{WL}	95.78	96.24	5.04	96.63	96.35	4.65	91.20	91.91	1.81
	GAT_{WL}	91.40	88.57	3.73	87.92	84.69	1.54	81.49	93.17	0.53
	GS_{WL}	97.84	97.02	15.74	97.84	96.62	13.59	93.19	91.97	3.04
	CARE	99.02	95.75	28.85	99.16	97.29	26.07	97.25	92.28	13.14
	PC	99.56	97.96	7.41	99.52	97.65	9.30	98.21	95.22	4.32
	BitcoNN	**99.83**	**99.71**	**83.19**	**99.84**	**99.73**	**76.63**	**99.32**	**98.85**	**57.61**
Inductive	GCN_{WL}	95.77	97.26	4.35	96.48	95.67	5.13	88.99	85.77	2.26
	GAT_{WL}	89.32	84.07	3.89	90.18	88.98	1.55	81.22	91.24	0.65
	GS_{WL}	97.61	96.54	15.28	97.55	96.16	11.98	89.70	81.84	6.80
	CARE	98.95	95.11	30.78	99.27	95.38	27.31	97.45	89.95	15.41
	PC	99.43	97.32	9.04	99.47	97.45	10.93	95.84	90.22	5.36
	BitcoNN	**99.80**	**99.64**	**82.87**	**99.60**	**98.98**	**75.58**	**97.83**	**94.79**	**62.69**

algorithms were originally designed for fake review detection, and positive rates in those tasks are around 10%. However, the positive rates for our work are less than 0.5%. Under these conditions, even if the model can filter out 99% of the true negatives, the precision will not be able to reach 50%. Therefore, achieving high precision in severe class imbalance problems is very challenging.

Another explanation is that those algorithms rely on the rich context of node functions, such as user profiles or text NLP. Both are designed to select neighbors with higher node similarity preferentially. However, Fig. 4 shows that there are a lot of benign nodes with similar characteristics to mixing nodes. Therefore, the existing models cannot perform at their best in our tasks.

In Table 2, we present the mixing detection capability of BitcoNN. In particular, the precision is outstanding. BitcoNN shows much greater than 70% in precision with the 2014 and 2015 data. The result of the second-best model is no more than 30%. We analyze how each component of BitcoNN benefits the performance in the next section.

3.4 Ablation Study

Next, we compare the variants of BitcoNN to demonstrate the effectiveness of each component. Table 3 shows the results of the inductive experiments because the differences in performance under the condition are more evident. We note that the experiments under the transductive condition show the same tendency. Since the variants in the first block evaluate the relational effects, all embedding methods are applied. *Homo* utilizes only the non-directional topology, and *Coop* adds the cooperative edges to it. *Direct* identifies the direction of Bitcoin transactions but does not utilize the cooperative edges. Adding the cooperative edges in *Homo* and separating the direction of all edges becomes BitcoNN. We observe that the inclusion of each relation component significantly improves the performance in terms of precision. In the 2014 and 2015 data, the direction of transactions plays an important role (*Direct*), whereas in the 2016 data, the

Table 3. Performance (%) comparison of variants of BitcoNN

	Dataset	Bitcoin 2014			Bitcoin 2015			Bitcoin 2016		
	Metric	AUC	Recall	Prec.	AUC	Recall	Prec.	AUC	Recall	Prec.
Rela-tional	*Homo*	99.04	98.48	37.89	99.42	98.01	17.31	96.08	92.13	9.06
	Coop	99.15	98.99	52.68	99.53	98.63	25.60	97.52	94.41	27.97
	Direct	99.78	**99.68**	75.42	99.58	98.85	57.44	96.04	92.55	11.48
Repr-esent.	*Base*	98.85	97.98	48.05	99.10	98.45	43.85	94.03	66.56	33.21
	Diff	99.72	99.54	80.56	99.58	98.25	**76.63**	97.31	**94.92**	40.65
	Edge	99.34	98.73	**83.37**	99.57	98.28	61.01	97.29	94.70	56.13
BitcoNN		**99.80**	99.64	82.87	**99.60**	**98.98**	75.58	**97.83**	94.79	**62.69**

cooperative edge shows more contributions (*Coop*). This is because the mixing service providers for the 2016 data differ from those for the 2014 and 2015 data. Most of the mixing nodes in the 2014 and 2015 data belonged to *BitcoinFog* (99.83% and 99.64%, respectively), while the majority of the 2016 data came from *HelixMixer* (95%). Thus, they have different mixing styles, and our results are a good reflection of that.

The second block examines the effectiveness of each representation embedding component. Base aggregates only information from neighbors. *Diff* adds the difference between center nodes and their neighbors to the message. In *Edge*, node representation and edge information are passed together. Combining the three embedding components yields BitcoNN. Comparing the *Diff* and *Edge*, the dissimilarity learning contributes more to financial anomaly detection than edge features. Also, *Diff* and *Edge* can improve the model performance independently, but the synergy between the two factors seems not as great as expected. BitcoNN does not consistently achieve the best performance for all metrics in Table 3. We assume this is because the edge information is already reflected in the node features. The synergy between the two components would be more significant if the node and edge features existed separately in graph data. In terms of AUC, nevertheless, BitcoNN shows the best results for all three data.

3.5 Adaptive Sampler Analysis

We devise the adaptive neighbor sampler that controls the sampling size for each relation by assessing relation scores. The upper row of Fig. 5 shows the change of relation scores according to the training epoch. We observe that outgoing scores are highest in the 2014 and 2015 data. In the 2016 data, the score of co-input rises rapidly. It is consistent with the previous experimental results, which distinguishing the direction of transactions further improves the mixing detection performance in the 2014 and 2015 data. It is also the same results in Table 3 that the cooperative edge appeared to be more critical. In addition, the co-output relation has a low score in all datasets, which is the relation with the lowest label similarity in Table 1. Therefore, as intended, we can consider that more important relations are awarded higher relation scores.

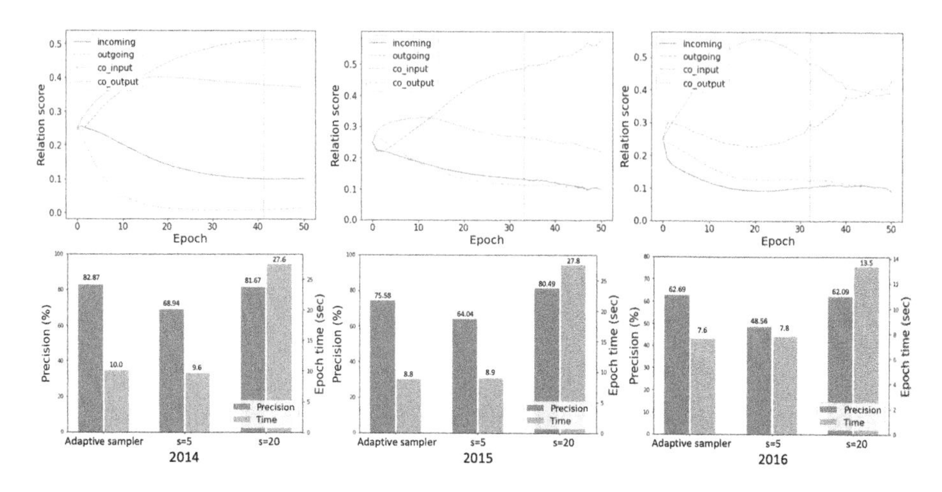

Fig. 5. (Upper row) Relation scores according to the training epoch. The gray dotted lines mark the point at which the model's performance has matured. (Bottom row) The adaptive neighbor sampler contributes to achieving competitive performance as high as referencing more neighbors while being as light as collecting fewer neighbors. (Color figure online)

The bottom row of Fig. 5 demonstrates the effectiveness of the adaptive neighbor sampler on large data sets. We compare the results of the experiments where the number of samples per relation is fixed at 5 and 20 and the adaptive sampler with the total sample size specified at 20. Comparing $s = 5$ and $s = 20$ settings, we observe that referencing more neighbors improves the precision. However, computational costs also increase along with the widened exploring area. Figure 5 shows that higher performance can be achieved by employing the adaptive neighbor sampler while keeping the total number of neighbors constant. The sampling sizes are determined according to the relation scores naturally obtained through the aggregation step, so no particular module is required. Therefore, it does not cause an increase in epoch time.

We highlight the significance of the adaptive sampler proposed in this study. So far, studies for dynamically determining the number of neighbors have been focused on the similarity between nodes [2,12]. Their interest has been to select more similar neighbors for the higher model performance rather than resource management. Therefore, their methods are expensive. By actively adjusting the number of neighbors from each relation, our framework achieves high fraud detection performance, such as referencing a larger number of neighbors, while containing the total sample size for lower computational cost simultaneously.

4 Conclusion

We present BitcoNN, a multi-relational GNN framework for financial anomaly detection. BitcoNN shows outstanding detection performance despite the challenges of lack of node features, extreme class imbalance, and large-scale cryptocurrency data. It utilizes a new type of temporal motifs, message aggregation methods, and edge features. In addition, the adaptive neighbor sampler provides the effect of referencing more neighbors efficiently while keeping the computational cost under control. This study is expected to be applied to actual investigation sites to contribute to anti-money laundering and a sound cryptocurrency ecosystem.

Limitations and Future Work. Our framework scores the relations in an attentive manner, but recently there has been a debate about whether attention truly reflects importance [5]. For more generalizability and explainability, we would further develop this approach and apply it to a broader range of problems.

Acknowledgement. This work was supported by Supreme Prosecutors' Office of the Republic of Korea grant funded by Ministry of Science and ICT(SPO2023-A1202digitalB).

References

1. Chainalysis: The 2022 Crypto Crime Report (2022)
2. Dou, Y., Liu, Z., Sun, L., Deng, Y., Peng, H., Yu, P.S.: Enhancing graph neural network-based fraud Detectors against camouflaged fraudsters. In: Proceedings of the International Conference on Information and Knowledge Management (2020)
3. Foley, S., Karlsen, J.R., Putniņš, T.J.: Sex, drugs, and bitcoin: how much illegal activity is financed through cryptocurrencies? Rev. Finan. Stud. **32**(5), 1798–1853 (2019)
4. Hamilton, W., Ying, Z., Leskovec, J.: Inductive representation learning on large graphs. In: Advances in Neural Information Processing Systems 30 (2017)
5. Jain, S., Wallace, B. C.: Attention is not explanation. In: Proceedings of the NAACL-HLT (2019)
6. Kipf, T.N., Welling, M.: Semi-supervised classification with graph convolutional networks. In: Proceedings of the International Conference on Learning Representations (2017)
7. Liu, Y., et al.: Pick and choose: a GNN-based imbalanced learning approach for fraud detection. In: Proceedings of the Web Conference (2021)
8. Liu, Z., Chen, C., Yang, X., Zhou, J., Li, X., Song, L.: Heterogeneous graph neural networks for malicious account detection. In: Proceedings of the International Conference on Information and Knowledge Management (2018)
9. Nakamoto, S.: Bitcoin: a peer-to-peer electronic cash system. Decentralized Business Review (2008)
10. Nguyen, V.H., Sugiyama, K., Nakov, P., Kan, M.Y.: FANG: leveraging social context for fake news detection using graph representation. In: Proceedings of the International Conference on Information and Knowledge Management (2020)
11. Paranjape, A., Benson, A.R., Leskovec, J.: Motifs in temporal networks. In: Proceedings of the International Conference on Web Search and Data Mining (2017)
12. Peng, H., Zhang, R., Dou, Y., Yang, R., Zhang, J., Yu, P.S.: Reinforced neighborhood selection guided multi-relational graph neural networks. ACM Trans. Inf. Syst. **40**, 4 (2022)

13. Veličković, P., Cucurull, G., Casanova, A., Romero, A., Lio, P., Bengio, Y.: Graph attention networks. In: Proceedings of the International Conference on Learning Representations (2018)
14. Wang, D., et al.: A semi-supervised graph attentive network for financial fraud detection. In: Proceedings of the IEEE International Conference on Data Mining (2019)
15. Wang, L., Li, P., Xiong, K., Zhao, J., Lin, R.: Modeling heterogeneous graph network on fraud detection: a community-based framework with attention mechanism. In: Proceedings of the International Conference on Information and Knowledge Management (2021)
16. Wu, J., Liu, J., Chen, W., Huang, H., Zheng, Z., Zhang, Y.: Detecting mixing services via mining bitcoin transaction network with hybrid motifs. IEEE Trans. Syst. Man Cybern. Syst. **52**(4), 2237–2249 (2021)
17. Zhong, Q., et al.: Financial defaulter detection on online credit payment via multi-view attributed heterogeneous information network. In: Proceedings of the Web Conference (2020)

Interpretability and Explainability

CeFlow: A Robust and Efficient Counterfactual Explanation Framework for Tabular Data Using Normalizing Flows

Tri Dung Duong[1], Qian Li[2], and Guandong Xu[1](✉)

[1] Faculty of Engineering and Information Technology, University of Technology Sydney, Ultimo, NSW, Australia
Guandong.Xu@uts.edu.au

[2] School of Electrical Engineering, Computing and Mathematical Sciences, Curtin University, Perth, WA, Australia

Abstract. Counterfactual explanation is a form of interpretable machine learning that generates perturbations on a sample to achieve the desired outcome. The generated samples can act as instructions to guide end users on how to observe the desired results by altering samples. Although state-of-the-art counterfactual explanation methods are proposed to use variational autoencoder (VAE) to achieve promising improvements, they suffer from two major limitations: 1) the counterfactuals generation is prohibitively slow, which prevents algorithms from being deployed in interactive environments; 2) the counterfactual explanation algorithms produce unstable results due to the randomness in the sampling procedure of variational autoencoder. In this work, to address the above limitations, we design a robust and efficient counterfactual explanation framework, namely CeFlow, which utilizes normalizing flows for the mixed-type of continuous and categorical features. Numerical experiments demonstrate that our technique compares favorably to state-of-the-art methods. We release our source code (https://github.com/tridungduong16/fairCE.git) for reproducing the results.

Keywords: Counterfactual explanation · Normalizing flow · Interpretable machine learning

1 Introduction

Machine learning (ML) has resulted in advancements in a variety of scientific and technical fields, including computer vision, natural language processing, and conversational assistants. Interpretable machine learning is a machine learning sub-field that aims to provide a collection of tools, methodologies, and algorithms capable of producing high-quality explanations for machine learning model judgments. A great deal of methods in interpretable ML methods has been proposed in recent years. Among these approaches, counterfactual explanation (CE) is the

H. Kashima et al. (Eds.): PAKDD 2023, LNAI 13936, pp. 133–144, 2023.
https://doi.org/10.1007/978-3-031-33377-4_11

prominent example-based method involved in how to alter features to change the model predictions and thus generates counterfactual samples for explaining and interpreting models [1,8,20,28,31]. An example is that for a customer A rejected by a loan application, counterfactual explanation algorithms aim to generate counterfactual samples such as "your loan would have been approved if your income was $51,000 more" which can act as a recommendation for a person to achieve the desired outcome. Providing counterfactual samples for black-box models has the capability to facilitate human-machine interaction, thus promoting the application of ML models in several fields.

The recent studies in counterfactual explanation utilize variational autoencoder (VAE) as a generative model to generate counterfactual sample [20,23]. Specifically, the authors first build an encoder and decoder model from the training data. Thereafter, the original input would go through the encoder model to obtain the latent representation. They make the perturbation into this representation and pass the perturbed vector to the encoder until getting the desired output. However, these approaches present some limitations. First, the latent representation which is sampled from the encoder model would be changed corresponding to different sampling times, leading to unstable counterfactual samples. Thus, the counterfactual explanation algorithm is not robust when deployed in real applications. Second, the process of making perturbation into latent representation is so prohibitively slow [20] since they need to add random vectors to the latent vector repeatedly; accordingly, the running time of algorithms grows significantly. Finally, the generated counterfactual samples are not closely connected to the density region, making generated explanations infeasible and non-actionable. To address all of these limitations, we propose a Flow-based counterfactual explanation framework (CeFlow) that integrates normalizing flow which is an invertible neural network as the generative model to generate counterfactual samples. Our contributions can be summarized as follows:

- We introduce CeFlow, an efficient and robust counterfactual explanation framework that leverages the power of normalizing flows in modeling data distributions to generate counterfactual samples. The usage of flow-based models enables to produce more robust counterfactual samples and reduce the algorithm running time.
- We construct a conditional normalizing flow model that can deal with tabular data consisting of continuous and categorical features by utilizing variational dequantization and Gaussian mixture models.
- The generated samples from CeFlow are close to and related to high-density regions of other data points with the desired class. This makes counterfactual samples likely reachable and therefore naturally follow the distribution of the dataset.

2 Related Works

An increasing number of methods have been proposed for the counterfactual explanation. The existing methods can be categorized into gradient-based methods [21,28], auto-encoder model [20], heuristic search methods [24,25] and integer linear optimization [15]. Regarding gradient-based methods, The authors in the study construct the cross-entropy loss between the desired class and counterfactual samples' prediction with the purpose of changing the model output. The created loss would then be minimized using gradient-descent optimization methods. In terms of auto-encoder model, generative models such as variational auto-encoder (VAE) is used to generate new samples in another line of research. The authors [23] first construct an encoder-decoder architecture. They then utilize the encoder to generate the latent representation, make some changes to it, and run it through the decoder until the prediction models achieve the goal class. However, VAE models which maximize the lower bound of the log-likelihood instead of measuring exact log-likelihood can produce unstable and unreliable results. On the other hand, there is an increasing number of counterfactual explanation methods based on heuristic search to select the best counterfactual samples such as Nelder-Mead [9], growing spheres [19], FISTA [4,27], or genetic algorithms [3,17]. Finally, the studies [26] propose to formulate the problem of finding counterfactual samples as a mixed-integer linear optimization problem and utilize some existing solvers [1,2] to obtain the optimal solution.

3 Preliminaries

Throughout the paper, lower-cased letters x and $\boldsymbol{x}$ denote the deterministic scalars and vectors, respectively. We consider a classifier $\mathcal{H} : \mathcal{X} \rightarrow \mathcal{Y}$ that has the input of feature space $\mathcal{X}$ and the output as $\mathcal{Y} = \{1...\mathcal{C}\}$ with $\mathcal{C}$ classes. Meanwhile, we denote a dataset $\mathcal{D} = \{\boldsymbol{x}_n, y_n\}_{n=1}^{N}$ consisting of N instances where $\boldsymbol{x}_n \in \mathcal{X}$ is a sample, $y_n \in \mathcal{Y}$ is the predicted label of individuals $\boldsymbol{x}_n$ from the classifier $\mathcal{H}$. Moreover, f_θ is denoted for a normalizing flow model parameterized by θ. Finally, we split the feature space into two disjoint feature subspaces of categorical features and continuous features represented by $\mathcal{X}^{\text{cat}}$ and $\mathcal{X}^{\text{con}}$ respectively such that $\mathcal{X} = \mathcal{X}_{\text{cat}} \times \mathcal{X}_{\text{con}}$ and $\boldsymbol{x} = (\boldsymbol{x}^{\text{cat}}, \boldsymbol{x}^{\text{con}})$, and $\boldsymbol{x}^{\text{cat}_j}$ and $\boldsymbol{x}^{\text{con}_j}$ is the corresponding j-th feature of $\boldsymbol{x}^{\text{cat}}$ and $\boldsymbol{x}^{\text{con}}$.

3.1 Counterfactual Explanation

With the original sample $\boldsymbol{x}_{\text{org}} \in \mathcal{X}$ and its predicted output $y_{\text{org}} \in \mathcal{Y}$, the counterfactual explanation aims to find the nearest counterfactual sample $\boldsymbol{x}_{\text{cf}}$ such that the outcome of classifier for $\boldsymbol{x}_{\text{cf}}$ is changed to desired output class y_{cf}. We aim to identify the perturbation $\boldsymbol{\delta}$ such that counterfactual instance $\boldsymbol{x}_{\text{cf}} = \boldsymbol{x}_{\text{org}} + \boldsymbol{\delta}$ is the solution of the following optimization problem:

$$\boldsymbol{x}_{\text{cf}} = \arg\min_{\boldsymbol{x}_{\text{cf}} \in \mathcal{X}} d(\boldsymbol{x}_{\text{cf}}, \boldsymbol{x}_{\text{org}}) \quad \text{subject to} \quad \mathcal{H}(\boldsymbol{x}_{\text{cf}}) = y_{\text{cf}} \tag{1}$$

where $d(\boldsymbol{x}_{\text{cf}}, \boldsymbol{x}_{\text{org}})$ is the function measuring the distance between $\boldsymbol{x}_{\text{org}}$ and $\boldsymbol{x}_{\text{cf}}$. Equation (1) demonstrates the optimization objective that minimizes the similarity of the counterfactual and original samples, as well as ensures to change the classifier to the desirable outputs. To make the counterfactual explanations plausible, they should only suggest minimal changes in features of the original sample [21].

3.2 Normalizing Flow

Normalizing flows (NF) [5] is the active research direction in generative models that aims at modeling the probability distribution of a given dataset. The study [6] first proposes a normalizing flow, which is an unsupervised density estimation model described as an invertible mapping $f_\theta : \mathcal{X} \rightarrow \mathcal{Z}$ from the data space $\mathcal{X}$ to the latent space $\mathcal{Z}$. Function f_θ can be designed as a neural network parametrized by θ with architecture that has to ensure invertibility and efficient computation of log-determinants. The data distribution is modeled as a transformation $f_\theta^{-1} : \mathcal{Z} \rightarrow \mathcal{X}$ applied to a random variable from the latent distribution $\boldsymbol{z} \sim p_{\mathcal{Z}}$, for which Gaussian distribution is chosen. The change of variables formula gives the density of the converted random variable $\boldsymbol{x} = f_\theta^{-1}(\boldsymbol{z})$ as follows:

$$\begin{aligned} p_{\mathcal{X}}(\boldsymbol{x}) &= p_{\mathcal{Z}}(f_\theta(\boldsymbol{x})) \cdot \left| \det \left(\frac{\partial f_\theta}{\partial \boldsymbol{x}} \right) \right| \\ &\propto \log \left(p_{\mathcal{Z}}(f_\theta(\boldsymbol{x})) \right) + \log \left(\left| \det \left(\frac{\partial f_\theta}{\partial \boldsymbol{x}} \right) \right| \right) \end{aligned} \tag{2}$$

With N training data points $\mathcal{D} = \{\boldsymbol{x}_n\}_{n=1}^{N}$, the model with respects to parameters θ can be trained by maximizing the likelihood in Eq. (3):

$$\theta = \arg\max_{\theta} \left(\prod_{n=1}^{N} \left(\log(p_{\mathcal{Z}}(f_\theta(\boldsymbol{x}_n))) + \log \left(\left| \det \left(\frac{\partial f_\theta(\boldsymbol{x}_n)}{\partial \boldsymbol{x}_n} \right) \right| \right) \right) \right) \tag{3}$$

4 Methodology

In this section, we illustrate our approach (CeFlow) which leverages the power of normalizing flow in generating counterfactuals. First, we define the general architecture of our framework in Sect. 4.1. Thereafter, Sect. 4.2 and 4.3 illustrate how to train and build the architecture of the invertible function f for tabular data, while Sect. 4.4 describes how to produce the counterfactual samples by adding the perturbed vector into the latent representation.

4.1 General Architecture of CeFlow

Figure 1 generally illustrates our framework. Let $\boldsymbol{x}_{\text{org}}$ be an original instance, and f_θ denote a pre-trained, invertible and differentiable normalizing flow model on the training data. In general, we first construct an invertible and differentiable function f_θ that converts the original instance $\boldsymbol{x}_{\text{org}}$ to the latent representation $\boldsymbol{z}_{\text{org}} = f(\boldsymbol{x}_{\text{org}})$. After that, we would find the scaled vector $\boldsymbol{\delta}_z$ as the perturbation and add to the latent representation $\boldsymbol{z}_{\text{org}}$ to get the perturbed representation

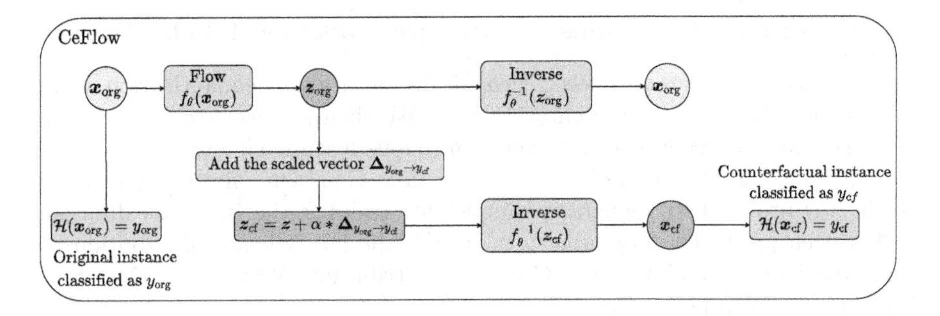

Fig. 1. Counterfactual explanation with normalizing flows (CeFlow).

z_{cf} which goes through the inverse function f_θ^{-1} to produce the counterfactual instance $\boldsymbol{x}_{\text{cf}}$. With the counterfactual instance $\boldsymbol{x}_{\text{cf}} = f_\theta^{-1}(\boldsymbol{z}_{\text{org}} + \boldsymbol{\delta}_z)$, we can re-write the objective function Eq. (1) into the following form:

$$\begin{cases} \boldsymbol{\delta}_z = \arg\min_{\boldsymbol{\delta}_z \in \mathcal{Z}} d(\boldsymbol{x}_{\text{org}}, \boldsymbol{\delta}_z) \\ \mathcal{H}(\boldsymbol{x}_{\text{cf}}) = y_{\text{cf}} \end{cases} \tag{4}$$

One of the biggest problems of deploying normalizing flow is how to handle mixed-type data which contains both continuous and categorical features. Categorical features are in discrete forms, which is challenging to model by the continuous distribution only [10]. Another challenge is to construct the objective function to learn the conditional distribution on the predicted labels [14,30]. In the next section, we will discuss how to construct the conditional normalizing flow f_θ for tabular data.

4.2 Normalizing Flows for Categorical Features

This section would discuss how to handle the categorical features. Let $\{\boldsymbol{z}^{\text{cat}_m}\}_{m=1}^{M}$ be the continuous representation of M categorical features $\{\boldsymbol{x}^{\text{cat}_m}\}_{m=1}^{M}$ for each $\boldsymbol{x}^{\text{cat}_m} \in \{0, 1, ..., K-1\}$ with $K > 1$. Follow by several studies in the literature [10, 12], we utilize variational dequantization to model the categorical features. The key idea of variational dequantization is to add noise $\boldsymbol{u}$ to the discrete values $\boldsymbol{x}^{\text{cat}}$ to convert the discrete distribution $p_{\mathcal{X}^{\text{cat}}}$ into a continuous distribution $p_{\phi_{\text{cat}}}$. With $\boldsymbol{z}^{\text{cat}} = \boldsymbol{x}^{\text{cat}} + \boldsymbol{u}_k$, ϕ_{cat} and θ_{cat} be models' parameters, we have following objective functions:

$$\begin{aligned} \log p_{\mathcal{X}^{\text{cat}}}(\boldsymbol{x}^{\text{cat}}) &\geq \int_{\boldsymbol{u}} \log \frac{p_{\phi_{\text{cat}}}(\boldsymbol{z}^{\text{cat}})}{q_{\theta_{\text{cat}}}(\boldsymbol{u}|\boldsymbol{x}^{\text{cat}})} du \\ &\approx \frac{1}{K} \sum_{k=1}^{K} \log \prod_{m=1}^{M} \frac{p_{\phi_{\text{cat}}}(\boldsymbol{x}^{\text{cat}_m} + \boldsymbol{u}_k)}{q_{\theta_{\text{cat}}}(\boldsymbol{u}_k|\boldsymbol{x}^{\text{cat}})} \end{aligned} \tag{5}$$

Followed the study [12], we choose Gaussian dequantization which is more powerful than the uniform dequantization as $q_{\theta_{\text{cat}}}(\boldsymbol{u}_k|\boldsymbol{x}^{\text{cat}}) = \text{sig}\left(\mathcal{N}\left(\boldsymbol{\mu}_{\theta_{\text{cat}}}, \boldsymbol{\Sigma}_{\theta_{\text{cat}}}\right)\right)$ with mean $\boldsymbol{\mu}_{\theta_{\text{cat}}}$, covariance $\boldsymbol{\Sigma}_{\theta_{\text{cat}}}$ and sigmoid function $\text{sig}(\cdot)$.

4.3 Conditional Flow Gaussian Mixture Model for Tabular Data

The categorical features $\boldsymbol{x}^{\text{cat}}$ going through the variational dequantization would convert into continuous representation $\boldsymbol{z}^{\text{cat}}$. We then perform merge operation on continuous representation $\boldsymbol{z}^{\text{cat}}$ and continuous feature $\boldsymbol{x}^{\text{con}}$ to obtain values $(\boldsymbol{z}^{\text{cat}}, \boldsymbol{x}^{\text{con}}) \mapsto \boldsymbol{x}^{\text{full}}$. Thereafter, we apply flow Gaussian mixture model [14] which is a probabilistic generative model for training the invertible function f_θ. For each predicted class label $y \in \{1...\mathcal{C}\}$, the latent space distribution $p_{\mathcal{Z}}$ conditioned on a label k is the Gaussian distribution $\mathcal{N}\left(\boldsymbol{z}^{\text{full}} \mid \boldsymbol{\mu}_k, \boldsymbol{\Sigma}_k\right)$ with mean $\boldsymbol{\mu}_k$ and covariance $\boldsymbol{\Sigma}_k$:

$$p_{\mathcal{Z}}(\boldsymbol{z}^{\text{full}} \mid y = k) = \mathcal{N}\left(\boldsymbol{z}^{\text{full}} \mid \boldsymbol{\mu}_k, \boldsymbol{\Sigma}_k\right) \tag{6}$$

As a result, we can have the marginal distribution of $\boldsymbol{z}^{\text{full}}$:

$$p_{\mathcal{Z}}(\boldsymbol{z}^{\text{full}}) = \frac{1}{\mathcal{C}} \sum_{k=1}^{\mathcal{C}} \mathcal{N}\left(\boldsymbol{z}^{\text{full}} \mid \boldsymbol{\mu}_k, \boldsymbol{\Sigma}_k\right) \tag{7}$$

The density of the transformed random variable $\boldsymbol{x}^{\text{full}} = f_\theta^{-1}(\boldsymbol{z}^{\text{full}})$ is given by:

$$p_{\mathcal{X}}(\boldsymbol{x}^{\text{full}}) = \log\left(p_{\mathcal{Z}}(f_\theta(\boldsymbol{x}^{\text{full}}))\right) + \log\left(\left|\det\left(\frac{\partial f_\theta}{\partial \boldsymbol{x}^{\text{full}}}\right)\right|\right) \tag{8}$$

Eq. (7) and Eq. (8) together lead to the likelihood for data as follows:

$$p_{\mathcal{X}}(\boldsymbol{x}^{\text{full}} \mid y = k) = \log\left(\mathcal{N}\left(f_\theta(\boldsymbol{x}^{\text{full}}) \mid \boldsymbol{\mu}_k, \boldsymbol{\Sigma}_k\right)\right) + \log\left(\left|\det\left(\frac{\partial f_\theta}{\partial \boldsymbol{x}^{\text{full}}}\right)\right|\right) \tag{9}$$

We can train the model by maximizing the joint likelihood of the categorical and continuous features on N training data points $\mathcal{D} = \{(\boldsymbol{x}_n^{\text{con}}, \boldsymbol{x}_n^{\text{cat}})\}_{n=1}^N$ by combining Eq. (5) and Eq. (9):

$$\begin{aligned} \theta^*, \phi^*_{\text{cat}}, \theta^*_{\text{cat}} &= \underset{\theta, \phi_{\text{cat}}, \theta_{\text{cat}}}{\arg\max} \prod_{n=1}^{N} \left(\prod_{\boldsymbol{x}_n^{\text{con}} \in \mathcal{X}^{\text{con}}} p_{\mathcal{X}}\left(\boldsymbol{x}_n^{\text{con}}\right) \prod_{\boldsymbol{x}_n^{\text{cat}} \in \mathcal{X}^{\text{cat}}} p_{\mathcal{X}}\left(\boldsymbol{x}_n^{\text{cat}}\right) \right) \\ &= \underset{\theta, \phi_{\text{cat}}, \theta_{\text{cat}}}{\arg\max} \prod_{n=1}^{N} \left(\log\left(\mathcal{N}\left(f_\theta(\boldsymbol{x}_n^{\text{full}}) \mid \boldsymbol{\mu}_k, \boldsymbol{\Sigma}_k\right)\right) + \log\left(\left|\det\left(\frac{\partial f_\theta}{\partial \boldsymbol{x}_n^{\text{full}}}\right)\right|\right)\right) \end{aligned} \tag{10}$$

4.4 Counterfactual Generation Step

In order to find counterfactual samples, the recent approaches [21,28] normally define the loss function and deploy some optimization algorithm such as gradient descent or heuristic search to find the perturbation. These approaches however demonstrates the prohibitively slow running time, which prevents from deploying in interactive environment [11]. Therefore, inspired by the study [13], we add the scaled vector as the perturbation from the original instance $\boldsymbol{x}_{\text{org}}$ to counterfactual one $\boldsymbol{x}_{\text{cf}}$. By Bayes rule, we notice that under a uniform prior distribution over labels $p(y = k) = \frac{1}{\mathcal{C}}$ for $\mathcal{C}$ classes, the log posterior probability becomes:

$$\log p_{\mathcal{X}}(y = k|\boldsymbol{x}) = \log \frac{p_{\mathcal{X}}(\boldsymbol{x}|y = k)}{\sum_{k=1}^{\mathcal{C}} p_{\mathcal{X}}(\boldsymbol{x}|y = k)} \propto ||f_\theta(\boldsymbol{x}) - \boldsymbol{\mu}_k||^2 \tag{11}$$

We observed from Eq. (11) that latent vector $\boldsymbol{z} = f_\theta(\boldsymbol{x})$ will be predicted from the class y with the closest model mean $\boldsymbol{\mu}_k$. For each predicted class $k \in \{1...\mathcal{C}\}$, we denote $\mathcal{G}_k = \{\boldsymbol{x}_m, y_m\}_{m=1}^{M}$ as a set of M instances with the same predicted class as $y_m = k$. We define the mean latent vector $\boldsymbol{\mu}_k$ corresponding to each class k such that:

$$\boldsymbol{\mu}_k = \frac{1}{M} \sum_{\boldsymbol{x}_m \in \mathcal{G}_k} f_\theta(\boldsymbol{x}_m) \tag{12}$$

Therefore, the scaled vector that moves the latent vector $\boldsymbol{z}_{\text{org}}$ to the decision boundary from the original class y_{org} to counterfactual class y_{cf} is defined as:

$$\boldsymbol{\Delta}_{y_{\text{org}} \rightarrow y_{\text{cf}}} = \left| \boldsymbol{\mu}_{y_{\text{org}}} - \boldsymbol{\mu}_{y_{\text{cf}}} \right| \tag{13}$$

The scaled vector $\boldsymbol{\Delta}_{y_{\text{org}} \rightarrow y_{\text{cf}}}$ is added to the original latent representation $\boldsymbol{z}_{\text{cf}} = f_\theta(\boldsymbol{x}_{\text{org}})$ to obtained the perturbed vector. The perturbed vector then goes through inverted function f_θ^{-1} to re-produce the counterfactual sample:

$$\boldsymbol{x}_{\text{cf}} = f_\theta^{-1}(f_\theta(\boldsymbol{x}_{\text{org}}) + \alpha \boldsymbol{\Delta}_{y_{\text{org}} \rightarrow y_{\text{cf}}}) \tag{14}$$

We note that the hyperparameter α needs to be optimized by searching in a range of values. The full algorithm is illustrated in Algorithm 1.

Algorithm 1. Counterfactual explanation flow (CeFlow)

Input: An original sample $\boldsymbol{x}_{\text{org}}$ with its prediction y_{org}, desired class y_{cf}, a provided machine learning classifier $\mathcal{H}$ and encoder model Q_ϕ.

1: Train the invertible function f_θ by maximizing the log-likelihood:

$$\theta^*, \phi^*_{\text{cat}}, \theta^*_{\text{cat}} = \underset{\theta, \phi_{\text{cat}}, \theta_{\text{cat}}}{\arg\max} \prod_{n=1}^{N} \left(\prod_{\boldsymbol{x}_n^{\text{con}} \in \mathcal{X}^{\text{con}}} p_{\mathcal{X}}\left(\boldsymbol{x}_n^{\text{con}}\right) \prod_{\boldsymbol{x}_n^{\text{cat}} \in \mathcal{X}^{\text{cat}}} p_{\mathcal{X}}\left(\boldsymbol{x}_n^{\text{cat}}\right) \right)$$

$$= \underset{\theta, \phi_{\text{cat}}, \theta_{\text{cat}}}{\arg\max} \prod_{n=1}^{N} \left(\log\left(\mathcal{N}\left(f_\theta(\boldsymbol{x}_n^{\text{full}}) \mid \boldsymbol{\mu}_k, \boldsymbol{\Sigma}_k \right) \right) + \log\left(\left| \det\left(\frac{\partial f_\theta}{\partial \boldsymbol{x}_n^{\text{full}}} \right) \right| \right) \right)$$

2: Compute mean latent vector $\boldsymbol{\mu}_k$ for each class k by $\boldsymbol{\mu}_k = \frac{1}{M} \sum_{\boldsymbol{x}_m \in \mathcal{G}_k} f(\boldsymbol{x}_m)$.

3: Compute the scaled vector $\boldsymbol{\Delta}_{y_{\text{org}} \rightarrow y_{\text{cf}}} = \left| \boldsymbol{\mu}_{y_{\text{org}}} - \boldsymbol{\mu}_{y_{\text{cf}}} \right|$.

4: Find the optimal hyperparameter α by searching a range of values.

5: Compute $\boldsymbol{x}_{\text{cf}} = f_\theta^{-1}(f_\theta(\boldsymbol{x}_{\text{org}}) + \alpha \boldsymbol{\Delta}_{y_{\text{org}} \rightarrow y_{\text{cf}}})$.

Output: $\boldsymbol{x}_{\text{cf}}$.

5 Experiments

We run experiments on three datasets to show that our method outperforms state-of-the-art approaches. The specification of hardware for the experiment is Python 3.8.5 with 64-bit Red Hat, Intel(R) Xeon(R) Gold 6238R CPU @ 2.20GHz. We implement our algorithm by using Pytorch library and adopt the RealNVP architecture [6]. During training progress, Gaussian mixture parameters are fixed: the means are initialized randomly from the standard normal distribution and the covariances are set to I. More details of implementation settings can be found in our code repository[1].

We evaluate our approach via three datasets: `Law` [29], `Compas` [16] and `Adult` [7]. `Law`[2] [29] dataset provides information of students with their features: their entrance exam scores (LSAT), grade-point average (GPA) and first-year average grade (FYA). `Compas`[3] [16] dataset contains information about 6,167 prisoners who have features including gender, race and other attributes related to prior conviction and age. `Adult`[4] [7] dataset is a real-world dataset consisting of both continuous and categorical features of a group of consumers who apply for a loan at a financial institution.

We compare our proposed method (CeFlow) with several state-to-the-art methods including Actionable Recourse (AR) [26], Growing Sphere (GS) [18], FACE [24], CERTIFAI [25], DiCE [21] and C-CHVAE [23]. Particularly, we implement the CERTIFAI with library PyGAD[5] and utilize the available source code[6] for implementation of DiCE, while other approaches are implemented with Carla library [22]. Finally, we report the results of our proposed model on a variety of metrics including success rate (success), l_1-norm (l_1), categorical proximity [21], continuous proximity [21] and mean log-density [1]. Note that for l_1-norm, we report mean and variance of l_1-norm corresponding to l_1-mean and l_1-variance. Lower l_1-variance aims to illustrate the algorithm's robustness.

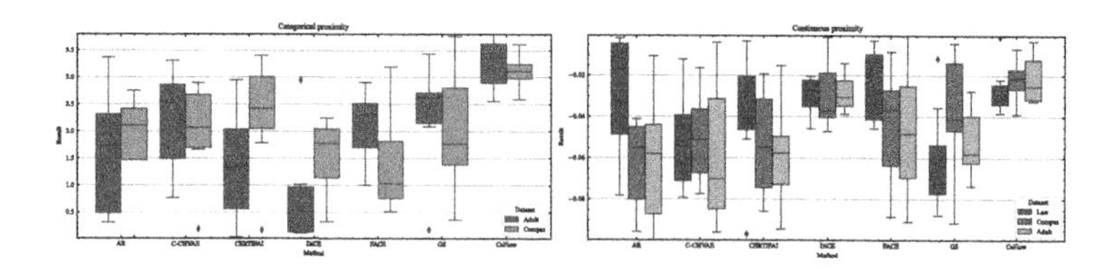

Fig. 2. Baseline results in terms of **Categorical proximity** and **Continuous proximity**. Higher continuous and categorical proximity are better.

[1] https://anonymous.4open.science/r/fairCE-538B.
[2] http://www.seaphe.org/databases.php.
[3] https://www.propublica.org.
[4] https://archive.ics.uci.edu/ml/datasets/adult.
[5] https://github.com/ahmedfgad/GeneticAlgorithmPython.
[6] https://github.com/divyat09/cf-feasibility.

Table 1. Performance of all methods on the classifier. We compute p-value by conducting a paired t-test between our approach (CeFlow) and baselines with 100 repeated experiments for each metric.

Dataset	Method	Performance				p-value		
		success	l_1-mean	l_1-var	log-density	success	l_1	log-density
Law	AR	98.00	3.518	2.0e-03	-0.730	0.041	0.020	0.022
	GS	100.00	3.600	2.6e-03	-0.716	0.025	0.048	0.016
	FACE	100.00	3.435	2.0e-03	-0.701	0.029	0.010	0.017
	CERTIFAI	100.00	3.541	2.0e-03	-0.689	0.029	0.017	0.036
	DiCE	94.00	**3.111**	2.0e-03	-0.721	0.018	0.035	0.048
	C-CHVAE	100.00	3.461	1.0e-03	-0.730	0.040	0.037	0.016
	CeFlow	**100.00**	3.228	**1.0e-05**	**-0.679**	-	-	-
Compas	AR	97.50	1.799	2.4e-03	-14.92	0.038	0.034	0.046
	GS	100.00	1.914	3.2e-03	-14.87	0.019	0.043	0.040
	FACE	98.50	1.800	4.8e-03	-15.59	0.036	0.024	0.035
	CERTIFAI	100.00	1.811	2.4e-03	-15.65	0.040	0.048	0.038
	DiCE	95.50	1.853	2.9e-03	-14.68	0.030	0.029	0.018
	C-CHVAE	100.00	1.878	1.1e-03	-13.97	0.026	0.015	0.027
	CeFlow	**100.00**	**1.787**	**1.8e-05**	**-13.62**	-	-	-
Adult	AR	100.00	3.101	7.8e-03	-25.68	0.044	0.037	0.018
	GS	100.00	3.021	2.4e-03	-26.55	0.026	0.049	0.028
	FACE	100.00	2.991	6.6e-03	-23.57	0.027	0.015	0.028
	CERTIFAI	93.00	3.001	4.1e-03	-25.55	0.028	0.022	0.016
	DiCE	96.00	2.999	9.1e-03	-24.33	0.046	0.045	0.045
	C-CHVAE	100.00	3.001	8.7e-03	-24.45	0.026	0.043	0.019
	CeFlow	**100.00**	**2.964**	**1.5e-05**	**-23.46**	-	-	-

Table 2. We report running time of different methods on three datasets.

Dataset	AR	GS	FACE	CERTIFAI	DiCE	C-CHVAE	CeFlow
Law	3.030 ± 0.105	7.126 ± 0.153	6.213 ± 0.007	6.522 ± 0.088	8.022 ± 0.014	9.022 ± 0.066	$\mathbf{0.850 \pm 0.055}$
Compas	5.125 ± 0.097	8.048 ± 0.176	7.688 ± 0.131	13.426 ± 0.158	7.810 ± 0.076	6.879 ± 0.044	$\mathbf{0.809 \pm 0.162}$
Adult	7.046 ± 0.151	6.472 ± 0.021	13.851 ± 0.001	7.943 ± 0.046	11.821 ± 0.162	12.132 ± 0.024	$\mathbf{0.837 \pm 0.026}$

The performance of different approaches regarding three metrics: l_1, success metrics and log-density are illustrated in Table 1. Regarding success rate, all three methods achieve competitive results, except the AR, DiCE and CERTIFAI performance in all datasets with around 90% of samples belonging to the target class. These results indicate that by integrating normalizing flows into counterfactuals generation, our proposed method can achieve the target of counterfactual explanation task for changing the models' decision. Apart from that, for l_1-mean, CeFlow is ranked second with 3.228 for `Law`, and is ranked first for `Compas` and `Adult` (1.787 and 2.964). Moreover, our proposed method generally achieves the best performance regarding l_1-variance on three datasets. CeFlow also demonstrates the lowest log-density metric in comparison with other approaches achieving at -0.679, -13.62 and -23.46 corresponding to `Law`, `Compas` and `Adult` dataset. This illustrates that the generated samples are more closely followed the distribution of data than other approaches. We furthermore perform a statistical significance test to gain more insights into the effectiveness of our proposed method in producing counterfactual samples compared with other approaches. Particularly, we conduct the paired t-test between our approach (CeFlow) and other methods on each dataset and each metric with the obtained results on 100 randomly repeated experiments and report the result of p-value in Table 1. We discover that our model is statistically significant with $p < 0.05$, proving CeFlow's effectiveness in counterfactual samples generation tasks. Meanwhile, Table 2 shows the running time of different approaches. Our approach achieves outstanding performance with the running time demonstrating around 90% reduction compared with other approaches. Finally, as expected, by using normalizing flows, CeFlow produces more robust counterfactual samples with the lowest l_1-variance and demonstrates an effective running time in comparison with other approaches.

Figure 2 illustrates the categorical and continuous proximity. In terms of categorical proximity, our approach achieves the second-best performance with lowest variation in comparison with other approaches. The heuristic search based algorithm such as FACE and GS demonstrate the best performance in terms of this metric. Meanwhile, DiCE produces the best performance for continuous proximity, whereas CeFlow is ranked second. In general, our approach (CeFlow) achieves competitive performance in terms of proximity metric and demonstrates the least variation in comparison with others. On the other hand, Fig. 3 shows the variation of our method's performance with the different values of α. We observed that the optimal values are achieved at 0.8, 0.9 and 0.3 for `Law`, `Compas` and `Adult` dataset, respectively.

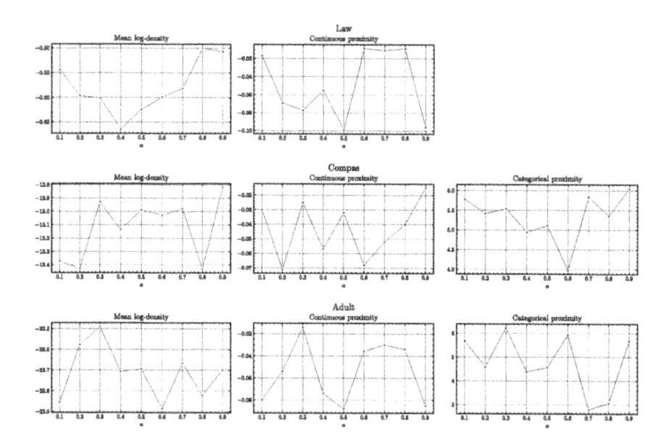

Fig. 3. Our performance under different values of hyperparameter α. Note that there are no categorical features in `Law` dataset.

6 Conclusion

In this paper, we introduced a robust and efficient counterfactual explanation framework called CeFlow that utilizes the capacity of normalizing flows in generating counterfactual samples. We observed that our approach produces more stable counterfactual samples and reduces counterfactual generation time significantly. The better performance witnessed is likely because that normalizing flows can get the exact representation of the input instance and also produce the counterfactual samples by using the inverse function. Numerous extensions to the current work can be investigated upon successful expansion of normalizing flow models in interpretable machine learning in general and counterfactual explanation in specific. One potential direction is to design a normalizing flow architecture to achieve counterfactual fairness in machine learning models.

Acknowledgement. This work is supported by the Australian Research Council (ARC) under Grant No. DP220103717, LE220100078, LP170100891, DP200101374 and National Science Foundation of China under Grant No. 62072257.

References

1. Artelt, A., Hammer, B.: Convex density constraints for computing plausible counterfactual explanations. arXiv preprint arXiv:2002.04862 (2020)
2. Bliek1ú, C., Bonami, P., Lodi, A.: Solving mixed-integer quadratic programming problems with ibm-cplex: a progress report. In: Proceedings of the Twenty-Sixth RAMP Symposium, pp. 16–17 (2014)
3. Dandl, S., Molnar, C., Binder, M., Bischl, B.: Multi-objective counterfactual explanations. arXiv preprint arXiv:2004.11165 (2020)
4. Dhurandhar, A., Pedapati, T., Balakrishnan, A., Chen, P.Y., Shanmugam, K., Puri, R.: Model agnostic contrastive explanations for structured data. arXiv preprint arXiv:1906.00117 (2019)
5. Dinh, L., Krueger, D., Bengio, Y.: Nice: non-linear independent components estimation. arXiv preprint arXiv:1410.8516 (2014)
6. Dinh, L., Sohl-Dickstein, J., Bengio, S.: Density estimation using real nvp. arXiv preprint arXiv:1605.08803 (2016)
7. Dua, D., Graff, C.: UCI machine learning repository (2017). http://archive.ics.uci.edu/ml
8. Grath, R.M., et al.: Interpretable Credit Application Predictions With Counterfactual Explanations. arXiv:1811.05245 [cs] (Nov 2018). arXiv: 1811.05245
9. Grath, R.M., et al.: Interpretable credit application predictions with counterfactual explanations. arXiv preprint arXiv:1811.05245 (2018)
10. Ho, J., Chen, X., Srinivas, A., Duan, Y., Abbeel, P.: Flow++: improving flow-based generative models with variational dequantization and architecture design. In: International Conference on Machine Learning, pp. 2722–2730. PMLR (2019)
11. Höltgen, B., Schut, L., Brauner, J.M., Gal, Y.: Deduce: generating counterfactual explanations at scale. In: eXplainable AI approaches for debugging and diagnosis (2021)
12. Hoogeboom, E., Cohen, T.S., Tomczak, J.M.: Learning discrete distributions by dequantization. arXiv preprint arXiv:2001.11235 (2020)
13. Hvilshøj, F., Iosifidis, A., Assent, I.: Ecinn: efficient counterfactuals from invertible neural networks. arXiv preprint arXiv:2103.13701 (2021)
14. Izmailov, P., Kirichenko, P., Finzi, M., Wilson, A.G.: Semi-supervised learning with normalizing flows. In: International Conference on Machine Learning, pp. 4615–4630. PMLR (2020)

15. Kanamori, K., Takagi, T., Kobayashi, K., Arimura, H.: Dace: Distribution-aware counterfactual explanation by mixed-integer linear optimization. In: Proceedings of the Twenty-Ninth International Joint Conference on Artificial Intelligence, IJCAI-20, Christian Bessiere (Ed.). International Joint Conferences on Artificial Intelligence Organization, pp. 2855–2862 (2020)
16. Larson, J., Mattu, S., Kirchner, L., Angwin, J.: How we analyzed the compas recidivism algorithm. ProPublica (5 2016) 9(1) (2016)
17. Lash, M.T., Lin, Q., Street, N., Robinson, J.G., Ohlmann, J.: Generalized inverse classification. In: Proceedings of the 2017 SIAM International Conference on Data Mining, pp. 162–170. SIAM (2017)
18. Laugel, T., Lesot, M.J., Marsala, C., Renard, X., Detyniecki, M.: Inverse classification for comparison-based interpretability in machine learning. arXiv preprint arXiv:1712.08443 (2017)
19. Laugel, T., Lesot, M.-J., Marsala, C., Renard, X., Detyniecki, M.: Comparison-based inverse classification for interpretability in machine learning. In: Medina, J., Ojeda-Aciego, M., Verdegay, J.L., Pelta, D.A., Cabrera, I.P., Bouchon-Meunier, B., Yager, R.R. (eds.) IPMU 2018. CCIS, vol. 853, pp. 100–111. Springer, Cham (2018). https://doi.org/10.1007/978-3-319-91473-2_9
20. Mahajan, D., Tan, C., Sharma, A.: Preserving causal constraints in counterfactual explanations for machine learning classifiers. arXiv preprint arXiv:1912.03277 (2019)
21. Mothilal, R.K., Sharma, A., Tan, C.: Explaining machine learning classifiers through diverse counterfactual explanations. In: Proceedings of the 2020 Conference on Fairness, Accountability, and Transparency, pp. 607–617 (2020)
22. Pawelczyk, M., Bielawski, S., Heuvel, J.v.d., Richter, T., Kasneci, G.: Carla: a python library to benchmark algorithmic recourse and counterfactual explanation algorithms. arXiv preprint arXiv:2108.00783 (2021)
23. Pawelczyk, M., Broelemann, K., Kasneci, G.: Learning model-agnostic counterfactual explanations for tabular data. In: Proceedings of the Web Conference 2020, pp. 3126–3132 (2020)
24. Poyiadzi, R., Sokol, K., Santos-Rodriguez, R., De Bie, T., Flach, P.: Face: feasible and actionable counterfactual explanations. In: Proceedings of the AAAI/ACM Conference on AI, Ethics, and Society, pp. 344–350 (2020)
25. Sharma, S., Henderson, J., Ghosh, J.: Certifai: a common framework to provide explanations and analyse the fairness and robustness of black-box models. In: Proceedings of the AAAI/ACM Conference on AI, Ethics, and Society, pp. 166–172 (2020)
26. Ustun, B., Spangher, A., Liu, Y.: Actionable recourse in linear classification. In: Proceedings of the Conference on Fairness, Accountability, and Transparency, pp. 10–19 (2019)
27. Van Looveren, A., Klaise, J.: Interpretable counterfactual explanations guided by prototypes. arXiv preprint arXiv:1907.02584 (2019)
28. Wachter, S., Mittelstadt, B., Russell, C.: Counterfactual explanations without opening the black box: automated decisions and the gdpr. Harv. JL Tech. **31**, 841 (2017)
29. Wightman, L.F.: Lsac national longitudinal bar passage study. lsac research report series. (1998)
30. Winkler, C., Worrall, D., Hoogeboom, E., Welling, M.: Learning likelihoods with conditional normalizing flows. arXiv preprint arXiv:1912.00042 (2019)
31. Xu, G., Duong, T.D., Li, Q., Liu, S., Wang, X.: Causality learning: a new perspective for interpretable machine learning. arXiv preprint arXiv:2006.16789 (2020)

Feedback Effect in User Interaction with Intelligent Assistants: Delayed Engagement, Adaption and Drop-out

Zidi Xiu[1(✉)], Kai-Chen Cheng[1], David Q. Sun[1(✉)], Jiannan Lu[1], Hadas Kotek[1], Yuhan Zhang[2], Paul McCarthy[1], Christopher Klein[1], Stephen Pulman[1], and Jason D. Williams[1]

[1] Apple, One Apple Park Way, Cupertino, CA 95014, USA
{z_xiu,dqs}@apple.com
[2] Department of Linguistics, Harvard University, Cambridge, MA 02138, USA

Abstract. With the growing popularity of intelligent assistants (IAs), evaluating IA quality becomes an increasingly active field of research. This paper identifies and quantifies the *feedback effect*, a novel component in IA-user interactions – how the capabilities and limitations of the IA influence user behavior over time. First, we demonstrate that unhelpful responses from the IA cause users to delay or reduce subsequent interactions in the short term via an observational study. Next, we expand the time horizon to examine behavior changes and show that as users discover the limitations of the IA's understanding and functional capabilities, they learn to adjust the scope and wording of their requests to increase the likelihood of receiving a helpful response from the IA. Our findings highlight the impact of the feedback effect at both the micro and meso levels. We further discuss its macro-level consequences: unsatisfactory interactions continuously reduce the likelihood and diversity of future user engagements in a feedback loop.

Keywords: Data Mining · Intelligent Assistant Evaluation

1 Introduction

Originated from spoken dialog systems (SDS), intelligent assistants (IAs) had rapid growth since the 1990s s [9], with both research prototypes and industry applications. As their capabilities grow with recent advancements in machine learning and increased adoption of smart devices, IAs are becoming increasingly popular in daily life [3,14]. Such IAs often offer a voice user interface, allowing users to fulfill everyday tasks, get answers to knowledge queries, or start casual social conversations, by simply speaking to their device [17,26]; that is, they take human voice as input, which they process in order to provide an appropriate response [29]. The evolution of these hands-free human-device interaction systems brings new challenges and opportunities to the data mining community.

IA systems often consist of several interconnected modules: Automated Speech Recognition (ASR), Natural Language Understanding (NLU), Response

Y. Zhang—Contributions made during the internship at Apple in the summer of 2022.

H. Kashima et al. (Eds.): PAKDD 2023, LNAI 13936, pp. 145–158, 2023.
https://doi.org/10.1007/978-3-031-33377-4_12

Generation & Text-to-Speech (TTS), Task Execution (*e.g.*, sending emails, setting alarms and playing songs), and Question Answering [9,12,22]. Many of the active developments in the field are formulated as supervised learning problems where the model predicts a target from an input, e.g., a piece of text from a speech audio input (ASR), a predefined language representation from a piece of text (NLU), or a clip of audio from a string of text (TTS). Naturally, the evaluation of these models often involves comparing model predictions to some ground-truth datasets.

When building such an evaluation dataset from real-world usage, we inevitably introduce user behavior into the measurement. User interactions with IA are likely to be influenced by the their pre-existing perception of IA's capabilities and limitations, therefore introducing a bias in the distribution of "chances of success" in logged user interactions – users are more likely to ask what they know the IA can handle. This hypothesis makes intuitive sense and has been partly suggested by an earlier study on vocabulary convergence in users learning to speak to an SDS [19].

In this context, we define *feedback effect* as the behavior pattern changes in users of an interactive intelligent system (*e.g.*, IA) that are attributable to their cumulative experiences with said system. Our contributions can be summarized as follows. First, we establish a causal link between IA performance and immediate subsequent user activities, and quantify its impact on users of a real-world IA. Second, we identify distinct dynamics of behavior change for a cohort of new users over a set period of time, demonstrating how users first explore the IA's capabilities before eventually adapting or quitting. Third, having examined the *feedback effect* and its impact in detail, we provide generalizable recommendations to mitigate its bias in IA evaluation.

2 Related Work

IA Evaluation Methods and Metrics. Many studies have been devoted to addressing the challenges in IA evaluation. Objective metrics like accuracy cannot present a comprehensive view of the system [8]. Human annotation is a crucial part of the process, but it incurs a high expense and is hard to scale [15]. Apart from human evaluation, *i.e.*, user self-reported scores or annotated scores, subjective metrics have been introduced. Jiang [12] designed a user satisfaction score prediction model based on user behavior patterns, ungrammatical sentence structures, and device features. Other implicit feedback from users (*e.g.*, acoustic features) are helpful to approximate success rates [16].

User Adaptation and Lexical Convergence. *Adaptation* (or *entrainment*) describes the phenomenon whereby the vocabulary and syntax used by speakers converge as they engage in a conversation over time [27]. Convergence can be measured by observing repetitive use of tokens in users' requests [6] and high frequency words [24]. Adaptation happens subconsciously and leads to more successful conversations [7]. In SDS, the speakers in a dialogue are the IA and the user. When the IA actively adapts to the user in the conversation, the quality of the generated IA responses increases substantially [30,31]. The phenomenon of lexical adaptation of users to the IA system has been investigated as well [19,25]. Currently, most IAs are built upon a limited domain with restricted vocabulary size [5]. Users' vocabulary variability tends to decrease as they engage with the IA over time. This naturally limits the linguistic diversity of user queries, although out-of-domain queries can happen from time to time [9].

3 Data Collection

We analyzed logged interactions (both user queries and the associated IA responses) from a real-world IA system. All data originate from users who have given consent to data collection, storage, and review. The data is associated with a *random, device-generated* identifier. Therefore, when we use the term 'user' in the context of the interaction data analysis, we are *actually* referring to this random identifier. While the identifier is a reasonable proxy of a user, we must recognize its limitations – in our analysis, we are unable to differentiate multiple users who share a single device to interact with the IA, nor to associate requests from a single user that were initiated on multiple devices.

The population of interest is US English-speaking smartphone users who interacted with the IA in 2021 and 2022. We randomly sampled interaction data from two distinct time periods *before* and *after* a special event in late 2021. This event entailed new software and hardware releases, potentially introducing nontrivial changes to user behavior and demographics, while simultaneously presenting unique opportunities for our particular investigation.

3.1 Study 1: Pre-event Control Period

To investigate the feedback effect on user engagement, we randomly sampled interaction data from a two-week period in August 2021. The choice of a relatively *short* time period *before* the special event helps us (i) directly control for seasonality and (ii) avoid the impact of the special event, where product releases and feature announcements usually stimulate user engagement and attract new users. (We return to a discussion of new users below.)

We further control for software and hardware versions, before taking a random sample of approximately 14,000 users who had at least one interaction with the IA during the study period. We then randomly sampled *one* interaction per user and used human-label review to determine whether the IA response was helpful. We additionally analyzed the frequency of interactions for the user in the 2 weeks prior to and 2 weeks following our causal analysis. In our sample, approximately 80% of the interactions were labeled as *helpful* to the user.[1] The results of this study are presented in Sect. 4.

3.2 Study 2: Post-event New User Period

To investigate language convergence among a new user cohort, we randomly sampled data from a six-month period immediately *after* the special event. With our interest in analyzing long-term behavior changes of a new user cohort, this

[1] This value is not necessarily a reflection of the aggregated or expected satisfaction metric, due to the sampling method and potential bias in the subpopulation of choice.

choice of sampling period has two interesting implications. First, special events often lead to a surge in new users of the IA. Second, feature announcements at the special event may cause some existing users to perceive the updated IA as "new" and explore it with a mindset akin to that of a new user. Given the challenges inherent to determining new user cohorts (to be further discussed in Sect. 5), these two factors are valuable as they collectively increase the share of new users, thus boosting the observability of the cohort. From this six-month period, we took a random sample of 5,000 users who used the new software version of the IA. For each user, all interactions with the IA in the full study period were used in our analysis. The study is described in Sect. 5.

4 Feedback Effect on Engagement

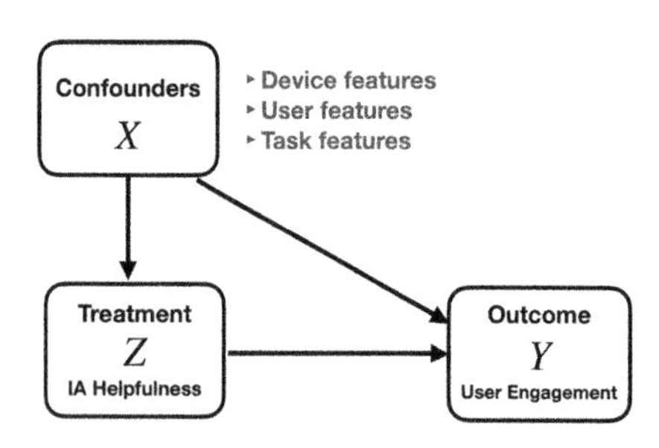

Fig. 1. Causal graph illustrating the observational study of IA feedback effect accounting for the existence of confounding factors.

Intuitively, unhelpful responses from an IA may discourage users from future interactions with the IA. Our work aims to empirically shed light on the relation between IA helpfulness[2] (*helpful* or *unhelpful* on a single interaction) and users' subsequent engagement patterns with the IA, as illustrated in Fig. 1. To establish such a causal relationship from IA performance to user engagement, we adopt an observational analysis framework with IA related features.

Given a dataset with N users, we denote unit i having (i) covariates $\mathbf{X}_i \in \mathbb{R}^p$, (ii) a treatment variable $Z_i \in \{0, 1\}$, indicating users experienced an *unhelpful* interaction with the IA or a *helpful* one respectively, and (iii) what would have happened if the unit is assigned to treatment and control, denoted by $Y_i(1)$ and $Y_i(0)$ respectively, according to the potential outcomes framework [10,28]. Consequently, the causal effect for unit i is defined as $\tau_i = Y_i(1) - Y_i(0)$, namely the difference between the outcomes if treated differently on the same user. However, the fundamental problem of causal inference is that only the potential outcome – the outcome in the group the subject was assigned to – can be observed, *i.e.*, $Y_i = Z_i Y(Z_i) + (1 - Z_i) Y(1 - Z_i)$. Individual level causal estimands, the contrast of values between the two potential outcomes, cannot be expressed with functions of observed data alone. Consequently, our primary focus is on population level causal effects like the Average Treatment Effect (ATE), $\tau = \mathbb{E}(\tau_i)$.

With a randomized controlled experiment, treated assignment mechanism is known and unconfounded, therefore we can directly and accurately estimate and infer causal effects (*e.g.*, ATE) from the observed data. However, in real-world scenarios, delicately designed experiments can be difficult or impossible to conduct. Instead, we must rely on observational techniques.

[2] The IA helpfulness of a given user request is defined as the user's satisfaction with the IA's response to the request, as determined by human annotators. .

4.1 Covariates and Outcome Variables

Observational studies are susceptible to *selection bias* due to confounding factors, which affect both the treatment **Z** and the outcome **Y**, as shown in Fig. 1. To address any confoundness to the best of our ability, we have collected rich sets of the following IA related features, and assuming that there are no unobserved observed confounders. (i) Device features: The type of device used to interact with the IA system, and the operating system version, (ii) Task features: The input sentence transcribed by the ASR system, the number of tokens in the input sentence, the word error rate (WER) of the transcription, and the domain that the IA executed with a confidence score provided by the NLU model (*e.g.*, weather, phone, etc.), (iii) User related features: Prior activity levels measured as the number of active days before the interaction, and temporal features including local day of the week and time of the day when the interaction happened.

To quantify user engagement after the annotated IA interaction, Sect. 4.3 focuses on time to next session ("immediate shock"), and Sect. 4.4 focuses on active day counts ("aftermath").

4.2 Observational Causal Methods

Matching Methods. Matching is a non-parametric method to alleviate the effects of confounding factors in observational studies. The goal is to obtain well-matched samples from different treatment groups, hoping to replicate randomized trial settings. The popular Coarsened Exact Matching (CEM) model is based on a monotonic imbalance reducing matching method at a pre-defined granularity with no assumption on assignment mechanisms [11].

Weighting Methods. Apart from matching covariates, weighting based methods use all of the high-dimensional data via summarizing scores, like the propensity score (PS). PS reflects the probability of assigned to treatment based on user's background attributes [20,28], $e(x) = P(Z_i = 1|X_i = x) = \mathbb{E}(\mathbf{Z}|\mathbf{X})$ Since the true PS is unknown, we adopt generalized linear regression models (GLMs) to estimate it, which are widely adopted by the scientific community.

With PS estimates available, the next question is how to leverage them. Li [20] proposed a family of balancing weights which enjoys balanced weighted distributions of covariates among treatment groups. Inverse-Probability Weights (IPW) are a special case of this family, shown in Eq. (1). As the name suggests, the weight is the inverse of the probability that a unit is assigned to the observed group, and the corresponding estimand is the ATE. However, IPW is very sensitive to outliers, *i.e.*, when PS scores approach 0 or 1. To mitigate this challenge, Overlap Weights (OW) which emphasize a target population with the most covariate overlap [20], shown in Eq. (2). More discussions are provided in the Supplemental Materials (SM)[3]

[3] Supplemental Materials: https://machinelearning.apple.com/research/feedback-effect.

$$\begin{cases} w_1^{\text{IPW}}(x) = \frac{1}{e(x)} \\ w_0^{\text{IPW}}(x) = \frac{1}{1-e(x)} \end{cases} \quad (1) \qquad \begin{cases} w_1^{\text{OW}}(x) = (1 - e(x)) \\ w_0^{\text{OW}}(x) = e(x), \end{cases} \quad (2)$$

where w_1 corresponds to the weight assigned to the treatment group, and w_0 to the control group, respectively. Then the population level causal estimands of interest, the Weighted Average Treatment Effect (WATE), are derived from the balanced weights. The target population varies with different weighting strategy. The causal estimand shown in Eq. (3) then becomes the average treatment effect for the overlap population.

$$\hat{\tau}^w = \frac{\sum_{i=1}^N w_1(\mathbf{x_i})\mathbf{Z_i}\mathbf{Y_i}}{\sum_{i=1}^N w_1(\mathbf{x_i})\mathbf{Z_i}} - \frac{\sum_{i=1}^N w_0(\mathbf{x_i})(\mathbf{1} - \mathbf{Z_i})\mathbf{Y_i}}{\sum_{i=1}^N w_0(\mathbf{x_i})(\mathbf{1} - \mathbf{Z_i})} \quad (3)$$

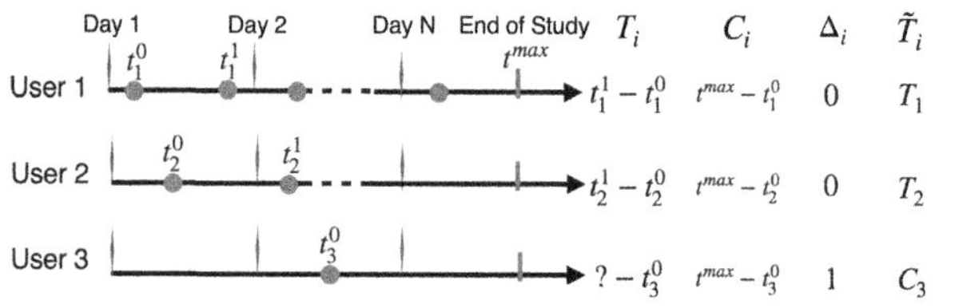

Fig. 2. Illustration of the users' engagement (blue dots) after the request was annotated at time t_i^0 for user i. We observed the time-to-next-engagement for user 1 and 2, but user 3 was censored (the next engagement was not observed). (Color figure online)

4.3 Time to Next Engagement

In this section, we establish causal links between interaction quality with the IA (as implied by the annotated helpfulness) and the user's time to next engagement. Specifically, our main hypothesis is that if a user has a helpful interaction with the IA, they are more likely to further engage with the IA in the future.

Unlike standard observational studies with well-defined and observable outcomes, time-to-event measures fall into the range of survival analyses, which focus on the length of time until the occurrence of a well-defined outcome [23,32]. A characteristic feature in the study of time-to-event distributions is the presence of *censored* instances: events that do *not* occur during the follow-up period of a subject. This can happen when the unit drops out during the study (right censoring), presenting challenges to standard statistical analysis tools.

As illustrated in Fig. 2, assume for user i: the time-to-next engagement is T_i with censoring time C_i, the observed outcome $\tilde{T}_i = T_i \wedge C_i$, and the censoring indicator $\Delta_i = \mathbb{1}\{T_i \leq C_i\}$. Under time-to-event settings, we observe a quadruplet $\{Z_i, \mathbf{X}_i, \tilde{T}_i, \Delta_i\}$ for each sample. Each user also has a set of potential outcomes, $\{T_i(1), T_i(0)\}$. Users may use the IA system at some point in our research and be assigned a *helpfulness* score, but not show up again before the data collection period ends (*e.g.*, User 3 illustrated in Fig. 2). This yields a censored time C_3 instead of a definite time-to-next-engagement outcome T_3 which is not observed within the study period.

Following Zeng [33], the causal estimand of interest is defined based on a function of the potential survival times, $\nu(T_i(z);t) = \mathbb{1}\{T_i(z) \geq t\}$. It can be interpreted as an at-risk function with the potential outcome $T_i(z)$. The expectation of the risk function corresponding to the potential survival function of user i, *i.e.*, the probability of no interaction with the IA until time t. Accordingly, the *re-engagement* probability for users in treatment group z within time t is therefore defined as Eq. (4).

$$\mathbb{E}[\nu(T_i(z);t)] = \mathbb{P}[T_i(z) \geq t] = \mathbb{S}_i(t;z) \quad (4) \qquad \mathbb{P}(t;z) = 1 - \mathbb{S}(t;z) \quad (5)$$

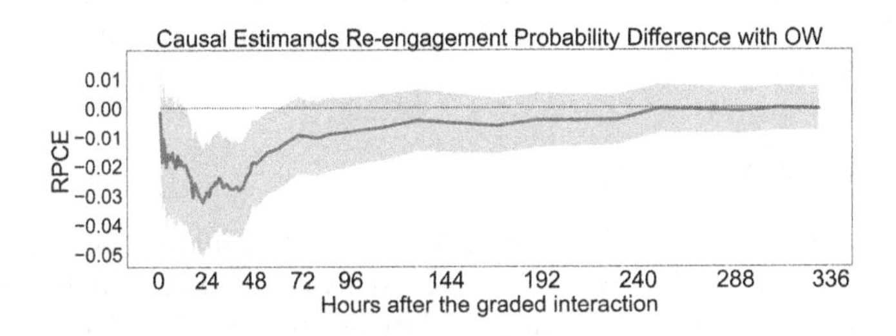

Fig. 3. The re-engagement probability causal estimands (RPCE) as a function of time after the annotated interaction, with associated 95% confidence interval (shaded gray). (Color figure online)

To properly apply balancing weights (2) with survival outcomes, right censoring needs to be accounted for. Pseudo-observation is therefore constructed based on re-sampling (a jack-knife statistic) and is interpreted as the individual contribution to the target estimate from a complete sample without censoring [2]. Given a time t, denote the expectation of the risk function at that time point , *i.e.*, $\mathbb{E}[\nu(T_i(z);t)]$ in Eq. (4), as $\theta(t)$, which is a population parameter. Without loss of generosity, we discuss the pseudo observation omitting the potential outcome notations. The pseudo-observation for each unit i can be specified as, $\hat{\theta}_i(t) = N\hat{\theta}(t) - (N-1)\hat{\theta}_{-i}(t)$, where $\hat{\theta}(t)$ is the Kaplan-Meier estimator of the population risk at time t, which is based on Δ_i and T_i. $\hat{\theta}_{-i}(t)$ is calculated without unit i. In this way, classic propensity score methods become applicable. Then the conditional causal effect averaged over a target population at time t is:

$$\begin{aligned}\hat{\tau}^w(t) &= \frac{\sum_{i=1}^{N} w_1(\mathbf{x_i})\mathbf{Z_i}\hat{\theta}_\mathbf{i}(\mathbf{t})}{\sum_{i=1}^{N} w_1(\mathbf{x_i})\mathbf{Z_i}} - \frac{\sum_{i=1}^{N} w_0(\mathbf{x_i})(\mathbf{1}-\mathbf{Z_i})\hat{\theta}_\mathbf{i}(\mathbf{t})}{\sum_{i=1}^{N} w_0(\mathbf{x_i})(\mathbf{1}-\mathbf{Z_i})} \\ &= (1-\hat{\mathbb{S}}^{w_0}(t;0)) - (1-\hat{\mathbb{S}}^{w_1}(t;1)) = \hat{\mathbb{P}}^{w_0}(t;0) - \hat{\mathbb{P}}^{w_1}(t;1) \end{aligned} \quad (6)$$

The estimator in Eq. (6) represents the survival probability causal effect, *i.e.*, the difference of the weighted *re-engagement* probability in the *Unhelpful* group and the *Helpful* group, or the *Re-engagement* Probability Causal Effect (RPCE). The results are shown in Fig. 3. The confidence interval is calculated based on the estimated standard error of SPCE [33].

The difference in estimated re-engagement probability is negative within the 336 h (two weeks) following the initial interaction, with a maximum causal difference of 3.2% at around 24 h (p-value is 0.007). The time window where the difference between the *Helpful* and *Unhelpful* groups is consistently statistically significant is between hours 8-65. The IPW result (provided in the SM) yields similar conclusions. Our main takeaway is that the inhibition effect of an unhelpful interaction reaches peak around 24 h after the interaction and then gradually weakening.

Specifically, we conclude the following, (i) An unhelpful interaction tends to have a stronger effect on whether the user wants to use the assistant again around the same hour on the next few days, perhaps affecting daily tasks like starting navigation to work, (ii) About one week later, the re-engagement probability difference becomes insignificant, as users' recollections of the unhelpful interaction fade away.

4.4 Number of Active Days

Section 4.3 established the immediate effect of IA helpfulness on time-to-next engagement. In this section, we widen the analysis window and focus on the number of active days after the annotated interaction. Let $A^{(k)}$ denote the number of active days within k-day window, $k \in \{3, 14\}$. The average treatment effect (ATE) is defined as $\mathbb{E}[A_i^{(k)}(1) - A_i^{(k)}(0)]$.

Table 1. Examples of low and high perplexity requests about weather.

Low Perplexity	**High Perplexity: syntactically complex sentences**	**High Perplexity: lexically diverse and rare topics**
What is the **weather**, 3.3	Show me hourly **weather** forecast, 17.1	What is the **UV index**, 11.8
	Could I have the **weather** for rest of the week in <Location> please, 20.8	Is there **tornado** nearby, 13.6
What is the **temperature**, 3.5	When is the **rain** supposed to start again, 19.5	How fast is the **wind** going, 15.6
Will it **rain** today, 5.9	When the **rain** going to stop, 21.8	When is the full **moon**, 15.7
Is it going to **snow** today, 7.6	How many inches of **snow** are we supposed to get, 20.4	What is the **barometric pressure** at <Location>, 27.1
	How tall will the **snow** get tonight, 27.6	

To estimate the causal effects with consistency, we applied four different statistical analysis tools at the two time windows respectively, belonging to two major branches of causal analysis. The first branch is weighting (IPW, entropy weights, overlap weights).[4] The corresponding WATE function is defined simi-

[4] Propensity weighting methods: https://cran.r-project.org/web/packages/PSweight.

larly as Eq. (3). The second branch is matching. Considering the dimensionality, we used the CEM method [11].

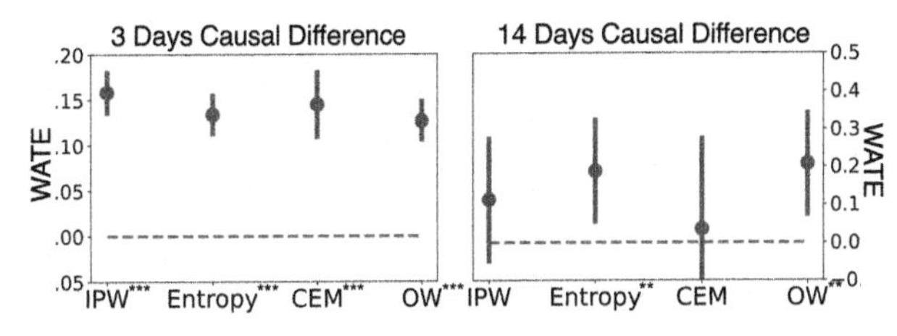

Fig. 4. Causal effect of an unhelpful IA interaction on activity levels. Bar length indicates 95% CI

In line with our previous findings, we observe statistically significant causal impacts on the activity level 3 days after the annotated IA interaction, shown in Fig. 4 (left). All four analysis tools yield p-values < 0.001. This also supports the finding that the inhibition effect of an unhelpful engagement fades in time. When we zoom out to a 14-day window, we observe that though the causal effects are not always significant, the directional consistency suggests a lessened effect of the unhelpful engagement compared to the 3-day window.

5 Language Convergence in New User Cohort

Having established the inhibition effect of an unhelpful interaction on a user's activity levels immediately following the interaction, we now expand both the scope and the time horizon of our analysis, to explore how prior engagements in turn shape users' linguistic choices over time.

5.1 New and Existing User Cohort Definition

Canonically, a *new user* to an IA is an individual who started using the IA for the first time in the observation window. As our data does not allow us to identify new users in this way, we rely instead on the following conservative, *necessary but insufficient*, condition for cohort determination: a user is assigned to the 'new user cohort' if they (i) had at least one interaction with the IA in the study period, and (ii) had no interaction with the IA in the first 60 days of the study period. By erring on the side of including existing users in the new user group, we can ensure that any patterns that remain are robust. Therefore, we argue that this determination method offers a reasonable (and likely inflated) approximation of the true new user cohort. In our dataset containing 6 months of interaction data, approximately 17% of all unique users were assigned to the new user cohort.

5.2 New User's Self-Selection: Drop-out or Adaption

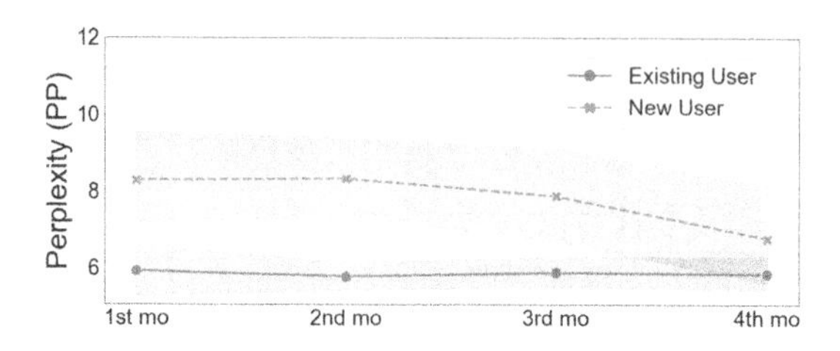

Fig. 5. New user cohort has a higher average perplexity in the beginning and converges toward existing user cohort in language perplexity over time.

We use a domain-specific language model-based perplexity (PP) score [18], which provides a comprehensive summary of the request's complexity characteristics [1]. PP score is defined as the inverse joint probability that a sentence belongs to the trained language model normalized by the number of tokens in the sentence [13], $PP(W) = \sqrt[N]{\frac{1}{\mathbb{P}(w_1 w_2 \ldots w_N)}}$, where W is the target sentence, w_k is individual token and N is the token count of the sentence. In our analysis, we adopted a tri-gram language model [4]. Table 1 presents examples of requests with perplexity scores. Here we use paraphrased variants rather than actual user data for illustration purposes. Higher perplexity correlates with more complex sentence structures, more diverse language representations and broader topics.

Intuitively, new users tend to explore the limits of the IA system, with broader vocabulary and diverse paraphrases of their requests. In this study, we track the average PP scores of new and existing user cohorts over a six month period. First, we empirically show that the *existing user* cohort has a lower and more stable perplexity score over time compared to the *new user* cohort (Figure 5). This result suggests that requests from existing users are more likely to conform to the typical wording of requests within a certain domain.

Second, we discover that the perplexity score in the new user cohort is 30% higher than in the existing user group in the first month, but it gradually converges to that of the existing user cohort. This trend suggests that new users are less familiar with the IA's capabilities and are more exploratory when they are first introduced to the system. Over time, they gradually familiarize themselves with it. Eventually, they adopt similar sentence structures and other linguistic characteristics to those used by existing users when expressing similar intents.

Next, within the new user cohort, we dive deeper into two subgroups: The *retained* group consists of users who were active for more than three out of the four-month follow-up period. The *dropout* group includes users who were active for no more than 30 days within the study period. Based on these criteria, the retained group has an average perplexity score of 7.5, while the dropout group has an average perplexity score of 10.6 and the difference has p-value < 0.001. That is, users who stop using the IA within the first month tend to have substantially higher perplexity scores than users who are retained. Further, we've shown in the SM that higher PP scores are closely related to higher unhelpful rates in the IA interactions. Following our findings from the previous sections, we expect unhelpful experiences to discourage users from continuing to engage with the IA.

In summary, we conclude that there are two plausible mechanisms that may explain the convergence of the perplexity score over time in the new user cohort:

1. **Dropout**: some new users who are either unfamiliar with the supported functionality or the language of the IA system suffer negative experiences. These high-perplexity language users stop using the system after a few tries.
2. **Adaptation**: despite some potential negative experiences in the beginning, some new users familiarize themselves with the system and adapt to its limitations. They continue to use the system after the first few months.

This represents a self-selection process among the users who choose to interact with the IA system: users adapt to the IA system in a way that lowers their language perplexity and consequently improves their experience, or they stop using it altogether. Crucially, as users adapt their behavior to the system over time, we expect to observe fewer and fewer requests that may lead to unhelpful interactions with the IA—as a result of the *feedback effect*. Consequently, we observe a bias that introduces a significant challenge to the meaningful offline evaluation of the IA system based on naive samples of the usage traffic.

6 Discussions

Evaluation of IA systems is an important yet challenging problem. On the one hand, the capabilities and limitations of IAs shape user behaviors (*e.g.*, delayed engagement, dropout, and adaption). On the other hand, these very user behavior shifts in turn influence data collection and consequently the assessment of the IAs' capabilities and limitations. To our knowledge, this two-sided problem has not been formally discussed in the literature, at least in the context of real-world IAs. To fill this gap, this paper empirically studied the "feedback effect" nature of IA evaluation. On the one hand, we demonstrated that unhelpful interactions with the IA led to delayed and reduced user engagements, both short-term and mid-term. On the other hand, we examined long-term user behaviors, which suggested that as users gradually learned the limitations of the IA, they either dropped out or adapted (*i.e.*, "gave in"), and consequently increased the likelihood of helpful interactions with the IA.

Beside raising awareness within the data mining community, this paper aims to equip researchers and practitioners with tools for trustworthy IA evaluations. First, in cases where randomized controlled experiments are infeasible, we offered best practices on properly employing observational causal inference methods, and constructing offline metrics that take the censoring of user engagements into account. Second, to reduce the *feedback loop* problem in data collection and sampling, it is important to gauge users' experience with the IA and control for confounding factors if possible. When not possible, researchers should consider stratified sampling or boosting the signals from more complex intents, or creating synthetic test data that varies in complexity, especially targeting more complex sentence structures and intent linguistic features which may be under-represented. Third, we have demonstrated that a key factor contributing to unsatisfactory IA experiences for new users is that the language they use is too complex in some way. We have also shown that users who fail to adapt by using simpler language often do not continue to use the IA. These insights immediately suggest growth opportunities to capitalize on. For example, multiple existing IAs offer a set of example conversations in different domains, in order to "train" new users to use the IA successfully right from the get go.

Our work implies multiple future directions, from both product and research perspectives. First, other than new user training (that might very well be skipped), what more can we do to convey the IA's capabilities and limitations, and help users engage more productively? Alternatively, how can we intervene early on and retain those "drop-outs," who provide invaluable feedback to help improve our system? Second, although we collected a rich set of covariates to ensure unconfoundedness, we can further assess the robustness of the established causal links, by leveraging classic sensitivity analysis techniques [21]. Third, while this paper focuses on off-line evaluation for IAs, it is possible to apply the proposed methodologies and recommendations in other settings (*e.g.*, on-line experimentation) and software products (*e.g.*, search engines).

Acknowledgements. This work was made possible by Zak Aldeneh, Russ Webb, Barry Theobald, Patrick Miller, Julia Lin, Tony Y. Li, Leneve Gorbaty, Jessica Maria Echterhof and many others at Apple. We also thank Ricardo Henao and Shuxi Zeng at Duke University for their support and feedback.

References

1. Adiwardana, D., et al.: Towards a human-like open-domain chatbot. arXiv preprint arXiv:2001.09977 (2020)
2. Andersen, P.K., Syriopoulou, E., Parner, E.T.: Causal inference in survival analysis using pseudo-observations. Stat. Med. **36**(17), 2669–2681 (2017)
3. de Barcelos Silva, A., et al.: Intelligent personal assistants: a systematic literature review. Expert Syst. Appl. **147**, 113193 (2020)
4. Bird, S., Klein, E., Loper, E.: Natural language processing with Python: analyzing text with the natural language toolkit. O'Reilly Media, Inc (2009)
5. Chattaraman, V., Kwon, W.S., Gilbert, J.E., Ross, K.: Should AI-based, conversational digital assistants employ social-or task-oriented interaction style? A task-competency and reciprocity perspective for older adults. Comput. Hum. Behav. **90**, 315–330 (2019)
6. Duplessis, G., Clavel, C., Landragin, F.: Automatic measures to characterise verbal alignment in human-agent interaction. In: SIGdial (2017)
7. Friedberg, H., Litman, D., Paletz, S.B.: Lexical entrainment and success in student engineering groups. In: SLT (2012)

8. Gao, J., Galley, M., Li, L.: Neural approaches to conversational AI. Found. Trends Inf. Retr. **13**(2–3), 127–298 (2019)
9. Glass, J.: Challenges for spoken dialogue systems. In: Proceedings of the 1999 IEEE ASRU Workshop, vol. 696 (1999)
10. Holland, P.W.: Statistics and causal inference. J. Am. Stat. Assoc. **81**(396), 945–960 (1986)
11. Iacus, S.M., King, G., Porro, G.: Causal inference without balance checking: Coarsened exact matching. Polit. Anal. **20**(1), 1–24 (2012)
12. Jiang, J., et al.: Automatic online evaluation of intelligent assistants. In: WWW (2015)
13. Jurafsky, D.: Speech & language processing. Pearson Education India (2000)
14. Kepuska, V., Bohouta, G.: Next-generation of virtual personal assistants (Microsoft Cortana, Apple Siri, Amazon Alexa and Google Home). In: IEEE CCWC (2018)
15. Kiseleva, J., et al.: Understanding user satisfaction with intelligent assistants. In: CHIIR (2016)
16. Komatani, K., Kawahara, T., Okuno, H.G.: Analyzing temporal transition of real user's behaviors in a spoken dialogue system. In: INTERSPEECH (2007)
17. Lee, D., et al.: A voice QR code for mobile devices. In: Lee, G.G., Kim, H.K., Jeong, M., Kim, J.-H. (eds.) Natural Language Dialog Systems and Intelligent Assistants, pp. 97–100. Springer, Cham (2015). https://doi.org/10.1007/978-3-319-19291-8_9
18. Lee, N., Bang, Y., Madotto, A., Khabsa, M., Fung, P.: Towards few-shot fact-checking via perplexity. In: NAACL (2021)
19. Levow, G.A.: Learning to speak to a spoken language system: vocabulary convergence in novice users. In: SIGDIAL Workshop of Discourse and Dialogue (2003)
20. Li, F., Morgan, K.L., Zaslavsky, A.M.: Balancing covariates via propensity score weighting. J. Am. Statist. Assoc. **113**(521), 1260466 (2018)
21. Liu, W., Kuramoto, S.J., Stuart, E.A.: An introduction to sensitivity analysis for unobserved confounding in nonexperimental prevention research. Prev. Sci. **14**, 570–580 (2013)
22. Lopatovska, I., et al.: Talk to me: exploring user interactions with the amazon Alexa. J. Librarianship Inf. Sci. **51**(4), 984–997 (2019)
23. Miller, R.G.: Survival analysis. John Wiley & Sons (2011)
24. Nenkova, A., Gravano, A., Hirschberg, J.: High frequency word entrainment in spoken dialogue. In: HLT, Short Papers (2008)
25. Parent, G., Eskenazi, M.: Lexical entrainment of real users in the let's go spoken dialog system. In: ISCA (2010)
26. Purington, A., Taft, J.G., Sannon, S., Bazarova, N.N., Taylor, S.H.: "Alexa is my new BFF" social roles, user satisfaction, and personification of the Amazon Echo. In: CHI (2017)
27. Reitter, D., Keller, F., Moore, J.D.: Computational modelling of structural priming in dialogue. In: HLT (2006)
28. Rubin, D.B.: Estimating causal effects of treatments in randomized and nonrandomized studies. J. Educ. Psychol. **66**, 688 (1974)
29. Santos, J., Rodrigues, J., Casal, J., Saleem, K., Denisov, V.: Intelligent personal assistants based on internet of things approaches. IEEE Syst. J. **12**(2), 1793–1802 (2016)
30. Walker, M.A., Stent, A., Mairesse, F., Prasad, R.: Individual and domain adaptation in sentence planning for dialogue. J. Artif. Intell. Res. **30**, 413–456 (2007)
31. Wen, T.H., Gasic, M., Mrksic, N., Su, P.H., Vandyke, D., Young, S.: Semantically conditioned LSTM-based natural language generation for spoken dialogue systems. In: EMNLP (2015)

32. Xiu, Z., Tao, C., Henao, R.: Variational learning of individual survival distributions. In: Proceedings of the ACM Conference on Health, Inference, and Learning (2020)
33. Zeng, S., Li, F., Hu, L.: Propensity score weighting analysis of survival outcomes using pseudo-observations. Stat. Sin. (2021). https://doi.org/10.5705/ss.202021.0175

Toward Interpretable Machine Learning: Constructing Polynomial Models Based on Feature Interaction Trees

Jisoo Jang(✉), Mina Kim, Tien-Cuong Bui, and Wen-Syan Li

Seoul National University, 1, Gwanak-ro Gwanak-gu, Seoul, Republic of Korea
{simonjisu,wensyanli}@snu.ac.kr

Abstract. As AI has been applied in many decision-making processes, ranging from loan application approval to predictive policing, the interpretability of machine learning models is increasingly important. Interpretable models and post-hoc explainability are two approaches in eXplainable AI (XAI). We follow the argument that transparent models should be used instead of black-box ones in real-world applications, especially regarding high-stakes decisions. In this paper, we propose PolyFIT to address two major issues in XAI: (1) bridging the gap between black-box and interpretable models and (2) experimentally validating the trade-off relationship between model performance and explainability. PolyFIT is a novel polynomial model construction method assisted by the knowledge of feature interactions in black-box models. PolyFIT uses extracted feature interaction knowledge to build interaction trees, which are then transformed into polynomial models. We evaluate the predictive performance of PolyFIT with baselines using four publicly available data sets, Titanic survival, Adult income, Boston house price, and California house price. Our method outperforms linear models by 5% and 56% in classification and regression tasks on average, respectively. We also conducted usability studies to derive the trade-off relationship between model performance and explainability. The studies validate our hypotheses about the actual relationship between model performance and explainability.

Keywords: eXplainable AI · transparent models · polynomial model · explainability evaluation

1 Introduction

AI is utilized in various decision-making processes, from loan application approval to predictive policing, making machine learning model interpretability increasingly important. Especially many of these models are considered black-boxes since it is difficult to review or explain their internal logic in a way humans can understand. Black-box models have made numerous serious mistakes in practice. The Pennsylvania Child Welfare Screening Tool [8] assists social workers in

H. Kashima et al. (Eds.): PAKDD 2023, LNAI 13936, pp. 159–170, 2023.
https://doi.org/10.1007/978-3-031-33377-4_13

determining which families should be investigated for child abuse and neglect. However, the predictive algorithm flagged a disproportionate number of black children for "mandatory" neglect investigations. Another example is the termination of Amazon's AI Recruitment System [14] due to discrimination against women candidates. Therefore, the difficulties of explaining "black-box" can hinder the adoption of machine learning in many real-world applications, especially regarding high-stakes decisions.

Much progress has been made in resolving the "black-box" issues in XAI. A post-hoc explainability is a popular approach that explains a model's predictions after training is done. It is based on the concept of *explanation by simplification*, also called *surrogate models*. LIME [18] is a popular tool for this approach. It simplifies the original model into a transparent model with input data perturbations. The *feature relevance explanation* describes the data instance by analyzing the contribution of each input feature to the target. In [19], Rudin argued that post-hoc explainability is often unreliable and misleading. For instance, the attention map from a CNN classifier is a computation result, not an explanation. Therefore, interpretable models must be used in high-stakes decisions since they provide explanations consistent with internal model computations and are more intelligible for humans. However, they usually suffer from poor performance with complex data [20].

In this paper, we propose PolyFIT, a **Poly**nomial model construction method based on **F**eature **I**nteraction **T**rees to address these two major issues in XAI:

1. Bridging the gap between black-box and interpretable models.
2. Experimentally validating the trade-off relationship between model performance and explainability.

To bridge the gap between black-box and interpretable models, since black-box models can capture complex patterns within input data, we can leverage their knowledge for constructing interpretable surrogate models. Surrogate models can achieve the performance level of black-box counterparts and, more importantly, are intelligible for humans. Specifically, we extract global feature interactions based on feature attribution methods like SHAP [13], as depicted in Fig. 1a. Based on these interactions, we construct polynomial models by iteratively adding the most relevant interaction terms. Finally, when the stopping criteria are met, we can get the polynomial model with the smallest performance gap with the black-box model and the Feature Interaction Tree as shown in Fig. 1b. The quantitative experimental results of our method showed that our proposed method is comparable to the existing approaches.

DARPA XAI Broad Agency Announcement [5] depicts a trade-off relationship between performance and interpretability that high-performance algorithms usually have lower explainability. This curve combines the performance and explainability of algorithms onto a single graph, allowing us to observe each algorithm's strengths and weaknesses with clarity. However, no quantification was in the trade-off curve, so we could not know how precise the graph was. To validate the "trade-off" relationship, we conducted a forward simulation test with actual users based on the concept proposed in [2] on four familiar data sets.

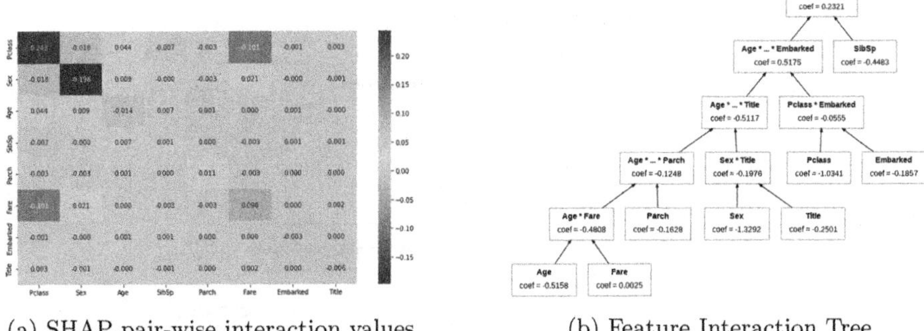

(a) SHAP pair-wise interaction values (b) Feature Interaction Tree

Fig. 1. (a) SHAP pair-wise interaction values from a sample in Titanic Survival data sets (ID=78). (b) Feature Interaction Tree generated from tree building process in Titanic Survival data sets

The results experimentally showed that the trade-off curves were similar for all four data sets but varied slightly depending on the data sets and the methods used to quantify explainability. For example, the explainability of linear models is not always superior to other models in our usability study.

The rest of the paper is organized as follows: Sect. 2 will introduce the background of knowledge used in our method and the evaluation of explainability. Also, it covers the previous works in building transparent models. Section 3 will introduce how PolyFIT creates a surrogate transparent model. Section 4 shows the results of the PolyFIT performance, the user subjective scores, and the explainability scores compared with the transparent models. Finally, in Sect. 5, we will summarize the findings and discuss the limitations and future of our methods and usability studies.

2 Related Work

2.1 SHAP and Pair-Wise Interaction Values

SHAP [13] explores each input feature's additive contribution to a machine learning model's output prediction. Given a model, SHAP uses a game-theoretical approach to generate Shapley values for each feature. The authors explain SHAP pair-wise interaction values in [12], which is an extension that can be obtained from the Shapley interaction index [4].

The result of SHAP pair-wise interaction values is shown in a symmetric matrix of size equal to the number of features for a single data instance, such as Fig. 1a. Other entries represent the interaction effect between the two features, while the diagonal values represent the main effects of the features.

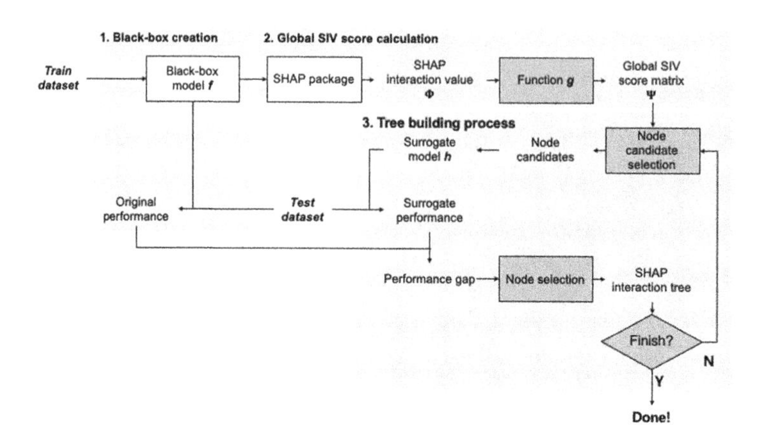

Fig. 2. The overall process of constructing polynomial models based on feature interaction trees. Details will be described in Sect. 3.

2.2 Polynomial Model and EBM

Polynomial models are useful when there exists a connection between two or more variables. It adds non-linearity to a simple linear model with interactions, which multiplies features to reflect relationships. Previous research compared a model's performance with and without a feature [16].

From this premise, we try to estimate the importance of feature interactions by altering the polynomial model to include only interaction terms in (1), where $p \neq q$ and g is the link function that adapts to either regression or classification:

$$g(h_{poly}(x)) = \beta_0 + \sum_i \beta_i x_i + \sum_{p,q} \beta_{p,q}(x_p \cdot x_q) \tag{1}$$

Explainable boosting machine (EBM) [15] is a tree-based cyclic gradient boosting Generalized Additive Model (GAM). EBM is completely interpretable and as accurate as state-of-the-art black-box models. However, it is slower to train than other contemporary algorithms [11]. EBM is a GAM of the form as follows, where g is the link function that adapts GAM to either regression or classification:

$$g(E[y]) = \beta_0 + \sum_i f_i(x_i) + \sum_{i,j} f_{i,j}(x_i, x_j) \tag{2}$$

3 Methodology

We intend to generate a transparent model h from a high-precision black-box model f. With Eq. (1), we can easily fit a polynomial model, but this method does not extract any valuable information from the black-box model. In addition, all conceivable combinations of features must also be included. This necessitates determining which interaction terms must be included in a linear model. To determine the importance of interaction terms, we constructed a hierarchical structure of feature interaction sets S, using the SHAP pair-wise interaction values. If the performance of h decreases when a certain interaction term is included, the term is deemed irrelevant. Based on this concept, we construct a transparent interaction model h as follows:

$$h(x;S) = \beta_0 + \sum_i \beta_i x_i + \sum_{s \in S} \beta_s \prod_k^s x_k \quad (3)$$

3.1 Black-box Model Creation

We split the data into train and test data sets, then train an XGBoost [1] model to get a black-box model f. Then we obtain the "original performance" by evaluating model f with the test data sets. This process is similar to most procedures in machine learning model training.

3.2 Global SHAP Interaction Value Score Calculation

Using the black-box model f and the training data, we calculate the SHAP pair-wise interaction values (SIV, Φ) using TreeSHAP technique [12]. SIV is a 3-dimensional matrix with a size of $N \times |M| \times |M|$, where N is the number of samples in the training data, and M is the entire feature set. We defined a global SIV as an average of absolute interaction values of the sample size. So, we process: $\Phi_{i,j} = \frac{1}{N}\sum_n^N |\Phi_{i,j}^n| \quad \forall i,j \in (1,\ldots,|M|)$. Next, we calculate the global SIV score matrix (Ψ) using one of the three g functions. We hypothesize g functions give different points of view when constructing hierarchical structures. (a) The absolute value (abs): only considers the size of interaction effects and the main effect, where $d(\phi_i) = |\phi_i|$, $d(\phi_{i,j}) = |\phi_{i,j}|$. (b) The absolute interaction value (abs_inter): ignores main effects and only considers the size of pair-wise interaction values, where $d(\phi_i) = 0$, $d(\phi_{i,j}) = |\phi_{i,j}|$. and (c) The interaction ratio (ratio): calculates the ratio of interaction values to the sum of main effects, where $d(\phi_i) = 0$, $d(\phi_{i,j}) = |\phi_{i,j}|/(|\phi_i| + |\phi_j|)$.

$$\Psi_{i,j} = \begin{cases} d(\phi_i) & \text{for } i = j \\ d(\phi_{i,j}) & \text{for } i \neq j \end{cases} \quad (4)$$

Let a node s be a container of a single feature or a hierarchical structure combination of features. Then "SIV score" for node s (refer as Ψ_s) is calculated given a global SIV score matrix Ψ, and will be used in tree-building process(3.3):

$$\Psi_s = \sum_{i \in \text{flatten}(s)} \Psi_{i,i} + \sum_{(p,q) \in \text{flatten}(s)} 2 \times \Psi_{p,q} \quad (5)$$

The SIV score matrix Ψ will always be symmetric since the SIV Φ is symmetric. "flatten(s)" is a flatten function that turns a hierarchical combination of feature sets into a single vector where $p \neq q$. The first term in Eq. (5) represents the summation of main effect scores. The second term is the summation of interaction scores.

3.3 Tree-building Process

The goal of the tree-building process is to find a node that leads to the minimal performance(refer as P) gap between the original model f and a surrogate polynomial model h at each step t:

$$\arg\min_{s} P_{gap} = P_f - P_h = L\big(y, f(x)\big) - L\big(y, h(x; S_t)\big) \tag{6}$$

where $S_t = \{S_{t-1}, s\}, (x, y) \in D^{train}$, L is a performance metric for a task, S_t represents the set of nodes with the lowest performance gap at step t, and D^{train} is a training data sets. At the beginning of the algorithm, we initialize nodes set S_0 with the summation of the SIV score matrix Ψ, which means the total effect of each feature in the feature sets M: $S_0 = \{i\colon \sum_j^{|M|}(\Psi_{i,j})|i \in M\}$

Construction of the Pair-Wise Interaction Node Candidates Set. The process starts with selecting a subset of nodes S_t' to reduce the searching space: $S_t' = \text{select}(S_{t-1}, N_s, \text{method})$

There are two methods to select a subset. The first is the random select number of nodes regardless of the scores in S_{t-1}. The other is the sort selection by top N_s nodes with scores in S_{t-1} in descending order. In our experiment, we examine which of two selection methods and what number of selected nodes N_s gains the best results with an increase of 10% on the number of selected nodes N_s. The minimum number of selected nodes is 2 due to the next step: constructing pair-wise node candidates set.

After selecting a subset of nodes, we construct a pair-wise interaction node candidates set C_t by calculating Ψ_s for each s in the combination of nodes in S_t': $C_t = \{s\colon \Psi_s | s \in \binom{S_t'}{2}\}$. Since there might be a lot of possible combinations in the early steps, to reduce the computational costs, we filter candidates to use as interaction terms for surrogate model h in Eq. (3): $C_t' = \text{filter}(C_t, N_f, \text{method})$. Same as selecting the subset of nodes process, there are two methods. We also examine which of the two methods and what number of filtered nodes N_f gains the best results.

Calculation of Performance Gap and Selection of Node. For each combination candidate in C_t', we construct a surrogate model h and calculate the performance gap P_{gap}^s with the performance of the original model f:

$$P_t = \left\{ s : P^s_{\text{gap}} = \begin{cases} P_f - P^s_h & \text{if task} = \text{classification} \\ P^s_h - P_f & \text{if task} = \text{regression} \end{cases} \right\} \quad \forall s \in C'_t \tag{7}$$

where $P^s_h = L\big(y, h(x; \{S_{t-1}, s\})\big)$

We select the node s with the lowest performance gap in P_t and form a new set of nodes $S_t = \{S_{t-1}, s\}$. In our experiment, we use the beam search algorithm to keep tracking K number of selected S_t to get better results for each step $H_t = \{S^k_t | k = (1, \ldots, K)\}$. Finally, Repeat the tree-building steps until meets the given finishing criteria.

4 Experiments

4.1 Model Performance

Configurations. We used four data sets for our study; two for binary classification problems: Titanic survival [10], and Adult income [3], and two for regression problems: Boston house price [6], and California house price [17]. We trained our XGBoost model using a learning rate of 0.1, a max tree depth of 6, and a seed of 7. We discovered the ideal number of boost rounds using three-fold cross-validation with 500 rounds. Table. 1 displays the final test data performance. In our experiment, PolyFIT performs better than LINEAR and POLY; the original model's interaction values helped boost the performance of constructing a polynomial model. The California and Titanic data sets performed better than the original model.

Table 1. Test performance of models by each task and data. Classification tasks use accuracy, and regression tasks use MSE. LINEAR is the baseline without interactions. POLY is the polynomial model with all feature interactions, and EBM refers to Explainable Boosting Machine.

Task	Dataset	XGBoost	PolyFIT	EBM	LINEAR	POLY
Classification↑	Adult	0.8732	0.8446	**0.8723**	0.8397	0.8403
	Titanic	0.8611	**0.8889**	0.8472	0.8056	0.8056
Regression↓	Boston	0.0164	0.0203	**0.0195**	0.0319	0.0266
	California	0.0838	**0.0339**	0.0787	0.1402	0.3571

Results About Parameters to Build PolyFIT. As explained in Sect. 3, we examine several parameters to build PolyFIT. Figure 3 shows the box plot of the performance gap about the number of the beam search, and Fig. 4 is the results of the number of selecting nodes and filtering candidates for reducing the search space. We calculated the algorithm's efficiency by dividing the exponential of the negative performance gap by the execution time(seconds). Other parameters did not show significant differences in the experiments.

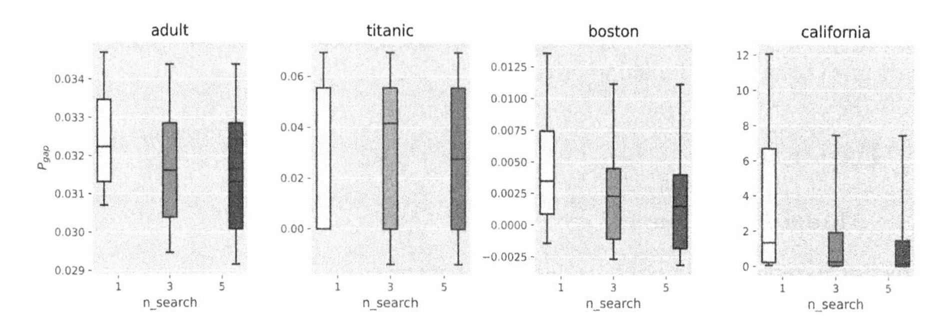

Fig. 3. The experiment about the number of hypotheses in the beam search(X-axis); if it equals one, refers to a greedy search. The Y-axis is the performance gap of the PolyFIT. Higher beam searches lead to a lower performance gap because the algorithm has a better chance of finding the best interaction combination.

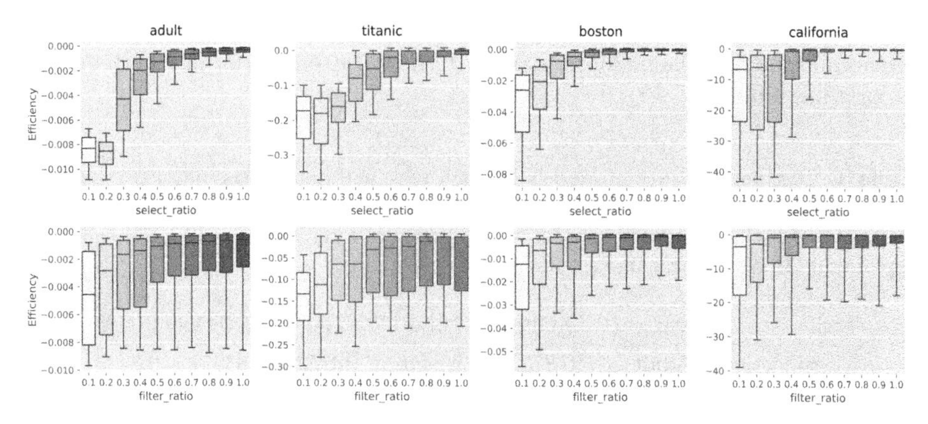

Fig. 4. The experiment about efficiency on the number of selecting nodes(first row) and filtering candidates(second row). The X-axis is the ratio of the number of selected nodes and filtered candidates. The higher ratio, the lower the efficiency, because more selection and filtering of nodes cause a higher execution time. From the results, 0.1∼0.2 ratio is the best efficiency for each data set.

4.2 Evaluating Interpretability

Similar to the studies in [7], we designed a forward simulation test that asks users to predict models' outcomes given inputs samples with and without explanations to evaluate the explainability in qualitative metrics. We also include a Likert-scaled subjective rating as an alternative measurement for explainability with the "helpfulness" and "easiness" of the explanation methods. Similar to [9], we administered subjective tests after respondents had completed all four tasks.

4.3 Usability Study

Our other goal is to quantify transparent models' explainability and user subjective ratings on easiness and helpfulness. We build a survey system to evaluate three transparent models and the global SHAP interaction matrix. The survey has two phases with four different data sets; the Pre-stage(pre-explanation phase) and the Post-stage(post-explanation phase). Explainability scores(δ) are differences in performance between the two stages that would be purely caused by the explanation method since it was the only difference made.

Survey Design. In the Pre-stage, a validation data set of 16 data points will be provided, including the feature values, labels, and model prediction results. The respondents are asked to estimate the model's prediction for 8 test data without explanation methods. They will fill them out one by one to be less affected by other questions.

The same validation data sets is kept visible in the Post-stage. The respondents do the same jobs for the same 8 test data in the Pre-stage with one of the explanation methods: linear model (LINEAR), global SHAP Interaction Value matrix (SIV), Explainable Boosting Machine (EBM), and our method (PolyFIT). Our system will randomly match four methods one-to-one to four data sets for each surveyor to ensure everyone can experience four explanation methods. Respondents can always see how the explanation tool works and how to interpret the results during the survey.

User Pool. We gathered 29 respondents that have basic knowledge of linear and polynomial models. They are either students in the graduate school of data science or workers in the field of data science. Respondents had difficulty with the regression tasks; we observed this phenomenon not only by their comments but also because they filled some of the problems with zeros. To control the survey's quality, we excluded eight regression task responses, therefore, received 928 classification and 736 regression responses.

Survey Results. Table. 2 shows the explainability scores(δ) in classification and regression tasks. From the results, at the significant level of 5%, SIV and LINEAR methods improved the performance in the California data set. Table. 3

Table 2. The average explainability scores(δ). CI gives the 95% confidence interval of Post-stage performance, calculated by bootstrap n responses. The p is the one-side Wilcoxon signed-rank test p-value to check if the Post-stage performance improves. The alternative hypothesis is the performance of the Post-stage is greater than the Pre-stage for classification tasks. For regression tasks, it will be the opposite.

	adult						titanic					
Method	N	Pre	δ	Post	CI	p	N	Pre	δ	Post	CI	p
SIV	32	0.719	−0.125	0.594	0.004	0.858	64	0.688	+0.063	0.750	0.004	0.207
LINEAR	72	0.556	+0.063	0.611	0.003	0.125	56	0.643	−0.036	0.607	0.005	0.760
EBM	48	0.646	+0.056	0.708	0.002	0.198	48	0.625	+0.125	0.750	0.004	0.065
PolyFIT	80	0.663	−0.013	0.650	0.005	0.500	64	0.665	−0.063	0.594	0.005	0.637
	Boston						California					
Method	N	Pre	δ	Post	CI	p	N	Pre	δ	Post	CI	p
SIV	40	0.118	−0.069	0.187	0.002	1.000	64	0.486	+0.191	0.295	0.002	0.039
LINEAR	40	0.174	−0.015	0.189	0.001	0.906	48	0.322	+0.063	0.258	0.003	0.047
EBM	64	0.170	−0.010	0.180	0.002	0.422	32	0.567	+0.069	0.498	0.007	0.313
PolyFIT	32	0.243	+0.141	0.102	0.002	0.063	48	0.333	−0.078	0.411	0.004	0.781

Table 3. The average helpfulness and easiness scores from the responses(29 respondents). The higher ratings infer the higher usefulness of explanations. Our method and LINEAR got high scores in the regression task.

		Average Helpfulness				Average Easiness			
Task	Dataset	SIV	LINEAR	EBM	PolyFIT	SIV	LINEAR	EBM	PolyFIT
Classification	Adult	2.750	2.889	3.000	2.500	3.000	3.444	2.667	2.600
	Titanic	4.000	3.000	2.833	3.250	3.875	3.857	2.500	3.250
Regression	Boston	2.143	3.600	2.667	2.800	2.429	4.200	2.750	4.200
	California	1.900	4.125	1.800	3.500	2.700	4.250	1.800	3.833

shows the helpfulness and easiness of four methods. Compared to the forward simulation, LINEAR and our method get a higher score in the regression task.

Figure 5 reveals the performance-explainability relationship using scores and user ratings. Our method outperforms linear models and is comparable to EBM. Explainability is not superior in classification tasks but performs better in regression tasks. Linear models' explainability isn't consistently better for all tasks, and users find them most useful except for the Titanic dataset(see subjective scores). Results differ in forward simulation.

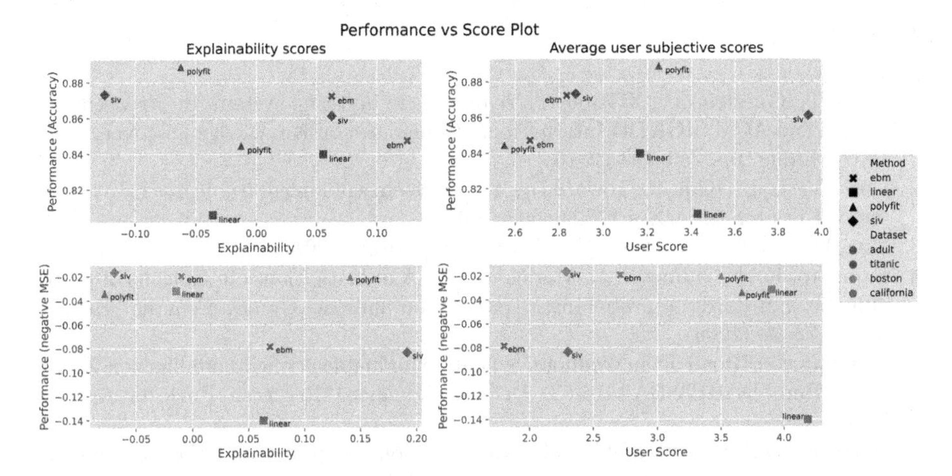

Fig. 5. The performance and explainability scores(δ), average user subjective scores plot. We plot the original model's score with the explainability of the global SIV matrix method to represent the black-box model. Classification tasks are measured by accuracy, while regression tasks use negative mean squared error. The user subjective ratings are calculated by the average of helpfulness and easiness.

5 Conclusion

We developed a novel polynomial model construction method, PolyFIT, leveraging feature interactions knowledge from black-box models, creating surrogate transparent models. Our method outperformed linear and naïve polynomial models and was comparable to EBM, an advanced polynomial construct method. We also address a crucial question in XAI regarding the trade-off relationship between model performance and explainability. Through carefully-designed user studies, we understood that the performance-explainability trade-off was not always accurate. The explainability of linear models is not always superior to other models for all tasks. The actual trade-off curve depends on the data sets and the method used to quantify explainability.

Even though PolyFIT can construct acceptable polynomial models, choosing the best one is still an open question. To address this limitation, we plan to evaluate the proposed technique with more real-world scenarios. Further work also includes measuring explainability in multiple aspects and domains.

Acknowledgements. This work was supported by the National Research Foundation of Korea's Brain Korea 21 FOUR Program and New Faculty Startup Fund from Seoul National University.

References

1. Chen, T., Guestrin, C.: XGBoost: a scalable tree boosting system. In: Proceedings of the 22nd ACM SIGKDD International Conference on Knowledge Discovery And Data Mining. pp. 785–794 (2016)
2. Doshi-Velez, F., Kim, B.: Towards a rigorous science of interpretable machine learning. arXiv preprint arXiv:1702.08608 (2017)
3. Dua, D., Graff, C.: UCI machine learning repository (2017)
4. Fujimoto, K., Kojadinovic, I., Marichal, J.L.: Axiomatic characterizations of probabilistic and cardinal-probabilistic interaction indices. Games Econom. Behav. **55**(1), 72–99 (2006)
5. Gunning, D.: Broad agency announcement, explainable artificial intelligence(XAI). Tech. Rep. DARPA-BAA-16-53, Defense Advanced Research Projects Agency (2016)
6. Harrison, D., Jr., Rubinfeld, D.L.: Hedonic housing prices and the demand for clean air. J. Environ. Econ. Manag. **5**(1), 81–102 (1978)
7. Hase, P., Bansal, M.: Evaluating explainable AI: which algorithmic explanations help users predict model behavior? arXiv preprint arXiv:2005.01831 (2020)
8. Ho, S., Burke, G.: Oregon dropping AI tool used in child abuse cases. The Associated Press (2022)
9. Hutton, A., Liu, A., Martin, C.: Crowdsourcing evaluations of classifier interpretability. In: 2012 AAAI Spring Symposium Series (2012)
10. kaggle: Titanic - machine learning from disaster. Accessed 05 July 2022
11. Lou, Y., Caruana, R., Gehrke, J., Hooker, G.: Accurate intelligible models with pairwise interactions. In: Proceedings of the 19th ACM SIGKDD International Conference on Knowledge Discovery and Data Mining, pp. 623–631 (2013)
12. Lundberg, S.M., Erion, G.G., Lee, S.I.: Consistent individualized feature attribution for tree ensembles. arXiv preprint arXiv:1802.03888 (2018)
13. Lundberg, S.M., Lee, S.I.: A unified approach to interpreting model predictions. In: Guyon, I., et al. (eds.) Advances in Neural Information Processing Systems 30, pp. 4765–4774. Curran Associates, Inc. (2017)
14. Meyer, D.: Amazon reportedly killed an AI recruitment system because it couldn't stop the tool from discriminating against Women. Fortune Media IP Limited (2018)
15. Nori, H., Jenkins, S., Koch, P., Caruana, R.: InterpretML: Aaunified framework for machine learning interpretability. arXiv preprint arXiv:1909.09223 (2019)
16. Oh, S.: Feature interaction in terms of prediction performance. Appl. Sci. **9**(23), 5191 (2019)
17. Pace, R.K., Barry, R.: Sparse spatial autoregressions. Statist. Prob. Lett. **33**(3), 291–297 (1997)
18. Ribeiro, M.T., Singh, S., Guestrin, C.: "why should i trust you?" explaining the predictions of any classifier. In: Proceedings of the 22nd ACM SIGKDD International Conference on Knowledge Discovery and Data Mining, pp. 1135–1144 (2016)
19. Rudin, C.: Stop explaining black box machine learning models for high stakes decisions and use interpretable models instead. Nat. Mach. Intell. **1**(5), 206–215 (2019)
20. Speith, T.: A review of taxonomies of explainable artificial intelligence (XAI) methods. In: 2022 ACM Conference on Fairness, Accountability, and Transparency, pp. 2239–2250 (2022)

Kernel Methods

BioSequence2Vec: Efficient Embedding Generation for Biological Sequences

Sarwan Ali[1], Usama Sardar[2], Murray Patterson[1(✉)], and Imdad Ullah Khan[2(✉)]

[1] Georgia State University, Atlanta, GA, USA
sali85@student.gsu.edu, mpatterson30@gsu.edu
[2] Lahore University of Management Sciences, Lahore, Pakistan
imdad.khan@lums.edu.pk

Abstract. Representation learning is an important step in the machine learning pipeline. Given the current biological sequencing data volume, learning an explicit representation is prohibitive due to the dimensionality of the resulting feature vectors. Kernel-based methods, *e.g.*, SVM, are a proven efficient and useful alternative for several machine learning (ML) tasks such as sequence classification. Three challenges with kernel methods are (i) the computation time, (ii) the memory usage (storing an $n \times n$ matrix), and (iii) the usage of kernel matrices limited to kernel-based ML methods (difficult to generalize on non-kernel classifiers). While (i) can be solved using *approximate* methods, challenge (ii) remains for typical kernel methods. Similarly, although non-kernel-based ML methods can be applied to kernel matrices by extracting principal components (kernel PCA), it may result in information loss, while being computationally expensive. In this paper, we propose a general-purpose representation learning approach that embodies kernel methods' qualities while avoiding computation, memory, and generalizability challenges. This involves computing a low-dimensional embedding of each sequence, using random projections of its k-mer frequency vectors, significantly reducing the computation needed to compute the dot product and the memory needed to store the resulting representation. Our proposed fast and alignment-free embedding method can be used as input to any distance (*e.g.*, k nearest neighbors) and non-distance (*e.g.*, decision tree) based ML method for classification and clustering tasks. Using different forms of biological sequences as input, we perform a variety of real-world classification tasks, such as SARS-CoV-2 lineage and gene family classification, outperforming several state-of-the-art embedding and kernel methods in predictive performance.

Keywords: Representation Learning · Sequence Classification · Kernel Function

M. Patterson and I. U. Khan — Joint Last Authors.

H. Kashima et al. (Eds.): PAKDD 2023, LNAI 13936, pp. 173–185, 2023.
https://doi.org/10.1007/978-3-031-33377-4_14

1 Introduction

The rate at which biological sequence data is being generated and stored currently outpaces Moore's law growth, even under fairly conservative estimates [33]. In the past decade, the amount of biological sequence data has already reached a level that automated — machine learning (ML) and deep learning (DL) — algorithms are able to learn from and perform analysis. A notable example is the AlphaFold framework [26] for structure prediction from a protein sequence. Another example is the Pangolin tool [28] for predicting lineage, *e.g.*, B.1.1.7 [29], from an assembled SARS-CoV-2 genomic sequence. The number of SARS-CoV-2 genomic sequences which are publicly available on databases such as GISAID[1] is more than 15 million and counting.

While approaches such as AlphaFold and Pangolin have outperformed the previous state-of-the-art by a significant margin, or can scale to orders of magnitude more in number of sequences, these learning approaches can be and will need to be optimized, as the amount of data grows to strain even these approaches. Since the majority of ML models deal with fixed-length numerical vectors, biological sequences cannot be used directly as input, making efficient embedding design an important step in such ML-based pipelines [15,24].

Sequence alignment is another important factor to be considered while performing sequence analysis. Although embeddings that require the sequences to be aligned, such as one-hot encoding (OHE) [27], are proven to be efficient in terms of predictive performance, researchers are interested in exploring alignment-free methods to avoid the expensive multiple sequence alignment operations as a preprocessing step [1,3,16,17,34]. Most alignment-free methods compute some form of a sketch of a sequence from short substrings, such as k-mers to generate a spectrum [7]. Although the existing alignment-free embedding methods yield promising predictive results; they produce vectors of high dimensionality, especially for very long sequences. An alternative to the traditional *feature engineering* methods is using deep learning (DL) models [4]. However, DL methods have not seen much success in the classification of tabular datasets [8,12].

Using a kernel (Gram) matrix for sequence classification and kernel-based ML classifiers, such as SVM, shows promising results [6,19]. Kernel-based methods outperform feature engineering-based methods [7]. These methods work by computing kernel (similarity) values between pairs of sequences based on the number of matches and mismatches between their k-mers [6]. The resultant kernel matrix can then be used to classify the sequences using SVM. However, there are serious challenges to the scalability of kernel-based methods to large datasets:

- Evaluating the kernel value between a pair of sequences takes time proportional to $|\Sigma|^k$, where Σ is the alphabet of sequences;
- Storing the $n \times n$ kernel matrix is memory intensive when n is very large; and
- Kernel matrices are limited to kernel-based machine learning models (such as SVM) downstream.

[1] https://www.gisaid.org/.

The first challenge of kernel evaluation can be overcome with the so-called *kernel trick* and approximating kernel values with quality guarantees [19]. To use more general classifiers like decision trees, one can compute principal components of the kernel matrix using kernel PCA [23], which can act as the embedding representation to tackle the third challenge. However, this process results in a loss of information and is computationally expensive. In general, the second challenge of the need to store an $n \times n$ kernel matrix in memory remains unaddressed.

In this paper, we propose a random projection-based sequence representation called BioSequence2Vec, which has the qualities of kernel methods in terms of efficiently computing pairwise similarity between sequences (overcoming the first challenge) while also addressing the memory overhead. Given a (biological) sequence as input, the BioSequence2Vec embedding projects frequency vectors of all k-mers in a sequence in "random directions" and uses these projections to represent the sequence. BioSequence2Vec computes the projections in one linear scan of the sequence (rather than explicitly computing the frequency of each of the $|\Sigma|^k$ k-mers in the sequence). Since our method computes the representation of a sequence in linear time (linear in the length of the sequence), it easily scales to a large number of sequences. The generated embeddings are low-dimensional (user-controlled), hence BioSequence2Vec overcomes the memory usage problem. The Euclidean (and cosine) similarity between a pair of embeddings is closely related to the kernel similarity of the pair, hence our method incorporates the benefits of kernel-based methods. BioSequence2Vec is efficient, does not require sequences to be aligned, and the embeddings can be used as input to any distance (*e.g.*, k nearest neighbors) and non-distance (*e.g.*, decision tree) based ML methods for the supervised tasks, solving the third problem.

In summary, our contributions are the following:

1. We propose a fast, alignment-free, and efficient embedding method, BioSequence2Vec, which quickly computes a low dimensional numerical embedding for biological sequences. It has the quality of kernel-based methods in terms of computing pair-wise similarity values between sequences while also addressing the memory usage issue of kernel methods — allowing it to scale to many more sequences.
2. We show that the proposed method can be generalized on different types of real-world biological sequences. It outperforms both alignment-based and alignment-free SOTA methods for predictive performance on different datasets.
3. Our method eliminates the expensive multiple sequence alignment step from the classification pipeline, hence making it a fast and scalable approach.

The rest of the paper is organized as follows: The literature for biological sequence analysis is given in Sect. 2. Section 3 outlines the details of the proposed model. The description of the dataset and experimental setup are given in Sect. 4. The empirical results are provided in Sect. 5. Section 6 concludes the paper and discusses future prospects.

2 Related Work

Designing numerical embeddings is an important step in the ML pipeline for supervised analysis [9,35]. The feature engineering-based methods, such as Spike2Vec [5] and PWM2Vec [2], which are based on the idea of using k-mers achieve reasonable predictive performance. However, they still face the problem of *curse of dimensionality*. As we increase the value of k, the spectrum (frequency count vector) becomes sparse (contains small k-mers counts). Hence the likelihood of observing a specific k-mer again decreases. To solve this sparse vector problem, authors in [20] propose the idea of using gapped/spaced k-mer. The use of k-mers counts for phylogenetic applications was first explored in [11], which constructed accurate phylogenetic trees from coding and non-coding nDNA sequences. Although phylogenetic-based methods are useful for sequence analysis, they are computationally expensive, hence cannot be scaled on bigger datasets.

Computing the pair-wise similarity between sequences by computing kernel/gram matrix is a well-studied problem in ML domain [7]. Since computing the pair-wise similarities could be expensive to compute, authors in [19] proposed an approximate method to improve the kernel computation time by computing the dot product between the spectrum of two sequences. The resultant kernel matrix can be used as input for kernel classifiers such as SVM or non-kernel classifiers [7] using kernel PCA [23] for classification purposes.

Authors in [30] propose a neural network-based model to extract the features using the Wasserstein distance. A ResNet model for the purpose of classification is proposed in [36]. However, DL methods, in general, do not show promising results when applied to tabular data [31]. Using pre-trained models is also explored in the literature for biological sequence analysis [13,22]. However, since those models are usually trained on a specific type of biological sequence, they cannot easily be generalized on different types of data.

3 Proposed Approach

In this section, we describe the details of our sequence representation, BioSequence2Vec. We also analyze the space and time complexity of computing the representations. As outlined above, sequences generally have varying lengths, and even when the lengths are the same, the sequences may not be aligned. Thus, they cannot be treated as vectors. Though in aligned sequences of equal length, a one-to-one correspondence between elements is established, treating them as vectors ignores the order of elements and their contiguity. In one of the most successful approaches that cater to all of the above issues, sequences are represented by fixed-dimensional feature vectors. The feature vectors are the spectra, or counts, of all k-mers appearing in the sequences [5].

Suppose we are given a set $\mathcal{S}$ of sequences (of, *e.g.*, nucleotides A, C, G, T). For a fixed integer k, let Σ^k be the set of all strings of length k made from characters in Σ (all possible k-mers) and let $s = |\Sigma|^k$. The spectrum $\Phi_k(X)$ of a sequence $X \in \mathcal{S}$ is a s-dimensional vector of the counts of each possible k-mer occurring in X. More formally,

$$\Phi_k(X) = (\Phi_k(X)[\gamma])_{\gamma \in \Sigma^k} = \left(\sum_{\alpha \in X} I(\alpha, \gamma) \right)_{\gamma \in \Sigma^k}, \tag{1}$$

where

$$I_k(\alpha, \gamma) = \begin{cases} 1, & \text{if } \alpha = \gamma \\ 0, & \text{otherwise} \end{cases} \tag{2}$$

Note that $\Phi_k(X)$ is a $s = |\Sigma|^k$-dimensional vector where the coordinate $\gamma \in \Sigma^k$ has a value equal to the frequency of γ in X. Since this dimensionality grows quickly for modest-sized alphabets — it is exponential in k — the space complexity of representing sequences can quickly become prohibitive.

In kernel-based machine learning, a *kernel function* computes a real-valued similarity score for a pair of feature vectors. The kernel function is typically the inner product of the respective spectra.

$$\begin{aligned} K(i,j) &= K(X_i, X_j) = \langle \Phi_k(X_i), \Phi_k(X_j) \rangle \\ &= \Phi(X_i) \cdot \Phi(X_j) = \sum_{\gamma \in \Sigma^k} \Phi_k(X_i)[\gamma] \times \Phi_k(X_j)[\gamma] \end{aligned} \tag{3}$$

The kernel matrix (of pairwise similarity scores) is input to a standard support vector machine (SVM) [18] classifier resulting in excellent classification performance in many applications [19]. Although, in the so-called *kernel trick*, the explicit computation of feature vectors are avoided, with quadratic space required to store the kernel matrix, even this approach does not scale to real-world sequences datasets. There are three challenges to overcome: (i) Explicit representation is prohibitive due to the dimensions of the feature vectors, (ii) Although explicit computation is avoided using the kernel trick [19], the storage complexity of the kernel matrix is too large, and (iii) Kernel methods do not allow non-kernel-based machine learning methods. In the following, we propose a representation learning approach, namely BioSequence2Vec, that encompasses the benefits of the kernels and allows employing both kernel-based and general purpose machine learning methods.

3.1 BioSequence2Vec Representation

The BioSequence2Vec representation, $\hat{\mathbf{x}}$ for a sequence X represents X by the random projections of $\Phi_k(X)$ on the "discrete approximations" of random directions. It allows the application of vector space-based machine learning methods. We show that the Euclidean distance between a pair of vectors in BioSequence2Vec representation is closely related to the kernel-based proximity measure between the corresponding sequences. We use 4-wise independent hash functions to compute $\Phi'(\cdot)$. Note that the definition of our representation of (6) is inspired by the work of AMS [10] to estimate the frequency moments of a stream.

Definition 1 (4-wise Independent hash function). *A family $\mathcal{H}$ of functions of the form $h : \Sigma^k \mapsto \{-1, 1\}$ is called 4-wise independent, or 4-universal, if a randomly chosen $h \in \mathcal{H}$ has the following properties:*

1. *for any $\alpha \in \Sigma^k$, $h(\alpha)$ is equally likely to be -1 or 1.*
2. *for any distinct $\alpha_i \in \Sigma^k$, and $m_i \in \{-1, 1\}$ $(1 \leq i \leq 4)$,*

$$Pr[h(\alpha_1) = m_1 \wedge \ldots \wedge h(\alpha_4) = m_4] = 1/2^4$$

Next, we give a construction of a 4-wise independent family of hash functions due to Carter and Wegman [14]

Definition 2. *Let p be a large prime number. For integers a_0, a_1, a_2, a_3, such that $0 \leq a_i \leq p - 1$, and $\alpha \in \Sigma^k$ (represented as integer base $|\Sigma|$), the hash function $h_{a_0,a_1,a_2,a_3} : \Sigma^k \mapsto \{-1, 1\}$ is defined as*

$$h_{a_0,a_1,a_2,a_3}(\alpha) = \begin{cases} -1 & \textit{if } g(\alpha) = 0 \\ 1 & \textit{if } g(\alpha) = 1 \end{cases} \tag{4}$$

where

$$g(\alpha) = \left(a_0 + a_1\alpha + a_2\alpha^2 + a_3\alpha^3 \mod p\right) \mod 2 \tag{5}$$

It is well-known that the family $\mathcal{H} = \{h_{a_0,a_1,a_2,a_3} : 0 \leq a_i < p\}$ is 4-universal. Choosing a random function from this family amounts to choosing four random coefficients of polynomial, and the hash value for a k-mer α is the value of the polynomial (with random coefficients) at α modulo the prime p and modulo 2.

We use the following property of any randomly chosen function h from $\mathcal{H}$ that directly follows from the definition.

Fact 1 *For any distinct $\alpha_1, \alpha_2 \in \Sigma^k$, $E[h(\alpha_2)h(\alpha_2)] = 0$*

The property of 4-wise independence is used to derive a bound on the variance of the inner product. Let t be a fixed positive integer (a user-specified quality parameter). For $1 \leq i \leq t$, let $h^{(i)} = h^{(i)}_{a_0,a_1,a_2,a_3}$ be t randomly and independently chosen functions from $\mathcal{H}$ (corresponds to choosing t sets of 4 integers modulo p).

The ith coordinate of our representation, $\hat{\mathbf{x}}$ of a sequence X is given by

$$\hat{\mathbf{x}}_i = \frac{1}{\sqrt{t}} \sum_{\alpha \in X} h^{(i)}(\alpha). \tag{6}$$

In other words, The ith coordinate is the projection on the random vector in $\mathbb{R}^{|\Sigma|^k}$, a corner of the $|\Sigma|^k$-dimensional hypercube. More precisely, $\hat{\mathbf{x}}$ is a t-dimensional vector, where the value at the ith coordinate is the (normalized) dot-product of $\Phi_k(X)$ with the vector in $\{-1, 1\}^{|\Sigma|^k}$ given by $h^{(i)}$.

Next, we show that the dot-product between the BioSequence2Vec representations $\hat{\mathbf{x}}$ and $\hat{\mathbf{y}}$ of a pair of sequences X and Y closely approximates the kernel similarity value given in (3). We are going to show that for any pair of sequences X and Y, $\hat{\mathbf{x}} \cdot \hat{\mathbf{y}} \simeq \Phi_k(X) \cdot \Phi_k(Y)$. For notational convenience let $\mathbf{x} = \Phi_k(X)$ and $\mathbf{y} = \Phi_k(Y)$, we show that $\hat{\mathbf{x}} \cdot \hat{\mathbf{y}} \simeq \mathbf{x} \cdot \mathbf{y}$.

Theorem 1. *For any $0 < \epsilon, \delta < 1$, if $t \geq 2/\epsilon^2 \log(1/\delta)$, then*

1. $E\big[\hat{\mathbf{x}} \cdot \hat{\mathbf{y}}\big] = \mathbf{x} \cdot \mathbf{y}$
2. $Pr\big[|\hat{\mathbf{x}} \cdot \hat{\mathbf{y}} - \mathbf{x} \cdot \mathbf{y}| \leq \epsilon \|\mathbf{x}\| \|\mathbf{y}\|\big] \geq 1 - \delta$

The proof of 1. will be provided in the full version of the paper.

The proof of 2. follows from a standard application of Hoeffding's inequality. First note that by construction for $1 \leq i \leq t$, we have

$$-\|\mathbf{x}\|/\sqrt{t} \leq \hat{\mathbf{x}}_i \leq \|\mathbf{x}\|/\sqrt{t}$$

Similar bounds hold on each coordinate of $\hat{\mathbf{y}}$. Also note that $\|\mathbf{x}\| = \|\Phi_k(X)\|$ is the number of k-mers in X. These bounds implies that

$$-\|\mathbf{x}\|\|\mathbf{y}\|/t \leq \hat{\mathbf{x}}_i \times \hat{\mathbf{y}}_i \leq \|\mathbf{x}\|\|\mathbf{y}\|/t$$

Using these bounds in Hoeffding's inequality, we get that

$$Pr\big[|\hat{\mathbf{x}} \cdot \hat{\mathbf{y}} - \mathbf{x} \cdot \mathbf{y}| \geq \epsilon \|\mathbf{x}\| \|\mathbf{y}\|\big] \leq e^{-t\epsilon^2/2}.$$

Substituting the value of t we get the desired probabilistic guarantee on the quality of our estimate. □

Note that the upper bound on the error is very loose, in practice we get far better estimates of the inner product.

Remark 1. The runtime of computing $\hat{\mathbf{x}}$ is tn_x, where n_x is the number of characters in X. The space complexity of saving $\hat{\mathbf{x}}$ is $2/\epsilon^2 \log(1/\delta)$, where both ϵ and δ are user-controlled parameters. In the error term, $\|\mathbf{x}\| = n_x - k + 1$, when $\mathbf{x} = \Phi_k(X)$.

Next, we show that the ℓ_2-distance between any two vectors, which is usually employed in vector-space machine learning methods (e.g. k-NN classification) is closely related to their inner product. The inner product of the BioSequence2Vec representations of two sequences closely approximate the kernel similarity score between two sequences, see Eq. (3). Thus, BioSequence2Vec achieves the benefits of kernel-based learning while avoiding the time complexity of kernel computation and the space complexity of storing the kernel matrix.

Suppose we scale the BioSequence2Vec representation $\hat{\mathbf{x}}$ of the sequence X by $\|\hat{\mathbf{x}}\|_2$ (ℓ_2 norm of $\hat{\mathbf{x}}$, to make them unit vectors). Then, by definition, we get the following relation between ℓ_2-distance and inner product between $\hat{\mathbf{x}}$ and $\hat{\mathbf{y}}$.

$$\begin{aligned} d^2(\hat{\mathbf{x}}, \hat{\mathbf{y}}) &= \sum_{i=1}^{t} (\hat{\mathbf{x}}_i - \hat{\mathbf{y}}_i) = \sum_{i=1}^{t} \hat{\mathbf{x}}_i^2 + \sum_{i=1}^{t} \hat{\mathbf{y}}_i^2 - 2\sum_{i=1}^{t} \hat{\mathbf{x}}_i \hat{\mathbf{y}}_i \\ &= 1 + 1 - 2(\hat{\mathbf{x}} \cdot \hat{\mathbf{y}}) = 2 - 2\ Cos\ \theta_{\hat{x}\hat{y}} \end{aligned}$$

where θ_{uv} is the angle between the $\mathbf{u}$ and $\mathbf{v}$ in $\mathbb{R}^t$. Thus, both the "Euclidean and cosine similarities" between two BioSequence2Vec vectors are proportional to the "kernel similarity" between the corresponding sequences.

The pseudocode of BioSequence2Vec is given in Algorithm 1. Our algorithm takes as input a set $\mathcal{S}$ of biological sequences, integers k, p, alphabet Σ, and the number of hash functions t. It outputs the embedding R, which is the t dimensional fixed-length numerical representation corresponding to set $\mathcal{S}$ of sequences.

Algorithm 1. BioSequence2Vec Computation

```
1: Input: Set S of sequences, integers k, p, Σ,t
2: Output: Embedding R
3: function ComputeEmbedding(S, k, p, Σ,t)
4:     R = []
5:     for X ∈ S do                                          ▷ for each sequence
6:         x̂ = []
7:         for i = 1 to t do                                 ▷ # of hash functions
8:             a0, a1, a2, a3 ← random(0, p-1)
9:                           ▷ Four random integers for coefficients of polynomial
10:            for j ∈ |X| − k + 1 do                         ▷ scan sequence
11:                α ← X[j : j + k]                           ▷ k-mer
12:                h ← a0 + a1αΣ + a2αΣ^2 + a3αΣ^3
13:                                     ▷ αΣ is numerical version of α base |Σ|
14:                h ← (h mod p) mod 2
15:                if h = 0 then
16:                    x̂[i] ← x̂[i] - 1                       ▷ Eq. (4)
17:                else
18:                    x̂[i] ← x̂[i] + 1                       ▷ Eq. (4)
19:            x̂[i] = 1/√t × x̂[i]                             ▷ Eq. (6)
20:        R.append(x̂)
21:    return R
```

4 Experimental Evaluation

This section discusses datasets and state-of-the-art (SOTA) methods for comparing results. All experiments are performed on a core i5 system (with a 2.40 GHz processor) having windows 10 OS and 32 GB memory. For experiments, we use 70-30% split for training and testing (held out) sets, respectively, and repeat experiments 5 times to report average and standard deviation (SD) results. To evaluate the proposed method, we use aligned and unaligned biological sequence datasets (see Table 1). For classification, we use SVM, Naive Bayes (NB), Multi-Layer Perceptron (MLP), KNN, Random Forest (RF), Logistic Regression (LR), and Decision Tree (DT). We use eight SOTA methods (both alignment-free and alignment-based) to compare results. The detail of SOTA methods and a brief comparison with the proposed model are given in Table 2.

5 Results and Discussion

In this section, we report the classification results for BioSequence2Vec using different datasets and compare the results with SOTA methods. A comparison of BioSequence2Vec with SOTA algorithms on Spike7k and the Human DNA dataset is shown in Table 3. We report the BioSequence2Vec results for $t = 1000$ and $k = 3$.

Table 1. Dataset Statistics.

Dataset	Detail	Source	Total Sequences	Total classes	Sequence Length		
					Min	Max	Average
Spike7k	Aligned spike protein sequences to classify lineages of coronavirus in humans	[21]	7000	22	1274	1274	1274.00
Human DNA	Unaligned nucleotide sequences to classify gene family to which humans belong	[25]	4380	7	5	18921	1263.59

Table 2. Different methods (ours and SOTA) description.

Method	Category	Detail	Source	Alignment Free	Computationally Efficient	Space Efficient	Low Dim. Embedding
Spike2Vec	Feature Engineering	Take biological sequence as input and design fixed-length numerical embeddings	[5]	✓	✓	✓	✗
Spaced k-mers			[32]	✓	✓	✓	✗
PWM2Vec			[2]	✗	✓	✓	✓
WDGRL	Neural Network (NN)	Take one-hot representation of biological sequence as input and design NN-based embedding method by minimizing loss	[30]	✗	✗	✓	✓
AutoEncoder			[37]	✗	✗	✓	✓
String Kernel	Kernel Matrix	Designs $n \times n$ kernel matrix that can be used with kernel classifiers or with kernel PCA to get feature vector based on principal components	[19]	✓	✗	✗	✓
SeqVec	Pretrained Language Model	Takes biological sequences as input and fine-tunes the weights based on a pre-trained model to get final embedding	[22]	✓	✗	✓	✓
ProteinBERT	Pretrained Transformer	A pretrained protein sequence model to classify the given biological sequence using Transformer/Bert	[13]	✓	✗	✓	✓
BioSequence2Vec (ours)	Hashing	Takes biological sequence as input and design embeddings based on the kernel property of preserving pairwise distance	-	✓	✓	✓	✓

For the aligned Spike7k protein sequence dataset, we can observe that the proposed BioSequence2Vec with random forest classifier outperforms all the SOTA methods for all but one evaluation metric. In the case of training runtime, WDGRL performs the best because of having the smallest size embedding.

For the unaligned Human DNA (nucleotide) data, we can observe in Table 3 that the random forest classifier with BioSequence2Vec outperforms all SOTA methods in all evaluation metrics except the classification training runtime. An important point to note here is that all alignment-free methods (i.e., Spike2Vec, Spaced k-mers, String kernel, and BioSequence2Vec) generally show better predictive performance as compared to the alignment-based methods such as PWM2Vec, WDGRL, AE. Among alignment-free methods, the proposed method performs the best (hence showing the generalizability property), showing that we can completely eliminate the multiple sequence alignment from the pipeline (an NP-hard step). Moreover, using pre-trained language models such as SeqVec and ProteinBert also did not improve the predictive performance.

Table 3. Classification results (averaged over 5 runs) on **Spike7k** and **Human DNA** datasets for different evaluation metrics. Best values are shown in bold.

Embeddings	Algo.	Spike7k							Human DNA						
		Acc. ↑	Prec. ↑	Recall ↑	F1 (Weig.) ↑	F1 (Macro) ↑	ROC AUC ↑	Train Time (sec.) ↓	Acc. ↑	Prec. ↑	Recall ↑	F1 (Weig.) ↑	F1 (Macro) ↑	ROC AUC ↑	Train Time (sec.) ↓
Spike2Vec	SVM	0.855	0.853	0.855	0.843	0.689	0.843	61.112	0.597	0.602	0.597	0.589	0.563	0.733	4.612
	NB	0.476	0.716	0.476	0.535	0.459	0.726	13.292	0.175	0.143	0.175	0.106	0.128	0.532	0.039
	MLP	0.803	0.803	0.803	0.797	0.596	0.797	127.066	0.618	0.618	0.618	0.613	0.573	0.747	22.292
	KNN	0.812	0.815	0.812	0.805	0.608	0.794	15.970	0.640	0.653	0.640	0.642	0.608	0.772	0.561
	RF	0.856	0.854	0.856	0.844	0.683	0.839	21.141	0.752	0.773	0.752	0.749	0.736	0.824	2.558
	LR	0.859	0.852	0.859	0.844	0.690	0.842	64.027	0.569	0.570	0.569	0.555	0.525	0.710	2.074
	DT	0.849	0.849	0.849	0.839	0.677	0.837	4.286	0.621	0.624	0.621	0.621	0.594	0.765	0.275
PWM2Vec	SVM	0.818	0.820	0.818	0.810	0.606	0.807	22.710	0.302	0.241	0.302	0.165	0.091	0.505	10011.3
	NB	0.610	0.667	0.610	0.607	0.218	0.631	1.456	0.084	0.442	0.084	0.063	0.066	0.511	4.565
	MLP	0.812	0.792	0.812	0.794	0.530	0.770	35.197	0.310	0.350	0.310	0.175	0.107	0.510	320.555
	KNN	0.767	0.790	0.767	0.760	0.565	0.773	1.033	0.121	0.337	0.121	0.093	0.077	0.509	2.193
	RF	0.824	0.843	0.824	0.813	0.616	0.803	8.290	0.309	0.332	0.309	0.181	0.110	0.510	65.250
	LR	0.822	0.813	0.822	0.811	0.605	0.802	471.659	0.304	0.257	0.304	0.167	0.094	0.506	23.651
	DT	0.803	0.800	0.803	0.795	0.581	0.791	4.100	0.306	0.284	0.306	0.181	0.111	0.509	1.861
String Kernel	SVM	0.845	0.833	0.846	0.821	0.631	0.812	7.350	0.618	0.617	0.618	0.613	0.588	0.753	39.791
	NB	0.753	0.821	0.755	0.774	0.602	0.825	0.178	0.338	0.452	0.338	0.347	0.333	0.617	0.276
	MLP	0.831	0.829	0.838	0.823	0.624	0.818	12.652	0.597	0.595	0.597	0.593	0.549	0.737	331.068
	KNN	0.829	0.822	0.827	0.827	0.623	0.791	0.326	0.645	0.657	0.645	0.646	0.612	0.774	1.274
	RF	0.847	0.844	0.841	0.835	0.666	0.824	1.464	0.731	0.776	0.731	0.729	0.723	0.808	12.673
	LR	0.845	0.843	0.843	0.826	0.628	0.812	1.869	0.571	0.570	0.571	0.558	0.532	0.716	2.995
	DT	0.822	0.829	0.824	0.829	0.631	0.826	0.243	0.630	0.631	0.630	0.630	0.598	0.767	2.682
WDGRL	SVM	0.792	0.769	0.792	0.772	0.455	0.736	0.335	0.318	0.101	0.318	0.154	0.069	0.500	0.751
	NB	0.724	0.755	0.724	0.726	0.434	0.727	0.018	0.232	0.214	0.232	0.196	0.138	0.517	**0.004**
	MLP	0.799	0.779	0.799	0.784	0.505	0.755	7.348	0.326	0.286	0.326	0.263	0.186	0.535	8.613
	KNN	0.800	0.799	0.800	0.792	0.546	0.766	0.094	0.317	0.317	0.317	0.315	0.266	0.574	0.092
	RF	0.796	0.793	0.796	0.789	0.560	0.776	0.393	0.453	0.501	0.453	0.430	0.389	0.625	1.124
	LR	0.752	0.693	0.752	0.716	0.262	0.648	0.091	0.323	0.279	0.323	0.177	0.095	0.507	0.041
	DT	0.790	0.799	0.790	0.788	0.557	0.768	**0.009**	0.368	0.372	0.368	0.369	0.328	0.610	0.047
Spaced *k*-mers	SVM	0.852	0.841	0.852	0.836	0.678	0.840	2218.347	0.746	0.749	0.746	0.746	0.728	0.844	26.957
	NB	0.655	0.742	0.655	0.658	0.481	0.749	267.243	0.177	0.233	0.177	0.122	0.142	0.533	0.467
	MLP	0.809	0.810	0.809	0.802	0.608	0.812	2072.029	0.722	0.723	0.722	0.720	0.689	0.817	126.584
	KNN	0.821	0.810	0.821	0.805	0.591	0.788	55.140	0.699	0.704	0.699	0.698	0.667	0.804	1.407
	RF	0.851	0.842	0.851	0.834	0.665	0.833	646.557	0.784	0.814	0.784	0.782	0.773	0.843	13.397
	LR	0.855	0.848	0.855	0.840	0.682	0.840	200.477	0.712	0.712	0.712	0.709	0.693	0.812	37.756
	DT	0.853	0.850	0.853	0.841	0.685	0.842	98.089	0.656	0.658	0.656	0.656	0.626	0.784	2.985
Auto-Encoder	SVM	0.699	0.720	0.699	0.678	0.243	0.627	4018.028	0.621	0.638	0.621	0.624	0.593	0.769	22.230
	NB	0.490	0.533	0.490	0.481	0.123	0.620	24.6372	0.260	0.426	0.260	0.247	0.268	0.583	0.287
	MLP	0.663	0.633	0.663	0.632	0.161	0.589	87.4913	0.621	0.624	0.621	0.620	0.578	0.756	111.809
	KNN	0.782	0.791	0.782	0.776	0.535	0.761	24.5597	0.565	0.577	0.565	0.568	0.547	0.732	1.208
	RF	0.814	0.803	0.814	0.802	0.593	0.793	46.583	0.689	0.738	0.689	0.683	0.668	0.774	20.131
	LR	0.761	0.755	0.761	0.735	0.408	0.705	11769.02	0.692	0.700	0.692	0.693	0.672	0.799	58.369
	DT	0.803	0.792	0.803	0.792	0.546	0.779	102.185	0.543	0.546	0.543	0.543	0.515	0.718	10.616
SeqVec	SVM	0.796	0.768	0.796	0.770	0.479	0.747	1.0996	0.656	0.661	0.656	0.652	0.611	0.791	0.891
	NB	0.686	0.703	0.686	0.686	0.351	0.694	0.0146	0.324	0.445	0.312	0.295	0.282	0.624	0.036
	MLP	0.796	0.771	0.796	0.771	0.510	0.762	13.172	0.657	0.633	0.653	0.646	0.616	0.783	12.432
	KNN	0.790	0.787	0.790	0.786	0.561	0.768	0.6463	0.592	0.606	0.592	0.591	0.552	0.717	0.571
	RF	0.793	0.788	0.793	0.786	0.557	0.769	1.8241	0.713	0.724	0.701	0.702	0.693	0.752	2.164
	LR	0.785	0.763	0.785	0.761	0.459	0.740	1.7535	0.725	0.715	0.726	0.725	0.685	0.784	1.209
	DT	0.757	0.756	0.757	0.755	0.521	0.760	0.1308	0.586	0.553	0.585	0.577	0.557	0.736	0.24
Protein Bert	-	0.836	0.828	0.836	0.814	0.570	0.792	14163.52	0.542	0.580	0.542	0.514	0.447	0.675	58681.57
BioSequence2Vec (ours)	SVM	0.848	0.858	0.848	0.841	0.681	0.848	9.801	0.555	0.554	0.555	0.543	0.497	0.700	13.251
	NB	0.732	0.776	0.732	0.741	0.555	0.771	1.440	0.263	0.518	0.263	0.244	0.239	0.572	0.095
	MLP	0.835	0.825	0.835	0.825	0.622	0.819	13.893	0.583	0.598	0.583	0.571	0.541	0.717	70.463
	KNN	0.821	0.818	0.821	0.811	0.616	0.803	1.472	0.613	0.625	0.613	0.615	0.565	0.748	0.313
	RF	**0.863**	**0.867**	**0.863**	**0.854**	**0.703**	**0.851**	2.627	**0.786**	**0.816**	**0.786**	**0.787**	**0.779**	**0.846**	1.544
	LR	0.500	0.264	0.500	0.333	0.031	0.500	11.907	0.527	0.522	0.527	0.501	0.457	0.674	29.029
	DT	0.845	0.856	0.845	0.841	0.683	0.839	0.956	0.663	0.666	0.663	0.664	0.639	0.795	4.064

For ProteinBert, the main reason for its comparable performance to BioSequence2Vec on Spike7k data while bad performance on Human DNA data is because it is pretrained on protein sequences in the original study, hence performing badly on Human DNA Nucleotide data (poor generalizability). Although SeqVec is also pretrained on protein sequences (in the original study), its comparatively better performance on nucleotide data is because we use it to design the embeddings and then use ML classifiers for the prediction, which performs better for tabular data compared to DL models [31]. To check if the computed results are statistically significant, we used the student t-test and observed the p-values using average and standard deviations (SD) of 5 runs. We noted that p-values were < 0.05 in the majority of the cases (because SD values are very low), confirming the statistical significance of the results.

6 Conclusion

In this paper, we propose an efficient and alignment-free method, called BioSequence2Vec, to generate embeddings for biological sequences using the idea of hashing. We show that BioSequence2Vec has the qualities of kernel methods while being memory efficient. We performed extensive experiments on real-world biological sequence data to validate the proposed model using different evaluation metrics. BioSequence2Vec outperforms the SOTA models in terms of predictive accuracy. Future work involves evaluating BioSequence2Vec on millions of sequences and other virus data. Applying this method to other domains (*e.g.*, music or video) would also be an interesting future extension.

References

1. Ali, S.: Evaluating covid-19 sequence data using nearest-neighbors based network model. In: 2022 IEEE International Conference on Big Data (Big Data), pp. 5182–5188. Osaka, Japan (2022). https://doi.org/10.1109/BigData55660.2022.10020653
2. Ali, S., Bello, B., Chourasia, P., Punathil, R.T., Zhou, Y., Patterson, M.: PWM2Vec: an efficient embedding approach for viral host specification from coronavirus spike sequences. Biology **11**(3), 418 (2022)
3. Ali, S., Bello, B., Tayebi, Z., Patterson, M.: Characterizing sars-cov-2 spike sequences based on geographical location. J. Comput. Biol. **30**, 0391 (2023)
4. Ali, S., Murad, T., Chourasia, P., Patterson, M.: Spike2signal: classifying coronavirus spike sequences with deep learning. In: 2022 IEEE Eighth International Conference on Big Data Computing Service and Applications (BigDataService), pp. 81–88 (2022)
5. Ali, S., Patterson, M.: Spike2vec: an efficient and scalable embedding approach for COVID-19 spike sequences. In: IEEE Big Data, pp. 1533–1540 (2021)
6. Ali, S., Sahoo, B., Khan, M.A., Zelikovsky, A., Khan, I.U., Patterson, M.: Efficient approximate kernel based spike sequence classification. IEEE/ACM Transactions on Computational Biology and Bioinformatics (2022)
7. Ali, S., Sahoo, B., Ullah, N., Zelikovskiy, A., Patterson, M., Khan, I.: A k-mer based approach for sars-cov-2 variant identification. In: International Symposium on Bioinformatics Research and Applications, pp. 153–164 (2021)
8. Ali, S., Sahoo, B., Zelikovsky, A., Chen, P.Y., Patterson, M.: Benchmarking machine learning robustness in COVID-19 genome sequence classification. Sci. Rep. **13**(1), 4154 (2023)
9. Ali, S., Zhou, Y., Patterson, M.: Efficient analysis of COVID-19 clinical data using machine learning models. Med. Biol. Eng. Comput. **60**(7), 1881–1896 (2022)
10. Alon, N., Matias, Y., Szegedy, M.: The space complexity of approximating the frequency moments. In: Symposium on Theory of computing, pp. 20–29 (1996)

11. Blaisdell, B.: A measure of the similarity of sets of sequences not requiring sequence alignment. Proc. Natl. Acad. Sci. **83**, 5155–5159 (1986)
12. Borisov, V., et al.: Deep neural networks and tabular data: a survey. arXiv preprint arXiv:2110.01889 (2021)
13. Brandes, N., Ofer, D., Peleg, Y., Rappoport, N., Linial, M.: ProteinBERT: a universal deep-learning model of protein sequence and function. Bioinformatics **38**(8), 2102–2110 (2022)
14. Carter, J.L., Wegman, M.N.: Universal classes of hash functions. In: ACM symposium on Theory of computing, pp. 106–112 (1979)
15. Chourasia, P., Ali, S., Ciccolella, S., Della Vedova, G., Patterson, M.: Clustering sars-cov-2 variants from raw high-throughput sequencing reads data. In: Computational Advances in Bio and Medical Sciences (ICCABS), pp. 133–148 (2022)
16. Chourasia, P., Ali, S., Patterson, M.: Informative initialization and kernel selection improves t-SNE for biological sequences. In: 2022 IEEE International Conference on Big Data (Big Data), pp. 101–106. Osaka, Japan (2022). https://doi.org/10.1109/BigData55660.2022.10020217
17. Chowdhury, B., Garai, G.: A review on multiple sequence alignment from the perspective of genetic algorithm. Genomics **109**(5–6), 419–431 (2017)
18. Cristianini, N., Shawe-Taylor, J., et al.: An introduction to support vector machines and other Kernel-based learning methods. Cambridge University Press (2000)
19. Farhan, M., Tariq, J., Zaman, A., Shabbir, M., Khan, I.U.: Efficient approximation algorithms for strings Kernel based sequence classification. In: NeurIPS, pp. 6935–6945 (2017)
20. Ghandi, M., Noori, M., Beer, M.: Robust k k-mer frequency estimation using gapped k-mers. J. Math. Biol. **69**(2), 469–500 (2014)
21. GISAID. https://www.gisaid.org/ (2022). Accessed 04 Dec 2022
22. Heinzinger, M., et al.: Modeling aspects of the language of life through transfer-learning protein sequences. BMC Bioinform. **20**(1), 1–17 (2019)
23. Hoffmann, H.: Kernel PCA for novelty detection. Pattern Recogn. **40**(3), 863–874 (2007)
24. Hu, W., Bansal, R., Cao, K., Rao, N., Subbian, K., Leskovec, J.: Learning backward compatible embeddings. arXiv preprint arXiv:2206.03040 (2022)
25. Human DNA. https://www.kaggle.com/code/nageshsingh/demystify-dna-sequencing-with-machine-learning/data. Accessed 10 Oct 2022
26. Jumper, J., et al.: Highly accurate protein structure prediction with AlphaFold. Nature **596**(7873), 583–589 (2021)
27. Kuzmin, K., et al.: Machine learning methods accurately predict host specificity of coronaviruses based on spike sequences alone. Biochem. Biophys. Res. Comm. **533**(3), 553–558 (2020)
28. O'Toole, A., et al.: Assignment of epidemiological lineages in an emerging pandemic using the pangolin tool. Virus Evol. **7**(2), veab064 (2021)
29. Rambaut, A., et al.: A dynamic nomenclature proposal for SARS-CoV-2 lineages to assist genomic epidemiology. Nature Microbiol. **5**, 1403–1407 (2020)
30. Shen, J., Qu, Y., Zhang, W., Yu, Y.: Wasserstein distance guided representation learning for domain adaptation. In: AAAI conference on A.I (2018)
31. Shwartz-Ziv, R., Armon, A.: Tabular data: deep learning is not all you need. Inf. Fusion **81**, 84–90 (2022)
32. Singh, R., Sekhon, A., et al.: GakCo: a fast gapped k-mer string kernel using counting. In: Joint ECML and Knowledge Discovery in Databases, pp. 356–373 (2017)

33. Stephens, Z.D., et al.: Big data: astronomical or genomical? PLoS Biol. **13**, e1002195 (2015)
34. Tayebi, Z., Ali, S., Patterson, M.: Robust representation and efficient feature selection allows for effective clustering of SARS-CoV-2 variants. Algorithms **14**(12), 348 (2021)
35. Ullah, A., Ali, S., Khan, I., Khan, M.A., Faizullah, S.: Effect of analysis window and feature selection on classification of hand movements using EMG signal. In: SAI Intelligent Systems Conference (IntelliSys), pp. 400–415 (2020)
36. Wang, Z., Yan, W., Oates, T.: Time series classification from scratch with deep neural networks: a strong baseline. In: IJCNN, pp. 1578–1585 (2017)
37. Xie, J., Girshick, R., Farhadi, A.: Unsupervised deep embedding for clustering analysis. In: International Conference on Machine Learning, pp. 478–487 (2016)

Matrices and Tensors

Relations Between Adjacency and Modularity Graph Partitioning

Hansi Jiang[1(✉)] and Carl Meyer[2]

[1] IoT Division, SAS Institute Inc., Cary, NC 27513, USA
Hansi.Jiang@sas.com
[2] Department of Mathematics, North Carolina State University, Raleigh, NC 27695, USA
meyer@ncsu.edu

Abstract. This paper develops the exact linear relationship between the leading eigenvector of the unnormalized modularity matrix and the eigenvectors of the adjacency matrix. We propose a method for approximating the leading eigenvector of the modularity matrix, and we derive the error of the approximation. There is also a complete proof of the equivalence between normalized adjacency clustering and normalized modularity clustering. Numerical experiments show that normalized adjacency clustering can be as twice efficient as normalized modulairty clustering.

Keywords: Spectral clustering · Graph partitioning · Adjacency matrix · Modularity matrix

1 Introduction

Graph partitioning is the process of breaking a graph into smaller components so the components can be characterized by specific properties. The problem, also known as clustering or community detection, is of high interest in both academia and industry. For example, Pothen [14] applies graph partitioning in scientific computing. Olson et al. [13] uses the concept of robotics. Tolliver and Miller [17] discusses the possibility of using graph partitioning for image segmentation. Recently, the scientific interest in graph partitioning has centered on dividing large graphs into smaller components by matching their size. This is done by minimizing the number of edges that are cut during the process [18].

A number of algorithms have been developed to solve problems related to graph partitioning. Among the many clustering methods, two spectral techniques that rely on adjacency matrices of graphs are widely used and extensively researched. Fiedler [5] develops the spectral clustering method, while Newman and Girvan [11] develop the modularity clustering method. As discussed in [5], the eigenvalue corresponding to the second smallest eigenvector of a graph adjacency matrix is closely related to the graph's structure. It is suggested in [6] to partition a graph based on the signs of eigenvector entries of its adjacency matrix. Newman [10] describes modularity clustering in detail. As with Fiedler's

H. Kashima et al. (Eds.): PAKDD 2023, LNAI 13936, pp. 189–200, 2023.
https://doi.org/10.1007/978-3-031-33377-4_15

spectral clustering method, the modularity clustering method uses entries in the eigenvector that correspond to a modularity matrix's eigenvalue.

There are some modified versions of the spectral clustering and modularity clustering methods. Chung [4] analyzes the properties of scaled Laplacian matrices. By utilizing normalized spectral clustering, Shi and Malik [16] provides a method to develop normalized Laplacian matrices and use them to segment images. In [12], another version of normalized spectral clustering is discussed. The Laplacian matrix is scaled on one side by the researchers in their method. In [1], a normalized version of modularity clustering is examined.

Since modularity matrices are derived from adjacency matrices, it would be interesting to see if similar clustering results can be obtained from the two kinds of matrices. One main contribution of this paper is to describe the relation between clustering results using modularity matrices and adjacency matrices, and to show that using normalized modularity matrices and normalized adjacency matrices will produce the same clustering results. As a practical motivation, this paper demonstrates that clustering can be sped up by using normalized adjacency matrices rather than normalized modularity matrices.

As follows is the organization of the paper. Section 2 contains some preliminary mathematical notations. Section 3 describes how to approximate the leading eigenvector of the modularity matrix with eigenvectors of the adjacency matrix. The equivalence between normalized adjacency clustering and normalized modularity clustering is presented in Sect. 4. Section 5 provides experimental results and discussions. Section 6 contains the conclusions.

2 Preliminaries

Throughout the paper, we assume $G(V, E)$ to be a connected simple graph with $m = |E|$ edges and $n = |V|$ vertices. Unless otherwise stated, $\mathbf{A}$ is assumed to represent an adjacency matrix, i.e.

$$\mathbf{A}_{ij} = \begin{cases} 1 \text{ if nodes } i \text{ and } j \text{ are adjacent} \\ 0 \text{ if otherwise.} \end{cases}$$

A vertex's degree is defined as

$$d_i = \sum_{i=1}^{n} a_{ji},$$

and

$$\mathbf{D} = \text{diag}(d_1, d_2, \cdots, d_n)$$

is a degree matrix containing the degrees of the vertices in a graph. In this paper, the number of clusters is always fixed at two. Clustering methods can be applied recursively if more clusters are needed, in which case a hierarchy is built to get the desired number of clusters. It is worth noting that this approach will result in a greedy algorithm which may lead to unsatisfactory results because of poor partitioning in the first stages.

Partitioning the graph is based on the signs of the entries in the eigenvectors. In real cases, the cases where zero entries emerge are rare, so it is assumed that there are no zero entries in the eigenvectors. Although the results are presented in this paper using adjacency matrices, it is also possible to extend the results to use similarity matrices. A graph Laplacian is defined as

$$\mathbf{L} = \mathbf{D} - \mathbf{A}, \tag{2.1}$$

and a modularity matrix defined as

$$\mathbf{M} = \mathbf{A} - \frac{\mathbf{d}\mathbf{d}^T}{2m}, \tag{2.2}$$

where

$$\mathbf{d} = \begin{pmatrix} d_1 \; d_2 \; \cdots \; d_n \end{pmatrix}^T \tag{2.3}$$

is the vector containing the degrees of the nodes. The normalized versions of the graph Laplacian and the modularity matrix are

$$\mathbf{L}_{sym} = \mathbf{D}^{-\frac{1}{2}}\mathbf{L}\mathbf{D}^{-\frac{1}{2}} \tag{2.4}$$

and

$$\mathbf{M}_{sym} = \mathbf{D}^{-\frac{1}{2}}\mathbf{M}\mathbf{D}^{-\frac{1}{2}}, \tag{2.5}$$

respectively. With $\mathbf{e}$ a vector that contains all 1's with proper dimension, it can be seen that $(0, \mathbf{e})$ is an eigenpair of $\mathbf{L}$ and $\mathbf{M}$, and $(0, \mathbf{D}^{\frac{1}{2}}\mathbf{e})$ is an eigenpair of $\mathbf{L}_{sym}$ and $\mathbf{M}_{sym}$.

3 Dominant Eigenvectors of Modularity and Adjacency Matrices

As a linear combination of the eigenvectors of the corresponding adjacency matrix, the eigenvector corresponding to the largest eigenvalue of a modularity matrix is written in this section. To begin with, we state a theorem from [2] regarding the interlacing property of a diagonal matrix and its rank-one modification, and how to calculate the eigenvectors of a diagonal plus rank one (DPR1) matrix [9]. The theorem is also discussed in [19]. We will use these results in our analysis.

Theorem 1. *Let* $\mathbf{P} = \mathbf{S} + \alpha\mathbf{u}\mathbf{u}^T$*, where* $\mathbf{S}$ *is diagonal,* $\|\mathbf{u}\|_2 = 1$*. Let* $s_1 \leq s_2 \leq \cdots \leq s_n$ *be the eigenvalues of* $\mathbf{S}$*, and let* $\tilde{s}_1 \leq \tilde{s}_2 \leq \cdots \leq \tilde{s}_n$ *be the eigenvalues of* $\mathbf{P}$*. Then* $\tilde{s}_1 \leq s_1 \leq \tilde{s}_2 \leq s_2 \leq \cdots \leq \tilde{s}_n \leq s_n$ *if* $\alpha < 0$*. If the* s_i *are distinct and all the elements of* $\mathbf{u}$ *are nonzero, then the eigenvalues of* $\mathbf{P}$ *strictly separate those of* $\mathbf{S}$*.*

Corollary 1. *By using the notations in Theorem 1, the eigenvector of* $\mathbf{P}$ *associated with eigenvalue* $\tilde{s}_i$ *can be calculated by*

$$(\mathbf{S} - \tilde{s}_i\mathbf{I})^{-1}\mathbf{u}. \tag{3.1}$$

By Theorem 1, we know that the eigenvalues of a DPR1 matrix interlace with the eigenvalues of the original diagonal matrix. A linear combination of the eigenvectors of the corresponding adjacency matrix is then used to compute the eigenvector representing the largest eigenvalue of a modularity matrix.

According to the notation in Sect. 1, because $\mathbf{A}$ is an adjacency matrix, it is symmetric and is therefore orthogonally similar to a diagonal matrix. It follows that there exists an orthogonal matrix $\mathbf{U}$ and a diagonal matrix $\mathbf{\Sigma_A}$ such that

$$\mathbf{A} = \mathbf{U}\mathbf{\Sigma_A}\mathbf{U}^T.$$

Suppose the rows and columns of $\mathbf{A}$ are ordered such that

$$\mathbf{\Sigma_A} = \text{diag}(\sigma_1, \sigma_2, \cdots, \sigma_n),$$

where $\sigma_1 \geq \sigma_2 \geq \cdots \geq \sigma_n$. Let $\mathbf{U} = \begin{pmatrix}\mathbf{u}_1 \ \mathbf{u}_2 \cdots \mathbf{u}_n\end{pmatrix}$. Similarly, since a modularity matrix $\mathbf{M}$ is symmetric, it is orthogonally similar to a diagonal matrix. Suppose the eigenvalues of $\mathbf{M}$ are $\beta_1 \geq \beta_2 \geq \cdots \geq \beta_n$.

Theorem 2. *Suppose $\beta_1 \neq \sigma_1$, $\beta_1 \neq \sigma_2$, and $|\beta_1 - \sigma_2| = \Delta$. The eigenvector corresponding to the largest eigenvalue of $\mathbf{M}$ is given by*

$$\frac{1}{\|\mathbf{U}^T\mathbf{d}\|_2} \sum_{i=1}^{n} \frac{\mathbf{u}_i^T\mathbf{d}}{\sigma_i - (\sigma_2 + \Delta)} \mathbf{u}_i, \tag{3.2}$$

where $\mathbf{d}$ is defined in Eq. 2.3.

Proof. Since $\mathbf{M} = \mathbf{A} - \mathbf{d}\mathbf{d}^T/(2m)$, we have

$$\begin{aligned}\mathbf{M} &= \mathbf{A} - \frac{\mathbf{d}\mathbf{d}^T}{2m} \\ &= \mathbf{U}\mathbf{\Sigma_A}\mathbf{U}^T - \frac{\mathbf{d}\mathbf{d}^T}{2m} \\ &= \mathbf{U}(\mathbf{\Sigma_A} + \rho\mathbf{y}\mathbf{y}^T)\mathbf{U}^T,\end{aligned} \tag{3.3}$$

where

$$\mathbf{y} = \frac{\mathbf{U}^T\mathbf{d}}{\|\mathbf{U}^T\mathbf{d}\|_2}$$

and

$$\rho = -\frac{\|\mathbf{U}^T\mathbf{d}\|_2^2}{2m}.$$

Since $\mathbf{\Sigma_A} + \rho\mathbf{y}\mathbf{y}^T$ is also symmetric, it is orthogonally similar to a diagonal matrix. So we have

$$\mathbf{M} = \mathbf{U}\mathbf{V}\mathbf{\Sigma_M}\mathbf{V}^T\mathbf{U}^T,$$

where $\mathbf{V}$ is orthogonal and $\mathbf{\Sigma_M}$ is diagonal. Since $\mathbf{\Sigma_A} + \rho\mathbf{y}\mathbf{y}^T$ is a DPR1 matrix, $\rho < 0$ and $\|\mathbf{y}\|_2 = 1$, the interlacing theorem applies to the eigenvalues of $\mathbf{A}$ and $\mathbf{M}$. More specifically, we have

$$\beta_n \leq \sigma_n \leq \beta_{n-1} \leq \sigma_{n-1} \leq \cdots \leq \beta_2 \leq \sigma_2 < \beta_1 < \sigma_1. \tag{3.4}$$

The strict inequalities hold because $\beta_1 \neq \sigma_1$ and $\beta_1 \neq \sigma_2$. Thus $|\beta_1 - \sigma_2| = \Delta$ implies $\beta_1 - \sigma_2 = \Delta$. Next, let

$$\mathbf{M}_1 = \mathbf{\Sigma}_\mathbf{A} + \rho \mathbf{y}\mathbf{y}^T.$$

Since $\mathbf{M} = \mathbf{U}\mathbf{M}_1\mathbf{U}^T$, we have $\mathbf{M}\mathbf{U} = \mathbf{U}\mathbf{M}_1$. Suppose $(\lambda, \mathbf{v})$ is an eigenpair of $\mathbf{M}_1$, then

$$\mathbf{M}\mathbf{U}\mathbf{v} = \mathbf{U}\mathbf{M}_1\mathbf{v} = \lambda\mathbf{U}\mathbf{v}$$

implies that $(\lambda, \mathbf{v})$ is an eigenpair of $\mathbf{M}_1$ if and only if $(\lambda, \mathbf{U}\mathbf{v})$ is an eigenpair of $\mathbf{M}$. By Corollary 1, the eigenvector of $\mathbf{M}_1$ corresponding to β_1 is given by

$$\begin{aligned} \mathbf{v}_1 &= (\mathbf{\Sigma}_\mathbf{A} - \beta_1\mathbf{I})^{-1}\mathbf{y} \\ &= (\mathbf{\Sigma}_\mathbf{A} - (\sigma_2 + \Delta)\mathbf{I})^{-1}\frac{\mathbf{U}^T\mathbf{d}}{\|\mathbf{U}^T\mathbf{d}\|_2}, \end{aligned} \tag{3.5}$$

and hence the eigenvector of $\mathbf{M}$ corresponding to β_1 is given by

$$\begin{aligned} \mathbf{m}_1 &= \mathbf{U}\mathbf{v}_1 \\ &= \mathbf{U}(\mathbf{\Sigma}_\mathbf{A} - (\sigma_2 + \Delta)\mathbf{I})^{-1}\frac{\mathbf{U}^T\mathbf{d}}{\|\mathbf{U}^T\mathbf{d}\|_2} \\ &= \frac{1}{\|\mathbf{U}^T\mathbf{d}\|_2}\sum_{i=1}^{n}\frac{\mathbf{u}_i^T\mathbf{d}}{\sigma_i - (\sigma_2 + \Delta)}\mathbf{u}_i. \end{aligned} \tag{3.6}$$

The aim of Theorem 2 is to demonstrate that the vector $\mathbf{b}_1$ is a linear combination of the $\mathbf{u}_i$. Let

$$\gamma_i = \frac{\mathbf{u}_i^T\mathbf{d}}{(\sigma_i - \beta_1)\|\mathbf{U}^T\mathbf{d}\|_2}, \tag{3.7}$$

the next theorem is intended to approximate $\mathbf{m}_1$, the eigenvector corresponding to the largest eigenvalue of $\mathbf{M}$, by a linear combination of $\mathbf{u}_i$ that has the largest $|\gamma_i|$, and to measure how good the approximation is by calculating the norm between $\mathbf{m}_1$ and its approximation.

Theorem 3. *With the notations and assumptions in Theorem 2 , and let γ_i has the expression in Eq. 3.7. Suppose $i_k \in \{1, 2, \cdots, n\}$, and γ_i are reordered such that*

$$|\gamma_{i_n}| \leq |\gamma_{i_{n-1}}| \leq \cdots \leq |\gamma_{i_1}|.$$

Then given $p \in \{1, 2, \cdots, n\}$, $\mathbf{m}_1$ can be approximated by

$$\hat{\mathbf{m}}_1 = \sum_{j=1}^{p}\gamma_{i_j}\mathbf{u}_{i_j},$$

with relative error

$$\frac{1}{q}\Big(\sum_{j=p+1}^{n}\gamma_{i_j}^2\Big)^{\frac{1}{2}},$$

where q is the 2-norm of the vector $\mathbf{m}_1$.

Proof. Since

$$\gamma_i = \frac{\mathbf{u}_i^T \mathbf{d}}{(\sigma_i - \beta_1)\|\mathbf{U}^T \mathbf{d}\|_2},$$

the vector $\mathbf{m}_1$ can be written as

$$\mathbf{m}_1 = \sum_{i=1}^{n} \gamma_i \mathbf{u}_i = \sum_{j=1}^{n} \gamma_{i_j} \mathbf{u}_{i_j}.$$

So if

$$\hat{\mathbf{m}}_1 = \sum_{j=1}^{p} \gamma_{i_j} \mathbf{u}_{i_j}, p \leq n$$

is an approximation of $\mathbf{m}_1$, then the difference between $\mathbf{m}_1$ and its approximation is

$$\mathbf{m}_1 - \hat{\mathbf{m}}_1 = \sum_{j=p+1}^{n} \gamma_{i_j} \mathbf{u}_{i_j},$$

and the 2-norm of $\mathbf{m}_1 - \hat{\mathbf{m}}_1$ is

$$\|\mathbf{m}_1 - \hat{\mathbf{m}}_1\|_2 = \left\| \sum_{j=p+1}^{n} \gamma_{i_j} \mathbf{u}_{i_j} \right\|_2 = \Big(\sum_{j=p+1}^{n} \gamma_{i_j}^2 \Big)^{\frac{1}{2}},$$

because the $\mathbf{u}_i$ are orthonormal. So if q is the 2-norm of the vector $\mathbf{m}_1$, then the relative error of the approximation is

$$\frac{\|\mathbf{m}_1 - \hat{\mathbf{m}}_1\|_2}{\|\mathbf{m}_1\|} = \frac{1}{q} \Big(\sum_{j=p+1}^{n} \gamma_{i_j}^2 \Big)^{\frac{1}{2}}.$$

We can use the error provided in Theorem 3 to gauge the number of terms we will need to approximate the dominant eigenvector of the modularity matrix with eigenvectors of the adjacency matrix to achieve a given level of accuracy.

4 Normalized Adjacency and Modularity Clustering

In parallel to the previous analysis, we will show that the eigenvectors corresponding to the largest eigenvalues of a normalized adjacency matrix and a normalized modularity matrix will produce the same clustering results. Bolla [1] mentions a similar statement without a complete proof, but Yu and Ding [20] consider it from a different angle.

Suppose $\mathbf{A}$ is an adjacency matrix, and

$$\mathbf{A}_{sym} = \mathbf{D}^{-\frac{1}{2}} \mathbf{A} \mathbf{D}^{-\frac{1}{2}}$$

is the corresponding normalized adjacency matrix. Let

$$\mathbf{L} = \mathbf{D} - \mathbf{A}$$

be the unnormalized Laplacian matrix and

$$\mathbf{L}_{sym} = \mathbf{D}^{-\frac{1}{2}}\mathbf{L}\mathbf{D}^{-\frac{1}{2}} = \mathbf{I} - \mathbf{A}_{sym}$$

be the normalized Laplacian matrix. Finally let $\mathbf{M}$ be the unnormalized modularity matrix defined in Sect. 1,

$$\mathbf{P} = \frac{\mathbf{d}\mathbf{d}^T}{2m},$$

and

$$\mathbf{M}_{sym} = \mathbf{D}^{-\frac{1}{2}}\mathbf{M}\mathbf{D}^{-\frac{1}{2}}$$

be the normalized modularity matrix. A theorem is first stated, followed by its proof.

Theorem 4. *Suppose that zero is a simple eigenvalue of* $\mathbf{M}_{sym}$*, and one is a simple eigenvalue of* $\mathbf{A}_{sym}$*. If* $\lambda \neq 0$ *and* $\lambda \neq 1$*, then* $(\lambda, \mathbf{u})$ *is an eigenpair of* $\mathbf{A}_{sym}$ *if and only if* $(\lambda, \mathbf{u})$ *is an eigenpair of* $\mathbf{M}_{sym}$*.*

This theorem may be proven by combining the following two observations. As the second observation requires more lines of explanation, we write it as a lemma.

Observation 5 $(\lambda, \mathbf{u})$ *is an eigenpair of* $\mathbf{L}_{sym}$ *if and only if* $(1 - \lambda, \mathbf{u})$ *is an eigenpair of* $\mathbf{A}_{sym}$ *because*

$$\begin{aligned} &\mathbf{L}_{sym}\mathbf{u} = \lambda\mathbf{u} \\ \iff &(\mathbf{I} - \mathbf{A}_{sym})\mathbf{u} = \lambda\mathbf{u} \\ \iff &\mathbf{A}_{sym}\mathbf{u} = (1 - \lambda)\mathbf{u}. \end{aligned}$$

Lemma 1. *Suppose that* 0 *is a simple eigenvalue of both* $\mathbf{L}_{sym}$ *and* $\mathbf{M}_{sym}$*. It follows that if* $\lambda \neq 0$ *and* $(\lambda, \mathbf{u})$ *is an eigenpair of* $\mathbf{L}_{sym}$*, then* $(1 - \lambda, \mathbf{u})$ *is an eigenpair of* $\mathbf{M}_{sym}$*. If* $\alpha \neq 0$ *and* $(\alpha, \mathbf{v})$ *is an eigenpair of* $\mathbf{M}_{sym}$*, then* $(1 - \alpha, \mathbf{v})$ *is an eigenpair of* $\mathbf{L}_{sym}$*.*

Proof. For $\mathbf{P} = \mathbf{d}\mathbf{d}^T/(2m)$, it is easy to observe that

$$\begin{aligned} \mathbf{M}_{sym} + \mathbf{L}_{sym} &= \mathbf{D}^{-\frac{1}{2}}(\mathbf{A} - \mathbf{P} + \mathbf{D} - \mathbf{A})\mathbf{D}^{-\frac{1}{2}} \\ &= \mathbf{I} - \mathbf{D}^{-\frac{1}{2}}\mathbf{P}\mathbf{D}^{-\frac{1}{2}}. \end{aligned} \tag{4.1}$$

Let $\mathbf{E} = \mathbf{D}^{-\frac{1}{2}}\mathbf{P}\mathbf{D}^{-\frac{1}{2}}$. If $(\lambda, \mathbf{u})$ is an eigenpair of $\mathbf{L}_{sym}$, we have

$$\begin{aligned} &\lambda\mathbf{u} = \mathbf{L}_{sym}\mathbf{u} \\ \Longrightarrow &\lambda\mathbf{u} = (\mathbf{I} - \mathbf{M}_{sym} - \mathbf{E})\mathbf{u} \\ \Longrightarrow &(1 - \lambda)\mathbf{u} = \mathbf{M}_{sym}\mathbf{u} + \mathbf{E}\mathbf{u}. \end{aligned}$$

Note that $\mathbf{P}$ is an outer product and $\mathbf{P} \neq \mathbf{0}$, so rank($\mathbf{P}$)=1. Since $\mathbf{E}$ is congruent to $\mathbf{P}$, $\mathbf{E}$ and $\mathbf{P}$ have the same number of positive, negative and zero eigenvalues by Sylvester's law [9]. Therefore

$$\text{rank}(\mathbf{E}) = \text{rank}(\mathbf{P}) = 1.$$

To prove $\mathbf{Eu} = \mathbf{0}$, it is sufficient to prove $\mathbf{u}$ is in the nullspace of $\mathbf{E}$. Let $\mathbf{e}$ be the vector such that all its entries are one. Observe that

$$\begin{aligned}\mathbf{E} \cdot \mathbf{D}^{\frac{1}{2}}\mathbf{e} &= \mathbf{D}^{-\frac{1}{2}}\mathbf{P}\mathbf{D}^{-\frac{1}{2}}\mathbf{D}^{\frac{1}{2}}\mathbf{e} \\ &= \mathbf{D}^{-\frac{1}{2}}\frac{\mathbf{d}\mathbf{d}^T}{2m}\mathbf{e} \\ &= \frac{\mathbf{d}^T\mathbf{e}}{2m}(\mathbf{D}^{-\frac{1}{2}}\mathbf{d}) \\ &= \mathbf{D}^{-\frac{1}{2}}\mathbf{d},\end{aligned} \tag{4.2}$$

because

$$\mathbf{d}^T\mathbf{e} = \sum_{i=1}^{n} d_i = 2m$$

is the sum of the degrees of all the nodes in the graph. Moreover, because

$$\mathbf{D}^{-\frac{1}{2}}\mathbf{d} = \mathbf{D}^{\frac{1}{2}}\mathbf{e},$$

$(1, \mathbf{D}^{\frac{1}{2}}\mathbf{e})$ is an eigenpair of $\mathbf{E}$. Also observe that

$$\begin{aligned}\mathbf{L}_{sym} \cdot \mathbf{D}^{\frac{1}{2}}\mathbf{e} &= \mathbf{D}^{-\frac{1}{2}}(\mathbf{D} - \mathbf{A})\mathbf{D}^{-\frac{1}{2}}\mathbf{D}^{\frac{1}{2}}\mathbf{e} \\ &= \mathbf{D}^{-\frac{1}{2}}\mathbf{L}\mathbf{e} = \mathbf{0}.\end{aligned} \tag{4.3}$$

Therefore, $(0, \mathbf{D}^{\frac{1}{2}}\mathbf{e})$ is an eigenpair of $\mathbf{L}_{sym}$. Since $\mathbf{u}$ is an eigenvector of $\mathbf{L}_{sym}$ corresponding to a nonzero eigenvalue λ, we have $\mathbf{u} \perp \mathbf{D}^{\frac{1}{2}}\mathbf{e}$, so $\mathbf{u}$ is in the nullspace of $\mathbf{E}$. This gives $\mathbf{Eu} = \mathbf{0}$ and thus $(1 - \lambda)\mathbf{u} = \mathbf{M}_{sym}\mathbf{u}$. Therefore $\lambda\mathbf{u} = \mathbf{L}_{sym}\mathbf{u} \Rightarrow (1 - \lambda)\mathbf{u} = \mathbf{M}_{sym}\mathbf{u}$.

On the other hand, if $(\alpha, \mathbf{v})$ is an eigenpair of $\mathbf{M}_{sym}$, then we have

$$\begin{aligned}&\alpha\mathbf{v} = \mathbf{M}_{sym}\mathbf{v} \\ \Longrightarrow\ &\alpha\mathbf{v} = (\mathbf{I} - \mathbf{L}_{sym} - \mathbf{E})\mathbf{v} \\ \Longrightarrow\ &\mathbf{L}_{sym}\mathbf{v} + \mathbf{E}\mathbf{v} = (1 - \alpha)\mathbf{v}.\end{aligned}$$

Observe that

$$\begin{aligned}\mathbf{M}_{sym} \cdot \mathbf{D}^{\frac{1}{2}}\mathbf{e} &= \mathbf{D}^{-\frac{1}{2}}\mathbf{M}\mathbf{D}^{-\frac{1}{2}}\mathbf{D}^{\frac{1}{2}}\mathbf{e} \\ &= \mathbf{D}^{-\frac{1}{2}}\mathbf{M}\mathbf{e} = \mathbf{0}\end{aligned} \tag{4.4}$$

because the row sums of $\mathbf{M}$ are all zeros. Therefore, $(0, \mathbf{D}^{\frac{1}{2}}\mathbf{e})$ is an eigenpair of $\mathbf{M}_{sym}$. Since $\mathbf{v}$ is an eigenvector of $\mathbf{M}_{sym}$ corresponding to a nonzero eigenvalue α, we have $\mathbf{v} \perp \mathbf{D}^{\frac{1}{2}}\mathbf{e}$, so $\mathbf{v}$ is in the nullspace of $\mathbf{E}$. This gives $\mathbf{Ev} = \mathbf{0}$ and thus $(1 - \alpha)\mathbf{v} = \mathbf{L}_{sym}\mathbf{v}$. Therefore $\alpha\mathbf{v} = \mathbf{M}_{sym}\mathbf{v} \Rightarrow (1 - \alpha)\mathbf{v} = \mathbf{L}_{sym}\mathbf{v}$.

As a result of Theorem 4, a bijection from the nonzero eigenvalues of $\mathbf{M}_{sym}$ to the nonzero eigenvalues of $\mathbf{A}_{sym}$ can be established, and the order of these eigenvalues is maintained. As zero is always an eigenvalue of $\mathbf{M}_{sym}$, the largest eigenvalue of $\mathbf{B}_{sym}$ is always nonnegative. Newman [10] discusses when $\mathbf{B}$ can have a zero largest eigenvalue. The congruence of $\mathbf{M}_{sym}$ and $\mathbf{M}$ logically implies that if zero is the largest Eigenvalue for $\mathbf{M}$, then it is also the largest Eigenvalue for $\mathbf{B}_{sym}$. Since $(0, \mathbf{D}^{\frac{1}{2}}\mathbf{e})$ is an eigenpair of $\mathbf{M}_{sym}$ and all entries in the vector $\mathbf{D}^{\frac{1}{2}}\mathbf{e}$ are greater than zero, all nodes in the graph will be put into one cluster. We prove below that, for nontrivial cases (i.e. when the largest eigenvalue of $\mathbf{M}$ is not zero), the eigenvectors for the largest eigenvalues of both a normalized adjacency matrix and a normalized modularity matrix are the same, so in nontrivial cases they will give the same clustering results.

Theorem 6. *With the assumptions in Theorem 4, and given that zero is not the largest eigenvalue of* $\mathbf{M}_{sym}$*, the eigenvector corresponding to the largest eigenvalue of* $\mathbf{M}_{sym}$ *and the eigenvector corresponding to the second largest eigenvalue of* $\mathbf{A}_{sym}$ *are identical.*

Proof. Due to the fact that $\mathbf{L}_{sym}$ is positive semi-definite [18], zero is the smallest eigenvalue of $\mathbf{L}_{sym}$. Then by Observation 5, one is the largest eigenvalue of $\mathbf{A}_{sym}$. Since all eigenvalues of $\mathbf{A}_{sym}$ that are not equal to one are also the eigenvalues of $\mathbf{M}_{sym}$, it follows that if the simple zero eigenvalue is not the largest eigenvalue of $\mathbf{M}_{sym}$, then the largest eigenvalue of $\mathbf{A}_{sym}$ is the second largest eigenvalue of $\mathbf{M}_{sym}$ and they have the same eigenvectors by Theorem 4.

Both adjacency clustering and modularity clustering involve calculation of all entries in the adjacency matrices, so they have the same time complexity of $\mathcal{O}(n^2)$. However, as shown in the next section, normalized adjacency clustering can be twice as effective as normalized modularity clustering.

5 Experiments

In this section, synthetic and practical data sets are used to corroborate the theoretical findings presented in the previous sections. Since normalized adjacency clustering and normalized modularity clustering provides the same eigenvalues and eigenvectors, only efficiency is compared in the experiments.

5.1 Synthetic Data Sets

Synthetic data sets with observations from 100 to $10,000$ are created, and for each of the data sets, the number of features is 10. The experimental results are shown in Fig. 1.

From Fig. 1, it can be seen that normalized adjacency clustering (the blue line) is about twice efficient as normalized modularity clustering (the orange line).

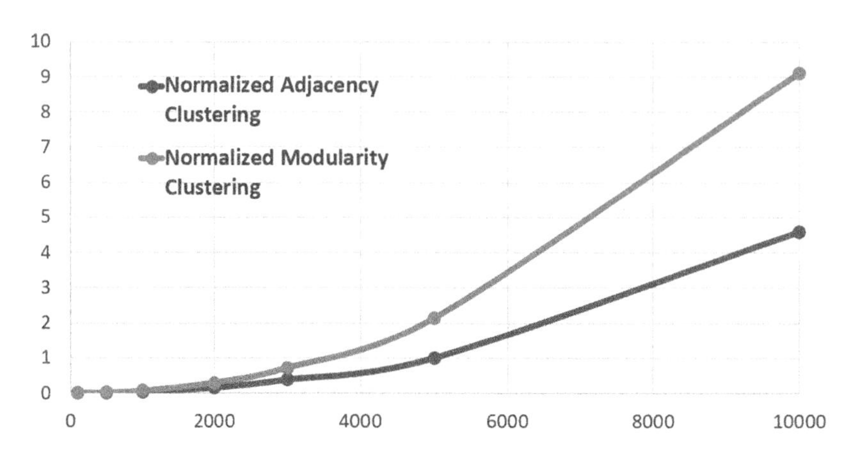

Fig. 1. The plot of run-time recordings of normalized adjacency clustering and normalized modularity clustering. X-axis is the number of observations, and y-axis is run-time in seconds.

5.2 PenDigit Data Sets from MNIST Database

The PenDigit database is a subset of the MNIST data set [3,7,8,15,21]. A training set of 60,000 handwritten digits from 44 writers is contained in the original data. Each data point is a row vector derived from a grayscale image. The images each have 28 pixels in height and 28 pixels in width, which makes 784 pixels in total. The row vectors contain the label of each digit as well as the lightness of each pixel. A pixel's lightness is represented by a number between 0 and 255 inclusively, with smaller numbers representing lighter pixels. The experiments were conducted using three subsets consisting of 1&7, 2&3, and 5&6. The experimental results are listed in Table 1.

Table 1. The plot of run-time recordings (in seconds) of normalized adjacency clustering and normalized modularity clustering on subsets of MNIST data set

Data	#data points	$\mathbf{A}_{sym}$	$\mathbf{M}_{sym}$
Digit1&7	9085	4.0920	9.1306
Digit2&3	8528	3.5197	7.0120
Digit5&6	7932	3.0505	6.5147

From Table 1, it can be seen that the experimental results from real data sets are similar to the ones from synthetic data sets in that normalized adjacency clustering as around twice efficient as normalized modularity clustering.

6 Conclusion

In this article, the exact linear relationship between the leading eigenvector of the unnormalized modularity matrix and the eigenvectors of the adjacency matrix is established. This paper demonstrates that the leading eigenvector of a modularity matrix can be written as a linear combination of the eigenvectors of an adjacency matrix, and the coefficients in the linear combination are deduced. An approximation method for the leading eigenvector of the modularity matrix is then given, along with a calculated relative error. Additionally, when the largest eigenvalue of the modularity matrix is nonzero, the normalized modularity clustering method will give the same results as using the eigenvector corresponding to the smallest eigenvalue of the normalized adjacency matrix. Experimental results indicate that using normalized adjacency clustering can be as twice efficient as normalized modularity clustering.

References

1. Bolla, M.: Penalized versions of the Newman-Girvan modularity and their relation to normalized cuts and k-means clustering. Phys. Rev. E **84**(1), 016108 (2011)
2. Bunch, J.R., Nielsen, C.P., Sorensen, D.C.: Rank-one modification of the symmetric eigenproblem. Numer. Math. **31**(1), 31–48 (1978)
3. Chitta, R., Jin, R., Jain, A.K.: Efficient kernel clustering using random Fourier features. In: 2012 IEEE 12th International Conference on Data Mining (ICDM), pp. 161–170. IEEE (2012)
4. Chung, F.R.: Spectral graph theory, vol. 92. American Mathematical Soc. (1997)
5. Fiedler, M.: Algebraic connectivity of graphs. Czechoslov. Math. J. **23**(2), 298–305 (1973)
6. Fiedler, M.: A property of eigenvectors of nonnegative symmetric matrices and its application to graph theory. Czechoslov. Math. J. **25**(4), 619–633 (1975)
7. Hertz, T., Bar-Hillel, A., Weinshall, D.: Boosting margin based distance functions for clustering. In: Proceedings of the Twenty-First International Conference On Machine Learning, p. 50. ACM (2004)
8. LeCun, Y., Bottou, L., Bengio, Y., Haffner, P.: Gradient-based learning applied to document recognition. Proc. IEEE **86**(11), 2278–2324 (1998)
9. Meyer, C.D.: Matrix analysis and applied linear algebra. SIAM (2000)
10. Newman, M.E.: Modularity and community structure in networks. Proc. Natl. Acad. Sci. **103**(23), 8577–8582 (2006)
11. Newman, M.E., Girvan, M.: Finding and evaluating community structure in networks. Phys. Rev. E **69**(2), 026113 (2004)
12. Ng, A.Y., Jordan, M.I., Weiss, Y., et al.: On spectral clustering: Analysis and an algorithm. Adv. Neural. Inf. Process. Syst. **2**, 849–856 (2002)
13. Olson, E., Walter, M.R., Teller, S.J., Leonard, J.J.: Single-cluster spectral graph partitioning for robotics applications. In: Robotics: Science and Systems, pp. 265–272 (2005)
14. Pothen, A.: Graph partitioning algorithms with applications to scientific computing. In: Keyes, D.E., Sameh, A., Venkatakrishnan, V. (eds.) Parallel Numerical Algorithms. ICASE/LaRC Interdisciplinary Series in Science and Engineering, vol. 4, pp. 323–368. Springer, Dordrecht (1997). https://doi.org/10.1007/978-94-011-5412-3_12

15. Race, S.L., Meyer, C., Valakuzhy, K.: Determining the number of clusters via iterative consensus clustering. In: Proceedings of the SIAM Conference on Data Mining (SDM), pp. 94–102. SIAM (2013)
16. Shi, J., Malik, J.: Normalized cuts and image segmentation. IEEE Trans. Pattern Anal. Mach. Intell. **22**(8), 888–905 (2000)
17. Tolliver, D.A., Miller, G.L.: Graph partitioning by spectral rounding: applications in image segmentation and clustering. In: 2006 IEEE Computer Society Conference on Computer Vision and Pattern Recognition (CVPR2006), vol. 1, pp. 1053–1060. IEEE (2006)
18. Von Luxburg, U.: A tutorial on spectral clustering. Stat. Comput. **17**(4), 395–416 (2007)
19. Wilkinson, J.H., Wilkinson, J.H., Wilkinson, J.H.: The algebraic eigenvalue problem, vol. 87. Clarendon Press, Oxford (1965)
20. Yu, L., Ding, C.: Network community discovery: solving modularity clustering via normalized cut. In: Proceedings of the Eighth Workshop on Mining and Learning with Graphs, pp. 34–36. ACM (2010)
21. Zhang, R., Rudnicky, A.I.: A large scale clustering scheme for kernel k-means. In: 2002 Proceedings. 16th International Conference on Pattern Recognition, vol. 4, pp. 289–292. IEEE (2002)

Model Selection and Evaluation

Bayesian Optimization over Mixed Type Inputs with Encoding Methods

Zhihao Liu, Weiming Ou, and Songhao Wang(✉)

SUSTech Business School, Southern University of Science and Technology, Shenzhen 518055, China
{11930201,12133007}@mail.sustech.edu.cn, wangsh2021@sustech.edu.cn

Abstract. Traditional Bayesian Optimization (BO) algorithms assume that the objective function is defined over numeric input space. To generalize BO for mixed numeric and categorical inputs, existing approaches mainly model or optimize them separately and thus cannot fully capture the relationship among different types of inputs. The complexity incurred by additional operations for the categorical inputs in these approaches can also reduce the efficiency of BO, especially when facing high-cardinality inputs. In this paper, we revisit the encoding approaches, which transfer categorical inputs to numerical ones to form a concise and easy-to-use BO framework. Specifically, we propose the target mean encoding BO (TmBO) and aggregate encoding BO (AggBO), where TmBO transfers each value of a categorical input based on the outputs corresponding to this value, and AggBO encodes multiple choices of a categorical input through several distinct ranks. Different from the prominent one-hot encoding, both approaches transfer each categorical input into exactly one numerical input and thus avoid severely increasing the dimension of the input space. We demonstrate that TmBO and AggBO are more efficient than existing approaches on several synthetic and real-world optimization tasks.

1 Introduction

Black-box function optimization refers to problems with no closed-form objective functions and some favorable conditions, such as convexity guarantee and gradient information. Instead, the function value can be evaluated at every query input point. These problems exist in many fields, such as material design [6] and autonomous planning [1].

BO is a popular sequential optimization approach to efficiently optimize black box function, especially when the function evaluation is costly and the computation budget is limited. Most of the current research work for BO focuses on optimizing objective functions over continuous input space. However, in many real-world applications, the decision variables involve mixed numerical and categorical inputs, where the inadequacy of BO for categorical variables can hinder its broader usage. For example, in tuning the hyper-parameters of machine learning models, *e.g.*, in XGBoost, some hyper-parameters are categorical, such

H. Kashima et al. (Eds.): PAKDD 2023, LNAI 13936, pp. 203–215, 2023.
https://doi.org/10.1007/978-3-031-33377-4_16

as the booster type and growth policies. In contrast, the remaining ones are continuous variables, including the learning rate and the regularisation weights. In material design, categorical parameters are often considered when selecting the best design choices, such as the catalysts and solvents in functional molecules and advanced materials [6]. In these applications, categorical variables represent non-numeric data, like a performance assessment ('low', 'medium', 'high') or the choice of material ('steel', 'titanium', 'aluminum'). In the former case, the three choices have an intrinsic order relationship, and the variable is called an ordered variable. In contrast, no explicit order relationship exists in the second case, and the variable is called a nominal variable.

Typically, BO iteratively fits a Gaussian process (GP) model. The GP model is designed for continuous input variables, making the traditional BO difficult to involve categorical inputs. The most straightforward way to solve this is to convert the categorical input using one-hot encoding and feed it to a GP model. However, the high cardinality of the categorical variable may cause the search space to explode exponentially. Another line of research tries to fix this issue by replacing the GP model with other surrogate models, such as the tree-based model in SMAC [7] and kernel density model in TPE [2]. These models are more friendly to mixed inputs, but their predictions are not consistently reliable and may lead to an inferior selection of new points. Recently, some hierarchical approaches handled categorical variables with multi-armed bandits (MAB) and the remaining continuous variables with GP, respectively. The separate operations between categorical and continuous variables may keep BO from fully considering their relationships when selecting new query points. Additionally, the extra computation time of these ad-hoc methods needs to be better considered.

We propose two simple and efficient approaches, TmBO and AggBO, based on encoding methods to address these challenging yet essential problems. TmBO employs target encoding that encodes a specific value $\hat{c_{i,j}}$ of a categorical input h_i based on the outputs at the evaluated points whose input h_i takes the value $c_{i,j}$. This method has shown competitive performance in several machine learning tasks involving mixed inputs [11,15]. In contrast, AggBO encodes the multiple values of a categorical input, trying to recover their intrinsic order. Specifically, we use different possible ranks to encode the multiple values of a categorical variable, and each rank is used in an individual GP model. AggBO then uses the weighted sum of the several GPs to select query points. If the categorical input is ordered, we could naturally use their intrinsic order. If, however, no natural order exists like in nominal variables, AggBO aggregates several possible distinct ranks to make it more robust to handle such variables. These approaches maintain the efficient and elegant GP-based BO framework without increasing the dimension of the search space. They typically achieve competitive performances with less computational time than current approaches.

Our contributions are as follows:

- We propose a target mean encoding **TmBO** method by adapting the original target mean encoding method (Sect. 4.1).

- We propose a novel **AggBO** method which utilizes the order information of categorical variables to improve optimization efficiency (Sect. 4.2).
- We demonstrate that our approaches are more effective and computationally efficient over several synthetic functions and real-world problems (Sect. 5).

2 Related Work

2.1 BO for Categorical and Continuous Inputs

The typical way to deal with categorical variables is to convert them to a one-hot encoding representation. However, the search space dimension will increase severely, especially when the cardinality is very high. This further exacerbates the curse of the dimension and challenges the acquisition function optimization when selecting the new query point. The tree-based model SMAC [7] uses the random forests (RF) [3] model as the underlying probabilistic surrogate model that can naturally handle categorical inputs. The mean and uncertainty of predictive distribution are obtained by the empirical mean and variance over all the decision trees' predictions for that test data. Nevertheless, SMAC may suffer from the unreliable estimation of randomness and easily become overfitting when the number of trees is not chosen properly [8]. Another related work is Tree Parzen Estimator (TPE) [2], which separates the training set into good and bad samples to fit two kernel density estimator (KDE) [18] and the acquisition function is correlated with the ratio of the two estimators. The drawback of TPE is that it requires more initial points in the early stage of modeling the KDE. HybridBO [4] utilizes the Gaussian kernel as the solution of a diffusion equation with the Laplacian operator to define a diffusion kernel directly used in mixed space. Nevertheless, it can only be valid for the Gaussian kernel. Another line of research connects the multi-armed bandits (MAB) with BO for mixed inputs [14,17]. The intuition is to use MAB to select the categorical components and BO to select the continuous components. For example, CoCaBO [17] uses one of MAB algorithms EXP3 to decide multiple categorical variables in BO and proposes a sum kernel (summing a Hamming kernel over categorical subspace and a Matern 5/2 kernel over continuous subspace) to capture the relationship of categorical variables and continuous variables. In CoCaBO, each agent maps to a categorical variable and decides the choice of this variable. However, each agent makes decisions independently, and only when these categorical values are fixed can CoCaBO select continuous values. Therefore, even though the summing kernel can partially model the relations between different types of inputs, the values of these inputs are chosen separately when selecting the query points.

2.2 Encoding Methods

Transferring categorical variables into numerical ones through encoding methods is a common practice for machine learning modeling, as some models only accept numerical values similar to the GP model. It has been shown that the one-hot encoding-based BO algorithm could provide satisfactory performance in certain circumstances [5] with a small number of categorical variables and low cardinality. We here provide an overview of different encoding methods. Typically, the encoding techniques are quite computationally efficient; thus, we may explore their applications in BO without too much overhead.

Generally, the encoding techniques can be categorized into two categories [16]: target encoding and target-agnostic encoding, see Table 1.

Table 1. Encoders

Target-based	Target-agnostic
Quantile Encoder	Binary Encode
Summary Encoder	Ordinal Encoder
Target mean Encoder	Hashing Encoder
M-estimate Encoder	One Hot Encoder
Weight of Evidence Encoder	Regular Simplex Encoder

Target-Based Encoding. The target encoding (TE) encodes each choice of a categorical variable to a numerical representation by incorporating information from target values. These target values generally correspond to the data points whose inputs contain this specific choice. We can then select the mean [11], quantile [13], or other statistics of these target values as the encoder of this choice. In addition, some regularization methods, such as k-fold and weighting, are usually used in TE to avoid overfitting. See [16] for a more detailed discussion.

Target-Agnostic Encoding. In contrast to TE, target-agnostic encoding is constructed without the knowledge of target values. The target-agnostic encoding methods include ordinal encoding (label encoding), which uses a single column of integers to represent different choices; Regular Simplex encoding [10], which encodes each choice of a k-value categorical input with a distinct vertex of a regular simplex in $k-1$ dimensional space, one-hot encoding, which transforms each categorical variable with N_j choices into N_j binary variables, etc.

TmBO uses target-based encoding in this work, while AggBO uses target-agnostic encoding.

3 Background

3.1 Problem Statement

We consider the problem of optimizing an expensive black-box function f defined over a mixed input type space, $\mathcal{X} \triangleq \mathcal{H} \times \mathcal{Z}$, where $\mathcal{H}$ is the categorical space and

$\mathcal{Z}$ is the continuous space. Each element $x \in \mathcal{X}$ has hybrid structure $[h, z]$, and $h \in \mathcal{H}$, where $h = [h_1, ..., h_c]$ are the categorical variables and each h_j has N_j choices, *i.e.*, its cardinality is N_j. Moreover, $z \in \mathcal{Z}$ stands for a d-dimensional continuous input. Formally, we aim to solve

$$x^* = [h^*, z^*] = \arg\max_{x \in \mathcal{X}} f(x). \quad (1)$$

3.2 Bayesian Optimization

Bayesian optimization (BO) is an efficient framework for solving expensive global optimization problems. It iteratively fits a surrogate model with all previous query points evaluated and then selects a new query point based on the model. In each iteration, BO consists of the following two main stages:

Surrogate Modeling. Typically, f is assumed to be a smooth function modeled by a Gaussian Process (GP) model. The GP model $f(x) \sim GP(m(x), k(x, x'))$ is characterized by a mean function $m(x)$ and a kernel function $k(x, x')$. The kernel function depicts how large the two values $f(x)$ and $f(x')$ are correlated. One commonly used kernel is the Matern 5/2 kernel:

$$k(x, x') = \sigma_f^2 \exp(-\sqrt{5}r)\left(1 + \sqrt{5}r + \frac{5}{3}r^2\right),$$

where $r^2 = (\mathbf{x} - \mathbf{x}')^T \boldsymbol{\Lambda} (\mathbf{x} - \mathbf{x}')$, $\boldsymbol{\Lambda}$ denotes a diagonal matrix contains a vector $\boldsymbol{\ell} = [\ell_1, ..., \ell_d]$ of lengthscale ℓ_i which control the correlation along the i-th dimension. From now on, the vector of the model parameters will be jointly denoted by $\boldsymbol{\theta} = (\boldsymbol{\ell}, \sigma_f)$.

Denote $\mathcal{D} = \{(x_i, y_i)\}_{i=1}^N$, $y_i = f(x_i) + \epsilon_i$, $\epsilon_i \sim \mathcal{N}(0, \sigma_\epsilon^2)$ as the evaluated points set. By fitting the observed data into the GP model, we obtain the *predictive Gaussian distribution* of $f(x)$ at any point x in the search space. The mean and variance are given by

$$\mu(x) = \mathbf{k}^T \left(\mathbf{K} + \sigma_\epsilon^2 \mathbf{I}\right)^{-1} \mathbf{y}, \quad \sigma^2(x) = k(x, x) - \mathbf{k}^T \left(\mathbf{K} + \sigma_\epsilon^2 \mathbf{I}\right)^{-1} \mathbf{k}, \quad (2)$$

where $\mathbf{y} = (y_1, \ldots, y_N)$, $\mathbf{k} = [k(x_i, x)]_{\forall x_i \in D}$ is the covariance between the new point x and all other observed points x_i, $\mathbf{K} = [k(x_i, x_j)]_{\forall x_i, x_j \in D}$ is the covariance matrix among all the training data points, and $\mathbf{I}$ is an identity matrix. The posterior mean and variance, calculated from Eq. 2, are used to define an acquisition function $\alpha(x)$ to guide the search of BO.

Acquisition Function. An acquisition function $\alpha(x)$ decides the next point evaluation. It is constructed with the GP model to balance the exploitation of the current optimal region and to explore potential promising regions that are less explored so far. Some commonly used acquisition functions include EI (expected improvement) [12] and UCB (upper confidence bound) [19].

4 The Proposed Framework

Rather than devising an ad-hoc method to separately select the optimal categorical variables or replace GP with another surrogate model, the proposed approaches adopt encoding methods to transfer categorical variables and select the different types of inputs simultaneously within the framework of GP-based BO. The two new approaches, TmBO and AggBO, are discussed in the following two subsections.

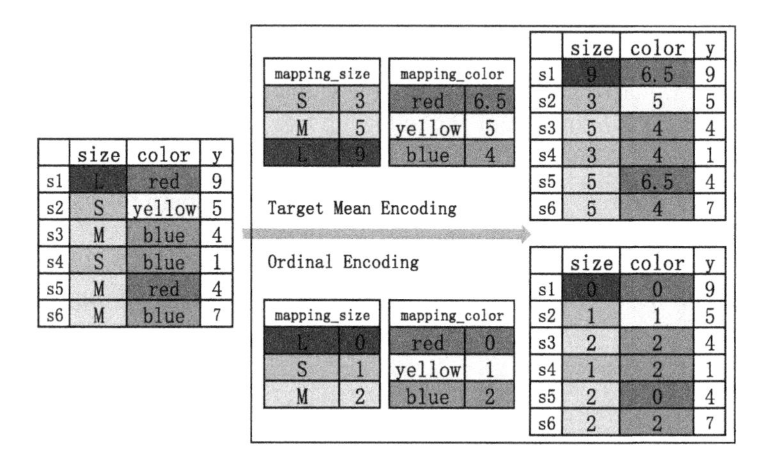

Fig. 1. Target Mean Encoding and Ordinal Encoding

4.1 Target Mean Encoding BO

Target Mean Encoding. As mentioned, the target-based encoding method encodes the categorical variable into a numerical variable by incorporating the output information. Here, we adopt the target mean method as it has shown to perform well in many machine learning tasks [9,11,16]. Given a dataset $\mathcal{D} = \{(x_i, y_i)\}_{i=1}^n$, for the i-th sample x_i, denote $h_{i,j}$ as its j-th categorical variable. We further denote the N_j choices of the j-th categorical variable h_j as $c_{j,1}, ..., c_{j,N_j}$. We then encode $c_{j,k}$ as:

$$\hat{c_{j,k}} = \frac{1}{s_{j,k}} \sum_{i=1}^{n} y_i \cdot \mathrm{I}\{h_{i,j} = c_{j,k}\}, \tag{3}$$

where $\mathrm{I}\{\cdot\}$ is the indicator function and $s_{j,k} = \sum_{i=1}^{n} \mathrm{I}\{h_{i,j} = c_{j,k}\}$ is the frequency of the choice $c_{j,k}$ appearing in $\mathcal{D}$. See the example in Fig. 1; each choice of the two categorical variables, color, and size, is encoded in the two tables *mapping_size* and *mapping_color*, respectively.

It has been observed that the query points selected by BO often cluster, typically around several local optimal points or points that perform reasonably well. Thus, the target values corresponding to different categorical choices can be quite close, leading to similar or extremely close encoding values for these choices. If this happens, we will get some query points that can have very close values in several dimensions, which may result in an ill-conditioned covariance matrix $\mathbf{K}$ in deriving the GP model. To ensure that the target mean encoding BO algorithm is more robust, we add a small noise to the encoding value to avoid singular matrix error, see in Eq. 4, where σ_e is the magnitude of the noise.

$$\hat{c_{j,k}} = \frac{1}{s_{j,k}} \sum_{i=1}^{n} y_i \cdot \mathrm{I}\{h_{i,j} = c_{j,k}\} + \epsilon, \quad \epsilon \sim \mathcal{N}(0, \sigma_e^2). \tag{4}$$

Algorithm 1. Target Mean Encoding BO

Input: budget T
Output: the best sample (x^*, y^*)
1: Initialize data $\mathcal{D}_0 = \{X_0, Y_0\}$ and design domain Q_0
2: **for** $t \leftarrow 0, T$ **do**
3: Fit the target mean encoder E
4: Get the encoding input $\hat{X}_t$, $\hat{Q}_t$ with E
5: Fit the GP model with $\{\hat{X}_t, Y_t\}$
6: Find $x_{t+1} = \arg\max_{x \in \mathcal{X}} UCB(\hat{Q_t})$ and its observation y_{t+1}
7: Update the data $\mathcal{D}_{t+1} = \mathcal{D}_t \cup \{(x_{t+1}, y_{t+1})\}$
8: **end for**

We next develop a target mean encoding BO based on this robust target mean encoding method, see in Algorithm 1. The TmBO algorithm uses the inputs whose categorical variables are encoded by the target mean encoder E_t (see in Eq. (4)) to fit the GP model. Once the acquisition function selects the next query point in the form of numerical values, we decode the component corresponding to the categorical inputs back to their original space to do the evaluation.

4.2 Aggregate Ordinal Encoding BO

The ordinal encoding method is one of the target-agnostic encoding methods, directly encoding choices as integers, 0, 1, 2, ..., in the order in which they appear (see the example in Fig. 1). Note that this order may not necessarily reveal their intrinsic relationships. Here, we propose AggBO (see in Algorithm 2), a more robust way to construct the order among choices. Specifically, we use m distinct ranks. Each rank can be considered an ordinal encoding of the choices and is used to derive a separate GP model. Therefore, in each iteration of AggBO, we get m different GPs, and we then use a weighted sum of them to build a final GP (notice that a linear combination of the GP models is also a GP model). This aggregate GP model will then be adopted to define the acquisition function. we decide the weights of the i-th GP model, w_i as follows. For each rank, we fit a predictive model on encoded train data, a subset of all existing observations, and then test it on the remaining test data. The MSE of the m predictive models on the test set, denoted as $M_1, ..., M_m$, will be used to compute w_i as in Eq. 5. A smaller MSE value leads to a larger weight of the model.

$$w_i = \frac{e^{-M_i}}{\sum_{j=1}^{m} e^{-M_j}}. \tag{5}$$

Algorithm 2. Aggregate Ordinal Encoding BO

Input: budget T, the number of GP model m
Output: the best sample (x^*, y^*)

1: Initialize data $\mathcal{D}_0 = \{X_0, Y_0\}$ and design domain Q_0
2: **for** $t \leftarrow 0, T$ **do**
3: Select m orders of the categorical variables as encoders $E_1, E_2, ..., E_m$
4: Split the data, $\{D^{train}, D^{test}\} \leftarrow \mathcal{D}_t$
5: **for** $j \leftarrow 0, m$ **do**
6: Get the encoding input $\hat{X}_t$, $\hat{Q}_t$, $\{D^{\hat{train}}, D^{\hat{test}}\}$ with E_j
7: Fit the GP model G_j with $\hat{\mathcal{D}}_t = \{\hat{X}_t, Y_t\}$
8: Obtain the test error M_j of $D^{\hat{test}}$
9: **end for**
10: Get the weighted GP model $\mathbf{WGP} = \sum_j^m w_j G_j$
11: Find $x_{t+1} = \arg\max_{x \in \mathcal{X}} UCB(\hat{Q}_t)$ of $\mathbf{WGP}$ and get its observation y_{t+1}
12: Update the data $\mathcal{D}_{t+1} = \mathcal{D}_t \cup \{(x_{t+1}, y_{t+1})\}$
13: **end for**

We give higher weights to the ranks resulting in more accurate predictions. If the real choices are intrinsically ordered, our approach can be treated as finding the proper rank to fit the model better. If, however, the choices are not inherently ordered like nominal variables, our approach provides a more robust solution than traditional ordinal encoding, as we maintain several different ranks, each of which provides a different perspective to view the relation between x and y. If any rank beats the other with a higher weight, this may indicate that such a rank is more beneficial to surrogate modeling fitting, even though the choices are not ordered or the order is inexplicit.

5 Experiments

In this section, we evaluate the performance of the two proposed algorithms (code available at https://github.com/ZhihaoLiu-git/Encoding_BO) and several benchmark approaches on synthetic functions and hyper-parameter tuning of machine learning algorithms.

5.1 Baseline Method and Evaluation Measures

The benchmark approaches are TPE[1], SMAC[2], CoCaBO[3], one-hot encoding BO[4] and RandomOrderBO. Here, RandomOrderBO randomly uses one ordinal encoding for categorical variables in each optimization iteration. For GP-based BO, we use Matern 5/2 kernel and UCB acquisition function. In CoCaBO, we set the mixed kernel parameter $\lambda = 0.5$, as suggested by the authors [17]. And we set $m = 6$ in AggBO. We start all optimization trials with 24 initial points.

A summary of the benchmark problems is shown in Table 2; for the synthetic functions, we convert some continuous dimensions of the Ackley-5C, HartmannSix-6C, and Michalewicz-4C into categorical variables with 17 choices each, as CoCaBO does [17]. For the hyper-parameter tuning task, SVM-Diabetes and MLP-Diabetes output the negative MSE of the SVM[5] regression model and Multi-layer Perceptron[6] on Diabetes dataset[7], respectively and XGBoost-MNIST returns the accuracy of the XGBoost model[8] on MNIST[9].

5.2 Performance and Computation Time

We ran each algorithm 20 trials for the synthetic functions and 10 trials for the hyper-parameter tuning problems. The mean and standard error of the current best function value with respect to the iteration number is illustrated in Fig. 2 where TmBO and AggBO perform quite well in all problems. We see that in all three synthetic functions, TmBO is much better than all the other approaches; in the three real-world problems, TmBO also achieves the best performance. This illustrates the benefit of transferring categorical variables through their performances in the function values. For AggBO, it reaches the second best results, just inferior to TmBO, almost in every problem except the MLP problem, where CoCaBO performs better. Note that it is better than the other two target-agnostic encoding approaches, one-hot and RandomOrderBO, which shows its robustness by synthesizing different encoding approaches. We can also observe that RandomOrderBO, although worse than AggBO, performs reasonably well in SVM and XGBoost. This demonstrates its potential in solving complex mixed input problems. We may develop more smart ways to choose the ordinal encoding in each iteration, instead of just random selection as here so that the algorithm can reach the performance of AggBO while maintaining the computational efficiency of RandomOrderBO.

[1] https://github.com/hyperopt/hyperopt.
[2] https://github.com/SheffieldML/GPyOpt.
[3] https://github.com/rubinxin/CoCaBO_code.
[4] https://github.com/scikit-learn-contrib/category_encoders.
[5] https://scikit-learn.org/stable/modules/generated/sklearn.svm.SVR.html?highlight=svr#sklearn.svm.SVR.
[6] https://scikit-learn.org/stable/modules/generated/sklearn.neural_network.MLPRegressor.html?highlight=mlp#sklearn.neural_network.MLPRegressor.
[7] https://www4.stat.ncsu.edu/~boos/var.select/diabetes.html.
[8] https://github.com/dmlc/xgboost.
[9] http://yann.lecun.com/exdb/mnist/.

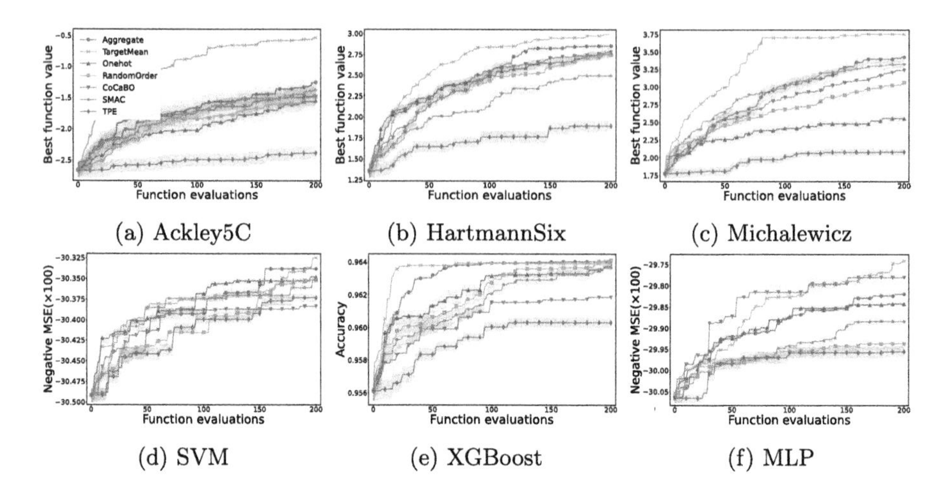

(a) Ackley5C (b) HartmannSix (c) Michalewicz

(d) SVM (e) XGBoost (f) MLP

Fig. 2. Experiments

Table 2. Optimization Problems

Problem	Input $x = [h, z]$	Input values
Ackley-5C $c = 5, d = 1$	$h = [17, ..., 17]^5$	$-1 + 1/8 \times h_i$
	$z = [z_1]$	$[-1, 1]$
HartmannSix-6C $c = 4, d = 2$	$h = [17, ..., 17]^4$	$1/16 \times h_i$
	$z = [z_1, z_2]$	$[0, 1]^2$
Michalewicz-4C $c = 4, d = 1$	$h = [17, ..., 17]^4$	$1/16 \times \pi \times h_i$
	$z = [z_1]$	$[0, \pi]$
SVM-Diabetes $c = 1, d = 2$	h_1 *kernel*	$\{poly, rbf, sigmoid\}$
	z_1 C	[1,50]
	z_2 *epsilon*	[0,1]
XGBoost-MNIST $c = 3, d = 5$	h_1 *booster*	$\{gbtree, dart\}$
	h_2 *grow_policy*	$\{depthwise, lossguide\}$
	h_3 *objective*	$\{softmax, softprob\}$
	z_1 *learning_rate*	[0,1]
	z_2 *max_depth*	[1,2,...,10]
	z_3 *gamma*	[0,10]
	z_4 *subsample*	[0.001,1]
	z_5 *reg_lambda*	[0,5]
MLP-Diabetes $c = 4, d = 3$	h_1 *activation*	$\{logistic, tanh, relu\}$
	h_2 *learning_rate*	$\{constant, invscaling, adaptive\}$
	h_3 *solver*	$\{sgd, adam\}$
	h_4 *early_stopping*	$\{True, False\}$
	z_1 *hidden_layer_sizes*	[1,200]
	z_2 *alpha*	[0.0001,1]
	z_3 *tol*	[0.00001,1]

Table 3. Computation Time

Method & Time(s)	Ackley5C	HartmannSix	Michalewicz	XGBoost	SVM	MLP
AggBO	142	152	143	2592	181	257
TmBO	63	48	60	2197	28	97
One-hot BO	101	65	93	2431	31	106
RandomOrderBO	33	30	26	2158	32	96
CoCaBO	307	279	267	2245	207	497
SMAC	217	188	171	2383	162	319
TPE	2	2	2	2101	5	69

We record the runtime of these approaches on an AMD Ryzen 7 5800 8-Core Processor machine and NVIDIA GeForce RTX 3060 (12 GB) for XGBoost acceleration. Table 3 shows the average one-trial total time for each experiment (in seconds). Note that the computation time of TPE is independent of the dimension of the problem, and it is the fastest one. However, it always gets poor performance. Except for TPE, TmBO and RandomOrderBO have the shortest running time as they adopt the efficient BO framework with one GP model. For AggBO, The increase in computation time compared with TmBO is mainly caused by the model fitting of the m GP model. However, it is still faster than CoCaBO, which generally takes longer due to the more complex operations to deal with the categorical variables, especially when the cardinality is high. We thus can find that TmBO is both effective and efficient, while AggBO provides reasonably good performances with longer running time (but the running time is still comparable with other well-adopted approaches). In these experiments, we only use 6 ranks in AggBO. It is expected that while more possible ranks may perform better. We leave it as a future direction to select better choices of ranks in AggBO.

6 Conclusion

We revisit encoding methods with BO to harness mixed categorical and numerical inputs and develop an easy-to-use encoding BO framework with target-based and target-agnostic encoding methods. Specifically, We propose a TmBO, which combines the target encoding method, and a novel AggBO which combines the ordinal encoding method with multiple ranks to achieve a more robust approach. Experiments demonstrate that TmBO and AggBO are more effective and efficient than some existing approaches, and TmBO generally takes less computational time than other approaches.

Acknowledgement. This work is supported by the National Natural Science Foundation (NNSF) of China under Grant 72101106 and the Shenzhen Science and Technology Program under Grant No. RCBS20210609103119020.

References

1. Abdelkhalik, O.: Autonomous planning of multigravity-assist trajectories with deep space maneuvers using a differential evolution approach. Int. J. Aerosp. Eng. **2013** (2013)
2. Bergstra, J., Bardenet, R., Bengio, Y., Kégl, B.: Algorithms for hyper-parameter optimization. In: Advances in Neural Information Processing Systems, vol. 24 (2011)
3. Breiman, L.: Random forests. Mach. Learn. **45**(1), 5–32 (2001)
4. Deshwal, A., Belakaria, S., Doppa, J.R.: Bayesian optimization over hybrid spaces. In: International Conference on Machine Learning, pp. 2632–2643. PMLR (2021)
5. González, J., Dai, Z.: GPYOPT: a Bayesian optimization framework in Python. Accessed (2016)
6. Häse, F., Aldeghi, M., Hickman, R.J., Roch, L.M., Aspuru-Guzik, A.: GRYFFIN: An algorithm for Bayesian optimization of categorical variables informed by expert knowledge. Appl. Phys. Rev. **8**(3), 031406 (2021)
7. Hutter, F., Hoos, H.H., Leyton-Brown, K.: Sequential model-based optimization for general algorithm configuration. In: Coello, C.A.C. (ed.) LION 2011. LNCS, vol. 6683, pp. 507–523. Springer, Heidelberg (2011). https://doi.org/10.1007/978-3-642-25566-3_40
8. Lakshminarayanan, B., Roy, D.M., Teh, Y.W.: Mondrian forests for large-scale regression when uncertainty matters. In: Artificial Intelligence and Statistics, pp. 1478–1487. PMLR (2016)
9. Larionov, M.: Sampling techniques in Bayesian target encoding. arXiv preprint arXiv:2006.01317 (2020)
10. McCane, B., Albert, M.: Distance functions for categorical and mixed variables. Pattern Recogn. Lett. **29**(7), 986–993 (2008)
11. Micci-Barreca, D.: A preprocessing scheme for high-cardinality categorical attributes in classification and prediction problems. ACM SIGKDD Explor. Newsl. **3**(1), 27–32 (2001)
12. Mockus, J., Tiesis, V., Zilinskas, A.: The application of Bayesian methods for seeking the extremum. Towards Glob. Optim. **2**(117–129), 2 (1978)
13. Mougan, C., Masip, D., Nin, J., Pujol, O.: Quantile encoder: tackling high cardinality categorical features in regression problems. In: Torra, V., Narukawa, Y. (eds.) MDAI 2021. LNCS (LNAI), vol. 12898, pp. 168–180. Springer, Cham (2021). https://doi.org/10.1007/978-3-030-85529-1_14
14. Nguyen, D., Gupta, S., Rana, S., Shilton, A., Venkatesh, S.: Bayesian optimization for categorical and category-specific continuous inputs. In: Proceedings of the AAAI Conference on Artificial Intelligence, vol. 34, pp. 5256–5263 (2020)
15. Pargent, F., Bischl, B., Thomas, J.: A benchmark experiment on how to encode categorical features in predictive modeling. Ph.D. thesis, Master Thesis in Statistics, Ludwig-Maximilians-Universität München ... (2019)
16. Pargent, F., Pfisterer, F., Thomas, J., Bischl, B.: Regularized target encoding outperforms traditional methods in supervised machine learning with high cardinality features. arXiv preprint arXiv:2104.00629 (2021)
17. Ru, B., Alvi, A., Nguyen, V., Osborne, M.A., Roberts, S.: Bayesian optimisation over multiple continuous and categorical inputs. In: International Conference on Machine Learning, pp. 8276–8285. PMLR (2020)

18. Sheather, S.J., Jones, M.C.: A reliable data-based bandwidth selection method for kernel density estimation. J. Roy. Stat. Soc.: Ser. B (Methodol.) **53**(3), 683–690 (1991)
19. Srinivas, N., Krause, A., Kakade, S.M., Seeger, M.: Gaussian process optimization in the bandit setting: no regret and experimental design. arXiv preprint arXiv:0912.3995 (2009)

Online and Streaming Algorithms

Using Flexible Memories to Reduce Catastrophic Forgetting

Wernsen Wong(✉), Yun Sing Koh, and Gillian Dobbie

School of Computer Science, The University of Auckland, Auckland, New Zealand
wwon129@aucklanduni.ac.nz, {ykoh,gill}@cs.auckland.ac.nz

Abstract. In continual learning, a primary factor of catastrophic forgetting is task-recency bias, which arises when a model is trained on an imbalanced set of new and old task instances. Recent studies have shown the effectiveness of rehearsal-based continual learning methods; however, a major drawback of these methods is the loss of accuracy on older tasks when training is biased towards newer tasks. To bridge this gap, we propose a λ Stability Wrapper (λSW), where the learner uses a task-based policy to adjust the probability of when instances are replaced in memory to account for task-recency bias to alleviate catastrophic forgetting. The policy results in an increased number of instances seen from older tasks. By construction, λSW can be applied with other rehearsal-based continual learning algorithms. We validate the effectiveness of λSW with three well known baseline methods: Gradient-based Sample Selection, Experience Replay, and Maximally Interfered Retrieval. Our experimental results show significant gains in accuracy on eleven out of twelve of our experiments across four datasets.

Keywords: Continual learning · Catastrophic forgetting · Task-recency bias

1 Introduction

Current continual learning (CL) research aims to incrementally learn a sequence of tasks [8]. Each task t consists of y_t ($y_t \geq 1$) classes to be learned. Once a task is learned, its training data is often no longer accessible. However, when a model learns a new task, it may be prone to catastrophically forget previous tasks. In a catastrophic forgetting event, the model suffers a significant drop in accuracy as the parameters learned for the previous tasks are modified while learning a new task [19]. An effective CL system must optimize for two conflicting goals. First, when the model encounters previous knowledge, it should be able to remember it accurately. Second, the maintenance of old knowledge in the model should not inhibit learning new knowledge. Maintaining these two simultaneous conditions represents the challenge known as the stability-plasticity dilemma [11]. A primary source of catastrophic forgetting is *task-recency bias* [17]. An example

Supplementary Information The online version contains supplementary material available at https://doi.org/10.1007/978-3-031-33377-4_17.

H. Kashima et al. (Eds.): PAKDD 2023, LNAI 13936, pp. 219–230, 2023.
https://doi.org/10.1007/978-3-031-33377-4_17

of *task-recency bias* is when a model with a softmax classifier has a strong bias towards the most recent task due to the imbalance of new and old tasks.

This paper focuses on a particular setting of CL, class incremental learning (Class-IL). In Class-IL, the model incrementally learns classes from a sequence of tasks. At the test time, the learned model classifies a test case without access to the task-id. In Task incremental learning (Task-IL), a model is constructed for each task in training. In testing, the task id is supplied for each test case.

This paper identifies an issue with rehearsal-based methods, which deal with catastrophic forgetting by adapting the model to previous tasks by storing a small number of task instances in a replay buffer [2,16,23] and replay them as required during training on new task instances [1]. To address the issue of *which* instances are replaced in the replay buffer, some approaches use a reservoir sampling method or a diversity measure like cosine similarity. However, these methods do not explicitly balance the number of new and older instances in the replay buffer, leading to a bias towards the new task and reduced accuracy on older tasks.

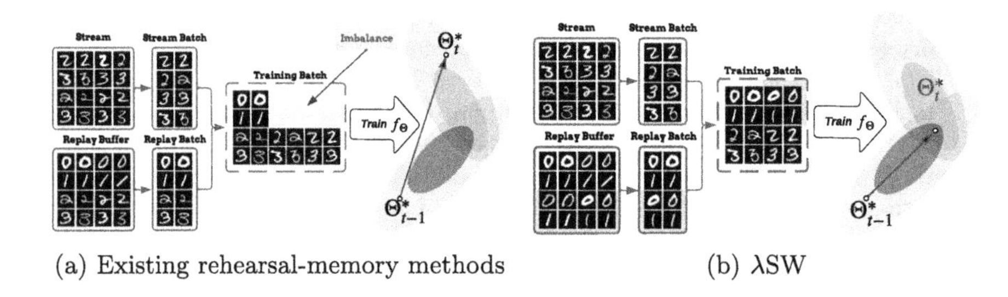

Fig. 1. Illustration of the gap between existing rehearsal-based methods and λSW.

To alleviate this problem, we developed a rehearsal-based CL framework called λ Stability Wrapper (λSW), which explicitly adjusts the replacement rate in the replay buffer over time. Our λSW does this by using a Stability Γ Policy, which we will discuss further in Sect. 4. Our framework is used in conjunction with other current state-of-the-art rehearsal-based methods, Gradient-based Sample Selection (GSS) [2], Experience Replay (ER) [24], and Maximally Interfered Retrieval (MIR) [1]. Figure 1(a) shows a classic rehearsal-based approach [23,24], which can result in an imbalanced training batch where the majority of the instances come from the new task, leading to suboptimal performance. In contrast, our method balances the old and new task instances in the replay buffer, rebalancing the training batch as shown in yellow in Fig. 1(b). The proposed λSW method determines whether to store a new task instance by comparing the diversity of the instances in the stream batch with the instances in the replay buffer. We evaluated the performance of λSW in a set of experiments on four benchmark datasets using three common rehearsal-based methods. The results showed that combining λSW with the existing rehearsal-based method improved accuracy in 11 out of 12 cases. We further investigate the conditions under which λSW can be applied in Sect. 5.

Our **main contributions** are two-fold. Firstly, we propose a novel method λSW, a framework to adjust the replacement rate over time to reduce the effects of *task-recency bias* that results in the method being more effective at alleviating catastrophic forgetting, which we empirically show through experiments. Secondly, we demonstrate the applicability of our λSW approach on two different types of rehearsal-based methods, specifically using reservoir sampling (*i.e.*, ER, MIR) [1,24] and cosine similarity (*i.e.*, GSS) [2].

2 Related Work

Recently, there has been significant progress in CL to mitigate the phenomenon of catastrophic forgetting [19] where the model "forgets" knowledge from past tasks when exposed to new ones. These methods can be divided into different categories, including regularization-based, knowledge distillation-based, rehearsal-based, and parameter isolation-based methods. To reduce this effect, several approaches have been investigated, such as regularization [9,29] that modify the model parameters with additional regularization constraints, separating parameters for old and new tasks [18,28], replaying instances from memory [7,21,23]. In this work, we build on rehearsal-based approaches, which have achieved higher accuracy in the CL setting.

Implicit Task-Recency Bias Mitigation. Implicit *task-recency bias* mitigation methods include adjusting the learning rate and adapting the $\mathcal{L}_2$ regularization. Jastrzebski et al. [13] proposed using an adaptive learning rate by applying large updates to the network weights, which enables the network to be plastic but at the cost of the stability of previous tasks. Rannen et al. [22] introduced $\mathcal{L}_2$ regularization to prevent network weights from deviating from the optimal performance of the previous task which enables the stability for specific tasks compared to others. Another approach is to use rehearsal-based methods, such as iCaRL [23], which equally divides the replay buffer between all tasks and randomly replay instances selected through a herding-based strategy. Gradient-based Memory Editing (GMED) [14] is a framework that edits stored instances in continuous input space via gradient updates to create more hard-to-learn instances for replay. All of these methods attempt to address task-recency bias implicitly. However, a major drawback is that the number of instances stored in the replay buffer is irrespective of when the task occurs. Thus, the distribution of replay instances may be biased towards the new task. In contrast, our λSW approach is used in conjunction with rehearsal-based approaches and explicitly adjusts the distribution of the replay buffer, which increases the probability of replay for earlier tasks.

Explicit Task-Recency Bias Mitigation. Explicit *task-recency bias* mitigation methods include replaying instances from the replay buffer multiple times and adding a correction layer in the network. Hou et al. [12] observed that the classifier norm is larger for new tasks than for older tasks, and the classifier bias favours the newest task. One effective approach proposed by [6] proposes an additional step, called *balanced training* at the end of each training session. In this step, an equal number of instances from all tasks are replayed for a limited number of iterations. However, this comes at the cost of overfitting the stored instances when these do not represent the distribution. Another approach proposed by Belouadah et al. [4] adds a layer that uses saved certainty statistics of classes from old tasks to adjust the network prediction if the outcome is predicted to be the new task. However, a major gap is that these methods require additional computational steps and may overfit the instances in the replay buffer. In contrast, our λSW approach adjusts the replay buffer distribution without modification to the amount of replay or the underlying network structure.

3 The Continual Learning Problem

We consider learning a model over a sequence of tasks $\mathcal{T} = \{\mathcal{T}_1, \ldots, \mathcal{T}_T\}$, where each task is composed of independently and identically distributed data points and their labels, such that task $\mathcal{T}_t$ includes $\mathcal{D}_t = \{x_{t,n}, y_{t,n}\}_{n=1}^{N_t} \sim \mathcal{X}_t \times \mathcal{Y}_t$, where N_t is the total number of instances for the t-th task, and $\mathcal{X}_t \times \mathcal{Y}_t$ is an unknown data generating distribution. We assume that an arbitrary set of labels for task $\mathcal{T}_t, y_t = \{y_{t,n}\}_{n=1}^{N_t}$ has unique classes. In a standard CL scenario, the model learns a corresponding task at each step, and the t-th task is accessible at step t only, but a small number of instances can be stored in a replay buffer. Let the neural network $f_\Theta : \mathcal{X}_{1:T} \rightarrow \mathcal{Y}_{1:T}$ be parameterized by a set of weights $\Theta = \{\theta_l\}_{l=1}^{L}$, where L is the number of layers in the neural network, We define the training objective at step t as follows:

$$\min_{\Theta} \sum_{n=1}^{N_t} \ell(f_\Theta(x_{t,n}), y_{t,n}) \tag{1}$$

where $\ell(.)$ is any standard loss function. The goal is to avoid forgetting past tasks while utilizing the limited stored instances when trained on new tasks.

λSW Objective. The most crucial step in the λSW approach is identifying "when" earlier task instances in the replay buffer should be replaced with later task instances. Consider $(x_{t,n}, y_{t,n})$ is a labelled instance received by the model for the t-th task, the goal is to design a suitable replacement function ϕ to determine if the instance $(x_{t,n}, y_{t,n})$ should replace an existing instance (x_i, y_i) in the replay buffer $\mathcal{M}$, where $\phi(x_{t,n})$ determines the replacement outcome. This is similar to replacement functions in prior work, *i.e.*, GSS [2], that uses cosine similarity to calculate a diversity score $c_{t,n}$ for instance $(x_{t,n}, y_{t,n})$ with memory instance (x_m, y_m), a replacement is made if $c_{t,n} > 0$.

4 The λ Stability Wrapper (λSW) for Replay Buffer Replacements

Rehearsal-based methods store and continually maintain a set of diverse instances in a fixed-size replay buffer. These instances are then replayed as a replay buffer batch during training to prevent forgetting of previous tasks. For example, Experience Replay (ER) [24] uses Reservoir Sampling [27] to select instances, while Gradient-based Sample Selection (GSS) [2] tries to diversify the gradient directions of the instances in the replay buffer. Figure 2 illustrates a brief architecture of λSW, which we describe in detail below.

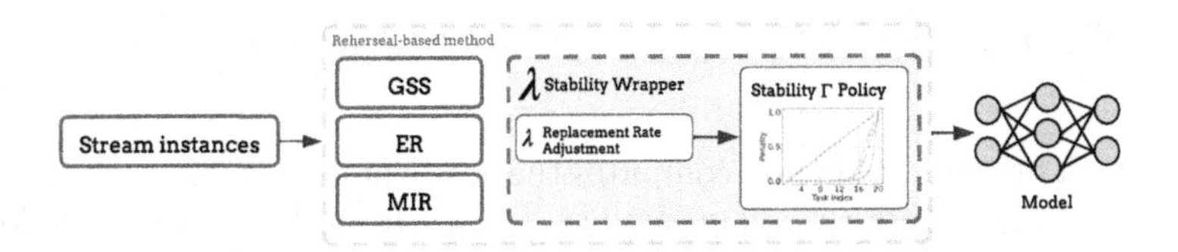

Fig. 2. The overall λ Stability Wrapper architecture in conjunction with existing rehearsal-based methods.

Our λSW approach adjusts the number of instances stored per task over time. The adjustment rate can be guided by a task-based replacement objective consisting of a dual-optimization function for balancing the instances of new and old tasks shown in Eq. 2, to ensure the model's accuracy for the tasks is maintained.

$$\lambda \sim f(x_{t,n}, y_{t,n}, \mathcal{M}) \quad (2)$$

where $f(.)$ is a selection function defined in Eq. 4 for cosine similarity and Eq. 5 for weighted reservoir sampling, $\mathcal{M}$ is the replay buffer, $(x_{t,n}, y_{t,n})$ is an instance from a stream of the t-th task. Intuitively this involves adjusting the replay buffer of new task instances to favour old task instances, thus increasing the replay of instances for older tasks. Our λSW approach consists of two components a λ parameter that controls the balance between the new and old tasks and a Stability Γ Policy that controls the replacement rate in the replay buffer.

Stability Γ Policy. The Stability Γ Policy controls the replacement rate in the replay buffer, and its effect on the replacement function depends on the rehearsal-based method. For example, with reservoir sampling, instances are replaced at random, but with a policy, the replacement rate can vary as more tasks are encountered. On the other hand, cosine similarity aims to maintain a diverse set of instances in the replay buffer, and when used in conjunction with a policy, it can further vary the diversity threshold over time. These different variations of the Stability Γ Policy allow for greater flexibility and adaptability. We propose three different Γ policies, Linear (*Lin*) $\Gamma(t) = t/|\mathcal{T}| - 1$, Exponential (*Exp*) $\Gamma(t) = e^t/e^{(|\mathcal{T}|-1)}$, Default $\Gamma(t) = 1$ where $|\mathcal{T}|$ is the total number of tasks and t is the index of the task. The goal of these policies is to allow different balancing to occur depending on the replay buffer replacement method. By adjusting the Γ policy, we can adapt the λSW approach for different settings.

Applying λSW with GSS. Figure 3 provides a schematic of the steps involved in GSS+λSW. When we receive the first task, the replay buffer $\mathcal{M} = \{(x_0, y_0), \ldots, (x_m, y_m)\}$ is initialized with incoming stream instances. When a

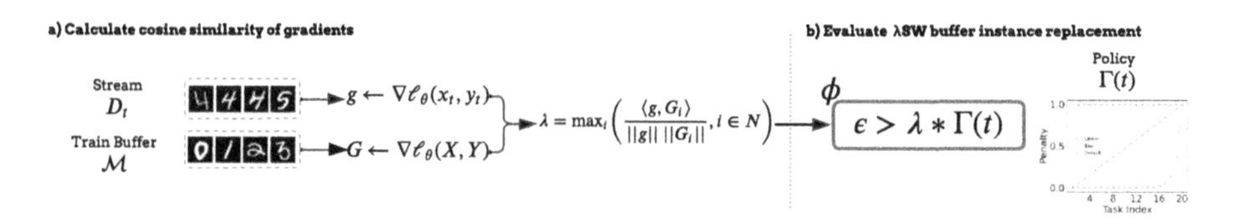

Fig. 3. The overall GSS+λSW architecture for the replay buffer.

new t-th task arrives, the model computes the gradient directions on the received instance $(x_{t,n}, y_{t,n})$ and a random instance (x_i, y_i) from $\mathcal{M}$, using the following equations:

$$g \leftarrow \nabla \ell_\theta(x_{t,n}, y_{t,n}); G_i \leftarrow \nabla \ell_\theta(x_i, y_i) \tag{3}$$

where g and G_i are the gradient changes for the stream and memory instances, respectively. Multiple gradients of the network are calculated from the replay buffer, and G_i represents a single instance from the replay buffer. We then calculate a "diversity" score (*i.e.,* cosine similarity of the gradient), as shown in Eq. 2 by substituting Eqs. 3 into it, the diversity score is calculated by:

$$f(.) = \max_i \left(\frac{\langle g, G_i \rangle}{||g|| \, ||G_i||}, i \in N \right) \tag{4}$$

where $\lambda = f(.)$ results in the "diversity" score. To determine if a buffer replacement occurs, we evaluate $\phi(x_{t,n}) = (\epsilon > \lambda + \Gamma(t))$ where ϕ is a replacement function to determine if an instance in the replay buffer should be replaced, $\Gamma(t)$ is the stability policy rate given the t-th task, ϵ is the threshold assigned to control the stability of the policy, and $\Gamma(t)$ is bounded between $[0, 1]$ and λ controls the replacement level in the replay buffer at task $\mathcal{T}_t$. Given the nature of cosine similarity, λ is bounded between $[-1, 1]$. In general, Γ policies increase at varying rates over time to account for various *task-recency bias* effects.

Applying λSW with ER and MIR. Since λSW can be applied to replacement functions, we integrate it with a range of existing rehearsal-based CL algorithms, specifically, ER [24] and MIR [1]. For integration with ER, we replace reservoir sampling with weighted reservoir sampling [10] as the replacement function. To enable λSW to adapt the λ parameter based on the changing needs of the model, we introduce the concept of a validation buffer. This buffer stores a subset of previously seen instances that were not used in the training of the current task, allowing us to calculate the model's performance on these instances and use this information to adjust the amount of older instances in the replay buffer. This allows us to calculate the validation accuracy, which is used as a proxy for the λ parameter to balance the amount of old and new instances in the replay buffer.

Figure 4 summarizes the process when a new task is seen. At the t-th task, the model receives an instance $(x_{t,n}, y_{t,n})$ from the training stream D_t. We first initialise a replay buffer $\mathcal{M} = \{(x_0, y_0), \ldots, (x_m, y_m)\}$, a validation buffer $\mathcal{V} = \{(x_0, y_0), \ldots, (x_v, y_v)\}$, and a weights buffer $\mathcal{P} = \{w_0, \ldots, w_m\}$ where m is the size of the replay buffer and v is the size of the validation buffer. Each instance

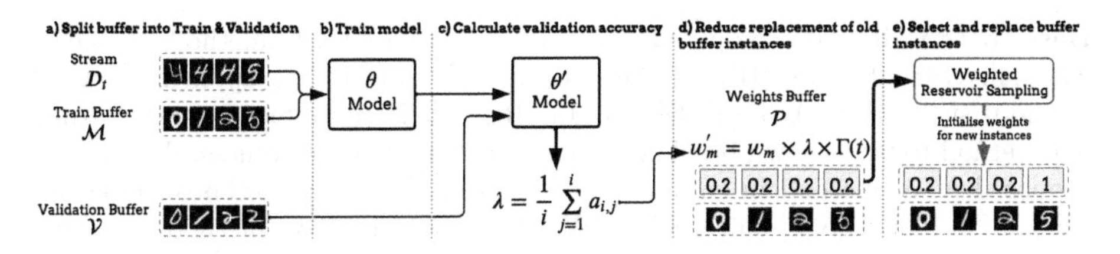

Fig. 4. The overall ER+λSW and MIR+λSW architecture for the replay buffer.

in the replay buffer (x_i, y_i) has an associated weight w_i. The initial weight is initialized with $w_i = 1$, the weight w_i controls the probability of replacement. To determine what is replaced in the replay buffer, we use weighted reservoir sampling to calculate an instance's replacement probability to be selected by using the following calculation $p_i = w_i / \sum_{j=0}^{m} w_j$ where p_i is the probability of replacement, for instance, i.

When a new task is seen, we calculate a penalty score, using Eq. 2 by substituting Eq. 5 into it. The penalty score is calculated as:

$$f(.) = \frac{1}{i} \sum_{j=1}^{i} v_{i,j} \tag{5}$$

where $\lambda = f(.)$ is calculated by the validation accuracy of the validation set. We define $v_{i,j}$ as the validation accuracy evaluated on the validation set $\mathcal{V}$ of task j after training the network from task i through to t. To reduce the bias of calculating the validation accuracy, the validation buffer instances are selected randomly and cannot be selected for replay during training of the task. The weight buffer scores $\mathcal{P}$ are then adjusted by $\lambda \times \Gamma(t)$ to reduce the probability of older buffer instance replacement.

5 Experimental Results

Our experiments address the following research questions: (i) what are the accuracy gains achieved from integrating λSW with existing rehearsal-based CL algorithms? (ii) how does the λ hyper-parameter and stability Γ policy affect the performance of reservoir sampling-based methods? (iii) how does the λ hyper-parameter perform under various task instance lengths and replay buffer batch sizes?

Datasets. We use four public CL datasets in our experiments. COIL 100 [20], which contains 100 classes, is split into 20 tasks, each of which has 5 classes; CIFAR10 [15], which contains 10 classes, is split into 5 tasks, each of which has 2 classes; CIFAR100 [15], which contains 100 classes, is split into 20 tasks, each of which has 5 classes; mini-ImageNet [26], which contains 100 classes, is split into 20 tasks, each of which has 5 classes.

Base Methods. We compared three rehearsal-based CL methods namely, ER [24], GSS [2], and MIR [1]. Our framework can also be applied to other methods such as ER-ACE [5], RM [3], InfoRS [25]. Here we focus on comparing our method to well-known baselines to demonstrate its effectiveness. We use our proposed λSW method in conjunction with three baseline methods, noted as ER+λSW, MIR+λSW, and GSS+λSW. The * next to λSW methods indicate significant accuracy improvement over the counterparts without λSW, with a p-values less than 0.05 in single-tailed paired t-tests.

Table 1. Current Task: Classification results showing the mean and standard deviation of accuracy (%) for λSW against baseline methods (Base).

	GSS			ER		MIR	
	Base	λSW(*Lin*)	λSW(*Exp*)	Base	λSW	Base	λSW
COIL-100							
M=100	36.7 ± 3.9	37.2 ± 4.8	40.8 ± 4.3*	47.8 ± 7.4	53.7 ± 5.0*	56.0 ± 5.0	**60.5 ± 4.8***
M=200	48.5 ± 6.7	46.2 ± 6.7	53.1 ± 4.9*	58.9 ± 7.6	62.5 ± 6.5*	68.9 ± 5.0	**70.9 ± 5.9***
M=500	47.2 ± 10.5	48.7 ± 7.0	53.4 ± 6.9*	65.5 ± 8.2	67.9 ± 7.7*	78.0 ± 6.4	**79.5 ± 5.7***
CIFAR-10							
M=100	18.8 ± 0.3	18.8 ± 0.9	19.1 ± 0.5*	19.7 ± 0.9	20.4 ± 0.9*	19.6 ± 0.5	**20.4 ± 0.8***
M=200	19.4 ± 0.9	20.7 ± 1.4*	20.3 ± 0.5*	21.0 ± 0.8	22.3 ± 1.3*	20.9 ± 1.0	**23.2 ± 1.7***
M=500	23.4 ± 1.4	26.0 ± 2.3	25.6 ± 1.9*	25.5 ± 2.1	29.5 ± 2.7*	26.9 ± 1.7	**30.2 ± 1.9***
CIFAR-100							
M=1K	8.3 ± 0.7	6.7 ± 0.4	8.0 ± 0.6	8.3 ± 0.8	9.8 ± 0.9*	8.0 ± 0.7	**8.8 ± 0.7***
M=5K	15.3 ± 1.3	12.3 ± 1.0	15.2 ± 1.1	17.6 ± 1.5	18.4 ± 1.5*	18.6 ± 1.1	**19.6 ± 1.5***
M=10K	15.1 ± 0.9	12.1 ± 1.1	14.9 ± 1.1	18.4 ± 1.8	19.1 ± 2.0*	18.6 ± 1.8	**19.2 ± 2.1***
mini-ImageNet							
M=1K	7.3 ± 1.2	6.1 ± 1.0	7.9 ± 1.0*	8.5 ± 1.1	**10.1 ± 0.9***	8.4 ± 1.1	8.9 ± 0.8*
M=5K	13.4 ± 2.0	13.7 ± 2.1	14.4 ± 1.9*	14.2 ± 1.8	14.9 ± 1.7*	14.8 ± 2.1	**15.6 ± 1.6***
M=10K	13.7 ± 1.2	12.4 ± 2.3	14.3 ± 1.4*	14.2 ± 1.8	14.7 ± 2.0*	14.0 ± 1.7	**14.8 ± 2.0***

Parameter Settings. We set the size of replay buffer as {1000, 5000, 10000} for CIFAR-100 and mini-ImageNet, and {100, 200, 500} for COIL-100 and CIFAR-10. We selected the policy based on the rehearsal-based method, for cosine similarity-based methods, *i.e.*, for GSS we selected {*Lin*, *Exp*} and for weighted reservoir sampling, *i.e.*, for ER and MIR we selected the *Default* policy. We adopt the hyperparameter settings used by Mai et al. [17] for the baselines and evaluate them using the original code with modifications for λSW. For λSW+GSS, we conduct a grid search to select the optimal ϵ hyper-parameter ($\epsilon = 0.125$) and report the best results. For λSW+ER and λSW+MIR we set the validation set size equal to 20% of the buffer size. All experiment results were produced after 20 runs for GSS and GSS+λSW, and 30 runs for ER, ER+λSW, MIR, and MIR+λSW.

Performance When Varying Datasets. We summarize the results obtained by integrating λSW with different CL algorithms. We compare the final accuracy at the end of all tasks, shown in Table 1. From the results, it is observed the reservoir sampling approaches ER+λSW and MIR+λSW achieves significant accuracy gains against the compared baselines with varying replay buffer sizes. For GSS that uses a diversity approach while GSS+λSW(*Exp*) have significant accuracy gains over three of the four datasets apart from CIFAR-100 while GSS+λSW(*Lin*) did not perform well as a constant penalty rate may have balanced too strongly towards the older tasks. We further investigated the accuracy performance of GSS+λSW(*Exp*) on CIFAR-100 and found that GSS already weighted the replay buffer towards older tasks, resulting in minimal effects when using our method. This suggests that the λSW approach may not be as effective when applied to methods that already incorporate weighting towards older tasks in their replay buffer. However, it can still improve the performance of other rehearsal-based methods that do not have this weighting towards older tasks, as shown by the results on the other datasets.

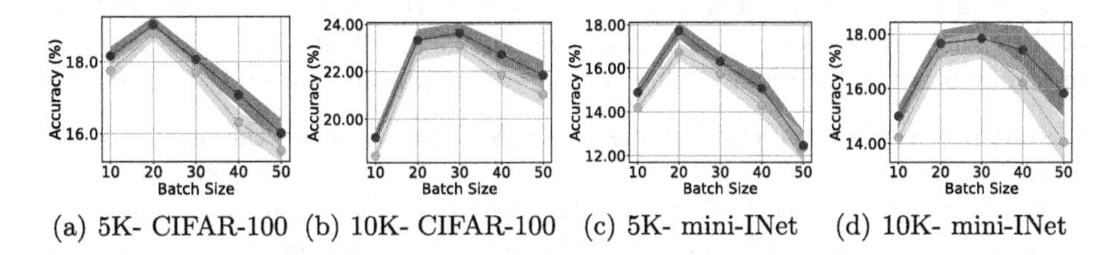

Fig. 5. Batch retrieval size of ER and ER+λSW with varying replay buffer sizes for CIFAR-100 and mini-ImageNet (mini-INet).

Performance When Varying Replay Buffer Batch Size. In this experiment, we investigate the relationship between the size of the replay buffer and the balance between old and new task instances. We vary the replay buffer batch size and observe the trade-off between the model's ability to memorize instances in the replay buffer and reducing the imbalance by increasing the replay of old instances, as shown in Fig. 5. Our results show that in both CIFAR-100 and mini-ImageNet, using a replay buffer size of 5000 with a batch size of 20 leads to better accuracy than the original setting of a batch size of 10. However, when the batch size is increased, accuracy decreases as the model may memorize the instances in the replay buffer. Additionally, the 10000 replay buffer for both datasets achieves the highest accuracy with a batch size of 30, which may be due to the larger size of the replay buffer allowing for less memorization.

Performance When Varying Training Instances Per Task. In this experiment, we vary the number of training instances per task to investigate the relationship between the final task accuracy and the effects of λ on the replay buffer by setting λ manually. This allows us to understand how the λ parameter affects the balance of old and new task instances in the replay buffer, and how this balance impacts the final task accuracy. The final task is the last task to be trained on in the dataset. Figure 6 shows the accuracy for the final task using ER and ER+λSW on CIFAR-10, which has 10000 training instances per task, while CIFAR-100 has 2500 training instances per task. On CIFAR-10, the final

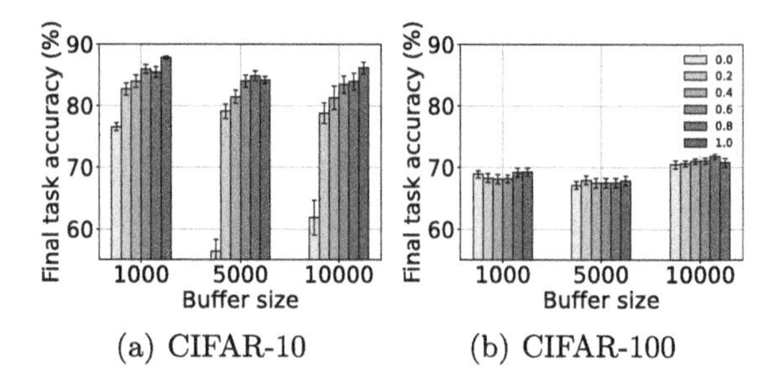

(a) CIFAR-10 (b) CIFAR-100

Fig. 6. Mean and standard error of final task accuracy (%) for ER versus ER+λSW.

task accuracy is significantly lower when λ is 0. However, for CIFAR-100 the final task accuracy does not show a significant change. Thus, forgetting within a task may occur if the replay buffer of a large training instance task is removed.

Visual Inspection for the Effects of λSW. Figure 7 shows the impact of the λSW on new tasks when the number of training instances per task is high. In this experiment, we used CIFAR-10 with 10000 training instances per task with a buffer size of 5000 on ER and ER+λSW for a single run and show the top five instances from three classes in the test dataset ranked by the difference between the first and second logit prediction scores. To understand the effects of the knowledge that is captured by λSW, we analyze the accuracy of the predictions based on the test data after training on all tasks. The test data is used as a proxy as it is the same dataset used across both methods. We used correctly classified instances from ER as a test case for ER+λSW, such that ER+λSW only sees the correct instances from the ER. We count the correctly classified instances on ER but misclassified by ER+λSW and call them *ER memorable*. Similarly, we count the instances that are correctly classified on ER+λSW but misclassified by ER and call them *ER+λSW memorable*. We count the instances that are both correctly classified on ER and ER+λSW and we call them *Both memorable*. We determine the loss that is incurred when using λSW on ER. For example, the test instances per task were set to 2000. We highlight that Task 0 contained classes of birds and boats where *Both memorable* is 628, *ER memorable* is 72 instances out of 700 that ER correctly classified, and *ER+λSW memorable* is 420 instances out of 1048 that ER+λSW correctly classified. From this, we can see that ER+λSW is correctly classifying more instances on this task.

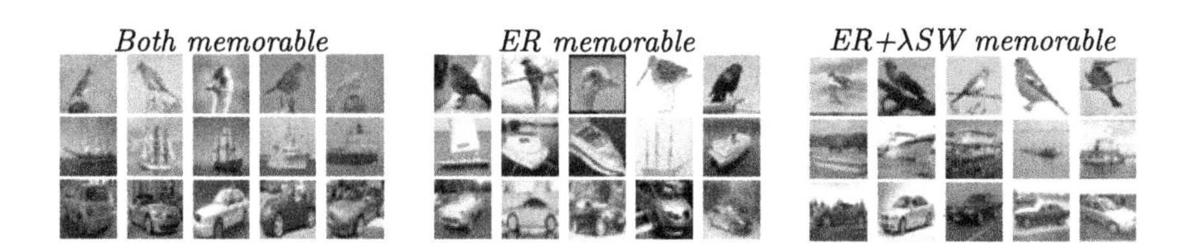

Fig. 7. Visual inspection on training instances in a single run for ER versus ER+λSW.

6 Conclusion

We proposed a λ Stability Wrapper (λSW) approach, where the learner uses a task-based policy to adjust the probability of when instances are replaced in memory to account for *task-recency bias.* Our method addresses the current gap in the field where current rehearsal-based methods implicitly deal with the imbalanced set of old and new task instances in the replay buffer while we explicitly control this balance over time. In our experiments, we showed that the proposed λSW method has significant gains in accuracy over the benchmark methods in eleven out of twelve cases. We observed that there are two factors affecting overall accuracy. First, the replay buffer batch size has a trade-off between memory memorization and the reduction of imbalance to gain higher accuracy by replaying more older task instances. Second, a new task with a high amount of training instances may be affected by forgetting within the task if the replay buffer is fully balanced towards the older task. In comparison, the model may still remember tasks with a small number of training instances.

References

1. Aljundi, R., et al.: Online continual learning with maximal interfered retrieval. In: Advances in Neural Information Processing Systems, vol. 32 (2019)
2. Aljundi, R., Lin, M., Goujaud, B., Bengio, Y.: Gradient based sample selection for online continual learning. In: Advances in Neural Information Processing Systems, vol. 32 (2019)
3. Bang, J., Kim, H., Yoo, Y., Ha, J.W., Choi, J.: Rainbow memory: continual learning with a memory of diverse samples. In: Proceedings of the IEEE/CVF Conference on Computer Vision and Pattern Recognition, pp. 8218–8227 (2021)
4. Belouadah, E., Popescu, A.: IL2M: class incremental learning with dual memory. In: Proceedings of the IEEE/CVF International Conference on Computer Vision, pp. 583–592 (2019)
5. Caccia, L., Aljundi, R., Asadi, N., Tuytelaars, T., Pineau, J., Belilovsky, E.: New insights on reducing abrupt representation change in online continual learning. In: International Conference on Learning Representations (2022). https://openreview.net/forum?id=N8MaByOzUfb
6. Castro, F.M., Marín-Jiménez, M.J., Guil, N., Schmid, C., Alahari, K.: End-to-end incremental learning. In: Ferrari, V., Hebert, M., Sminchisescu, C., Weiss, Y. (eds.) ECCV 2018. LNCS, vol. 11216, pp. 241–257. Springer, Cham (2018). https://doi.org/10.1007/978-3-030-01258-8_15
7. Chaudhry, A., et al.: On tiny episodic memories in continual learning. arXiv preprint arXiv:1902.10486 (2019)
8. Chen, Z., Liu, B.: Lifelong machine learning. Synth. Lect. Artif. Intell. Mach. Learn. **12**(3), 1–207 (2018)
9. Ebrahimi, S., Elhoseiny, M., Darrell, T., Rohrbach, M.: Uncertainty-guided continual learning in Bayesian neural networks, pp. 75–78 (2019)
10. Efraimidis, P.S., Spirakis, P.G.: Weighted random sampling with a reservoir. Inf. Process. Lett. **97**(5), 181–185 (2006)
11. Grossberg, S.: How does a brain build a cognitive code? In: Grossberg, S. (ed.) Studies of Mind and Brain, pp. 1–52. Springer, Dordrecht (1982). https://doi.org/10.1007/978-94-009-7758-7_1

12. Hou, S., Pan, X., Loy, C.C., Wang, Z., Lin, D.: Learning a unified classifier incrementally via rebalancing. In: Proceedings of the IEEE/CVF Conference on Computer Vision and Pattern Recognition, pp. 831–839 (2019)
13. Jastrzebski, S., et al.: The break-even point on optimization trajectories of deep neural networks. In: International Conference on Learning Representations (2019)
14. Jin, X., Sadhu, A., Du, J., Ren, X.: Gradient-based editing of memory examples for online task-free continual learning. Adv. Neural. Inf. Process. Syst. **34**, 29193–29205 (2021)
15. Krizhevsky, A., Hinton, G., et al.: Learning multiple layers of features from tiny images. Citeseer (2009)
16. Lopez-Paz, D., Ranzato, M.: Gradient episodic memory for continual learning. In: Advances in Neural Information Processing Systems, vol. 30 (2017)
17. Mai, Z., Li, R., Jeong, J., Quispe, D., Kim, H., Sanner, S.: Online continual learning in image classification: an empirical survey. Neurocomputing **469**, 28–51 (2022)
18. Mallya, A., Lazebnik, S.: Packnet: adding multiple tasks to a single network by iterative pruning. In: Proceedings of the IEEE Conference on Computer Vision and Pattern Recognition, pp. 7765–7773 (2018)
19. McCloskey, M., Cohen, N.J.: Catastrophic interference in connectionist networks: the sequential learning problem. In: Psychology of Learning and Motivation, vol. 24, pp. 109–165. Elsevier (1989)
20. Nene, S.A., Nayar, S.K., Murase, H., et al.: Columbia object image library (coil-100). Citeseer (1996)
21. Pan, P., Swaroop, S., Immer, A., Eschenhagen, R., Turner, R., Khan, M.E.E.: Continual deep learning by functional regularisation of memorable past. Adv. Neural. Inf. Process. Syst. **33**, 4453–4464 (2020)
22. Rannen, A., Aljundi, R., Blaschko, M.B., Tuytelaars, T.: Encoder based lifelong learning. In: Proceedings of the IEEE International Conference on Computer Vision, pp. 1320–1328 (2017)
23. Rebuffi, S.A., Kolesnikov, A., Sperl, G., Lampert, C.H.: ICARL: incremental classifier and representation learning. In: Proceedings of the IEEE Conference on Computer Vision and Pattern Recognition, pp. 2001–2010 (2017)
24. Rolnick, D., Ahuja, A., Schwarz, J., Lillicrap, T., Wayne, G.: Experience replay for continual learning. In: Advances in Neural Information Processing Systems, pp. 350–360 (2019)
25. Sun, S., Calandriello, D., Hu, H., Li, A., Titsias, M.: Information-theoretic online memory selection for continual learning. In: International Conference on Learning Representations (2022). https://openreview.net/forum?id=IpctgL7khPp
26. Vinyals, O., Blundell, C., Lillicrap, T., Wierstra, D., et al.: Matching networks for one shot learning. In: Advances in Neural Information Processing Systems, vol. 29 (2016)
27. Vitter, J.S.: Random sampling with a reservoir. ACM Trans. Math. Softw. (TOMS) **11**(1), 37–57 (1985)
28. Yan, S., Xie, J., He, X.: Der: Dynamically expandable representation for class incremental learning. In: Proceedings of the IEEE/CVF Conference on Computer Vision and Pattern Recognition, pp. 3014–3023 (2021)
29. Zeng, G., Chen, Y., Cui, B., Yu, S.: Continual learning of context-dependent processing in neural networks. Nat. Mach. Intell. **1**(8), 364–372 (2019)

Fair Healthcare Rationing to Maximize Dynamic Utilities

Aadityan Ganesh[1], Pratik Ghosal[1], Vishwa Prakash HV[1], and Prajakta Nimbhorkar[1,2](✉)

[1] Chennai Mathematical Institute, Chennai, India
{aadityanganesh,pratik,vishwa,prajakta}@cmi.ac.in
[2] UMI ReLaX, Chennai, India

Abstract. Allocation of scarce healthcare resources under limited logistic and infrastructural facilities is a major issue in the modern society. We consider the problem of allocation of healthcare resources like vaccines to people or hospital beds to patients in an online manner. Our model takes into account the arrival of resources on a day-to-day basis, different categories of agents, the possible unavailability of agents on certain days, and the utility associated with each allotment as well as its variation over time.

We propose a model where priorities for various categories are modelled in terms of utilities of agents. We give online and offline algorithms to compute an allocation that respects eligibility of agents into different categories, and incentivizes agents not to hide their eligibility for some category. The offline algorithm gives an optimal allocation while the online algorithm gives an approximation to the optimal allocation in terms of total utility. Our algorithms are efficient, and maintain fairness among different categories of agents. Our models have applications in other areas like refugee settlement and visa allocation. We evaluate the performance of our algorithms on real-life and synthetic datasets. The experimental results show that the online algorithm is fast and performs better than the given theoretical bound in terms of total utility. Moreover, the experimental results confirm that our utility-based model correctly captures the priorities of categories.

1 Introduction

Healthcare rationing has become an important issue in the world amidst the COVID-19 pandemic. At certain times, the scarcity of medical resources like vaccines, hospital beds, ventilators, medicines especially in developing countries raised the question of fair and efficient distribution of these resources. A New York Times article has mentioned this as one of the hardest decisions for health organizations [12]. A natural approach is to define priority groups [23,29]. However, allocating resources within the groups in a transparent manner is still challenging [11,30], including the concern for racial equity [5].

The authors contributed equally to this work and are listed in alphabetical order.

H. Kashima et al. (Eds.): PAKDD 2023, LNAI 13936, pp. 231–242, 2023.
https://doi.org/10.1007/978-3-031-33377-4_18

The healthcare rationing problem has been recently addressed by market designers. In [22], the problem was framed as a two-sided matching problem (see e.g. [26]), with reserve categories having their own priority ordering of people. This ordering is based on the policy decisions made according to various ethical guidelines. It is shown in [22] that running the Deferred Acceptance algorithm of Gale and Shapley [13] has desired properties like eligibility compliance, non-wastefulness and respect to priorities. This approach of [22] has been recommended or adopted by organizations like the NASEM (National Academies of Sciences, Engineering, and Medicine) [15]. It has also been recognized in medical literature [23,28], and is mentioned by the Washington Post [8]. The Smart Reserves algorithm of [22] gives a maximum matching satisfying the desired properties mentioned earlier. However, it assumes a global priority ordering on people. In a follow-up work, [2] generalize this to the case where categories are allowed to have heterogeneous priorities. Their Reverse Rejecting (REV) rule, and its extension to Smart Reverse Rejecting (S-REV) rule are shown to satisfy the goals like eligibility compliance, respect to priorities, maximum size, non-wastefulness, and strategyproofness.

However, the allocation of healthcare resources is an ongoing process. On a day-to-day basis, new units arrive in the market and they need to be allocated to people, depending on their availability, avoiding wastage. The previous models do not encompass this dynamic nature of resources. While priority groups or categories aim to model the urgency to allocate a resource to individuals by defining a priority order on people, defining a strict ordering is not practically possible nor desirable. Even if categories are allowed to have ties in their ordering, the ordering still provides only an ordinal ranking.

Our model provides the flexibility to have cardinal rankings in terms of prioritizing people by associating a *utility value* for each individual. The goal is to find an allocation with maximum total utility while respecting category quotas. However, utilities can change over time. We model this through *dynamic utilities*, that diminish over time by a multiplicative *discounting factor* $0 < \delta < 1$. Such exponential discounting is commonly used in economics literature [25,27]. Our utility maximization objective can thus be seen as maximization of *social welfare*. Our algorithms to find a maximum utility allocation are based on network flows. They adhere to the following important ethical principles which were introduced by Aziz et al. in [2]:

1. complies with the eligibility requirements
2. is strategyproof (does not incentivize agents to under-report the categories they qualify for or days what they are available on),
3. is non-wasteful (no unit is unused but could be used by some eligible agent)

Using category quotas and utility values, we provide a framework in which more vulnerable populations can be prioritized while maintaining a balance among the people vaccinated through each category on a day-to-day basis. Our models and algorithms are also applicable in other settings like school admissions [1], refugee settlement [3,10,10], visa allocation [4,7,9,20], hospital residents problem [2,18,19,21] etc. A detailed discussion about the related work can be found in the full version [14].

1.1 Our Models

We define our model below and then define its extension. Throughout this paper, we consider vaccines as the medical resource to be allocated. People are referred to as agents.

Model 1: Our model consists of a set of agents A, a set of categories C, and a set of days D. Vaccine shots available on day $d_j \in D$, called *daily supply*, is denoted by s_j. For each category $c_i \in C$, and each day $d_j \in D$, we define a *daily quota* q_{ij} which is the upper bound on the number of vaccines that can be allocated to c_i on day d_j. Each agent a_k has a priority factor α_k. Let $\alpha_{\max}$, and $\alpha_{\min}$ be the maximum and minimum priority factors. Utilities have a multiplicative *discount factor* $\delta \in (0,1)$. If agent a_k is vaccinated on day d_j, the utility obtained is $\alpha_k \cdot \delta^j$. Each agent a_k has an *availability vector* $v_k \in \{0,1\}^{|D|}$. The jth entry of v_k is 1 if and only if a_k is available for vaccination on day d_j.

Model 2: Model 2 is an extension of Model 1 in the following way. The sets A, C, D and the daily supply and daily quotas are the same as those in model 1. Apart from the daily quota, each category c_i also has an *overall quota* q_i that denotes the maximum total number of vaccines that can be allocated for category c_i over all the days. Note that overall quota is also an essential quantity in applications like visa allocation and refugee settlement.

In both the models, a matching $M : A \to (C \times D) \cup \{\emptyset\}$ is a function denoting the day on which a person is vaccinated and the category through which it is done, such that the category quota(s) and daily supply values do not exceed on any day. Thus if we define variables x_{ijk} such that $x_{ijk} = 1$ if $M(a_k) = (c_i, d_j)$ and $x_{ijk} = 0$ if $M(a_k) = \emptyset$, then we have $\sum_{i,j} x_{ijk} \leq 1$ for each k, $\sum_{k,j} x_{ijk} \leq q_i$ for each i, $\sum_k x_{ijk} \leq q_{ij}$ for each i, j, and $\sum_{i,k} x_{ijk} \leq s_j$ for each j. Here $1 \leq i \leq |C|, 1 \leq j \leq |D|, 1 \leq k \leq |A|$. If $M(a_k) = \emptyset$ for some $a_k \in A$, it means the person could not be vaccinated through our algorithm within $|D|$ days.

In both the models, the utility associated with a_k is $\alpha_k \cdot \delta^{j-1}$ where $M(a_k) = (c_i, d_j)$. The goal is to find a matching that maximizes the total utility.

1.2 Our Contributions

The utilities α_k and discounting factor δ have some desirable properties. If agent a_k is to be given a higher priority over agent a_ℓ, then we set $\alpha_k > \alpha_\ell$. On any day d_j, $\alpha_k \cdot \delta^j > \alpha_\ell \cdot \delta^j$. Moreover, the difference in the utilities of the two agents diminishes over time i.e. if $j < j'$ then $(\alpha_k - \alpha_\ell)\delta^j > (\alpha_k - \alpha_\ell)\delta^{j'}$. Thus the utility maximization objective across all days vaccinates a_k earlier than a_ℓ.

We consider both online and offline settings. The offline setting relies on the knowledge about availability of agents on all days. This works well in a system where agents are required to fill up their availability in advance e.g. in case of planned surgeries, and visa allocations. The online setting involves knowing the availability of all the agents only on the current day as in a *walk-in* setting. Thus the availability of an agent on a day in future is not known.

Following are our results:

Theorem 1. *There is a polynomial-time algorithm that computes an optimal solution for any instance of Model* 1 *in the offline setting.*

We give theoretical guarantees on the performance of online algorithms in terms of their *competitive ratio* in comparison with the utility of an offline optimal solution.

Theorem 2. *There is an online algorithm (Algorithm 1) that gives a competitive ratio of (i)* $1 + \delta$ *for Model* 1 *and (ii) of* $1 + \delta + (\alpha_{\max}/\alpha_{\min})\delta$ *for Model* 2 *when* δ *is the common discounting factor for all agents. The algorithm runs in polynomial time.*

The details omitted due to space constraints are available in the full version [14].

Strategy-proofness: It is a natural question whether agents benefit by hiding their availability on some days. We show that the online algorithm is strategy-proof by exhibiting that the offline setting has a *pure Nash equilibrium* that corresponds to the solution output by the online algorithm.

Theorem 3. *Let an offline optimal solution that breaks ties according to a random permutation* π *match agent* a_i *on day* d_i. *Then for each agent* a_i, *reporting availability exactly on day* d_i *(unmatched agents mark all days as unavailable) is a pure Nash equilibrium. Moreover, the Nash equilibrium corresponds to a solution output by the online algorithm.*

Experimental Results: We also give experimental results in Sect. 5 using real-world datasets. Apart from maximization of utilities, we also consider the number of days taken by the online algorithm for vaccinating high priority people. Our experiments show that the online algorithm almost matches the offline algorithm in terms of both of these criteria.

Selection of utility values: An important aspect of our model is that the choice of utility values does not affect the outcome as long as the utility values have the same numerical order as the order of priorities among agents. Thus the output of online as well as offline algorithm remains the same as long as $\alpha_k > \alpha_\ell$ whenever agent a_k has a higher priority over agent a_ℓ.

2 Algorithms for Model 1

We present two algorithms for this model - an optimal offline algorithm and an online algorithm. The offline algorithm is based on flow and runs in polynomial-time [14]. The online algorithm we present achieves a tight competitive ratio of $1 + \delta$, where δ is the discounting factor of the agents.

2.1 Online Algorithm for Model 1

We present an online algorithm which greedily maximizes utility on each day. We show that this algorithm indeed achieves a competitive ratio of $1 + \delta$.

Outline of the Algorithm: On each day d_i, starting from day d_1, we construct a bipartite graph $H_i = (A_i \cup C, E_i, w_i)$ where A_i is the set of agents who are available on day d_i and are not vaccinated earlier than day d_i. Let the weight of the edge $(a_j, c_k) \in E_i$ be $w_i(a_j, c_k) = \alpha_j.\delta^{i-1}$. We define capacity of the category $c_k \in C$ as $b'_{i,k}$. In this graph, our algorithm finds a maximum weighted b-matching of size not more than the daily supply value s_i.

Algorithm 1. Online Algorithm for Vaccine Allocation

Input: An instance I of Model 1
Output: A matching $M : A \to (C \times D) \cup \{\varnothing\}$

1: Let D, A, C be the set of Days, Agents and Categories respectively.
2: $M(a_j) \leftarrow \varnothing$ for each $a_j \in A$
3: **for** day d_i in D **do**
4: $\quad A_i \leftarrow \{a_j \in A \mid a_j$ is available on d_i and a_j is not vaccinated$\}$
5: $\quad E_i \leftarrow \{(a_j, c_k) \in A_i \times C \mid a_j$is eligible to be vaccinated under category c_k $\}$
6: $\quad$ **for** (a_j, c_k) in E_i **do**
7: $\quad\quad$ Let $w_i(a_j, c_k) \leftarrow \alpha_j \delta^{i-1}$
8: $\quad$ **end for**
9: $\quad$ Construct weighted bipartite graph $H_i = (A_i \cup C, E_i, w_i)$.
10: $\quad$ **for** c_k in C **do**
11: $\quad\quad$ $b'_{i,k} \leftarrow q_{ik}$ {Where q_{ik} is the daily quota}
12: $\quad$ **end for**
13: $\quad$ Find maximum weight b-matching M_i in H_i of size at most s_i. {Where s_i is the daily supply}
14: $\quad$ **for** each edge (a_j, c_k) in M_i **do**
15: $\quad\quad$ $M(a_j) \leftarrow (c_k, d_i)$ {Mark a_j as vaccinated on day d_i under category c_k}
16: $\quad$ **end for**
17: **end for**
18: **return** M

The following lemma shows that the maximum weight b-matching computed in Algorithm 1 is also a maximum size b-matching of size at most s_i.

Lemma 1. *The maximum weight b-matching in H_i of size at most s_i is also a maximum size b-matching of size at most s_i.*

Proof. We prove that applying an augmenting path in H_i increases the weight of the matching. Consider a matching M_i in H_i such that M_i is not of maximum size and $|M_i| < s_i$. Let $\rho = (a_1, c_1, a_2, c_2, \cdots, a_k, c_k)$ be an M_i-augmenting path in H_i. We know that every edge incident to an agent has the same weight in H_i. If we apply the augmenting path ρ, the weight of the matching increases by the weight of the edge (a_1, c_1). This proves that a maximum weight matching in H_i of size at most s_i is also a maximum size b-matching of size at most s_i.

2.2 Charging Scheme

We compare the solution obtained by Algorithm 1 with the optimal offline solution to get the worst-case competitive ratio for Algorithm 1. Let M be the output of Algorithm 1 and N be an optimal offline solution. To compare M and N, we devise a *charging scheme* by which, each agent a_p matched in N *charges* a unique agent a_q matched in M. The amount charged, referred to as the *charging factor* here is the ratio of utilities obtained by matching a_p and a_q in M and N respectively.

Properties of the charging scheme:

1. Each agent matched in N charges exactly one agent matched in M,
2. Each agent a_q matched in M is charged by at most two agents matched in N, with charging factors at most 1 and δ. This implies that the utility of N is at most $(1 + \delta)$ times the utility of M.

We divide the agents matched in N into two types. Type 1 agents are those which are matched in M on an earlier day compared to that in N. Thus $a_p \in A$ is a Type 1 agent if a_p is matched on day d_i in M and on day d_j in N, such that $i < j$. The remaining agents are called Type 2 agents. Our charging scheme is as follows:

1. Each Type 1 agent a_p charges themselves with a charging factor δ, since the utility associated with them in N is at most δ times that in M.
2. Here onwards, we consider only Type 2 agents and discuss the charging scheme associated with them.
   Let X_i be the set of Type 2 agents matched on day d_i in N, and let Y_i be the set of agents matched on day d_i in M. Since Algorithm 1 greedily finds a maximum size b-matching of size at most s_i, and as each edge in the b-matching corresponds to a unique agent, we show the following lemma holds:

Lemma 2. *For each $d_i \in D$, the set $|X_i| \leq |Y_i|$.*

Proof. Since X_i contains only Type 2 agents matched in N_i, the agents in X_i are not matched by M until day $i - 1$. Therefore $X_i \subseteq A_i$, where A_i is defined in Algorithm 1. The daily quota and the daily supply available for computation of N_i and M_i is the same i.e. $q_{i,k}$, and s_i respectively.

By construction, M_i is a matching that matches maximum number of agents in A_i, up to an upper limit of s_i, $|X_i| \leq |Y_i|$.

To obtain the desired competitive ratio we design an injective mapping according to which, each agent a_p in X_i can uniquely charge an agent a_q in Y_i such that $\alpha_p \leq \alpha_q$. The following lemma shows that such an injective mapping always exists.

Lemma 3. *There exists an injective mapping $f : X_i \to Y_i$ such that if $f(a_p) = a_q$, then $\alpha_p \leq \alpha_q$.*

Order of Charging Among Type 2 Agents: First, every agent who has both M_i and N_i edges indecent on it, charges herself. Next every agent who is an end-point of an even-length path charges the agent represented by the other end-point. The rest of the agents are end-points of an odd-length path matched in N_i. We proved that the edges incident on these agents have a weight smaller than every edge in M_i. They can charge any agent of M_i who has not been charged yet by any agent of N_i, as stated above.

Proof (of Theorem 2 (i)). Let a_q be an agent who is vaccinated by the online matching M on day i. Then a_q can be charged by at most two agents matched in N. Suppose a_q is vaccinated by the optimal matching N on some day $i' > i$. Assume that the agent a_p of type 2 who also charges a_q. If the priority factor of a_q and a_p are α_q and α_p respectively, then

$$\frac{\alpha_p.\delta^i + \alpha_q.\delta^{i'}}{\alpha_q.\delta^i} = \left(\frac{\alpha_p}{\alpha_q}\right)^i + \delta^{i'-i} \leq 1 + \delta.$$

The last inequality follows as $0 < \alpha_p \leq \alpha_q < 1$, and $i' > i$. Therefore the utility obtained by a_p and a_q in M_i is atmost $1 + \delta$ times the the utility of a_q in M_i. Therefore the competitive ratio of Algorithm 1 is at most $1 + \delta$.

2.3 Tight Example for the Online Algorithm

The example in Fig. 1 shows that the competitive ratio of Algorithm 1 is tight. Let $A = \{a_1, a_2\}$ be the set of agents and $C = \{c_1, c_2\}$ be the set of categories. Agent a_1 is eligible under $\{c_1, c_2\}$ and agent a_2 is eligible only under $\{c_2\}$. The daily supply: $s_1 = 1$ and $s_2 = 1$. The daily quota of each category on each day is set to 1. Both agents have equal priority. An optimal allocation scheme vaccinates all agents (highlighted in green), whereas the online allocation does not vaccinate agent a_2 if agent a_1 is vaccinated on day 1 under the category c_1 (highlighted in pink). Therefore the competitive ratio is $1 + \delta$, which is tight.

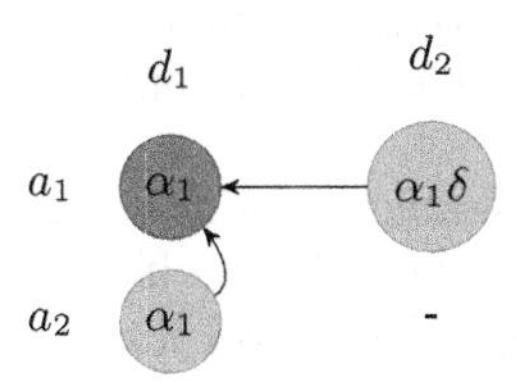

Fig. 1. A tight example with competitive ratio $1 + \delta$. Online allocation indicated in red, Optimal allocation indicated in green and arrows indicate charging (Color figure online)

3 Online Algorithm for Model 2

We present an online algorithm which greedily maximizes utility on each day. We assume that the discounting factor of the agents is δ. Moreover each agent a_k has a priority factor α_k. Let $\alpha_{\max} = \max_i\{\alpha_i \mid \alpha_i$ is the priority factor of agent $a_i\}$ and $\alpha_{\min} = \min_i\{\alpha_i \mid \alpha_i$ is the priority factor of agent $a_i\}$. We show that this algorithm indeed achieves a competitive ratio of $1 + \delta + \frac{\alpha_{\max}}{\alpha_{\min}}\delta$.

3.1 Outline of the Charging Scheme

We compare the solution obtained by the online algorithm for Model 2 with the optimal offline solution to get the worst-case competitive ratio for the online algorithm. Let M be the output of the Algorithm and N be an optimal offline solution. To compare M and N, we devise a *charging scheme* similar to that in Sect. 2.2, by which each agent a matched in N *charges* a unique agent a' matched in M. The amount charged, referred to as the *charging factor* here is the ratio of utilities obtained by matching a and a' in M and N respectively.

Properties of the charging scheme:

1. Each agent matched in N charges exactly one agent matched in M,
2. Each agent matched in M is charged by at most three agents matched in N, with charging factors at most $1, \delta$ and $\frac{\alpha_{\max}}{\alpha_{\min}}\delta$. This implies that the utility of N is at most $(1 + \delta + \frac{\alpha_{\max}}{\alpha_{\min}}\delta)$ times the utility of M.

Using analogous reasoning to that presented in Sect. 2.3, we demonstrate that the competitive ratio of Model 2 is also tight. For a more comprehensive description of this algorithm, please refer to [14].

4 Strategy-Proofness of the Online Algorithm

In the previous sections, we considered the case when agents truthfully report their availability and categories. In this section, we look into the scenario when agents are strategic, and might misreport their availability and/or eligible categories. Agents have an incentive in doing so because they prefer obtaining the healthcare resource over not getting it. Further, an agent would prefer to obtain the resource as early as possible. Offline optimal algorithm is prone to such strategic manipulations.

The online algorithm is strategy-proof because, on any day d_i, it does not consider availability on future days. Hence if an agent hides his availability on day d_i, the agent does not benefit. We show that there is a pure Nash equilibrium in the offline setting such that the utility of the Nash equilibrium has the same value as that of the solution obtained by the online algorithm. Thus, in the presence of strategic agents, the online algorithm has the advantage over the offline algorithm that it avoids Nash equilibria with very low utility value.

5 Experimental Evaluation

In Sect. 2 we prove worst-case guarantees for the online algorithm. We also give a tight example instance achieving a competitive ratio of $1+2\delta$. Here, we experimentally evaluate the performance of the online algorithm and compare it with the worst-case guarantees on a real-life dataset. For finding the optimal allocation that maximizes utility, we solve the network flow linear program with the additional constraint for overall quota $\sum_{i \in A, k \in D} x_{ijk} \leq q_j \quad \forall c_j \in C$. This LP is described in [14]. The code and datasets for the experiments can be found at [16]

5.1 Methodology

All experiments run on a 64-bit Ubuntu 20.04 desktop of 2.10 GHz * 4 Intel Core i3 CPU with 8 GB memory.

The proposed online approximation algorithm runs in polynomial time. In contrast, the optimal offline algorithm solves an integer linear program which might take exponential time depending on the integrality of the polytope. We relax the integrality constraints to achieve an upper bound on the optimal allocation. For comparing the performance of the online Algorithm 1 and the offline Algorithm, we use vaccination data of 24 hospitals in Chennai, India for the month of May 2022. We use small datasets with varying instance sizes for evaluating the running times of the algorithms. We use large datasets of smaller instance sizes for evaluating competitive ratios.

All the programs used for the simulation are written in Python language. For solving LP, ILP, and LPR, we use the general mathematical programming solver COIN-OR Branch and Cut solver MILP (Version: 2.10.3) [6] on PuLP (Version 2.6) framework [24]. When measuring the running time, we consider the time taken to solve the LP.

5.2 Datasets

Our dataset can be divided into two parts.

Supply: We consider vaccination data of twenty-four hospitals in Chennai, India for the month of May 2022. This data is obtained from the official COVID portal of India using the API's provided. The dataset consists of details such as daily vaccination availability, type of vaccines, age limit, hospital ID, hospital zip code, etc. for each hospital.

Demand: Using the Google Maps API [17], we consider the road network for these 24 hospitals in our dataset. From this data, we construct a complete graph with hospitals as vertices and edge weights as the shortest distance between any two hospitals. For each hospital $h \in H$, we consider the cluster $C(h)$ as the set of hospitals which are at most five kilometres away from h. We consider these clusters as our categories. Now, we consider 10000 agents who are to be vaccinated. For each agent a, we pick a hospital h uniformly at random. The agent a belongs to every hospital in the cluster $C(h)$. Each agent's availability over 30 days is independently sampled from the uniform distribution. Now, we consider the age-wise population distribution of the city. For each agent, we assign an age sampled from this distribution. Now, we partition the set of agents as agents of age 18–45 years, 45–60 years and 60+. We assign α-values $0.96, 0.97$ and 0.99 respectively. We also consider the same dataset with α-values $0.1, 0.5$ and 0.9 respectively. We set the discounting factor δ to 0.95.

For analyzing the running time of our algorithms, we use synthetically generated datasets with a varying number of instance sizes ranging from 100 agents to 20000 agents. Each agent's availability and categories are chosen randomly from a uniform distribution.

5.3 Results and Discussions

We show that the online algorithm runs significantly faster than the offline algorithm while achieving almost similar results. We give a detailed empirical evaluation of the running times in [14].

To compare the performance of the online Algorithm 1 against the offline algorithm we define a notion of *remaining fraction of un-vaccinated agents.* That is, on a given day d_i, we take the set of agents P_{d_i} who satisfy both of the following conditions:

1. Agent a is available on some day d_j on or before day d_i.
2. Agent a belongs to some hospital h and h has non-zero capacity on day d_j

P_{d_i} is the set of agents who could have been vaccinated without violating any constraints. Let $\gamma_i = |P_{d_i}|$.

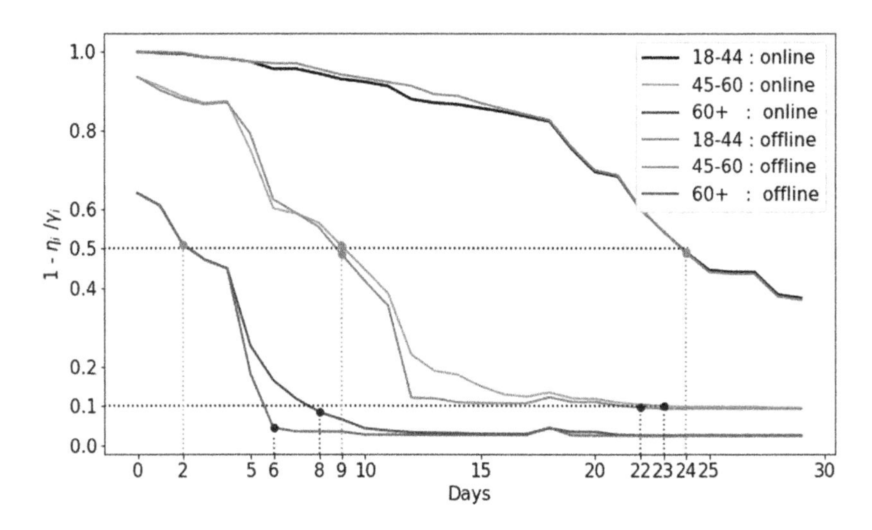

Fig. 2. The $1 - \eta_i/\gamma_i$ value achieved by the online algorithm is very similar to that of the offline algorithm across age groups. Both algorithms vaccinate achieves vaccinate 90% of the most vulnerable group within 8 days.

Let V_{d_i} be the set of agents who are vaccinated by the algorithm on or before day d_i. Let $\eta_i = |V_{d_i}|$. Now, $1 - \eta_i/\gamma_i$ represents the fraction of unvaccinated agents. In Fig. 2 we compare the age-wise $1-\eta_i/\gamma_i$ of both of our online and offline algorithms. We note that the vaccination priorities given to vulnerable groups by the online approximation algorithm are very close to that of the offline optimal algorithm. In both the algorithms, By the end of day 2, 50% of $1 - \eta_i/\gamma_i$ was achieved for agents of the 60+ age group. By the end of day 8, only 10% of the most vulnerable group remained unvaccinated.

6 Conclusion

We investigate the problem of dynamically allocating perishable healthcare goods to agents arriving over a period of time. We capture various constraints while allocating a scarce resource to a large population, like production constraint on the resource, infrastructure and constraints. While we give an offline optimal algorithm for Model 1, getting one for Model 2 or showing NP-hardness remains open.

We also propose an online algorithm approximating welfare that elicits information every day and makes an immediate decision. The online algorithm does not require a foresight and hence has a practical appeal. Our experiments show that the online algorithm generates a utility roughly equal to the utility of the offline algorithm while achieving very little to no wastage.

References

1. Abdulkadiroğlu, A., Sönmez, T.: School choice: a mechanism design approach. Am. Econ. Rev. **93**(3), 729–747 (2003)
2. Aziz, H., Brandl, F.: Efficient, Fair, and Incentive-Compatible Healthcare Rationing, pp. 103–104. Association for Computing Machinery, New York (2021)
3. Bansak, K., et al.: Improving refugee integration through data-driven algorithmic assignment. Science **359**(6373), 325–329 (2018)
4. Borjas, G.J., et al.: Immigration policy, national origin, and immigrant skills: a comparison of Canada and the united states. Small differences that matter: labor markets and income maintenance in Canada and the United States, pp. 21–43 (1993)
5. Bruce, L., Tallman, R.: Promoting racial equity in COVID-19 resource allocation. J. Med. Ethics **47**(4), 208–212 (2021)
6. COIN-OR branch and cut solver (2022). https://coin-or.github.io/Cbc/intro.html. Accessed 15 Aug 2022
7. Clark, X., Hatton, T.J., Williamson, J.G.: Explaining us immigration, 1971–1998. Rev. Econ. Stat. **89**(2), 359–373 (2007)
8. https://www.covid19reservesystem.org/media (2022). Accessed 14 Jan 2022
9. Delacrétaz, D.: Processing reserves simultaneously. In: Proceedings of the 22nd ACM Conference on Economics and Computation, pp. 345–346 (2021)
10. Delacrétaz, D., Kominers, S.D., Teytelboym, A.: Refugee Resettlement. University of Oxford, Department of Economics Working Paper (2016)

11. Emanuel, E.J., et al.: Fair allocation of scarce medical resources in the time of COVID-19. N. Engl. J. Med. **382**(21), 2049–2055 (2020)
12. Fink, S.: The hardest questions doctors may face: who will be saved? Who won't? The New York Times, March 2020
13. Gale, D., Shapley, L.S.: College admissions and the stability of marriage. Am. Math. Mon. **69**(1), 9–15 (1962)
14. Ganesh, A., Nimbhorkar, P., Ghosal, P., HV, V.P.: Fair healthcare rationing to maximize dynamic utilities (2023). https://arxiv.org/abs/2303.11053
15. Gayle, H., Foege, W., Brown, L., Kahn, B. (eds.): Framework for Equitable Allocation of COVID-19 Vaccine. National Academies of Sciences Engineering and Medicine (2020)
16. GitHub Repository of experimental evaluation: Fair_healthcare_rationing. https://github.com/severus-tux/Fair_Healthcare_Rationing
17. Google maps platform (2022). https://developers.google.com/maps. Accessed 15 Aug 2022
18. Kamada, Y., Kojima, F.: Efficient matching under distributional constraints: theory and applications. Am. Econ. Rev. **105**(1), 67–99 (2015)
19. Kamada, Y., Kojima, F.: Recent developments in matching with constraints. Am. Econ. Rev. **107**(5), 200–204 (2017)
20. Kato, T., Sparber, C.: Quotas and quality: the effect of h-1b visa restrictions on the pool of prospective undergraduate students from abroad. Rev. Econ. Stat. **95**(1), 109–126 (2013)
21. Kojima, F.: New Directions of Study in Matching with Constraints, pp. 479–482. Future of Economic Design, November 2019
22. Pathak, P.A., Sönmez, T., Ünver, M.U., Yenmez, M.B.: Fair Allocation of Vaccines, Ventilators and Antiviral Treatments: Leaving No Ethical Value Behind in Health Care Rationing. Boston College Working Papers in Economics 1015, Boston College Department of Economics, July 2020
23. Persad, G., Peek, M.E., Emanuel, E.J.: Fairly prioritizing groups for access to COVID-19 vaccines. JAMA **324**(16), 1601–1602 (2020)
24. PuLP optimization (2022). https://coin-or.github.io/pulp/. Accessed 14 Jan 2022
25. Ramsey, F.P.: A mathematical theory of saving. Econ. J. **38**(152), 543–559 (1928)
26. Roth, A.E., Sotomayor, M.A.O.: Two-Sided Matching: A Study in Game-Theoretic Modeling and Analysis. Econometric Society Monographs. Cambridge University Press, Cambridge (1990)
27. Samuelson, P.A.: A note on measurement of utility. Rev. Econ. Stud. **4**(2), 155–161 (1937)
28. Sönmez, T., Pathak, P.A., Ünver, M.U., Persad, G., Truog, R.D., White, D.B.: Categorized priority systems: a new tool for fairly allocating scarce medical resources in the face of profound social inequities. Chest **159**(3), 1294–1299 (2021)
29. Truog, R.D., Mitchell, C., Daley, G.Q.: The toughest triage - allocating ventilators in a pandemic. N. Engl. J. Med. **382**(21), 1973–1975 (2020)
30. WHO: A global framework to ensure equitable and fair allocation of COVID-19 products and potential implications for COVID-19 vaccines. Technical report, World Health Organization (2020)

A Multi-player MAB Approach for Distributed Selection Problems

Jinyu Mo[1] and Hong Xie[2(✉)]

[1] College of Computer Science, Chongqing University, Chongqing, China
374151683@qq.com

[2] Chongqing Institute of Green and Intelligent Technology, Chinese Academy of Sciences, Beijing, China
xiehong2018@foxmail.com

Abstract. Motivated by distributed selection problems, we formulate a new variant of multi-player multi-armed bandit (MAB) model, which captures stochastic arrival of requests to each arm and the policy of allocating requests to players. The challenge is how to design a distributed learning algorithm such that players select arms according to the optimal arm pulling profile without communicating to each other. We first design a greedy algorithm, which locates one of the optimal arm pulling profiles with a polynomial computational complexity. We also design an iterative distributed algorithm for players to commit to an optimal arm pulling profile with a constant number of rounds in expectation. We apply the explore then commit (ETC) framework to address the online setting when model parameters are unknown. We design an exploration strategy for players to estimate the optimal arm pulling profile. Since such estimates can be different across different players, it is challenging for players to commit. We then design an iterative distributed algorithm, which guarantees that players can arrive at a consensus on the optimal arm pulling profile in only M rounds. We conduct experiments to validate our algorithm.

1 Introduction

Multi-player MAB has attracted extensive attentions [1,3,4,6,12,17,19]. The canonical multi-player MAB model [1,4] was motivated from the channel access problem in cognitive radio applications. In this channel access problem, multiple secondary users (modeled as players) compete for multiple channels (modeled as arms). In each decision round, each player can select one arm. When collision happens (i.e., multiple players selecting the same arm), all players in the collision receive no reward. Players can not communicate with each other and they are aware of whether they encounter a collision or not. The objective is to maximize the total reward of all players. A number of algorithms were proposed [1,4,6,12, 17,19]. Various extensions of the canonical model were studied [6–9,15] and one can refer to Sect. 2 for more details.

Supplementary Information The online version contains supplementary material available at https://doi.org/10.1007/978-3-031-33377-4_19.

H. Kashima et al. (Eds.): PAKDD 2023, LNAI 13936, pp. 243–254, 2023.
https://doi.org/10.1007/978-3-031-33377-4_19

Existing multi-player MAB models are mainly built on the reward model that in each round either only one reward is generated from an arm [1,4,17], or multiple rewards are generated but the number of rewards equals the number of players [8,10,14]. And they are based on the collision assumption that when collision happens either all players in the collision receive no reward (one reward is generated), or each player receives an independent reward. The reward model and collision model make existing multi-player models not a satisfactory model for solving distributed selection problems arise from ridesharing applications like Uber food delivery applications like DoorDash, etc. For those applications, an arm can model a riding pickup location or food pickup port. A player can model a driver or a delivery driver. The ridesharing request or food delivery request arrives at an arm in a stochastic manner, which is independent of the number of players who will select this arm. In case of collision each player can serve at most one request in the manner that if the number of requests exceeds the number of players in the collision, each player serves one request and the remaining requests are unserved, and on the contrary the remaining players are idle serving no requests.

To model and design efficient algorithms for allocating requests to players, we formulate a new variant of the multi-player MAB model to address the above distributed selection problem. Our model consists of $M \in \mathbb{N}_+$ arms and $K \in \mathbb{N}_+$ players. For each arm, the number of requests across different decision rounds are independent and identically distributed (IID) samples from a probability distribution (called request distribution) and the reward of different requests are IID samples from another probability distribution (called reward distribution). The request distribution and reward distribution across different arms can be different. Each player is allowed to serve at most one request. In the request assigning process, there is no differentiation between players or between requests. Players can not communicate with each other. When a decision round ends, the platform makes the number of requests and the number of players on each arm public to all players. The objective is to maximize the total reward of all players without knowing the request distribution and reward distribution.

Example 1. Consider $M = 3$ arms and $K = 2$ players. For simplicity, in each time slot, two/two/one ride-sharing requests arrive at arm 1/2/3. Each request in arm 1/2/3 is associated with a reward of 0.2/0.2/0.3. Let $n_{t,1}, n_{t,2}, n_{t,3}$ denote the number of players who pull arm (or go to pickup location) 1, 2 and 3 respective. All possible arm pulling profiles, i.e., $(n_{t,1}, n_{t,2}, n_{t,3})$, can be expressed as $\{(2,0,0),(0,2,0),(0,0,2),(1,1,0),(0,1,1),(1,0,1)\}$. The total number of arm pulling profiles can be calculated as $\binom{M+K-1}{M-1} = \binom{4}{2} = 6$. Among them, the arm pulling profiles (0,1,1) and (1,0,1) achieve the highest total reward, i.e., $0.3 + 0.2 = 0.5$. Players can not communicate with each other in pulling arms. In the first time slot, the arm pulling profile can be $(0, 0, 2)$, i.e., both players pull arm 3. Then, one player serves a request and receive a reward of 0.3, and the other one gets no requests (the allocation of requests to players is specified by the application itself). In the second time slot, the arm pulling profile can be $(1, 1, 0)$, then each player gets a reward of 0.2.

Example 1 illustrates that under the simplified setting where reward and arrival are deterministic and given, the number of all possible arm pulling profiles is

$\binom{M+K-1}{M-1}$. Exhaustive search is computationally infeasible when the number of arms and players are large. In practice, both reward and arrival are stochastic capturing uncertainty in real-world applications, which further complicates the problem. *How to design computationally efficient searching algorithms to locate the optimal arm pulling profile under the offline setting?* Once the searching algorithm is developed, players can use it locally to locate an optimal arm pulling profile. As illustrated in Example 1, there can be multiple optimal arm pulling profiles. Players cannot communicate with each other on which one they should commit to. *How to design a distributed algorithm such that players will commit to an optimal arm pulling profile?* After we address the aforementioned two challenges, we turn our attention to address the online setting via the ETC framework, which was also used in previous works [5,17]. In the exploration phase, players estimate the request distribution and reward distribution. Using these estimates of distributions in the searching algorithm developed in the offline setting, each player can obtain an estimate of the optimal arm pulling profile. When there are multiple optimal arm pulling profiles, these estimates may classify some optimal arm pulling profiles as suboptimal ones. Different players can have different estimates of distributions and therefore have different estimates of arm pulling profiles. What makes it challenging is that the estimate of the optimal arm pulling profile at one player may be classified as a suboptimal one at some other players. Increasing the length of exploration phase can reduce this uncertainty but it may induce a larger regret. *How to determine the exploration length? How to make players commit to an optimal arm pulling profile?*

We address them and our contributions are: (1) In the offline setting with model given, we design a greedy algorithm which can locate one of the optimal arm pulling profiles with a computational complexity of $O(KM)$. We also design an iterative distributed algorithm for players to commit to a unique optimal arm pulling profile with a constant number of rounds in expectation. When there are multiple optimal arm pulling profiles, by using the same deterministic tie breaking rule in the greedy algorithm, all players can locate the same optimal arm pulling profile and then our committing algorithm can be applied. (2) In the online setting with unknown model, we design an exploration strategy with a length such that each player can estimate one of the optimal arm pulling profiles with high probability. These estimates of the optimal arm pulling profile can be different across different players. We design an iterative distributed algorithm, which guarantees that players reach a consensus on the optimal arm pulling profile with only M rounds. Players can then run the commit algorithm developed in the offline setting to commit to this consensus. Putting them together, we obtain an algorithm with a logarithmic regret. We conduct experiments to validate the efficiency of our algorithms.

2 Related Work

Stochastic Multi-player MAB with Collision. The literature on multi-player MAB starts from a static (i.e., the number of players is fixed) and informed collision (in each round, players know whether they encounter a collision or not) setting. In this setting, Liu *et al.* [11] proposed a time-division fair sharing algorithm, which attains a logarithmic total regret in the asymptotic sense. Anandkumar *et al.* [1] proposed an algorithm with a logarithmic total regret in the finite number of rounds sense. Rosenski *et al.* [17] proposed a communication-free algorithm with constant regret in the high probability sense. Besson *et al.* [4] improved the regret lower bound, and proposed *RandTopM* and *MCTopM* which outperform existing algorithms empirically. The regret of these algorithms depends on the gaps of reward means. Lugosi *et al.* [12] suggested the idea of using collision information as a way to communicate and they gave the first square-root regret bounds that do not depend on the gaps of reward means. Boursier *et al.* [6] further explored the idea of using collision information as a way to communicate and they proposed the SIC-MMAB algorithm, which attains the same performance as a centralized one. Inspired SIC-MMAB algorithm, Shi *et al.* [18] proposed the error correction synchronization involving communication algorithm, which attains the regret of a centralized one. Hanawal *et al.* [9] proposed the leader-follower framework and they developed a trekking approach, which attains a constant regret. Inspired by the leader-follower framework, Wang *et al.* [19] proposed the DPE1 algorithm which attains the same asymptotic regret as that obtained by an optimal centralized algorithm. A number of algorithms were proposed for the static but unknown collision setting. In particular, Besson *et al.* [4] proposed a heuristic with nice empirical performance. Lugosi *et al.* [12] developed the first algorithm with theoretical guarantees, i.e., logarithmic regret, and an algorithm with a square-root regret type that does not depend on the gaps of the reward means. Shi *et al.* [18] identified a connection between communication phase without collision information and Z-channel model in information theory. Bubeck *et al.* [7] proposed an algorithm with near-optimal regret $O(\sqrt{T\log(T)})$. They argued that the logarithmic term $\sqrt{\log(T)}$ is necessary. A number of algorithms were proposed for the dynamic (i.e., players can join or leave) and informed collision setting. In particular, Avner *et al.* [2] proposed an algorithm with a regret of $O(T^{2/3})$. Rosenski *et al.* [17] proposed the first communication-free algorithms which attains a regret of $O(\sqrt{T})$. Boursier [6] proposed a SYN-MMAB algorithm with the logarithmic growth of the regret. Hanawal *et al.* [9] proposed an algorithm based on a trekking approach. All the above works considered a homogeneous setting, i.e., all players have the same reward mean over the same arm. A number of works studied the heterogeneous setting, i.e., different players may have different reward mean over the same arm. Bistritz *et al.* [5] proposed the first algorithm which attains a total regret of order $O(\ln^2 T)$. Magesh *et al.* [13] proposed an algorithm which attains a total regret of $O(\log T)$. Mehrabian *et al.* [15] proposed an algorithm which attains a total regret of $O(\ln(T))$ and it solved an open question in [5].

Stochastic Multi-player MAB Without Collision. The typical setting is that when collision happens, each player in the collision obtain an independent reward. Players can share their reward information using a communication graph. In this setting, Landgren *et al.* [10] proposed a decentralized algorithm which utilizes a running consensus algorithm for agents to share reward information. Martínez-Rubio *et al.* [14] proposed a DD-UCB algorithm which utilizes a consensus procedure to estimate reward mean. Wang *et al.* [19] proposed DPE2 algorithm which is optimal in the symptotic sense and it outperforms DD- UCB [14]. Dubey *et al.* [8] proposed MP-UCB to handle heavy tail reward.

Summary of Difference. Different from the above works, we formulate a new variant of multi-player multi-armed bandit (MAB) model to address the distributed selection problems. From a modeling perspective, our model captures stochastic arrival of request and request allocation policy of these applications. From an algorithmic perspective, our proposed algorithms presents new ideas in searching the optimal pulling profile, committing to optimal pulling profile, achieving consensus when different players have different estimates on the optimal arm pulling profile, etc.

3 Platform Model and Problem Formulation

The Platform Model. We consider a platform composed of requests, players and a platform operator. We use a discrete time system indexed by $t \in \{1, \ldots, T\}$, where $T \in \mathbb{N}_+$, to model this platform. The arrival of requests is modeled by a finite set of arms denoted by $\mathcal{M} \triangleq \{1, \ldots, M\}$, where $M \in \mathbb{N}_+$. Each arm can be mapped as a pickup location of ride sharing applications or a pickup port of food delivery applications. Each arm $m \in \mathcal{M}$ is characterized by a pair of random vectors $(\boldsymbol{D}_m, \boldsymbol{R}_m)$, where $\boldsymbol{D}_m \triangleq [D_{t,m} : t = 1, \ldots, T]$ and $\boldsymbol{R}_m \triangleq [R_{t,m} : t = 1, \ldots, T]$ model the stochastic request and reward of arm m across time slots $1, \ldots, T$. More concretely, the random variable $D_{t,m}$ models the number of requests arrived at arm m in time slot t, and the support of $D_{t,m}$ is $\mathcal{D} \triangleq \{1, \ldots, d_{\max}\}$, where $d_{\max} \in \mathbb{N}_+$. Each request can be mapped as a ride sharing request or a food delivering request. We consider a stationary arrival of requests, i.e., $D_{1,m}, \ldots, D_{t,m}$ are independent and identically distributed (IID) random variables. Note that in each time slot unserved requests will be dropped. This captures the property of ride sharing like applications that a customer may not wait until he is served, but instead he will cancel the ride sharing request and try other alternatives such as buses if he is not severed in a time slot. Let $\boldsymbol{p}_m \triangleq [p_{m,d} : \forall d \in \mathcal{D}]$ denote the probability mass function (pmf) of these IID random variables $D_{1,m}, \ldots, D_{t,m}$, formally $p_{m,d} = \mathbb{P}[D_{t,m} = d], \forall d \in \mathcal{D}, m \in \mathcal{M}$. In time slot t, the rewards of $D_{t,m}$ requests are IID samples of the random variable $R_{t,m}$. Without loss of generality we assume the support of $R_{t,m}$ is $[0, 1]$. The rewards $R_{1,m}, \ldots, R_{t,m}$ are IID random variables. We denote the mean of these IID random variables $R_{1,m}, \ldots, R_{t,m}$ as $\mu_m = \mathbb{E}[R_{t,m}], \forall m \in \mathcal{M}$. For simplicity, we denote the reward mean vector as $\boldsymbol{\mu} \triangleq [\mu_m : \forall m \in \mathcal{M}]$ and probability mass matrix as $\boldsymbol{P} \triangleq [\boldsymbol{p}_1, \ldots, \boldsymbol{p}_M]^{\mathrm{T}}$. Both $\boldsymbol{P}$ and $\boldsymbol{\mu}$ are unknown to players.

We consider a finite set of players denoted by $\mathcal{K} \triangleq \{1, \ldots, K\}$, where $K \in \mathbb{N}_+$. Each player can be mapped as a driver in ride sharing applications, or a deliverer in food deliverer applications. In each time slot t, each player is allowed to pull only one arm. Let $a_{t,k} \in \mathcal{M}$ denote the action (i.e., the arm pulled by player k) of player k in time slot t. We consider a distributed setting that players can not communicated with each other. Let $n_{t,m} \triangleq \sum_{k\in\mathcal{K}} \mathbb{1}_{\{a_{t,k}=m\}}$ denote the number of players who pull arm m. The $n_{t,m}$ satisfies that $\sum_{m\in\mathcal{M}} n_{t,m} = K$. Namely, all players are assigned to arms. Recall that in round t, the number of requests arrived at arm m is $D_{t,m}$. These $D_{t,m}$ requests will be allocated to $n_{t,m}$ players randomly (our algorithm can be applied to other assignment policies also). Regardless of the details of the allocation policy, two desired properties of the allocation is: (1) if the number of requests is larger than the number of players, i.e., $D_{t,m} \geq n_{t,m}$, then each player serves one request and $(D_{t,m} - n_{t,m})$ request remains unserved, (2) if the number of requests is smaller than the number of players, i.e., $D_{t,m} \leq n_{t,m}$, then only $D_{t,m}$ players can serve requests (one player per request) and $(n_{t,m} - D_{t,m})$ remains idle.

Let $\boldsymbol{n}_t \triangleq [n_{t,m} : \forall m \in \mathcal{M}]$ denote the action profile in time slot t. Let $\widetilde{\boldsymbol{D}}_t \triangleq [D_{t,m} : \forall m \in \mathcal{M}]$ denote the request arrival profile in time slot t. At the end the each time slot, the platform operator makes $\boldsymbol{n}_t$ and $\widetilde{\boldsymbol{D}}_t$ public to all players. The platform operator ensures that each player is allowed to serve at most one request. When the number of requests exceeds the number of players who pull the arm, each player serves one request and the remaining requests are unserved, and on the contrary, the remaining players will be idle.

Online Learning Problem. Let $U_m(n_{t,m}, \boldsymbol{p}_m, \mu_m)$ denote the total reward earned by $n_{t,m}$ players pull arm m. Then, it can be expressed as: $U_m(n_{t,m}, \boldsymbol{p}_m, \mu_m) = \mu_m \mathbb{E}[\min\{n_{t,m}, D_{t,m}\}]$. Denote the total reward for earned by all players in time slot t as $U(\boldsymbol{n}_t, \boldsymbol{P}, \boldsymbol{\mu}) \triangleq \sum_{m\in\mathcal{M}} U_m(n_{t,m}, \boldsymbol{p}_m, \mu_m)$. The objective of players is to maximize the total reward across T time slots, i.e., $\sum_{t=1}^{T} U(\boldsymbol{n}_t, \boldsymbol{P}, \boldsymbol{\mu})$. The optimal arm pulling profile of players is $\boldsymbol{n}^* \in \arg\max_{\boldsymbol{n}\in\mathcal{A}} U(\boldsymbol{n}, \boldsymbol{P}, \boldsymbol{\mu})$, where $\mathcal{A} \triangleq \{(n_1, \ldots, n_M) | n_m \in \mathcal{K} \cup \{0\}, \sum_{m\in\mathcal{M}} n_m = K\}$ is defined as a set of all arm pulling profiles. There are $|\mathcal{A}| = \binom{M+K-1}{M-1}$ possible arm pulling profiles, which poses a computational challenge in searching the optimal arm pulling profile. Furthermore, both the probability mass matrix $\boldsymbol{P}$ and the reward mean vector $\boldsymbol{\mu}$ are unknown to players and players can not communicate to each other. Denote $\mathcal{H}_{t,k} \triangleq (a_{1,k}, X_{1,k}, \boldsymbol{n}_1, \widetilde{\boldsymbol{D}}_1, \ldots, a_{t,k}, X_{t,k}, \boldsymbol{n}_t, \widetilde{\boldsymbol{D}}_t)$ as the historical data available to player k up to time slot t. Each player has access to his own action history and reward history as well as the arm pulling profile history and request arrival profile which are made public by the platform operator. Denote the regret as $R_T \triangleq \sum_{t=1}^{T}(U(\boldsymbol{n}^*, \boldsymbol{P}, \boldsymbol{\mu}) - \mathbb{E}[U(\boldsymbol{n}_t, \boldsymbol{P}, \boldsymbol{\mu})])$.

4 The Offline Optimization Problem

Searching the Optimal Arm Pulling Profile. We define the marginal reward gain function as: $\Delta_m(n) \triangleq U_m(n+1, \boldsymbol{p}_m, \mu_m) - U_m(n, \boldsymbol{p}_m, \mu_m)$.

Lemma 1. *The* $\Delta_m(n) = \mu_m P_{m,n+1}$*, where* $P_{m,n+1} \triangleq \sum_{d=n+1}^{d_{\max}} p_{m,d}$*. Furthermore,* $\Delta_m(n+1) \leq \Delta_m(n)$*.*

Based on Lemma 1, Algorithm 1 searches the optimal arm pulling profile by sequentially adding K players one by one to pull arms. More specifically, players are added to arms sequentially according to their index in ascending order. Player with index k, is added to the arm with the largest marginal reward gain given the assignment of players indexed by $1, \dots, k-1$. When all players are added to arms, the resulting arm pulling profile is returned as an output. For simplicity, denote $\widetilde{\boldsymbol{P}} = [\widetilde{\boldsymbol{P}}_1, \dots, \widetilde{\boldsymbol{P}}_M]^T$, where $\widetilde{\boldsymbol{P}}_m = [P_{m,k} : \forall k \in \mathcal{K}]$.

Algorithm 1. `OptArmPulProfile` $(\boldsymbol{\mu}, \widetilde{\boldsymbol{P}})$

1: $\Delta_m \leftarrow \mu_m, \forall m \in \mathcal{M}, \quad n_{\text{greedy},m} \leftarrow 0, \forall m \in \mathcal{M}$
2: **for** $k = 1, \dots, K$ **do**
3: $\quad i = \arg\max_{m \in \mathcal{M}} \Delta_m$ (if there is a tie, breaking it arbitrarily)
4: $\quad n_{\text{greedy},i} \leftarrow n_{\text{greedy},i} + 1, \quad \Delta_i \leftarrow \mu_i P_{i,n_{\text{greedy},i}}$
5: **end for**
6: **return** $\boldsymbol{n}_{\text{greedy}} = [n_{\text{greedy},m} : \forall m \in \mathcal{M}]$

Theorem 1. *The output of Algorithm 1 satisfies $\boldsymbol{n}_{greedy} \in \arg\max_{\boldsymbol{n} \in \mathcal{A}} U(\boldsymbol{n}, \boldsymbol{P}, \boldsymbol{\mu})$. The computational complexity of $\boldsymbol{n}_{greedy}$ of Algorithm 1 is $O(KM)$.*

Committing to Optimal Arm Pulling Profile. We focus on the case with a unique optimal arm pulling profile. Due to page limit, the case with multiple optimal pulling profiles is handled in our technical report [16]. Each player first applies Algorithm 1 to locate the unique optimal pulling profile $\boldsymbol{n}^*$. In each round t, player k selects arm based on $\boldsymbol{n}^*$ and $\boldsymbol{n}_1, \dots, \boldsymbol{n}_{t-1}$. Our objective is that n^*_m players commit to arm m. Let $c_{t,k} \in \{0\} \cup \mathcal{M}$ denote the index of the arm that player k commits to. The $c_{t,k}$ is calculated from $\boldsymbol{n}_1, \dots, \boldsymbol{n}_t$ and $\boldsymbol{n}^*$ as follows. Initially, player k sets $c_{0,k} = 0$ representing that he has not committed to any arm yet. In each time slot t, after $\boldsymbol{n}_t$ is published, each player k calculates $c_{t,k}$ based on $\boldsymbol{n}_t$, $\boldsymbol{n}^*$ and $c_{t-1,k}$ as follows:

$$c_{t,k} = \mathbb{1}_{\{c_{t-1,k} \neq 0\}} c_{t-1,k} + \mathbb{1}_{\{c_{t-1,k}=0\}} \mathbb{1}_{\{n_{t,a_{t,k}} \leq n^*_{a_{t,k}}\}} a_{t,k}. \tag{1}$$

Equation (1) states that if player k has committed to an arm, i.e., $c_{t-1,k} \neq 0$, this player will stay committed to this arm. In other words, once a player commits to an arm, he will keep pulling it in all remaining time slots. If player k has not committed to any arm yet, i.e., $c_{t-1,k} = 0$, player k commits to the arm he pulls $a_{t,k}$, only if the number of players pull the same arm does not exceed the number of players required by this arm, i.e., $n_{t,a_{t,k}} \leq n^*_{a_{t,k}}$. In each round only the players who have not committed to any arm need to selecting different arms.

To assist players who have not committed to any arm selecting arms, for each arm, each player keeps a track of the number of players who have committed to it. Let $n^+_{t,m}$ denote the number of players committing to arm m up to time slot t. Initially, no players commit to each arm, i.e., $n^+_{0,m} = 0, \forall m \in \mathcal{M}$. After round t, each player uses the following rule to update $n^+_{t,m}$:

$$n^+_{t,m} = \mathbb{1}_{\{n_{t,m} \leq n^*_m\}} n_{t,m} + \mathbb{1}_{\{n_{t,m} > n^*_m\}} n^+_{t-1,m}. \tag{2}$$

Equation (2) states an update rule that is consistent with Eq. (1). More concretely, if the number of players $n_{t,m}$ pull arm m does not exceed the optimal number of players n_m^* for arm m, then all these $n_{t,m}$ players commit to arm m. Otherwise, as $n_{t,m} > n_m^*$, it is difficult for players to decide who needs to commit to arm m without communication. According to Eq. (1), the commitment status of players pulling arm m does not update, resulting that the number of player commit to arm m remains unchanged. Equation (2) implies that $n_{t,m}^+ \leq n_m^*$.

In time slot t, each player k selects arm based on $c_{t-1,k}$ and $\boldsymbol{n}_{t-1}^+ \triangleq [n_{t-1,m}^+ : \forall m \in \mathcal{M}]$ as follows. If player k has committed to an arm, i.e., $c_{t-1,k} \neq 0$, this player sticks to the arm that he commits to. Otherwise, player k has not committed to any arm yet, and he needs to select an arm. To achieve this, player k first calculates the number of players that each arm m lacks, which is denoted by $n_{t-1,m}^- \triangleq n_m^* - n_{t-1,m}^+$. Note that $n_m^* - n_{t-1,m}^+ \geq 0$ because Eq. (2) implies that $n_{t-1,m}^+ \leq n_m^*$. Then, player k selects an arm with a probability proportional to the number of players that the arm lacks, i.e., selects arm m with probability $n_{t-1,m}^-/N_{t-1}^-$, where $N_{t-1}^- \triangleq \sum_{m\in\mathcal{M}} n_{t-1,m}^-$. We summarize the arm selection strategy as follows: $\mathbb{P}[a_{t,k} = m] = \mathbb{1}_{\{c_{t-1,k}\neq 0\}}\mathbb{1}_{\{c_{t-1,k}=m\}} + \mathbb{1}_{\{c_{t-1,k}=0\}} n_{t-1,m}^-/N_{t-1}^-$. Players use the same committing strategy, Algorithm 2 uses player k as an example to outline the above committing strategy, where T_{start} denotes the index of the time slot that players start committing. We prove that Algorithm 2 terminates in a constant number of rounds in expectation. Due to page limit, we present them in our technical report [16].

Algorithm 2. `CommitOptArmPulProfile` $(k, \boldsymbol{n}^*, T_{\text{start}})$

1: $c_k \leftarrow 0$, $n_{t,m}^- \leftarrow n_m^*, \forall m \in \mathcal{M}$, $n_{t,m}^+ \leftarrow 0, \forall m \in \mathcal{M}$
2: **for** $t = T_{\text{start}}, \ldots, T$ **do**
3:     **if** $c_k \neq 0$ **then**
4:         $a_{t+1,k} \leftarrow c_k$
5:     **else if** $n_{a_{t,k}}^* \geq n_{t,a_{t,k}}$ **then**
6:         $c_k \leftarrow a_{t,k}$, $a_{t+1,k} \leftarrow a_{t,k}$
7:     **else**
8:         **for** $m = 1, \ldots, M$ **do**
9:             **if** $n_m^* \geq n_{t,m}$ **then**
10:                 $n_{t,m}^+ \leftarrow n_{t,m}$, $n_{t,m}^- \leftarrow n_m^* - n_{t,m}^+$
11:             **end if**
12:         **end for**
13:         $N_t^- = \sum_{m\in\mathcal{M}} n_{t,m}^-$. $a_{t+1,k} \leftarrow m$, w.p. $n_{t,m}^-/N_t^-$
14:     **end if**
15:     player k pulls arm $a_{t+1,k}$, player k receives $\boldsymbol{n}_{t+1}$
16: **end for**

5 Online Learning Algorithm

We address the online setting with unknown $\boldsymbol{\mu}$ and $\boldsymbol{P}$ via the ETC framework.

Exploration Phase. In the exploration, each player aims to estimate the probability $\widetilde{\boldsymbol{P}}$ and the mean vector $\boldsymbol{\mu}$. We consider a random exploring strategy that in each decision round, each player randomly selects an arm. Let T_0 be the length of the exploration phase and $X_{t,k}$ be the reward the player k gets at round t. Each player uses the same exploration strategy, and Algorithm 3 uses player $k \in \mathcal{K}$ as an example to illustrate our exploration strategy. We quantify the impact of the length of exploration phase on the accuracy of estimating $\boldsymbol{\mu}$ and $\widetilde{\boldsymbol{P}}$. We present them in our technical report [16].

Algorithm 3. Explore(k, T_0)

1: $S_m \leftarrow 0$, $\widehat{P}^{(k)}_{m,d} \leftarrow 0$, $\widetilde{c}_m \leftarrow 0, \forall m \in \mathcal{M}$
2: **for** $t = 1, \ldots, T_0$ **do**
3: $\quad a_{t,k} \sim U(1, \ldots, M)$. Player k receives $\widetilde{\boldsymbol{D}}_t$ and $X_{t,k}$
4: $\quad$**if** $X_{t,k} \neq$ null **then**
5: $\quad\quad S_{a_{t,k}} \leftarrow S_{a_{t,k}} + X_{t,k}, \quad \widetilde{c}_{a_{t,k}} \leftarrow \widetilde{c}_{a_{t,k}} + 1$
6: $\quad$**end if**
7: $\quad \widehat{P}^{(k)}_{m,d} \leftarrow \widehat{P}^{(k)}_{m,d} + \mathbb{1}_{\{\widetilde{D}_{t,m} \geq d\}}, \forall d \in \mathcal{D}, m \in \mathcal{M}$
8: **end for**
9: $\widehat{\mu}^{(k)}_m \leftarrow S_m/\widetilde{c}_m, \forall m \in \mathcal{M}, \quad \widehat{P}^{(k)}_{m,d} \leftarrow \widehat{P}^{(k)}_{m,d}/T_0, \forall d, m$
10: **return** $\widehat{\boldsymbol{\mu}}^{(k)} = [\widehat{\mu}^{(k)}_m : \forall m]$, $\widehat{\boldsymbol{P}}^{(k)} = [\widehat{P}^{(k)}_{m,d} : \forall m, d]$

Committing Phase. When the optimal arm pulling profile is not unique, it is highly likely that players have different estimates on the optimal arm pulling profile, i.e., $\exists k, \tilde{k}$ such that $\widehat{\boldsymbol{n}}^*(k) \neq \widehat{\boldsymbol{n}}^*(\tilde{k})$. This creates a challenge in committing to the optimal arm pulling profile. To address this challenge, we make the following observations. An element $\Delta_m(n)$ is a borderline element if $\Delta_m(n) = \Delta^{(K)}$.

Lemma 2. *Suppose $p_{m,d} > 0$ holds for all $m \in \mathcal{M}$ and $d \in \mathcal{D}$. Suppose $\boldsymbol{n}^*$ and $\widetilde{\boldsymbol{n}}^*$ denote to optimal arm pulling profiles. Then, $|n^*_m - \widetilde{n}^*_m| \leq 1, \forall m \in \mathcal{M}$ and if $n^*_m \neq \widetilde{n}^*_m$, $\Delta_m(\max\{n^*_m, \widetilde{n}^*_m\})$ is a borderline element.*

Based on Lemma 2, Algorithm 4 uses player k as an example to illustrate our consensus algorithm, which enables players to reach a consensus on the optimal arm pulling profile. Note that Algorithm 4 focuses on the case that each player has an accurate estimate of the optimal arm pulling profile, i.e., $\widehat{\boldsymbol{n}}^*(k) \in \arg\max_{\boldsymbol{n} \in \mathcal{A}} U(\boldsymbol{n}, \boldsymbol{P}, \boldsymbol{\mu})$, but their estimates can be different. First, all players run M rounds to identify disagreements in their estimates of optimal arm pulling profiles (step 2 to 11). In each round, they check disagreements on one arm, if they identify one disagreement, each player records the corresponding borderline elements. After this phase, each player eliminates the identified borderline elements from its estimate on the optimal arm pulling profile (step 12 to 16). After this elimination, players agree on their remaining arm pulling profile, but this arm pulling profile only involves a number of players less than K. Thus, finally, each player adding the same number of borderline elements as the number of eliminated ones back, using the same rule, i.e., guaranteeing that all players add the same borderline elements back (step 17 and 18).

Algorithm 4. $\texttt{Consensus}(k, \widehat{\boldsymbol{n}}^*(k))$

1: $\boldsymbol{v}_{\text{board},k} \leftarrow \boldsymbol{0}$, $\mathcal{V}_{\text{board},k} \leftarrow \emptyset$, Num$\leftarrow$0
2: **for** $t = T_0 + 1, \dots, T_0 + M$ **do**
3: player k pulls arm $(\widehat{n}^*_{t-T_0}(k) \mod M + 1)$
4: **if** $\max\{m|n_{t,m} > 0\} - \min\{m|n_{t,m} > 0\} == 1$ **then**
5: Update borderline elem. $\boldsymbol{v}_{\text{board},k} \leftarrow (\boldsymbol{v}_{\text{board},k}, t - T_0)$
6: $\mathcal{V}_{\text{board},k} \leftarrow \mathcal{V}_{\text{board},k} \cup \{(t - T_0, \widehat{n}^*_{t-T_0}(k) + \mathbb{1}_{\{\widehat{n}^*_{t-T_0}(k) \mod M+1=\min\{m|n_{t,m}>0\}\}})\}\}$
7: **else if** $\max\{m|n_{t,m} > 0\} - \min\{m|n_{t,m} > 0\} > 1$ **then**
8: $\boldsymbol{v}_{\text{board},k} \leftarrow (\boldsymbol{v}_{\text{board},k}, t - T_0)$
9: $\mathcal{V}_{\text{board},k} \leftarrow \mathcal{V}_{\text{board},k} \cup \{(t - T_0, \widehat{n}^*(k) + \mathbb{1}_{\{\widehat{n}^*_{t-T_0}(k) \mod M+1=\max\{m|n_{t,m}>0\}\}})\}$
10: **end if**
11: **end for**
12: **for** $m = 1, \dots, M$ **do**
13: **if** $\{(m, \widehat{n}^*_m(k))\} \in \mathcal{V}_{\text{board},n}$ **then**
14: $\widehat{n}^*_m(k) \leftarrow \widehat{n}^*_m(k) - 1$, Num$\leftarrow$Num $+1$
15: **end if**
16: **end for**
17: $\boldsymbol{v}_{\text{board},n} \leftarrow \boldsymbol{v}_{\text{board},n}$ sorted in descending order
18: $\widehat{n}^*_{\boldsymbol{v}_{\text{board},k}(i)}(k) \leftarrow \widehat{n}^*_{\boldsymbol{v}_{\text{board},k}(i)}(k) + 1, \forall i = 1, \dots, \text{Num}$
19: **return** $\widehat{\boldsymbol{n}}^*(k) = [\widehat{n}^*_m(k) : \forall m \in \mathcal{M}]$

Theorem 2. *Suppose* $\widehat{\boldsymbol{n}}^*(k) \in \arg\max_{\boldsymbol{n} \in \mathcal{A}} U(\boldsymbol{n}, \boldsymbol{P}, \boldsymbol{\mu}), \forall k \in \mathcal{K}$, *and* $M \geq 3$. *Algorithm 4 reaches a consensus.*

Putting Them Together and Regret Analysis. Puting all previous algorithms together, Algorithm 5 outlines our algorithm for the online setting. Due to page limit, we present the regret analysis of Algorithm 5 in our technical report [16].

Algorithm 5. Distributed learning algorithm for multi-player MAB with stochastic requests

---

1: Exploration: $(\widehat{\boldsymbol{\mu}}^{(k)}, \widehat{\boldsymbol{P}}^{(k)}) \leftarrow \texttt{Explore}(k, T_0)$
2: Estimate optimal arm pulling profile:
$\boldsymbol{n}_{\text{greedy}}(k) \leftarrow \texttt{OptArmPulProfile}(\widehat{\boldsymbol{\mu}}^{(k)}, \widehat{\boldsymbol{P}}^{(k)})$
3: Consensus: $\widetilde{\boldsymbol{n}}^* \leftarrow \texttt{Consensus}(k, \boldsymbol{n}_{\text{greedy}}(k))$
4: Committing to consensus:
$\texttt{CommitOptArmPulProfile}(k, \widetilde{\boldsymbol{n}}^*, T_0 + M)$
5: Sticking to the committed consensus

---

6 Experiment

Experiment Setup. Unless we vary them explicitly, we consider the following default parameters: $T = 10^4$, $K = 150$, $M = 50$, each arm's reward have same standard deviation $\sigma = 0.1$ and $d_{\max} = 50$. The mean reward of each arm is drawn from $[0, 1]$ uniformly at random. We generate the reward of each request via a normal distribution. For each arm, we first generate $d_{\max}$ numbers from $[0, 1]$ uniformly at random. Then we normalize these number such that their sum equals one. We use these normalized numbers as the probability mass of one arm. Repeating this process for all arms we obtain the probability mass matrix. We consider the following two baselines. (1) **MaxAvgReard**, where each player pulls arm with the largest average reward estimated from the collected historical rewards. (2) **SofMaxReward**, where each player selects arm m with probability proportional to the exponential of the average reward estimated from the collected historical rewards, i.e., softmax of average reward. We consider two metrics, i.e., regret and total reward.

Experimental Results. Figure 1(a) shows the regret of Algorithm 5 as we vary the length of exploration from T_0=0.01T to T_0=0.2T. One can observe that the regret curve first increases sharply in the exploration phase, and then becomes flat in the committing phase. This verifies that Algorithm 5 has a nice convergence property. Figure 1(b) shows that when T_0=0.1T the reward curve of Algorithm 5 lies in the top. Namely, Algorithm 5 has a larger reward than two comparison baselines. This statement also holds when the length of exploration increases as shown in Fig. 1(c) and 1(d). More experiment results are in our technical report [16].

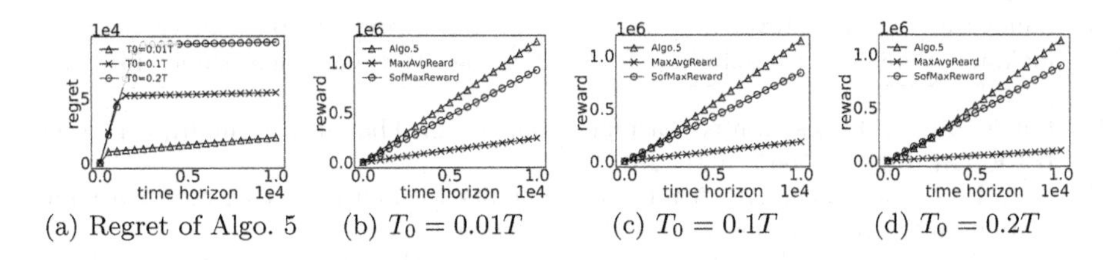

(a) Regret of Algo. 5 (b) $T_0 = 0.01T$ (c) $T_0 = 0.1T$ (d) $T_0 = 0.2T$

Fig. 1. Impact of length of exploration.

7 Conclusion

This paper formulates a new variant of multi-player MAB model for distributed selection problems. We designed a computational efficient greedy algorithm, to located one of the optimal arm pulling profiles. We also designed an iterative distributed algorithm for players to commit to an optimal arm pulling profile with a constant number of rounds in expectation. We designed an exploration strategy with a length such that each player can have an accurate estimate on the optimal arm pulling profile with high probability. Such estimates can be different across different players. We designed an iterative distributed algorithm, which guarantees that players arrive at a consensus on the optimal arm pulling profile. We conduct experiments to validate our algorithms.

Acknowledgment. This work was supported in part by Chongqing Talents: Exceptional Young Talents Project (cstc2021ycjhbgzxm0195), the Chinese Academy of Sciences "Light of West China" Program, the Key Cooperation Project of Chongqing Municipal Education Commission (HZ2021008, HZ2021017), and the "Fertilizer Robot" project of Chongqing Committee on Agriculture and Rural Affairs.

References

1. Anandkumar, A., Michael, N., Tang, A.K., Swami, A.: Distributed algorithms for learning and cognitive medium access with logarithmic regret. IEEE JSAC **29**(4), 731–745 (2011)
2. Avner, O., Mannor, S.: Concurrent bandits and cognitive radio networks. In: ECML (2014)
3. Awerbuch, B., Kleinberg, R.: Competitive collaborative learning. JCSS **74**(8), 1271–1288 (2008)
4. Besson, L., Kaufmann, E.: Multi-player bandits revisited. In: ALT (2018)
5. Bistritz, I., Leshem, A.: Distributed multi-player bandits-a game of thrones approach. In: NeurIPS (2018)
6. Boursier, E., Perchet, V.: Sic-mmab: Synchronisation involves communication in multiplayer multi-armed bandits. In: NeurIPS (2019)
7. Bubeck, S., Budzinski, T.: Coordination without communication: optimal regret in two players multi-armed bandits. In: COLT (2020)
8. Dubey, A., et al.: Cooperative multi-agent bandits with heavy tails. In: ICML (2020)
9. Hanawal, M.K., Darak, S.: Multi-player bandits: a trekking approach. IEEE TAC (2021)
10. Landgren, P., Srivastava, V., Leonard, N.E.: Distributed cooperative decision-making in multiarmed bandits: frequentist and Bayesian algorithms. In: CDC (2016)
11. Liu, K., Zhao, Q.: Distributed learning in multi-armed bandit with multiple players. IEEE TSP **58**(11), 5667–5681 (2010)
12. Lugosi, G., Mehrabian, A.: Multiplayer bandits without observing collision information. arXiv preprint arXiv:1808.08416 (2018)
13. Magesh, A., Veeravalli, V.V.: Multi-user mabs with user dependent rewards for uncoordinated spectrum access. In: CSSC (2019)
14. Martínez-Rubio, D., Kanade, V., Rebeschini, P.: Decentralized cooperative stochastic bandits. In: NeurIPS (2019)
15. Mehrabian, A., Boursier, E., Kaufmann, E., Perchet, V.: A practical algorithm for multiplayer bandits when arm means vary among players. In: AISTAT (2020)
16. Mo, J., Xie, H.: A Multi-player MAB Approach for Distributed Selection Problems (2023). https://www.dropbox.com/s/oidjkrbeukzd7pd/supplementary.pdf?dl=0
17. Rosenski, J., Shamir, O., Szlak, L.: Multi-player bandits-a musical chairs approach. In: ICML (2016)
18. Shi, C., Xiong, W., Shen, C., Yang, J.: Decentralized multi-player multi-armed bandits with no collision information. In: AISTAT (2020)
19. WANG, P.A., Proutiere, A., Ariu, K., Jedra, Y., Russo, A.: Optimal algorithms for multiplayer multi-armed bandits. In: AISTAT (2020)

A Thompson Sampling Approach to Unifying Causal Inference and Bandit Learning

Hanxuan Xu[1] and Hong Xie[2](✉)

[1] College of Computer Science, Chongqing University, Chongqing, China
[2] Chongqing Institute of Green and Intelligent Technology, Chinese Academy of Sciences, Beijing, China
xiehong2018@foxmail.com

Abstract. Offline logged data is quite common in many web applications such as recommendation, Internet advertising, etc., which offers great potentials to improve online decision making. It is a non-trivial task to utilize offline logged data for online decision making, because the offline logged data is observational and it may mislead online decision making. The VirUCB is one of the latest notable algorithmic frameworks in this research line. This paper studies how to extend VirUCB from upper confidence bound (UCB) based online decision making to Thompson sampling based online decision making, for the purpose of improving the online decision accuracy. We first show that naively applying Thompson sampling to the VirUCB framework is not effective and we reveal fundamental insights on why it is not effective. Based on these insights, we design a filtering algorithm to filter out the logged data corresponding to the optimal arm. To address the challenge that the optimal arm is unknown, we estimate it through the posterior of the reward mean. Putting them together, we obtain our VirTS-DF algorithm. Extensive experiments on two real-world datasets validate the superior performance of VirTS-DF.

1 Introduction

Offline logged data is quite common in many web applications such as recommendation, Internet advertising, etc., which offers great potential to improve online decision making [1]. It is a non-trivial task to utilize offline logged data for online decision making, because the offline logged data is observational and it may mislead online decision making [1,2]. A number of works investigated the problem of applying offline logged data to improve the accuracy of online decision making [1,3–6]. Interested readers can refer to Sect. 6 for more details of these works. The VirUCB [1] is one of the latest notable framework to utilize offline logged data to improve sequential decision making. Essentially, it unifies offline causal inference and online bandit learning through virtual play. The virtual play refers to that the feedback or reward of the online bandit learning is synthesized from the offline logged data via causal inference techniques. The bandit learning algorithm does not distinguish virtual play or true play. This property makes

H. Kashima et al. (Eds.): PAKDD 2023, LNAI 13936, pp. 255–266, 2023.
https://doi.org/10.1007/978-3-031-33377-4_20

the VirUCB framework has nice theoretical guarantee on the regret upper and lower bound.

The VirUCB framework uses UCB based algorithm for bandit learning or online decision making. Thompson sampling is a notable alternative for the UCB based algorithms in bandit learning problems, and it is shown to have superior performance over UCB based algorithms in many scenarios. This paper studies how to extend VirUCB from upper confidence bound (UCB) based online decision making to Thompson sampling based online decision making, for the purpose of improving the online decision accuracy. Through this, we aim to deliver fundamental insights on applying Thompson sampling to unify offline causal inference and online bandit learning.

It is non-trivial to extend VirUCB from upper confidence bound (UCB) based online decision making to Thompson sampling based online decision making. Our fist contribution is showing that naively applying Thompson sampling to the VirUCB framework is not effective. We also use numerical simulations to show when and why naively applying Thompson sampling to the VirUCB framework has a poor online decision accuracy. Our results show that the logged data on the optimal arm may make the bandit learning algorithm miss the optimal. Our second contribution is designing a data filtering algorithm to address this limitation. In particular, we design a filtering algorithm to filter out the logged data corresponds to the optimal arm. To address the challenge that the optimal arm is unknown, we estimate it through the posterior of the reward mean. Putting them together, we obtain our VirTS-DF algorithm. Our last contribution is that we conduct extensive experiments on two real-world datasets to validate the superior performance of VirTS-DF. We believe that our work reveals important insights on unifying offline causal inference and online bandit learning.

2 Model

2.1 The Bandit Learning Model

For the simplicity of presentation, we consider the contextual multi-armed bandit learning model. We consider one decision maker and a finite number of $T \in \mathbb{N}_+$ decision rounds indexed by $t \in [T] \triangleq \{1, \ldots, T\}$. Let $\mathcal{A} \subset \mathbb{N}_+$ denote the arm set, where $|\mathcal{A}|=K<\infty$. Let $\boldsymbol{x}_t \in \mathbb{R}^d$ denote the context vector in round t. We consider the case the context vector $\boldsymbol{x}_t$ is arbitrarily generated. Let $R_{a,t}$ denote the reward of pulling $a \in \mathcal{A}$ in round t, formally

$$R_{a,t} \triangleq \mu_{a,t} + \epsilon_{a,t}, \tag{1}$$

where $\mu_{a,t} \in \mathbb{R}$ represents the mean and $\epsilon_{a,t} \in \mathbb{R}$ is a random variable representing the stochastic noise satisfying $\mathbb{E}[\epsilon_{a,t}] = 0$. The reward mean $\mu_{a,t}$ is uniquely determined by the context vector, i.e.,

$$\boldsymbol{x}_t = \boldsymbol{x}_{t'} \Rightarrow \mu_{a,t} = \mu_{a,t'}. \tag{2}$$

We do not assume any further parametric form of dependency, e.g., linear, non-linear, etc., between the context and the reward mean. For simplicity, denote the reward mean vector as $\boldsymbol{\mu}_t \triangleq [\mu_{a,t} : a \in \mathcal{A}]$. Let $a_t \in \mathcal{A}$ denote the arm selected by the decision maker. Only the reward $R_{a_t,t}$ is revealed to the decision maker in round t. Note that the reward means $\mu_{a,t}, \forall a, t$, is unknown to the decision maker. The objective is to design an arm selection algorithm to maximize the cumulative reward $\sum_{t=1}^{T} R_{a_t,t}$.

2.2 The Data Model

The Offline Logged Data. We consider a finite set of $I \in \mathbb{N}_+$ offline logged data tuples. For presentation convenience, we index these logged data tuples using the set $[-I] \triangleq \{-I, \ldots, -1\}$ to highlight that they are collected before the first online decision making round. Let $(i, a_i, \boldsymbol{x}_i, y_i)$ denote the i-th logged data tuple, where $i \in [-I]$, $a_i \in \mathcal{A}$ denotes an arm, $y_i \in \mathcal{Y} \subseteq \mathbb{R}$ denotes the outcome (or reward) and $\boldsymbol{x}_i$ represents the context (or feature) associated with data item $i \in [-I]$. More precisely, $\boldsymbol{x}_i \triangleq [x_{i,1}, \ldots, x_{i,d}] \in \mathcal{X}$ where $d \in \mathbb{N}_+$ and $\mathcal{X} \in \mathbb{R}^d$. The reward y_i satisfies that $y_i = \mu_{a_i,i} + \epsilon_{a_i,i}$, where $\mu_{a_i,i}$ and $\epsilon_{a_i,i}$ have the same meaning as Eq. (1). Furthermore, $\mu_{a_i,i}$ satisfies Eq. (2) for all $i \in [-I] \cup [T]$. Let

$$\mathcal{L} \triangleq \{(i, a_i, \boldsymbol{x}_i, y_i) | i \in [-I]\}$$

denote a set of all the offline logged data.

The offline logged data set $\mathcal{L}$ is observational and the arm a_i can be correlated with the outcome y_i [2], where $i \in [-I]$. Similar with previous works [1,6], we apply the potential outcome framework to characterize such correlations [2]. The feature $\boldsymbol{x}_i$ is referred to as "observed confounder" in the causal inference literature [2]. We do not assume any probability law that generates the feature vector, i.e., we consider the general case that $\{x_i\}, \forall i \in [-I]$ can be non-random. Let random variable $Y_i(a)$ denote the potential outcome of arm $a \in \mathcal{A}$ instead of a_i being selected for data item $i \in [-I]$. Similar with previous works [1,6], we induce the following two conventional assumptions on the logged data.

Assumption 1. *The potential outcome $Y_i(a)$ satisfies:*

$$\mathbb{P}[Y_i(a){=}y | A_i, A_j] = \mathbb{P}[Y_i(a){=}y | A_i], \qquad \forall y{\in}\mathcal{Y}, j{\neq}i, a{\in}\mathcal{A}.$$

where A_i denotes a random variable whose realization is a_i.

Assumption 1 states the independence property of the offline logged dataset that the potential outcome associated with data item is independent of the arms associated with other data items.

Assumption 2. *Given a context vector $\boldsymbol{x}_i$, the outcome $Y_i(a), \forall a \in \mathcal{A}$ satisfies:*

$$[Y_i(a) : a{\in}\mathcal{A}] \perp A_i | \boldsymbol{x}_i, \quad \forall i{\in}[-I],$$

where A_i is defined in Assumption 1.

Assumption 2 states the incorrigibility property of the logged data set that the observed arm of each data item is independent of the outcome. In other words, there is no unobserved or latent confounders in the logged dataset. Extending our work to the setting with unobserved confounders is an interesting future work.

2.3 Problem Formulation

We aim to learn a context dependent optimal arm, i.e.,

$$a_t^* = \arg\max_{a \in \mathcal{A}} \mu_{a,t}. \tag{3}$$

For simplicity, let

$$\mathcal{H}_t \triangleq \{(i, a_i, \boldsymbol{x}_i, R_{a_i,i}) | i \in [t]\},$$

denote the online feedback history up to time slot t, where $[t] \triangleq \{1, ..., t\}$. Let $\mathcal{S}_t \subseteq \mathcal{L}$ denote a set of selected offline logged data up to time slot t. Let $\mathbb{A}$ denote an arm selection algorithm, i.e., $a_t = \mathbb{A}(\mathcal{H}_{t-1}, \mathcal{S}_t, \boldsymbol{x}_t)$. Following the convention of bandit learning, we quantify the performance of $\mathbb{A}$ via pseudo-regret:

$$\text{Reg}(T, \mathbb{A}) \triangleq \mathbb{E}\left[\sum_{t=1}^{T} (\mu_{a_t^*,t} - \mu_{a_t,t}) \middle| a_t = \mathbb{A}(\mathcal{H}_{t-1}, \mathcal{S}_t, \boldsymbol{x}_t)\right]$$

We aim to utilize offline data to assist arm selection so as to minimize the regret.

3 Limitations of Naively Applying Thompson Sampling

3.1 VirTS: Naively Applying Thompson Sampling

We first consider a naive extension of the virtual play UCB framework, i.e., VirUCB [1]. This naive extension is obtained by replacing the UCB based online decision oracle of VirUCB by the Thompson sampling based online decision oracle, which is outlined in Algorithm 1. For simplicity of presentation, Algorithm 1 combines `EM`, i.e., exact matching in causal inference, with `TS`, i.e., Thompson sampling based online decision making algorithm. In the Thompson sampling algorithm `TS` $(\mathcal{A}, \mathcal{H}, \tilde{\mathcal{S}}, \boldsymbol{x}_t, \{\mathbb{P}_0(\cdot; a, \boldsymbol{x}_t), \forall a\})$, the decision maker needs to specify a prior on the reward distribution denoted by $\{\mathbb{P}_0(\cdot; a, \boldsymbol{x}_t), \forall a\}$, and then calculate the posterior of the reward distribution based on the feedback associated with this arm, denoted by $\mathbb{P}_a(\cdot|\mathcal{H}_{a,\boldsymbol{x}_t})$. For each arm a, one sample is generated from its posterior $\mathbb{P}_a(\cdot|\mathcal{H}_{a,\boldsymbol{x}_t})$, and then use the generated sample to estimate the reward mean of the arm. Finally, the arm with the largest estimated reward mean is selected. Note that VirTS is a generic framework and can be applied to combine a broad class of causal inference algorithms with a broad class of sequential decision making algorithms [1].

Algorithm 1: VirTS

1 **Init:** $\mathcal{S}_0 \leftarrow \emptyset, \mathcal{H}_0 \leftarrow \emptyset$
2 **for** $t = 1$ *to* T **do**
3   Observe context $\boldsymbol{x}_t$, Set $\mathcal{S} \leftarrow \emptyset$
4   **while** *true* **do**
5         $a \leftarrow \texttt{TS}(\mathcal{A}, \mathcal{H}_{t-1}, \mathcal{S}_{t-1} \cup \mathcal{S}, \boldsymbol{x}_t)$
6   SelectedData $\leftarrow$ $\texttt{EM}$ $(\mathcal{L}, \mathcal{S}_{t-1} \cup \mathcal{S}, \boldsymbol{x}_t, a)$
7   **if** *SelectedData* $\neq$ *Null* **then**
8             $\mathcal{S} \leftarrow \mathcal{S} \cup \{\text{SelectedData}\}$
9   **else**
10   **break**
11     $\mathcal{S}_t \leftarrow \mathcal{S}_{t-1} \cup \mathcal{S}$, $a_t \leftarrow \texttt{TS}$ $(\mathcal{A}, \mathcal{H}_{t-1}, \mathcal{S}_t, \boldsymbol{x}_t)$
12   Observe y_t, $\mathcal{H}_t \leftarrow \mathcal{H}_{t-1} \cup (t, a_t, \boldsymbol{x}_t, y_t)$
13 ---
14 **Function** *TS* $(\mathcal{A}, \mathcal{H}, \tilde{\mathcal{S}}, \boldsymbol{x}_t, \{\mathbb{P}_0(\cdot; a, \boldsymbol{x}_t), \forall a\})$ **:**
15     $\mathcal{W} \leftarrow \tilde{\mathcal{S}} \cup \mathcal{H}$
16   **foreach** $a \in \mathcal{A}$ **do**
17         $\mathcal{H}_{a,\boldsymbol{x}_t} \leftarrow \{(i, a_i, \boldsymbol{x}_i, y_i) | (i, a_i, \boldsymbol{x}_i, y_i) \in \mathcal{W}, a_i = a, \boldsymbol{x}_i = \boldsymbol{x}_t\}$
18   Calculate the posterior $\mathbb{P}_a(\cdot | \mathcal{H}_{a,\boldsymbol{x}_t})$ based on $\mathbb{P}_0(\cdot; a, \boldsymbol{x}_t)$ and $\mathcal{H}_{a,\boldsymbol{x}_t}$
19         $\boldsymbol{p}_a \sim \mathbb{P}_a(\cdot | \mathcal{H}_{a,\boldsymbol{x}_t})$, $r_a \leftarrow \int r d\boldsymbol{p}_a(r)$
20   **return** $\arg\max_{a \in \mathcal{A}} r_a$
21 **Function** *EM* $(\mathcal{L}, \tilde{\mathcal{S}}, \boldsymbol{x}_t, a)$**:**
22     $\mathcal{C} \leftarrow \{i | \boldsymbol{x}_i = \boldsymbol{x}_t, a_i = a, (i, a_i, \boldsymbol{x}_i, y_i) \in \mathcal{L} \setminus \tilde{\mathcal{S}}\}$
23   **if** $\mathcal{C} \neq \emptyset$ **then**
24         $i \leftarrow$ a random sample from $\mathcal{C}$
25   **return** $(i, a_i, \boldsymbol{x}_i, y_i)$
26   **else**
27   **return** *Null*

3.2 Limitations of VirTS

To illustrate the limitation of naively applying the Thompson sampling and reveal fundamental insights on the limitation, we consider a simplified setting with only one context and two arms indexed by $\{1, 2\}$. The reward of arm 1 and 1 follows a multi-nominal distribution $\boldsymbol{p}_1 = (p_{1,1}, \ldots, p_{1,5})$ and $\boldsymbol{p}_2 = (p_{2,1}, \ldots, p_{2,5})$ respectively. Without loss of generality, we assume that the reward mean of arm 1 is large than that of arm 2, i.e., $\sum_{i=1}^{5} i p_{1,i} > \sum_{i=1}^{5} i p_{2,i}$. We set $T = 1000$ and consider the same offline logged datasets as follows:

- **Balanced:** it consists of 50 offline logged data tuples. Half (The other half) of them are IID samples from $\boldsymbol{p}_1$ ($\boldsymbol{p}_2$).
- **Imbalanced toward optimal arm:** it consists of 50 offline logged data items. If $\sum_{i=1}^{5} i p_{1,i} > \sum_{i=1}^{5} i p_{2,i}$, all of them are IID samples from $\boldsymbol{p}_1$, otherwise all of them are IID samples from $\boldsymbol{p}_2$.

- **Imbalanced toward suboptimal arm:** it consistes of 50 offline logged data items. If $\sum_{i=1}^{5} ip_{1,i} > \sum_{i=1}^{5} ip_{2,i}$, all of them are IID samples from a $\boldsymbol{p}_2$, otherwise all of them are IID samples from $\boldsymbol{p}_1$.

To make our numerical results convincing, we repeat our algorithm for 100 times and compute the average regret. In each repeat, we generate $\boldsymbol{p}_1$ and $\boldsymbol{p}_2$ independently from the Dirichlet distribution Dir(1, 1, 1, 1, 1). Consider VirTS-Bal, VirTS-Opt, VirTS-Sub and VirTS-No, which denotes VirTS with balanced, imbalanced toward optimal arm, imbalanced toward suboptimal arm and without offline logged data respectively.

Figure 1(a) shows that VirTS-Bal and VirTS-Sub have a smaller regret than VirTS-No, while VirTS-Opt has a larger regret than VirTS-No. Namely, when the offline logged data is imbalanced, VirTS owns a learning speed slower than the baseline algorithm without offline logged data. Figure 1(b) shows that the slow learning speed is caused by consuming 50 offline logged data items. This implies that the VirTS algorithm is not robust and it is not efficient in utilizing offline logged data. Improving the utilization efficiency is important as in many applications collecting the offline logged data is associated with certain cost.

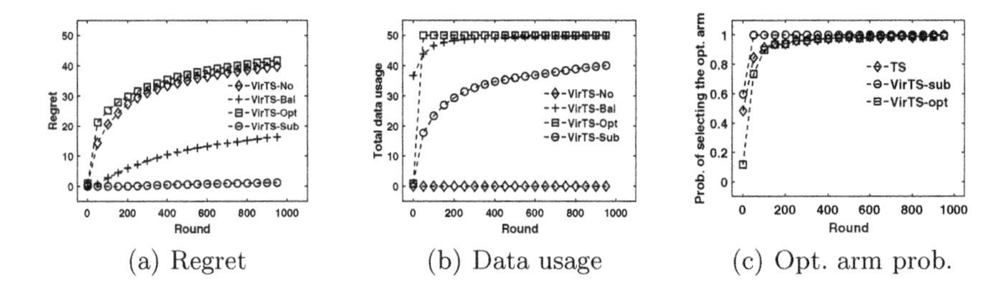

(a) Regret (b) Data usage (c) Opt. arm prob.

Fig. 1. Impact of offline logged data on VirTS

Figure 1(c) shows the probability of selecting the optimal arm under the VirTS-No, VirTS-Opt and VirTS-Sub algorithm respectively. When the logged data is imbalanced, VirTS-Sub performs better than TS and VirTS-Opt for its high possibility of choosing the optimal arm. This shows that the logged data may cause the decision maker to miss the optimal arm.

4 VirTS-DF: Improving VirTS via Offline Data Filtering

In this section, we present the design of VirTS-DF, which addresses the limitation of VirTS via filtering out some logged data that may lead the decision maker to miss the optimal arm. In particular, instead of greedily utilize the logged data as VirTS, we aim to filter out the logged data corresponding to the optimal arm. Though this, only logged data corresponding to sub-optimal arms will be utilized. One challenge is that the optimal arm is unknown. To address this challenge, we estimate the optimal arm via the posterior means of the reward means of arms. More specifically, we estimate the reward mean of each arm via its posterior mean, and the arm with the largest posterior mean is estimated as the optimal arm. Algorithm 2 summarizes the above ideas, where $\widehat{\psi}_t(i, a_i, \boldsymbol{x}_i, y_i, \tilde{\mathcal{S}}, \{\mathbb{P}_0(\cdot; a, \boldsymbol{x}_t), \forall a\})$ is the data filtering algorithm. It calculates the posterior of the reward means via first generating a sample of the reward distribution from its posterior and then using the reward distribution sample to calculate a reward mean.

Algorithm 2: VirTS-DF

```
1  S_0 ← ∅, H_0 ← ∅
2  for t = 1 to T do
3      Observe context x_t, set S ← ∅
4      while true do
5          Obtain SelectedData by step 5-6 of Algorithm 1
6          if SelectedData ≠ ∅ and ψ̂_t(SelectedData, S_{t-1} ∪ S) == 1 then
7              S ← S ∪ {SelectedData}
8          else
9              break
10     Execute step 11-12 in Algorithm 1
11 Function ψ̂_t(i, a_i, x_i, y_i, S̃, {P_0(·; a, x_t), ∀a}) :
12     W ← S̃ ∪ H_{t-1}
13     foreach a ∈ A do
14         H_{a,x_t} ← {(i, a_i, x_i, y_i)|(i, a_i, x_i, y_i) ∈ W, a_i=a, x_i = x_t}
15         Calculate the posterior P_a(·|H_{a,x_t}) based on P_0(·; a, x_t) and H_{a,x_t}
16         p_a ~ P_a(·|H_{a,x_t}), r_a ← ∫ r dp_a(r)
17     â*_t ← arg max_{a∈A} r_a
18     if x_i = x_t and a_i ≠ â*_t then
19         return 1
20     else
21         return 0
```

5 Experiments on Real-world Data

5.1 Experimental Settings

We conduct experiments on two datasets from Amazon[1] and MovieLens[2]. These two datasets consists of ratings of movies. The rating is cardinal of five levels ranging from 1 to 5. For fair comparison with the baseline in [6], we process the data following the same procedures as that of [6]. In the end, we process

[1] https://snap.stanford.edu/data/web-Movies.html.
[2] https://grouplens.org/datasets/movielens/.

the data to a setting with 20 arms and two contexts. Every decision round, we choose one of the two contexts uniformly at random. We compare our VirTS-DF algorithm with three baselines: (1) **TS** which is the Thompson sampling without logged data; (2) **VirTS**; (3) **EffVirUCB** which is one of the latest algorithm that unifies causal inference and online bandit learning [6]. Similar with [6], we use the vector $[0.1, 1, ..., K-1]^{-\alpha}$ to characterize the distribution of the number of offline logged data across arms, where $\alpha \in \mathbb{R}$. In particular, the arm whose ground truth mean ranked k-th has a number of offline logged data proportional to the k-the element of the vector $[0.1, 1, ..., K-1]^{-\alpha}$.

5.2 Experiment Results

Extremely Imbalanced Offline Data. Figure 2 shows the regret and logged data usage of four algorithms under extremely imbalanced offline data, i.e., $\alpha = 3$. From Fig. 2(a) and (c) , one can observe that the regret curve of VirTS-DF lies below that of VirTS on both the Amazon and Movielens dataset. In other words, VirTS-DF has a smaller regret than VirTS, which implies that it has a faster learn speed than VirTS. It shows the merit and necessity of data filtering. Furthermore, VirTS-DF has a significant smaller regret than the EffVirUCB algorithm. This shows the merit of Thompson sampling based approach over the UCB based approach. On the Amazon dataset, the regret curve of TS lies in the bottom. This shows that the logged data on the optimal arm disturbs the learning speed. Furthermore, there is a room for improving our data filtering algorithm on the Amazon dataset. However, our data filtering algorithm performs quite well on the Movielens dataset. From Fig. 2(b) and (d), one can observe that logged data usage curve of VirTS-DF also lies in the bottom, while the logged data usage curve of VirTS lies in the top. This shows that VirTS-DF is highly efficiently in utilizing the logged data. The data usage curve of VirTS-DF overlaps with that of the EffVirUCB. This implies that VirTS-DF utilizes nearly the same amount of logged data as EffVirUCB.

Imbalanced Offline Logged Data. Figure 3 shows the regret and logged data usage of four algorithms under imbalanced offline logged data. From Fig. 3(a) and (c), one can observe that the regret curve of VirTS lies above the regret curve of TS under Amazon and MovieLens dataset respectively. This implies that VirTS has a larger regret than that of TS algorithm under the real-world datasets. In other words, under real-world datasets, imbalanced offline logged data makes the learning speed of VirTS algorithm slower than that of TS which does not have offline logged data. Furthermore, VirTS-DF has a significant smaller regret than the EffVirUCB algorithm. This shows the merit of Thompson sampling based approach over the UCB based approach. On the Movielens dataset, the regret curve of TS lies in the bottom. This shows that the logged data on the optimal arm disturbs the learning speed. Furthermore, there is a room for improving our data filtering algorithm on the MovieLens dataset. However, our data filtering algorithm performs quite well on the Amazon dataset. From Fig. 3(b) and (d), one can observe that logged data usage curve of VirTS-DF also lies in the bottom,

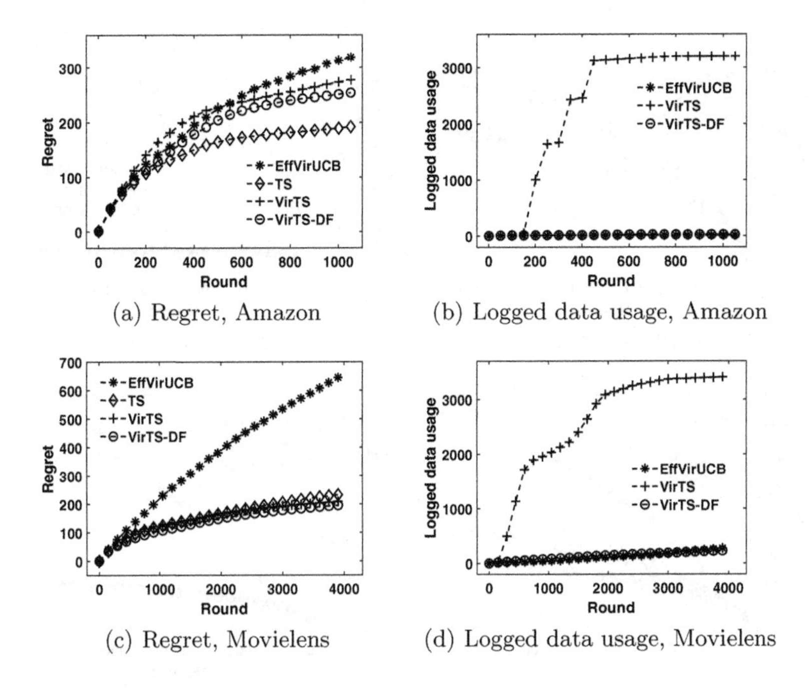

(a) Regret, Amazon (b) Logged data usage, Amazon

(c) Regret, Movielens (d) Logged data usage, Movielens

Fig. 2. Extremely imbalanced offline logged data ($\alpha = 3$).

while the logged data usage curve of VirTS lies in the top. This shows that VirTS-DF is highly efficiently in utilizing the logged data. The data usage curve of VirTS-DF overlaps with that of the EffVirUCB. This implies that VirTS-DF utilizes nearly the same amount of logged data as EffVirUCB.

Relatively Balanced Offline Data. Figure 4 shows the regret and logged data usage of four algorithms under relatively balanced offline data. From Fig. 4(a) and (c), one can observe that the regret curves of VirTS-DF lies nearly in the bottom. This implies a fast learning speed of VirTS-DF. one can observe that the regret curve of VirTS-DF lies below that of VirTS on both the Amazon and Movielens dataset. In other words, VirTS-DF has a smaller regret than VirTS, which implies that it has a faster learn speed than VirTS. It shows the merit and necessity of data filtering. From Fig. 4(b) and (d), one can observe that logged data usage curve of VirTS-DF also lies in the bottom, while the logged data usage curve of VirTS lies in the top. This shows that VirTS-DF is highly efficient in utilizing the logged data.

6 Related Work

Multi armed bandit [7] is a fundamental tool to study the exploration vs. exploitation tradeoff in online decision making. Two typical algorithmic frameworks of MAB learning include: (1) upper confidence bound (UCB) based algo-

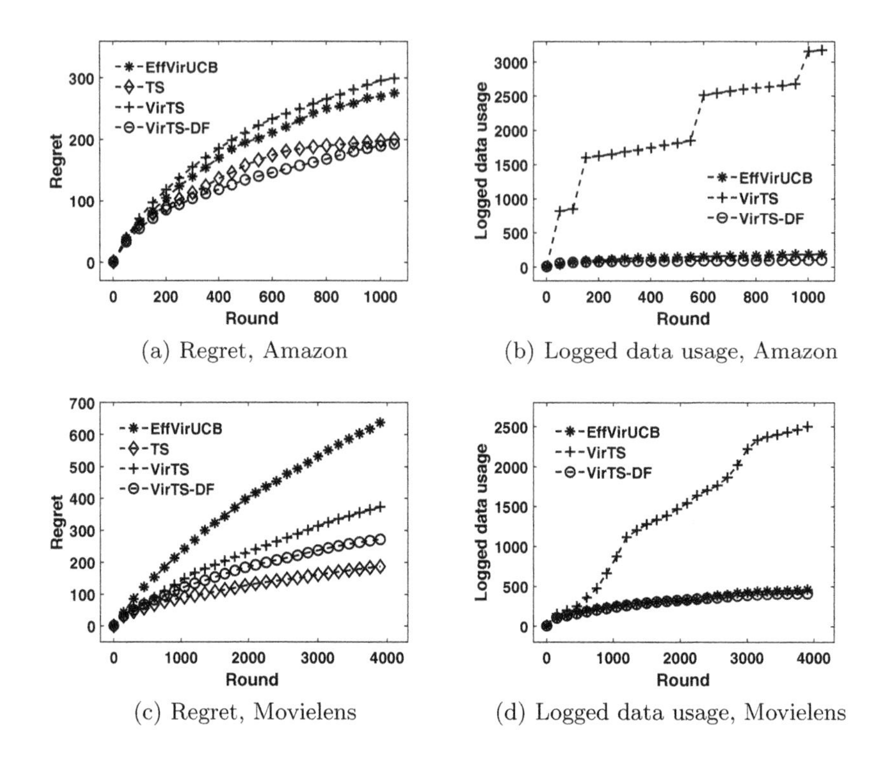

(a) Regret, Amazon (b) Logged data usage, Amazon

(c) Regret, Movielens (d) Logged data usage, Movielens

Fig. 3. Imbalanced offline logged data ($\alpha = 1.5$).

rithmic frameworks [8], and (2) Thompson sampling based algorithmic frameworks [9]. Each algorithmic framework has its own merits and limitations. Our work applies Thompson sampling to unify offline causal inference and bandit learning. Many variants of MAB has been proposed from a modeling perspective such as combinatorial MAB [10], multi-player MAB [11], and MAB with additional information [3], etc. Our work essentially falls into the research line MAB with additional information. In particular, the additional information refers to logged data. Notable works in this research include the following. Wang *et al.* [12] modeled the additional information as a random variable and the random variable reveals side information on the rewards of arms, which can be used to improve the estimation of arm rewards. Sharma *et al.* [13] modeled the additional information as confidence bounds on the mean of each arm and proposed algorithm to utilize them. Yun *et al.* [14] modeled side information as additional feedbacks on arms that are not pulled. These feedbacks are assumed to be IID. Zuo *et al.* [4] studied a similar additional feedback model, but they study the problem under the multi-player MAB setting. Shivaswamy *et al.* [3] modeled side information as IID rewards on arms. They identified sufficient conditions such that offline logged data can reduce the regret MAB to a constant. Li *et al.* [1] treated the additional information as the observational logged data. The observational data rise a new challenge to debias the data. They proposed a virtual

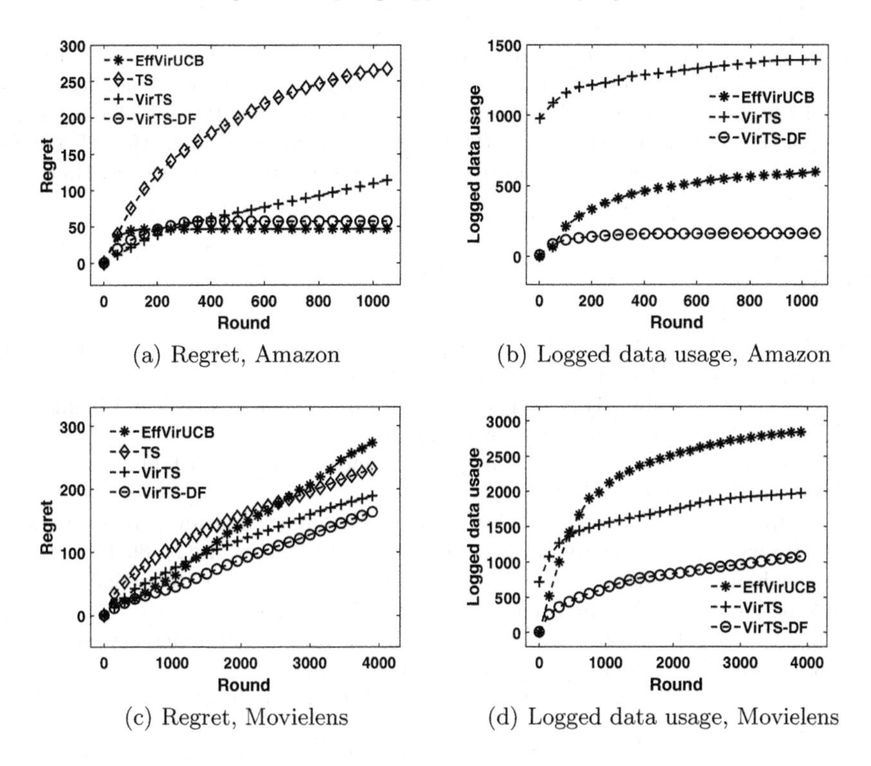

(a) Regret, Amazon

(b) Logged data usage, Amazon

(c) Regret, Movielens

(d) Logged data usage, Movielens

Fig. 4. Relatively balanced logged data ($\alpha = 0.5$).

play framework to utilize the logged data. Tang *et al.* [6] improved the work of [1] to be more robust and efficient in utilizing logged data. Different from the above works, our work explores Thompson sampling for unifying causal inference and bandit learning. We show that naively applying the Thompson sampling is not efficient and we design a data filtering algorithm to improve the efficiency.

7 Conclusion

This paper presents a Thompson sampling approach to unify offline causal inference and online bandit learning. In particular, we extend VirUCB from UCB based online decision making to Thompson sampling based online decision making, which leads to improved accuracy. We first show that naively applying Thompson sampling to the VirUCB framework is not effective and we reveal fundamental insights on why it is not effective. Based on these insights, we design a filtering algorithm to filter out the logged data corresponding to the optimal arm. To address the challenge that the optimal arm is unknown, we estimate it through the posterior of the reward mean. Putting them together, we obtain our VirTS-DF algorithm. Extensive experiments on two real-world datasets validate the superior performance of VirTS-DF.

Acknowledgment. This work was supported in part by Chongqing Talents: Exceptional Young Talents Project (cstc2021ycjhbgzxm0195), the Chinese Academy of Sciences "Light of West China" Program, the Key Cooperation Project of Chongqing Municipal Education Commission (HZ2021008, HZ2021017), and the "Fertilizer Robot" project of Chongqing Committee on Agriculture and Rural Affairs. Hong Xie is the corresponding author.

References

1. Li, Y., Xie, H., Lin, Y., Lui, J.C.: Unifying offline causal inference and online bandit learning for data driven decision. In: WWW (2021)
2. Imbens, G.W., Rubin, D.B.: Causal inference in statistics, social, and biomedical sciences. Cambridge University Press (2015)
3. Shivaswamy, P., Joachims, Y.: Multi-armed bandit problems with history. In: AISTAT, pp. 1046–1054 (2012)
4. Zuo, J., Zhang, X., Joe-Wong, C.: Observe before play: multi-armed bandit with pre-observations. ACM SIGMETRICS Perform. Eval. Rev. **46**(2), 89–90 (2019)
5. Tennenholtz, G., Shalit, U., Mannor, S., Efroni, Y.: Bandits with partially observable offline data. arXiv preprint arXiv:2006.06731 (2020)
6. Tang, Q., Xie, H.: A robust algorithm to unifying offline causal inference and online multi-armed bandit learning. In: IEEE ICDM (2021)
7. Lattimore, T., Szepesvári, C.: Bandit algorithms. Cambridge University Press (2020)
8. Auer, P., Cesa-Bianchi, N., Fischer, P.: Finite-time analysis of the multiarmed bandit problem. Mach. Learn. **47**(2–3), 235–256 (2002)
9. Russo, D., Van Roy, B., Kazerouni, A., Osband, I., Wen, Z.: A tutorial on Thompson sampling. arXiv preprint arXiv:1707.02038 (2017)
10. Chen, W., Wang, Y., Yuan, Y.: Combinatorial multi-armed bandit: general framework and applications. In: ICML, pp. 151–159 (2013)
11. Bistritz, I., Leshem, A.: Distributed multi-player bandits-a game of thrones approach. In: NIPS (2018)
12. Wang, C.-C., Kulkarni, S.R., Poor, H.V.: Bandit problems with side observations. IEEE TAC **50**(3), 338–355 (2005)
13. Sharma, N., Basu, S., Shanmugam, K., Shakkottai, S.: Warm starting bandits with side information from confounded data. arXiv preprint arXiv:2002.08405 (2020)
14. Yun, D., Proutiere, A., Ahn, S., Shin, J., Yi, Y.: Multi-armed bandit with additional observations. Proceed. ACM Measure. Anal. Comput. Syst. **2**(1), 1–22 (2018)

Parallel and Distributed Mining

Maverick Matters: Client Contribution and Selection in Federated Learning

Jiyue Huang[1], Chi Hong[1], Yang Liu[2], Lydia Y. Chen[1(✉)], and Stefanie Roos[1(✉)]

[1] Delft University of Technology, Delft, The Netherlands
{j.huang-4,c.hong,y.chen-10,s.roos}@tudelft.nl
[2] AI Industry Research (AIR), Tsinghua University, Beijing, China
liuy03@air.tsinghua.edu.cn

Abstract. Federated learning (FL) enables collaborative learning between parties, called clients, without sharing the original and potentially sensitive data. To ensure fast convergence in the presence of such heterogeneous clients, it is imperative to timely select clients who can effectively contribute to learning. A realistic but overlooked case of heterogeneous clients are Mavericks, who monopolize the possession of certain data types, e.g., children hospitals possess most of the data on pediatric cardiology. In this paper, we address the importance and tackle the challenges of Mavericks by exploring two types of client selection strategies. First, we show theoretically and through simulations that the common contribution-based approach, *Shapley Value*, underestimates the contribution of Mavericks and is hence not effective as a measure to select clients. Then, we propose FedEMD, an adaptive strategy with competitive overhead based on the Wasserstein distance, supported by a proven convergence bound. As FedEMD adapts the selection probability such that Mavericks are preferably selected when the model benefits from improvement on rare classes, it consistently ensures the fast convergence in the presence of different types of Mavericks. Compared to existing strategies, including *Shapley Value*-based ones, FedEMD improves the convergence speed of neural network classifiers with FedAvg aggregation by 26.9% and its performance is consistent across various levels of heterogeneity.

Keywords: Federated learning · data heterogeneity · client selection · shapley value · wasserstein distance

1 Introduction

Federated Learning (FL) enables clients (either individuals or institutes who own data) to collaboratively train a global machine learning models by exchanging locally trained models instead of data [16,18]. Thus, Federated Learning allows the training of models when data cannot be transferred to a central server and is hence often a suitable alternative for medical research and other domains, such as finance, with high privacy requirements. The effectiveness of FL, in terms of accuracy and convergence, highly depends on how the local models are selected and aggregated.

Supplementary Information The online version contains supplementary material available at https://doi.org/10.1007/978-3-031-33377-4_21.

H. Kashima et al. (Eds.): PAKDD 2023, LNAI 13936, pp. 269–282, 2023.
https://doi.org/10.1007/978-3-031-33377-4_21

In FL, clients tend to own heterogeneous datasets [14] rather than identically and independent distributed (*i.i.d.*) ones. The prior art has recently addressed the challenge of heterogeneity from either the perspective of skewed distribution [28] or skewed quantity [23] among all clients. However, a common real-world scenario, where one or a small group of clients monopolize the possession of a certain class, is universally overlooked. For example, in the widely used image classification benchmark, Cifar-10 [12], most people can contribute images of cats and dogs. However, deer images are bound to be owned by comparably few clients. We call these types of clients *Mavericks*. Another relevant example, shown in Fig. 1, arises from learning predictive medicine from clinics who specialize in different conditions, e.g., AIDS and Amyotrophic Lateral Sclerosis, and own data of exclusive disease types. Without involving Mavericks into the training, it is impossible to achieve high accuracy on the classes for which they own the majority of all training data, e.g., rare diseases.

Given its importance, it is not well understood when to best involve Mavericks in FL training, because the effectiveness of FL, in terms of accuracy and convergence, highly depends on how those local models are selected and aggregated. The existing client selection[1] considers either the contribution of local models [3] or difference of data distributions [19]. The contribution-based approaches select clients based on contribution scores preferring clients with higher scores [7], whereas the distance-based methods choose clients based on the pairwise feature distance. Both types of selection methodologies have their suitable application scenarios and it is hard to weigh the benefits of one over the other in general.

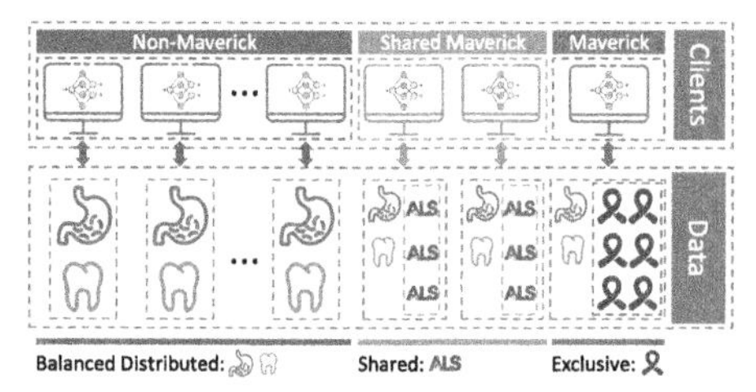

Fig. 1. Illustration of Mavericks.

In this paper, we aim to effectively select Mavericks in FL so that users are able to collaboratively train an accurate model in a low number of communication rounds. We first explore *Shapley Value* as a contribution metric for client selection. Although *Shapley Value* is shown to be effective in measuring contribution for the *i.i.d.* case, it is unknown if it can assess the contribution of Mavericks and effectively involve them via the selection strategy. Moreover, we propose FedEMD, which selects clients based on Wasserstein distance [2] of the global distribution and current distribution. As FedEMD adapts the selection probability such that Mavericks are preferably selected when the model benefits from improvement on rare classes, it consistently ensures the fast convergence in the presence of different types of Mavericks.

Our main **contributions** for this work can be summarized as follows. *i)* We explore the effectiveness of both contribution-based and distance-based selection strategies for Mavericks. *ii)* Both our theoretical and empirical results show that the contribution of clients with skewed data or very large data quantity is measured below average by *Shapley Value*. *iii)* We propose FedEMD, a novel adaptive client selection based on the Wasserstein distance, derive a convergence bound, and show that it significantly outperforms SOTA selection methods in terms of convergence speed across different scenarios of Mavericks.

[1] Note that here we only discuss selection on statistical challenges, the selections considering system resources, e.g., unreliable networks are left for other works.

2 Related Studies

Contribution Measurement. Although the self-reported contribution evaluation [7] is easy to implement, it is fragile too dishonest parties. Besides, existing work on contribution measurement can be categorized into two classes: *i)* local approach: clients exchange the local updates, i.e., model weights or gradients, and measure the contribution of each other, e.g., by creating a reputation system [11], and *ii)* global approach: all clients send all their model updates to the *federator* who in turn aggregates and computes the contribution via the marginal loss [1,25]. Prevailing examples of globally measuring contribution are Influence [1] and *Shapley Value* [22,25]. The prior art demonstrates that *Shapley Value* can effectively measure the client's contribution for the case when client data is *i.i.d.* or of biased quantity [22]. A work [24] has proposed federated *Shapley Value* to capture the effect of participation order on data value. The experimental results indicate that *Shapley Value* is less accurate in estimating the contribution of heterogeneous clients than for *i.i.d.* cases. However, there is no rigorous analysis on whether *Shapley Value* can effectively evaluate the contribution from heterogeneous users with skewed data distributions.

Client Selection. Selecting clients within a heterogeneous group of potential clients is key to enabling fast and accurate learning based on high data quality. The state-of-the-art client selection strategies focus on the resource heterogeneity [10,21] or data heterogeneity [3,4,14]. In case of data heterogeneity, which is the focus of our work, selection strategies [3,4,8] gain insights on the distribution of clients' data and then select them in specific manners. Goetz et. al [8] apply active sampling and Cho et. al [4] use Power-of-Choice to favor clients with higher local loss. TiFL [3] considers both resource and data heterogeneity to mitigate the impact of stragglers and skewed distributions. TiFL applies a contribution-based client selection by evaluating the accuracy of selected participants each round and chooses clients of higher accuracy. FedFast [19] chooses classes based on clustering and achieves fast convergence for recommendation systems. One recently work [17] focuses on reduce wall-clock time for convergence under high degrees of system and statistical heterogeneity. However, there is no selection strategy that addresses the Maverick scenario.

3 Federated Learning with Mavericks

In this section, we first formalize a Federated Learning framework with Mavericks. Then we rigorously analyze the contribution of clients based on *Shapley Value* and argue that the contribution of Mavericks is underestimated by the *Shapley Value*, which leads to a severe selection bias and a suboptimal integration of Mavericks into the learning process.

Suppose there are a total of N clients in a federated learning system. We denote the set of possible inputs as $\mathcal{X}$ and the set of L class labels as $\mathcal{Y} = \{1, 2, ..., L\}$. Let $f\colon \mathcal{X} \rightarrow \mathcal{P}$ be a prediction function and ω be the learnable weights of the machine learning tasks, the objective is then defined as: $\min \mathcal{L}(\boldsymbol{\omega}) = \min \sum_{l=1}^{L} p(y = l)\mathbb{E}_{\boldsymbol{x}|y=l}\left[\log f_l(\boldsymbol{x}, \boldsymbol{\omega})\right]$.

The training process of a FL system has the following steps[2]: *i)* INITIALIZATION. Initialize global model ω_0 and distribute it to the available clients, i.e., a set $\mathcal{C}$ of N clients. *ii)* CLIENT SELECTION. Enumerate the K clients $\mathcal{C}(\pi, \omega_r)$, selected in round r with selection strategy π, by $C_1, \ldots, C_K$. *iii)* UPDATE AND UPLOAD. Each client C_k selected in round r computes local updates ω_r^k and the *federator* aggregates the results. Concretely, with η being the learning rate, C_k updates their weights in the r-th global round by: $\omega_r^k = \omega_{r-1} - \eta \sum_{l=1}^{L} p^k(y=l) \nabla_\omega \mathbb{E}_{x|y=l} \left[\log f_l(x, \omega_{r-1})\right]$. *iv)* AGGREGATION. Client updates are aggregated to one global update. The most common aggregation method is quantity-aware FedAvg, defined as follows with n^k indicating the data quantity of C_k: $\omega_r = \sum_{k=1}^{K} \frac{n^k}{\sum_{k=1}^{K} n^k} \omega_r^k$. To facilitate our discussions, we also define the following:

Local Distribution: The array of all L class quantities $\mathcal{D}^i(y=l), l \in \{1, .., L\}$ owned by client C_i.

Global Distribution: The quantity of all clients' data by class as $\mathcal{D}_g = \sum_{i=1}^{N} \mathcal{D}^i(y = l), l \in \{1, .., L\}$.

Current Distribution at R**:** By summing up the class quantity of all clients' data reported, which have been chosen up to round R as: $\mathcal{D}_c{}^R = \sum_{t=1}^{R} \sum_{C_k \in \mathcal{K}^t} \mathcal{D}^{C_k}$.

Definition 1 (Maverick). *Let Y_{Mav} be the set of class labels that are primarily owned by Mavericks. An exclusive Maverick is one client that owns one or more classes exclusively. A shared Maverick is a small group of clients who jointly own one class exclusively. That is:*

$$D_i = \begin{cases} \{\{x_l, y_l\}^i_{l \in Y_{Mav}}, \{x_l, y_l\}^i_{l \notin Y_{Mav}}\}, \textit{if } C_i \textit{ is a Maverick} \\ \{x_l, y_l\}^i_{l \notin Y_{Mav}}, \textit{if } C_i \textit{ is not a Maverick}, \end{cases} \tag{1}$$

where D_i denotes the dataset for C_i, $\{x_l, y_l\}^i$ denotes the dataset in C_i with label l.

In the rest of the paper, we assume the global distribution organized by the server's preprocessing has high similarity with the real-world (test dataset) distribution, which is balanced, so that data $\{x_l, y_l\}_{l \notin Y_{Mav}}$ are evenly distributed across all parties, whereas $\{x_l, y_l\}_{l \in Y_{Mav}}$ either belong to one exclusive Maverick or are evenly distributed across all shared Maverick parties. We focus our analysis on exclusive Mavericks since shared Maverick are a straightforward extension. Based on the assumptions above, we obtain the following properties for Mavericks.

Property 1. Because the data distribution is balanced, Mavericks have a larger data quantity than non-Mavericks. Concretely, let n^n be the data quantity of a non-Maverick. Let n^m be the quantity for Mavericks, then $n^m = ((N/m - 1) \times Y_{Mav} + L) \times n^n$, where m is the number of Mavericks.

Property 2. Assume $N > 2$, the KL divergence of a Maverick's data to the normalized global distribution is expected to be larger than for a non-Maverick due to their

[2] Here we assume all the clients are honest. Since we focus on the statistical challenge, the impact of unreliable networking and insufficient computation resources is ignored.

specific distribution, i.e., $D_{KL}(\mathcal{P}_g||\mathcal{P}_m)$ ¿ $D_{KL}(\mathcal{P}_g||\mathcal{P}_n)$, where $\mathcal{P}_m$, $\mathcal{P}_n$ are the data distribution with class labels for Maverick and non-Maverick, where $\mathcal{P}_g$ denotes for global distribution.

3.1 *Shapley Value* for Mavericks

Definition 2 (*Shapley Value*). *Let $\mathcal{K} = \mathcal{C}(\pi, \omega_r)$ denote the set of clients selected in a round including C_k, $\mathcal{K} \setminus \{C_k\}$ denote the set $\mathcal{K}$ without C_k.* Shapley Value *of C_k is:*

$$SV(C_k) = \sum_{S \subseteq \mathcal{K} \setminus \{C_k\}} \frac{|S|!(|\mathcal{K}| - |S| - 1)!}{|\mathcal{K}|!} \delta C_k(\mathcal{S}). \tag{2}$$

Here we let $\delta C_k(S)$ be the Influence [1]. Influence can be defined on loss, accuracy, etc., here we apply the most commonly used loss-based Influence written as $Infs_S(C_k)$ for set C_k.

Lemma 1. *Based on Shapley Value in Eq. 2, the difference of Maverick C_m's and non-Maverick C_n's Shapley Value is:*

$$\begin{aligned} SV(C_m) - SV(C_n) = \frac{1}{|\mathcal{K}|!}\Big(& (|\mathcal{K}| - 1)!(\mathcal{L}(C_m) - \mathcal{L}(C_n)) \\ & + \sum_{S \subseteq S_-} |S|!(|\mathcal{K}| - |S| - 1)!(Infs_S(C_m) - Infs_S(C_n)) \\ & + \sum_{S \subseteq S_+} |S|!(|\mathcal{K}| - |S| - 1)!(Infs_S(C_m) - Infs_S(C_n))\Big), \end{aligned} \tag{3}$$

with $S_- = \mathcal{K} \setminus \{C_n, C_m\}$, $S_+ = \mathcal{K} \setminus \{C_n, C_m\} \cup C_M$, $C_M \in \{C_n, C_m\}$. Note that we simplify $Infs_{S \cup C_i}(C_i)$ as $Infs_S(C_i)$ for readability.

Comparison of *Shapley Value* and Influence: Rather than considering Influence for the complete set of K clients, Eq. 3 only considers Influence on a subset S. However, our derivations for Influence are independent from the number of selected clients and remain applicable for subsets S, meaning that indeed the second and the third term of Eq. 3 are negative. Similarly, the first term is negative as the loss for clients only owning one class is higher. However, *Shapley Value* obtains higher values for *i.i.d.* clients with large data sets than Influence since $\mathcal{L}(C_m) - \mathcal{L}(C_n)$ increases if the distance between C_m's distribution and the global distribution is small, in line with a previous work [9].

Property 3. Shapley Value and Influence share the same trend in contribution measurement for Mavericks.

Theorem 1. *Let C_m and C_n be a Maverick and a non-Maverick client, respectively, and denote by $SV_t(C_k)$ the Shapley value of C_k in round r. Then $SV_1(C_m) < SV_1(C_n)$ and $SV_t(C_m)$ converges towards $SV_t(C_n)$.*

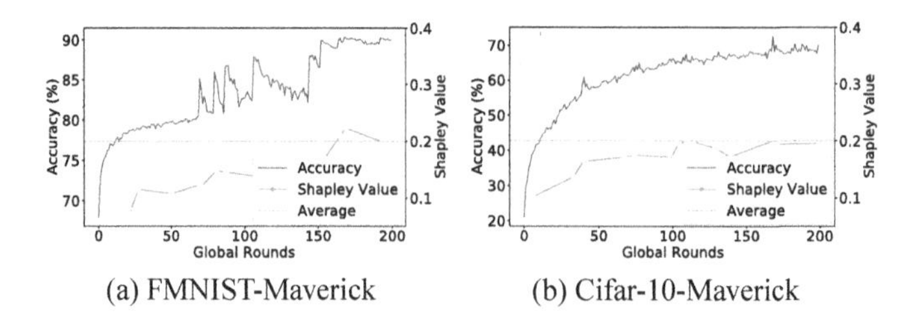

(a) FMNIST-Maverick (b) Cifar-10-Maverick

Fig. 2. Relative *Shapley Value* during training under multiple exclusive and shared Mavericks.

We present the empirical evidences of how one or multiple Mavericks are measured by *Shapley Value*. We here focus on single exclusive Mavericks and leave multiple Mavericks, shared and exclusive, for our in-depth experimental evaluation in the supplementary material. We use Fashion-MNIST (Fig. 2a) and Cifar-10 (Fig. 2b) as learning scenarios and use random client selection with FedAvg.

Figure 2 shows the global accuracy and the relative *Shapley Value* during training, with the average relative *Shapley Value* of the 5 selected clients out of 50 indicated by the dotted line. The contribution is only evaluated when a Maverick is selected. Looking at Fig.(2a, b), The *Shapley Value* of the Maverick indeed increases over time but remains below average until round 160, providing concrete evidence of **Theorem** 1. Furthermore, the accuracy increases when a Maverick is selected, indicating that Mavericks contribute highly to improving the model. Thus, assigning Mavericks a lower contribution measure is unreasonable, especially in the early stage of the learning process. All of the empirical results are consistent with our theoretical analysis.

4 *FEDEMD*

In this section, we propose a novel adaptive client selection algorithm FEDEMD, which enables FL systems with Mavericks to achieve faster convergence compared with SOTA methods, including *Shapley Value*-based ones. The key idea is to assign a higher probability for selecting Maverick clients initially to accelerate convergence; later we reduce the selection probability to avoid skewing the distribution towards Maverick classes. To measure the differences in data distributions, we adopt Wasserstein distance (EMD) [2], which is used to characterize weight divergence in FL [27]. The Wasserstein distance (EMD) is defined as:

$$\mathrm{EMD}\left(P_r, P_\theta\right)=\inf_{\gamma \in \Pi} \sum_{x, y}\|x-y\| \gamma(x, y)=\inf_{\gamma \in \Pi} \mathbb{E}_{(x, y) \sim \gamma}\|x-y\|, \tag{4}$$

where $\Pi(P_r, P_\theta)$ represents the set of all possible joint probability distributions of P_r, P_θ. $\gamma(x, y)$ represents the probability that x appears in P_r and y appears in P_θ.

Overview. The complete algorithm is shown in Algorithm 1, we here summarize the different components that make up the algorithm. ***i) Data Reporting and Initialization*** (Line 1–3): Clients report their data quantity so that the *federator* is able to compute the global data size array $\mathcal{D}_g$ and initialize the current size array $\mathcal{D}_c^1$.

ii) Dynamic Weights Calculation (Line 4–11): In this key step, we utilize a light-weight measure based on EMD to calculate dynamic selection probabilities over time, which achieve faster convergence, yet avoid overfitting, concretely we compute

$$Proba^r = softmax(\widetilde{emd}_g - t\beta\widetilde{emd}_c^r) \tag{5}$$

Algorithm 1: FEDEMD Clients Selection

Data: $\mathcal{D}^i$ for $i \in 1, 2, ..., N$.
Result: $\mathcal{K}$: selected participants.
1 **Set:** distance coefficient $\beta > 0$;
2 initialize probability $Proba^1$;
3 initialize current distribution $\mathcal{D}_c^1$;
4 $\mathcal{D}_g \leftarrow \sum_{i=1}^{N} \mathcal{D}^i$;
5 calculate $\widetilde{emd}_g$ by Eq. 6;
6 **for** *round r = 1, 2, ..., R* **do**
7 $\quad \mathcal{K}^r = rand(K, \mathcal{C}, Proba^r)$
8 $\quad \mathcal{D}_c^{r+1} \leftarrow \mathcal{D}_c^r + \sum_{C_k \in \mathcal{K}}^{r} \mathcal{D}^{C_k}$;
9 $\quad$ calculate $\widetilde{emd}_c^r$ by Eq. 7;
10 $\quad$ **for** *client i = 1, ..., N* **do**
11 $\quad\quad$ update $Proba^{r+1}$ by Eq. 5

where $Proba_i^r$ is the probability for selecting C_i in round r. β is a coefficient to weigh the global and current distance and shall be adapted for different initial distributions, i.e., different dataset and distribution rules. $\widetilde{emd}_g$ and $\widetilde{emd}_c^r$ are the normalized EMDs between the global/current and local distributions (Line 5, 9), namely

$$\widetilde{emd}_g = Norm([EMD(\mathcal{D}_g, \mathcal{D}^i)|_{i\in\{1,...,N\}}]), \tag{6}$$

which is constant through the learning process as long as the local distribution of clients stays the same. The larger $\widetilde{emd}_g$ is, the higher the probability $Proba_i^r$ that a client C_i is selected to increase model accuracy (Line 11), since C_i brings more distribution information to train ω_r. However, for convergence, a smaller $\widetilde{emd}_c$ is preferred in selection, so that $\widetilde{emd}_c$ depends on the round r:

$$\widetilde{emd}_c^r = Norm([EMD(\mathcal{D}_c^r, \mathcal{D}^i)|_{i\in\{1,...,N\}}]), \tag{7}$$

where $\mathcal{D}_c^r$ is the accumulated $\mathcal{D}^i$ of selected clients over rounds (Line 8). Let l denote one class randomly chosen by the *federator* except for the Maverick class from $\mathcal{D}$, here we apply normalization: $Norm(emd, \mathcal{D}) = \frac{emd}{\sum_{i=1}^{N} \mathcal{D}^i(y=l)/N}$.

iii) Weighted Random Client Selection (Line 7): At each round r, we select clients based on a probability distribution characterized by the dynamic weights [6] $Proba^r$:

$$\mathcal{K}^r = rand(K, \mathcal{C}, Proba^r). \tag{8}$$

Sampling K out of N clients based on $Proba^r$ has a complexity of $O(K \log(N/K))$, so comparably low. Thus, Mavericks with larger global distance and smaller current distance initially are preferred to be selected. The decrease of probability for selecting Mavericks elaborates based on the global and current distances changes over the learning procedure. As r increases, so does the impact of the current distance based on Eq. 5, reducing the probability to select a Maverick, as intended.

Convergence Analysis: To derive the convergence bound, we follow the setting of [15]. We let F_k be the local objective of client C_k and define $F(\boldsymbol{\omega}) \triangleq \sum_{k=1}^{N} p_k F_k(\boldsymbol{\omega})$, where p_k is the weight of client C_k when doing the aggregation. We have the FL optimization framework $\min_\omega F(x) = \min_\omega \sum_{k=1}^{N} p_k F_k(\boldsymbol{\omega})$. We make the *L-smooth* and *μ-strongly convex* assumptions on the functions $F_1, ..., F_N$ [15,20]. Let T be the total number of SGDs in a client, E be the number of local iterations of each client in each round. t is used to index the SGDs in each client. Thus, the relationship between E, t and global round r is $r = \lfloor t/E \rfloor$. F^* and F_k^* are the minimum values of F and F_k. $\Gamma = F^* - \sum_{k=1}^{N} p_k F_k^*$ is used to represent the degree of heterogeneity. We obtain:

Theorem 2. Let ξ_t^k be a sample chosen from the local data of each client. For $k \in [N]$, assume that:

$$\mathbb{E}\left\|\nabla F_k(\boldsymbol{\omega}_t^k, \xi_t^k) - F_k(\boldsymbol{\omega}_t^k)\right\|_2^2 \leq \sigma_k^2, \tag{9}$$

and

$$\mathbb{E}\left\|F_k(\boldsymbol{\omega}_t^k, \xi_t^k)\right\|_2^2 \leq G^2. \tag{10}$$

Then let $\epsilon = \frac{L}{\mu}$, $\gamma = \max\{8\epsilon, E\}$ and the learning rate $\eta_t = \frac{2}{\mu(\gamma+t)}$. We have the following convergence guarantee for Algorithm 1.

$$\mathbb{E}[F(\boldsymbol{\omega}_T)] - F^* \leq \frac{\epsilon}{\gamma + T - 1}\left(\frac{2(\Psi + \Phi)}{\mu} + \frac{\mu\gamma}{2}\mathbb{E}\left\|\boldsymbol{\omega}_1 - \boldsymbol{\omega}^*\right\|_2^2\right),$$

where $\Psi = \sum_{k=1}^{N} (Proba_k^{\lfloor T/E \rfloor})^2 {\sigma_k}^2 + 6L\Gamma + 8(E-1)^2 G^2$ and $\Phi = \frac{4}{K} E^2 G^2$.

Since all the notations except T in Expression (2) are constants, we have $O(\frac{1}{T})$ convergence rate for the algorithm where $\lim_{T\to\infty} \mathbb{E}[F(\boldsymbol{\omega}_T)] - F^* = 0$.

5 Experimental Evaluation

In this section, we comprehensively evaluate the effectiveness and convergence of FEDEMD in comparison to *Shapley Value*-based selection and SOTA baselines. The evaluation considers both exclusive and shared Mavericks.

Datasets and Classifier Networks. We use public image datasets: *i)* Fashion-MNIST [26] for bi-level image classification; *ii)* MNIST [13] for simple and fast tasks that require a low amount of data; *iii)* Cifar-10 [12] for more complex task such as colored image classification; *iv)* STL-10 [5] for applications with small amounts of local data for all clients. We note that light-weight neural networks are more applicable for FL scenarios, where clients typically have limited computation and communication resources [19]. Thus, here we apply light-weight CNNs for all datasets.

Federated Learning System. The system considered has 50 participants with homogeneous computation and communication resources and 1 *federator*. At each round, the *federator* selects 10% of clients using different client selection algorithms. The *federator* uses average or quantity-aware aggregation to aggregate local models from selected clients. We set one local epoch for both aggregations to enable a fair comparison of the two aggregation approaches. Two types of Mavericks are considered: exclusive and shared Mavericks with up to 3 Mavericks. We demonstrate the case of single Maverick owning an entire class of data in most of our experiments.

Evaluation Metrics. *i)* Global test accuracy for all classes; *ii)* Source recall for classes owned by Mavericks exclusively; *iii)* $R@99$: the number of communication rounds required to reach 99% of test accuracy of random selection results; *iv)* Normalized *Shapley Value* ranging between $[0, 1]$ to measure the contribution of Mavericks.

Baselines. We consider four selection strategies: Random [18], *Shapley Value*-based, FedFast [19], and TiFL [3][3] under both average and quantity-aware aggregation methods. Further, in order to compare with state-of-the-art solutions for heterogeneous FL that focus on the optimizer, we evaluate FedProx [14] as one of the baselines.

5.1 *FedEMD* Is Effective for Client Selection

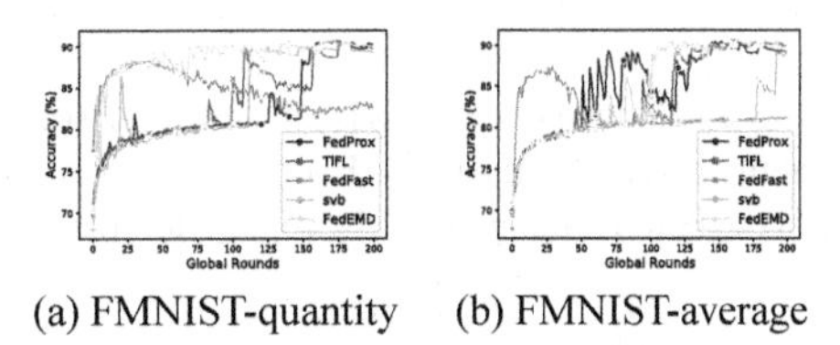

(a) FMNIST-quantity (b) FMNIST-average

Fig. 3. Comparison on FedEMD with baselines.

Figure (3a, b) show global accuracy over rounds. First we focus on the comparison between the contribution-based SVB and our proposed distance-based FedEMD. FedEMD achieves an accuracy close to the maximum almost immediately for FedAvg while SVB requires about 100 rounds (72 and 104 rounds for $R@99$ for SVB and FedEMD). For average aggregation, both client selection methods have a slower convergence but FedEMD still only requires about half the number of rounds to achieve the same high accuracy as SVB. Indeed, SVB fails in reaching $R@99$ within 200 rounds. The reason is that SVB rarely selects the Maverick in the early phase, as the Maverick has a below-average *Shapley Value*. We can also see the superiority of FedEMD among results presented for the baselines in the figures. The detailed analysis will be discussed together with Table 1 below.

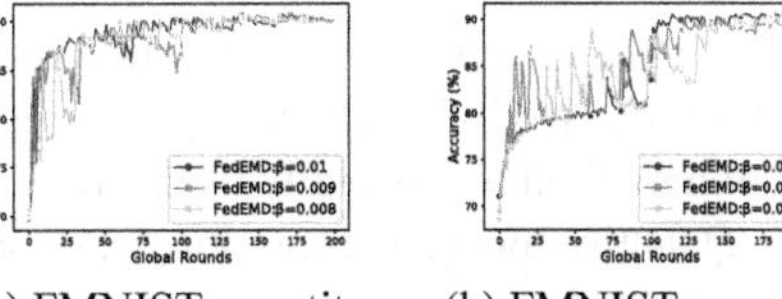

(a) FMNIST-quantity (b) FMNIST-average

Fig. 4. Comparison on FedEMD over different β.

We evaluate the effects of the hyperparameter β in Fig. (4a, b). The server can apply a preliminary client selection simulation before training based on the self-reported data size array. FedEMD works best when the average probability of selecting Maverick is within $[1/N - \epsilon, 1/N + \epsilon]$ based on our observation experiments, where $\epsilon > 0$ is a task-aware

[3] We focus on their client selection and leave out other features, e.g., communication acceleration in TiFL. We apply distribution mean clustering for FedFast following the setting in their paper.

small value. In our example with Fashion-MNIST, we choose β equal to 0.008, 0.009 and 0.01, with the results displayed in Fig. 4. These three values all satisfy the average probability above with $\epsilon \geq 0.002$. The results shows that all of the 3 numbers work for Fashion-MNIST, verifying the effectiveness of FEDEMD for various values of the hyper-parameter. However, there are also values of β that are not suitable, e.g., $\beta = 0.1$ for which the Maverick is selected too rarely.

Comparison with Baselines. We summarize the comparison with the state-of-the-art methodologies in Table 1. The reported $R@99$ is averaged over three replications. Note that we run each simulation for 200 rounds, which is mostly enough to see the convergence statistics for these lightweight networks. The rare exceptions when 99% maximal accuracy is not achieved for random selection are indicated by > 200.

Due to its distance-based weights, FEDEMD almost consistently achieves faster convergence than all other algorithms. The reason for this result is that FEDEMD enhances the participation of the Maverick during the early training period, speeding up learning of the global distribution. For most settings, the difference in convergence rounds is considerable and clearly visible.

Table 1. Convergence rounds of selection strategies in $R@99$ Accuracy, under average and quantity-aware aggregation (Every result is averaged over three runs and is marked with standard deviation among all of the replication results).

Dataset	Average Aggregation					
	Random	**FedProx**	**TiFL**	**FedFast**	**SVB**	**FEDEMD**
MNIST	133 ± 44.47	118 ± 8.50	111 ± 21.66	>200 ± NA	147 ± 52.50	**<u>99</u>** ± 24.70
Fashion-MNIST	144 ± 51.47	135 ± 20.59	140 ± 8.62	>200 ± NA	**<u>103</u>** ± 56.00	131 ± 37.29
Cifar-10	141 ± 6.11	164 ± 15.00	147 ± 10.97	>200 ± NA	184 ± 9.24	**<u>140</u>** ± 15.13
STL-10	122 ± 49.94	186 ± 4.36	125 ± 57.50	171 ± 16.74	190 ± 3.06	**<u>96</u>** ± 4.93
Dataset	Quantity-aware Aggregation					
	Random	**FedProx**	**TiFL**	**FedFast**	**SVB**	**FEDEMD**
MNIST	72 ± 29.26	51 ± 8.19	84 ± 37.99	>200 ± NA	49 ± 2.52	**<u>40</u>** ± 5.57
Fashion-MNIST	111 ± 37.75	92 ± 12.12	146 ± 38.18	>200 ± NA	80 ± 40.13	**<u>80</u>** ± 10.79
Cifar-10	143 ± 26.29	144 ± 39.46	120 ± 9.45	174 ± 9.50	132 ± 26.50	**<u>107</u>** ± 10.58
STL-10	180 ± 0.58	179 ± 6.24	>200 ± NA	153 ± 34.88	181 ± 10.97	**<u>95</u>** ± 2.65

The only exception are easy tasks with simple averaging rather than weighted, e.g., Fashion-MNIST with average aggregation, which indicates our distribution-based selection method is especially useful for data size-aware aggregation and more complex tasks. Quantity-aware aggregation nearly always outperforms plain average aggregation as its weighted averaging assigns more impact to the Maverick. While such an increased weight caused by larger data size can lead to a decrease in accuracy in the latter phase of training, Mavericks are rarely selected in the latter phase by FEDEMD, which successfully mitigates the effect and achieves a faster convergence.

In order to demonstrate the comparison of FEDEMD and SVB across multiple datasets, here we also provide the experimental results with MNIST and Cifar-10, which is inline with our conclusion of Fashion-MNIST in Fig. 4 for better convergence performance of FEDEMD.

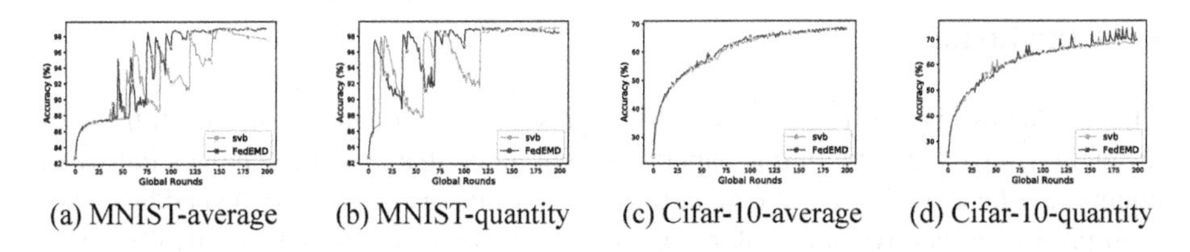

(a) MNIST-average (b) MNIST-quantity (c) Cifar-10-average (d) Cifar-10-quantity

Fig. 5. Comparison on FEDEMD with SVB.

5.2 *FEDEMD* Works for Multiple Mavericks

We explore the effectiveness of FEDEMD on both types of Mavericks: exclusive and shared Mavericks.

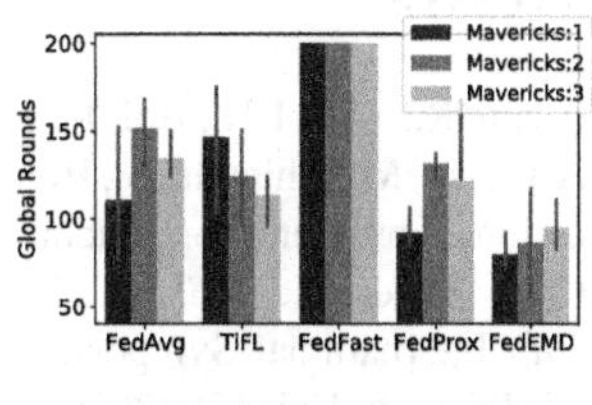

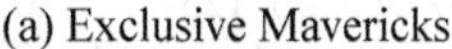

(a) Exclusive Mavericks

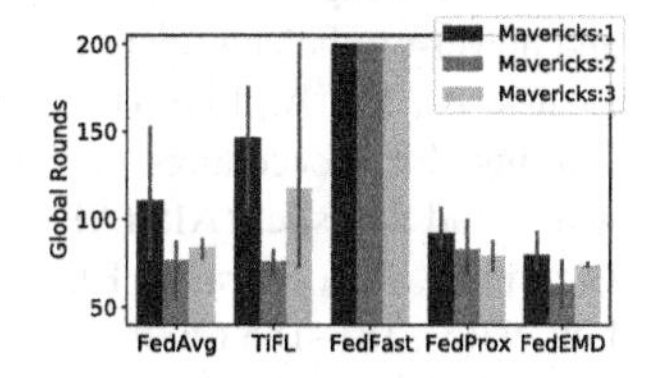

(b) Shared Mavericks

Fig. 6. Convergence rounds $R@99$ for multiple Mavericks.

We vary the number of Mavericks between one and three and use the Fashion-MNIST dataset. The Maverick classes are 'T-shirt', 'Trouser', and 'Pullover'. Results are shown with respect to $R@99$.

Figure (6a) illustrates the case of multiple exclusive Mavericks. For exclusive Mavericks, the data distribution becomes more skewed as more classes are exclusively owned by Mavericks. FEDEMD always achieves the fastest convergence, though its convergence rounds increase slightly as the number of Mavericks increases, reflecting the increased difficulty of learning in the presence of skewed data distribution. FedFast's K-mean clustering typically results in a cluster of Mavericks and then always includes at least one Maverick. In some initial experiments, we found that constantly including a Maverick hinders convergence, which is also reflected in FedFast's results. TiFL outperforms FedAvg with random selection for multiple Mavericks. However, TiFL's results differ drastically over runs due to the random factor in its local computations. Thus, TiFL is not a reliable choice for Mavericks. Comparably, FedProx tends to achieve the best performance among the SOTA algorithms but still exhibits slower convergence than FEDEMD as higher weight divergence entails higher penalty on the loss function.

For shared Mavericks, a higher number of Mavericks indicates a more balanced distribution. Similar to the exclusive case, FEDEMD has the fastest convergence and FedFast again trails the others. The improvement of FEDEMD over the other methods is less visible due to the limited advantage of FEDEMD on balanced data. A higher number of Mavericks resembles the case of *i.i.d.*. Random performs the most similar to FEDEMD for shared Mavericks, as random selection is best for *i.i.d.* scenarios. Note that the standard deviation of FEDEMD is smaller, implying a better stability.

6 Conclusion

Client selection is key to successful FL as it enables maximizing the usefulness of different diverse datasets. In this paper, we highlighted that existing schemes fail when clients have heterogeneous data, in particular if one class is exclusively owned by one or multiple Mavericks. We first explore *Shapley Value*-based selection, theoretically showing its limitations in addressing Mavericks. We then propose FEDEMD that encourages the selection of diverse clients at the opportune moment of the training process, with guaranteed convergence. Evaluation results on multiple datasets across different scenarios of Mavericks show that FEDEMD reduces the communication rounds needed for convergence by 26.9% compared to the state-of-the-art client selection methods.

References

1. Adam, R., Aris, F.R., Boi, F.: Rewarding high-quality data via influence functions (2019)
2. Arjovsky, M., Chintala, S., Bottou, L.: Wasserstein generative adversarial networks. In: International conference on Machine Learning (ICML), pp. 214–223. PMLR (2017)
3. Chai, Z., et al.: TiFL: a tier-based federated learning system. In: Proceedings of the 29th International Symposium on High-Performance Parallel and Distributed Computing (HPDC), pp. 125–136 (2020)
4. Cho, Y.J., Wang, J., Joshi, G.: Client selection in federated learning: convergence analysis and power-of-choice selection strategies. arXiv preprint arXiv:2010.01243 (2020)
5. Coates, A., Ng, A., Lee, H.: An analysis of single-layer networks in unsupervised feature learning. In: Proceedings of the Fourteenth International Conference on Artificial Intelligence Aad Statistics (AISTATS). JMLR Workshop and Conference Proceedings (2011)
6. Efraimidis, P.S., Spirakis, P.G.: Weighted random sampling with a reservoir. Inf. Process. Lett. **97**(5), 181–185 (2006)
7. Feng, S., Niyato, D., Wang, P., Kim, D.I., Liang, Y.: Joint service pricing and cooperative relay communication for federated learning. In: 2019 IEEE iThings and GreenCom and IEEE Cyber and CPSCom and SmartData, iThings/GreenCom/CPSCom/SmartData 2019, pp. 815–820. Atlanta, GA, USA, 14–17 July 2019. IEEE (2019)
8. Goetz, J., Malik, K., Bui, D., Moon, S., Liu, H., Kumar, A.: Active federated learning. arXiv preprint arXiv:1909.12641 (2019)
9. Huang, J., Talbi, R., Zhao, Z., Boucchenak, S., Chen, L.Y., Roos, S.: An exploratory analysis on users' contributions in federated learning. arXiv preprint arXiv:2011.06830 (2020)
10. Huang, T., Lin, W., Wu, W., He, L., Li, K., Zomaya, A.: An efficiency-boosting client selection scheme for federated learning with fairness guarantee. IEEE Transactions on Parallel and Distributed Systems (2020)
11. Kang, J., Xiong, Z., Niyato, D., Xie, S., Zhang, J.: Incentive mechanism for reliable federated learning: a joint optimization approach to combining reputation and contract theory. IEEE Internet Things J. **6**(6), 10700–10714 (2019)
12. Krizhevsky, A., Hinton, G., et al.: Learning multiple layers of features from tiny images (2009)
13. LeCun, Y., Bottou, L., Bengio, Y., Haffner, P.: Gradient-based learning applied to document recognition. Proc. IEEE **86**(11), 2278–2324 (1998)
14. Li, T., Sahu, A.K., Zaheer, M., Sanjabi, M., Talwalkar, A., Smith, V.: Federated optimization in heterogeneous networks. In: Proceedings of Machine Learning and Systems (MLsys) (2020)

15. Li, X., Huang, K., Yang, W., Wang, S., Zhang, Z.: On the convergence of FedAvg on non-IID data. ICLR (2020)
16. Liu, S., Feng, X., Zheng, H.: Overcoming forgetting in local adaptation of federated learning model. In: Gama, J., Li, T., Yu, Y., Chen, E., Zheng, Y., Teng, F. (eds.) Advances in Knowledge Discovery and Data Mining. PAKDD 2022. LNCS, vol. 13280, pp. 613–625. Springer, Cham (2022). https://doi.org/10.1007/978-3-031-05933-9_48
17. Luo, B., Xiao, W., Wang, S., Huang, J., Tassiulas, L.: Tackling system and statistical heterogeneity for federated learning with adaptive client sampling. In: IEEE INFOCOM 2022 - IEEE Conference on Computer Communications, London, United Kingdom, 2–5 May 2022, pp. 1739–1748. IEEE (2022)
18. McMahan, B., Moore, E., Ramage, D., Hampson, S., y Arcas, B.A.: Communication-efficient learning of deep networks from decentralized data. In: Proceedings of Artificial Intelligence and Statistics (AISTATS), pp. 1273–1282 (2017)
19. Muhammad, K., et al.: FedFast: going beyond average for faster training of federated recommender systems. In: Proceedings of the 26th ACM International Conference on Knowledge Discovery & Data Mining, pp. 1234–1242 (2020)
20. Nguyen, H.T., Sehwag, V., Hosseinalipour, S., Brinton, C.G., Chiang, M., Poor, H.V.: Fast-convergent federated learning. IEEE J. Sel. Areas Commun. **39**(1), 201–218 (2020)
21. Nishio, T., Yonetani, R.: Client selection for federated learning with heterogeneous resources in mobile edge. In: IEEE International Conference on Communications (ICC), pp. 1–7 (2019)
22. Sim, R.H.L., Zhang, Y., Chan, M.C., Low, B.K.H.: Collaborative machine learning with incentive-aware model rewards. In: International Conference on Machine Learning (ICML), pp. 8927–8936. PMLR (2020)
23. Wang, L., Xu, S., Wang, X., Zhu, Q.: Addressing class imbalance in federated learning. In: Thirty-Fifth AAAI Conference on Artificial Intelligence, AAAI, IAAI, EAAI, pp. 10165–10173. AAAI Press (2021)
24. Wang, T., Rausch, J., Zhang, C., Jia, R., Song, D.: A principled approach to data valuation for federated learning. In: Yang, Q., Fan, L., Yu, H. (eds.) Federated Learning. LNCS (LNAI), vol. 12500, pp. 153–167. Springer, Cham (2020). https://doi.org/10.1007/978-3-030-63076-8_11
25. Wei, S., Tong, Y., Zhou, Z., Song, T.: Efficient and fair data valuation for horizontal federated learning. In: Federated Learning, pp. 139–152 (2020)
26. Xiao, H., Rasul, K., Vollgraf, R.: Fashion-MNIST: a novel image dataset for benchmarking machine learning algorithms (2017)
27. Zhao, Y., Li, M., Lai, L., Suda, N., Civin, D., Chandra, V.: Federated learning with non-IID data. arXiv:1806.00582 (2018)
28. Zhu, Z., Hong, J., Zhou, J.: Data-free knowledge distillation for heterogeneous federated learning. In: Meila, M., Zhang, T. (eds.) Proceedings of the 38th International Conference on Machine Learning, ICML 2021, 18–24 July 2021, Virtual Event. Proceedings of Machine Learning Research, vol. 139, pp. 12878–12889. PMLR (2021)

pFedV: Mitigating Feature Distribution Skewness via Personalized Federated Learning with Variational Distribution Constraints

Yongli Mou[1], Jiahui Geng[2], Feng Zhou[3](✉), Oya Beyan[4], Chunming Rong[2], and Stefan Decker[1,5]

[1] Chair of Computer Science 5, RWTH Aachen University, Ahornstr. 55, 52074 Aachen, Germany
{mou,decker}@dbis.rwth-aachen.de

[2] Faculty of Science and Technology, Department of Electrical Engineering and Computer Science, University of Stavanger, Stavanger, Norway
{jiahui.geng,chunming.rong}@uis.no

[3] Center for Applied Statistics and School of Statistics, Renmin University of China, Beijing, China
feng.zhou@ruc.edu.cn

[4] Institute for Medical Informatics, Faculty of Medicine and University Hospital Cologne, University of Cologne, Cologne, Germany
oya.beyan@uni-koeln.de

[5] Fraunhofer Institute for Applied Information Technology, Sankt Augustin, Germany
stefan.decker@fit.fraunhofer.de

Abstract. Statistical heterogeneity, especially feature distribution skewness, among the distributed data is a common phenomenon in practice, which is a challenging problem in federated learning that can lead to a degradation in the performance of the aggregated global model. In this paper, we introduce pFedV, a novel approach that leverages a variational inference perspective by incorporating a variational distribution into neural networks. During training, we add the KL-divergence term to the loss function to constrain the output distribution of layers for feature extraction and personalize the final layer of models. The experimental results demonstrate the effectiveness of our approaches in mitigating the distribution shift in feature space in federated learning.

Keywords: federated learning · statistical heterogeneity · variational inference

1 Introduction

Despite the impressive results that deep learning-based approaches have achieved in recent decades, training deep learning models is data-driven and intensively depends on the availability and accessibility of high-quality data. Conventionally,

H. Kashima et al. (Eds.): PAKDD 2023, LNAI 13936, pp. 283–294, 2023.
https://doi.org/10.1007/978-3-031-33377-4_22

data is brought to the computation by following a data centralization approach, leading to privacy breaches and the loss of data sovereignty. As the related issues are increasingly aware, data protection legislation has emerged worldwide in the last few years, e.g., the General Data Protection Regulation (GDPR) in the European Union explicitly prohibits organizations from exchanging data without clear consent from users. Besides, commercial competition and complicated administrative procedures also hinder data integration and data sharing, which makes data exist in the form of isolated islands [22]. As a promising paradigm to provide privacy protection in machine learning, federated learning [16] has been widely adopted in academia and industry. Federated learning enables the participating clients collaboratively train a global machine learning model without revealing local private data. Due to its privacy-preserving characteristics, federated learning is increasingly drawing attention from a wide range of applications and domains such as healthcare [18], finance [22,23], and IoT [8].

Despite federated learning's benefits, its continued popularity is usually accompanied by new emerging problems [6,11], such as the lack of trust among participants, the vulnerability exposed to privacy inferences, the limited or unreliable connectivity, etc. Among these, statistical heterogeneity is considered to be the most challenging problem. It is also called the non-IID problem, where data are not independent and identically distributed across clients. For example, medical radiology images in different hospitals are acquired by different devices using disparate standards [14]. Studies have shown that non-IID data can lead to poor accuracy and slow convergence, sometimes even divergence, if without appropriate optimization algorithms [13]. In practice, the non-IID scenarios are complicated to be categorized, but statistical heterogeneities with regard to label distribution, feature distribution and quantity are mainly being studied. To tackle the aforementioned challenges, it is necessary to adopt appropriate optimization algorithms for federated learning.

In this work, we mainly focus on the feature distribution skewness problem. The main contributions of our paper could be summarized as follows: **(1)** we propose a novel FL training strategy, called pFedV to mitigate the covariance shift, i.e., one of the major problems of statistical heterogeneity. The last layer for feature extraction is modified before the classification layers in the neural networks, instead of compressing the input into the hidden feature space, that layer generates the variational distribution of the feature maps. A regularization term is added in the loss function for the local training in federated learning, i.e., the KL-divergence term makes the variational distribution of the local model close to the output distribution of the global model or a certain pre-defined distribution. We design two variational distribution models, a strong restricted one using zero-mean, unit-variance Gaussian for all clients and another one using the distribution in the global model. **(2)** Furthermore, we adopt the idea of FedBN [14] to train the last classification layer individually at each client, as a personalized technique for federated learning. **(3)** Finally, we evaluate our proposed approaches on five related but heterogeneous data sets and our empirical studies validate pFedV's superior performance on non-IID data.

2 Related Work and Background

2.1 Federated Learning

Unlike conventional machine learning where training is centralized and the data is collected from different sites and stored in central storage [1], federated learning is a distributed machine learning paradigm and trains a global model across data generated from distributed clients participating in each communication round. A typical federated learning system consists of a server and clients, where the server orchestrates the training process by repeating the steps including client selection, model distribution, client training and model aggregation [6], and the clients train the global model with local data. The server aggregates the collected client models according to a specified strategy and the aggregated global model is expected to surpass the performance of independently trained client models. Considering multi-class classification problem, given K clients with client i holding a dataset $\mathcal{D}_i := \{(\mathbf{x}_i^{(n)}, y_i^{(n)})\}_{n=1}^{N_i}$, where $\mathbf{x}_i^{(n)} \in \mathcal{X} \subseteq \mathbb{R}^D$ and $y_i^{(n)} \in \{1, 2, \cdots, C\}$, N_i is the number of data on client i, D is the number of input dimension and C is the number of classes, federated learning can basically be formalized as an optimization problem to minimize the objective function $\min \mathcal{F}(\theta) = \sum_{i=1}^{K} \pi_i \mathcal{F}_i(\theta)$, where θ, π_i and $\mathcal{F}_i$ are the global model, the relative impact and the local objective function $\mathcal{F}_i(\theta) = \frac{1}{N_i} \sum_{n=1}^{n=N_i} \mathcal{L}(\theta, \mathbf{x}_i^{(n)}, y_i^{(n)})$ for client i, respectively. The relative impact π_i can be user-defined with $\sum_{i=1}^{K} \pi_i = 1$ normally as N_i/N, where $N = \sum_{i=1}^{K} N_i$ is the total number of samples. FedSGD [16] used stochastic gradient decent as the optimizer and updated the model on the server for each local training step. However, this approach has a main obstacle i.e., high communication cost. and potential risk of data leakage from gradients [5]. To reduce the communication cost and prevent privacy leakage, FedAvg [16], instead of the one-step gradient descent scheme, is an aggregation strategy that updates models with multiple steps.

2.2 Statistical Heterogeneity

The local objective function $\mathcal{F}_i$ is often defined as the empirical risk over local data and is the same across all clients, while the local data distribution $P_i(X, Y)$ often varies among different clients capturing data heterogeneity. The joint distribution $P_i(X, Y)$ can be rewritten as $P_i(X|Y)P_i(Y)$ and $P_i(Y|X)P_i(X)$ and Kairous et al. simplified the non-identical distributions into five categories, namely (1) covariate shift as feature distribution skew, (2) prior probability shift as label distribution skew, (3) concept shifts including same label- but different feature distributions and same feature- but different label distributions, and (4) quantity skew [6]. In practice, the non-identical distribution can be combined and even more complicated. Studies [13] show that the performance on the convergence rate and the accuracy of FedAvg on heterogeneous data are significantly reduced, compared to the results on homogeneous data. Empirical works address non-IID issues by modifying operations in different steps. For example, FedProx [12] used a proximal term in the local training stage as a regularization term to suppress the divergence of model updates. FedNova [21] improved the aggregation stage by considering different parties may conduct different numbers of local steps. Li et al. [10] proposed comprehensive data partitioning strategies to cover the typical non-IID data cases. To mitigate such performance degradation, FedBN [14] is designed to alleviate the feature shift before averaging models via local batch normalization. Anit et al. [20] chose to add a proximal item to reduce the difference between the global model and the local model parameters, avoiding the failure of convergence during training. Mou et al. [17] demonstrated that additional small balanced datasets can be used to overcome model differences caused by class imbalance. Sai et al. [7] proposed SCAFFOLD that uses a control variable (variance reduction) to correct for client drift in local updates, which is claimed to reduce the number of communication rounds required for training and the impact due to data heterogeneity or client sampling. Recently, a lot of work apply the Bayesian framework to federated learning. Instead of maximizing the log-likelihood $\log p(\mathcal{D}|\theta)$, the Bayesian framework is to find the posterior of model parameters as $p(\theta|\mathcal{D}) = \frac{p(\mathcal{D}|\theta)p(\theta)}{p(\mathcal{D})}$, where $p(\theta)$ is the prior of model parameters, $p(\mathcal{D}|\theta)$ is the likelihood. FedBE [4] adopted Bayesian inference to achieve robust aggregation of local models through Bayesian model ensemble. It uses Gaussian or Dirichlet distributions and Monte to efficiently model data distributions. FOLA [15] proposed to approximate the client and server posteriors using online Laplacian approximation, and employed a multivariate Gaussian on the server side to construct and maximize the global posterior, thereby reducing aggregation errors and local forgetting due to large model differences. pFedBayes [24] introduced the uncertainty of weights, i.e., Bayesian neural networks (BNNs) [3], into the federated learning system. Each client achieves personalization by balancing between the construction error of its own private data and the KL divergence with the global model.

2.3 Variational Inference

Variational autoencoder (VAE) [9] is a generative model that consists of an encoder yielding approximate posterior distribution $q_\theta(z|x)$ and a decoder yielding approximate likelihood distribution $p_\phi(x|z)$. The objective of VAE is to minimize the KL-divergence between approximate posterior and real posterior as shown in Eq. 1.

$$D_{KL}(q_\theta(z|x)||p(z|x)) = -\int q_\theta(z|x)\log(\frac{p(z|x)}{q_\theta(z|x)})\mathrm{d}z \tag{1}$$

The evidence lower bound (ELBO) is defined as the boxed part on the right-hand side in Eq. 2. We note that the log probability of the data on the left-hand side in Eq. 2 is a constant, therefore maximizing the ELBO is equal to minimizing the KL-divergence.

$$\log p(x) = D_{KL}(q_\theta(z|x)||p(z|x)) + \boxed{\int q_\theta(z|x)\log(\frac{p_\phi(x|z)p(z)}{q_\theta(z|x)})\mathrm{d}z} \tag{2}$$

The ELBO can be derived into two terms, namely, the KL-divergence term and the reconstruction term as shown in Eq. 3. The KL-divergence term is a constraint on the form of the approximate posterior as a regularizer while the reconstruction term is a measure of the likelihood of reconstructed data output at the decoder. The detailed derivation is available in [19].

$$\text{ELBO} = \int q_\theta(z|x)\log(\frac{p(z)}{q_\theta(z|x)})\mathrm{d}z + \int q_\theta(z|x)\log(p_\phi(x|z))\mathrm{d}z \tag{3}$$

$$= D_{KL}(q_\theta(z|x)||p(z)) + \mathbb{E}_{z\sim q_\theta(z|x)[\log p_\phi(x|z)]} \tag{4}$$

3 Methodology

3.1 Problem Formulation

As mentioned above, we consider the horizontal federated learning scenario (i.e., each client shares the same feature space but differs in sample ID space) with a supervised learning task (e.g., multi-class classification). We use neural networks for the task and formalize as a function $f(\mathbf{x}) = h(g(\mathbf{x}))$ consisting of two parts, i.e., $g(\cdot)$ is the encoder function parameterized by θ_g that extracts input features and the $h(\cdot)$ is the classifier function parameterized by θ_h that classifies the extracted features. We write $\mathbf{z} = g(\mathbf{x})$ and $y = h(\mathbf{z})$, where $\mathbf{z} \in \mathbb{R}^M$ and M is the dimension of the latent representations. Usually, deep neural networks are formed by stacking layer upon layer. Therefore, the parameters of the encoder and classifier can be further formulated as $\theta_g = (\theta_g^{(1)}, \theta_g^{(2)}, \cdots, \theta_g^{(G)})$ and $\theta_h = (\theta_h^{(1)}, \theta_h^{(2)}, \cdots, \theta_h^{(H)})$, where G and H are the number of layers in the encoder and classifier, and $\theta_g^{(i)}$ and $\theta_h^{(j)}$ denote the parameters of i-th and j-th layer in the encoder and classifier, respectively. For the statistical heterogeneity, we focus on feature distribution skewness, i.e., for two clients, their corresponding joint distributions vary due to the covariate shift, i.e., $P_i(X,Y) \neq P_j(X,Y), \forall i \neq j$ since $P_i(X) \neq P_j(X), \forall i \neq j$, assuming the conditional distribution $P(Y|X)$ is shared across clients.

3.2 Derivation of Variational Distribution Constraints

In our model, we denote the input and output of the neural networks as $\mathbf{x}$ and y and the latent representation as $\mathbf{z}$. We aim to learn the true posterior distribution $p(\mathbf{z}|y)$ for a given label y, which ensures that the learned latent representation is informative about the label and can be used to make accurate predictions on new data. In general, it is difficult to infer the posterior of latent variable $\mathbf{z}$ for a given label y when the likelihood is non-conjugated to the prior. To circumvent this issue, we resort to the variational inference [2] which uses a variational distribution to approximate the true posterior. Following the standard variational inference, the objective is to minimize the KL divergence between the variational distribution $q_\theta(\mathbf{z})$[1] and the true posterior (as shown in Eq. 5) to learn a variational distribution that is as close as possible to the true posterior, which is equivalent to maximizing the evidence lower bound (ELBO).

$$D_{KL}(q_\theta(\mathbf{z})||p(\mathbf{z}|y)) = -\int q_\theta(\mathbf{z}) \log(\frac{p(\mathbf{z}|y)}{q_\theta(\mathbf{z})})\mathrm{d}z \tag{5}$$

Similar to the derivation of the ELBO of VAE, we derive the ELBO[2] as in Eq. 6. Basically, ELBO consists of two parts: on the one hand, it enforces the model to fit the data better with the log-likelihood term; and on the other hand, it makes the variational distribution $q_\theta(\mathbf{z})$ as close as possible to the prior $p(\mathbf{z})$ by using Kull-back-Leibler (KL) divergence.

$$\text{ELBO} = \mathbb{E}_{\mathbf{z}\sim q_\theta(\mathbf{z})} \log p(y|\mathbf{z}) - D_{KL}(q_\theta(\mathbf{z})||p(\mathbf{z})), \tag{6}$$

Our goal is to find the optimal variational distribution of the latent representation $\mathbf{z}$. Specifically, we assume the variational distribution of $\mathbf{z}$ is a Gaussian distribution $q_\theta(\mathbf{z}) = \mathcal{N}(\mathbf{z}|\boldsymbol{\mu}_{\theta_g}(\mathbf{x}), \text{diag}(\boldsymbol{\sigma}^2_{\theta_g}(\mathbf{x})))$ where $\text{diag}(\cdot)$ denotes the diagonalization of a vector. The mean and variance are modeled by an encoder whose parameters are denoted as θ_g. After drawing a $\mathbf{z}$ from the corresponding approximate variational distribution from $q_\theta(\mathbf{z})$, known as the reparameterization trick [9], we can classify the current sample with the help of a classifier constructed by another neural network $\hat{y} = h(\mathbf{z})$ where $\hat{y}$ is the predicted class label and $h(\cdot)$ denotes the classifier parameterized by θ_h. We replace the log-likelihood term in 6 by the cross entropy loss in our case and finally obtain the following objective for our model, where CE is the cross entropy loss:

$$\theta_g^*, \theta_h^* = \underset{\theta_g,\theta_h}{\text{argmin}}\ \mathbb{E}_{q_{\theta_g}(\mathbf{z})}\text{CE}(\hat{y}_{\theta_h}(\mathbf{z}), y) + \alpha D_{KL}(q_{\theta_g}(\mathbf{z})||p(\mathbf{z})), \tag{7}$$

Comparing to conventional classification model training, a KL divergence term is added to the objective function as shown above. We add a weight factor α to the KL term, which is a hyperparameter, to adjust the strength of the penalty. In our case, we set it to 0.5 in all experiments related to variational distribution.

3.3 Personalized Federated Learning with Variational Distribution Constraints

In this section, we present our proposed approach, personalized federated learning with variational distribution constraints (pFedV). Figure 1 gives the overview of pFedV.

[1] The variational distribution is the output of the encoder parameterized by θ, which is equivalent to θ_g in the previous section.

[2] We omitted x in the formula since all distributions are given the condition of $\mathbf{x}$, e.g., $q_\theta(\mathbf{z}) = q_\theta(\mathbf{z}|\mathbf{x})$.

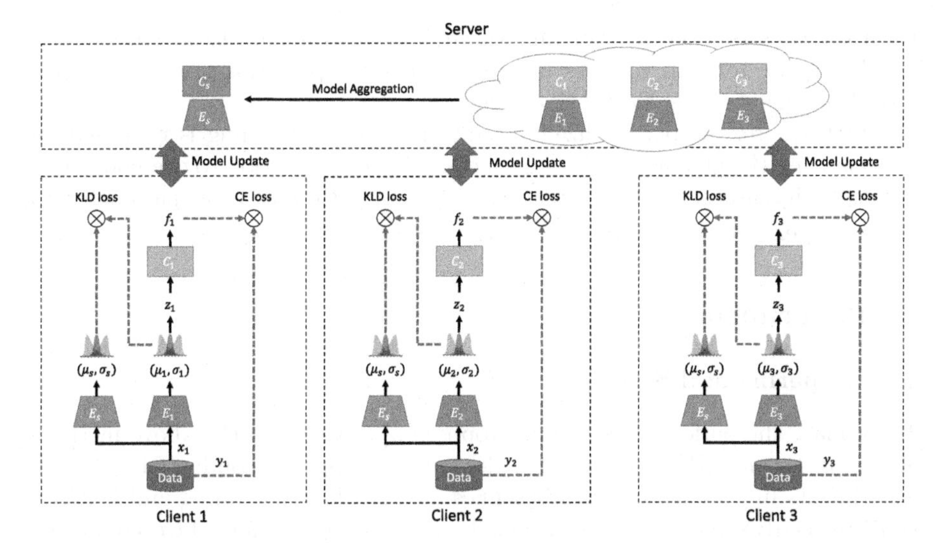

Fig. 1. Overview of pFedV: At each communication round, the server sends the global model to the clients participating in the local training; During the local training, models are trained with the above-mentioned loss function; After training for a given number of epochs, model updates are sent back to the server for the model aggregation, in which the last layer for classification is reserved at each client if the personalized setting is chosen

Like conventional federated learning systems, a server is employed for orchestrating the federated learning process repeating the steps of model update and aggregation. The blue bidirectional arrows between the server and clients indicate the communication for the model update. The server sends the global model to the clients at the beginning of each communication round and the clients send local models to the server after the local training.

The variational distribution constraints and loss functions described above are applied during the local training. We make an assumption for the variational distribution, i.e., the Gaussian distribution. The encoder of the neural network is modified to output the mean and standard deviation (for the non-negativity guarantee of standard deviation, we use log variance instead).

For the construction of the prior, we utilize two different strategies: **(1)** a fixed prior distribution like the classical variational inference and **(2)** continuous update. For the fixed prior solution, we use strong prior constraints, i.e., zero-mean, unit-variance Gaussian distribution for all clients. For the continuous update solution, we abstract the aggregated knowledge into the prior distribution and use the output of the variational distribution of the global model, i.e., the prior is constantly updated as the server communicates with clients in our federated learning framework. Specifically, we assume the prior of $\mathbf{z}$ is also a Gaussian distribution $p(\mathbf{z}) = \mathcal{N}(\mathbf{z}|\boldsymbol{\mu}_{\theta_s}(\mathbf{x}), \mathrm{diag}(\boldsymbol{\sigma}^2_{\theta_s}(\mathbf{x})))$. The mean and variance are modelled by another encoder whose parameter is denoted as θ_s.

With the construction of prior and variational posterior by two encoders, the KL divergence term in the loss function makes the posterior of extracted features from clients close to the global one.

Furthermore, the personalized variant of federated learning is proposed for personalizing the global model for each client in the federation to overcome data heterogeneity issues. In our approaches, we propose to reserve the parameters of the last layer of the classifier $\theta_h^{(H)}$ to achieve personalization.

4 Experiments

4.1 Experimental Settings

To evaluate the performance of our proposed approaches in the above methodologies in the non-IID scenario of federated learning. we conducted extensive experiments in comparison with baselines, i.e., single-site training and FedAvg, FedProx, FedBN and FOLA. Additionally, we report the results of the conducted experiments and analyze the effect of variational distribution constraints.

Datasets. To demonstrate the feature distribution skewness problem, we conduct all experiments on Digits-Five dataset, namely MNIST, SVHN, USPS, Synthetic Digits and MNIST-M. They all contain digit images and are for the multi-class classification task. Figure 2 shows some sample images of the Digits-Five dataset, from which we can observe the non-IID phenomenon in feature space, i.e., the digits from different datasets vary considerably.

Model. For all experiments presented in this section, we implement a simple convolutional neural network model for classification with three convolutional layers with 5×5 kernel (the first and the second with 64 channels and the last with 128 channels, each followed by batch normalization, 2×2-max pooling and ReLU activation) and three fully connected layers with batch normalization followed by ReLU activation (the first with 2048 units, the second with 512 units and the last with 10 units a.k.a. logits). In between, the extracted feature maps by convolutional neural networks are flattened into a 6272-dimensional vector. For variational distribution, we doubled the channels of the third convolutional layer, that the first half represents the mean and the second half represents the variance of the encoder output, and by using the reparameterization technique draw the feature maps following corresponding distributions.

Setups. MNIST, SVHN, USPS, Synthetic digits, and MNIST-M consist of the training sets of 60000, 73257, 7291, 479400, and 60000 examples and test sets of 10000, 26032, 2007, 9553, 10000, respectively. In our experiments, we set the quality of data at each client to 7291 and models evaluate models on the original test sets. The image size and the number of channels of images are different from each dataset. We resize all data into the size of 28×28 and the number of channels of input data is set to 3. For single-site training, models are trained for 50 epochs, while in federated settings the number of communication rounds is set to 50 and at each communication round, models are trained for one epoch at each client. All experiments adopt the stochastic gradient descent (SGD) optimizer with a learning rate of 0.01 and batch size of 32. For FOLA, the weight factor of prior task loss is set to 0.5 (a.k.a., CSD importance) and same for the weight factor of KL divergence term our proposed pFedV. Since the classes are relatively balanced, accuracy (in percentage) is the only metric we used to measure and compare the performance of models trained in different ways.

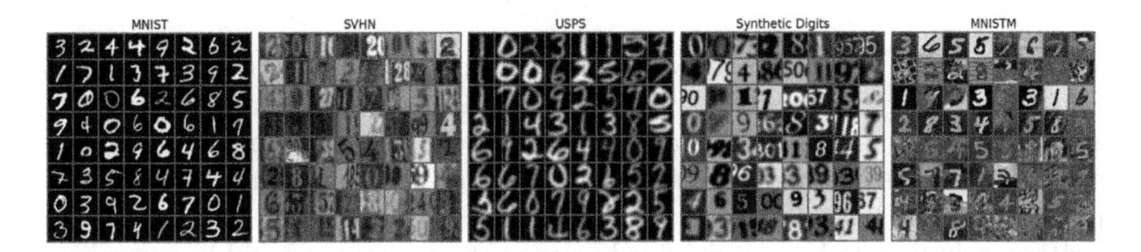

Fig. 2. Example images of datasets used for feature shift (Non-IID) experiments.

4.2 Results

We conduct experiments of single-site training, i.e., models are trained on each client individually and tested on the test sets of MNIST, SVHN, USPS, Synthetic Digits and MNIST-M. The results of the accuracy of single-site trained models are illustrated in Table 1. Each row represents a model trained on the corresponding dataset individually. We can observe that the high-performance values always occur on the diagonal, i.e., models fit well on the test set of the dataset that is the same as that used for training. Of course, there is the possibility of overfitting due to small data sets. We also found that feature complexity is also one of the factors to influence model performance. For example, MNIST and USPS are two datasets with relatively simple features, while SVHN is much more complex as it often occurs more than one digit in one single picture obtained from street view. The interesting result in this table is that the MNIST accuracy of the model trained on MNIST-M is even higher than MNIST-M itself since MNIST-M extended MNIST dataset with randomly extracted patch background. The model trained on MNIST-M has learned the basic features of MNIST with additional generalized feature abstraction and thus works even better on MNIST. However, single-site trained models are overall poor in generalization to other datasets. For example, the second column shows that models trained on other datasets can hardly perform well on SVHN test set, e.g., only 7.95% by the model trained on USPS.

To evaluate the contribution of our approaches to overcoming the non-IID problem in federated learning setting, we compare the results with baselines such

Table 1. Results of models via single site training on test sets of MNIST, SVHN, USPS, Synthetic Digits and MNIST-M

Model (trained on)	MNIST	SVHN	USPS	Synthetic Digits	MNIST-M
MNIST	**98.72**	19.73	28.50	14.92	37.28
SVHN	51.48	**85.18**	64.52	81.43	37.11
USPS	24.41	7.95	**97.11**	23.76	18.60
Synth	82.63	77.97	84.26	**95.04**	54.19
MNIST-M	96.63	30.17	56.05	41.94	**93.62**

as FedAvg, FedProx and FedBN, as well as one of the other Bayesian methods FOLA, as illustrated in Table 2. In general, we can see the effectiveness of variational distribution constraints as the results of FedV that without the personalized layer is also improved on all test sets, which also shows the generalization property of the variational distribution constraints. However, compared with continuously updated prior, the fixed prior does not provide a stable generalization guarantee, for example, it is even worst than FedAvg on SVHN. Overall, our pFedV outperforms others as it achieved 2.36%, 1.05%, 1.74% 1.79% improvement on SVHN, USPS, Synthetic Digits and MNIST-M and slight improvement on MNIST in comparison with FedAvg.

Table 2. Results of methods on test sets of MNIST, SVHN, USPS, Synthetic Digits and MNIST-M in the federated setting

Methods	MNIST	SVHN	USPS	Synthetic Digits	MNIST-M
FedAvg	98.86	83.23	96.16	93.43	90.56
FedProx	98.61	83.36	96.01	93.66	90.59
FedBN	98.67	**86.58**	**97.21**	94.06	91.79
FOLA (CSD 0.5)	98.83	86.46	96.86	94.67	90.50
FedV	98.74	84.80	96.71	94.14	90.93
FedV (Gaussian prior)	98.60	83.04	96.51	93.54	90.46
pFedV	**98.91**	85.99	**97.21**	**95.17**	**92.35**
pFedV (Gaussian prior)	98.86	83.65	**97.21**	94.69	91.74

5 Conclusion

In this paper, we propose a novel federated learning training strategy pFedV to tackle the non-IID problem in federated learning, in particular the covariate shift, a.k.a. feature distribution skewness. Through empirical results, we demonstrate that the proposed approaches vastly improved the federated learning accuracy performance under the scenario of non-IID problem where feature distributions vary across the clients and the results are comparable to state-of-the-art methods like FedBN and FOLA. We have shown the generalization capability of variational distribution in federated learning and the advance of it combined with personalization. For future work, it deserves further investigation of the impact of the combination of multiple variational distribution constraint layers, since the framework is scalable. Besides, it will be interesting to explore more non-IID scenarios and extend to more general settings in addition to feature distribution skewness.

Acknowledgements. This work was supported by the German Ministry for Research and Education (BMBF) projects CORD_MI, POLAR_MI, Leuko-Expert and WestAI (Grant no. No. 01ZZ1911M, 01ZZ1910E, ZMVI1-2520DAT94C and 01IS22094D, respectively), CLARIFY Project (Marie Skłodowska-Curie under Grant no. 860627), and by National Natural Science Foundation of China (NSFC) Project (No. 62106121). This research was supported by Public Computing Cloud, Renmin University of China.

References

1. Banabilah, S., Aloqaily, M., Alsayed, E., Malik, N., Jararweh, Y.: Federated learning review: fundamentals, enabling technologies, and future applications. Inf. Process. Manage. **59**(6), 103061 (2022)
2. Blei, D.M., Kucukelbir, A., McAuliffe, J.D.: Variational inference: a review for statisticians. J. Am. Stat. Assoc. **112**(518), 859–877 (2017)
3. Blundell, C., Cornebise, J., Kavukcuoglu, K., Wierstra, D.: Weight uncertainty in neural network. In: International Conference on Machine Learning, pp. 1613–1622. PMLR (2015)
4. Chen, H.Y., Chao, W.L.: FedBE: Making Bayesian model ensemble applicable to federated learning. arXiv preprint arXiv:2009.01974 (2020)
5. Geng, J., et al.: Towards general deep leakage in federated learning. arXiv preprint arXiv:2110.09074 (2021)
6. Kairouz, P., et al.: Advances and open problems in federated learning. Found. Trends® Mach. Learn. **14**(1–2), 1–210 (2021)
7. Karimireddy, S.P., Kale, S., Mohri, M., Reddi, S., Stich, S., Suresh, A.T.: Scaffold: Stochastic controlled averaging for federated learning. In: International Conference on Machine Learning, pp. 5132–5143. PMLR (2020)
8. Khan, L.U., Saad, W., Han, Z., Hossain, E., Hong, C.S.: Federated learning for internet of things: recent advances, taxonomy, and open challenges. IEEE Commun. Surv. Tutorials **PP**, 1 (2021)
9. Kingma, D.P., Welling, M.: Auto-encoding variational Bayes. arXiv preprint arXiv:1312.6114 (2013)
10. Li, Q., Diao, Y., Chen, Q., He, B.: Federated learning on non-IID data silos: an experimental study. In: 2022 IEEE 38th International Conference on Data Engineering (ICDE), pp. 965–978. IEEE (2022)
11. Li, T., Sahu, A.K., Talwalkar, A., Smith, V.: Federated learning: Challenges, methods, and future directions. IEEE Signal Process. Mag. **37**(3), 50–60 (2020)
12. Li, T., Sahu, A.K., Zaheer, M., Sanjabi, M., Talwalkar, A., Smith, V.: Federated optimization in heterogeneous networks. Proceed. Mach. Learn. Syst. **2**, 429–450 (2020)

13. Li, X., Huang, K., Yang, W., Wang, S., Zhang, Z.: On the convergence of FedAvg on non-IID data. arXiv preprint arXiv:1907.02189 (2019)
14. Li, X., Jiang, M., Zhang, X., Kamp, M., Dou, Q.: FedBN: Federated learning on non-IID features via local batch normalization. arXiv preprint arXiv:2102.07623 (2021)
15. Liu, L., Zheng, F., Chen, H., Qi, G.J., Huang, H., Shao, L.: A bayesian federated learning framework with online Laplace approximation. arXiv preprint arXiv:2102.01936 (2021)
16. McMahan, B., Moore, E., Ramage, D., Hampson, S., Arcas, B.A.: Communication-efficient learning of deep networks from decentralized data. In: Artificial intelligence and statistics, pp. 1273–1282. PMLR (2017)
17. Mou, Y., Geng, J., Welten, S., Rong, C., Decker, S., Beyan, O.: Optimized federated learning on class-biased distributed data sources. In: Machine Learning and Principles and Practice of Knowledge Discovery in Databases. ECML PKDD 2021. Communications in Computer and Information Science, vol. 1524, pp. 146–158. Springer, Cham (2021). https://doi.org/10.1007/978-3-030-93736-2_13
18. Nguyen, D.C., et al.: Federated learning for smart healthcare: a survey. ACM Comput. Surv. (CSUR) **55**(3), 1–37 (2022)
19. Odaibo, S.: Tutorial: Deriving the standard variational autoencoder (VAE) loss function. arXiv preprint arXiv:1907.08956 (2019)
20. Sahu, A.K., Li, T., Sanjabi, M., Zaheer, M., Talwalkar, A., Smith, V.: On the convergence of federated optimization in heterogeneous networks. arXiv preprint arXiv:1812.06127 3, 3 (2018)
21. Wang, J., Liu, Q., Liang, H., Joshi, G., Poor, H.V.: Tackling the objective inconsistency problem in heterogeneous federated optimization. Adv. Neural. Inf. Process. Syst. **33**, 7611–7623 (2020)
22. Yang, Q., Liu, Y., Chen, T., Tong, Y.: Federated machine learning: concept and applications. ACM Trans. Intell. Syst. Technol. (TIST) **10**(2), 1–19 (2019)
23. Yang, W., Zhang, Y., Ye, K., Li, L., Xu, C.-Z.: FFD: a federated learning based method for credit card fraud detection. In: Chen, K., Seshadri, S., Zhang, L.-J. (eds.) BIGDATA 2019. LNCS, vol. 11514, pp. 18–32. Springer, Cham (2019). https://doi.org/10.1007/978-3-030-23551-2_2
24. Zhang, X., Li, Y., Li, W., Guo, K., Shao, Y.: Personalized federated learning via variational Bayesian inference. In: International Conference on Machine Learning, pp. 26293–26310. PMLR (2022)

Probabilistic Models and Statistical Inference

Inverse Problem of Censored Markov Chain: Estimating Markov Chain Parameters from Censored Transition Data

Masahiro Kohjima(✉), Takeshi Kurashima, and Hiroyuki Toda

NTT Human Informatics Laboratories, NTT Corporation, Kanagawa, Japan
{masahiro.kohjima.ev,takeshi.kurashima.uf}@hco.ntt.co.jp,
hirotoda@acm.org

Abstract. Due to the difficulty of collecting comprehensive data sets (factors include limited GPS coverage and the existence of competitors), most transition data collected from, for example, people and automobiles, have been censored, and so record only the visits of known observable states (locations). In this paper, we tackle the problem of estimating Markov chain parameters from censored transition data. Our parameter estimation method utilizes the theory of the censored Markov chain, the Markov chain that has unobservable states. Our problem formulation can be seen as the inverse of existing studies that construct censored Markov chains from (original) Markov chains and unobservable states. We confirm the effectiveness of the proposal by experiments on synthetic and real data sets.

Keywords: probabilistic models · markov chain · censored transition data · inverse problem

1 Introduction

Markov chain (MC) is a versatile tool for modeling dynamic systems. Markov chain variants have been used to analyze city systems such as land cover use, traffic, and people flow [1–3] in addition to application to queuing systems [4], marketing [5] and biology [6].

Since the MC parameters, initial state probability and transition probability, are unknown in practice, the parameters need to be estimated from observed transition data. In the ideal scenario, where the visits of all states (locations) are observable, shown in Fig. 1a, the parameters can be estimated directly from the number of transitions between the states [7]. Unfortunately, comprehensive data collection is rare because of factors such as limited GPS coverage and the existence of competitors. Thus the data to be analyzed is, in practice, degraded from the ideal data; the actual data takes the form of *censored transition data* where only the visits of known observable states are recorded.

An example of censored transition data are the GPS-based trajectories provided by a taxi company. Figure 1b shows example trajectories between streets

H. Kashima et al. (Eds.): PAKDD 2023, LNAI 13936, pp. 297–308, 2023.
https://doi.org/10.1007/978-3-031-33377-4_23

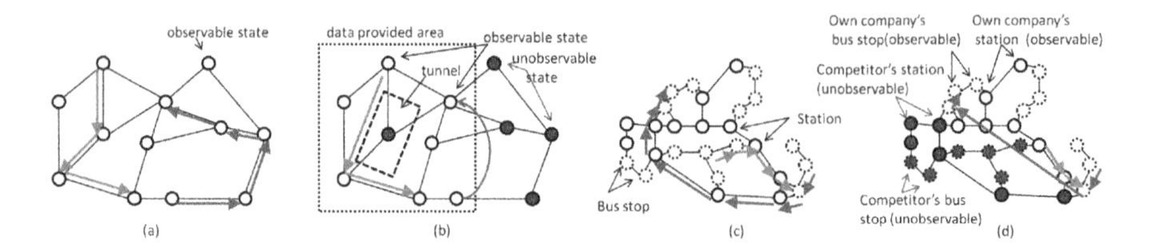

Fig. 1. Example of trajectory in (a) ideal setting and (b) real world setting and trips recorded in (c) ideal setting and (d) real world setting. Green and blue arrows indicate observed transition data in each setting. Since the visits of the *unobservable states* (red circle) are censored (not observed) in (b)(d), we call the transition data in (b)(d) *censored transition data.* (Color figure online)

or areas in a city made by applying map-matching to raw GPS data [8]. The data contains only trajectories from data-provided areas, which may correspond to prefectures or countries. Note that it does not contain trajectories from GPS-blind areas such as tunnels and building shadows. Therefore, the visits to *unobservable states*, states that lie outside the observable area, are not recorded; this yields censored transition data that consists of only the transitions between observable states. Similarly, passenger trip data provided from a train company, e.g., entrance-exit history recorded in a smart card, are also a form of censored transition data (Fig. 1cd). Because the trips made on the other transportation modes (competitors) such as buses, taxis, etc. are not recorded in smart card, the card history cannot cover entire movements. Trajectories made from person re-identification [9] are also censored transition data since only visits to areas hosting cameras are observed. Thus many examples of censored transition data exist, and the importance of analyzing this data is increasing.

Existing methods for estimating MC parameters from randomly missing transition data cannot work well for censored transition data since the nature of the censored transition data is completely different; unobservable states never appear and the number of steps censored between transitions is unknown (in fact, infinitely many number of steps may be censored) in the censored transition data. These difficulties make it impossible to estimate transition probability from and to unobservable states and to adopt the EM algorithm for estimating missing values [10] since we can never know how many latent variables need to be placed between the observed transitions.

In this paper, we tackle the problem of estimating Markov chain parameters from censored transition data. The key to our approach is its use of the theory of the censored Markov chain (CMC) [11–13], which can handle Markov chains with unobservable states; this allows us to model the generative process that yields censored transition data. We can design the loss function of the proposed method by using the relation between the parameters of the original MC and those of CMC. By estimating the parameters of the original MC, we can know, for example, the preference of people for a certain path which may contain unobservable states. Our problem can be seen as the *inverse* problem of existing works, which construct CMC from an original MC. Experiments are conducted on synthetic and real car probe data captured in the greater Tokyo area to confirm the effectiveness of the proposed method.

As far as we aware, this is the first study in the data mining and machine learning community to use CMC for modeling data. Moreover, our method can be seen as versatile since it can adopt (almost) any probabilistic model including log-linear model and deep neural networks which can represent transition probability and initial state probability of MC. We believe this study has great potential to become the foundation of censored transition data analysis.

2 Related Works

Our study follows recent work that investigated new problem formulations for the Markov chain. For example, Morimura et al. and Kumar et al. tackle the problem of estimating the MC parameters from a steady state distribution [14, 15]. The problem of parameter estimation from node-level aggregated traffic [16] and from snapshots of population [17,18] have also been investigated. However, existing studies do not consider the use of censored transition data; we succeeds in identifying a new problem and introducing a new formulation.

The theory of CMC is the main key to our study. CMC is constructed from the original MC and a set of observable states. References [19,20] state that this construction was first shown by Paul Lévy [11–13]. The various properties of CMC have already been clarified [19,21] and used, among other applications, to obtain the steady state probability of an infinite state MC [22]. We also use one of the results; the relation between the transition probability of the original MC and that of CMC. However, our problem can be seen as the *inverse* problem of existing works; we estimate the original MC from (transition data on) the CMC.

3 Markov Chain and Censored Markov Chain

Let $\mathcal{X} = \{1, 2, \cdots, |\mathcal{X}|\}$ be a finite state space. A homogeneous and discrete time Markov chain (MC) on $\mathcal{X}$ is a stochastic process $\{X_t; t = 0, 1, 2, \cdots\}$ that satisfies the Markov property: $Pr(X_{t+1} = x_{t+1}|X_k = x_k; k = 0, \cdots, t) = Pr(X_{t+1} = x_{t+1}|X_t = x_t)$. MC is thus defined by triplet $\{\mathcal{X}, \mathcal{P}, q\}$, where $\mathcal{P} : \mathcal{X} \times \mathcal{X} \to [0, 1]$ is the transition probability and $q : \mathcal{X} \to [0, 1]$ is the initial state probability, i.e., $\mathcal{P}(x_{next}|x) = Pr(X_{t+1} = x_{next}|X_t = x)$ and $q(x_0) = Pr(X_0 = x_0)$. Throughout this paper, we consider only the *irreducible* Markov chain, i.e., any state is reachable from any state. We also denote adjacency information by $\varGamma = \{\varGamma_i\}_{i \in \mathcal{X}}$ where $\varGamma_i$ denotes the set of reachable states from state i by one-step transitions.

We also define the *censored Markov chain* (CMC), which is also called the *watched Markov chain* or *induced chain* [20,21,23], as it will be used in the next section. Let $\mathcal{O}$ be a subset of state space $\mathcal{X}$, $\mathcal{O} \subseteq \mathcal{X}$. $\mathcal{O}$ indicates the set of *observable states*. We also denote the set of *unobservable states* as $\mathcal{U}$. CMC is constructed from the original MC $\{X_t; t = 0, 1, 2, \cdots\}$ and observable states $\mathcal{O}$. CMC is a stochastic process $\{X^c_t; t = 0, 1, 2, \cdots\}$; the state at time t is the state of the MC at time σ_t which is the time of the t-th visit to the observable states $\mathcal{O}$, i.e., $X_t^c := X_{\sigma_t}$. Intuitively, CMC is the MC watched only when it stays in states in $\mathcal{O}$. The formal definition is given as follows:

Definition 1. *(censored Markov chain) Let $\{\sigma_t; t = 0, 1, 2, \cdots\}$ be a sequence such that $\sigma_0 = 0$ if $X_0 \in \mathcal{O}$, otherwise $\sigma_0 = \inf\{m \geq 1 : X_m \in \mathcal{O}\}$, and $\sigma_t = \inf\{m > \sigma_{t-1} : X_m \in \mathcal{O}\}$. Censored Markov chain $\{X_t^c; t = 0, 1, 2, \cdots\}$ is defined as a stochastic process such that $X_t^c := X_{\sigma_t}$.*

Without loss of generality, we consider that the states are re-ordered so that the matrix representation of the transition probability, $\boldsymbol{P}$, $(\boldsymbol{P})_{xx'} = \mathcal{P}(x'|x)$, and the vector representation of the initial probability, $\boldsymbol{q}$, $(\boldsymbol{q})_x = q(x)$, can be written as follows:

$$\boldsymbol{P} = \begin{array}{c} \\ \mathcal{O} \\ \mathcal{U} \end{array} \!\!\begin{array}{c} \begin{array}{cc} \mathcal{O} & \mathcal{U} \end{array} \\ \begin{pmatrix} \boldsymbol{P}_{oo} & \boldsymbol{P}_{ou} \\ \boldsymbol{P}_{uo} & \boldsymbol{P}_{uu} \end{pmatrix} \end{array}, \quad \boldsymbol{q} = \begin{array}{c} \begin{array}{cc} \mathcal{O} & \mathcal{U} \end{array} \\ \begin{pmatrix} \boldsymbol{q}_o & \boldsymbol{q}_u \end{pmatrix} \end{array}, \tag{1}$$

where the sizes of $\boldsymbol{P}_{oo}, \boldsymbol{P}_{ou}, \boldsymbol{P}_{uo}, \boldsymbol{P}_{uu}$ are given by $|\mathcal{O}| \times |\mathcal{O}|$, $|\mathcal{O}| \times |\mathcal{U}|$, $|\mathcal{U}| \times |\mathcal{O}|$, $|\mathcal{U}| \times |\mathcal{U}|$, respectively.

The following property of CMC has been clarified.

Theorem 1. (e.g., Lemma 6-6 [21]) *A censored Markov chain is a Markov chain with transition probability $\boldsymbol{R} = \boldsymbol{P}_{oo} + \boldsymbol{P}_{ou}(\boldsymbol{I} - \boldsymbol{P}_{uu})^{-1}\boldsymbol{P}_{uo}$.*

Note that $(\boldsymbol{I} - \boldsymbol{P}_{uu})^{-1}$ comes from the infinite sum of the power of $\boldsymbol{P}_{uu}$, $\sum_{\ell=0}^{\infty}(\boldsymbol{P}_{uu})^{\ell}$; the inverse matrix exists for the irreducible Markov chain. $\boldsymbol{R}$ is constructed from the transition probability between observable states $\boldsymbol{P}_{oo}$ plus the probability of moving from observable states to unobservable states ($\boldsymbol{P}_{ou}$) and of entering observable states ($\boldsymbol{P}_{uo}$) after (arbitrary length) occupancy in unobservable states ($\sum_{\ell=0}^{\infty}(\boldsymbol{P}_{uu})^{\ell}$). We can derive the following result for the initial probability in an analogous manner.

Theorem 2. *The initial state probability of CMC is $\boldsymbol{s} = \boldsymbol{q}_o + \boldsymbol{q}_u(\boldsymbol{I} - \boldsymbol{P}_{uu})^{-1}\boldsymbol{P}_{uo}$.*

Theorems 1 and 2 state that the censored Markov chain induced from the Markov chain $\{\mathcal{X}, \mathcal{P}, q\}$ and set of observable states $\mathcal{O}$ is the Markov chain $\{\mathcal{O}, \mathcal{R}, s\}$.

4 Estimating Markov Chain Parameters from Censored Transition Data

This section describes our problem, estimating MC parameters from censored transition data. Figure 2 shows our problem formulation. Since our purpose is to recover the original MC from (transition data of) the CMC, this problem can be seen as the inverse of the existing study shown in the previous section, which constructs a CMC from the original MC.

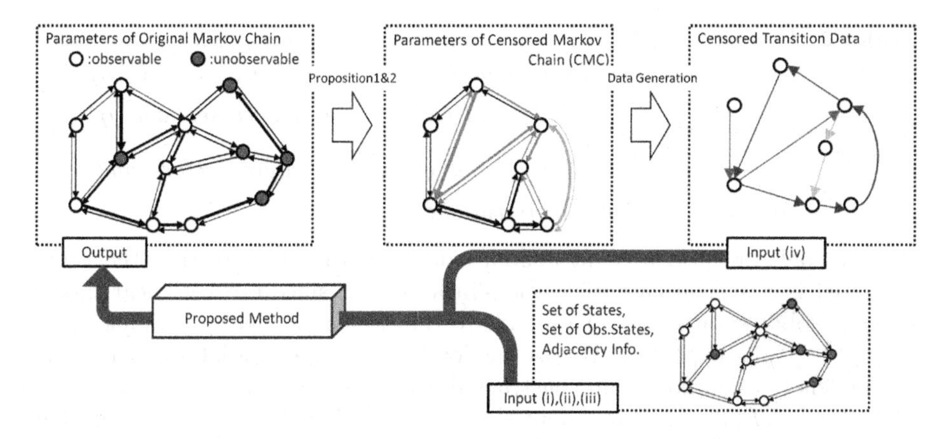

Fig. 2. Our problem formulation. The proposed method estimates the original Markov chain parameters from observed censored transition data. This can be seen as the inverse problem of the existing study shown in Theorem 1 which constructs a censored Markov chain from the original Markov chain and observable states.

4.1 Problem Formulation

Input data of our problem consists of (i) a set of states of original Markov chain, $\mathcal{X}$, (ii) adjacency information, $\Gamma = \{\Gamma_i\}_{i \in \mathcal{X}}$, (iii) set of observable states, $\mathcal{O}$, (iv) censored transition data. We consider that the censored transition data are summarized as $\mathcal{D}$, which is defined as $\mathcal{D} = \{N_{ij}\}_{ij \in \mathcal{O}} \cup \{N_k^{ini}\}_{k \in \mathcal{O}}$, where N_{ij} is the number of transitions from observable state $i \in \mathcal{O}$ to state $j \in \mathcal{O}$ and N_k^{ini} is the number of counts of state $k \in \mathcal{O}$ in the initial state.

In the example shown in Fig. 1bd, a set of states $\mathcal{X}$ and adjacency information Γ can be obtained from the street/traffic networks that are publicly available. The set of observable states $\mathcal{O}$ can be extracted from censored transition data. In addition, for the example of citywide level analysis, we do not require person-level transition data, only the crowd-level censored transition data made by aggregating the person-level data. To ensure privacy protection, the scenario considered here assumes that actual time information (representing when a user visits a state) is not provided/available and that only the order of states visited can be used.

Input model is a model of the transition probability and the initial state probability of the (original) Markov chain with parameters $\theta = (\nu, \lambda)$. To emphasize parameter dependency, we denote the model of the transition probability and the initial probability as P^{ν}, q^{λ}. Given a model and parameters, a state transition and initial state probability on the original Markov chain are modeled as $Pr(X_{t+1}{=}j|X_t{=}i, \theta) = (\boldsymbol{P}^{\nu})_{ij}$, $Pr(X_0{=}k|\theta) = (\boldsymbol{q}^{\lambda})_k$. Popular examples of the model used for MC are the tabular model and the log-linear model.

Model 1 (Tabular Model). *Let us define model parameters ν and λ as $\nu = \{\{a_{ij}\}_{j \in \Gamma_i}\}_{i \in \mathcal{X}}$ and $\lambda = \{b_i\}_{i \in \mathcal{X}}$ where $\sum_j a_{ij} = 1$ for all i and $\sum_i b_i = 1$. The tabular model can be defined as $(\boldsymbol{P}^{\nu})_{ij} = a_{ij}$ if $j \in \Gamma_i$, and $(\boldsymbol{P}^{\nu})_{ij} = 0$ otherwise, and $(\boldsymbol{q}^{\lambda})_k = b_k$.*

Model 2 (Log-Linear Model). *The transition probability and initial state probability using the log-linear model can be defined as* $(\boldsymbol{P}^\nu)_{ij} = \frac{\exp\{v_{ij}^{loc}+\phi(i,j)^T \boldsymbol{v}^{glo}\}}{\sum_{k\in\Gamma_i}\exp\{v_{ik}^{loc}+\phi(i,k)^T \boldsymbol{v}^{glo}\}}$ *if* $j \in \Gamma_i$ *and* $(\boldsymbol{P}^\nu)_{ij} = 0$ *otherwise, and* $(\boldsymbol{q}^\lambda)_i = \frac{\exp\{w_i^{loc}+\psi(i)^T \boldsymbol{w}^{glo}\}}{\sum_k \exp\{w_k^{loc}+\psi(k)^T \boldsymbol{w}^{glo}\}}$, *where* $\phi(i,j)$ *and* $\psi(k)$ *are feature vectors and* $\nu = \{\boldsymbol{v}^{loc}, \boldsymbol{v}^{glo}\}$ *and* $\lambda = \{\boldsymbol{w}^{loc}, \boldsymbol{w}^{glo}\}$ *are parameters.* [1]

These models are also used in related works [14,15,17]. Note that the tabular model is a special case of the log-linear model since the log-linear model without global parameters corresponds to the tabular model by defining $a_{ij} = \exp\{v_{ij}^{\rm loc}\}/\sum_{k\in\Gamma_i}\exp\{v_{ik}^{\rm loc}\}$. We use the log-linear model in a later experiment.

We denote a Markov chain constructed using the model with parameter θ as $\mathcal{M}(\boldsymbol{\theta}) = \{\mathcal{X}, P^\nu, q^\lambda\}$. Throughout this paper, similar to [24], we assume that Markov chain $\mathcal{M}(\boldsymbol{\theta})$ is *irreducible* for any parameter θ, and a model of transition probability P^ν and that of initial state probability q^λ are differentiable everywhere w.r.t. parameter $\theta = (\nu, \lambda)$. We emphasize that our method can handle any probabilistic model that satisfies the above assumptions and that can represent the transition probability and initial state probability of MC; for example, if the feature vector in the log-linear model is a high-dimensional vector, the deep neural network architecture can be used. Similar to Eq. (1), without loss of generality, we assume that the states are re-ordered so that the matrix representation of the model can be written as $\boldsymbol{P}^\nu = (\boldsymbol{P}^\nu_{oo}, \boldsymbol{P}^\nu_{ou}; \boldsymbol{P}^\nu_{uo}, \boldsymbol{P}^\nu_{uu})$ and $\boldsymbol{q}^\lambda = (\boldsymbol{q}^\lambda_o, \boldsymbol{q}^\lambda_u)$.

Output of the proposed method consists of Markov chain parameters $\theta = (\nu, \lambda)$. Namely, the entire transition probability represented by $|\mathcal{X}| \times |\mathcal{X}|$ matrix P^ν and $|\mathcal{X}|$-dimension vector q^λ are estimated. Note that if all states are observable, $\mathcal{X}=\mathcal{O}$, we get the standard setting of parameter estimation for MC [7].

4.2 Difficulty of Our Problem

The nature of the censored transition data creates two key problem difficulties: (i) unobservable states never appear and (ii) the number of steps censored between observations is unknown. See Fig. 3a. In this example, state-3 never appears and we are unable to know the number of steps that have been censored (two steps are censored between X_2^c and X_3^c). From characteristic (i), when we adopt standard parameter estimation for MC (e.g., [7]), the learned transition probability to unobservable states is zero, i.e. biased, and we cannot estimate the probability of moving from unobservable states. Even if the true parameter is known, due to characteristic (ii), the time-steps of original data and those of censored data do not correspond; consequently, we cannot adopt the EM using latent variables for missing information [10] since we can never know how many latent variables need to be placed between the observed states. Accordingly, we construct a new method using CMC.

[1] $\phi(i,j)$ may be the (inverse) distance between states i and j and $\psi(k)$ may be the attribute information about state k. If no such information is available, the term related to the feature and global parameters $\boldsymbol{v}^{\rm glo}, \boldsymbol{w}^{\rm glo}$ can be excluded from the model.

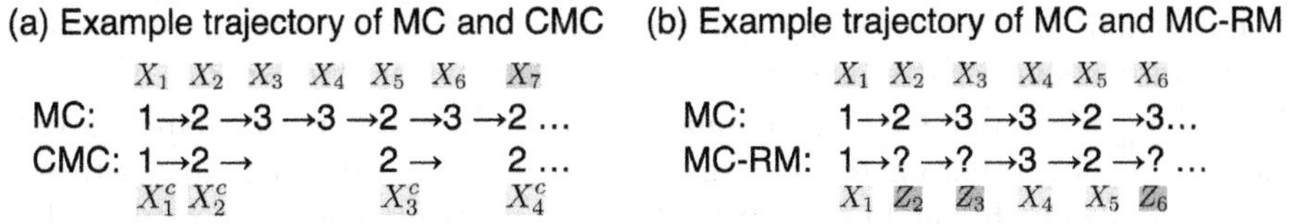

(a) Example trajectory of MC and CMC

	X_1	X_2	X_3	X_4	X_5	X_6	X_7
MC:	1→	2 →	3 →	3 →	2 →	3 →	2 ...
CMC:	1→	2 →			2 →		2 ...
	X_1^c	X_2^c			X_3^c		X_4^c

(b) Example trajectory of MC and MC-RM

	X_1	X_2	X_3	X_4	X_5	X_6
MC:	1→	2 →	3 →	3 →	2 →	3...
MC-RM:	1→	? →	? →	3 →	2 →	? ...
	X_1	Z_2	Z_3	X_4	X_5	Z_6

Fig. 3. Example trajectories of (a) CMC (state-3 is unobservable) and (b) MC with randomly-missing data (MC-RM) in 3 state MC.

Remark: The above difficulties do not exist in handling MC with randomly omitted data (MC-RM) shown in Fig. 3b since it is usually assumed that all states are observed at least once and that the time-steps of the original data and missing data correspond one to one. Thus, it can be solved by the EM algorithm using latent variables (Z in Fig. 3b) for the missing states and applying the Viterbi algorithm to estimate the latent variables similar to the hidden Markov model. However, this cannot not be applied to our problem.

4.3 Parameter Estimation via Divergence Minimization

Our proposed method estimates output parameters by optimizing a loss function. Although various type of loss functions such as L_2 divergence and Kullback Leibler (KL) divergence can be adopted, here we focus on KL divergence, which is defined as $\mathrm{KL}(q||p) = \sum_j q_j \log q_j - q_j \log p_j$, where $q = \{q_j\}$ and $p = \{p_j\}$ are (discrete) probability distributions. The use of KL yields an objective function that corresponds to the negative log likelihood function.

The theory of CMC has an important role in designing the loss function. As shown in Fig. 2, the censored transition data $\mathcal{D}$ can be regarded as the empirical transition data on CMC $\{\mathcal{O}, \boldsymbol{R}^*, \boldsymbol{s}^*\}$ where $\boldsymbol{R}^*$ and $\boldsymbol{s}^*$ are unknown true parameters. From Theorem 1 and 2, given Markov chain $\{\mathcal{X}, P^\nu, q^\lambda\}$ and a set of observable states $\mathcal{O}$, we can construct censored Markov chain $\{\mathcal{O}, \boldsymbol{R}^\nu, \boldsymbol{s}^{\nu,\lambda}\}$ where $\boldsymbol{R}^\nu$ and $\boldsymbol{s}^{\nu,\lambda}$ can be defined as

$$\boldsymbol{R}^\nu = \boldsymbol{P}^\nu_{oo} + \boldsymbol{P}^\nu_{ou}(\boldsymbol{I} - \boldsymbol{P}^\nu_{uu})^{-1}\boldsymbol{P}^\nu_{uo}, \quad \boldsymbol{s}^{\nu,\lambda} = \boldsymbol{q}^\lambda_o + \boldsymbol{q}^\lambda_u(\boldsymbol{I} - \boldsymbol{P}^\nu_{uu})^{-1}\boldsymbol{P}^\nu_{uo}. \tag{2}$$

Note that $(\boldsymbol{I} - \boldsymbol{P}^\nu_{uu})^{-1}$ exists for *irreducible* MC and the elements of $\boldsymbol{R}^\nu$ and those of $\boldsymbol{s}^{\nu,\lambda}$ correspond to the following probability:

$$Pr(X^c_{t+1}{=}j|X^c_t{=}i, \theta) = (\boldsymbol{R}^\nu)_{ij}, \; Pr(X^c_0{=}k|\theta) = (\boldsymbol{s}^{\nu,\lambda})_k.$$

Therefore, KL divergence between $\boldsymbol{R}^\nu$ and $\boldsymbol{R}^*$, $\boldsymbol{s}^{\nu,\lambda}$ and $\boldsymbol{s}^*$, is given by $L^{\mathrm{KL}}_R(\boldsymbol{R}^*||\boldsymbol{R}^\nu) = \sum_{i,j\in\mathcal{O}}(\boldsymbol{R}^*)_{ij}\log(\boldsymbol{R}^*)_{ij} - (\boldsymbol{R}^*)_{ij}\log(\boldsymbol{R}^\nu)_{ij}$, $L^{\mathrm{KL}}_s(\boldsymbol{s}^*||\boldsymbol{s}^{\nu,\lambda}) = \sum_{i\in\mathcal{O}}(\boldsymbol{s}^*)_i\log(\boldsymbol{s}^*)_i - (\boldsymbol{s}^*)_i\log(\boldsymbol{s}^{\nu,\lambda})_i$. The objective function is defined as the sum of the above two KL divergence and regularization terms. Taking sample average and removing constant terms, the objective is given by

$$\mathcal{L}(\theta) = -\,\mathcal{Z}^{-1}\sum_{i,j\in\mathcal{O}} N_{ij}\log(\boldsymbol{R}^\nu)_{ij} - \mathcal{Z}^{-1}_{ini}\sum_{k\in\mathcal{O}} N^{ini}_k \log(\boldsymbol{s}^{\nu,\lambda})_k + \Omega(\theta),$$

where $\mathcal{Z}$ and $\mathcal{Z}_{ini}$ are normalizing factors, and $\Omega(\theta)$ is the (differentiable) regularization term or negative logarithm of prior distribution. We use $\mathcal{Z} = \sum_{i,j} N_{ij}$, $\mathcal{Z}_{ini} = \sum_k N_k^{ini}$ and L_2 norm regularization $\Omega(\theta) = \alpha\|\theta\|^2$ in a later experiment. α is a hyperparameter. Use of the regularization term contributes to not only avoiding overfitting but also handling redundant parameters as explained in next subsection. Estimated parameter $\hat{\theta}$ is obtained by optimizing this objective, $\hat{\theta} = \arg\min_\theta \mathcal{L}(\theta)$. We could derive the objective function in an analogous manner when the other divergence is adopted. For minimizing the objective function, we could arbitrarily use any optimization method such as gradient descent and L-BFGS [25]. ∇_θ is the partial derivative operator w.r.t. θ. The property of this algorithm depends on the model chosen; it may reach a local optimum or a stationary point for "large complex" models such as the deep neural network. In a later experiment, we use the BFGS method.

5 Experiment

This section confirms that the proposed method well estimates the parameters of the original MC from censored transition data. We prepared training data and *true-test* data in two experiment settings: synthetic chain and greater Tokyo. True-test data are the non-censored transition data such as that shown in Fig. 1a and training data are the censored transition data such as Fig. 1b. The performance of the proposed method is evaluated using the true-test data. In order to further check whether the proposed method can well estimate CMC (if the parameter of original MC is well estimated, it will also well recover CMC), we also prepared *censored-test* data that has the form of censored transition data.

5.1 Setting

Synthetic Data: In the synthetic-chain experiment, we set the number of states to 100 and randomly generated chain edges following [26]. The true transition probability was set using the log-linear model (Model 2), where parameters ν^*, λ^* and feature ϕ were generated using a standard normal distribution (feature ψ was not used). We also added symmetric Dirichlet noise with parameter 0.3 to the transition probability by taking the weight sum with $\beta = 0.1$ for the noise term and $1 - \beta = 0.9$ for the transition probability. We generated training and two type of test data by generating episodes (sequences of states from initial state) with 20 steps in common. The number of training and test episodes were 100 and 1000, respectively. We randomly selected observable states while varying the ratio of observable states $|\mathcal{O}|/|\mathcal{X}|$; training and *censored-test* data were made by censoring the episodes by excluding unobservable states. We prepared 5 sets of training, *true-test* and *censored-test* data.

Real Data: In the greater Tokyo experiment, we used car probe data provided by NAVITIME JAPAN Co, Ltd. The dataset is a collection of GPS trajectories of car navigation users in the greater Tokyo area, Japan. We divided the region using an approximately 5km $\times$ 5km grid mesh; the total number of mesh cells was approximately 150. We used the data recorded during the period between 2015.4.13 to 2015.4.17 (5 working days in total) in the morning (6:00 am to 10:59 am). The number of unique users per day was, on average, approximately 8000. The data were made by converting the GPS trajectories into sequences of visited mesh cells (states). We excluded the states that appeared less than 20 times per day on average and only used the episodes containing more than 2 steps. Here we did not use feature ϕ or ψ. Training and test data were made from the logs of one day and that of the next day, respectively. Training data and *censored-test* data were censored in a manner analogous to the synthetic data. We prepared 4 sets of training, *true-test* and *censored-test* data.

Evaluation Measure: As the performance metric, we used the *true* negative test log likelihood, which was computed using the *true-test* data explained at the beginning of this section. The *true* negative test log likelihood is defined as $(1/\mathcal{T}_{tt})\sum_{i,j\in\mathcal{X}} -n_{ij}^{tt}\log\hat{p}_{ij}$, where $\mathcal{T}_{tt}$ is the number of total transitions and n_{ij}^{tt} indicates the number of transitions from state i to state j in the *true-test* data. $\hat{p}_{ij}$ is the estimated transition probability in the original MC. A lower *true* negative test log likelihood indicates the method estimates the original MC more precisely. To also investigate whether the proposed method can well estimate CMC, we also computed the *censored* negative test log likelihood, using *censored-test* data, which is defined as $(1/\mathcal{T}_{ct})\sum_{i,j\in\mathcal{O}} -n_{ij}^{ct}\log\hat{p}_{ij}$, where $\mathcal{T}_{ct}$ is the number of total transitions and n_{ij}^{ct} indicates the number of transitions from state i to state j in *censored-test* data. $\hat{p}_{ij}$ is the estimated transition probability in CMC. A lower *censored* negative test log likelihood indicates the method estimates the CMC more precisely.

Baseline Methods: Since our proposal is the first method that can estimate parameters from censored transition data, there are no "strong" baselines[2]. Thus, we compared the performance of the proposed method with two simple and reasonable baseline methods based on the standard Markov model. The baseline methods are constructed by mixing uniform distributions since the result of applying the standard Markov model to censored transition data seems to be, from Theorems 1 and 2, an acceptable approximation of the original MC when the number of unobservable states is small. The result of the uniform distribution is also a compromise solution when the number of unobservable states is large; we expect that mixing these two yields useful benchmarks. Baseline1 (Base1) and Baseline2 (Base2) are constructed as $\hat{p}_{ij}^{\mathrm{Base1}}=(1-\eta)P_{ij}^{M1}+\eta P_{ij}^{U1}$ and $\hat{p}_{ij}^{\mathrm{Base2}}=(1-\xi)P_{ij}^{M2}+\xi P_{ij}^{U2}$, where $P_{ij}^{M1}=N_{ij}/Z_i$ if $j\in\Gamma_i$ and $P_{ij}^{M1}=0$ otherwise, $P_{ij}^{U1}=1/|\Gamma_i|$ if $j\in\Gamma_i$ and $P_{ij}^{U1}=0$ otherwise, $P_{ij}^{M2}=N_{ij}/Z_i'$ and $P_{ij}^{U2}=1/|\mathcal{X}|$. Z_i and Z_i' are the normalizing constants and η and ξ are the hyperparameters. Baseline1 and 2 are used for computing *true* negative test log likelihood and *censored* negative test log likelihood, respectively. We selected the baselines' hyperparameters, η and ξ, from $\{10^{-3},10^{-2},10^{-1},0.3,0.5,0.7\}$; top-3 results are shown in next subsection. The proposed method uses $\boldsymbol{P}^{\nu}$ for *true* negative test log likelihood, $\hat{p}_{ij}=(\boldsymbol{P}^{\nu})_{ij}$, and $\boldsymbol{R}^{\nu}$ for *censored* negative test log

[2] No existing model, including the hidden Markov model, suits this problem as explained in Subect. 4.2.

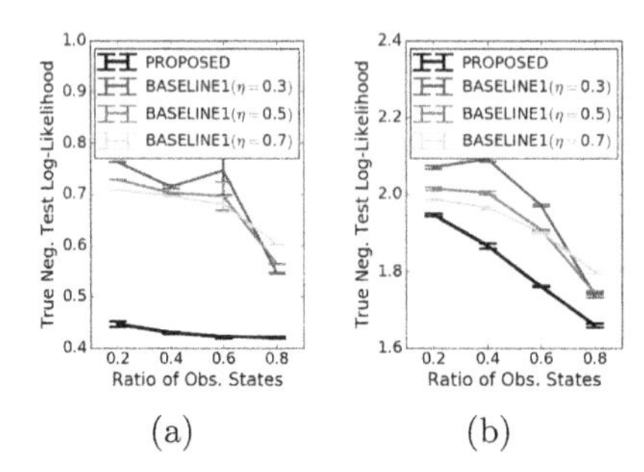

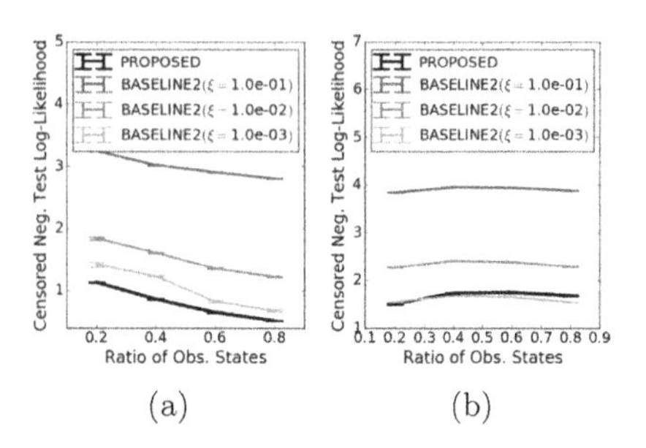

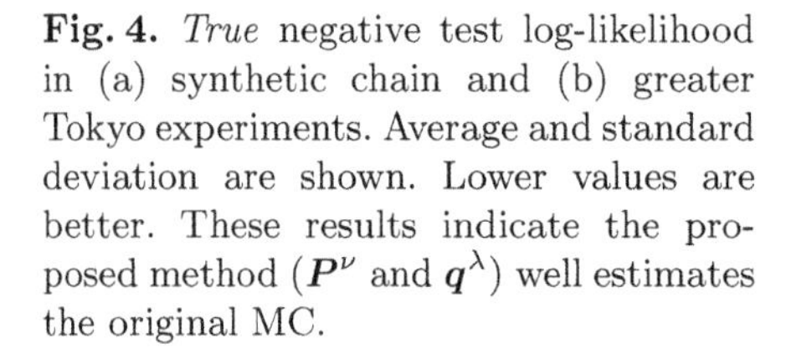

Fig. 4. *True* negative test log-likelihood in (a) synthetic chain and (b) greater Tokyo experiments. Average and standard deviation are shown. Lower values are better. These results indicate the proposed method ($\boldsymbol{P}^{\nu}$ and $\boldsymbol{q}^{\lambda}$) well estimates the original MC.

Fig. 5. *Censored* negative test log-likelihood in (a) synthetic chain and (b) greater Tokyo. Average and standard deviation are shown. Lower values are better. These results indicate the proposed method ($\boldsymbol{R}^{\nu}$ and $s^{\nu,\lambda}$) has performance comparable with the baseline for estimating CMC.

likelihood, $\hat{p}_{ij} = (\boldsymbol{R}^{\nu})_{ij}$. We set the hyperparameter $\alpha = 10^{-3}$ in all experiments since this yielded stable performance in a preliminary experiment[3].

5.2 Results

Quantitative Evaluation (Original MC Estimation): Figure 4a and b shows the *true* negative test log-likelihood results. We can confirm that the proposed method outperforms baseline methods regardless of the ratio of the observable states. These results validate the effectiveness of our method for estimating the parameters of the original MC.

Quantitative Evaluation (CMC Estimation): Figure 5 shows the results of *censored* negative test log-likelihood. We can confirm that the proposed method matches the CMC estimation performance of the baseline method regardless of the ratio of the observable states. Although the proposed method is designed to recover not CMC but original MC, this result means that the proposed method can well estimate CMC. Therefore, from Figs. 4 and 5, we can say that the proposed method well estimates the parameters without sacrificing either original MC estimation or CMC estimation performance.

Qualitative Evaluation (Original MC Estimation): Figure 6 illustrates the true and estimated transition probability in the synthetic experiment. It shows the estimated probability output by the proposed method is close to the true probability. This also implies that our method well estimates the original MC.

[3] The use of cross-validation is not appropriate given our problem since it selects the best one in terms of *censored* test log-likelihood (in asymptotic limit), not *true* test log-likelihood; the result also shows the baselines with small hyperparameter values provide better censored log-likelihood and worse true log-likelihood, and vice versa.

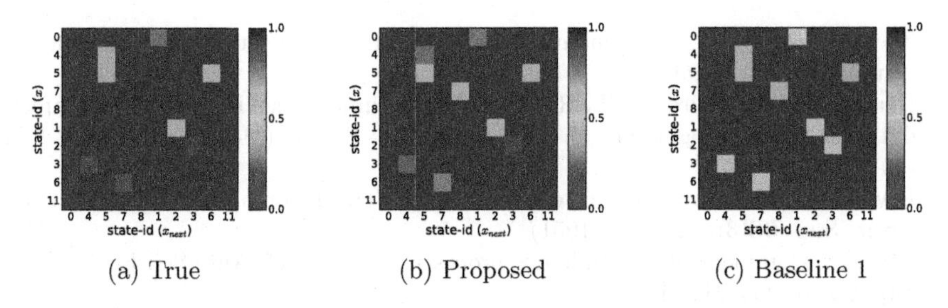

(a) True (b) Proposed (c) Baseline 1

Fig. 6. (a) True and estimated transition probability yielded by (b) Proposed method, (c) Baseline1 in synthetic chain experiment where the ratio of observed states is 0.5. The probability between ten extracted states are shown. First 5 states (0, 4, 5, 7, 8) are observable; remaining states are unobservable.

6 Conclusion

This paper tackled the problem of estimating Markov chain parameters from censored transition data. By formulating it as the inverse problem of CMC, we developed a new estimation method on the theory of CMC. We confirmed the effectiveness of the proposed method by experiments on both synthetic and real car probe data.

Remaining future work is to construct a method for hyperparameter optimization and to provide a theoretical analysis of the proposed method, especially the dependency of the performance on the ratio of observable states. Investigation of the performance using other types of divergence for the loss function such as L_2 divergence and using more complex models, e.g., deep neural networks and Markov mixtures [27,28], is also a promising future direction.

References

1. Muller, M.R., Middleton, J.: A Markov model of land-use change dynamics in the Niagara region, Ontario, Canada. Landscape Ecol. **9**(2), 151–157 (1994)
2. Crisostomi, E., Kirkland, S., Shorten, R.: A google-like model of road network dynamics and its application to regulation and control. Int. J. Control **84**(3), 633–651 (2011)
3. Fan, Z., Song, X., Shibasaki, R., Adachi, R.: CityMomentum: an online approach for crowd behavior prediction at a citywide level. In: International Joint Conference on Pervasive and Ubiquitous Computing, pp. 559–569 (2015)
4. Neuts, M.F.: Matrix-geometric solutions in stochastic models: an algorithmic approach. Johns Hopkins University Press (1981)
5. Pfeifer, P.E., Carraway, R.L.: Modeling customer relationships as Markov chains. J. Interact. Mark. **14**(2), 43–55 (2000)
6. Nowak, M.A.: Evolutionary dynamics. Harvard University Press (2006)
7. Billingsley, P.: Statistical inference for Markov processes. University of Chicago Press (1961)
8. Zheng, Y., Zhou, X.: Computing with spatial trajectories. Springer Science & Business Media (2011). https://doi.org/10.1007/978-1-4614-1629-6

9. Zheng, L., Yang, Y., Hauptmann, A.G.: Person re-identification: past, present and future. arXiv preprint arXiv:1610.02984 (2016)
10. Dempster, A.P., Laird, N.M., Rubin, D.B.: Maximum likelihood from incomplete data via the EM algorithm. J. Roy. Stat. Soc.: Ser. B (Methodol.) **39**(1), 1–22 (1977)
11. Lévy, P.: Systèmes markoviens et stationnaires. cas dénombrable. Ann. Sci. École Norm. Sup. **68**(3), 327–381 (1951)
12. Lévy, P.: Complément à l'étude des processus de Markoff. Ann. Sci. École Norm. Sup. **69**(3), 203–212 (1952)
13. Lévy, P.: Processus markoviens et stationnaires. cas dénombrable. Ann. Inst. H. Poincaré **18**, 7–25 (1958)
14. Morimura, T., Osogami, T., Idé, T.: Solving inverse problem of Markov chain with partial observations. In: Advances in Neural Information Processing Systems, pp. 1655–1663 (2013)
15. Kumar, R., Tomkins, A., Vassilvitskii, S., Vee, E.: Inverting a steady-state. In: International Conference on Web Search and Data Mining, pp. 359–368 (2015)
16. Maystre, L., Grossglauser, M.: ChoiceRank: identifying preferences from node traffic in networks. In: International Conference on Machine Learning, pp. 2354–2362 (2017)
17. Iwata, T., Shimizu, H., Naya, F., Ueda, N.: Estimating people flow from spatiotemporal population data via collective graphical mixture models. ACM Trans. Spatial Alg. Syst. **3**(1), 2 (2017)
18. Akagi, Y., Nishimura, T., Kurashima, T., Toda, H.: A fast and accurate method for estimating people flow from spatiotemporal population data. In: International Joint Conference on Artificial Intelligence, pp. 3293–3300 (2018)
19. Freedman, D.: Approximating countable Markov chains. Springer-Verlag, New York (1983). https://doi.org/10.1007/978-1-4613-8230-0
20. Zhao, Y.Q., Liu, D.: The censored Markov chain and the best augmentation. J. Appl. Probab. **33**(3), 623–629 (1996)
21. Kemeny, J.G., Snell, J.L., Knapp, A.W.: Denumerable Markov chains. Springer-Verlag, New York (1976). https://doi.org/10.1007/978-1-4684-9455-6
22. Grassmann, W.K., Heyman, D.P.: Computation of steady-state probabilities for infinite-state Markov chains with repeating rows. ORSA J. Comput. **5**(3), 292–303 (1993)
23. Levin, D.A., Peres, Y.: Markov chains and mixing times. Amer. Math, Soc (2017)
24. Baxter, J., Bartlett, P.L.: Infinite-horizon policy-gradient estimation. J. Artif. Intell. Res. **15**, 319–350 (2001)
25. Liu, D.C., Nocedal, J.: On the limited memory BFGs method for large scale optimization. Math. Program. **45**(1–3), 503–528 (1989)
26. Morimura, T., Uchibe, E., Yoshimoto, J., Doya, K.: A generalized natural actor-critic algorithm. In: Advances in Neural Information Processing Systems, pp. 1312–1320 (2009)
27. Frydman, H., Schuermann, T.: Credit rating dynamics and Markov mixture models. J. Banking Finance **32**(6), 1062–1075 (2008)
28. Gupta, R., Kumar, R., Vassilvitskii, S.: On mixtures of Markov chains. In: Advances in Neural Information Processing Systems, pp. 3441–3449 (2016)

Parameter-Free Bayesian Decision Trees for Uplift Modeling

Mina Rafla[1,2](✉), Nicolas Voisine[1], and Bruno Crémilleux[2]

[1] Orange Labs, 22300 Lannion, France
{mina.rafla,nicolas.voisine}@orange.com
[2] UNICAEN, ENSICAEN, CNRS - UMR GREYC, Normandie Univ, 14000 Caen, France
bruno.cremilleux@unicaen.fr

Abstract. Uplift modeling aims to estimate the incremental impact of a treatment, such as a marketing campaign or a drug, on an individual's behavior. These approaches are very useful in several applications such as personalized medicine and advertising, as it allows targeting the specific proportion of a population on which the treatment will have the greatest impact. Uplift modeling is a challenging task because data are partially known (for an individual, responses to alternative treatments cannot be observed). In this paper, we present a new tree algorithm named UB-DT designed for uplift modeling. We propose a Bayesian evaluation criterion for uplift decision trees T by defining the posterior probability of T given uplift data. We transform the learning problem into an optimization one to search for the uplift tree model leading to the best evaluation of the criterion. A search algorithm is then presented as well as an extension for random forests. Large scale experiments on real and synthetic datasets show the efficiency of our methods over other state-of-art uplift modeling approaches.

Keywords: Uplift Modeling · Decision trees · Random Forests · Bayesian methods · Machine Learning · Treatment Effect Estimation

1 Introduction

Uplift modeling aims to estimate the incremental impact of a treatment, such as a marketing campaign or a drug, on an individual's behavior. These approaches are very useful in several applications such as personalized medicine and advertising, as it allows targeting the specific proportion of a population on which the treatment will have the greatest impact. Uplift estimation is based on groups of people who have received different treatments. A major difficulty is that data are only partially known: it is impossible to know for an individual whether the chosen treatment is optimal because their responses to alternative treatments cannot be observed. Several works address challenges related to the uplift modeling, among which uplift decision tree algorithms became widely used [15,17].

H. Kashima et al. (Eds.): PAKDD 2023, LNAI 13936, pp. 309–321, 2023.
https://doi.org/10.1007/978-3-031-33377-4_24

Despite their usefulness, current uplift decision tree methods have limitations such as local splitting criteria. A split criterion decides whether to divide a terminal node. However these splits are independent to each other and a pruning step is then used to ensure generalization and avoid overfitting. Moreover, these methods require parameters to set. In this paper, we present UB-DT (Uplift Bayesian Decision Tree) a parameter-free method for uplift decision tree based on the Bayesian paradigm. Contrary to state-of-art uplift decision tree methods, we define a global criterion designed for an uplift decision tree. A major advantage of a global tree criterion is it allows to get rid of the pruning step, since it acts as a regularization to avoid overfitting. We transform the uplift tree learning problem to an optimization problem according to the criterion. Then a search algorithm is used to find the decision tree that optimizes the global criterion. Moreover our approach is easily extended to random forests and we propose UB-RF (Uplift Bayesian Random Forest). We evaluate both UB-DT and UB-RF to state-of-art uplift modeling approaches through a benchmarking study.

This paper is organized as follows. Section 2 introduces an overview of uplift modeling and related work. Section 3 presents UB-DT. We conduct experiments in Sect. 4 and conclude in Sect. 5.

2 Context and Literature Overview

2.1 Uplift Problem Formulation

Uplift is a notion introduced by Radcliffe and Surry [11] and defined in Rubin's causal inference models [14] as the *Individual Treatment effect (ITE)*.

We now outline the notion of uplift and its modeling. Let X be a group of N individuals indexed by $i : 1 \ldots N$ where each individual is described by a set of variables $\mathbb{K}$. X_i denotes the set of values of $\mathbb{K}$ for the individual i. Let Z be a variable indicating whether or not an individual has received a treatment. Uplift modeling is based on two groups: the individuals having received a treatment (denoted $Z = 1$) and those without treatment (denoted $Z = 0$). Let Y be the outcome variable (for instance, the purchase or not of a product). We note $Y_i(Z = 1)$ the outcome of an individual i when he received a treatment and $Y_i(Z = 0)$ his outcome without treatment. The uplift of an individual i, denoted by τ_i, is defined as: $\tau_i = Y_i(Z = 1) - Y_i(Z = 0)$.

In practice, we will never observe both $Y_i(Z = 1)$ and $Y_i(Z = 0)$ for a same individual and thus τ_i cannot be directly calculated. However, uplift can be empirically estimated by considering two groups: a treatment group (individual with a treatment) and a control group (without treatment). The estimated uplift of an individual i denoted by $\hat{\tau}_i$ is then computed by using the CATE (Conditional Average Treatment Effect) [14]:

$$CATE : \hat{\tau}_i = \mathbb{E}[Y_i(Z = 1)|X_i] - \mathbb{E}[Y_i(Z = 0)|X_i] \quad (1)$$

As the real value of τ_i cannot be observed, it is impossible to directly use machine learning algorithms such as regression to infer a model to predict τ_i. The next section describes how uplift is modeled in the literature.

2.2 Related Work

Uplift modeling approaches Uplift modeling approaches are divided into two categories. The first one (called *metalearners*) is made up of methods that take advantage of usual machine learning algorithms to estimate the CATE. One of the most intuitive approaches is the *two-model approach.* It consists of fitting two independent classification models, one for the treated group and another for the control group. The estimated uplift is then the difference between the estimations of the two classification models. While this approach is simple, intuitive and allows the usage of any machine learning algorithm, it has also known weaknesses with particular patterns [12]. The causal inference community has also proposed other metalearners such as X-learner [8], R-Learner and DR-learner [7].

The second category is closer to our work. This category gathers tailored methods for uplift modeling such as tree-based algorithms. Trees are built using recursive partitioning to split the root node to child nodes according to a splitting criterion. [15] defines a splitting criterion that compares the probability distributions of the outcome variable in each of the treatment groups using weighted divergence measures like the Kullback-Leibler (KL), the squared euclidean distance (ED) and the chi-squared divergence. [17] proposes the Contextual Treatment Selection algorithm (CTS) where a splitting criterion directly maximizes a performance measure called the expected performance. Causal machine learning algorithms were also developed such as the Causal Trees algorithm [1] and the Causal Forests [2].

Uplift tree splitting criterion and Bayesian approaches. Building an uplift tree requires to discretize variables to detect areas with homogeneous treatment effects. The global criterion of UB-DT to select a variable on a node takes advantage of on a univariate parameter-free Bayesian approach for density estimation through discretization called UMODL [13]. More precisely, UMODL applies a Bayesian approach to select the most probable uplift discretization model M given the data. This implies finding the model M that maximizes the posterior probability $P(M|Data)$, hence maximizing $P(M) \times P(Data|M)$. Finally, a global criterion within the Bayesian framework for decision trees is given in [16] but it does not deal with uplift.

3 UB-DT: Uplift Decision Tree Approach

UB-DT is made up of two ingredients: a global criterion $C(T)$ for a binary uplift decision tree T and a tree search algorithm to find the most probable optimal tree. We start by presenting the structure of an uplift tree model. Then we describe the new global criterion for an uplift decision tree and the algorithm to give the best tree. Finally we show how the approach is straightforwardly extended to random forests.

3.1 Parameters of an Uplift Tree Model T

We define a binary uplift decision tree model T by its structure and the distribution of instances and class values in this structure. The structure of T consists

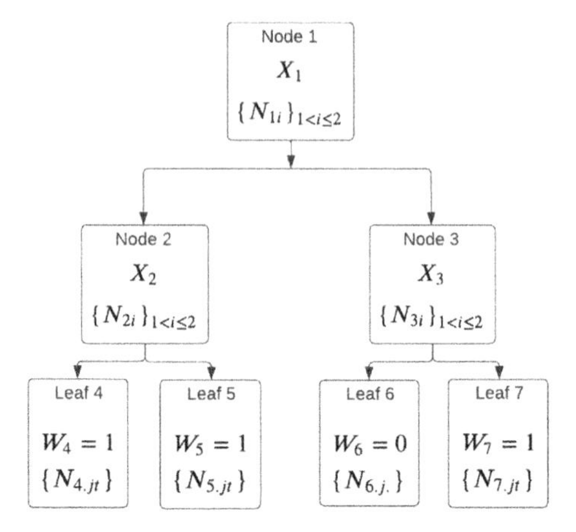

Fig. 1. Example of an uplift tree model. Internal nodes are described by the segmentation variable X_s and the distribution of instances in each of the two children $\{N_{si}\}$. Leaf nodes containing a treatment effect (i.e. $W_l = 1$) are described by the class distribution for each treatment. This applies to leaves 4, 5 and 7. Leaf nodes containing no treatment effect (i.e. $W_l = 0$) are only described by the class distribution (this is the case of leaf 6).

of the set of internal nodes $\mathbb{S}_T$ and the set of leaf nodes $\mathbb{L}_T$. The distribution of the instances in this structure is described by the partition of the segmentation variable X_s for each internal node s, the class frequency in each leaf node where there is no treatment effect, and the class frequency on each treatment in the leaf nodes with a treatment effect. More precisely, T is defined by:

- the subset of variables $\mathbb{K}_T$ used by model T. This includes the number of the selected variables K_T and their choice among a set of $\mathbb{K}$ variables provided in a dataset, we note $K = |\mathbb{K}|$.
- a binary variable I_n indicating the choice of whether each node n is an internal node ($I_n = 1$) or a leaf node ($I_n = 0$).
- the distribution of instances in each internal node s, which is described by the segmentation variable X_s of the node s and how the instances of s are distributed on its two child nodes.
- a binary variable W_l indicating for each leaf node l if there is a treatment effect ($W_l = 1$) or not ($W_l = 0$). If $W_l = 0$, l is described by the distribution of the output values $\{N_{l.j.}\}_{1\leq j\leq J}$, where $N_{l.j.}$ is the number of instances of output value j in leaf l. If $W_l = 1$, l is described by the distribution of the class values per treatment $\{N_{l.jt}\}_{1\leq j\leq J, 1\leq t\leq 2}$, where $N_{l.jt}$ is the number of instances of output value j and treatment t in leaf l.

These parameters are automatically optimized by the search algorithm (presented in Sect. 3.4) and not fixed by the user. In the rest of the paper, the following notations $N_{s.}$, $N_{si.}$, $N_{l.}$ and $N_{l..t}$ will additionally be used to respectively designate the number of instances in node s, in the ith child of node s, in the leaf l and treatment t in leaf l.

3.2 Uplift Tree Evaluation Criterion

We now present the new global criterion $C(T)$ which is an uplift tree model evaluation criterion. UB-DT applies a Bayesian approach to select the most probable uplift tree model T that maximizes the posterior probability $P(T|Data)$. This is equivalent to maximizing the product of the prior and the likelihood i.e. $P(T) \times P(Data|T)$. Taking the negative log turns the maximization problem into a minimization one: $C(T) = -\log(P(T) \times P(Data|T))$, $C(T)$ is the cost of the uplift tree model T. T is optimal if $C(T)$ is minimal. By exploiting the hierarchy of the presented uplift tree parameters and assuming a uniform prior, we express $C(T)$ as follows (cf. Eq. 2):

$$
\begin{aligned}
C(T) = & \underbrace{\log(K+1) + \log\binom{K+K_T-1}{K_T}}_{\text{Variable selection}} \\
& + \underbrace{\sum_{s \in \mathbb{S}_{T_n}} \log 2 + \log K_T + \log(N_{s.}+1) +}_{\text{Prior of internal nodes}} \underbrace{\sum_{l \in \mathbb{L}_T} \log 2}_{\text{Treatment effect W}} \\
& + \underbrace{\sum_{l \in \mathbb{L}_T} \log 2 + \sum_{l \in \mathbb{L}_T} (1-W_l) \log\binom{N_{l.}+J-1}{J-1} + \sum_{l \in \mathbb{L}_T} W_l \sum_t \log\binom{N_{l..t}+J-1}{J-1}}_{\text{Prior of leaf nodes}} \\
& + \underbrace{\sum_{l \in \mathbb{L}_T} (1-W_l) \log \frac{N_{l.}!}{N_{l.1.}! N_{l.2.}! \ldots N_{l.J.}!} + \sum_{l \in \mathbb{L}_T} W_l \sum_t \log \frac{N_{l..t}!}{N_{l.1t}! .. N_{l.Jt}!}}_{\text{Tree Likelihood}}
\end{aligned} \quad (2)
$$

The next section demonstrates Eq. 2.

3.3 $C(T)$: Proof of Equation 2

We express the prior and the likelihood of a tree model, resp. $P(T)$ and $P(Data|T)$ according to the hierarchy of the uplift tree parameters. Assuming the independence between all the nodes, the prior probability of an uplift decision tree is thus defined as:

$$
\begin{aligned}
& P(T) = P(\mathbb{K}_T) \times \\
& \prod_{s \in \mathbb{S}_T} P(I_s) P(X_s \mid \mathbb{K}_T) P(N_{si.} \mid \mathbb{K}_T, X_s, N_{s.}, I_s) \times \\
& P(\{W_l\}) \times \prod_{l \in \mathbb{L}_T} P(I_l) \left[(1-W_l) \times p(\{N_{l.j}\} \mid \mathbb{K}_T, N_{l.}) + W_l \times \prod_t P(\{N_{l.jt}\} \mid \mathbb{K}_T, N_{l..t}) \right]
\end{aligned} \quad (3)
$$

The first line is the prior probability of the variable selection, the second line the prior of internal nodes and the third line the prior of the leaf nodes.

Variable Selection Probability. A hierarichal prior is chosen: first the choice of the number of selected variables K_T, then the choice of the subset $\mathbb{K}_T$ among $\mathbb{K}$ variables. By using a uniform prior the number K_T can have any value between 0 and K in an equiprobable manner. For the choice of the subset $\mathbb{K}_T$, we assume that every subset has the same probability. Then the prior of the variable selection can be defined as:

$$P(\mathbb{K}_T) = \frac{1}{K+1} \frac{1}{\binom{K+K_T-1}{K_T}}$$

Prior of Internal Nodes. Each node can either be an internal node or a leaf node with equal probability. This implies that: $P(I_s) = \frac{1}{2}$

The choice of the segmentation variable is equiprobable between 1 and K_T. We obtain:

$$P(X_s|\mathbb{K}_T) = \frac{1}{K_T}$$

All splits of an internal node s to two intervals are equiprobable. We then obtain:

$$P\left(N_{si.} \mid \mathbb{K}_T, X_s, N_{s.}, I_s\right) = \frac{1}{N_s+1}$$

Prior of Leaf Nodes. Similar to the prior of internal nodes, each node can either be internal or a leaf node with equal probability leading to $P(I_l) = \frac{1}{2}$. For each leaf node, we assume that a treatment can have an effect or not, with equal probability, we get:

$$P(\{W_l\}) = \prod_l \frac{1}{2}$$

In the case of a leaf node l where there is not effect of the treatment ($W_l = 0$), UB-DT describes one unique distribution of the class variable. Assuming that each of the class distributions is equiprobable, we end up also with a combinatorial problem:

$$P\left(\{N_{l.j}\} \mid \mathbb{K}_T, N_{l.}\right) = \frac{1}{\binom{N_{l.}+J-1}{J-1}}$$

In a leaf node with an effect of the treatment ($W_i = 1$), UB-DT describes two distributions of the outcome variable, with and without the treatment. Given a leaf l and a treatment t, we know the number of instances $N_{l..t}$ Assuming that each of the distributions of class values is equiprobable, we get:

$$P\left(\{N_{l.jt}\} \mid \mathbb{K}_T, N_{l..t}\right) = \frac{1}{\binom{N_{l..t}+J-1}{J-1}}$$

Tree Likelihood. After defining the tree's prior probability, we establish the likelihood probability of the data given the tree model. The class distributions depend only of the leaf nodes. For each multinomial distribution of the outcome variable (a single or two distinct distributions per leaf depending on whether the treatement has an effect or not), we assume that all possible observed data D_l consistent with the multinomial model are equiprobable. Using multinomial terms, we end up with:

$$P(Data \mid T) = \prod_{l \in L} P(D_l|M)$$

$$\prod_{l \in L} \left[(1 - W_l) \times \frac{1}{N_{l.}!/N_{l.1.}!N_{l.2.}! \dots N_{l.J.}!} + W_l \times \prod_t \frac{1}{(N_{l..t}!/N_{i.1t}!..N_{i.Jt}!)} \right] \quad (4)$$

By combining the prior and the likelihood (resp. Eq. 3 and 4) and by taking their negative log, we obtain $C(T)$ and thus Eq. 2 is proved.

3.4 Search Algorithm

The induction of an optimal uplift decision tree from a data set is NP-hard [10]. Thus, learning the optimal decision tree requires exhaustive search and is limited to very small data sets. As a result, heuristic methods are required to build uplift decision trees. Algorithm 1 (see below) selects the best tree according to the global criterion. Algorithm 1 chooses a split among all possible splits in all terminal nodes only if it minimizes the global criterion of the tree. The algorithm continues as long as the global criterion is improved. Since a decision tree is a partitioning of the feature space, a prediction for a future instance is then the average uplift in its corresponding leaf. This algorithm is deterministic and thus it always leads to the same local optimum. Experiments show the quality of the building trees.

3.5 UB-RF

UB-DT is easily extended to random forests. For that purpose, a split is randomly chosen among all possible splits that improve the global criterion. The number of trees is set by the analyst and the prediction of a forest is the average predictions of all the trees.

4 Experiments

We experimentally evaluate the quality of UB-DT as an uplift estimator and compare UB-DT and UB-RF versus state-of-art uplift modeling approaches[1].

[1] Code, datasets and complementary results are at https://github.com/MinaWagdi/UB-DT.

We use the following state-of-art methods: (1) metalearners: two-model approach (2M), X-Learner and R-Learner, each with Xgboost; (2) uplift trees: CTS-DT,KL-DT, Chi-DT, ED-DT; (3) uplift random forests: CTS-RF,KL-RF, Chi-RF, ED-RF [15]; (4) and causal forests (all forest methods were used with 10 trees).

4.1 Is UB-DT a Good Uplift Estimator?

To be able to measure the estimated uplift we need to know the real uplift and therefore we use synthetic data. Figure 2 depicts two synthetic uplift patterns where $P(Y = 1|X, T = 1)$ and $P(Y = 1|X, T = 0)$ are identified for each instance. The grid pattern can be considered as a tree-friendly pattern whereas the continuous pattern is much more difficult. We generated several datasets according to these patterns with several different numbers of instances (also called data size) ranging from 100 to 100,000 instances. Uplift models were built using 10-fold stratified cross validation and the RMSE (Root Mean Squared Error) was used to evaluate the performance of the models.

Results: Figure 3 gives the RMSE for the two synthetic patterns according to the data size for different uplift methods. We see that UB-DT is a good estimator for uplift. With UB-DT, RMSE decreases and converges to zero when data sizes increase both for the grid and continuous patterns. This is the expected behavior of a good uplift estimator. This also means that UB-DT, thanks to its global criterion, avoids overfitting of uplift trees. The two-model approach with decision trees also shows competitive performance. UB-DT clearly outperforms the other tree-based methods, these latter having similar performances. With the continuous pattern, KL-DT, Chi-DT, ED-DT and CTS-DT approaches have

Algorithm 1: UB-DT algorithm

input : T the root tree
output: the tree T^* which minimizes the proposed criterion
$T^* \leftarrow T$
while $C(T^*)$ *decreases*:
    $T' \leftarrow T^*$
    for *leaf* l **in** $\mathbb{L}_T$:
        for X **in** $\mathbb{K}$:
            Get the best Split $S_X(l)$ according to UMODL
            $T_X \leftarrow T^* + S_X(l)$
            if $C(T_X) < C(T')$:
                $T' \leftarrow T_X$
    if $C(T') < C(T^*)$:
        $T^* \leftarrow T'$
Prediction: The output of a tree is a partition of the feature space. The predicted uplift for each instance is the average uplift of its leaf node.

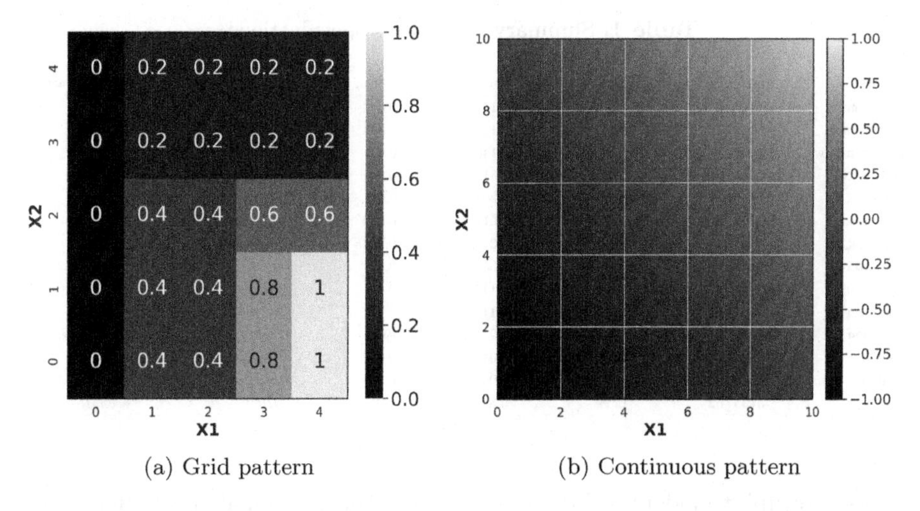

(a) Grid pattern

(b) Continuous pattern

Fig. 2. Uplift for 2 synthetic patterns. Figure 2a (grid pattern): uplift values for each cell. Figure 2b (continuous pattern): uplift values are $P(Y|T = 0, x1, x2) = 1 - (x1 + x2)/20$ while $P(Y|T = 1, x1, x2) = (x1 + x2)/20$.

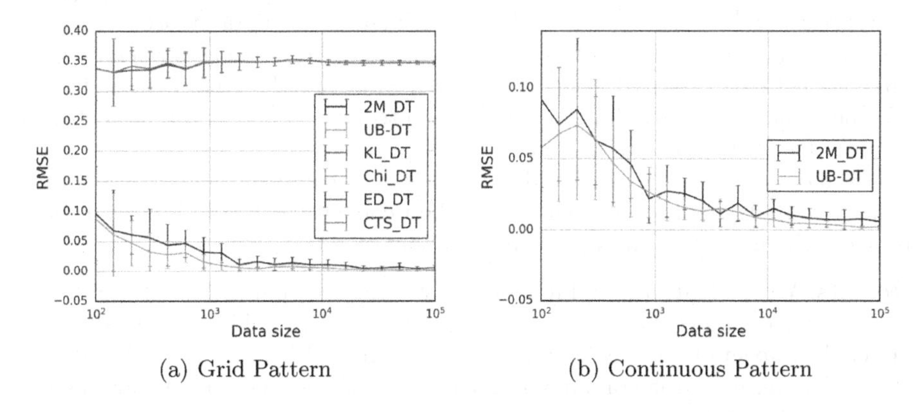

(a) Grid Pattern

(b) Continuous Pattern

Fig. 3. The RMSE of tree-based approaches according to data size

lower performances (their RMSE are around 0.5). To avoid a cluttered visualisation, their performances are not included in Fig. 3b.

4.2 UB-DT and UB-RF Versus State of the Art Methods

Datasets. We conducted experiments on 8 real and synthetic datasets widely used in the uplift modeling community: (1) *Hillstrom*[2] (a classical dataset for uplift modeling with data of customers who either received emails featuring men's/ women's products, or received no emails); (2) *Criteo* [5] (a marketing

[2] http://blog.minethatdata.com/2008/03/minethatdata-e-mail-analytics-and-data.html.

Table 1. Summary of datasets specifications

Dataset	No. Rows	No. Columns	Treatment ratio	Outcome Ratio	Average Uplift	Treatment variable	Outcome variable
Hillstrom-m	42,613	10	0.5	0.145	0.076	*'mens'*	*'visit'*
Hillstrom-w	42,693	10	0.5	0.128	0.045	*'womens'*	*'visit'*
Hillstrom-mw	64,000	10	0.67	0.146	0.06	*'mens' & 'womens'*	*'visit'*
Gerber-N	229,444	16	0.166	0.31	0.081	*'neighbour'*	*'voted'*
Geber-S	229,461	16	0.166	0.304	0.04	*'self'*	*'voted'*
Starbucks	84,534	9	0.5	0.012	0.009	*'promotion'*	*'purchase'*
Information	20,000	69	0.5	0.2	0.0018	*'treatment'*	*'purchase'*
Bank-tel	15,926	17	0.18	0.05	0.09	*'telephone'*	*'Y'*
Bank-cell	42,305	17	0.6	0.115	0.11	*'cellular'*	*'Y'*
Bank-tel-cel	45,211	17	0.71	0.116	0.107	*'telephone'&'cellular'*	*'Y'*
Megafon	600,000	52	0.5	0.2	-0.18	*'treatment'*	*'conversion'*
Criteo-v	13,979,592	12	0.85	0.047	0.68	*'treatment'*	*'visit'*
Criteo-c	13,979,592	12	0.85	0.0029	0.37	*'treatment'*	*'conversion'*
RHC	5735	62	0.38	0.35	-0.05	*'RHC'*	*'swang1'*

dataset for uplift modeling) (3) *Bank* [9] (a marketing campaign conducted by a bank) (4) *Information*[3] (a marketing dataset in the insurance domain, a part of the Information R package); (5) *Megafon*[4] (a synthetic dataset generated by a telecom company); (6) *Starbucks*[5] (an advertising promotion tested to improve customers purchases); (7) *Gerber* [6] (a policy-relevant dataset used to study the effect of social pressure on voter turnout); (8) *Right Heart Catheterization (RHC)* [3] (a real dataset from the medical domain, the treatment indicates whether a patient received a RHC and the outcome is whether the patient died at any time up to 180 d after admission to the study).

Each dataset was used with different settings of treatment and outcome variables. For all datasets, each treatment and outcome variables are binary. Table 1 provides the most relevant specifications about the data sets.

Results. We evaluate the uplift models by using the qini metric [4]. Qini is a variant of the Gini coefficient. Its values are in $[-1, 1]$, the higher the value, the larger the impact of the predicted optimal treatment. Figure 4a (resp. Figure 4b) shows the overall average ranking of tree based methods (resp. meta-learners and forest-based methods) according to its qini performance against each dataset. Compared to other tree-based methods and to the two-model approach with decision trees, Fig. 4a shows that UB-DT achieves the best performance. Table 2 reports the results of the experiment for the qini metric. This table shows that UB-DT is also a good estimator of the uplift on real data. Figure 4b shows that both UB-RF and 2M have the best rank. Table 3 indicates that the random forest strategy improves the performance of the uplift models (qini values are higher with UB-RF than UB-DT). UB-RF has the best performance on 4 out the 14 experiments.

[3] https://cran.r-project.org/web/packages/Information/index.html.
[4] https://ods.ai/tracks/df21-megafon/competitions/megafon-df21-comp/data.
[5] https://github.com/joshxinjie/Data_Scientist_Nanodegree/tree/master/starbucks_portfolio_exercisejoshxinjie.

Table 2. Average qini values and standard deviation (multiplied by 100). The best qini value for each dataset is marked in bold.

Dataset	2M_DT	KL_DT	Chi_DT	ED_DT	CTS_DT	UB-DT
Hillstrom-m	0.3(1.0)	1.1(1.9)	1.0(1.9)	0.0(1.4)	0.2(1.0)	**1.6(1.6)**
Hillstrom-w	0.8(1.6)	5.2(2.5)	5.2(2.6)	**6.4(1.2)**	−0.4(2.0)	4.8(2.3)
Hillstrom-mw	−0.6(0.8)	−0.1(1.2)	−0.8(1.1)	**4.4(2.7)**	−0.0(1.0)	−0.4(1.4)
Gerber-n	**5.6(0.8)**	1.3(0.8)	1.2(0.8)	1.1(0.6)	1.3(0.8)	1.9(0.6)
Gerber-s	**5.5(1.1)**	0.4(0.5)	0.4(0.6)	0.5(0.3)	0.4(0.4)	0.8(0.6)
Criteo-c	8.0(1.5)	4.1(1.4)	4.8(1.5)	**15.2(0.3)**	1.7(0.3)	13.7(3.2)
Criteo-v	0.4(0.3)	−1.2(0.2)	−1.1(0.3)	−1.3(0.3)	0.4(1.1)	**3.6(1.2)**
Megafon	5.1(0.6)	4.5(0.9)	4.7(0.9)	4.7(0.9)	4.9(0.8)	**7.8(0.8)**
Bank-tel	5.4(7.6)	−12.5(2.8)	−10.8(7.0)	−10.2(7.8)	−12.8(2.9)	**12.8(8.0)**
Bank-cell	11.1(3.0)	−2.0(1.5)	−1.4(2.5)	−2.2(1.5)	−3.7(1.5)	**38.4(3.4)**
Bank-tel-cell	10.3(1.6)	−1.9(1.2)	−1.2(2.1)	−1.8(1.2)	−3.4(1.4)	**37.1(2.6)**
Information	4.6(3.4)	−6.3(2.8)	−6.3(2.8)	−2.8(1.5)	−5.4(1.5)	**11.8(2.4)**
Starbucks	1.4(1.4)	20.1(3.0)	18.3(3.4)	19.9(3.2)	13.9(3.9)	**20.2(3.5)**
RHC	12.8(1.9)	18.4(3.8)	19.9(4.2)	18.4(3.8)	16.7(2.5)	**20.7(5.0)**

Table 3. Average qini values and standard deviation (multiplied by 100) across datasets and uplift approaches. In bold, the best value for each dataset

Dataset	XLearner	RLearner	DR	2M	KL_RF	Chi_RF	ED_RF	CTS_RF	UB-RF	CausalForest
Hillstrom-m	0.3(2.3)	0.3(1.8)	1.2(1.6)	0.7(2.3)	−0.0(2.1)	−0.9(1.5)	0.7(1.5)	1.1(1.9)	**1.8(1.6)**	−0.2(1.6)
Hillstrom-w	6.2(1.7)	6.2(1.4)	6.0(1.4)	4.9(1.1)	6.2(1.1)	**7.0(1.0)**	6.2(1.1)	5.7(1.3)	6.7(1.1)	2.1(1.9)
Hillstrom-mw	3.7(2.3)	**3.9(2.7)**	3.8(2.8)	3.0(2.0)	3.0(1.3)	2.8(1.5)	3.6(2.5)	2.3(2.4)	3.1(1.7)	0.1(1.7)
Gerber-n	**3.7(0.6)**	1.9(0.7)	0.5(0.9)	3.1(0.6)	1.8(1.0)	2.1(1.1)	1.9(0.5)	1.4(1.0)	2.7(0.7)	2.9(1.0)
Gerber-s	2.4(0.9)	1.7(0.7)	0.6(0.9)	2.2(0.8)	1.3(1.0)	1.4(0.6)	1.6(0.8)	1.4(0.7)	1.8(0.8)	**3.1(0.5)**
Criteo-c	**22.3(1.8)**	19.4(1.0)	20.0(0.6)	19.5(1.6)	14.6(3.5)	12.4(4.3)	21.1(2.3)	7.3(3.9)	18.7(1.5)	10.9(2.4)
Criteo-v	0.3(0.8)	5.3(0.5)	4.8(1.5)	3.9(0.5)	5.4(1.2)	4.8(1.7)	**6.1(1.0)**	2.4(0.8)	5.7(0.7)	0.4(0.4)
Megafon	**18.2(0.6)**	2.6(0.5)	2.2(0.9)	16.6(0.9)	11.2(0.7)	11.0(1.2)	10.8(0.8)	9.2(1.1)	12.8(1.0)	9.7(0.7)
Bank-tel	14.5(7.6)	2.8(8.8)	16.0(9.0)	21.1(11.6)	−15.5(6.3)	−6.1(12.6)	−15.8(5.6)	−18.7(2.9)	**26.7(7.2)**	25.4(5.3)
Bank-cell	18.8(4.7)	23.3(3.6)	17.4(6.5)	31.0(3.9)	0.4(2.3)	1.5(2.5)	-2.5(2.6)	-1.0(1.9)	**45.5(2.7)**	20.8(2.6)
Bank-tel-cell	16.2(5.6)	23.8(2.5)	17.0(3.4)	30.5(2.7)	1.4(3.4)	−0.4(5.7)	−1.7(3.1)	−0.5(2.3)	**46.1(2.1)**	23.5(2.9)
Information	**14.9(3.3)**	10.0(3.1)	4.1(2.3)	13.7(4.1)	9.6(2.0)	9.7(3.1)	11.2(2.9)	10.6(2.9)	12.0(3.1)	10.5(3.2)
Starbucks	22.3(4.5)	22.4(3.9)	22.4(3.7)	22.7(4.1)	22.4(2.1)	21.4(3.4)	**23.4(3.2)**	20.8(3.1)	20.2(3.3)	8.1(3.7)
RHC	32.4(3.5)	31.3(4.3)	30.3(5.0)	**34.6(4.3)**	29.6(4.2)	29.7(5.0)	30.0(4.1)	29.1(3.7)	27.2(5.0)	27.6(4.5)

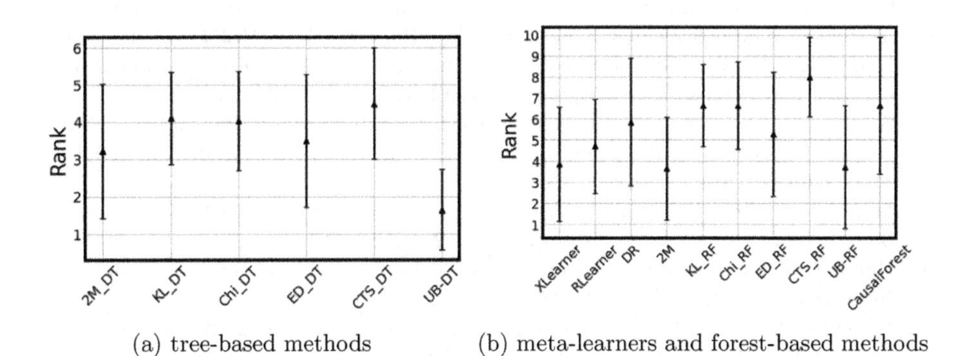

(a) tree-based methods (b) meta-learners and forest-based methods

Fig. 4. Overall average ranking of the uplift approaches

5 Conclusion and Perspectives

In this paper, we presented a new parameter-free method called UB-DT for uplift decision trees. We have designed a Bayesian approach to select the most probable uplift tree model T that maximizes the posterior probability $P(T|Data)$. Contrary to state-of-art uplift decision tree approaches, UB-DT is characterized by a global criterion to build a tree, so the splits in one node depend on the splits in the other nodes. This approach avoids overfitting and the need for a pruning step. A search algorithm finds the tree that optimizes this criterion. We showed that our approach is easily extended to random forests and we defined UB-RF. Evaluations on real and synthetic data sets show that UB-DT is a good uplift estimator and our tree and forests methods perform competitively with state-of-art uplift modeling approaches including non tree methods.

This work opens several perspectives. Studies on *general* trees (with more than two child nodes) is promising. In addition, studies with multiple treatments are still open work in uplift modeling. Moreover, the search algorithm leads to a local optimum and may create under-fitted uplift trees. To go above this horizon effect, it would be interesting to use a post-pruning algorithm [16].

References

1. Athey, S., Imbens, G.: Recursive partitioning for heterogeneous causal effects. Proc. Nat. Acad. Sci. **113**(27), 7353–7360 (2016)
2. Athey, S., Tibshirani, J., Wager, S.: Generalized random forests. Ann. Stat. **47**(2), 1148–1178 (2019)
3. Connors, A.F., et al.: The effectiveness of right heart catheterization in the initial care of critically ill patients. support investigators. JAMA **276** 11, 889–97 (1996)
4. Devriendt, F., Van Belle, J., Guns, T., Verbeke, W.: Learning to rank for uplift modeling. IEEE Trans. Knowl. Data Eng., pp. 1–1 (2020)
5. Diemert, E., Betlei, A., Renaudin, C., Amini, M.R.: A Large Scale Benchmark for Uplift Modeling. In: KDD. London, United Kingdom (2018)
6. Gerber, A.S., Green, D.P., Larimer, C.W.: Social pressure and voter turnout: Evidence from a large-scale field experiment. Am. Polit. Sci. Rev. **102**(1), 33–48 (2008)
7. Kennedy, E.H.: Towards optimal doubly robust estimation of heterogeneous causal effects (2020). https://arxiv.org/abs/2004.14497
8. Künzel, S.R., Sekhon, J.S., Bickel, P.J., Yu, B.: Meta-learners for estimating heterogeneous treatment effects using machine learning. Proc. Nat. Acad. Sci. **116**(10), 4156–4165 (2019)
9. Moro, S., Cortez, P., Rita, P.: A data-driven approach to predict the success of bank telemarketing. Decis. Support Syst. **62**, 22–31 (2014)
10. Naumov, G.: Np-completeness of problems of construction of optimal decision trees. In: Soviet Physics Doklady. 36, 270 (1991)
11. Radcliffe, N., Surry, P.: Differential response analysis: modeling true responses by isolating the effect of a single action. Credit Scoring and Credit Control IV (1999)
12. Radcliffe, N.J., Surry, P.D.: Real-world uplift modelling with significance-based uplift trees. Stochastic Solutions (2011)

13. Rafla, M., Voisine, N., Crémilleux, B., Boullé, M.: A non-parametric bayesian approach for uplift discretization and feature selection. In: ECML PKDD (2022) https://doi.org/10.1007/978-3-031-26419-1_15
14. Rubin, D.B.: Estimating causal effects of treatments in randomized and nonrandomized studies. J. Educ. Psychol. **66**, 688–701 (1974)
15. Rzepakowski, P., Jaroszewicz, S.: Decision trees for uplift modeling with single and multiple treatments. Knowl. Inf. Syst. **32**(2), 303–327 (2012)
16. Voisine, N., Boullé, M., Hue, C.: A bayes evaluation criterion for decision trees. In: Advances in knowledge discovery and management, pp. 21–38. Springer (2009) https://doi.org/10.1007/978-3-642-00580-0_2
17. Zhao, Y., Fang, X., Simchi-Levi, D.: Uplift modeling with multiple treatments and general response types. In: SIAM Int. Conf. on Data Mining. SIAM (2017)

Interpretability Meets Generalizability: A Hybrid Machine Learning System to Identify Nonlinear Granger Causality in Global Stock Indices

Yixiao Lu(✉), Yokiu Lee, Haoran Feng, Johnathan Leung, Alvin Cheung, Katharina Dost, Katerina Taskova, and Thomas Lacombe

University of Auckland, Auckland 1010, New Zealand
{ylu306,ylee942,hfen962,jleu075,ache706,kdos481}@aucklanduni.ac.nz,
{katerina.taskova,thomas.lacombe}@auckland.ac.nz

Abstract. Globalization has posed challenges to financial risk management, connecting markets with each other, and making it more difficult to diversify the portfolio to uncorrelated markets than ever in history. In light of this growing complexity of causal relationships between global stock markets, nonlinear Granger causality has superseded its linear counterpart in providing quantitative evidence for these relationships. In this paper, we propose a hybrid system that extends existing nonlinear Granger causality frameworks using machine learning-based time series prediction models. We improve the accuracy of identifying nonlinear Granger causality by combining p-values of causality statistics from individual machine learning models. By adjusting a model independence coefficient, our model is generalized to datasets where the strength of causality varies. Meanwhile, the causality statistic is still interpretable, because the distribution and critical value are known. Our findings challenge the current understanding that the United States market has the dominating influence and show that Asian markets play a significant role in spreading financial risk worldwide.

Keywords: Granger causality · machine learning · significance test · financial risk

1 Introduction

Alongside its contributions to economic growth, globalization is a conduit for risk transmission among international stock markets [2,7,20]. Besides financial connectedness inside the market, global events outside stock markets, such as COVID-19, cryptocurrency, and regional wars, also have an impact on stock markets worldwide. To address the complex financial risk transmission network worldwide and reduce the risk on stock markets, diversifying the portfolio to uncorrelated markets is a widely accepted approach [15].

Supplementary Information The online version contains supplementary material available at https://doi.org/10.1007/978-3-031-33377-4_25.

H. Kashima et al. (Eds.): PAKDD 2023, LNAI 13936, pp. 322–334, 2023.
https://doi.org/10.1007/978-3-031-33377-4_25

Conditional Granger causality [9] is based on the idea that the past values of a time series can help predict the future value of another time series. To calculate the causality statistic, they train a pair of linear autoregression models with and without the causing series and find their residuals. The limitation is that it only identifies linear causality, while causality between stock indices in the real world is nonlinear [22]. To fill the gap between the linear causality model and nonlinear causal relationships in the real world, researchers often use machine learning models to make the prediction and identify nonlinear Granger causality, such as *importance causal analysis* (ICA) [17], *group lasso* [8], and *cLSTM neural network* [24]. These methods expand the scope of Granger causality.

However, these individual models investigate one particular type of causality and are specialized to one scenario. The causality statistics and hypothesis tests are different among these models, making it difficult to combine these models. To tackle this issue, we propose a unified general framework integrating several machine learning methods. Our framework is powerful enough to identify Granger causality from different perspectives, thus increasing the robustness and reliability of the results.

In this paper, we use a hybrid framework to identify nonlinear Granger causality between stock indices. This framework combines p-values of causality statistics from different machine learning methods, searches a model independence coefficient to find critical values, and combines the significance test results. Our main contributions are:

- We enhance the generalizability of nonlinear Granger causality identifiers by using a hybrid strategy to combine individual models.
- We integrate p-values from individual models on the supervised dataset, where the hybrid causality statistic is interpretable in hypothesis testing.
- Any time series prediction model, even if it is not specially designed for identifying Granger causality, is compatible with our framework. This allows us to combine all learnings from individual methods based on the stacking method, which increases robustness.

The remainder of this paper is organized as follows. Section 2 introduces the background of our work, Sect. 3 reviews related literature, Sect. 4 proposes our hybrid machine learning system, Sect. 5 presents experimental results, and Sect. 6 concludes our findings and suggestions on future work.

2 Linear Granger Causality

Conditional Granger causality is a method to identify causal relationships in a group of time series. Let $\mathbf{y} = \{y_1, ..., y_N\}$ be a group of time series that are sampled from the same time period and frequency. Each sequence y_j of length T has observations $\{y_{j,1}, ..., y_{j,T}\}$. Each time series is stationary and is significant on lags of length m. For each pair of causal relationship from y_i to y_j, there is an *unrestricted* (UR) model and a *restricted* (R) model as follows:

$$\begin{aligned} \text{UR}: y_{j,t} &= \beta_0 + \sum_{k \in \{1:N\}} \boldsymbol{\beta}_k \mathbf{y}_{k,t-m:t-1} + \epsilon_{j,t}, \quad && t = m+1, ..., T \\ \text{R}: y_{j,t} &= \beta_0 + \sum_{k \in \{1:N\} \setminus i} \boldsymbol{\beta}_k \mathbf{y}_{k,t-m:t-1} + \eta_{j,t}, \quad && t = m+1, ..., T, \end{aligned} \tag{1}$$

where $\{1 : N\} \setminus i$ is the sequence from 1 to N excluding i, random variables $\epsilon_t, \boldsymbol{\eta}_t$ are the residuals, and $\boldsymbol{\beta}$ are coefficients of the linear autoregression model.

Both equations are fitted using the linear autoregression model. The Granger causality statistic is defined as

$$F_{i,j} = \frac{(T - 2m - 1) \left(\sum_{t=m+1}^{T} \hat{\eta}_{j,t}^2 - \sum_{t=m+1}^{T} \hat{\epsilon}_{j,t}^2 \right)}{m \sum_{t=m+1}^{T} \hat{\epsilon}_{j,t}^2}. \tag{2}$$

The null hypothesis of the causality statistic is "time series y_i has no causal effect on time series y_j". The alternative hypothesis is "time series y_i has a causal effect on time series y_j". The null distribution fits $F(m, T - 2m - 1)$. For a confidence level $\alpha = 0.01$, there is causality from y_i to y_j if $p_{i,j} < \alpha$.

We traverse each pair of (i, j) in all N time series $\{y_1, ..., y_N\}$ except (i, i). Then, the p-values of causality statistics form a matrix $\mathbf{p}$ of size $N \times N$ with an empty diagonal. Therefore, the adjacency matrix of the Granger causality network $\mathbf{P}$ is $I[\mathbf{p} < \alpha]$, where $I[\cdot]$ is the Iverson bracket.

3 Related Work

The Granger causality matrix $\mathbf{p}$ introduced above can be learned using machine learning models. In this section, we review existing attempts, as summarized in Table 1. Subsequently, we discuss the applications and results of Granger causality in global stock markets, including linear and nonlinear models.

3.1 Machine Learning Methods

Marinazzo et al. [18] use a *Support Vector Machine* (SVM), which prevents overfitting by using a filtered Granger causality index to test the probability of false positive predictions. However, the statistic is still linear and based on Pearson's correlation coefficients. Zheng and Song [26] use a *Generalized Radial Basis Function* (GRBF) neural network. Their method has an advantage when there is a large number of time series, and each has few observations as time points, but they cannot account for multiplicative relationships between time series. Gao and Yang [8] solve two problems: firstly, they apply group Lasso regression to overcome the sparsity in high-dimensional regression when the lag items were long; secondly, they transform the regression coefficients from a constant to a function related to time. Thus, the model can discover causality changing over time. Tank et al. [24] propose

Table 1. Overview of ML-based Granger causality detection method. Framework means whether their modeling objective is the same as Eq. 3.

Article	Model	Tackled issues	Limitation	Significance test	Framework
Marinazzo et al. [18]	SVM	overfitting	Significance on linear basis	Transformed Pearson correlation coefficient	No
Zheng and Song [26]	GRBF	high-dimension sparsity	Ignore multiplicative relationships	F-test	Yes
Gao and Yang [8]	Group Lasso	high-dimension sparsity, causality changes over time	Not mentioned	Regularization (nonzero implies significance)	Yes
Tank et al. [24]	cLSTM, cMLP	automatically lag selection, long-range dependencies	No critical value of significance test	Regularization (nonzero implies significance)	No
Leng et al. [17]	Random forest	dynamic causality, indirect causality	Not mentioned	T-test	No
Rosol et al. [21]	GRU, LSTM, MLP	dynamic causality, program implementation	Not mentioned	Wilcoxon signed-rank test	Yes

constraint neural networks, which automatically select the length of lag between time series. Their method makes a prediction for each time series independently, then results in the causality statistic based on the weights in previous prediction models. Tank et al.'s and Gao and Yang's works use the regularization method, while not carrying out the significance test. Leng et al. [17] are inspired by *decision trees* and propose the *Importance Causal Analysis* (ICA) method. Their method helps distinguish direct and indirect causality, and is applicable to real-time problems where causality is dynamic. Rosol et al. [21] integrate neural network-based nonlinear Granger causality models into a Python package. Their method allows for identifying causality that changes over time.

In summary, machine learning models in these works usually do not use UR and R models, and their causality statistics are not based on residuals as in linear cases. They have two problems in common: generalizability and interpretability. For generalizability, each of their causality statistics is only based on one specific machine learning model and is not compatible with others. When the model fails on other datasets, we can neither reuse their framework by choosing another machine learning method, nor integrate these methods into a hybrid framework. For interpretability, each causality statistic they proposed has different meanings and leads to different distributions, some even do not use a statistic. If we integrate or compare these methods, it is hard to explain the results.

In contrast, our approach not only bridges the causality statistic and machine learning methods, but also represents linear and nonlinear cases in the same framework. We also propose a method to combine the results of different models.

3.2 Granger Causality in Finance

Linear Granger causality is widely used in identifying Granger causality in global stock markets. Tang et al. [23] and Zheng & Song [26] showed that United States indices were in a pivotal position of the causal network, while Asian and Pacific indices were less important. With these models proposed in recent years for identifying nonlinear Granger causality, some research investigated nonlinear Granger causality in global stock markets. Al-Yahyaee et al. [1] attempted several statistical methods to detect nonlinear Granger causality and found asymmetric volatility transmission and asymmetric causalities. Their results showed a dominating influence of the US over Europe.

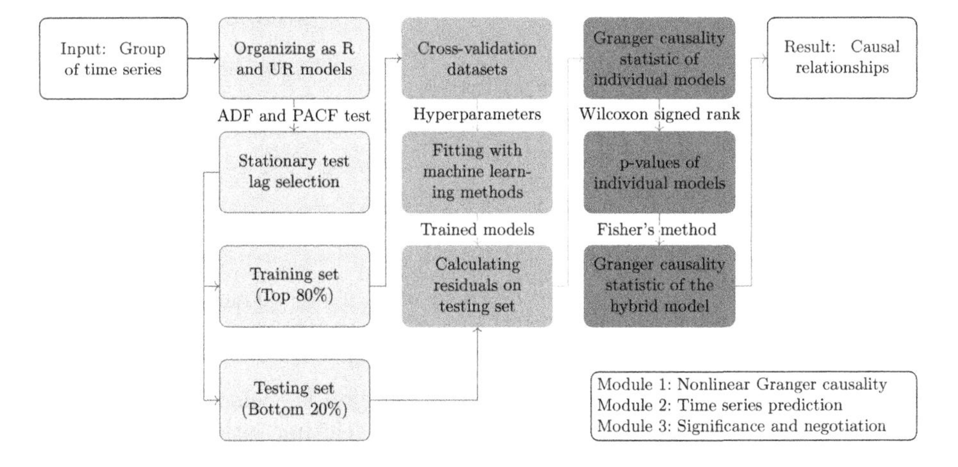

Fig. 1. Our proposed framework

The different opening and closing hours of the stock market worldwide cause a non-synchronous trading effect. Without adjustment, the conclusions tend to underestimate the causality from the United States to Asian and Pacific markets. Baumohl and Vyrost [3] accounted for this effect by adjusting the time scale of the causing series based on the effect series. Before adjusting, their results agreed with Tang et al. [23] and Zheng & Song [26]. But after adjusting, they found that Asian indices linearly caused others, and the European market caused the United States market. To the best of our knowledge, it was the only paper that found Asian indices causing others but not United States ones. Their limitation was that they only included five indices in the experiment.

4 Methodology

In this section, we identify the shortcomings in existing approaches, such as the lack of significance test and the incompatibility with the framework presented in Eq. 3. To address these limitations, we propose a three-module framework. See Fig. 1 for an overview.

The first module transforms the dataset into the form of nonlinear Granger causality in Eq. 3. Then, it tests if each time series is stationary and selects a suitable length of lag. We then do a train-test split with the processed dataset. This module bridges the causality statistic and the residuals of time series prediction models. It works with machine learning models, and also degenerates to linear Granger causality if we still use linear autoregression.

The second module trains different machine learning methods on the transformed data. We calculate the residuals for both UR and R models and for each machine learning method. This module does not require additional statistics from models, so it is compatible with any machine learning methods.

The third module calculates the Wilcoxon signed-rank statistic based on the residuals. With this causality statistic, we establish the p-value matrix for each individual model. We use Fisher's method to combine these p-value matrices as a new causality statistic. By using a model independence coefficient, this module is generalized to different strengths of causality and still uses hypothesis testing to justify the final results. We discuss all modules in depth below[1].

4.1 Module 1: Nonlinear Granger Causality

As mentioned in Sect. 2, the conditional Granger causality statistic is calculated via the residuals of the *unrestricted* (UR) and the *restricted* (R) model. In the nonlinear case, these two models are

$$\begin{aligned} \text{UR}: y_{j,t} &= f(\mathbf{y}_{1:N,t-m:t-1}) + \epsilon_{j,t}, & t &= m+1,...,T \\ \text{R}: y_{j,t} &= f(\left[\mathbf{y}_{\{1:N\}\setminus i,t-m:t-1}\ \tilde{\mathbf{y}}_{i,t-m:t-1}\right]) + \eta_{j,t}, & t &= m+1,...,T, \end{aligned} \tag{3}$$

where $f(\cdot)$ is any machine learning method with the same hyperparameter and training process in both equations.

In the linear case, the loss function is convex, and a redundant variable will not increase the sum squared residual. However, the loss function of the machine learning model is not guaranteed to be convex. The contribution of causality may be weaker than the interference of the redundant dimension, making it hard to identify the existence of causality. We introduce the bootstrap method to tackle this issue: we shuffle the series $\mathbf{y}_{i,t-m:t-1}$ and include the shuffled series $\tilde{\mathbf{y}}_{i,t-k}, t = m+1,...,T$ in the R model, instead of removing y_i. This method eliminates the effect of dimension differences between the UR and R models.

We split the dataset for each of the UR and R models into a training set with 80% time points and a test set with 20%, such that the time of the test set is after that of the training set. In the training set, we firstly carry out *Augmented Dickey-Fuller* (ADF) test to check if the time series is stationary, which is a requirement of time series prediction models. The length of lag m is automatically chosen by the *Akaike Information Criterion* (AIC). Also, we use the *Partial Auto-Correlation Function* (PACF) test to determine the most

[1] Our code as well as supplementary materials are provided at https://github.com/cloudy-sfu/GC-significance-test.

suitable length of lag items. The restricted and unrestricted models are built and fitted on the processed dataset, using the same hyperparameters, structures, and training methods.

4.2 Module 2: Time Series Prediction

In this module, we use machine learning-based time series prediction models to fit UR and R models on the training set, and calculate the residuals on the test set. The models we use are SVM, an L2-regularized *Multi-Layer Perceptron* (MLP), and an L2-regularized *Long-Short Term Memory* (LSTM) neural network. We tune the hyperparameters for each model using 5-fold cross-validation and grid search. Please see our supplementary materials for details on the tuning and the model architectures.

When there are N time series, the total number of predictions is $N(N-1)$. We evaluate the accuracy of these predictions by counting the number of correct predictions when comparing the matrix $I[\mathbf{p} < \alpha]$ with the ground truth, normalized by the total number of predictions.

4.3 Module 3: Significance and Negotiation

With machine learning models, the causality statistic in Eq. 2 is not guaranteed to fit the F distribution. To find its null distribution, some research [12, 19] simulates the unrestricted model many times and estimates the empirical distribution of $\sum_{t=m+1}^{T} \hat{\epsilon}_{j,t}^2$. The limitation of the simulation method is computational cost as many machine learning models need to be trained, in proportion to the number of time series.

As an alternative, our causality statistic is the Wilcoxon signed-rank statistic

$$\mathcal{T}_{i,j} = \sum_{p=m+1}^{T} \left(\operatorname{sgn}(r_p) \sum_{q=m+1}^{T} I[r_p \leq r_q] \right), \tag{4}$$

where $\operatorname{sgn}(r_p)$ is the sign of $r_p = \hat{\epsilon}_{j,t}^2 - \hat{\eta}_{j,t}^2$. It measures whether the distribution of r is symmetric around 0. See Kolassa [13] for the distribution of $\mathcal{T}_{i,j}$.

Each machine learning model provides a p-value of the Wilcoxon signed-rank test. If there are k methods to identify the causality from y_i to y_j, we obtain p-values $p_1, ..., p_k$. We have to negotiate and find a joint p-value. This value should still be reliable when some methods fail and output inaccurate p-values, such that a hybrid system involving many machine learning models remains generalizable.

One approach to combine all p-values from y_i to y_j into one final output is *stacking* [25]. This method combines the individual values using a regression that can be trained via cross-validation. The disadvantage of stacking is that there is a risk that if one of the individual p-values is insignificant, the combined p-value will also be insignificant, which puts too much emphasis on a single result. The UR and R models are treated independently, and the difference between these two groups is not taken into account.

Another approach to finding a combined p-value is calculating a joint p-value by Fisher's method [4]. If the machine learning methods are independent, then

$$\chi_0 = -2 \sum_{i=1}^{k} \ln(p_i) \sim \chi^2(2k). \tag{5}$$

There is causality from y_i to y_j if $\chi_0 > \chi^2_{1-\alpha}(2k)$. However, we cannot guarantee that the machine learning models are independent, so the null distribution of χ_0 is no longer $\chi^2(2k)$.

Brown [5] and Kost [14] have proposed an extension of Fisher's method for dependent p-values. If the distribution of the individual p-values is known, then $c\chi_0 \sim \chi^2(2k)$, and c has a closed form. Unfortunately, we cannot make this assumption for p-values from machine learning models in general.

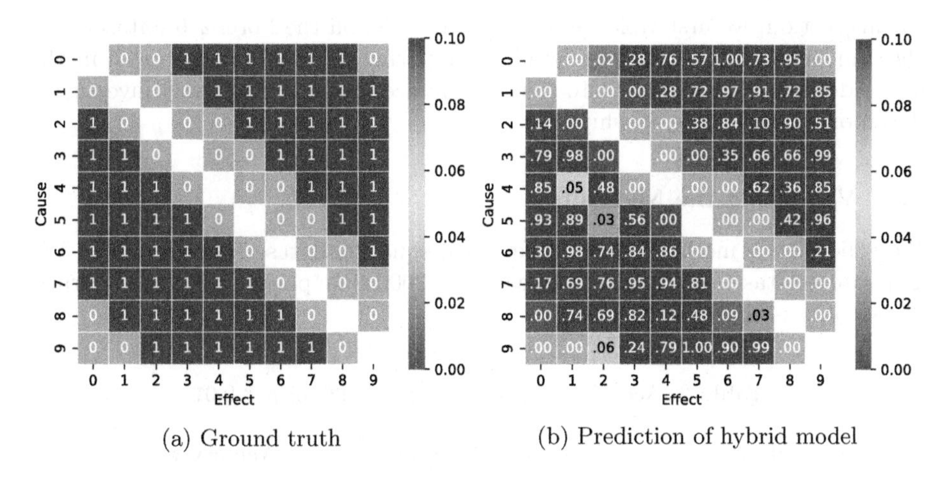

(a) Ground truth (b) Prediction of hybrid model

Fig. 2. Causality in the Lorenz96 system

To overcome this, we should provide a numerical estimation of c, the model independence coefficient. The causality network is entirely disjoint when $c = 0$ and is fully connected when $c \to +\infty$, but neither would provide meaningful results. The most appropriate c should be insensitive to the change of topology of the causality network. The node connectivity serves as a heuristic proxy to find the right balance. We search for that value of c that maximizes the slope of the corresponding node connectivity, as a jump in node connectivity indicates a stable topology.

Denote the node connectivity of a causality graph $\mathbf{P}$ as $\kappa(\mathbf{P})$, which represents the minimum edges removed from $\mathbf{P}$ to make it disconnected. Note that $\mathbf{P}$ depends on c and α, and could therefore be written as $\mathbf{P}(c, \alpha)$. We formally calculate c as

$$c^* = \max_c \frac{\partial \kappa(\mathbf{P}(c, \alpha))}{\partial c}. \tag{6}$$

The derivative is estimated numerically to obtain a smoother result [6].

After picking the value of c, we obtain a causality statistic and corresponding p-value per time series tuple (y_i, y_j). These p-values can be assembled into the matrix $\mathbf{p}$, which is the final output of our hybrid system.

5 Experiments and Results

In this section, we first validate our hybrid model on the Lorenz96 dataset (see the supplementary material for details), comparing its accuracy with other methods. Additionally, we apply it to the stock indices and quantitatively investigate its reported causal relationships.

5.1 Validating the Methods

We validate our method on the differential equation dataset Lorenz96 [11]. We generate a dataset of 10 series, and each has 2000 time points. The ground truth of causality is shown in Fig. 2a.

Table 2. Accuracy of models on Lorenz96 for $p < 0.01$

Name	**Accuracy**	**Name**	**Accuracy**
Linear Regression	0.711	SVM	0.900
LSTM	0.744	LSTM-L2	0.800
MLP	0.833	MLP-L2	0.844
cLSTM	0.967	**Hybrid (ours)**	**0.978**

We apply the linear autoregression, SVM, Tank et al.'s method [24], as well as MLP and LSTM with and without L2-regularization to the Lorenz96 dataset, calculating the accuracy score for each of the methods. For each individual machine learning model, the Wilcoxon signed-rank test results in a p-value for each pair of time series. For the hybrid system, we search for the best c and build the Fisher statistic $c\chi_0$, and its quantile in $\chi^2(2k)$ is the p-value. For the competitors, we follow the significance tests in the original papers and calculate the p-value. We conclude there is a causality if $p < 0.01$, and there is no causality otherwise. By comparing it with the ground truth, we obtain an accuracy score for each method, as shown in Table 2.

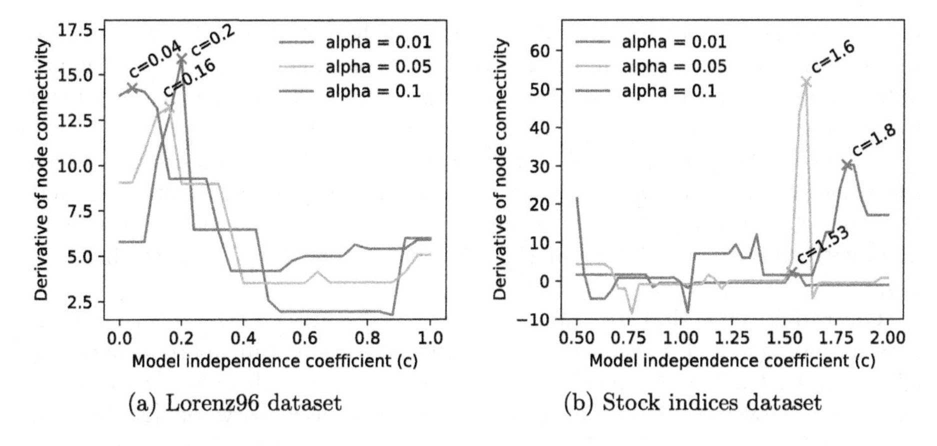

(a) Lorenz96 dataset (b) Stock indices dataset

Fig. 3. Picking c by node connectivity of causality network

The results show that unregularized individual models (SVM, LSTM, MLP) have an advantage in identifying Granger causality in nonlinear systems over the linear Granger causality model. L2-regularization increases the accuracy and reduces the number of false positive predictions. Moreover, the hybrid model beats any of its base models and exceeds the performance of the state-of-the-art cLSTM model, as shown in Fig. 2b. Though it does not outperform cLSTM a lot, it can figure out p-values while cLSTM cannot. The corresponding model independent coefficient is $c^* = 0.2$, as shown in Fig. 3a.

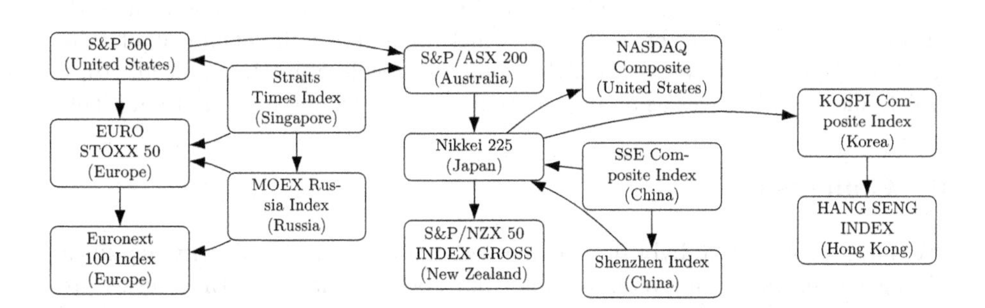

Fig. 4. Causal network of stock indices

5.2 Application of the Stock Indices

Our stock indices dataset[2] that we used in the experiments contains the properly adjusted closing price of stock indices from the start of 2016 to the end of 2019

[2] Source: https://finance.yahoo.com/world-indices, extracted using the yfinance PyPI package.

in 14 stock markets. The dataset skips worldwide holidays and fills other missing values with their previous value. Following the common practice in the finance industry, we use the logarithmic rate of return, standardizing the mean to 0 and the standard deviation to 1.

We apply our hybrid model to the stock indices. The relationship between the value of c and the derivative of $\kappa(\mathbf{P})$, along with the effect of confidence level, are shown in Fig. 3b. The optimal value is $c^* = 1.53$, which results in 17 pairs with causal relationships at the confidence level of $p \leq 0.01$, as shown in Fig. 4.

Our results agree that Australia and Japan act as agencies that transmit risks between markets [26]. Specifically, the United States and Singapore have causal effects on Australia and Japan, which then transmit such effects to New Zealand and Korea. However, we challenge the results from previous works [23,26] that claim the United States has the dominating influence while Asian markets have little to no influence.

Some of our results could be explained by the impact of the *financial sector*, which is a major contributor to systemic risk [26]. Therefore, the reason for the non-causality from NASDAQ to other markets could be that NASDAQ only has 5% of *financial sector* holdings while the S&P 500 has 18%. And our result of Singapore showing dominance with causal effects on many other countries could be because the Straits Times Index has over 40% holdings in the Banking sector.

The causal effect from China to Japan and through Japan to other markets such as NASDAQ could be explained by the escalating trade war between the United States and China which evolved into a tech war in late 2018, as both Japan and NASDAQ are heavily tech-weighted [16]. Looking within the Chinese markets, our result of casual effect from Shanghai to Shenzhen contradicts the linear result from Tang et al. [23] in which Shanghai is independent of other markets. However, we agree with the nonlinear result from Huang et al. [10] in which Shanghai behaves as an information source for Chinese domestic markets.

6 Conclusion

We have proposed a hybrid machine learning system to identify nonlinear Granger causality. Our system enables not specially designed time series prediction models to identify nonlinear causality while guaranteeing a robust and consistent result. By this means, different nonlinear Granger causality models can be integrated together, thus being generalized to different scenarios. As an example, we integrate SVM, LSTM-L2, and MLP-L2 models, and the significance of the causality statistic can be tested by Wilcoxon signed-rank test and negotiated by Fisher's method. Experiments on the Lorenz96 datasets indicate that our model performs better than existing ones.

Our hybrid model can be generalized to real-world stock indices, finding plausible results the previous model missed. Meanwhile, the significance of the causality statistic is still interpretable by probability-based hypothesis testing methods. By investigating the global stock indices from 2016–2019, we find the

United States stock market no longer dominates, and Singapore has a more significant influence on the global stock markets than expected. Also, Japan plays the role of connecting Asia, the Pacific Ocean, and the United States. It is also impacted by the Shanghai market, which we believe is regulated and not in free circulation.

Future work is expected to be conducted in three aspects. First, we plan to investigate further individual models in our ensemble and select those that integrate best into our framework. Second, we expect to extend the weights in neural networks to a function dependent on time, tackling the time-varying causality. Third, we aim to find a theoretically supported rather than heuristic closed form for c.

References

1. Al-Yahyaee, K.H., Mensi, W., Al-Jarrah, I.M.W., Tiwari, A.K.: Testing for the Granger-causality between returns in the US and GIPSI stock markets. Physica A **531**, 120950 (2019)
2. Balcilar, M., Elsayed, A., Hammoudeh, S.: Financial connectedness and risk transmission among MENA countries: Evidence from connectedness network and clustering analysis. J. Int. Financial Markets, Inst. Money, p. 101656 (2022)
3. Baumohl, E., Vyrost, T.: Stock market integration: granger causality testing with respect to nonsynchronous trading effects. Finance a Uver: Czech J. Econ. Finance **61**(1) (2011)
4. Birnbaum, A.: Combining independent tests of significance. J. Am. Stat. Assoc. **49**(267), 559–574 (1954)
5. Brown, M.B.: 400: A method for combining non-independent, one-sided tests of significance. Biometrics, pp. 987–992 (1975)
6. Chartrand, R.: Numerical differentiation of noisy, nonsmooth data. International Scholarly Research Notices 2011 (2011)
7. Choi, S.Y.: Credit risk interdependence in global financial markets: evidence from three regions using multiple and partial wavelet approaches. J. Int. Finan. Markets. Inst. Money **80**, 101636 (2022)
8. Gao, W., Yang, H.: Time-varying group lasso Granger causality graph for high dimensional dynamic system. Pattern Recognition, p. 108789 (2022)
9. Granger, C.W.: Investigating causal relations by econometric models and cross-spectral methods. Econometrica: J. Econometric Soc., 424–438 (1969)
10. Huang, W., Lai, P.C., Bessler, D.A.: On the changing structure among Chinese equity markets: Hong Kong, Shanghai, and Shenzhen. Eur. J. Oper. Res. **264**(3), 1020–1032 (2018)
11. Karimi, A., Paul, M.R.: Extensive chaos in the Lorenz-96 model. Chaos Interdisciplinary J. Nonlinear Sci. **20**(4), 043105 (2010)
12. Ko, H.H., Ogaki, M.: Granger causality from exchange rates to fundamentals: what does the bootstrap test show us? Int. Rev. Econ. Finance **38**, 198–206 (2015)
13. Kolassa, J.E.: Edgeworth approximations for rank sum test statistics. Stat. Probability Lett. **24**(2), 169–171 (1995)
14. Kost, J.T., McDermott, M.P.: Combining dependent p-values. Stat. Probability Lett. **60**(2), 183–190 (2002)
15. Koumou, G.B.: Diversification and portfolio theory: a review. Fin. Markets. Portfolio Mgmt. **34**(3), 267–312 (2020). https://doi.org/10.1007/s11408-020-00352-6

16. Kwan, C.H.: The China-US trade war: deep-rooted causes, shifting focus and uncertain prospects. Asian Econ. Policy Rev. **15**(1), 55–72 (2020)
17. Leng, S., Xu, Z., Ma, H.: Reconstructing directional causal networks with random forest: Causality meeting machine learning. Chaos: Interdisciplinary J. Nonlinear Sci. **29**(9), 093130 (2019)
18. Marinazzo, D., Pellicoro, M., Stramaglia, S.: Kernel method for nonlinear Granger causality. Phys. Rev. Lett. **100**(14), 144103 (2008)
19. Marques, A.M., Lima, G.T.: Testing for Granger causality in quantiles between the wage share in income and productive capacity utilization. Structural Change and Economic Dynamics (2022)
20. McLemore, P., Mihov, A., Sanz, L.: Global banks and systemic risk: the dark side of country financial connectedness. J. Int. Money Financ. **129**, 102734 (2022)
21. Rosoł, M., Młyńczak, M., Cybulski, G.: Granger causality test with nonlinear neural-network-based methods: Python package and simulation study. Comput. Methods Programs Biomed. **216**, 106669 (2022)
22. Shojaie, A., Fox, E.B.: Granger causality: a review and recent advances. Annual Rev. Stat. Appl. **9**, 289–319 (2022)
23. Tang, Y., Xiong, J.J., Luo, Y., Zhang, Y.C.: How do the global stock markets influence one another? evidence from finance big data and Granger causality directed network. Int. J. Electron. Commer. **23**(1), 85–109 (2019)
24. Tank, A., Covert, I., Foti, N., Shojaie, A., Fox, E.B.: Neural Granger causality. IEEE Trans. Pattern Anal. Mach. Intell. **44**(8), 4267–4279 (2021)
25. Wolpert, D.H.: Stacked generalization. Neural Networks **5**(2), 241–259 (1992)
26. Zheng, Q., Song, L.: Dynamic contagion of systemic risks on global main equity markets based on Granger causality networks. Discrete Dynamics in Nature and Society 2018 (2018)

Reinforcement Learning

A Dynamic and Task-Independent Reward Shaping Approach for Discrete Partially Observable Markov Decision Processes

Sepideh Nahali[1(✉)], Hajer Ayadi[1], Jimmy X. Huang[1], Esmat Pakizeh[3], Mir Mohsen Pedram[3], and Leila Safari[2]

[1] Information Retrieval and Knowledge Management Research Lab, York University, Toronto, Canada
{sepidnah,hajayal,jhuang}@yorku.ca
[2] University of Zanjan, Zanjan, Iran
[3] Kharazmi University, Tehran, Iran

Abstract. Agents often need a long time to explore state-action space in order to learn how to act expectedly in Partially Observable Markov Decision Processes (POMDPs). With the reward shaping method, real-time POMDP planning can be guided both in terms of reliability and speed. In this paper, we propose Low Dimensional Policy Graph (LDPG), a new reward shaping method for reducing the dimension of the value function to extract the best state-action pairs. The reward function is then shaped using these key pairs. For accelerating learning speed, we analyze the Transition Function graph to discover significant paths to the learning agent's goal. Direct comparison on five standard testbeds indicates LDPG brings about the deterministic finding of optimal actions faster regardless of the task type. Our method is shown to reach the goals more quickly (by 41.48 % improvement) and performed 61.57 % better in receiving rewards in the $4 \times 5 \times 2$ domain.

Keywords: Dynamic reward shaping · Markov decision making · Planning · Dimension reduction · Reinforcement Learning

1 Introduction

Reward shaping in reinforcement learning accelerates the discovery of optimal solutions to complex problems by adding a supplementary reward signal to the environment reward [25]. This is especially useful in complex and uncertain environments like POMDPs. However, it can be challenging to apply reward shaping in POMDPs due to the need for expert knowledge and the importance of defining the appropriate reward function. An incorrect reward function can result in sub-optimal behavior, particularly in online tasks with limited planning time. This paper addresses the limitations of existing reward shaping approaches for discrete POMDPs by using our Low Dimensional Policy Graph (LDPG) method. To reduce POMDP complexity, we applied Isometric Feature Mapping (ISOMAP),

H. Kashima et al. (Eds.): PAKDD 2023, LNAI 13936, pp. 337–348, 2023.
https://doi.org/10.1007/978-3-031-33377-4_26

an advanced method of dimension reduction [24], on value function vectors while iterating through the horizons (the number of time steps considered in solving the problem). ISOMAP preserves the relation between state-action values by considering the point's neighbors. It extends MDS to maintain distance proportions between states in a low-dimensional space, and its output is the value function vectors of state-actions for the optimal solution. During each iteration of the value iteration process, the LDPG method dynamically identifies the sub-goals. Next, LDPG rewards states located on the path of the sub-goal (sub-paths) to induce the agent to follow sub-path. Through this strategy, the algorithm convergence is sped up, and the final Expected Reward (ER) and Average Cumulative Reward (ACR) are increased. To the best of our knowledge, there are no dynamic reward shaping methods for POMDPs without human supervision.

Our contributions to this work are as follows. First, we propose and deploy the first dynamic reward shaping approach for discrete POMDPs. Second, our method can extend to the POMDP problems that are solvable by a point-based value iteration algorithm. Third, we experimentally demonstrate the utility of our approach, employing it for complex domains (larger state space). Fourth, we conducted experimental results on five POMDP tasks to demonstrate the effectiveness of LDPG in terms of speed of ACR, final ER, and convergence speed.

2 Background and Notation

Markov decision process (MDP) is used for sequential decision-making in observable environments with uncertain system dynamics [17]. **Partially Observable Markov decision process (POMDP)** is similar to MDP, but it deals with unfamiliar and partially observable environments [29]. In POMDP, a set of observations is added to the model to indirectly provide information about the states to the agent. Observations are represented with a probability function indicating the likelihood of each observation for each state in the model.

A POMDP is formally defined by a tuple $< S, A, T, R, Z, O, \gamma >$. In this tuple, S is a finite set of latent states s, A is the set of action a, $T(s_0|s,a)$ is the transition probability function, $R(s,a) \in [0,1]$ is the reward function which is a real-time reward that is received when the agent performs action a in state s, Z is a set of observations, $O(z|s_0,a)$ is the observation probability function, and $\gamma \in [0,1]$ is a discount factor that gives a lower weight to further rewards in future.

By introducing additional rewards (F) to the main reward function (R), reward shaping can accelerate the planning process for POMDPs. This approach is particularly beneficial as it provides valuable and informative feedback to agents based on previous knowledge, which can influence the optimal policy. However, if the reward function is not appropriate, it can mislead the agent and cause divergence [21]. The concept of **Potential-based reward shaping (PBRS)** function was introduced to address this problem. PBRS is a technique designed to ensure that a policy remains optimal during the learning process. This is accomplished through the use of a potential function ϕ, which maps states to real numbers [19]. PBRS assumes shaping rewards are not defined for all goal states since no actions are needed in those states. But if there are multiple goals and the potential function of the goal states are not zero, it can impact the optimal policy [9].

3 Related Work

A diverse range of tasks have been subject to experimentation and development of reward shaping techniques. Some of these methods use deep learning [11] and generative models for shaping state-action potential reward functions statically [26] and some of them designed a probabilistic learning paradigm to learn reward functions for RL problems [28]. These methods are designed for MDPs in discrete domains and require human supervision [20]. However, Dong et al. proposed a static reward shaping method for MDPs, performing well in continuous and discrete spaces [6]. The existing reward shaping methods for POMDPs are very limited. These limited number of reward shaping methods are either static [7] or use human demonstration [1]. In some cases, the solutions hardly depend on a specific task and cannot generalize to new situations [18]. Information Particle Filter Tree (IPFT) is a dynamic method designed for continuous $\rho POMDP$ that utilizes Monte Carlo Tree Search [8]. Also, a reward shaping method is designed for DQN, an RNN-based reinforcement learning (RL) method in POMDPs [27]. This reward shaping method is developed to deal with limited training data, and the issue of the sparse reward of DQNs [12]. It is worth mentioning that sparse rewards in long-horizon tasks can lead to low efficiency in the RL problem [10]. On the other hand, existing deep learning approaches for reward shaping rely on RNNs, which often cause sub-optimality in complex tasks (i.e., high dimensional and continuous environments). Besides, RNNs are prone to vanishing gradients for long paths and have slow training procedures. Moreover, utilizing deep learning models for shaping in POMDPs requires large training patterns which makes RNN-based reward shaping a data-extensive and time-consuming process [9]. The referenced works indicate the benefits of utilizing a reward shaping approach for POMDPs to addresses the above limitations.

4 LDPG: Our Proposed Reward Shaping Method

The LDPG method is an online algorithm for POMDPs based on detecting the paths leading to the sub-goals (sub-paths). To discover an appropriate path, the agent uses the value function vectors as we iterate through the horizons to mark specific states as high-priority states. Hence, the agent's focus becomes more concentrated on finding the optimal path, and it drew to this sub-path in the future, resulting in the best route being found. To be specific, in LDPG, the agent looks for the sub-goals at each iteration of the value iteration process. Then, if the sub-goals were experienced before, a subset of state-action pairs to reach the sub-goals are extracted. This subset of state-action pairs includes states that have appeared more frequently on the route to reach the sub-goals. A reward equivalent to the value of the extracted states is then added to their immediate reward. The augmented reward allows the agent to reach the goal in a shorter time by avoiding unnecessary states while preserving the policy's optimality. When there are no sub-goals or the agent hasn't experienced any sub-goals, a standard RL algorithm will be used. In the initial steps of the algorithm, the agent does not have experience, so it learns without reward shaping and uses the original rewards received from the environment. In later steps, the reward function shapes dynamically as the agent's history expands. Furthermore, the state space is not changed because we do not attempt to limit it for the agent. As a result, the agent can still visit disregarded states and will not behave biased.

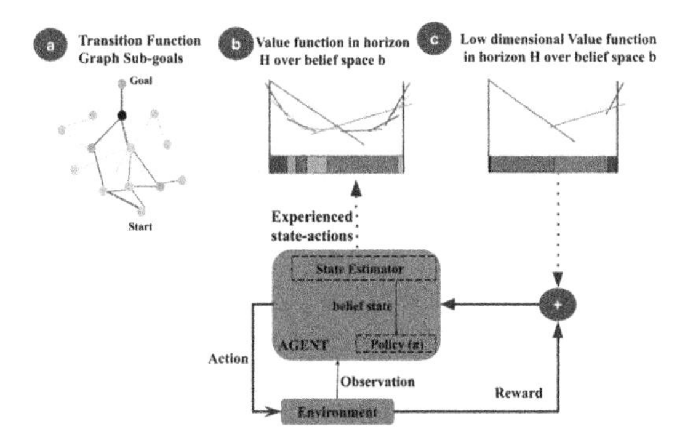

Fig. 1. An overview of our LDPG method

Figure 1 shows the procedure of the LDPG. Each color region in the value function represents all the belief states where the action with the same color is the best strategy to use. This belief state refers to a probability distribution over the underlying states of the system, which the agent updates based on the observations and actions it takes. It is important to note that each color in parts b and c correspond to the same colored line in the value function and not all of the transformed lines are useful for presenting the maximal value. The following steps describe each step in detail.

Step 1: Compute Betweenness Centrality. We represent the Transition Function Graph (Policy Graphs) of actions and the states with a directed graph. The node of the policy graph determines which action to take. Next, the observation received is used to find the next node. The edges in the policy graph indicate the previous vectors used to construct the current one. By using the policy graph, which is actually a finite state controller, we can execute the optimal policy instead of tracking the belief state [14]. To create a policy graph for infinite horizon POMDP domains, value iteration can be used by computing a sequence of value functions incrementally. The relative importance of nodes can be measured using vertex centrality [4]. Here, for finding the sub-goals of each goal from a graph, we use the Betweenness Centrality of the node u, denoted by $BC(u)$, which is defined as the frequency that a node lies on the shortest path connecting two distinct nodes. The BC is specified by the Eq. 1.

$$BC(u) = \sum_{u \neq s \neq t \in V} \frac{\sigma_{st}(u)}{\sigma_{st}} \tag{1}$$

where σ_{st} is the total number of shortest paths from node s to node t and $\sigma_{st}(u)$ is the number of those paths that pass through u [2]. The node with the greatest BC is the sub-goal of G.

Step 2: Search for Value Function Vectors of Sub-goals. Once sub-goals are identified, the agent checks if it has encountered them before by inspecting its history for their corresponding value function vectors. Value function vectors represent the value function, considered piecewise linear and convex (PWLC). For each action, there is a vector comprised of the coefficients of a hyperplane. These hyperplanes represent one side of the value function and pass through the origin. The length of the vectors is equal to the number of states in the POMDP [22]. The value function for an environment with k actions and N states is a set of N-dimensional alpha vectors, where each row corresponds to an action and each column corresponds to a state. The best action for each belief state, given the PWLC value function, is the one with the highest dot product between its alpha vector coefficients and the belief state probabilities.

Step 3: Apply ISOMAP on Value Function Vectors of Sub-goals. The value function vectors of experienced sub-goals are then fed to the ISOMAP to find the "best" state-action pair for reaching the sub-goals. By reducing the dimension of the value function vectors we make the value of actions zero for some states. Hence, it is easier and faster to calculate the value of each action over a belief state. In other words, these pairs have the most influential role in achieving the sub-goals and discovering strongly connected regions in the policy graph.

Step 4: Assign Rewards to Value Function Vectors' States. To encourage the agent to pass through the experienced successful states extracted in Step 3 and appear in the path of the sub-goal, the values of these states are used from the output vectors of the value function. Then these values are added to the original rewards of these states with a discount factor γ. Specifically, we set the reward shaping function as specified by Eq. 2.

$$\phi(S_i) = V(S_i) \tag{2}$$

where V is the value of state S_i, and $\phi(S_t) = 0$, where $S_t \in$ goal and sub-goal states and all other states which have not appeared in the low dimensional value function vectors of the sub-goals. Therefore, the agent tends to pass through experienced successful states to get the specified rewards. We use potential-based reward shaping by learning a potential function ϕ that shapes the reward for each state. The potential function is set to zero for terminal state s_T to ensure policy invariance in finite-horizon environments. However, a terminal state in one path may be non-terminal in another path, in which case the potential of that state will be used. Setting $\phi(s_T) = 0$ balances positive and negative shaping rewards to maintain optimality of the original policy by compensating for any potentials accumulated before visiting s_T.

Algorithm 1. Computing an optimal policy using LDPG

1: Transition function $\psi(b, a, b')$, Value function in iteration t V_t, policy ϕ, policy graph G, α-vectors of state s Ψ_s, # Actions A, Belief state b, Terminal States Tr
2: **Begin**
3: $S \leftarrow \{S_0, .., S_k\}$, $t := 1$, $V_1(s) := 0 \forall s$
4: **while** $sup_b|v_t(b) - v_{t-1}(b)| < \epsilon$ **do**
5: $\quad t := t + l$
6: $\quad$ **for** $a \in A$ **do**
7: $\quad\quad G =: ConstructPolicyGraph(\phi, \psi(b, a, b'))$
8: $\quad\quad$ BC $= \sum_{u \neq s \neq t \in A} \frac{\rho_{s\,t}(u)}{\rho_{s\,t}}$
9: $\quad\quad SubGoals \leftarrow \{SUB_0, .., SUB_p\}$
10: $\quad\quad$ for all $sub \in SubGoals$
11: $\quad\quad$ **if** $\Psi_{sub_b} \in V_t$ **then**
12: $\quad\quad\quad \Psi_{sub_b}^{D \times K} =$ ISOMAP$(\Psi_{sub_b}^{A \times K})$
13: $\quad\quad\quad$ **for** $S_i \mid \Psi_{Si} \in \Psi_{sub_b}^{D \times K}$ & $S_i \notin Tr$ **do** ▷ For all possible actions in state
14: $\quad\quad\quad\quad Reward(S_i) := |V_t(S_i)|$
15: $\quad\quad\quad\quad R := R(S_i, a) + Reward(S_i)$
16: $\quad\quad\quad$ **end for**
17: $\quad\quad$ **end if**
18: $\quad\quad Q_t^a := IncrementalPrunning(V_{t-1}, a, R)$ ▷ value of starting in state s in t
19: $\quad$ **end for**
20: $\quad$ prune $\cup_a Q_t^a$ to get V_t
21: **end while**
22: **End**

The pseudo-code in Algorithm 1 outlines our reward shaping approach. At iteration t, the algorithm has the optimal t-step value function. Separate Q-functions for each action are represented by sets of policy graphs within the value-iteration loop, and are obtained by calling the Incremental Pruning algorithm using the previous iteration's value function. The union of these sets forms the optimal value function, which may contain extraneous policy trees that are pruned to yield the useful set of r-step policy trees V_t. To demonstrate the LDPG algorithm's convergence towards global optimality, we must establish that it satisfies the conditions of the Policy Iteration Theorem [23]. This theorem guarantees convergence to a globally optimal policy. In order to confirm that the LDPG Algorithm satisfies these conditions, we must demonstrate that it repeatedly applies the policy improvement and policy evaluation operators [13]. The LDPG Algorithm applies the policy improvement operator in the for-loop beginning at line 6, where it constructs a policy graph using the transition

function for each action, and then prunes the policy graph using incremental pruning to obtain optimal Q value for each action. The LDPG Algorithm then applies the policy evaluation operator in the for-loop beginning at line 9, where it computes the reward for each state using the current value function and the updated value function obtained from incremental pruning. The LDPG Algorithm then prunes the value function using incremental pruning to obtain the updated value function. This process of iteratively applying the policy improvement and policy evaluation operators continues until the stopping criterion at line 4 is met. Since the LDPG Algorithm satisfies the conditions of the Policy Iteration Theorem, we can conclude that the sequence of policies generated by the LDPG Algorithm will converge to a unique optimal policy, which achieves global optimality.
To prove that LDPG uses a potential-based reward function, we need to show that the reward shaping function used in the algorithm depends only on the potential function. Looking at the algorithm, the reward shaping function is defined as Eq. 2. Since the value function is a measure of the expected return from that state on-wards, this reward shaping function is encouraging the agent to move towards states with higher value functions. Now, if we consider the Bellman equation for the value function $V(s)$ as specified in Eq. 3.

$$V(s) = \max_{a \in A} \sum_{s' \in S} \psi(s, a, s')[R(s, a, s') + \gamma V(s')] \tag{3}$$

We can see that the value function is a function of the potential function, since the transition function ψ and the reward function R are both functions of the potential function. Therefore, since the reward shaping function used in LDPG is a function of the value function, we can conclude that LDPG uses a potential-based reward function.

The algorithm for convergence involves constructing a policy graph, calculating betweenness centrality, ISOMAP dimension reduction, computing rewards, and pruning the value function, with a time complexity of $O(t*S^2*A+K^2 \log K + t*S*E+t*S^2)$. The formula involves t iterations, S states, A actions, and K nearest neighbors in ISOMAP. The algorithm's space complexity is $O(|S|)$.

5 Experimental Results

To evaluate the LDPG, experiments were conducted on five goal-conditioned testbeds: Hallway, Shuttle, ALOHA, BULKHEAD, and $4 \times 5 \times 2$ [14]. In goal-conditioned environments, the agent is responsible for achieving specific goals or sets of goals [5]. Using a goal-conditioned approach provides an advantage, as it allows the agent to learn a more general policy that can be applied to various goal states, rather than being limited to a specific set of states [3,16].

In the routing problem of navigating the **Hallway**, a robot equipped with sensors moves through an office. The sensors may produce inaccurate readings with probabilities, leading to errors. The robot can detect only four things: walls, doors, open space, and undetermined areas. If the robot mistakenly identifies its location as the goal, it receives a penalty of minus one. The **Shuttle** is a simulated docking task with random actions and imprecise sensors. The objective is to move supplies between two space stations, with a bonus of +10 given for each successful action at the station with fewer visits. The agent incurs a penalty of −3 for colliding with the station and no bonus otherwise. The agent can disconnect, slow down, throw, approach, or collide depending on the state. Backtracking has a 0.3 probability of throwing the agent into space and an 0.8 probability of approaching a station from space. Lastly, the agent fails to load 30% of the time.

ALOHA is a packet-switched network that shares a channel between packets to increase efficiency, but only one packet can be transmitted at a time. The probability of a packet waiting to be sent is $a = 1/S_b$, where S_b is the number of accumulated packets. The set of actions is defined as $A = \{S_b | S_b = 1, 2, \ldots, M\}$, where M is the maximum number of accumulated packets allowed. The reward is zero at the highest number of packets and +1 for every number less than the highest. In **BULKHEAD**, a titanium aircraft engine's chipping and blade inspection process is modeled. In each step, one of two processing or inspection techniques is selected. Agent observations are processing actions or inspections. Process actions observations include information on whether something has caused a part to be stressed. These events are also divided into three states. The transition states in this environment are defined according to a uniform probability function by applying inspection. **4 × 5 × 2** is a maze world problem that consists of two floors, each with a 4×5 maze. The agent intends to move towards the goal placed in one of the states. In this problem, observations include nothing if the agent didn't observe the target in the current state, landmark-level-lower if in upper floor states, landmark-level-upper if the agent is on the lower floor, and goal when in the goal position. Table 1 offers supplementary details regarding these environments. Specifically, multiple sub-goals listed in the table share the same rank and had the highest BC factor.

The simulations were conducted on a computer with an Intel Core i7 CPU and 8GB of memory. POMDP-solve, which uses the Perseus algorithm and dynamic programming with backward recursion, was used to develop the method [22]. ISOMAP was used with two coordinates for the manifold and one neighbor for each point. Finally, for the eigenvalue decomposition, the 'auto' or 'dense' modes were used depending on the Transition Function graph for each domain.

Table 1. The details of the POMDP domains used in Experiment 1.

Name	States	Actions	obs	Sub-goal states	Goal states
$4 \times 5 \times 2$	39	4	4	'38'	'0'
Hallway	60	5	21	'56' to '59'	'52' to '55'
ALOHA	30	9	3	'0' to '26'	'29'
Shuttle	8	3	5	'0'	'3'
BULKHEAD	10	6	6	'6'	'7'

5.1 Results

This section compares POMDP agent performance across environments using three standard criteria: Average Cumulative Reward (ACR), Expected Reward (ER), and speed of convergence. A higher ACR indicates accurate learning, a higher expected reward indicates success in achieving goals [15], and faster convergence indicates passing more states towards the goal in 1000 stages. Each experiment was performed five times and results were averaged. The Q function was initialized with a value of zero. The discount factor γ is 0.95 ± 0.5. The algorithm stops if the execution time is more than 1000 time units. Also, if the absolute difference between the two values is less than 10^{-7}, the algorithm finishes. Experiment 1 compares the performance of the POMDP agent in environments with standard rewards and environments with reward shaping (LDPG method). Figure 2 depicts ACR results for standard RL and LDPG methods. The bar chart indicates improvement percentage, based on the final value of the ACR for each domain. Results are summarized in Table 2. The results from experiment 1 establish the potential for using LDPG method.

Experiment 2 examines how complexity and size of the environment affect the performance of LDPG. For this aim, the number of states of environments has been increased to 10 times while maintaining the same goals. This expansion has also led to the development of the transfer function. However, the observation space is still kept as in the first experiment. Table 3 shows the results of this experiment. Figure 3 shows the ACR in extended environments for the standard RL and LDPG method for this experiment. In both experiments, BULKHEAD is the most complex problem in terms of complexity as the agent requires more time to achieve positive rewards. It is also confirmed by the steep slope of the ACR in Fig. 2 and Fig. 3. As the number of states in all of the environments (except $4 \times 5 \times 2$) is increased in Experiment 2, LDPG improves less than Experiment 1, but is still superior to standard RL without reward shaping. Tables 2 and 3 show that both experiments were satisfactory in all environments since values related to both criteria were improved. As LDPG is based on the paths leading to the sub-goals of each goal, the results for the domains with a more robust set of sub-goals are expected to be more satisfactory. According to the results in Table 3, LDPG still performs well when the size of the environments increases to 10 times wider. Moreover, with Wilcoxon signed-rank test, performance differences in speed and expected reward are considered statistically significant over learning without shaping when the p-value < 0.05.

Table 2. LDPG performance in speed and reward across different environments.

	Speed (LDPG)	Speed (no shaping)	Improved (%)	ER (LDPG)	ER (no shaping)	Improved (%)
$4 \times 5 \times 2$	135	79	41.48	16.11	6.19	61.57
HALLWAY	52	48	7.69	14.48	9.29	35.84
SHUTTLE	271	269	0.73	18.66	9.517	48.99
BULKHEAD	17194	15768	8.29	341933197.5	242344682 .2	29.12
ALOHA	1819	1599	12.09	828.87	786.446	5.11

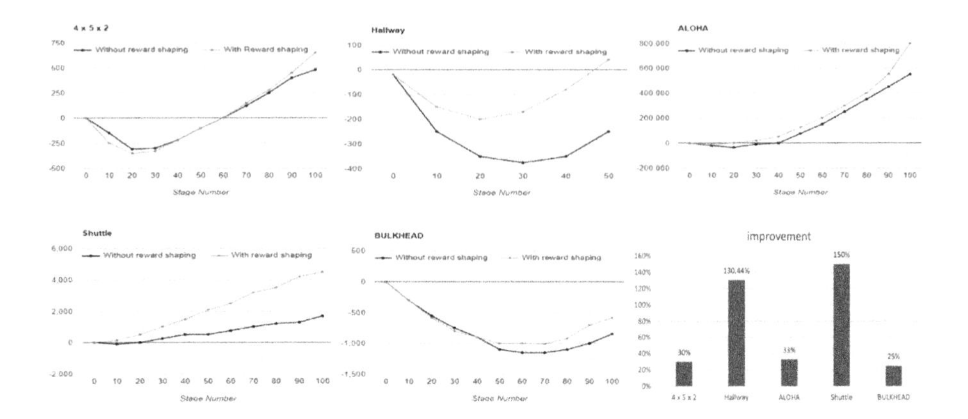

Fig. 2. LDPG performance in ACR across different environments.

Table 3. LDPG performance in speed and reward across different environments with increased size.

	Speed (LDPG)	Speed (no shaping)	Improved (%)	ER (LDPG)	ER (no shaping)	Improved (%)
$4 \times 5 \times 2$	103	76	26.21	16.07	6.19	86.12
HALLWAY	37.39	35	6.83	3113.21	9.29	34.24
SHUTTLE	271	269	0.7	18.66	9.517	48.98
BULKHEAD	3543	2842	19.78	144183608.0	114283101.3	2.73
ALOHA	2497	2269	9.13	890.390	885.442	0.55

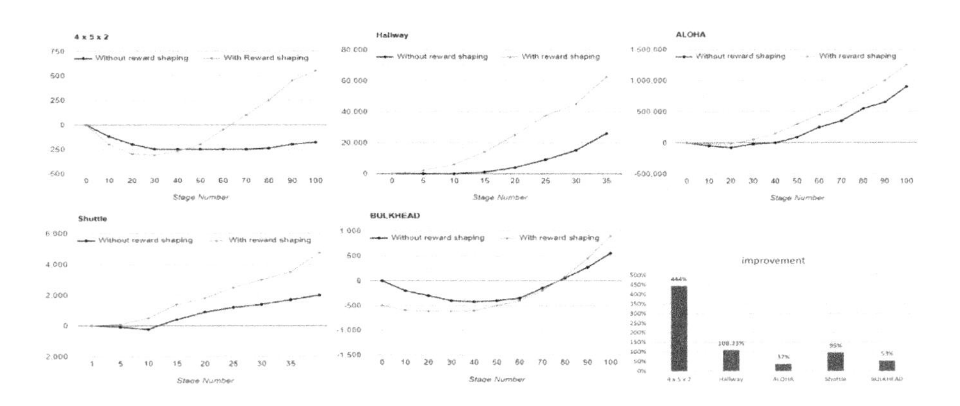

Fig. 3. LDPG performance in ACR across environments with increased size.

6 Conclusion and Future Work

Planning in POMDPs is challenging in problems where the immediate reward is obtained after long action sequences. In this work, we are the first to propose a dynamic and task-independent reward shaping method for discrete POMDPs. We use a non-linear dimension reduction method on the Transition Function to help the agent decide between a set of pruned state-actions and choose the most optimal in a complex task. LDPG dynamically shapes the reward function during the value iteration algorithm by considering the distance between states. Our evaluation reveals that breaking down the task into shorter overlapping sub-paths of the main goal's path is highly promising for POMDPs, and LDPG improves performance measures. In our future research, we will examine the impact of the LDPG on continuous domains. Additionally, using neural networks is another direction for future research.

Acknowledgements. This research is supported by the research grants from Natural Sciences and Engineering Research Council (NSERC) of Canada. We thank four anonymous reviewers for their thorough review comments on this paper.

References

1. Abbeel, P., Ng, A.Y.: Exploration and apprenticeship learning in reinforcement learning. In: Proceedings of the 22nd International Conference on ML, ICML 2005, pp. 1–8, New York, NY, USA (2005)
2. Brandes, U.: A faster algorithm for betweenness centrality. J. Math. Sociol. **25**(2), 163–177 (2001)
3. Chane-Sane, E., Schmid, C., Laptev, I.: Goal-conditioned reinforcement learning with imagined subgoals. In: Meila, M., Zhang, T. (eds.) Proceedings of the 38th International Conference on Machine Learning, ICML 2021. Proceedings of Machine Learning Research, vol. 139, pp. 1430–1440. PMLR (2021)
4. Chehreghani, M.H., Bifet, A., Abdessalem, T.: Efficient exact and approximate algorithms for computing betweenness centrality in directed graphs. In: Advances in Knowledge Discovery and Data Mining, PAKDD, pp. 752–764 (2018)
5. Colas, C., Karch, T., Sigaud, O., Oudeyer, P.: Autotelic agents with intrinsically motivated goal-conditioned reinforcement learning: a short survey. J. Artif. Intell. Res. **74**, 1159–1199 (2022)
6. Dong, Y., Tang, X., Yuan, Y.: Principled reward shaping for reinforcement learning via lyapunov stability theory. Neurocomputing **393**, 83–90 (2020)
7. Eck, A., Soh, L.K., Devlin, S., Kudenko, D.: Potential-based reward shaping for finite horizon online pomdp planning. Auton. Agent. Multi-Agent Syst. **30**, 403–445 (2015)
8. Fischer, J., Ömer Sahin Tas: Information particle filter tree: an online algorithm for pomdps with belief-based rewards on continuous domains. In: Proceedings of the 37th International Conference on ML, ICML. vol. 119, pp. 3177–3187 (2020)
9. Grzeundefined, M.: Reward shaping in episodic reinforcement learning. In: Proceedings of the 16th Conference on Autonomous Agents and MultiAgent Systems, AAMAS 2017, pp. 565–573. International Foundation for Autonomous Agents and Multiagent Systems, Richland, SC (2017)
10. Guo, Y., Wu, Q., Honglak, L.: Learning action translator for meta reinforcement learning on sparse-reward tasks. In: Proceedings of the AAAI Conference on Artificial Intelligence, vol. 36(6), pp. 6792–6800 (2022)

11. Hussein, A., Elyan, E., Gaber, M.M., Jayne, C.: Deep reward shaping from demonstrations. In: International Joint Conference on Neural Networks (IJCNN), pp. 510–517 (2017)
12. Hausknecht, M., Stone, P.: Deep recurrent q-learning for partially observable mdps. In: AAAI Fall Symposium on Sequential Decision Making for Intelligent Agents (AAAI-SDMIA15) (2015)
13. Howard, R.A.: Dynamic programming and Markov processes. MIT Press (1960)
14. Kaelbling, L.P., Cassandra, A.R.: Exact and approximate algorithms for partially observable Markov decision processes. In: Proceedings of the 13th Conference on Uncertainty in Artificial Intelligence, pp. 374–381. Morgan Kaufmann Publishers Inc. (1998)
15. Kalra, B., Munnangi, S.K., Majmundar, K., Manwani, N., Paruchuri, P.: Cooperative monitoring of malicious activity in stock exchanges. In: Trends and Applications in Knowledge Discovery and Data Mining. PAKDD, pp. 121–132 (2021)
16. Kim, J., Seo, Y., Shin, J.: Landmark-guided subgoal generation in hierarchical reinforcement learning. In: Advances in Neural Information Processing Systems, vol. 34, pp. 28336–28349. Curran Associates, Inc. (2021)
17. Liu, S., Krishnan, R., Brunskill, E., Ni, L.M.: Modeling social information learning among taxi drivers. In: Advances in Knowledge Discovery and Data Mining, PAKDD, pp. 73–84. Berlin (2013)
18. Mafi, N., Abtahi, F., Fasel, I.: Information theoretic reward shaping for curiosity driven learning in pomdps. In: Proceedings of the 2011 IEEE International Conference on Development and Learning (ICDL), vol. 2, pp. 1–7 (2011)
19. Ng, A.Y., Harada, D., Russell, S.: Policy invariance under reward transformations: theory and application to reward shaping. In: In Proceedings of the Sixteenth International Conference on ML, pp. 278–287. Morgan Kaufmann (1999)
20. Nourozzadeh: Shaping Methods to Accelerate Reinforcement Learning: From Easy to Challenging Tasks. Master's thesis, Delft University of Technology (2010)
21. Snel, M., Whiteson, S.: Multi-task reinforcement learning: Shaping and feature selection. In: Proceedings of the 9th European Conference on Recent Advances in Reinforcement Learning, EWRL 2011, pp. 237–248. Springer, Berlin (2011)
22. Spaan, M.T.J., Vlassis, N.: Perseus: randomized point-based value iteration for pomdps. J. Artif. Int. Res. **24**(1), 195–220 (2005)
23. Sutton, R.S., Barto, A.G.: Reinforcement learning: An introduction. MIT press (2018)
24. Tenenbaum, J.B., Silva, V.d., Langford, J.C.: A global geometric framework for nonlinear dimensionality reduction. Science **290**(5500), 2319–2323 (2000)
25. Wang, P., Fan, Y., Xia, L., Zhao, W.X., Niu, S., Huang, J.X.: KERL: a knowledge-guided reinforcement learning model for sequential recommendation. In: Proceedings of the 43rd International ACM SIGIR conference on research and development in Information Retrieval SIGIR, China, pp. 209–218 (2020)
26. Yuchen Wu, M.M., Shkurti, F.: Shaping rewards for reinforcement learning with imperfect demonstrations using generative models. In: 2021 IEEE International Conference on Robotics and Automation (ICRA), pp. 6628–6634 (2020)
27. Zhanhong J., Michael J. Risbeck, V.R.S.M.J.A.C.Z.Y.M.L., Drees, K.H.: Building hvac control with reinforcement learning for reduction of energy cost and demand charge. Energy Buildings **239**, 110833 (2021)
28. Zhou, W., Li, W.: Programmatic reward design by example. In: Proceedings of the AAAI Conference on Artificial Intelligence 36(8), pp. 9233–9241 (2022)
29. Åström, K.: Optimal control of Markov processes with incomplete state information. J. Math. Anal. Appl. **10**(1), 174–205 (1965)

Multi-Agent Meta-Reinforcement Learning with Coordination and Reward Shaping for Traffic Signal Control

Xin Du, Jiahai Wang(✉), and Siyuan Chen

School of Computer Science and Engineering, Sun Yat-sen University, Guangzhou, China
{duxin23,chensy47}@mail2.sysu.edu.cn, wangjiah@mail.sysu.edu.cn

Abstract. Traffic signal control (TSC) plays an important role in alleviating heavy traffic congestion problem. It is helpful to provide an effective transportation system by optimizing traffic signals intelligently. Recently, many deep reinforcement learning methods are proposed to solve TSC. However, most of these methods are trained and tested in a fixed roadnet with the same traffic flow environment. They can not adapt to new environments. Some meta-reinforcement learning methods are proposed to solve this problem, but they can not properly decide when to coordinate traffic signals. This paper proposes a multi-agent meta-reinforcement learning method with coordination and reward shaping to solve TSC. The proposed method combines independent learning and neighbor-aware learning to adapt to different TSC environments. Besides, the proposed method constructs a novel reward shaping. The reward shaping can enhance traffic efficiency by encouraging adjacent intersections to generate more green waves. Based on green waves, vehicles can go straight through multiple intersections without stopping. Experimental results demonstrate that the proposed method achieves the state-of-the-art generalization performance on synthetic and real-world datasets.

Keywords: Traffic signal control · Multi-agent meta-reinforcement learning · Coordination · Reward shaping · Green wave

1 Introduction

Along with the increasing urbanization and the latest advance in transportation, the modern cities suffer from heavy traffic congestions. It results in several negative problems such as air pollution, fuel consumption and economic losses. To solve these problems, it is necessary to construct an intelligent transportation system to control traffic signals. The application of traffic signal control (TSC) to Hefei city brain is a good example. The city brain totally monitors 2417 intersections, 1017 roads and 50 traffic grids. In 2020, It reduced average vehicle queue length by 13% and average travel time by 20% at the intersections around Wuhu avenue in Hefei. Therefore, the city brain increased traffic

H. Kashima et al. (Eds.): PAKDD 2023, LNAI 13936, pp. 349–360, 2023.
https://doi.org/10.1007/978-3-031-33377-4_27

efficiency by controlling traffic signals intelligently. Moreover, some researchers pay attention to design an adaptive TSC system [2,6,7]. The system can relieve traffic congestions by adapting to different traffic flows.

Recently, more and more deep reinforcement learning (RL) methods [1,4,10, 16–18] are proposed to solve TSC. These RL methods use different strategies to coordinate traffic signals. Some methods [4,10,16] learn the interactions among agents by graph neural networks. Besides, some methods [17,18] design the reward functions based on traffic pressure. Their reward functions consider the vehicle queue lengths in multiple intersections rather than a single intersection. These methods can well coordinate traffic signals from different perspectives, but their coordination strategies can only adapt to a fixed traffic roadnet. Therefore, some meta-reinforcement learning based methods [19,21,22] are proposed to solve this problem. These methods use meta-reinforcement learning to train their models in one traffic flow environment and test in other environments. They can handle multiple patterns of traffic flows. However, these methods can not learn a general coordination strategy to adapt to various complex traffic environments. In these environments, some intersections have the low traffic pressures. It is not necessary to cost much time to coordinate these intersections with low traffic pressures, but the meta-reinforcement learning based methods can not properly decide when to coordinate. Besides, the reward functions of these methods do not consider the correlations of traffic signal phases between two adjacent intersections. More related works for TSC are shown in supplementary material[1].

To address the problems mentioned above, this paper designs a new coordination strategy. This coordination strategy combines independent learning and neighbor-aware learning to adapt to light and heavy traffic conditions, respectively. The independent learning aims to learn a TSC policy for a single intersection. The neighbor-aware learning can learn a general coordination strategy between two adjacent intersections. Besides, this paper designs a reward shaping to encourage adjacent intersections to generate more green waves. Based on green waves, vehicles can go straight through multiple intersections without stopping. Based on the coordination strategy and the reward shaping, this paper proposes a multi-agent **Meta**-reinforcement learning method with **C**oordination and reward **S**haping (Meta-CSLight) to solve TSC. Meta-CSLight is composed of two stages: the meta-training stage and the meta-testing stage. In the meta-training stage, Meta-CSLight constructs the independent learning model and the neighbor-aware learning model. The independent learning model is trained by using the data from light traffic intersections, while the neighbor-aware learning model is from heavy traffic intersections. To distinguish light from heavy cases, Meta-CSLight defines the concept of traffic pressure based on the theory of max pressure [12,13,17]. In the meta-testing stage, Meta-CSLight uses the trained models from the meta-training stage to adapt to new traffic environments. The main contributions of this paper are threefold:

- A multi-agent meta-reinforcement learning method with coordination and reward shaping, named Meta-CSLight, is proposed to adapt to multiple traffic

[1] https://github.com/08doudou/Meta-CSLight-Appendix.

flow environments in different cities. Meta-CSLight combines independent learning and neighbor-aware learning to coordinate traffic signals according to traffic pressures. This coordination strategy can adapt to various complex traffic environments.

- Meta-CSLight designs the reward shaping based on green wave. The reward shaping correlates the traffic signals between downstream and upstream intersections to allow vehicles to go through these intersections without stopping. Therefore, it can reduce traffic delay by encouraging adjacent intersections to generate more green waves.
- Meta-CSLight is trained on Hangzhou dataset and evaluated on four different datasets. The experimental results show that Meta-CSLight achieves the superior performance on generalizing to different traffic flows.

2 Problem Definition

Definition 1 (Traffic signal phase): A traffic signal phase can be denoted by the combination of movement signals. There are eight valid paired-signal phases [15,23] in TSC as shown in Fig. 1. These phases are generated by pairing all the non-conflicting movement signals.

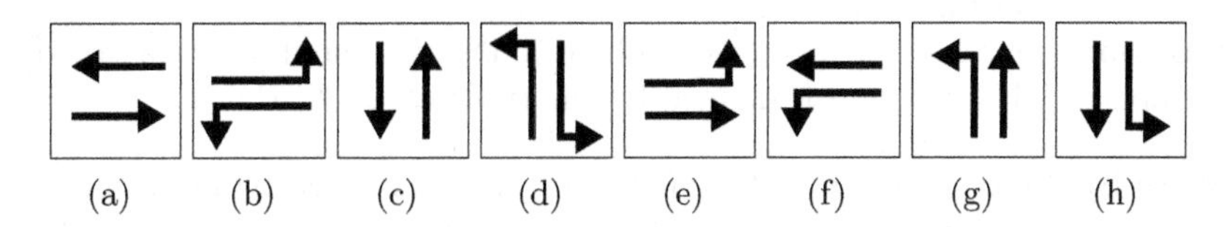

Fig. 1. Traffic signal phases. (a) Going straight from west and east. (b) Turning left from west and east. (c) Going straight from north and south. (d) Turning left from north and south. (e) Going straight and turning left from west. (f) Going straight and turning left from east. (g) Going straight and turning left from south. (h) Going straight and turning left from north.

Definition 2 (Traffic movement): A traffic movement can be defined as traveling from an incoming lane to an outgoing lane. Therefore, the traffic movement from lane l to lane n can be denoted as (l, n), where l and n are connected with the same intersection.

Definition 3 (Traffic pressure): The traffic pressure of a traffic movement (l, n) can be denoted as the difference of vehicle density between lane l and lane n [12,13,17]. This paper simply defines the traffic pressure of (l, n) as the difference of vehicle number between lane l and lane n, because the traffic pressure is only used to distinguish whether the traffic conditions are light or heavy. The definition of traffic pressure of (l, n) is as follows:

$$m(l, n) = x(l) - x(n), \tag{1}$$

where $x(\cdot)$ denotes the vehicle number. Therefore, the traffic pressure of intersection i can be defined as follows:

$$P_i = \sum_{(l,n)\in i} m(l,n). \quad (2)$$

Definition 4 (Green wave): Green wave denotes that vehicles travel along a direction without stopping at any intersections [11,15]. To achieve green wave, the offset of the green signals between two adjacent intersections in the same direction can be calculated as follows:

$$\Delta t = \frac{DIS_{i,j}}{v}, \quad (3)$$

where $DIS_{i,j}$ is the distance between two adjacent intersections i and j; v is the average vehicle travel speed. Based on the definition of Δt, this paper defines the green wave between two adjacent intersections i and j as follows.

At time t, the agent i selects a traffic phase which can make vehicles go from intersection i to j. After Δt, these vehicles can go straight through intersection j without stopping. Consequently, there is a green wave occurring between intersection i and j from time t to $t + \Delta t$. Therefore, the number of the green waves between arbitrary two neighbor intersections over a period of time can reflect the degree of coordination.

Definition 5 (Traffic signal control problem): TSC is casted as a Markov decision problem. This problem is defined by the tuple $(\mathcal{S}, \mathcal{A}, \mathcal{P}, \mathcal{R}, \gamma)$:

- **State Space** $\mathcal{S}$: $\mathcal{S}$ denotes the joint state space observed by all agents. The state of each agent is defined as $s_i^t \in \mathcal{S}_\rangle$ at time t, where i is the index of agents. s_i^t includes the traffic signal phase and the vehicle number of the incoming lanes connected with the intersection i.
- **Action Space** $\mathcal{A}$: $\mathcal{A}$ denotes the joint action space of agents. Each agent can select actions from predefined eight valid traffic signal phases every a fixed time interval.
- **Transition Function** $\mathcal{P}$: $\mathcal{P}$ denotes the transition function. This function maps the state-action pair at time t to the state at time $t + 1$. Formally, the transition function $\mathcal{P}$ can be defined as $\mathcal{P}$: $\mathcal{S} \times \mathcal{A} \rightarrow P(\mathcal{S})$.
- **Reward** $r_i^t \in \mathcal{R}$: r_i^t denotes the reward of agent i for taking the action a_i^t at time t.
- **Discount Factor** γ: $\gamma \in [0, 1]$ is the discount factor for future rewards.

3 Method

The framework of Meta-CSLight is presented in Fig. 2. Meta-CSLight is composed of two stages: the meta-training stage and the meta-testing stage. In the meta-training stage, the coordination strategy and the reward are designed to

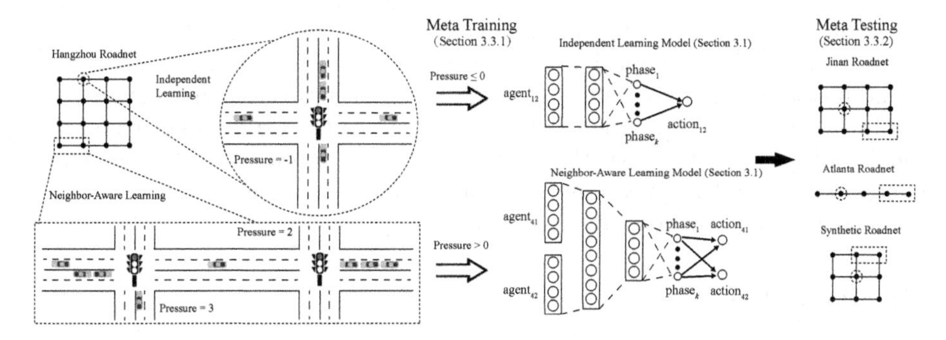

Fig. 2. The framework of Meta-CSLight. Meta-CSLight is composed of two stages: the meta-training stage and the meta-testing stage. In the meta-training stage, Meta-CSLight constructs the independent learning model and the neighbor-aware learning model to adapt to light and heavy traffic intersections, respectively. To distinguish light from heavy cases, Meta-CSLight uses the concept of traffic pressure. In the meta-testing stage, Meta-CSLight uses the trained models from the meta-training stage to adapt to new traffic environments.

enhance traffic efficiency. This section first introduces the coordination strategy and the reward design in Meta-CSLight. Then, the procedure of Meta-CSLight is described.

3.1 Coordination Strategy

The coordination strategy designed in Meta-CSLight uses two models to train the traffic flow environment V_e: the independent learning model $f_{\theta_e^{\text{indep}}}$ and the neighbor-aware learning model $f_{\theta_e^{\text{nei}}}$, where e is the index of traffic flow environments. $f_{\theta_e^{\text{indep}}}$ is trained by using the data from light traffic intersections with $P \leqslant 0$, while $f_{\theta_e^{\text{nei}}}$ is from heavy traffic intersections with $P > 0$. Both models use deep Q-network (DQN) [8] as the function approximator to estimate Q-values of state-action pairs, because compared with other RL algorithms, DQN is easier to handle the discrete action space in TSC. The two models are designed as follows:

$$\begin{aligned} q_{e,j}^t &= f_{\theta_e^{\text{indep}}}(s_{e,j}^t) && P_j \leqslant 0, \\ q_{e,k}^t, q_{e,l}^t &= f_{\theta_e^{\text{nei}}}([s_{e,k}^t, s_{e,l}^t]) && P_k > 0, l = \arg\max_l(P_l), l \in \mathcal{N}_k, \end{aligned} \tag{4}$$

where $f_{\theta_e^{\text{indep}}}$ and $f_{\theta_e^{\text{nei}}}$ are L-layer multilayer perceptrons with the activation function ReLU; $s_{e,j}^t$ is the state representation of agent j at time step t in traffic flow V_e; $[\cdot,\cdot]$ is the concatenation symbol; $q_{e,j}^t$ is the predicted Q-value vector of agent j; P_j is the traffic pressure of intersection j; $\mathcal{N}_k$ is the neighbor intersections of intersection k.

To prepare the training data of the two models for the traffic flow V_e, at each time step, the traffic pressures of intersections in V_e are first calculated.

Then, the intersection k with the maximum traffic pressure in V_e is selected. The intersection k will be coupled with a neighbor intersection with the maximum traffic pressure in $\mathcal{N}_k$ to train $f_{\theta_e^{\text{nei}}}$. The neighbor-aware learning model $f_{\theta_e^{\text{nei}}}$ can learn a better coordination strategy to alleviate the traffic pressures in the couple of intersections. Next, the operation above is repeated in the remaining intersections until there is no intersections with $P > 0$. At last, all the states of the remaining intersections are used to train $f_{\theta_e^{\text{indep}}}$.

3.2 Reward Design

The reward in Meta-CSLight includes the immediate reward ri_i^t and the reward shaping rs_i^t. The immediate reward ri_i^t uses the vehicle queue length in intersection i at time t, because the vehicle queue length can reflect the degree of the traffic congestion in intersection i. The reward shaping rs_i^t is designed based on green wave. It uses the amount of the green waves between intersection i and its neighbor intersections from time $t-T$ to t, where T is the width of the time window. Therefore, the immediate reward ri_i^t, the reward shaping rs_i^t and the total reward r_i^t of agent i are defined as follows:

$$
\begin{aligned}
ri_i^t &= -\sum_l L_{i,l}^t, \\
rs_i^t &= \sum_{j \in \mathcal{N}_i} (N_{i,j}^{[t-T,t]} - max(\lfloor \frac{T - \Delta t}{\Delta t_{\text{action}}} + 1 \rfloor, 0)), \\
r_i^t &= ri_i^t + \alpha \cdot rs_i^t,
\end{aligned}
\tag{5}
$$

where $L_{i,l}^t$ denotes the vehicle queue length at lane l; $\mathcal{N}_i$ denotes the neighbor intersections of intersection i; $N_{i,j}^{[t-T,t]}$ is the amount of the green waves between intersection i and j from time $t-T$ to t; Δt_{action} is the minimum time interval of changing traffic phases; Δt is the offset; $max(\lfloor \frac{T-\Delta t}{\Delta t_{\text{action}}} + 1 \rfloor, 0)$ denotes the maximum possible value of the green waves between two adjacent intersections in T seconds; $\alpha \in [0, 1]$ is the weight to balance the immediate reward and the reward shaping.

The reward shaping is designed to encourage more green waves between adjacent intersections. Green waves can reduce traffic delay by allowing vehicles to go through two adjacent intersections without stopping. Therefore, the reward shaping can enhance the cooperation of traffic signals to improve traffic efficiency.

The reward shaping is controlled by the weight α. If $\alpha = 0$, each agent only considers the immediate reward. The agent will greedily reduce the vehicle queue length in its own intersection without considering the neighbor intersections. If $\alpha = 1$, each agent may suffer from credit assignment problem [3,14]. It means that the TSC policy of each agent is easily affected by its neighbor agents. Therefore, this paper sets $0 < \alpha < 1$ to balance the immediate reward and the reward shaping. This setting adapts to all the agnets.

3.3 Procedure of Meta-CSLight

As mentioned above, Meta-CSLight is composed of the meta-training stage and the meta-testing stage. In the meta-training stage, Meta-CSLight constructs the independent learning model and the neighbor-aware learning model to adapt to light and heavy traffic intersections, respectively. To distinguish light from heavy cases, Meta-CSLight uses the concept of traffic pressure. In the meta-testing stage, Meta-CSLight uses the trained models from the meta-training stage to adapt to new traffic environments. The details of the procedure of Meta-CSLight are given as follows.

3.3.1 Meta-Training Stage

The meta-training stage aims to learn a global parameter initialization to adapt to new traffic environments. This stage follows the architecture of a first-order meta-learning method, Reptile [9]. Reptile is motivated by the model-agnostic meta-learning method (MAML) [5]. MAML is used as the meta-reinforcement learning architecture in MetaLight and GeneraLight for TSC. These methods will cost much time to train meta-learner in the outer loop. Compared with MAML, Reptile is simple to implement without a training-test split for each task. Therefore, Reptile is used as the meta-reinforcement learning architecture in Meta-CSLight.

The meta-training stage consists of the inner loop and the outer loop. In the inner loop, a global parameter initialization θ_0 of the meta-learner f_{θ_0} is provided. f_{θ_0} includes two models: the independent learning model $f_{\theta_0^{\text{indep}}}$ and the neighbor-aware learning model $f_{\theta_0^{\text{nei}}}$. Then, the parameter initializations θ_e^{indep} and θ_e^{nei} for each specific traffic flow environment V_e are set by θ_0^{indep} and θ_0^{nei}, respectively. Each V_e is used to train the corresponding parameter initializations θ_e^{indep} and θ_e^{nei}. In the outer loop, the global parameter initializations θ_0^{indep} and θ_0^{nei} are updated by aggregating the adaptations of θ_e^{indep} and θ_e^{nei}, respectively. The details of the meta-training stage are given as follows.

Inner Loop: In the inner loop, M different traffic flow environments are sampled to train the independent learning model $f_{\theta_e^{\text{indep}}}$ and the neighbor-aware learning model $f_{\theta_e^{\text{nei}}}$, where $e \in \{1, ..., M\}$. $f_{\theta_e^{\text{indep}}}$ and $f_{\theta_e^{\text{nei}}}$ are initialized by the meta-learner f_{θ_0}. The parameter updating process of the two models for each time step is formulated as follows:

$$\theta_e \leftarrow \theta_e - \beta_{\text{inner}} \nabla_{\theta_e} \mathcal{L}(f_{\theta_e}, D_e), \tag{6}$$

$$\mathcal{L}(f_{\theta_e}, D_e) = \mathbb{E}_{s^t, a^t, r^t, s^{t+1} \sim D_e}[r^t + \gamma \max_{a^{t+1}} f_{\theta'_e}(s^{t+1}, a^{t+1}) - f_{\theta_e}(s^t, a^t)], \tag{7}$$

where β_{inner} is the inner-loop step size; $\mathcal{L}$ is the loss function; D_e is the memory replay; γ is the discount factor; $f_{\theta'_e}$ is the target Q-network.

Outer Loop: The outer loop aims to update the global parameter initialization θ_0 by using θ_e. The parameter updating process in the outer loop for each time step is formulated as follows:

$$\theta_0 \leftarrow \theta_0 + \beta_{\text{outer}} \frac{1}{M} \sum_{e=1}^{M} (\theta_e - \theta_0), \tag{8}$$

where β_{outer} is the outer-loop step size; M is the number of environments in the inner loop. The global parameter initialization will be used to adapt to new traffic flow environments in the meta-testing stage. The algorithm of the meta-training stage is shown in supplementary material.

3.3.2 Meta-Testing Stage

This stage aims to use the trained meta-learner f_{θ_0} to adapt to new traffic flow environments W_e, where $e \in \{1, 2, ..., R\}$. In these environments, R_{adapt} ($R_{\text{adapt}} < R$) environments are selected to do fine-tuning based on θ_0. The fine-tuning process is formulated as follows:

$$\theta_e \leftarrow \theta_e - \beta_f \nabla_{\theta_e} \mathcal{L}(f_{\theta_e}, D_e), \tag{9}$$

where β_f is the fine-tuning step size. After fine-tuning, θ_e will be used to test the remaining $R - R_{\text{adapt}}$ environments. The algorithm of the meta-testing stage is shown in supplementary material.

4 Experiments

4.1 Datasets

This paper uses 3 public real-world datasets $Hangzhou_{4\times4}$, $Jinan_{3\times4}$, $Atlanta_{1\times5}$ and a synthetic dataset $Syn_{3\times3}$[2] [15,16,23]. These datasets are conducted on an open-source traffic simulator CityFlow [20]. However, these datasets are obviously insufficient to train a generalized model for any type of environments. Therefore, following the previous works [21], this paper uses Wasserstein generative adversarial network (WGAN) to generate more traffic flows based on the 4 datasets mentioned above. Trained on the generated datasets, Meta-CSLight can enhance the generalization ability of the meta-learner f_{θ_0}. Then, f_{θ_0} can adapt to different traffic flow environments. The details of the traffic flow datasets are shown in supplementary material.

4.2 Baseline Methods

The baseline methods compared with Meta-CSLight are described as follows:

- **GCN** [10]: This method uses GCN to fuse the traffic features of neighbor intersections with the same attention weights.

[2] https://traffic-signal-control.github.io/.

- **MADRL-STFF** [4]: MADRL-STFF proposes a spatio-temporal feature fusion method to solve TSC. This method uses self-attention and temporal convolutional network to capture temporal dependency. Besides, to capture spatial dependency, this method uses GAT to make the neighbor agents and the agents in the same subnetwork share their traffic features.
- **PressLight** [17]: PressLight proposes an RL method based on max pressure control. This method uses the traffic pressure as the reward to essentially evaluate the real-time traffic condition.
- **CoLight** [16]: CoLight uses GAT to extract and fuse traffic features of neighbor intersections to achieve cooperative traffic signal control.
- **CSLight**: CSLight is designed by removing the Reptile architecture from Meta-CSLight. It aims to show that Reptile can help Meta-CSLight to enhance the generalization ability compared with CSLight.
- **Reptile** [9]: Reptile is combined with PressLight and CoLight, respectively, to improve their generalization ability.
- **MetaLight** [19]: MetaLight designs a value-based meta-reinforcement learning framework by combining individual-level adaptation and global-level adaptation. Based on this framework, MetaLight can pay attention to finding a generalized model for any type of intersections and phase settings.
- **GeneraLight** [21]: GeneraLight is designed to learn a generalized model for traffic flow environments. GeneraLight clusters similar traffic environments based on average travel time. The traffic environments of the same cluster can be utilized to train a global parameter initialization. Then, a set of global parameter initializations will be tested in the meta-testing stage.

Following the existing methods [16,17,21], average travel time is selected as the performance metric in TSC. This metric is calculated by the average travel time of all the vehicles traveling in the roadnet (in seconds).

The parameter settings, the effect of independent learning and neighbor-aware learning, the effect of reward shaping and case study are shown in supplementary material. The experimental results demonstrate that the neighbor-aware learning model is more effective than the independent learning model. Besides, Meta-CSLight can converge to the best performance by setting $\alpha = 0.2$.

4.3 Comparison with Baseline Methods

This section compares Meta-CSLight with baseline methods on synthetic datasets and real-world datasets. To show the generalization ability of all the methods, the experiments are conducted with and without fine-tuning. The experimental results are shown in Table 1 and Table 2 with and without fine-tuning, respectively. The source code of Meta-CSLight is available on request. Several observations can be found as follows.

1) Meta-CSLight outperforms all the meta-reinforcement learning methods on all the datasets in terms of average travel time. On four test datasets, when the meta-testing stage includes the fine-tuning process, Meta-CSLight outperforms MetaLight and GeneraLight by 8.57% and 8.39% on average, respectively,

Table 1. Performance comparison on synthetic and real-world datasets with fine-tuning w.r.t average travel time (the lower the better).

	$Hangzhou_{4\times4,dis=0.1}$	$Jinan_{3\times4,dis=0}$	$Atlanta_{1\times5,dis=0}$	$Syn_{3\times3,dis=0}$
GCN [10]	694.76	733.15	314.65	961.41
CoLight [16]	657.98	503.99	279.47	820.72
CoLight+Reptile	640.32	448.64	273.82	655.43
MADRL-STFF [4]	499.66	394.89	268.72	675.95
PressLight [17]	481.24	345.56	247.84	167.50
PressLight+Reptile	442.67	304.08	229.36	166.11
MetaLight [19]	381.90	284.00	203.14	156.63
GeneraLight [21]	372.34	292.31	206.85	152.58
CSLight	367.45	271.29	246.03	145.16
Meta-CSLight	**355.30**	**262.25**	**178.46**	**144.96**

Table 2. Performance comparison on synthetic and real-world datasets without fine-tuning w.r.t average travel time (the lower the better).

	$Hangzhou_{4\times4,dis=0.1}$	$Jinan_{3\times4,dis=0}$	$Atlanta_{1\times5,dis=0}$	$Syn_{3\times3,dis=0}$
GCN [10]	717.38	953.84	364.84	1171.88
CoLight [16]	682.61	743.12	295.28	974.02
CoLight+Reptile	664.80	631.40	307.79	861.97
MADRL-STFF [4]	617.45	647.88	297.48	954.03
PressLight [17]	623.96	786.57	290.91	806.83
PressLight+Reptile	559.76	779.90	288.32	651.76
MetaLight [19]	422.37	535.24	276.91	446.73
GeneraLight [21]	415.38	727.99	270.38	478.47
CSLight	390.73	296.07	272.30	173.76
Meta-CSLight	**363.80**	**289.64**	**267.83**	**164.44**

and Meta-CSLight outperforms PressLight and CoLight both combined with Reptile by 17.11% and 49.69% on average, respectively. When the meta-testing stage removes the fine-tuning process, Meta-CSLight outperforms MetaLight and GeneraLight by 31.56% and 34.8% on average, respectively, and Meta-CSLight outperforms PressLight and CoLight both combined with Reptile by 44.94% and 48.33% on average, respectively. These meta-reinforcement learning baseline methods are designed for the traffic environments in fixed roadnets. In contrast, Meta-CSLight combines independent learning and neighbor-aware learning to coordinate traffic signals according to traffic pressures. This coordination strategy can adapt to various complex traffic environments. Besides, the independent learning model adapts to light traffic flows and the neighbor-aware learning model adapts to heavy traffic flows. The adaptation strategy ensures the stable training results, because the training data of two models come from the similar traffic conditions, respectively. Therefore, Meta-CSLight achieves the best generalization performance compared with other meta-reinforcement learning methods.

2) CSLight outperforms all the non-meta-learning methods on four datasets in terms of average travel time. On four test datasets, when the fine-tuning process is conducted, CSLight outperforms GCN, CoLight, MADRL-STFF and PressLight by 54.21%, 46.15%, 36.18% and 14.79% on average, respectively. When the fine-tuning process is removed, CSLight outperforms the four non-meta-learning baseline methods by 56.26%, 48.22%, 45.32% and 46.15% on average, respectively. For GCN, CoLight and MADRL-STFF, the coordination strategies are closely related to the traffic flow distributions in the training road-nets. Their coordination strategies are based on the interactions between the target intersection and its four neighbor intersections. Unlike these methods, CSLight learns a coordination strategy between arbitrary two adjacent intersections by neighbor-aware learning. This coordination strategy can easily adapt to the variations of traffic flow environments. For PressLight, the reward function is designed based on traffic pressure. It aims to reduce the traffic pressure in a single intersection. Instead, motivated by green wave, CSLight designs the reward shaping to coordinate traffic signals between two intersections. The reward shaping can encourage more green waves to generate between two adjacent intersections. Therefore, the traffic efficiency can be enhanced by allowing vehicles to go through these intersections without stopping.

5 Conclusion

This paper proposes Meta-CSLight to solve TSC. Meta-CSLight is composed of two stages: the meta-training stage and the meta-testing stage. In the meta-training stage, Meta-CSLight constructs the independent learning model and the neighbor-aware learning model. The independent learning model is trained by using the data from light traffic intersections, while the neighbor-aware learning model is from heavy traffic intersections. To distinguish light from heavy cases, Meta-CSLight defines the concept of traffic pressure. Besides, the reward shaping based on green wave is designed to enhance traffic efficiency. In the meta-testing stage, Meta-CSLight uses the trained models from the meta-training stage to adapt to new traffic environments. Experimental results demonstrate that Meta-CSLight achieves the best generalization performance in four test datasets.

In the future, the emergence mechanism of green wave can be investigated further. This mechanism can improve the results of Meta-CSLight.

Acknowledgements. This work is supported by the National Key R&D Program of China (2018AAA0101203), the National Natural Science Foundation of China (62072483), and the Guangdong Basic and Applied Basic Research Foundation (2022A1515011690, 2021A1515012298).

References

1. Bi, J., et al.: Learning generalizable models for vehicle routing problems via knowledge distillation. In: NeurIPS, vol. 35, pp. 31226–31238 (2022)

2. Chen, Y., et al.: Engineering a large-scale traffic signal control: A multi-agent reinforcement learning approach. In: IEEE INFOCOM WKSHPS, pp. 1–6 (2021)
3. Chu, T., et al.: Multi-agent deep reinforcement learning for large-scale traffic signal control. IEEE Trans. Intell. Transp. Syst. **21**(3), 1086–1095 (2020)
4. Du, X., et al.: Multi-agent deep reinforcement learning with spatio-temporal feature fusion for traffic signal control. In: ECML-PKDD 2021, pp. 470–485 (2021)
5. Finn, C., Abbeel, P., Levine, S.: Model-agnostic meta-learning for fast adaptation of deep networks. In: ICML 2017, pp. 1126–1135 (2017)
6. Haydari, A., Yılmaz, Y.: Deep reinforcement learning for intelligent transportation systems: A survey. IEEE Trans. Intell. Transp. Syst. **23**(1), 11–32 (2022)
7. Li, Z., Yu, H., Zhang, G., Dong, S., Xu, C.Z.: Network-wide traffic signal control optimization using a multi-agent deep reinforcement learning. Transport. Res. Part C: Emerg. Technol. **125**, 103059 (2021)
8. Mnih, V., et al.: Human-level control through deep reinforcement learning. Nature **518**(7540), 529–533 (2015)
9. Nichol, A., Achiam, J., Schulman, J.: On first-order meta-learning algorithms. arXiv preprint arXiv:1803.02999 (2018)
10. Nishi, T., Otaki, K., Hayakawa, K., Yoshimura, T.: Traffic signal control based on reinforcement learning with graph convolutional neural nets. In: 2018 21st International Conference on Intelligent Transportation Systems, pp. 877–883 (2018)
11. Roess, R.P., et al.: Traffic Engineering, 5th Edition. Pearson (2019)
12. Varaiya, P.: Max pressure control of a network of signalized intersections. Transp. Res. Part C: Emerg. Technol. **36**, 177–195 (2013)
13. Varaiya, P.: The max-pressure controller for arbitrary networks of signalized intersections. In: Advances in Dynamic Network Modeling in Complex Transportation Systems, pp. 27–66 (2013)
14. Wang, X., et al.: Large-scale traffic signal control using a novel multiagent reinforcement learning. IEEE Trans. Cybern. **21**(3), 1086–1095 (2020)
15. Wei, H., Zheng, G., Gayah, V., Li, Z.: A survey on traffic signal control methods. arXiv preprint arXiv:1904.08117 (2019)
16. Wei, H., et al.: CoLight: Learning network-level cooperation for traffic signal control. In: CIKM 2019, pp. 1913–1922 (2019)
17. Wei, H., et al.: PressLight: Learning max pressure control to coordinate traffic signals in arterial network. In: KDD 2019, pp. 1290–1298 (2019)
18. Wu, L., Wang, M., Wu, D., Wu, J.: DynSTGAT: Dynamic spatial-temporal graph attention network for traffic signal control. In: CIKM 2021, pp. 2150–2159 (2021)
19. Zang, X., et al.: MetaLight: Value-based meta-reinforcement learning for traffic signal control. In: AAAI 2020. vol. 34, pp. 1153–1160 (2020)
20. Zhang, H., et al.: CityFlow: A multi-agent reinforcement learning environment for large scale city traffic scenario. In: WWW 2019, pp. 3620–3624 (2019)
21. Zhang, H., et al.: GeneraLight: Improving environment generalization of traffic signal control via meta reinforcement learning. In: CIKM 2020, pp. 1783–1792 (2020)
22. Zhang, Z., et al.: Meta-learning-based deep reinforcement learning for multiobjective optimization problems. In: IEEE Transactions on Neural Networks and Learning Systems, pp. 1–14 (2022)
23. Zheng, G., et al.: Learning phase competition for traffic signal control. In: CIKM 2019, pp. 1963–1972 (2019)

Regularization of the Policy Updates for Stabilizing Mean Field Games

Talal Algumaei[1(✉)], Ruben Solozabal[1], Reda Alami[2], Hakim Hacid[2], Merouane Debbah[2], and Martin Takáč[1]

[1] Mohamed bin Zayed University of Artificial Intelligence, Masdar City, UAE
{talal.algumaei,ruben.solozabal,martin.takavc}@mbzuai.ac.ae
[2] Technology Innovation Institute, Masdar City, UAE
{reda.alami,hakim.hacid,merouane.debbah}@tii.ae

Abstract. This work studies non-cooperative Multi-Agent Reinforcement Learning (MARL) where multiple agents interact in the same environment and whose goal is to maximize the individual returns. Challenges arise when scaling up the number of agents due to the resultant non-stationarity that the many agents introduce. In order to address this issue, Mean Field Games (MFG) rely on the symmetry and homogeneity assumptions to approximate games with very large populations. Recently, deep Reinforcement Learning has been used to scale MFG to games with larger number of states. Current methods rely on smoothing techniques such as averaging the q-values or the updates on the mean-field distribution. This work presents a different approach to stabilize the learning based on proximal updates on the mean-field policy. We name our algorithm *Mean Field Proximal Policy Optimization (MF-PPO)*, and we empirically show the effectiveness of our method in the OpenSpiel framework.

Keywords: Reinforcement learning · mean-field games · proximal policy optimization

1 Introduction

Despite the recent success of Reinforcement Learning (RL) in learning strategies in games (e.g., the game of Go [1], Chess [2] or Starcraft [3]), learning in games with a large number of players is still challenging. Independent Learning leads to instabilities due to the fact that the environment becomes non-stationary. Alternatively, learning centralised policies can be applied to handle coordination problems and avoid the non-stationarity. However, centralised learning is hard to scale, as the joint action space grows exponentially with the number of agents. Many works in Multi-Agent Reinforcement Learning (MARL) have succeeded in decomposing the objective function into individual contributions [4], although this is also intractable when the number of agents is large. In this sense, mean field theory addresses large population games by approximating the distribution of the players. An infinite population of agents is represented by a continuous distribution of identical players that share the same behaviour. This reduces the learning problem to a representative player interacting with the representation of the whole population.

H. Kashima et al. (Eds.): PAKDD 2023, LNAI 13936, pp. 361–372, 2023.
https://doi.org/10.1007/978-3-031-33377-4_28

This work in particular focuses on learning in Mean Field Games (MFG), non-cooperative games in which many agents act independently to maximise their individual reward, and the goal is to reach the Mean Field Nash Equilibrium (MFNE). Learning in MFG is not an easy task as most of the problems do not have an analytical solution. Traditionally numerical methods have been used to address these problems [5]; nonetheless, these methods do not scale well. In this sense, numerous game theory approaches have been brought into MFG. A classical algorithm is the Banach-Picard (BP) [6] algorithm, which uses a fixed-point iteration method to interactively update the population's behaviour based on the best response of a single representative agent against the mean-field distribution. However, acting in a best response to other agents might cause the others to actuate in the same way, leading to instabilities in the learning (referred to as the *curse of many agents* in game-theory [7]). In practice, smoothing techniques derived from optimization theory are used to guarantee the convergence of these algorithms under reasonable assumptions [8].

More recently, deep RL has been introduced to scale MFG to games with larger state spaces [8]. Nevertheless, traditional approaches cannot be directly applied when using non-linear function approximators as neural networks to represent the objectives in the game. Traditional algorithms average the policy, the mean-field distribution, or both, in order to guarantee a theoretical convergence to the MFNE. This can be done in the case of games with small state spaces under linear or tabular policies, but it is not straightforward when using neural networks. Recent works [9] have derived deep learning algorithms based on value learning suitable for MFG. However, to the best of our knowledge, there is no approach based on policy optimization that addresses this issue.

The main contribution of this paper is bringing policy-based optimization into MFG. This is performed through developing an algorithm based on Proximal Policy Optimization (PPO) [10]. We refer to this algorithm as *Mean Field Proximal Policy Optimization (MF-PPO)*. Conducted experiments in the OpenSpiel framework [11] show better convergence performance of MF-PPO compared with current state-of-the-art methods for MFG. This validates our approach and broadens the spectrum of algorithms on MFG to policy-based methods, traditionally dominant in the literature on environments with large or continuous action spaces.

The remainder of this paper is organised as follows. In Sect. 2, we present the state-of-the-art related to solving the mean-field games. In Sect. 3, we provide a formal description of the problem formulation. Then, in Sect. 4 we present the designed algorithm MF-PPO, that we validate experimentally in Sect. 5. Finally, Sect. 6 concludes the paper. [1]

2 Related Works

In the literature, numerous RL approaches have been designed to address MFG. These can be classified based on the property used to represent the population

[1] Code available at: https://github.com/Optimization-and-Machine-Learning-Lab/open_spiel/tree/master/open_spiel/python/mfg.

Table 1. Summary on the RL literature for MFG.

	Setting	Learning	Requires Oracle	Best Response
Heinrich et al. [19]	General RL	Value-based	Yes	Yes
Laurière et al. [9]	General RL	Value-based	Yes	No
Koppel et al. [23]	General RL	Value-based	No	No
Xie et al. [24]	General RL	Value-based	No	No
Fu et al. [25]	LQR	Policy-based	No	Yes
Our Approach	General RL	Policy-based	Yes	No

into (i) mean-field action and (ii) mean-field state distribution. Examples of mean-field action can be found in [12], in these works the interaction within the population is done based on the average behaviour of the neighbours. A more common approach is using the mean-field state distribution [13]. This approach approximates the infinitum of agents by the state distribution or *distribution flow* of the population. In this case, each player is affected by other players through an aggregate population state. Also, regarding the problem setup, MFG can be classified as (i) stationary or (ii) non-stationary. In the stationary setup, the mean field distribution does not evolve during the episode [14]. A more realistic scenario, and the one discussed in this work, is the non-stationary [15]. In that case the mean-field state is influenced by the agents decisions.

The methodology to address MFG in the literature is also diverse. The classical method for learning the MFNE is the (BP) algorithm [6]. BP is a fixed point iteration method that iteratively computes the Best Response (BR) for updating the mean field distribution. The convergence of the BP algorithm is restrictive [16], and in practice, it might appear with oscillations. To address this issue, the Fictitious Play (FP) algorithm [17] averages the mean field distribution over the past iterations. This stabilizes the learning and improves the convergence properties of the algorithm [18]. Several attempts have been made in the literature to scale FP. For example, [19] proposed Neural Fictitious Self-play algorithm based on fitted Q-learning that learns from best response behaviours on previous experiences. Also, Deep Average Fictitious Play [9] presents a similar idea in a model-free version of FP in which the BR policy is learned though deep Q-learning. Although learning the best response using deep RL allows scaling this method to games with larger state spaces, in practice learning the BR policy is computationally inefficient. In this sense, algorithms based on policy iteration have been also applied to MFG [20]. These methods have proved to be more efficient [8] as they do not require the computation of the best response but they perform a policy update per evaluation step. An example is Online Mirror Descent (OMD) [21], which averages the evaluation of the Q-function from where it derives the mean-field policy. A deep learning variant of it is the Deep-Munchausen OMD (D-MOMD) [9]. This algorithm uses the Munchausen algorithm [22] to approximate the cumulative Q-function when parameterized using a neural network.

Last but not least, oracle-free methods [26] are complete model-free RL methods applied to MFG. Oracle-free algorithms do not require the model dynamics but they estimate the mean-field distribution induced by the population. In [23], the authors propose a two timescale approach with a Q-learning algorithm suitable for both cooperative and non-cooperative games that simultaneously updates the action-value function and the mean-field distribution in a two timescale setting.

Regardless of the numerous works on value-based learning, the attention to policy optimization methods in MFG has been limited. Related works cover the linear quadratic regulator setting [27] but not general RL, a summary can be observed in Table 1. Motivated by [28], work that emphasizes the effectiveness of PPO in multi-agent games, this paper brings PPO into MFG by presenting a solution to the stabilization issues based on proximal policy updates.

3 Problem Formulation

In Mean Field Games (MFG) the interaction between multiple agents is reduced to a uniform and homogeneous population represented by the mean-field distribution. This is the distribution over states that the continuum of agents define when following the mean-field policy. The way in which MFG addresses the problem is selecting a *representative player* that interacts with the mean-field distribution. This simplifies the problem and facilitates the computation of the equilibria.

More formally, we consider the non-stationary setting with a finite time horizon in which we denote by $n \in \{0, 1, ..., N_T\}$ the time steps in an episode. The state and actions of an agent at each time-step are denoted as $s_n \in \mathcal{S}$ and $a_n \in \mathcal{A}$, both finite in our setting. The mean-field state is represented by the distribution of the population states $\mu_n \in \Delta^{|\mathcal{S}|}$, where $\Delta^{|\mathcal{S}|}$ is the set of state probability distributions on $\mathcal{S}$. In the non-stationary setting, the mean field distribution μ_n evolves during the episode and it characterizes the model dynamics $P : \mathcal{S} \times \mathcal{A} \times \Delta^{|\mathcal{S}|} \to \Delta^{|\mathcal{S}|}$ and the reward function $R : \mathcal{S} \times \mathcal{A} \times \Delta^{|\mathcal{S}|} \to \mathbb{R}$. The policy of the agents depends on a prior on the mean-field distribution. Although, without loss of generality, we can define a time-dependent policy $\pi_n \in \Pi : \mathcal{S} \to \Delta^{|\mathcal{A}|}$ that independently reacts to the mean-field state at every step. The model dynamics are therefore expressed as

$$s_{n+1} \sim P(\cdot|s_n, a_n, \mu_n) \qquad a_n \sim \pi_n(\cdot|s_n). \tag{1}$$

We define the policy $\boldsymbol{\pi} := (\pi_n)_{n\geq 0}$ as the aggregated policy for every time-step, similarly the mean-field distribution $\boldsymbol{\mu} := (\mu_n)_{n\geq 0}$. The value function is calculated as $V^{\boldsymbol{\pi},\boldsymbol{\mu}}(s) := \mathbb{E}[\sum_{n=0}^{N_T} \gamma^n r(s_n, a_n, \mu_n)]$. Given a population distribution $\boldsymbol{\mu}$ the objective for the representative agent is to learn the policy $\boldsymbol{\pi}$ that maximizes the expected total reward,

$$J(\boldsymbol{\pi}, \boldsymbol{\mu}) = \mathbb{E}_{a_n \sim \pi_n(\cdot|s_n), s_{n+1} \sim P(\cdot|s_n, a_n, \mu_n)} \left[\sum_{n=0}^{N_T} \gamma^n R(s_n, a_n, \mu_n) \mid \mu_0 \sim m_0 \right] \tag{2}$$

where μ_0 is the initial mean-field state drawn from the initial distribution of the population m_0 and $0 < \gamma < 1$ denotes the discount factor.

Nash Equilibrium in MFG. The desired solution in games is computing the Nash Equilibrium. This is the set of policies that followed by all players maximize their individual reward such that no agent can unilaterally increase deviating from the Nash policy. Furthermore, in MFG the agents share the same interest and an extension of the Nash equilibrium is needed.

Definition 1. *A mean-field Nash equilibrium (MFNE) is defined as the pair* (π^*, μ^*) *that satisfies the rationality principle* $V^{\pi^*,\mu^*}(s) \geq V^{\pi,\mu^*}(s)\ \forall s, \pi$*; and the consistency principle,* μ^* *is the mean-field state distribution induced by all agents following optimal policy* π^**.*

Mean-Field Dynamics. This work relies on an *oracle* to derive the mean-field state. Given the initial mean-field distribution $\mu_0 = m_0$, the oracle uses the transition function P to compute the mean-field distribution induced by the policy π_n at each time step $n \in \{0, 1, ..., N_T\}$,

$$\mu_{n+1}(s') = \sum_{s,a \in \mathcal{S} \times \mathcal{A}} \mu_n(s)\pi_n(a|s)P(s'|s, a, \mu_n) \quad \forall s' \in \mathcal{S}. \tag{3}$$

In a similar way, the policy is evaluated analytically by computing the expected total costs of the policy $\boldsymbol{\pi}$ under the mean field $\boldsymbol{\mu}$ as follows:

$$J(\boldsymbol{\pi}, \boldsymbol{\mu}) = \sum_{n=0}^{N_t} \sum_{s,a \in \mathcal{S} \times \mathcal{A}} \mu_n(s)\pi_n(a|s)R(s, a, \mu_n). \tag{4}$$

Exploitability. The metric of choice for estimating the MFNE convergence is the exploitability. This metric is well known in game-theory [29,30] and it characterizes the maximum increase in the expected reward a representative player can obtain deviating from the policy the rest of the population adopted. The exploitability is obtained as follows:

$$\phi(\boldsymbol{\pi}, \boldsymbol{\mu}) = \max_{\boldsymbol{\pi}'} J(\boldsymbol{\pi}', \boldsymbol{\mu}) - J(\boldsymbol{\pi}, \boldsymbol{\mu}). \tag{5}$$

An interpretation of the exploitability is to consider it as a measure of how close the learned policy is to the MFNE. Small values of exploitability indicate less incentive for any agent to change its policy.

4 Proposal: Proximal Policy Updates for MFG

Learning in MFG is commonly achieved in the literature via fixed-point iteration [6], where the set $\{(\pi_k, \mu_k)\}_{k \geq 0}$ is recursively updated. Particularly, at iteration k the best response policy for the MDP induced by μ_k is computed and the mean-field is updated μ_{k+1} as a result of the many agent following π_k^{BR}. Under the assumptions discussed in [6], contraction mapping holds and the algorithm is proof to converge to a unique fixed point $\{(\pi^*, \mu^*)\}$. This problem corresponds to finding the optimal policy for an MDP induced by μ,

Algorithm 1. MF-PPO algorithm

Initial policy parameters θ, initial value function parameters ϕ, initialize mean-field policy parameters $\theta^0 \leftarrow \theta$, initial mean-field distribution $\mu_0 = m_0$
for iteration $k = 1, 2, ..., K$ **do**
  Compute the mean-field distribution μ^k induced by the policy $\pi_{\theta^{k-1}}$
  for epsiode $e = 1, 2, ..., E$ **do**
    Sample a minibatch of transitions: $\mathcal{D}_e = \{s_n, a_n, r_n, s_{n+1}\}$ by running the policy $\pi_n(\theta)$ on the game governed by the mean-field μ_n^k.
    Compute the advantage estimate: $\hat{A}_n = G_n - \hat{V}_\phi(s_n)$
    Update the policy network: $\theta^\star \leftarrow \arg\max_\theta \mathcal{L}^{\text{MF-PPO}}(\theta)$ (8)
    Update the value network: $\phi \leftarrow \arg\min_\phi \hat{E}_n[\ ||G_n - \hat{V}_\phi(s_n)||^2\ ]$
  end for
  Update the mean-field policy parameters: $\theta^k \leftarrow \theta$
end for
return μ^K, π_{θ^K}

$\text{MDP}_\mu := (\mathcal{S}, \mathcal{A}, P(\mu), R(\mu), \gamma)$. This can be solved using modern RL techniques that allow in practice to scale the method to large games.

However, solving the BR is demanding, and in practice, it leads to instabilities in learning. In this paper, we aim to provide a solution to these instabilities by regularizing the updates in the mean-field policy. To this end, we bring the proximal policy updates developed in PPO [10] into MFG.

Let start defining how PPO can be used to estimate the best response $\hat{\pi}_\mu^{BR}$ to the MDP_μ. Based on the trajectories collected during the iteration k, one can perform policy optimization on the following objective function

$$\mathcal{J}_\mu^{PPO}(\theta) = \hat{E}_n\left[\min(r_n\hat{A}_n,\ \text{clip}(r_n \pm \epsilon)\hat{A}_n)\right] \qquad r_n(\theta) = \frac{\pi(\cdot|s_n;\theta)}{\pi(\cdot|s_n;\theta_{\text{old}})} \tag{6}$$

where $\pi(a_n|s_n;\theta)$ is a stochastic policy, $\pi(a_n|s_n;\theta_{old})$ is the policy before the update and $\hat{A}_n$ is an estimator of the advantage function at timestep n. $\hat{E}$ is the empirical expectation based in Monte-carlo rollouts. The theory behind PPO suggests relaxing the update on the policy to prevent large destructive updates by using a clip function applied on the ratio between the old policy and the current one. PPO imposes this constraint, forcing the ratio on the policy update $r_n(\theta)$ to stay within a proximal interval. This is controlled with the clipping hyperparameter ϵ.

In this work, we extend the regularization of the policy updates to successive iterations on the MFG. We call the algorithm *Mean-Field Proximal Policy Optimization (MF-PPO)* and it combines a double proximal policy regularization for the intra- and inter-iteration policy updates. This prevents the mean-field policy from having a large update between iterations, obtaining a smoothing effect that has previously been reported beneficial in value-based algorithms for MFG [9]. We denote the probability ratios for the intra- and inter-iteration policy updates as

$$r_n^{\text{e}}(\theta) = \frac{\pi_n(a_n|s_n;\theta)}{\pi_n(a_n|s_n;\theta_{\text{old}}^e)} \qquad r_n^{\text{k}}(\theta) = \frac{\pi_n(a_n|s_n;\theta)}{\pi_n(a_n|s_n;\theta_{\text{old}}^k)} \tag{7}$$

where the superscript $k \in [1, K]$ refers to the iteration and the superscript $e \in [1, E]$ to the episode.

In order to derive an appropriate objective function for MFG we extend the objective function of the classical PPO by adding an additional term that limits the policy updates w.r.t. the previous iteration. We can think in this term as a proximal update that limits the divergence between iterations preventing the policy from reaching the BR at iteration k. The MF-PPO objective is therefore expressed as

$$\mathcal{L}^{\text{MF-PPO}}(\theta) = \hat{E}[\alpha \min(r_n^{\text{e}} \hat{A}_n, \text{clip}(r_n^{\text{e}} \pm \epsilon_e)\hat{A}_n) \\ + (1-\alpha)\min(r_n^k \hat{A}_n, \text{clip}(r_n^{\text{k}} \pm \epsilon_k)\hat{A}_n)] \quad (8)$$

where $0 < \alpha < 1$ balances the proximity of the policy between the inter and intra-iteration updates.

5 Experimentation

In this section, we describe the experiments conducted to validate the proposed MF-PPO algorithm. We analyze the hyper-parameter selection and finally, we present the numerical results obtained against the state-of-the-art algorithms namely Deep-Munchausen Online mirror decent (D-MOMD) and Deep Average-Network Fictitious Play (D-ANFP) [9].

5.1 Experimental Setup

We opted for the OpenSpiel suite [11] to benchmark the proposed algorithm in selected crowd modeling with congestion scenarios. Particularly the scenarios used for evaluation are:

Four-rooms. A simple setup on a four-room grid with 10×10 states and a time horizon of 40 steps. The agents receive a reward for navigating close to the goal located in the bottom right room while there also exists an adversion to crowded areas.

Maze. The maze is a more complex scenario with 20×20 states and a time horizon of 100 steps. In this setting, the agent must correctly steer through a complex maze to reach the goal while, similar to the previous case, evading congested areas.

In both environments the state-space is a two-dimension grid, where the state is represented by the agent's current position. Furthermore, the action space consists of five discrete actions: up, down, left, right, or nothing. Those actions are always valid if the agent is confined within the boundaries. Finally, the reward signal is defined as:

$$r(s, a, \mu) = r_{\text{pos}}(s) + r_{\text{move}}(a, \mu(s)) + r_{\text{pop}}(\mu(s)) \quad (9)$$

where the first term measures the distance to the target, the second penalizes movement, and the last term is a penalty which encourages the agents to avoid crowded areas, and is given by the inverse of the concentration of the distribution at a particular state.

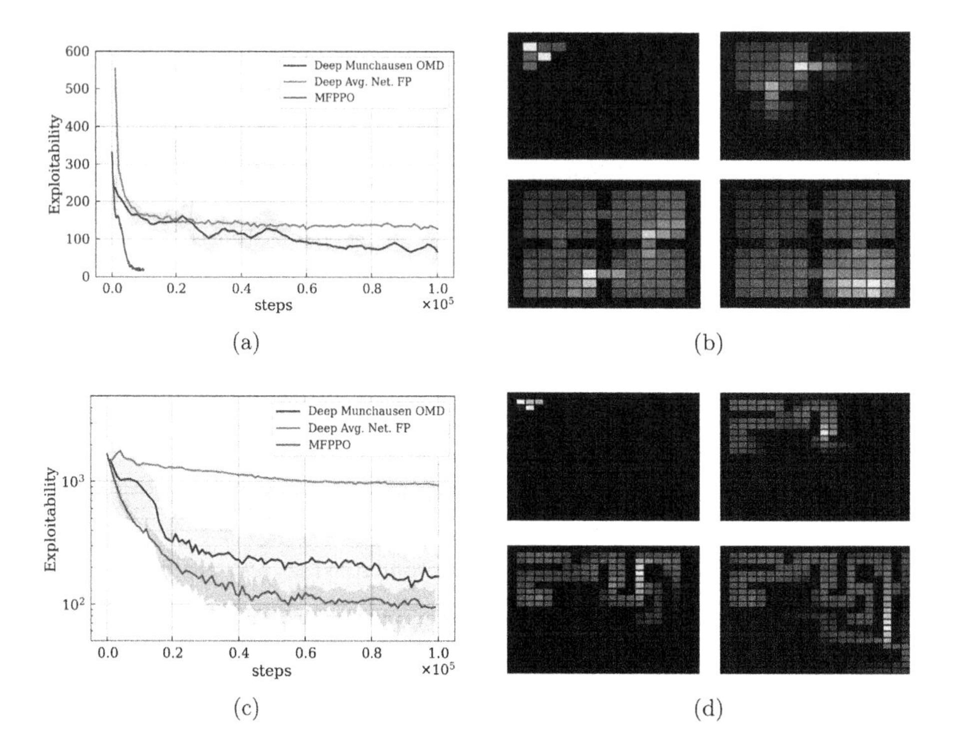

Fig. 1. On the left, the exploitability results obtained on the (a) four-room and (c) maze environments. Results are averaged over five seeds and the confidence interval corresponds to one standard deviation. On the right, the mean-field distribution of the agents generated by the MF-PPO policy on the (b) four-room and (d) maze environments.

5.2 Numerical Results

In this section, we present the results MF-PPO achieves in the selected scenarios. We compare our results with Deep-Munchausen Online Mirror Descent (D-MOMD) and Deep Average-Network Fictitious Play (D-ANFP) [9], both state-of-the-art algorithms in the selected settings. We report the exploitability metric, which is used in the literature as a proxy for quantifying convergence to the MFNE. The results are depicted in Fig. 1 and summarized in Table 2.

Four-Rooms. Obtained results show that MF-PPO outperforms D-NAFP and D-MOMD algorithms, not only by converging to a better ϵ-MFNE solution but, as depicted in Fig. 1, converging in a significantly fewer number of steps. We speculate that this can be credited to the fact our solution learns the optimal policy directly which in this situation is superior to learning the value function that the other methods use and then extract the optimal policy. Figure 1(b) shows the learned mean-field distribution learned using MF-PPO. The agents

Table 2. Comparison of the exploitability metric of the different algorithms. Results are averaged over five different seeds and reported as mean ± std.

Environment	Four Rooms	Maze
D-MOMD	64.41 ± 24.84	153.80 ± 93.05
D-ANFP	127.37 ± 15.19	929.54 ± 46.36
MF-PPO	**15.84** ± 1.95	**93.63** ± 38.11

Table 3. Comparison of the CPU execution time of the different algorithms. Results are averaged over five different seeds and reported as mean ± std.

Environment	Four Rooms	Maze
D-MOMD	3H48M ± 1.79 Min	7H35M ± 1.53 Min
D-ANFP	8H35M ± 56.36 Min	9H45M ± 2.24 Min
MF-PPO	**33M32S** ± 16.58 Sec	**5H36M** ± 3.37 Min

gather as expected around the goal state at the right-bottom room, reaching it by equally distributing over the two symmetric paths.

Maze. Similarly, on the Maze environment Fig. 1(c) shows that MF-PPO and D-MOMD converge to a favorable ϵ-MFNE solution, whereas D-ANFP does to a sub-optimal solution. Still, the policy learned by MF-PPO is closer to the MFNE, reported by a smaller exploitability. Finally, Fig. 1(d) corroborates that the flow of agents over the maze distribute around the goal located in the lower right part of the maze.

In Table 3 we present the CPU execution time of the tested algorithms. In all experiments we used AMD EPYC 7742 64-Core server processor to produce presented results. We note that the official implementation of D-ANFP and D-MOMD was used to reproduce previously presented results. MF-PPO coverages faster than both approaches, more notably, as evidenced by Fig. 1(a) in the four rooms case, MF-PPO converges within roughly 34 min compared to hours by the other two methods. We see a similar, although not as remarkable, trend in the maze as well, where MF-PPO converges in roughly five and half hours to a better MFNE point in comparison with the other techniques.

5.3 Analysis on the Hyper-parameters

This section investigates the influence on the hyper-parameter selection in the learning process. The experiments are conducted on the Maze environment. First, we focus on the configuration where $\alpha = 0$, i.e., we update the iteration policy only and neglect the episode updates entirely. The results are depicted in Fig. 2(a), we see no sign of convergence indicated by high exploitability throughout learning. Furthermore, as the value of α assigned to episode updates increases, we observe a significantly better convergence rate. Nevertheless, it introduces oscillations that impede good convergence on the MFNE. This could be explained by the following dilemma: at each iteration, the representative agent learns an policy far better

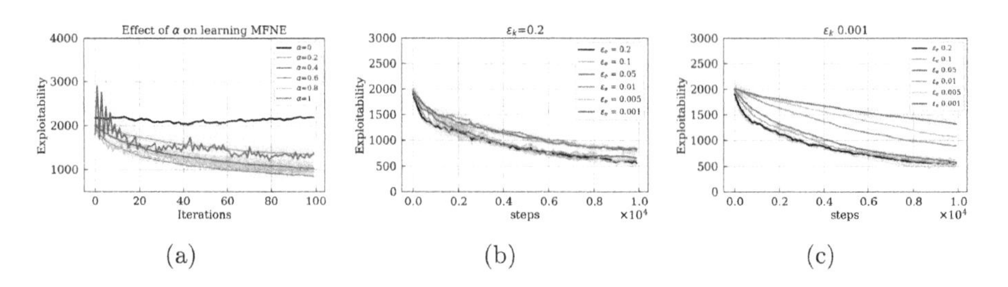

Fig. 2. (a) Study on the impact of the hyper-parameter α on the learning. Moreover, (b) and (c) show the iteration clipping factor ϵ_k contribution to the smoothness and convergence of the MF-PPO algorithm.

than what is available to the current population. Hence, agents have the incentive to deviate from the current policy resulting in an increment in the exploitability. Moreover, at the next iteration, the distribution is updated with such policy, resulting in a sharp decline in the exploitability. This phenomenon can be smoothed by reducing the rate at which the agent's policy updates with respect to the mean-field policy. Consequently, this is controlled using the parameter α as shown by the remaining curves in Fig. 2(a).

Table 4. Hyper-parameter selection.

Environment	Four Rooms	Maze
Input dimension	81	145
Critic/Actor network size	[32, 32]	[64, 64]
Critic output dimension	1	
Actor output dimension	5	
Activation function	ReLU	
Alpha α	0.5	0.6
Iteration ϵ_k	0.01	0.05
Episode ϵ_e	0.2	
Learning rate	1E-03	6E-04
Optimizer	Adam	
Update iteration	100	
Update episodes	20	200
Update epochs	5	
Batch size	200	500
Number of mini-batches	5	4
Gamma γ	0.99	0.9

Then we analyze the impact on both iteration ϵ_k and episode ϵ_e clipping factors. We consider two extreme cases for ϵ_k and different values of ϵ_e. In Fig. 2(b), we set $\epsilon_k = 0.2$, and compare for different ϵ_e values, we observe high variance in exploitability mainly due to more significant policy updates. On the other end, for $\epsilon_k = 0.001$, the curves look much smoother since the policy update is largely constrained; however, the drawback is a slower convergence rate as shown in Fig. 2(c). Finally, all the hyper-parameters used in the experiments are summarized in Table 4.

6 Conclusion

In this work, we propose the *Mean Field Proximal Policy Optimization (MF-PPO)* algorithm for mean field games (MFG). Opposed to current strategies for stabilizing MFG based on averaging the q-values or the mean-field distribution, this work constitutes the first attempt for regularizing the mean-field policy updates directly. Particularly, MF-PPO algorithm regularizes the updates between successive iterations in the mean-field policy updates using a proximal policy optimization strategy. Conducted experiments in the OpenSpiel framework show a faster convergence to the MFNE when compared to current state-of-the-art methods for MFG, namely the Deep Munchausen Online Mirror Descent and Deep Average-Network Fictitious Play.

As future work, the first track would be the investigation of the mathematical analysis of the MFNE reached by the MF-PPO algorithm. Second, investigating the optimization of the computation time of the proposed approach is of interest. Finally, the application of the approach on large-scale real cases would push the boundaries of the approach.

References

1. Silver, D., et al.: Mastering the game of go with deep neural networks and tree search. Nature **529**(7587), 484–489 (2016)
2. Schrittwieser, J., Antonoglou, I., Hubert, T., Simonyan, K., Sifre, L., Schmitt, S., Guez, A., Lockhart, E., Hassabis, D., Graepel, T., et al.: Mastering atari, go, chess and shogi by planning with a learned model. Nature **588**(7839), 604–609 (2020)
3. Vinyals, O., et al.: Grandmaster level in starcraft ii using multi-agent reinforcement learning. Nature **575**(7782), 350–354 (2019)
4. Son, K., Kim, D., Kang, W.J., Hostallero, D.E., Yi, Y.: Qtran: learning to factorize with transformation for cooperative multi-agent reinforcement learning. In: International Conference on Machine Learning, pp. 5887–5896. PMLR (2019)
5. Mathieu. Lauriere. Numerical methods for mean field games and mean field type control. Mean Field Games 78 p221 (2021)
6. Huang, M., et al.: Large population stochastic dynamic games: closed-loop mckean-vlasov systems and the nash certainty equivalence principle. Commun. Inf. Syst. **6**(3), 221–252 (2006)
7. Sonu, E., Chen, Y., Doshi, P.: Decision-theoretic planning under anonymity in agent populations. J. Artif. Intell. Res. **59**, 725–770 (2017)
8. Perolat, J., et al.: Scaling up mean field games with online mirror descent. arXiv:2103.00623 (2021)
9. Laurière, M., et al.: Scalable deep reinforcement learning algorithms for mean field games. arXiv:2203.11973 (2022)

10. Schulman, J., Wolski, F., Dhariwal, P., Radford, A., Klimov, O.: Proximal policy optimization algorithms. arXiv:1707.06347 (2017)
11. Lanctot, M., et al.: Openspiel: a framework for reinforcement learning in games. arXiv:1908.09453 (2019)
12. Subramanian, S.G., Taylor, M.E., Crowley, M., Poupart, P.: Partially observable mean field reinforcement learning. arXiv:2012.15791 (2020)
13. Angiuli, A., Fouque, J.-P., Laurière, M.: Unified reinforcement q-learning for mean field game and control problems. Mathematics of Control, Signals, and Systems, pp. 1–55 (2022)
14. Subramanian, J., Mahajan, A.: Reinforcement learning in stationary mean-field games. In: Proceedings of the 18th International Conference on Autonomous Agents and MultiAgent Systems, pp. 251–259 (2019)
15. Mishra, R.K., Vasal, D., Vishwanath, S.: Model-free reinforcement learning for non-stationary mean field games. In: 2020 59th IEEE Conference on Decision and Control (CDC), pp. 1032–1037. IEEE (2020)
16. Cui, K., Koeppl, H.: Approximately solving mean field games via entropy-regularized deep reinforcement learning. In: International Conference on Artificial Intelligence and Statistics, pp. 1909–1917. PMLR (2021)
17. Cardaliaguet, P., Hadikhanloo, S.: Learning in mean field games: the fictitious play. ESAIM: Control, Optimisation and Calculus of Variations, 23 (2017)
18. Perrin, S., Pérolat, J., Laurière, M., Geist, M., Elie, R., Pietquin, O.: Fictitious play for mean field games: Continuous time analysis and applications. Advances in Neural Information Processing Systems, 33 (2020)
19. Heinrich, J., Silver, D.: Deep reinforcement learning from self-play in imperfect-information games. arXiv:1603.01121 (2016)
20. Cacace, S., Camilli, F., Goffi, A.: A policy iteration method for mean field games. ESAIM: Control, Optimisation and Calculus of Variations, 27, p. 85 (2021)
21. Shalev-Shwartz, S., et al.: Online learning and online convex optimization. Found. Trends Mach. Learn. **4**(2), 107–194 (2012)
22. Vieillard, N., Pietquin, O., Geist, M.: Munchausen reinforcement learning. Adv. Neural. Inf. Process. Syst. **33**, 4235–4246 (2020)
23. Zaman, M.A.U., et al.: Oracle-free reinforcement learning in mean-field games along a single sample path. In: International Conference on Artificial Intelligence and Statistics, pp. 10178–10206. PMLR (2023)
24. Xie, Q., Yang, Z., Wang, Z., Minca, A.: Learning while playing in mean-field games: Convergence and optimality. In: International Conference on Machine Learning, pp. 11436–11447. PMLR (2021)
25. Fu, Z., Yang, Z., Chen, Y., Wang, Z.: Actor-critic provably finds nash equilibria of linear-quadratic mean-field games. arXiv:1910.07498 (2019)
26. Angiuli, A., Fouque, J.-P., Lauriere, M.: Reinforcement learning for mean field games, with applications to economics. arXiv:2106.13755 (2021)
27. Aneeq uz Zaman, M., Zhang, K., Miehling, E., Basar, T.: Reinforcement learning in non-stationary discrete-time linear-quadratic mean-field games. In: 2020 59th IEEE Conference on Decision and Control (CDC), pp. 2278–2284. IEEE (2020)
28. Yu, C., Velu, A., Vinitsky, E., Wang, Y., Bayen, A., Wu, Y.: The surprising effectiveness of ppo in cooperative, multi-agent games. arXiv:2103.01955 (2021)
29. Bowling, M., Burch, N., Johanson, M., Tammelin, O.: Heads-up limit hold'em poker is solved. Science **347**(6218), 145–149 (2015)
30. Lanctot, M., Waugh, K., Zinkevich, M., Bowling, M.: Monte carlo sampling for regret minimization in extensive games. Advances in neural information processing systems, 22 (2009)

Constrained Portfolio Management Using Action Space Decomposition for Reinforcement Learning

David Winkel[1,2](✉), Niklas Strauß[1,2], Matthias Schubert[1,2], Yunpu Ma[1,2], and Thomas Seidl[1,2]

[1] LMU Munich, Munich, Germany
{winkel,strauss,schubert,ma,seidl}@dbs.ifi.lmu.de
[2] Munich Center for Machine Learning (MCML), Munich, Germany

Abstract. Financial portfolio managers typically face multi-period optimization tasks such as short-selling or investing at least a particular portion of the portfolio in a specific industry sector. A common approach to tackle these problems is to use constrained Markov decision process (CMDP) methods, which may suffer from sample inefficiency, hyperparameter tuning, and lack of guarantees for constraint violations. In this paper, we propose Action Space Decomposition Based Optimization (ADBO) for optimizing a more straightforward surrogate task that allows actions to be mapped back to the original task. We examine our method on two real-world data portfolio construction tasks. The results show that our new approach consistently outperforms state-of-the-art benchmark approaches for general CMDPs.

Keywords: Reinforcement Learning · Constrained Action Space · Decomposition · CMDP · Portfolio Optimization

1 Introduction

Constrained portfolio optimization is an important problem in finance. A typical example is a portfolio that must have at least 40% of the total portfolio value invested in environmentally friendly companies at each time step of the investment horizon or a portfolio that is not permitted to invest more than 20% in a particular industry sector. Another example of an action constraint task is a 130-30 strategy, in which the portfolio manager bets on group A of (potentially) overperforming stocks against group B of (potentially) underperforming stocks. This strategy is carried out by short-selling stocks worth 30% of the investment budget from Group B and leveraging the investment into stocks worth 130% of the investment budget from Group A.

The action space for these tasks can be considered as a continuous distribution of weights for a given set of assets. Therefore, reinforcement learning (RL) with policy gradient [16] is well-suited for this task. Because the invested capital

H. Kashima et al. (Eds.): PAKDD 2023, LNAI 13936, pp. 373–385, 2023.
https://doi.org/10.1007/978-3-031-33377-4_29

totals 100%, the codomain of the policy function is typically assumed to be a standard simplex. Existing solutions model the policy using a softmax output layer [1] or based on the Dirichlet distribution [20]. However, the constraints mentioned above cause a change in the shape of the policy's codomain, making the standard solutions no longer directly applicable.

A way to optimize policies for tasks with constrained action spaces is by using approaches for CMDPs with constraints on the action spaces. However, state-of-the-art general approaches for CMDPs often have drawbacks such as expensive training loops, sample inefficiency, or only guarantees for asymptotical constraint compliance [2,4,10,19,21].

In this paper, we propose ADBO, a dedicated approach for dealing with the two important types of investment tasks mentioned previously: (a) investment tasks that invest *at least* or *at most* a certain percentage of a portfolio in a specific group of assets, and (b) short-selling tasks. ADBO can overcome the aforementioned shortcomings of general policy optimization methods for CMDPs. This is achieved by decomposing the non-standard-simplex action space into a surrogate action space. Solutions found in the surrogate action space can then be mapped back into the original constrained action space. In contrast to the non-standard-simplex action space, the surrogate action space is designed to be easily represented in the policy function approximator, allowing us to model the problem as a standard Markov decision process (MDP). Due to the lack of penalties and reward shaping, finding an optimal policy for an MDP is less complex than finding an optimal policy for a CMDP with constrained actions. Furthermore, ADBO ensures that the actions adhere to the constraints both during and after training.

In the experimental section, we demonstrate that the ADBO approach can handle two types of investment tasks using real-world financial data. The first task focuses on investing each time step at least a certain percentage of the portfolio in companies considered to be environmentally sustainable. The second task allows the agent to short-sell selected stocks, i.e., allowing for negative portfolio weights. Our proposed approach outperforms the state-of-the-art benchmark approaches for handling CMDPs on various criteria in both tasks.

2 Related Work

CMDPs were introduced by [3] to model constrained sequential decision tasks. constrained Reinforcement Learning (CRL) approaches for finding optimal policies for CMDPs have a wide range of applications, including finance [7,20], autonomous electric vehicle routing [14], network traffic [9], and robotics [2,8]. A **Trust Region**-based approach was introduced by [2] to find optimal policies for CMDPs that may still exhibit constraint violation due to approximation errors. Another approach proposed by [6] is based on **prior knowledge** and involves a one-time pretraining to predict simple one-step dynamics of the environment. **Lagrangian-based** approaches are another option for dealing with CMDPs. These approaches convert the original constraint optimization problem into an unconstrained optimization problem by applying a Lagrangian relaxation to the constraints. Lagrangian-based approaches can be classified into two types: The first type is **Primal-Dual algorithms**, in which the Lagrange multipliers for a saddle point problem are chosen *dynamically* [5,19]. The second type of Lagrangian-based approach employs **manually selected Lagrange multipliers**, which remain *static*, as shown in [13,17]. Instead of a *saddle point problem*, as in the first type, using a static Lagrange multiplier transforms the problem into a maximization problem, which is more stable and computationally less expensive to solve. Some approaches carefully select Lagrange multipliers to model preferences in a trade-off problem rather than as a means to enforce constraints in an optimization problem. This is commonly seen in risk-return trade-off settings, such as in [7,17,20].

The **factorization of high-dimensional action spaces** in RL, i.e., splitting action spaces into smaller sub-action spaces as a Cartesian product, is an active area of research that has resulted in improved scalability and training performance. In their work, [11] introduce the Sequential DQN approach, which trains the agent for a sequence of n 1-dimensional actions rather than training the agent for n-dimensional actions of the original action space, effectively factorizing the original action space. The approach by [18] introduces an action branching architecture, which models the policies for the sub-action spaces in parallel. Our approach, like theirs, uses a Cartesian product of sub-action spaces. However, the sub-action spaces in our new approach ADBO are the outcome of a decomposition based on the Minkowski sum, resulting in a surrogate action space rather than a factorization of the original action space.

3 Problem Setting

We consider an agent that needs to allocate wealth across N different assets over T time steps. The allowed actions of the agent are defined by the investor's investment task and are contained in the *constrained* **action space** $\mathcal{A}$. The investment task type $T1$ requires the investor to invest at least c_{T1} of the portfolio into assets from group V_{T1}. In practice, these group definitions are often linked to individual risk profiles, industry sectors, or features such as being an environmentally friendly investment. The action space for investment task type $T1$ is then defined as

$$\mathcal{A}_{T1} = \left\{ a \in \mathbb{R}^N : \sum_{i=0}^{N-1} a_i = 1 \; , \; \sum_{i \in V_{T1}} a_i \geq c_{T1} \; , \; a_i \geq 0, \; c_{T1} > 0 \right\}$$

and represents an N-dimensional convex polytope. Task type $T1$ also includes cases that require investing *at most* c_{T1} into assets in V_{T1} because this case is equivalent to investing *at least* $(1 - c_{T1})$ into the remaining assets a_i for $i \in I \setminus V_{T1}$.

The investment task type $T2$ represents investors who believe that a group of assets V_{T2} will underperform relatively compared to the rest of the investment universe I. The investor pays a borrowing fee to short-sell assets in group

V_{T2} worth $|c_{T2}|$ of his total portfolio value and then uses the freed-up cash to invest $1 + |c_{T2}|$ into assets of the other investment universe. The action space for investment task type $T2$ is defined as

$$\mathcal{A}_{T2} = \left\{ a \in \mathbb{R}^N : \sum_{i=0}^{N-1} a_i = 1, \sum_{j \in V_{T2}} a_j = c_{T2}, \ a_j \leq 0, \ a_k \geq 0 \, \forall k \in I \setminus V_{T2}, \ c_{T2} < 0 \right\}$$

and represents an N-dimensional convex polytope as well.

The **observation space** is defined as $\mathcal{O} = \mathcal{W} \times \mathcal{V} \times \mathcal{U}$ where $\mathcal{W} \subseteq \mathbb{R}^+$ is the current absolute wealth level, $\mathcal{V} \subseteq \mathbb{R}^N$ is the current relative portfolio weight of each of the N assets, and $\mathcal{U} \subseteq \mathbb{R}^N$ represents all the observed single asset returns from the previous time step.

The economic return of each asset is individually modeled for each time step by the random vector $\Theta = [\Theta_0, \ldots, \Theta_{N-1}] \in \mathcal{U}$. The portfolio return is then a random variable with an expected value denoted as $\mathbb{E}[\Theta_{PF}] = a^\intercal \mathbb{E}[\Theta]$ with the portfolio weights $a \in \mathcal{A}$. There are two potential sources of cost to consider for the agent: First, the transaction costs caused by changes in the portfolio weights a_t in time step t by the agent defined as $tc_t = (|a_t - v_t|)^\intercal c$, where $v_t \in \mathcal{V}$ and vector $c = [c_0, .., c_{N-1}]$ represents the asset-specific transaction costs caused by trading a specific asset. Second, borrowing fees in case the agent is allowed to short-sell assets. These costs occur every period as long as assets are short-sold, i.e., assigned to a negative portfolio weight. The borrowing fees are defined as $bf_t = (\mathbb{1}_{a_i<0} \circ a_t)^\intercal b$ where $\mathbb{1}_{a_i<0}$ is an indicator vector signaling for each individual asset a_i if the current portfolio weight is negative, $\circ$ is an operator for element-wise vector multiplication, and the vector $b = [b_0, .., b_{N-1}]$ represents asset-specific borrowing fees per time step.

The **reward** for the agent is a combination of transaction costs tc, borrowing fees bf, and a realization ϑ_{PF} of the random variable of the portfolio's economic return Θ_{PF}, i.e., $r = \vartheta_{PF} - tc - bf$. The agent's goal is to maximize the expected cumulative reward, which we will refer to as *total economic payoff*.

4 Solution as CMDP

A CMDP is an extension of an MDP and is defined as a tuple $(\mathcal{S}, \mathcal{A}, R, P, \gamma, \mathcal{C})$ where $\mathcal{S}$ is the set of states, $\mathcal{A}$ is the set of actions, R is the immediate reward function, which maps transition tuples to their respective expected reward, i.e., $R : \mathcal{S} \times \mathcal{A} \times \mathcal{S} \to \mathbb{R}$. P denotes the transition probability function, whereas $P(s_{t+1}|s_t, a_t)$ gives the probability of transitioning to state $s_{t+1} \in \mathcal{S}$ given state $s_t \in \mathcal{S}$ and action $a_t \in \mathcal{A}$. The parameter $\gamma \in [0, 1)$ represents a discount factor. $\mathcal{C} = \{C_0, \ldots, C_m\}$ is a set of immediate constraint functions $C_i : \mathcal{S} \times \mathcal{A} \times \mathcal{S} \to \mathbb{R}$ for $i \in \{0, \ldots, m\}$ that map transition tuples to the respective cost. We let $r_{t+1} := R(s_t, a_t, s_{t+1})$ and define the return for a trajectory τ as the observed discounted cumulative rewards. The objective function J is then defined as the expected return for a given policy π, i.e., $J(\pi) := \mathbb{E}_{\tau \sim P(\tau|\pi)} \left[\sum_{t=0}^{T-1} \gamma^t r_{t+1} \right]$.

The expected cumulative discounted immediate cost for constraint i under policy π is defined as $J_{C_i}(\pi) := \mathbb{E}_{\tau \sim P(\tau|\pi)} \left[\sum_{t=0}^{T-1} \gamma^t C_i(s_t, a_t, s_{t+1}) \right]$. We also define constant trajectory constraint limits $d_0, \ldots, d_m$. The optimization problem for the CMDP is then defined as:

$$\begin{aligned} \underset{\pi}{\text{maximize}}\ J(\pi) &= \underset{\tau \sim P(\tau|\pi)}{\mathbb{E}}(G) = \underset{\tau \sim P(\tau|\pi)}{\mathbb{E}} \left[\sum_{t=0}^{T-1} \gamma^t r_{t+1} \right] \\ \text{s.t.} \quad & J_{C_i}(\pi) \leq d_i \quad \forall i \end{aligned}$$

In the following, we will show how to formulate the tasks defined in Sect. 3 as a CMDP. In Sect. 3, we defined the observation space $\mathcal{O}$, the constrained action space $\mathcal{A}_i$, and the reward R. The transition function P and the state space $\mathcal{S}$ are unknown. However, we assume that we can sample transitions from an environment. Therefore, we can employ reinforcement learning based on a learned state representation function to learn effective policies. To address the action constraints of tasks $T1$ and $T2$, we define the following cost function for each respective task $i \in \{1, 2\}$: $C_{T_i}(s_t, a_t, s_{t+1}) = \mathbb{1}_{a_t \notin \mathcal{A}_{T_i}} \cdot \zeta$ where constant $\zeta > 0$ indicates the non-zero cost of a constraint violation. The respective constraint for each task is then defined as $J_{C_{T_i}}(\pi) \leq 0$.

5 Action Space Decomposition Based Optimization

We define a surrogate MDP $(\mathcal{S}, \tilde{\mathcal{A}}, R, P, \gamma)$ and ensure that there exists a surjective function $f : \tilde{\mathcal{A}} \rightarrow \mathcal{A}$ that allows reaching any $a \in \mathcal{A}$ from at least one $\tilde{a} \in \tilde{\mathcal{A}}$. For a formal description of our method, we first introduce the Minkowski sum:

Definition 1. *Given two sets A and B of vectors in n-dimensional Euclidean space, the* ***Minkowski sum*** *of A and B is generated by adding each vector in A to each vector in B, i.e., the set $A + B = \{a + b | a \in A, b \in B\}$ in which we refer to A and B as Minkowski summands.*

In our setting, the Minkowski sum describes how multiple decomposed action sets can be combined to reconstruct the original constraint action set. The masked scaled standard simplex (MSSS) describes a part of the original constrained action which can be described as a simplex:

Definition 2. *Let mask $M \subseteq \{0, \ldots, N-1\}$. MSSS is defined as:*

$$MSSS_{M,c} = \left\{ y \in \mathbb{R}^N : \sum_{j \in M} y_j = c, y_i = 0\ \forall i \in I \setminus M \right\}$$

with either $(c \geq 0 \wedge y_i \geq 0\ \forall i \in M)$ *or* $(c < 0 \wedge y_i \leq 0\ \forall i \in M)$.

The surrogate action space is modeled as the Cartesian product of independent sub-action spaces $\tilde{\mathcal{A}} = \tilde{\mathcal{A}}_1 \times \tilde{\mathcal{A}}_2$. The sub-action spaces $\tilde{\mathcal{A}}_i$ with $i \in \{1, 2\}$ are required to have the two properties: (a) being a decomposition of $\mathcal{A}$ in such a way that the Minkowski sum (see Definition 1) of all the sub-action spaces $\tilde{\mathcal{A}}_i$ is $\mathcal{A}$, i.e., $\mathcal{A} = \tilde{\mathcal{A}}_1 + \tilde{\mathcal{A}}_2$, and (b) being an MSSS as defined in Definition 2. Property (a) guarantees the existence of function f that can be defined as $f(\tilde{a}) = \tilde{a}_1 + \tilde{a}_2 = a$ with $\tilde{a} = [\tilde{a}_1, \tilde{a}_2] \in \tilde{\mathcal{A}} \subset \mathbb{R}^{2N}$ and $\tilde{a}_i \in MSSS_i \subset \mathbb{R}^N$ for $i \in \{1, 2\}$, i.e., a summation of vectors in a subspace of $\mathbb{R}^N$. Property (b) allows utilizing well-established RL methods for handling standard simplex action spaces with only minor modifications by adding a scaling and masking logic in order to model single $MSSS$ action spaces.

The following two theorems show that constrained action spaces as defined in Sect. 3 can be decomposed into two MSSS that satisfy both of the requirements mentioned above. Theorem 1 describes the decomposition for task $T1$:

Theorem 1. *Any convex polytope $P \neq \emptyset$ defined as*

$$\sum_{i \in I} x_i = 1, x_i \geq 0 \quad \forall i \in I, \sum_{i \in V_1} x_i \geq c_1$$

with $c_1 > 0$, $I = \{0, \ldots, N-1\}$ and $V_1 \subseteq I$ can be decomposed into two MSSSs:

$$MSSS_{S_1, z_1}, \text{ i.e. } \forall y_1 \in MSSS_{S_1, z_1} : \sum_{S_1} y_{i,1} = z_1 \quad \text{with } S_1 = V_1 \text{ and } z_1 = c_1$$

$$MSSS_{S_2, z_2}, \text{ i.e. } \forall y_2 \in MSSS_{S_2, z_2} : \sum_{S_2} y_{i,2} = z_2 \quad \text{with } S_2 = I \text{ and } z_2 = 1 - c_1$$

so that the Minkowski sum of the MSSSs equals the original polytope P.

Correspondingly, the following theorem formulates the decomposition of the action space in task $T2$:

Theorem 2. *Any convex polytope $P \neq \emptyset$ defined as*

$$\sum_{i \in I} x_i = 1, \sum_{i \in V_1} x_i = c_1, x_i \geq c_1 \; \forall i \in V_1, x_i \leq 0 \; \forall i \in V_1, x_i \geq 0 \; \forall i \in I \backslash V_1$$

with $c_1 < 0$, $I = \{0, \ldots, N-1\}$ and $V_1 \subseteq I$ can be decomposed into two MSSSs:

$$MSSS_{S_1, z_1}, \text{ i.e. } \forall y_1 \in MSSS_{S_1, z_1} : \sum_{S_1} y_{i,1} = z_1 \quad \text{with } S_1 = V_1 \text{ and } z_1 = c_1$$

$$MSSS_{S_2, z_2}, \text{ i.e. } \forall y_2 \in MSSS_{S_2, z_2} : \sum_{S_2} y_{i,2} = z_2 \quad \text{with } S_2 = I \backslash V_1 \text{ and } z_2 = 1 - c_1$$

so that the Minkowski sum of the MSSSs equals the original polytope P.

Theorem 1 and 2 can be proven by showing that the two sets of closed half-spaces, one describing the polytope P and the other describing the Minkowski sum of the two MSSSs, are equal resulting in the equality of the two polytopes.

ADBO is based on the PPO algorithm [15]. The agent's policy network is designed in such a way that the action representation is distributed across multiple *independent* segments, i.e., one head for each MSSS. A *shared* state encoder, on the other hand, provides a learned state representation to both heads. For the state encoder we use a neural network of four fully connected layers of size 1024, 512, 256, and 64 with ReLU activation functions followed by GTrXL element allowing to handle tasks requiring memory. The GTrXL element is based on [12]. The element is composed of a single *transformer unit* with a single encoder layer as well as a single decoder layer with four attention heads and an embedding size of 64. While the sub-actions in ADBO are stochastically independent, the parameters of the two distributions from which the sub-actions are drawn are partially coordinated, i.e., parts of the actions rely on the same shared latent state representation. To further ensure coordination between the sub-actions, the different sub-actions are all evaluated using a joint reward. This means that if a *joint action* performs poorly, all independent segments receive a poor reward signal, regardless of individual sub-action performance.

We use a Dirichlet distribution to model each MSSS in the architecture of the policy function approximator. The expected value of a random vector $X = [X_0, \ldots, X_{N-1}]$ following a Dirichlet distribution with a parameter vector of $\alpha = [\alpha_0, \ldots, \alpha_{N-1}]$ is defined as $\mathbb{E}[X_i] = \alpha_i \cdot \left(\sum_{n=0}^{N-1} \alpha_n\right)^{-1}$ with $\alpha_i > 0$ for $i \in \{0, \ldots, N-1\}$. By adjusting the parameter vector of a Dirichlet distribution and applying a linear scaling transformation, we can create a random variable with the set of all possible realizations equaling $MSSS_{M,c}$. The set M contains index values which we map to an N-dimensional indicator vector $\mathbb{1}_M$, with the vector's elements set to one if their respective index occurs in M and zero otherwise. The parameter vector passed to the Dirichlet distribution is calculated as $\alpha_{\mathbb{1}_M} = \max(\alpha \circ \mathbb{1}_M, \epsilon)$, where α is the initial parameter vector before applying the masking and $\epsilon > 0$ is an arbitrary small number. The operator $\circ$ represents element-wise multiplication for vectors. In the final step, a linear scaling transformation is applied, i.e., $Y = c \cdot X$ with $X \sim Dir(\alpha_{\mathbb{1}_M})$.

ADBO requires the uses of two MSSSs, i.e., $MSSS_{M_1,c_1}$ and $MSSS_{M_2,c_2}$. It should be noted that the gradient of the policy during training is based on a policy interacting with the surrogate action space $\tilde{\pi}(\cdot|s)$ rather than a policy interacting with the original constrained action space $\pi(\cdot|s)$. We only use f to convert $\tilde{a}$ into a representation a that can interact with the environment. Various inputs $\tilde{a}$ for f may sum to the same output value a, resulting in f being a many-to-one function. For some $\tilde{a} \in \tilde{\mathcal{A}}$, this results in $\mathbb{P}(\tilde{a}|s) \neq \mathbb{P}(a|s)$ with $a = f(\tilde{a})$. However, we argue that finding *one* possible representation for an action a belonging to an optimal policy for the original problem is sufficient from an optimization standpoint.

6 Experiments

The environment is based on [20], and uses the same real-world financial data from the Nasdaq-100 index that was fetched and processed using the qlib package.[1] The investment universe of the environment consists of 13 assets, one of

[1] https://github.com/microsoft/qlib/tree/main.

which is cash. The remaining 12 assets are chosen at random from a list of 35 stocks that remain after filtering the Nasdaq-100 data set for companies that have been part of the index since January 1, 2010 and have no missing data. The original Nasdaq-100 data set is supplemented with the Environmental Score Metric (ESM) assigned by financial data provider LSEG.[2] The score rates a company's environmental sustainability based on various evaluation categories, such as carbon emissions, willingness to innovate in this field, and transparency in reporting relevant information. The score ranges from 0 to 100, representing the percentiles of a ranking system.

We compare **ADBO** to three other state-of-the-art approaches for optimizing policies in CMDPs. **RCPO** is proposed by [19] and belongs to the class of Lagrangian-based approaches. The interior-point policy optimization approach **IPO** is introduced by [10]. **P3O** is proposed by [21] and uses a first-order optimization over an unconstrained objective with a penalty term equal to the original constraint objective. All benchmark approaches are implemented in the RLlib framework[3] based on their papers and publically available.[4]

Two experimental settings are examined: the **SUSTA setting** is based on task type $T1$. The investor must invest at least 40% of his capital in the top 20% of environmentally sustainable companies, i.e., companies with an ESM score of 80 or higher. A score of 80 or higher "indicates excellent relative [...] performance and a high degree of transparency in reporting material" by a company.[5] The **SHORT setting** is based on task type $T2$. It employs a **130-30 strategy**, a popular long/short equity strategy among investors to invest 130% of the available capital in stocks they believe will outperform and short-sell stocks worth 30% of the available capital they believe will underperform. In the experiments, we choose the companies Automatic Data Processing Inc., Paccar Inc., and Amgen Inc. to be sold short based on being the worst performers in 2020, the final year before the start of the backtesting period.

$\mathcal{A}_{short}$ is not a subset of the standard simplex because negative weights are permitted. As a result, the RCPO, IPO, and P3O approaches must be modified to be applicable to SHORT setting. The agent performed very poorly in initial tests using $\mathbb{R}^N$ as a base action space and applying constraints accordingly and was unable to learn meaningful policies. Instead, using a standard simplex as the base action space and applying action scaling produced better results. For *action scaling*, the agent uses the output of a Dirichlet distribution as an encoded action $\tilde{a} = [\tilde{a}_0, \ldots, \tilde{a}_{N-1}]$ that is then transformed, i.e., scaled into the final action $a = [a_0, \ldots, a_{N-1}]$: the cumulative weights of the stocks sold *short* and the cumulative weights of the stocks bought *long* are added up in their absolute values, resulting in a scaling factor $\alpha_{total} = |\alpha_{long}| + |\alpha_{short}|$. Then, for all elements i of the encoded action $a_i = \tilde{a}_i \cdot \alpha_{total}$ that are bought, a positive scaling factor is applied, and for all elements j of the encoded action $a_j = \tilde{a}_j \cdot (-\alpha_{total})$

[2] https://www.lseg.com/.
[3] https://docs.ray.io/en/master/rllib/index.html.
[4] https://github.com/DavWinkel/RL_ADBO.
[5] https://www.refinitiv.com/en/sustainable-finance/esg-scores.

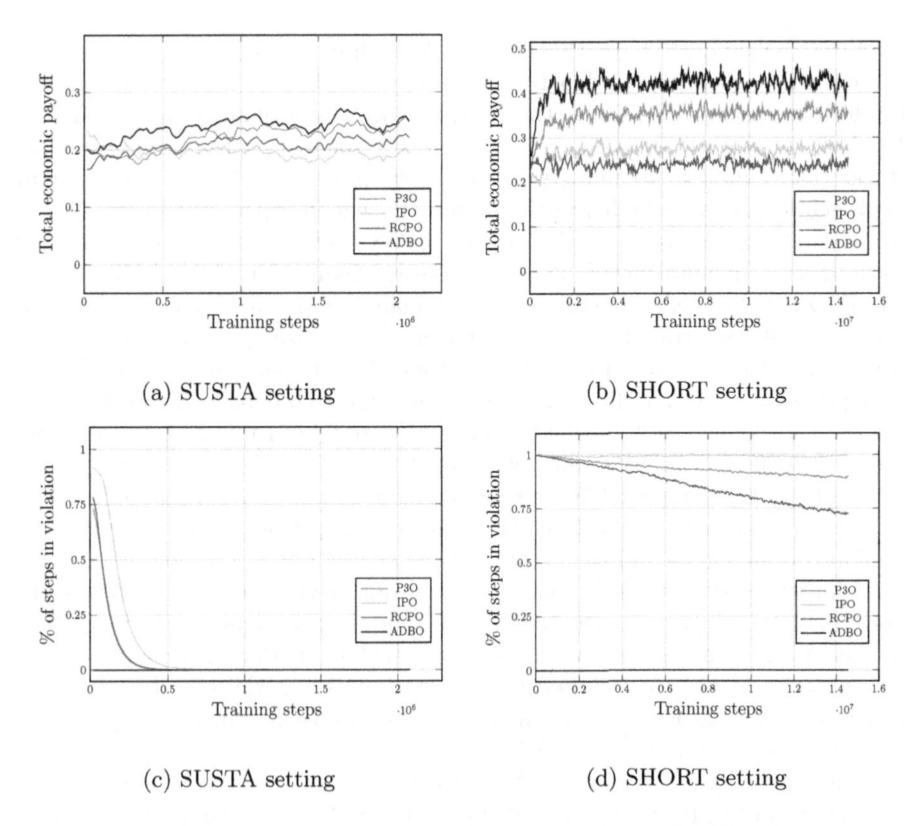

(a) SUSTA setting (b) SHORT setting

(c) SUSTA setting (d) SHORT setting

Fig. 1. Performance during training for all four approaches in the SUSTA setting and the SHORT setting regarding Total economic payoff and % of steps in violation.

that are sold short, a negative scaling factor is applied. It should be noted that actions generated as described above are no longer guaranteed to sum up to 1.0. Because IPO is a logarithmic barrier function-based approach that does not apply to equality constraints, we must additionally soften equality constraints of the form $x = c$ to inequality constraints that allow values in a α-neighborhood of x, i.e., $x \leq c + \alpha$ and $x \geq c + \alpha$.

To evaluate the four approaches, we will report performance during and after training for both the SUSTA setting and the SHORT setting. The total economic payoff defined in Sect. 3 is used to measure economic performance. The results of the SUSTA setting will be discussed first. The training in the SUSTA setting lasts 500 iterations and consists of approximately 2.1 million training steps. Figure 1a shows that ADBO and P3O perform best during training by steadily improving their total economic payoff. RCPO also shows improvements, although at a much slower rate. Table 1 shows the evaluation of economic performance following training completion in two setups: in the **(A) environment setup** 1000 trajectories are sampled from the same environment used for the training. ADBO generates the highest total economic payoff in the SUSTA setting, followed by

P3O, RCPO, and IPO. In the **(B) backtesting setup** a single trajectory, namely the real-world Nasdaq-100 trajectory in 2021, is used for evaluation. In the backtesting year 2021, the overall yearly performance of the Nasdaq-100 index was above average, returning 27.5%, indicating that the individual stocks that comprise the index were also performing well. As a result, the four approaches generated high returns in the (B) backtesting setup, with ADBO performing best, followed by P3O. In the SHORT setting, the training time had to be increased significantly. This increase was required because IPO, RCPO, and P3O failed to generate constraint-compliant actions satisfactorily. However, due to insufficient training progress, which will be discussed in detail later in this section, the training was eventually stopped after 3500 iterations, consisting of approximately 14.7 million training steps. Figure 1b depicts the evolution of the total economic payoff during training. After roughly 1 million training steps, the performance of ADBO converges to a level that it then maintains for the remainder of the training. P3O improves its performance over 3 million training steps until it reaches a stable level. IPO improves its performance during the first million training steps and then stabilizes, whereas RCPO does not show significant improvements in total economic payoff during training. Table 1 shows the performance evaluation in the SHORT setting after the training is completed. In the (A) environment setup, ADBO performs best, with an average total economic payoff of 42.72%, followed by P3O with 35.12%. ADBO outperforms its benchmark approaches by a wide margin in the (B) backtesting setup, achieving a total economic payoff of 102.05%.

The experiments show that violations of the action constraints occurred during the training of IPO, RCPO, and P3O in the SUSTA setting. Figure 1c shows that this is especially true at the beginning of the training phase, while the number of time steps with actions in violation decreases almost to zero later on. After completion of the training in the (A) environment setup, RCPO is the only approach generating actions in violations, as shown in Table 1. However, violations occur only on a small number of time steps, i.e., nine out of 12'000 time steps. All approaches are free of constraint violations in the (B) backtesting setup. For the SHORT setting, the majority of actions generated by the approaches IPO, RCPO, and P3O violated the constraints during training. However, as training time progresses, the number of actions in constraint violation decreases for RCPO and P3O. As a result, the training time was increased sevenfold when compared to SUSTA setting. Nevertheless, the training was eventually halted due to insufficient speed in reducing constraint violations. Figure 1d shows the best-performing variants of the agents after extensive tuning of their hyperparameters. Table 1 displays the evaluation results after the training in the SHORT setting was completed. In the SHORT setting, IPO, RCPO, and P3O fail to generate results free of constraint violations for both the (A) environment and (B) backtesting setups. ADBO, on the other hand, guarantees by design actions free of violations during and after training.

Table 1. Evaluation after training is completed. (A) environment setup has a total of 12'000 time steps (1000 trajectories), (B) backtesting setup has a single trajectory with 12 time steps.

	SUSTA setting		SHORT setting	
	Total econ. payoff (12 months)	Total violations	Total econ. payoff (12 months)	Total violations
(A) environment				
RCPO	0.2238	0	0.2418	8656
IPO	0.2013	0	0.2721	11943
P3O	0.2561	9	0.3512	10865
ADBO (Ours)	**0.2603**	0	**0.4272**	0
(B) backtesting				
RCPO	0.4640	0	0.5285	9
IPO	0.3499	0	0.6262	12
P3O	0.5475	0	0.7654	11
ADBO (Ours)	**0.5758**	0	**1.0205**	0

7 Conclusion

In this paper, we train agents to manage investment portfolios over multiple periods, given two types of tasks that are commonly encountered in practice. Task type $T1$ constrains the allocation of a particular group of assets, e.g., assets belonging to a specific industry sector. Task type $T2$ requires the investor to short-sell one group of assets while increasing the investment in another. We propose ADBO, which finds a performant policy for a surrogate MDP rather than for the more complex CMDP. The surrogate MDP is based on an action space decomposition of the original action space. We show that ADBO outperforms general CMDP approaches for both task types in experimental settings. For future work, we will examine extensions of action space decomposition based on the Minkowski sums to a broader group of convex polytopes.

References

1. Abrate, C., et al.: Continuous-action reinforcement learning for portfolio allocation of a life insurance company. In: Dong, Y., Kourtellis, N., Hammer, B., Lozano, J.A. (eds.) ECML PKDD 2021. LNCS (LNAI), vol. 12978, pp. 237–252. Springer, Cham (2021). https://doi.org/10.1007/978-3-030-86514-6_15
2. Achiam, J., Held, D., Tamar, A., Abbeel, P.: Constrained policy optimization. In: International Conference on Machine Learning, pp. 22–31. PMLR (2017)
3. Altman, E.: Constrained Markov decision processes: stochastic modeling. Routledge (1999)
4. Ammar, H.B., Tutunov, R., Eaton, E.: Safe policy search for lifelong reinforcement learning with sublinear regret. In: International Conference on Machine Learning, pp. 2361–2369. PMLR (2015)

5. Bhatnagar, S., Lakshmanan, K.: An online actor-critic algorithm with function approximation for constrained Markov decision processes. J. Optim. Theory Appl. **153**(3), 688–708 (2012)
6. Dalal, G., Dvijotham, K., Vecerik, M., Hester, T., Paduraru, C., Tassa, Y.: Safe exploration in continuous action spaces. arXiv preprint arXiv:1801.08757 (2018)
7. Di Castro, D., Tamar, A., Mannor, S.: Policy gradients with variance related risk criteria. arXiv preprint arXiv:1206.6404 (2012)
8. Gu, S., Holly, E., Lillicrap, T., Levine, S.: Deep reinforcement learning for robotic manipulation with asynchronous off-policy updates. In: 2017 IEEE International Conference on Robotics and Automation (ICRA), pp. 3389–3396. IEEE (2017)
9. Hou, C., Zhao, Q.: Optimization of web service-based control system for balance between network traffic and delay. IEEE Trans. Autom. Sci. Eng. **15**(3), 1152–1162 (2017)
10. Liu, Y., Ding, J., Liu, X.: Ipo: Interior-point policy optimization under constraints. In: Proceedings of the AAAI Conference on Artificial Intelligence, vol. 34, pp. 4940–4947 (2020)
11. Metz, L., Ibarz, J., Jaitly, N., Davidson, J.: Discrete sequential prediction of continuous actions for deep RL. arXiv preprint arXiv:1705.05035 (2017)
12. Parisotto, E., et al.: Stabilizing transformers for reinforcement learning. In: International Conference on Machine Learning, pp. 7487–7498. PMLR (2020)
13. Peng, X.B., Abbeel, P., Levine, S., Van de Panne, M.: DeepMimic: example-guided deep reinforcement learning of physics-based character skills. ACM Trans. Graph. (TOG) **37**(4), 1–14 (2018)
14. Qin, Z., Chen, Y., Fan, C.: Density constrained reinforcement learning. In: International Conference on Machine Learning, pp. 8682–8692. PMLR (2021)
15. Schulman, J., Wolski, F., Dhariwal, P., Radford, A., Klimov, O.: Proximal policy optimization algorithms. arXiv preprint arXiv:1707.06347 (2017)
16. Sutton, R.S., McAllester, D., Singh, S., Mansour, Y.: Policy gradient methods for reinforcement learning with function approximation. In: Advances in Neural Information Processing Systems 12 (1999)
17. Tamar, A., Mannor, S.: Variance adjusted actor critic algorithms. arXiv preprint arXiv:1310.3697 (2013)
18. Tavakoli, A., Pardo, F., Kormushev, P.: Action branching architectures for deep reinforcement learning. In: Proceedings of the AAAI Conference on Artificial Intelligence, vol. 32 (2018)
19. Tessler, C., Mankowitz, D.J., Mannor, S.: Reward constrained policy optimization. In: International Conference on Learning Representations (2018)
20. Winkel, D., Strauß, N., Schubert, M., Seidl, T.: Risk-aware reinforcement learning for multi-period portfolio selection. In: Amini, M.R., Canu, S., Fischer, A., Guns, T., Kralj Novak, P., Tsoumakas, G. (eds.) Machine Learning and Knowledge Discovery in Databases. ECML PKDD 2022. LNCS, vol. 13718. Springer, Cham (2022). https://doi.org/10.1007/978-3-031-26422-1_12
21. Zhang, L., et al.: Penalized proximal policy optimization for safe reinforcement learning. In: Proceedings of the Thirty-First International Joint Conference on Artificial Intelligence, IJCAI-22, pp. 3744–3750 (2022)

Transfer Reinforcement Learning Based Negotiating Agent Framework

Siqi Chen[1](✉), Tianpei Yang[2], Heng You[1], Jianing Zhao[1], Jianye Hao[1], and Gerhard Weiss[3]

[1] College of Intelligence and Computing, Tianjin University, Tianjin 300072, China
siqichen@tju.edu.cn

[2] University of Alberta, Edmonton, Canada

[3] Department of Advanced Computing Sciences, Maastricht University, Maastricht, The Netherlands

Abstract. While achieving tremendous success, there is still a major issue standing out in the domain of automated negotiation: it is inefficient for a negotiating agent to learn a strategy from scratch when being faced with an unknown opponent. Transfer learning can alleviate this problem by utilizing the knowledge of previously learned policies to accelerate the current task learning. This work presents a novel Transfer Learning based Negotiating Agent (TLNAgent) framework that allows a negotiating agent to transfer previous knowledge from source strategies optimized by deep reinforcement learning, to boost its performance in new tasks. TLNAgent comprises three key components: the negotiation module, the adaptation module and the transfer module. To be specific, the negotiation module is responsible for interacting with the other agent during negotiation. The adaptation module measures the helpfulness of each source policy based on a fusion of two selection mechanisms. The transfer module is based on lateral connections between source and target networks and accelerates the agent's training by transferring knowledge from the selected source strategy. Our comprehensive experiments clearly demonstrate that TL is effective in the context of automated negotiation, and TLNAgent outperforms state-of-the-art Automated Negotiating Agents Competition (ANAC) negotiating agents in various domains.

Keywords: Automated negotiation · Transfer learning · Reinforcement learning · Deep learning

1 Introduction

In the domain of automated negotiation, autonomous agents attempt to reach a joint agreement on behalf of human negotiators in a buyer-seller or consumer-provider setup. The biggest driving force behind research into automated negotiation is arguably augmentation of human negotiators' abilities as well as the broad spectrum of potential applications in industrial and commercial domains [2,6]. The interaction framework enforced in automated negotiation lends itself to the use of machine learning techniques for exploring effective

H. Kashima et al. (Eds.): PAKDD 2023, LNAI 13936, pp. 386–397, 2023.
https://doi.org/10.1007/978-3-031-33377-4_30

strategies. Inspired by advances in deep learning (DL) [8,11] and reinforcement learning (RL) [14,15], the application of DRL on negotiation has made significant success [1,3,4,7,9]. However, all these methods need to learn from scratch when faced with new opponents, which is inefficient and impractical.

The existing works mainly focus on how to use the gained experience to train an agent to deal with the encountered opponents [13]. In practice, the agent however may be faced with unfamiliar or unknown opponent strategies, in which its policy may be ineffective, and the agent thus needs to learn a new policy from scratch. Besides, in most negotiation settings, agents are required to negotiate with multiple types of opponents in turn which may be unknown. The problem behind it is that learning in such manner is time-costly and may also restrict its potential performance (e.g., ignoring all previous experience and learned policies that are relevant with the current task). So, a core question arises: how to accelerate the learning process of new opponent strategy, while improving the performance of the learned policy.

This paper describes an attempt to answer the question with transfer learning (TL), which has emerged as a promising technique to accelerate the learning process of the target task by leveraging prior knowledge. We propose a novel TL-based negotiating agent called TLNAgent, which is the first RL-based framework to apply TL in automated negotiation. It comprises three key components: the negotiation module, the adaptation module, and the transfer module. The negotiation module is responsible for interacting with other agents according to the current strategy represented by a deep RL policy and providing information for other modules. The adaptation module measures the helpfulness of the source task concurrently based on the two metrics: similarity between the source opponents and the current opponent, as well as the specific performance of the source policies on the target task. The transfer module is the core of our agent framework, which accelerates the agent's training utilizing the source policies that the adaptation module selects. The comprehensive experiments conducted in the work clearly demonstrate the effectiveness of TLNAgent. Precisely, the performance of TLNAgent is carefully studied from the following aspects:

- The performance of TLNAgent and baselines are compared under standard transfer settings.
- The tournament consisting of recent ANAC winning agents is run to investigate how well TLNAgent performs against state-of-the-art negotiating agents in a broad range of negotiation scenarios.

2 Preliminaries

2.1 Negotiation Settings

The negotiation settings consist of a **negotiation protocol** and a **negotiation environment** [5]. First, the negotiation protocol defines the rules and procedures in the negotiation process. This paper considers the stacked alternating offers protocol, which defines the negotiation as alternating between two agents who can choose to accept each other's offers or propose new offers in their rounds. The negotiation terminates when both parties agree with an agreement ω, or the allowed negotiation rounds run out. Second, the negotiation environment contains the opponents and domains that the agent interacts with in the negotiation process. The strategy of an opponent makes decision at each round. The negotiation domain is composed of multiple issues and preference profiles of both parties. The preference profiles define the relative importance that an agent assigns to each issue under negotiation, and each agent only knows its own preference profile. The outcome space Ω of the negotiation domain can be denoted by $\Omega = \{\omega_1, \cdots, \omega_n\}$, where ω_i represents different offers available in the i-th domain. The offer ω_i includes an arrangement between two negotiation agents for multiple issues of the domain.

2.2 Reinforcement Learning

Markov Decision Process We model the bilateral negotiation as a MDP represented by a $\langle \mathcal{T}, \mathcal{S}, \mathcal{A}, \mathcal{P}, \mathcal{R} \rangle$ tuple. In the negotiation setting of this paper, TLNAgent will be penalized if the negotiation is not finished before the allowed negotiation rounds run out. Therefore, time T which indicates negotiation rounds is an important factor affecting the negotiation. In addition, historical offer is also a key information that affects whether agents accept the last offer or make a new offer. In conclusion, we define the state at time t as

$$\begin{aligned} S_t =& \{t_r, U_o(\omega_o^{t-2}), U_s(\omega_s^{t-2}), U_o(\omega_o^{t-1}) \\ &, U_s(\omega_s^{t-1}), U_o(\omega_o^{t}), U_s(\omega_s^{t})\} \end{aligned} \tag{1}$$

where $t_r = \frac{t}{T}$ is the relative time denoting the progress of negotiation, and the ω_o and ω_s represent the offers made by the opponent and us at time t, respectively. Since the structures of offers are completely different in diverse environments and the number of offers is spacious, it's difficult to apply the offers directly to MDP modeling. Therefore, we introduce a utility function U to map the specific offer to a value between [0, 1]. This not only contributes to the modeling of the state space but also helps us to define the action space:

$$\begin{aligned} a_t = u_s^t, \quad u_s^t < 1 \\ U^{-1}(u_s^t) = \arg\max_{\omega}(U(\omega) - u_s^t), \quad \forall \omega \in \Omega \end{aligned} \tag{2}$$

where U^{-1} is an inverse utility function that maps the utility value to a real offer. The inverse utility function U^{-1} maps the action value given by our agent to an offer ω with the closest utility value in the offer space Ω. The agent receives only one reward during the whole negotiation process based on the negotiation result. If the negotiation results in an agreement ω, the agent receives the final reward corresponding to the utility value $U(\omega)$. Otherwise, if the negotiation fails, both parties receive the same reward -1. The reward function R is defined as follows:

$$R(s_t, a_t, s_{t+1}) = \begin{cases} U_s(\omega_a), & \text{if there is an agreement } \omega_a \\ -1, & \text{if no agreement in the end} \\ 0, & \text{otherwise} \end{cases}$$

3 Transfer Learning Based Agent

3.1 Framework Overview

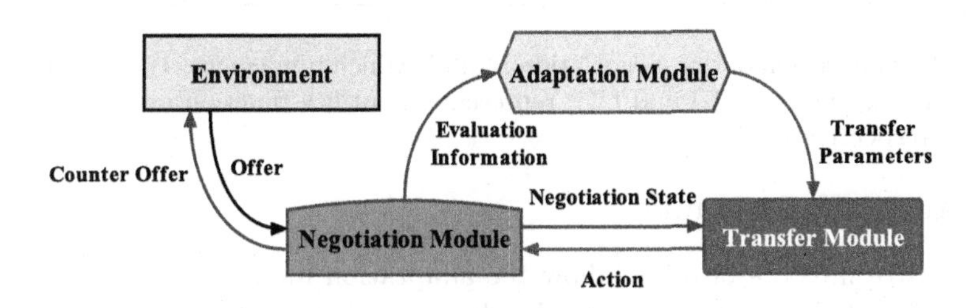

Fig. 1. An overview of our framework

To enable the agent to reuse the learned knowledge and learn how to deal with new opponents, we firstly propose the **Transfer Learning Based Agent For Automated Negotiation** framework (See Fig. 1). The framework is composed of three modules: negotiation module, adaptation module, and transfer module. Through the cooperation of three modules, the framework can accelerate the learning process when encountered a new opponent and improve the learned policy performance. Our framework performs much better than traditional methods based on RL, which will be validated in our experiments.

3.2 Negotiation Module

In this section, we introduce how the negotiation module helps the agent reaches an agreement in a negotiation process. As shown in Fig. 2, the module initializes the session information including the negotiation domain and agent preference in the beginning. Then, the negotiation module generates offers using information sent by transfer module, which implements the bidding policy. Specifically, the negotiation module passes the current state s_t according to Eq. (1) into the transfer module. Subsequently, the negotiation module utilizes Eq. (2) to convert the action a_t given by transfer module to an real offer.

When the agent receives an offer from the opponent, the negotiation module considers two actions: accept or make a counter offer. It first makes a new offer based on the present state. By comparing the utilities which are calculated by utility function $U(\cdot)$ between this offer and the received offer from opponent, the negotiation module decides whether to accept (i.e., accept when the counter offer is better than the new offer), which implements the acceptance policy.

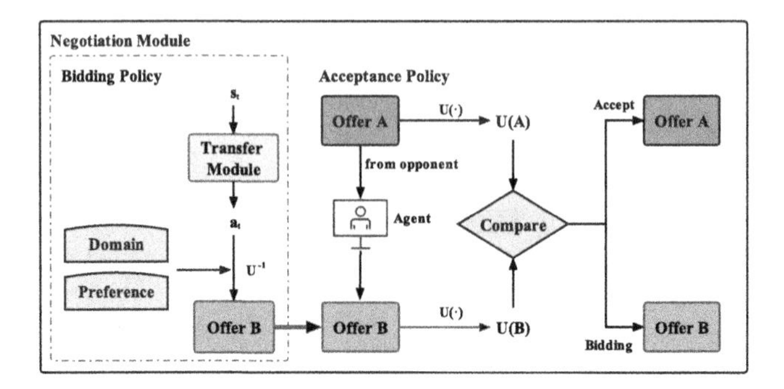

Fig. 2. An illustration of our Negotiation Module which implements the bidding policy and acceptance policy. $U(\cdot)$ and U^{-1} represent the utility function and inverse utility function respectively.

3.3 Adaptation Module

Now we dive into the of details how the adaptation module measures the helpfulness of multiple source policies. In the case of multiple source policies, the primary matter is how to transfer the most relevant knowledge to the target task under different negotiation environments. To solve this problem, we propose two evaluation metrics: performance metric and similarity metric.

As for the performance metric, it is a standard and intuitive approach to directly evaluate the average performance of each source policy when faced with the current opponent. In this work, we use the average utilities $U = \{U_1, \cdots, U_n\}$ of each source policy negotiating with the current opponent in random domains to evaluate them, where $U_n = \frac{1}{I}\sum\limits_{i=1} u_i^n$ and u_i^n denotes the reward value obtained by teacher n for the episode i of evaluation. To ensure the fairness of the negotiation, the evaluation process is only based on the mean results of different domains and is not dependent on the current environment. Subsequently, we pass U through the softmax function to get the weight: $P_{teachers} = \{p_1, \cdots, p_n\}$, where $p_i = \frac{e^{U_i}}{\sum_{i=1}^{n} e^{U_i}}$. The updating of the performance metric is performed continuously throughout the training process and soft changed to ensure the accuracy of the evaluation process and the overall training speed.

The performance metric can obtain the overall performance of source policy when faced with the current opponent. However, it is not rigorous enough to assess the source policy relying on this metric alone because only a part of information in source policies is useful and the performance metric is not fine-grained enough. Therefore, we introduce the Wasserstein distance [10] as our similarity metric to help evaluate the source policy, which compares the similarity between the opponent and the teacher library $O = \{o_1, \cdots, o_n\}$. Specifically, the teacher library contains the opponents used to train source policies. To compare the similarity of the library and the current opponent, we collected our agent's negotiation trajectories τ with different opponents under fixed episodes to calculate the Wasserstein distance. $l_\tau^o = \{H_\tau(\omega_1^o), \cdots, H_\tau(\omega_n^o)\}$ denotes the probability distribution of offers given by the opponent o in a negotiation trajectory τ, where $H(\cdot)$ is used to calculate the probability of the appearance of the corresponding offer. Then, $l_o = \{\frac{1}{k}\sum_{i=1}^{k} H_{\tau_i}(\omega_1^o), \cdots, \frac{1}{k}\sum_{i=1}^{k} H_{\tau_i}(\omega_n^o)\}$ denotes the average probability distribution of opponent's offers over the k trajectories. Similarly, we can obtain the distribution $L = \{l_1. \cdots, l_n\}$ for different opponents in the teacher library. By comparing the value of $\mathbb{W}$ in Eq. (3), we can get the similarity between the opponent which is used to train source policy and the current opponent.

$$\mathbb{W}(l_i, l_o) = \inf_{\gamma \in \Gamma(l_i, l_o)} \mathbb{E}_{(x,y)\sim\gamma}[\|x - y\|] \tag{3}$$

where $\Gamma(l_i, l_o)$ denotes the set of all joint distributions $\gamma(x, y)$ whose marginals are respectively l_i and l_o. A higher value of $\mathbb{W}(l_i, l_o)$ means that more knowledge in the corresponding source policy will be helpful to the current opponent. Then, the adaptation module takes the inverse of $\mathbb{W}(l_i, l_o)$ and passes it through the softmax function to get the weight $D_{teachers} = \{d_1, \cdots, d_n\}$, where

$$d_i = \frac{\exp(\mathbb{W}(l_i, l_o)^{-1})}{\sum_{i=1}^{n} \exp(\mathbb{W}(l_i, l_o)^{-1})}$$

As our agent's policy is constantly changing in the training process, the similarity metric $D_{teachers}$ will be soft updated every certain number of episodes throughout the negotiation process.

The weighted combination of $P_{teachers}$ and $D_{teachers}$ is used to comprehensively evaluate each source policy. To find the best performance combination, we conducted several experiments to determine the weighting factors μ and λ of the two evaluation metrics described above. The weighted factors are eventually determined as 0.5 and 0.5 for the two approaches based on multiple experiments.

$$W_{teachers} = \mu P_{teachers} + \lambda D_{teachers}$$

In conclusion, the adaptation module measures the helpfulness of each source policy by the two metrics. Then it selects the two most helpful source policies based on $W_{teachers}$ and utilizes their knowledge in the following transfer module.

3.4 Transfer Module

With the guidance of the weighting factors obtained from the adaptation module, the transfer module makes decisions by extracting suitable knowledge from multiple source policies. In the following, we will refer to these source policies as teachers and our agent as student for convenience. Inspired by prior work [12,16], we draw out knowledge directly from teachers' policies and state-value networks using the transfer method of lateral connections. We assume teachers and student have the same number of hidden layers in both the policy and value networks, where N_π and N_V denote the number of hidden layers in the policy networks and state-value networks of teachers and the student respectively. Teacher j's policy networks and state-value networks are represented by $\pi_{\phi_j'}$ and $V_{\psi_j'}$, where the parameters (ϕ'_j, ψ'_j) are fixed in the training process. In the same, the networks' trainable parameters of the student are represented by (ϕ, ψ).

In the negotiation, the student gets the current state s_t and pass it through teachers' networks to extract the pre-activation outputs of the i-th hidden layers of the j-th teachers' networks:

$$\{h^i_{\phi'_j},\ 1 \leq i \leq N_\pi,\ 1 \leq j \leq N\}$$
$$\{h^i_{\psi'_j},\ 1 \leq i \leq N_V,\ 1 \leq j \leq N\}$$

To obtain the i-th hidden layer outputs $\{h^i_{\pi_\phi}, h^i_{V_\psi}\}$ of student networks, we performed two weighted linear combinations for the pre-activations of student's networks with the pre-activations of teachers' networks [12,16]:

$$h^i_{\pi_\phi} = ph^i_\phi + (1-p)\sum_{j=1}^{N} w_j h^i_{\phi'_j}$$
$$h^i_{V_\psi} = ph^i_\psi + (1-p)\sum_{j=1}^{N} w_j h^i_{\psi'_j}$$

where $p \in [0, 1]$ is a weighted factor controlling the impact of source policies in the current environment which is increasing with training time. As p increases, source policies have a decreasing influence on our agent in the current environment to avoid the negative transfer. Besides, w_j represents the weight of source policy π_j obtained from the adaptation module. The higher the w_j, the greater the influence of the corresponding π_j on our agent in the current environment, which means the more valuable knowledge and the more helpful for forming our policy. In this way, our agent can leverage the knowledge of multiple source policies to learn a policy to deal with the current opponent.

4 Experiments

In this section, we conduct systematic studies to verify the capability of the TLNAgent compared with RL-based methods and other baselines.

Environments: We implemented 11 ANAC winning agents in our negotiation environment to evaluate the negotiation ability of our agent in different scenarios: Atlas3, ParsAgent, Caduceus, YXAgent, Ponpoko, CaduceusDC16, AgreeableAgent2018, Agent36, AlphaBIU, MatrixAlienAgent and TripleAgent. And we used all the 18 domains of ANAC2013 in the experiments.

Baselines: To demonstrate the advantages of using previous knowledge and the superiority of the transfer method when faced with new opponents, we consider the following two baselines in the experiment of Sect. 4.1: 1) Learn from scratch, which uses the standard DRL algorithm SAC and learns without prior knowledge in the new negotiation environment; 2) Learn from teachers, which is directly trained by the opponents that are used to train the source policies.

4.1 New Opponent Learning Task

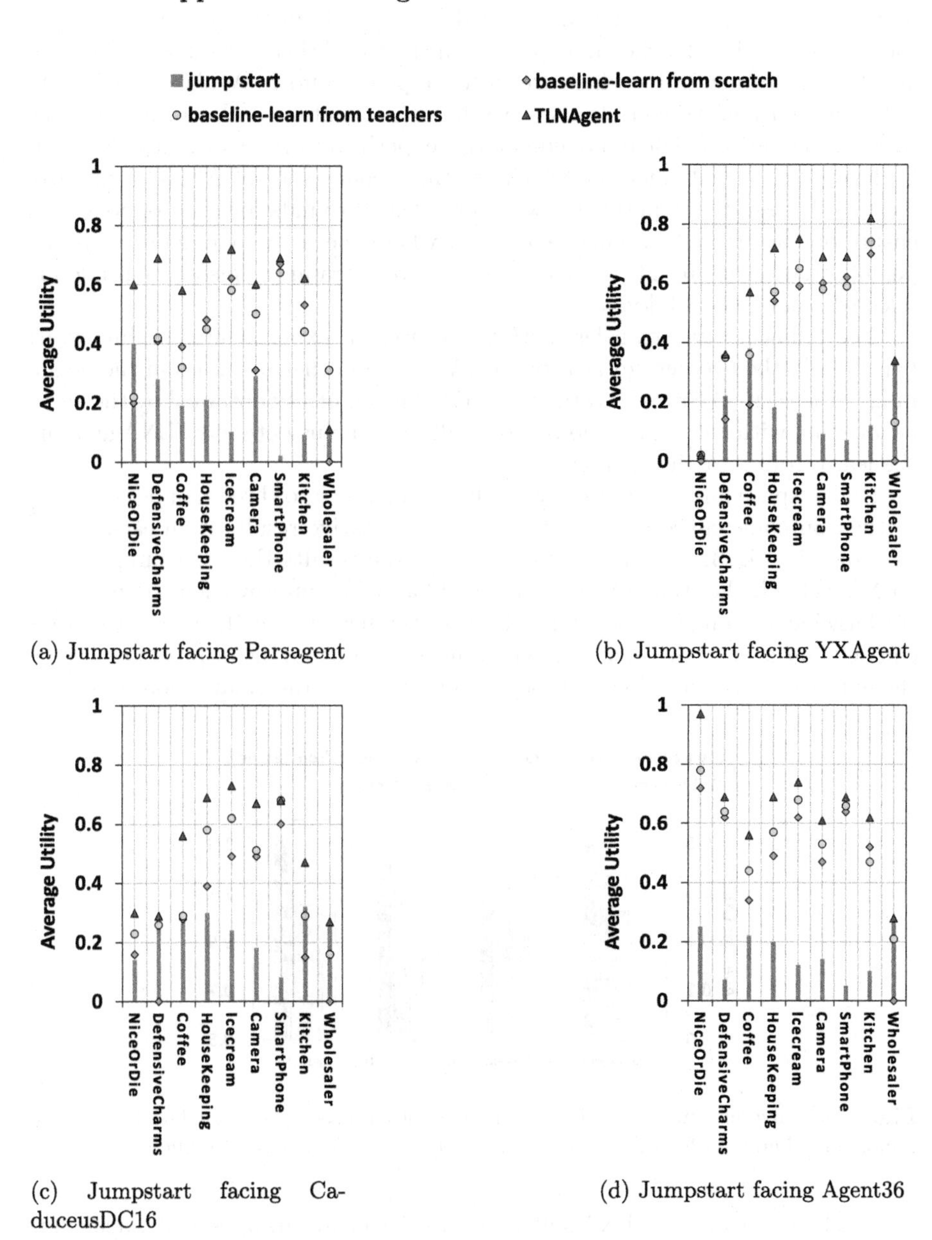

(a) Jumpstart facing Parsagent

(b) Jumpstart facing YXAgent

(c) Jumpstart facing CaduceusDC16

(d) Jumpstart facing Agent36

Fig. 3. The difference in starting rewards between TLNAgent and other baselines. The dots represent the jumpstarts of different agents. The rectangle represents the difference between TLNAgent and the learning from scratch baseline.

In this section, to verify the efficient learning ability of TLNAgent for previously unknown opponents, we evaluate the agent with multiple tasks consisting of different opponents and domains. Assume that TLNAgent is only equipped with 4 response policies that are trained by 4 agents in the teacher library as source polices. The teacher library is comprised of Atlas3, Caduceus, Ponpoko and AgreeableAgent2018, which are the champion agents of ANAC from 2015 to 2018. In addition, we consider two baselines (as mentioned above) in the same task for comparison. The opponents of this experiment are Parsagent, YXAgent, CaduceusDC16, and Agent36, which are the second place of ANAC from 2015 to 2018. During the experiment, we train 300,000 rounds for each opponent to ensure our agent and baselines converge, where the allowed number of round per negotiation is 30. The domain used in every training episode is randomly selected among the 18 domains.

The following two transfer metrics are used in experiments, 1) Jumpstart benchmark: the average rewards of TLNAgent and other baselines in the beginning of the task; 2) Transfer ratio: the ratio of mean utility obtained by the agent negotiating with a certain opponent over all 18 domains between TLNAgent and the learn from scratch baseline.

Due to space limitation, we divide all 18 domains into three groups according to their outcome space and select three representative results from each group, as shown in Fig. 3. It can be observed from the results that the jumpstart of TLNAgent is higher than two baselines and has a 50% improvement compared to the baseline learning from scratch. This result indicates that the transfer module can help our agent gain an advantage in the early stage of the negotiation, even if the improvement is not obvious in some scenarios (e.g., the SmartPhone domain).

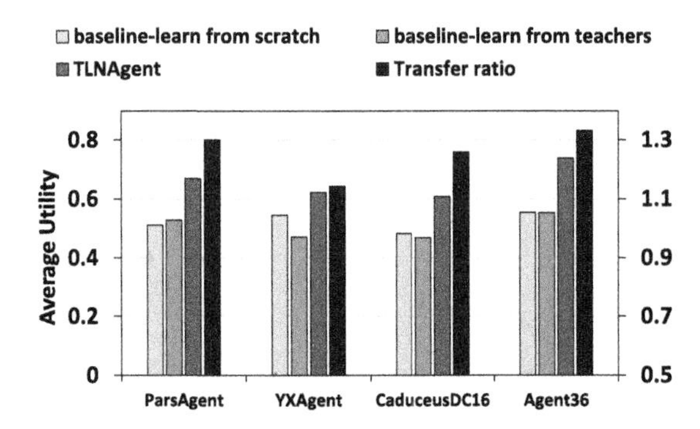

Fig. 4. The average utility of TLNAgent and other baselines when faced with new opponents. The transfer ratio is shown by green bar. (Color figure online)

As shown in Fig. 4, TLNAgent performs better for all opponents, achieving a 26% improvement in average utility compared to the two baselines. This is because TLNAgent transfers helpful knowledge from multiple source policies to the target task learning process through the transfer module. In addition, the adaptation module effectively selects the most appropriate combination of source policies in the current environment so that TLNAgent can decide when and which source policy is more valuable to conduct the adaptive transfer.

4.2 Performance Against ANAC Winning Agents

This section presents the experimental results of a tournament of our agent and 11 ANAC winning agents. To be specific, the experiment consider the top two agents from 2015 to 2018 competitions plus the top three in the 2021 competition[1]. In the tournament, every agent pair will perform a bilateral negotiation of 1000 episodes. The results are shown in Table 1, and the experiments use the following metrics, 1) Average utility benchmark: the mean utility obtained by the agent $p \in A$ when negotiating with every other agent $q \in A$ on all domains D, where A and D denote all the agents and all the domains used in the tournament, respectively; 2) Agreement rate benchmark: the agreement achievement rate between the agent and all others throughout the tournament.

Table 1. Comparison of our proposed TLAgent with 11 ANAC winning agents using average utility benchmark and average agreement achievement rate.

Agent	Average utility	95% confidence interval		Average agreement rate
		Lower Bound	Upper Bound	
Atlas3	0.513	0.487	0.539	0.53
ParsAgent	0.408	0.391	0.425	0.51
Caduceus	0.428	0.415	0.441	0.55
YXAgent	0.474	0.453	0.495	0.37
Ponpoko	0.393	0.382	0.404	0.44
CaduceusDC16	0.452	0.432	0.472	0.53
AgreeableAgent2018	0.533	0.512	0.554	0.79
Agent36	0.315	0.289	0.341	0.47
AlphaBIU	0.572	0.552	0.592	0.64
MatrixAlienAgent	0.558	0.534	0.582	0.59
TripleAgent	0.549	0.532	0.546	0.57
TLAgent	**0.626**	**0.619**	**0.633**	**0.82**

Table 1 shows the performance of our TLNAgent on the average utility benchmark with standard deviation, concurrently with average agreement achievement rate. Our TLNAgent outperforms all ANAC winning agents in the tournament, as exemplified by the higher average utility and higher agreement achievement rate. Without considering the advanced ANAC winning agents of 2021 who have access to past negotiation data, the average utility obtained by our agent is 40% higher than the average benchmark over all other ANAC winning agents. Even when 2021 ANAC winning agents are considered in the comparison, TLNAgent still manages to achieve around 30% advantage in the average utility benchmark. This means that when encountering a new opponent, the agent can utilize the

[1] Note that the themes of ANAC 2019 & 2020 are to elicit preference information from a user during the negotiation, which are different from our negotiation setting.

knowledge of source policies through the adaptation module and transfer module to enhance its negotiation performance rapidly facing the opponent. In addition, TLNAgent achieves the highest agreement rate in the tournament among all agents. The results together show the effectiveness of our framework.

5 Conclusion and Future Work

In this paper we introduced a novel transfer reinforcement learning based negotiating agent framework called TLNAgent for automated negotiation. The framework contains three components: the negotiation module, the adaptation module and the transfer module. Furthermore, the framework adopts the performance metric and the similarity metric to measure the transferbility of the source policies. The experimental results show a clear performance advantage of TLNAgent over state-of-the-art baselines in various aspects. In addition, an analysis was also performed from the transfer perspective.

TLNAgent opens several new research avenues, among which we consider the following as most promising. First, as opponent modeling is another helpful way to improve the efficiency of a negotiation, it's worthwhile investigating how to combine opponent modeling techniques with our framework. Also, it is very interesting to see how well TLNAgent performs against human negotiators. The third important avenue we see is to enlarge the scope of the proposed framework to other negotiation forms.

Acknowledgments. This study was supported by the National Natural Science Foundation of China (Grant No. 61602391).

References

1. Bagga, P., Paoletti, N., Alrayes, B., Stathis, K.: A deep reinforcement learning approach to concurrent bilateral negotiation. In: Proceddings of IJCAI-20 (2020)
2. Chen, S., Ammar, H.B., Tuyls, K., Weiss, G.: Using conditional restricted Boltzmann machine for highly competitive negotiation tasks. In: Proceedings of the 23th International Joint Conference on Artificial Intelligence, pp. 69–75. AAAI Press (2013)
3. Chen, S., Su, R.: An autonomous agent for negotiation with multiple communication channels using parametrized deep Q-network. Math. Biosci. Eng. **19**(8), 7933–7951 (2022). https://doi.org/10.3934/mbe.2022371
4. Chen, S., Sun, Q., Su, R.: An intelligent chatbot for negotiation dialogues. In: Proceedings of IEEE 20th International Conference on Ubiquitous Intelligence and Computing (UIC), pp. 68–73. IEEE (2022)
5. Chen, S., Weiss, G.: An intelligent agent for bilateral negotiation with unknown opponents in continuous-time domains. ACM Trans. Auton. Adapt. Syst. **9**(3), 1–24 (2014). https://doi.org/10.1145/2629577
6. Chen, S., Weiss, G.: An approach to complex agent-based negotiations via effectively modeling unknown opponents. Expert Syst. Appl. **42**(5), 2287–2304 (2015). https://doi.org/10.1016/j.eswa.2014.10.048

7. Chen, S., Yang, Y., Su, R.: Deep reinforcement learning with emergent communication for coalitional negotiation games. Math. Biosci. Eng. **19**(5), 4592–4609 (2022). https://doi.org/10.3934/mbe.2022212
8. Chen, S., Yang, Y., Zhou, H., Sun, Q., Su, R.: DNN-PNN: a parallel deep neural network model to improve anticancer drug sensitivity. Methods **209**, 1–9 (2023). https://doi.org/10.1016/j.ymeth.2022.11.002
9. Lillicrap, T.P., et al.: Continuous control with deep reinforcement learning. In: 4th International Conference on Learning Representations, ICLR 2016, Conference Track Proceedings (2016)
10. Ramdas, A., Trillos, N.G., Cuturi, M.: On Wasserstein two-sample testing and related families of nonparametric tests (2017)
11. Su, R., Yang, H., Wei, L., Chen, S., Zou, Q.: A multi-label learning model for predicting drug-induced pathology in multi-organ based on toxicogenomics data. PLoS Comput. Biol. **18**(9), e1010402 (2022). https://doi.org/10.1371/journal.pcbi.1010402
12. Wan, M., Gangwani, T., Peng, J.: Mutual information based knowledge transfer under state-action dimension mismatch. In: Proceedings of the Thirty-Sixth Conference on Uncertainty in Artificial Intelligence (2020)
13. Wu, L., Chen, S., Gao, X., Zheng, Y., Hao, J.: Detecting and learning against unknown opponents for automated negotiations. In: Pham, D.N., Theeramunkong, T., Governatori, G., Liu, F. (eds.) PRICAI 2021: Trends in Artificial Intelligence (2021)
14. Yang, T., Hao, J., Meng, Z., Zhang, C., Zheng, Y., Zheng, Z.: Towards efficient detection and optimal response against sophisticated opponents. In: Proceedings of the Twenty-Eighth International Joint Conference on Artificial Intelligence, pp. 623–629. ijcai.org (2019)
15. Ye, D., et al.: Towards playing full MOBA games with deep reinforcement learning. In: Proceedings of the 34th International Conference on Neural Information Processing Systems (2020)
16. You, H., Yang, T., Zheng, Y., Hao, J., Taylor, M.E.: Cross-domain adaptive transfer reinforcement learning based on state-action correspondence. In: Uncertainty in Artificial Intelligence, Proceedings of the Thirty-Eighth Conference on Uncertainty in Artificial Intelligence (2022)

Relational Learning

A Relational Instance-Based Clustering Method with Contrastive Learning for Open Relation Extraction

Xiaoge Li[1(✉)], Dayuan Guo[2], and Tiantian Wang[1]

[1] School of Computer Science and Technology, Xi 'an University of Posts and Telecommunications, Xi 'an 710121, China
lixg@xupt.edu.cn

[2] Shaanxi Key Laboratory of Network Data Analysis and Intelligent Processing, Institute, Xi' an 710121, China

Abstract. Unsupervised text representations significantly narrow the gap with supervised pretraining, and relation clustering has gradually become an important method of open relational extraction (OpenRE). However, different relational categories generally overlap in the high-dimensional representation space, and distance-based clustering is difficult to separate different categories. In this work, we propose a relational instance-based clustering method with contrastive learning (RICL) - a framework to leverage similarity distribution information and contrastive method to promote better aggregation and relational representation. Specifically, to enable the model to better represent relation instances with word-level features, we construct an augmented dataset using only standard dropout as noise and iteratively optimize the vector representation of relation instances by fully using self-supervised signals. Experiments on real-world datasets show that RICL can achieve excellent performance compared with previous state-of-the-art methods.

Keywords: Relation Extraction · Unsupervised Clustering · Contrastive learning

1 Introduction

Relation extraction is an important basic work for building large-scale knowledge bases such as semantic networks and knowledge graphs [1–3]. However, conventional relation extraction methods such as semi-supervision and distant supervision are generally used to deal with pre-defined relations and cannot well identify emerging relations in the real world.

Against this background, OpenRE has been widely studied for its ability to mine novel relation from massive text data. At present, OpenRE is mainly based on unsupervised methods, which can be divided into two categories. The first group is pattern extraction models [4–6], which usually uses sentence analysis tools, combined with linguistics and professional domain knowledge, to construct artificial rules based on

H. Kashima et al. (Eds.): PAKDD 2023, LNAI 13936, pp. 401–411, 2023.
https://doi.org/10.1007/978-3-031-33377-4_31

lexical, syntactic and semantic features. When performing relation extraction tasks, different relation types are obtained by matching rules with the preprocessed text. However, with the expansion of the relational model set, the complexity of the system is greatly increased, and it is difficult to be widely used in the open field. The second group is to discover various relation types through unsupervised methods [7–9]. This work optimizes the representation of relations to improve the accuracy of unsupervised clustering while overcoming the instability of unsupervised training. Recently, some RE methods work begin to study better utilization of hand-crafted features, which only use named entities to induce relation types [10]. The hierarchy information in relation types is further exploited for better novel relation extraction [11].

However, much research has shown that complex linguistic information requires high-dimensional embeddings so that the meaning of the text becomes clear [12]. This complex information may contain local syntactic [13] and semantic structures [14]. Therefore, the position and relative distance in the high-dimensional vector space is not completely consistent with the relational semantic similarity. Especially before model training starts, even with deep neural networks, different classes may still overlap in high-dimensional space [15].

We propose a relational instance-based clustering method with contrastive learning in this work. In order to make the model better mine the information of the relation instance itself to produce better clustering results, the nonlinear mapping is optimized by using the difference information of the constructed relation instance's comparative dataset and the distribution information of the original instance dataset. High-dimensional relational instance features of complex information are transformed into relation-oriented low-dimensional feature representations. Specifically, we pull together instances representing the same relationship while pushing apart those from different ones by jointly optimizing distribution loss and contrastive loss so that the learned representation is cluster-friendly. In addition, the proposed method obtains supervision from the data itself and its corresponding augmented dataset and iteratively learns better feature representations for relation classification tasks to improve the quality of supervision, which in turn improves cluster purity and separates distances between different clusters.

Overall, our work has the following contributions: (1) we propose a self-supervised framework which can fine-tune pretrained MLMs into capable universal relational encoders and extensively learn to cluster relational data; (2) we show how to use contrastive learning to learn and improve representations of relation instances in a self-supervised manner.

2 Related Work

Self-supervised learning has recently achieved excellent results on multiple tasks in the image and text domains, and many studies have been further developed thanks to its effectiveness in feature representation work. The quality of learned representations is assured by a theoretical framework based on contrast learning [16], which learns self-features from unlabeled data and formalize the concept of semantic similarity through latent classes to improve the performance of classification tasks. Hu et al. [9] propose adaptive clustering algorithms and uses pseudo-labels of relations as self-supervised signals to optimize their semantic representations. Recently, there has been an increasing interest in contrast learning using individual raw sentences based on PLMs [15, 17, 18].

Meanwhile, inspired by research related to contrast learning in computer vision [19, 20], we utilize "multi-view" contrastive learning for relation extraction. Previous work mainly uses sentences as the smallest unit of text input, builds enhanced datasets by randomly masking characters or replacing words, and uses semantic similarity as the goal of the measurement model. In contrast, our work takes entity word pairs as the minimum granularity of semantic representation, abstracts various types of relations, and obtains their vector representations with the help of the idea of clustering. It not only maintains the advantages of unsupervised learning, which can deal with deal with undefined relation types, but also exerts the advantages of supervised learning, which has a strong guiding ability for relational feature learning.

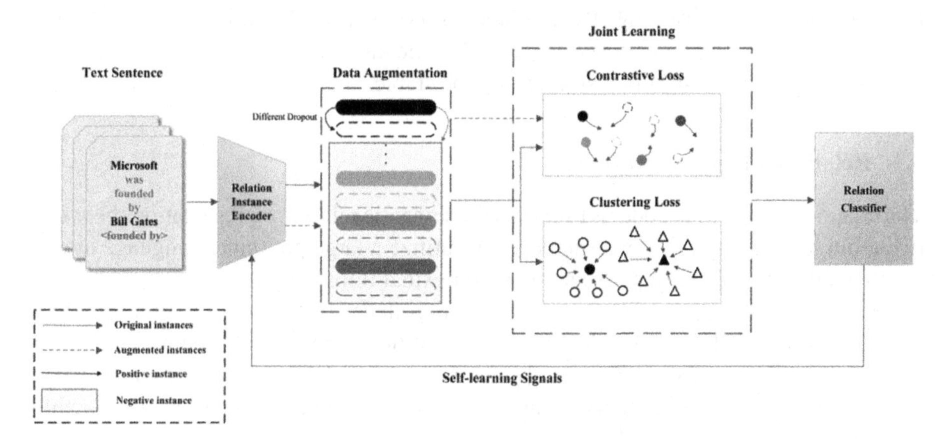

Fig. 1. Overall architecture of RICL

3 Methodology

In this work, we propose a simple and effective approach to relation clustering, which exploits relation instance distribution information in unlabeled data and semantic information from pretrained models, enabling the model to optimize the representation of relations.

In order to alleviate the overlap of different relation clusters in the representation space, we improve the clustering of unlabeled data by contrastive learning to promote better separation. The proposed method is shown in Fig. 1.

We build a "multi-view" of the training corpus, gradually optimize the representation of relation instances in a joint learning manner and aggregate to generate pseudo-labels, and fine-tune the pre-trained language model through the classification. As shown in Figure 1., we mainly iteratively perform the following steps:

(1) First, we use the pretrained BERT as the encoder of relational instances $\{h_i\}_{i=1,\ldots,N}$; each relational instance h_i is composed of an entity pair vector as the output

vector. However, high-dimensional representations of h contain too much information (structural features, semantic information, etc.), and the direct use of high-dimensional vectors for clustering cannot align well with the relationships corresponding to instances.

(2) In order to better reflect the semantic similarity between each other through the distance between the relation representation spaces, we transform the high-dimensional representations of relation instances h_i into low-dimensional representations $h_i^{'}$ through non-linear mapping g. However, the quality of pseudo-labels produced by direct clustering is not high, which is not conducive to downstream classification tasks.

(3) In order to reduce the negative impact of pseudo-label errors, we apply different dropouts under the same pre-training model to construct a positive set and other data under the same batch as a negative set. During the training process, aiming at the aggregation of clusters of similar relational instances and the separation of different instances, the representation of relation instances is optimized to improve the quality of pseudo-labels produced by clustering. Pseudo-labels serve as prior knowledge of the dataset and are finally used for supervised relation classification. The above steps are executed iteratively until the clustering result tends to be stable.

3.1 Relational Instance Encoder

The relational instance encoder is to extract the semantic relation representations between two arbitrary given entities in a sentence. We utilize a large pretrained language model to efficiently encode entity pairs and their contextual information.

For sentence $S = [s_1, \ldots, s_n]$, we introduce two pairs of special identifiers $[E1\backslash], [\backslash E1], [E2], [\backslash E2]$ to mark entities and inject them to $S = [s_1, \ldots, [E1\backslash], s_i, \ldots, s_k, [\backslash E1], \ldots, [E2\backslash], s_m, \ldots s_j, [\backslash E2], \ldots, s_n]$. We adopt BERT [21] as our encoder $l(\bullet)$ due to its strong performance and wide application in extracting semantic information. Formally:

$$v_1, \ldots, v_n = l(s_1, \ldots, s_n) \tag{1}$$

$$h = \left[v_{[E1\backslash]}, v_{[E2\backslash]}\right] \tag{2}$$

where v_i is a word vector generated by BERT, we use the outputs concatenated by $v_{[E1/]}$ and $v_{[E2/]}$ as the representation of the relational instance. This method of relational representation has been widely used in previous RE methods [9, 22, 23].

3.2 Instance-Relational Contrastive Learning

We use the distribution information of relation instances and their own feature information to build a joint model to achieve deep clustering. As shown in **Fig. 1**, our joint learning model is composed of two components, $f(\bullet)$ and $g(\bullet)$, using clustering loss and contrastive loss, respectively. We describe the specific structure of the model in Sect. 4.

Dropout Noise as Data Augmentation. We use different dropouts to obtain different vector representations of the same text. Specifically, for each batch $B = \{h_i\}_{i=1}^{M}$, we generate a new vector representation for each relation instance in B and then get an augmented batch $B^a = \{h_i, \tilde{h}_i\}_{i=1}^{M}$. The positive pair h_i, $\tilde{h}_i$ takes exactly the same relational instance, and their embeddings only differ in dropout masks, while treating the other $2M - 2$ instances as negative instances N of this positive pair. Here the dropout rate p is 0.1.

Given a batch of data B^a, τ denotes a temperature parameter. We leverage the standard InfoNCE loss [24] to aggregate the positive pairs together and separate the negative pairs in the embedding space:

$$L_a = -\sum\nolimits_{i=1}^{M} \log \frac{\exp(\cos(g(h_i), g(\tilde{h}_i))) / \tau}{\sum_{h_j \in N} \exp(\cos(g(h_i), g(h_i)))} \tag{3}$$

3.3 Clustering

Different from contrastive learning, clustering focuses on the similarity between different instances, encodes abstract semantic information into representations of relation instances, and finally aggregates instances of the same relation.

The known dataset consists of K relation classes. The centroid representation of each class denoted as u_k, $k \in \{1, ..., K\}$. We compute the probability of assigning h_i to the kth cluster by student's t-distribution [25]:

$$q_{ik} = \frac{(1 + ||h_i - u_k||_2^2 / \alpha)^{-\frac{\alpha+1}{2}}}{\sum_{k'=1}^{K} (1 + ||h_i - u_{k'}||_2^2 / \alpha)^{-\frac{\alpha+1}{2}}} \tag{4}$$

Here α denotes the degree of freedom of the student's t-distribution and q_{ik} can be regarded as the probability of the cluster assignment. In general, we follow Maaten and Hinton [25] by setting $\alpha = 1$.

A linear layer $f(\bullet)$ is used to fit the centroid of each relation cluster and then iteratively improve it by the auxiliary distribution proposed by Xie et al. [26] Concretely, defining p_{ik} as the auxiliary probability:

$$p_{ik} = \frac{q_{ik}^2 / f_k}{\sum_{k'} q_{ik}^2 / f_{k'}} \tag{5}$$

where $f_k = \sum_{i=1}^{M} q_{ik}$, $k = 1, \ldots, K$ is the cluster frequency within a batch, the purpose of this is to encourage learning from high-confidence cluster assignments while improving low-confidence tasks against biases caused by imbalanced clusters, resulting in better clustering performance.

We optimize the KL divergence loss between the cluster assignment probability and the target distribution:

$$L_b = KL(P||Q) = \sum\nolimits_{i} \sum\nolimits_{k} p_{ik} \log \frac{p_{ik}}{q_{ik}} \tag{6}$$

In conclusion, our overall objective is,

$$L = (1 - \varepsilon)L_a + \varepsilon L_b \tag{7}$$

ε balances between the clustering loss and the contrastive loss of RICL is set to 0.65. Note that L_b is only optimized on the initial data, and the parameters for $f(\bullet)$ and $g(\bullet)$ will be updated-parameters in the $l(\bullet)$ are not improved when minimizing L.

Finally, we obtain $\{h_i'\}_{i=1}^{M}$ using the optimized $g(\bullet)$ and $f(\bullet)$, and then generate pseudo-labels y' by k-means algorithm:

$$y' = Kmeans(h') \tag{8}$$

3.4 Relation Classification

Based on the pseudo-labels y' generated by clustering, we can use supervised learning to train the classifier and refine relational instance h to encode more relational semantic information:

$$l_n = \mu_\tau(l_\theta(S)) \tag{9}$$

$$L_C = \min_{\theta,\tau} \frac{1}{M} \sum\nolimits_{n=1}^{M} loss(l_n, one\_hot(y_n')) \tag{10}$$

where μ_τ denotes the relation classification module parameterized by τ and I_n is a probability distribution over K pseudo-labels for the original data. In order to find the best-performing parameters θ for Relational Instance Encoder and τ for Relation Classification, we optimize the above classification loss.

4 Experimental Setup

We first introduce publicly available datasets for training and evaluation. Then we briefly introduce the baseline models used for comparison. Finally, we elaborate on the hyperparameter configuration and implementation details of RICL.

4.1 Datasets

We conduct experiments and comparisons on three open-domain datasets.

FewRel. Few-Shot Relation Classification Dataset is derived from Wikipedia and annotated by humans [27]. FewRel contains 80 types of relations, each with 700 instances. Following the paper [7], we use all instances of 64 relations as training set, and the test set of FewRel, which randomly selects 16 relations with 1600 instances.

T-REx SPO and T-REx DS. They come from the T-Rex dataset [28], which is generated by aligning Wikipedia corpus with Wiki-data. At first, we need to preprocess each sentence in the dataset. If there are multiple entity pairs in a sentence, the sentence will be retained for the same number of times according to the number of occurrences of different entity pairs. And then, we built two datasets, T-REx SPO and T-REx DS, according to Hu et al. [9]. In both datasets, 80% of sentences will be used for model training, and the remaining 20% were set aside for validation, the rest for testing.

4.2 Baseline and Evaluation Metrics

We use standard unsupervised evaluation metrics for comparisons with the other six baseline algorithms. For all models, we assume the number of target relation classes is known in advance, but no human annotations are available to extract relations from the open-domain data. We set the number of clusters to the number of ground-truth classes and evaluate performance with B3, V-measure, and ARI [8, 9, 29]. To evaluate the effectiveness of our method, we select the following SOTA OpenRE models for comparison.

VAE [30] consists of a classifier that predicts relations and a factorization model which reconstructs arguments. The model is jointly optimized by reconstructing entities from pairing entities and predicted relations.

UIE [8] trains a discriminative relation extraction model by introducing a skewness loss and a distribution distance loss to make the model confidently predict each relation and encourage the average prediction of all relations.

SelfORE [9] uses an adaptive clustering algorithm to obtain relation sets based on a large pretrained language model and then uses the pseudo-labels of relations as self-supervised signals to optimize their semantic representations.

EI_ORE [29] conduct Element Intervention, which intervenes on the context and entities respectively to obtain the underlying causal effects of them, to address the spurious correlations from entities and context to the relation type.

RW-HAC [31] reconstructs word embeddings and uses single feature reduction to alleviate the feature sparsity problem for relation extraction through clustering.

Etype + [10] consists of two regularization methods and a link predictor and uses only named entity types to induce relation types.

4.3 Implementation Details

Follow the settings used in previous work [8, 9, 29, 30], at T-REx SPO and T-REx DS datasets, RICL are trained with 10 relation classes. Although it is lower than the number of real relationships in the dataset, it still reveals important insights due to the very imbalanced distribution of relationships on the 10 relation classes of data used for training and testing.

For Relational Instance Encoder, we use the default tokenizer in BERT to preprocess all datasets and set the max length of a sentence as 128. We use the BERT-base-uncased model to initialize parameters for $l(\bullet)$ and use BertAdam to optimize the loss.

For Instance-relational Contrastive Learning, we use an MLP $g(\bullet)$ with fully connected layers with the following dimensions $\mathbb{R}^d$-512–512-256. We randomly initialize weights following Xie et al. [26]. For Clustering, we use a linear layer $f(\bullet)$ of size $256 \times K$ with K indicating the number of clusters, and initialize the cluster centers by the Kmean algorithm.

For Relation Classification, we use a fully connected layer as μ_τ and set the dropout rate to 10%, the learning rate to $5e-5$, and the warm-up rate to 0.1. In the process of fine-tuning BERT, we freeze its first 8 layers. All experiments are conducted using an NVIDIA GeForce RTX 3090 with 24GB memory.

5 Results and Analysis

In this section, we present the experimental results of RICL on three open-domain datasets, and verify the rationality of the framework through ablation experiments. Finally, we prove its effectiveness by combining data characteristics and visual analysis.

Table 1. Main results on three relation extraction datasets.

Dataset	Model	B3			V-measure			ARI
		F1	Prec.	Rec.	F1	Hom.	Comp.	
T-Rex SPO	VAE [30]	24.8	20.6	31.3	23.6	19.1	30.6	12.6
	UIE-BERT [8]	38.1	30.7	50.3	39.1	37.6	40.8	23.5
	SelfORE [9]	41.0	39.4	42.8	41.1	40.3	42.5	33.7
	EI_ORE [29]	45.0	46.7	43.4	45.3	45.4	45.2	36.6
	Our	44.6	42.9	44.4	47.2	46.2	48.2	37.1
T-Rex DS	VAE [30]	9.0	6.4	15.5	5.7	4.5	7.9	1.9
	UIE-BERT [8]	22.4	17.6	30.8	31.2	26.3	38.3	12.3
	SelfORE [9]	32.9	29.7	36.8	32.4	30.1	35.1	20.1
	EI_ORE [29]	42.9	40.2	45.9	47.3	46.9	47.8	25.0
	Our	43.3	41.3	46.6	47.1	47.3	48.6	28.2
FewRel	VAE [30]	36.5	30.9	44.6	47.3	44.8	50.0	29.1
	RW-HAC [31]	33.7	25.6	49.2	43.3	39.1	48.5	25.0
	EType + [10]	31.9	23.8	48.5	40.8	36.4	46.3	24.9
	SelfORE [9]	51.2	50.8	51.6	58.3	57.9	58.8	34.7
	Our	53.9	50.9	57.4	65.3	63.2	67.6	47.3

5.1 Main Results

Table 1 reports model performances on T-Rex SPO, T-Rex DS, and FewRel dataset, which shows that the proposed method achieves state-of-the-art results on the OpenRE task. Benefiting from the rich information in the pre-trained model, RICL exploits the relation distribution in unlabeled data and optimizes the relation representation through the method of contrastive learning so as to achieve a better clustering effect, thus greatly surpassing previous cluster-based baselines.

5.2 Ablation Study

In order to study the effect of each algorithm in the proposed framework, we conduct ablation experiments on two datasets, respectively, and the results are presented in **Table 2**.

Table 2. Ablation results on T-Rex SPO and FewRel

Method	T-Rex SPO			FewRel		
	B3-F1	V-F1	ARI	B3-F1	V-F1	ARI
w/o contrastive learning	41.8	43.1	22.9	51.0	57.8	34.1
w/o clustering	39.9	39.8	19.6	47.5	52.0	34.9
w/o classification	42.9	46.6	32.0	51.9	59.2	45.5
RICL	44.6	47.2	37.1	53.9	65.3	47.3

The results show that the model performance is degraded if L_a is removed, indicating that Instance-relational Contrastive Learning can produce superior relation embeddings from either unlabeled data. It is worth noting that Clustering has an important role in RICL. It prevents the excessive separation of the same relation instance in the space, avoids the collapse of the relation semantic space. At the same time, it provides guidance for downstream relation classification and optimizes the representation of relation instances. In addition, joint optimizing on both the Clustering and the Contrastive Learning is also very important. While alleviating the overlap of different relation classes in the representation space, different instances under the same class are aggregated.

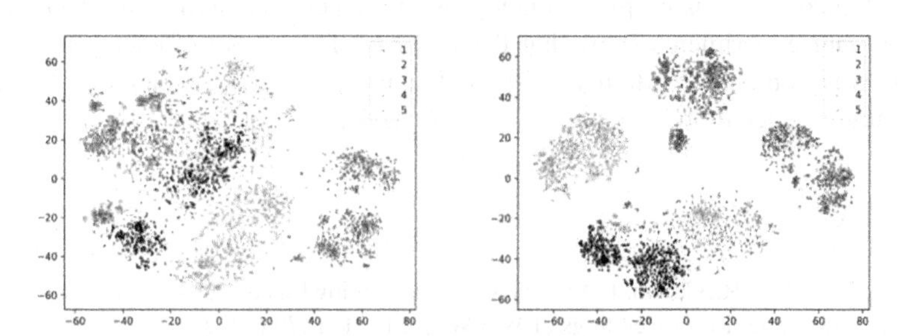

Fig. 2. Visualization of feature embeddings on FewRel-5

5.3 Visualization and Analysis

To further explore the performance of RICL and the rationality of its design, we randomly select 5 types of data in the FewRel dataset and visualize the embedded features from BERT-base-uncased (left) and RICL (right) with t-SNE in **Fig. 2**. It is convenient for us to observe the changes in class distribution.

In the initial distribution, we observe that classes 2, 3, 4 have high purity, but these classes are not highly clustered and have slight overlap at the boundaries. The relation instances of class 1 and 5 are heavily overlapped in space. Through the analysis of relationship classes and their instances, class 1 describes the "located in" relation between the airport and the place it belongs to, and class 5 describes the "located in" relation between the regional locality and the city or country. These two classes are affected by factors such as relational semantics and entity types [10], and some relation instances are spatially closely distributed.

From a global perspective, RICL achieves better separation of each class in space, solves the problem of blurred boundaries, ensures the overall consistency, and explores the possibility of further subdividing categories under the same class. While classes 2, 3, 4 are aggregated, they are separated from different class as much as possible in space to ensure semantic consistency. When dealing with class 1 and class 5 overlapping problems, RICL locally aggregates discretely distributed class 5 instances and separates them from class 1 while guaranteeing relational consistency, thereby improving class purity as much as possible.

6 Conclusions

In this paper, we propose a novel self-supervised learning framework for open-domain relation extraction, namely RICL. It aims to enable the neural network to obtain better relation-oriented representation encoding and how to better handle relational instances in the open domain in a self-supervised manner. We utilize instance distribution information and contrastive learning to promote better aggregation and relational representation, improving clustering accuracy and reducing error propagation, thus benefiting downstream classification tasks. Moreover, we iteratively improve the robustness of the neural encoder by using pseudo-labels as self-supervised signals for relation classification. Our experiments show that RICL can perform more efficient and accurate relation extraction on open-domain corpora than previous methods, and can construct a representation space more suitable for semantic tasks.

References

1. Xiong, C., Power, R., Callan, J.: Explicit semantic ranking for academic search via knowledge graph embedding. In: Proceedings of WWW, pp. 1271–1279 (2017)
2. Wang, Z., Zhang, J., Feng, J., Chen, Z.: Knowledge graph embedding by translating on hyperplanes. In: Proceedings of AAAI, pp. 1112–1119 (2014)
3. Dong, L., Wei, F. R., Zhou, M., Xu, K.: Question answering over freebase with multicolumn convolutional neural networks. In: Proceedings of ACL-IJCNLP, pp. 260–269 (2015)
4. Anthony, F., Stephen, S., Oren, E.: Identifying relations for open information extraction. In: Proceedings of EMNLP, pp. 1535–1545 (2011)
5. Jiang, M., Shang, J., Taylor, C., Ren, X., Lance, M., Timothy, P., Han, J.: Metapad: Meta pattern discovery from massive text corpora. In: Proceedings of KDD, pp. 877–886 (2017)
6. Zheng, S., et al.: DIAG-NRE: A neural pattern diagnosis framework for distantly supervised neural relation extraction. In: Proceedings of ACL, pp. 1419–1429. (2019)
7. Wu, R., et al.: Open relation extraction: Relational knowledge transfer from supervised data to unsupervised data. In: Proceedings of EMNLP-IJCNLP, pp. 219–228 (2019)
8. Étienne, S., Vincent, G., Benjamin, P.: Unsupervised information extraction: Regularizing discriminative approaches with relation distribution losses. In: Proceedings of ACL, pp. 1378–1387 (2019)
9. Hu, X., Wen, L., Xu, Y., Zhang, C., Philip Y.: SelfORE: self-supervised relational feature learning for open relation extraction. In: Proceedings of EMNLP, pp 3673–3682 (2020)

10. Tran, T., Le, P., Ananiadou, S.: Revisiting unsupervised relation extraction. In: Proceedings of ACL, pp. 7498–7505 (2020)
11. Zhang, K, et al.: Open Hierarchical Relation Extraction. In: Proceedings of ACL, pp. 5682–5693 (2021)
12. Choudhary, R., Doboli, S., Minai, A.: A Comparative Study of Methods for Visualizable Semantic Embedding of Small Text Corpora. In: Proceedings of IJCNN, pp. 1–8 (2021)
13. Hewitt, J., Manning, C.: A structural probe for finding syntax in word representations. In: Proceedings of NAACL, pp. 4129–4138 (2019)
14. Richie, R., White, B., Bhatia, S., Hout, M.C.: The spatial arrangement method of measuring similarity can capture high-dimensional semantic structures. Behav. Res. Methods **52**(5), 1906–1928 (2020). https://doi.org/10.3758/s13428-020-01362-y
15. Zhang, D., Nan, F., Wei, X., et al.: Supporting clustering with contrastive learning. In: Proceedings of ACL, pp. 5419–5430 (2021)
16. Arora, S., Khandeparkar, H., Khodak, M., Plevrakis, O., Saunshi, N.: A Theoretical Analysis of Contrastive Unsupervised Representation Learning. arXiv preprint arXiv: 1902.09229 (2019)
17. Liu, F., Vulić, I., Korhonen, A., et al.: Fast, effective, and self-supervised: Transforming masked language models into universal lexical and sentence encoders. In: Proceedings of EMNLP, pp. 1442–1459 (2021)
18. Gao, T., Yao, X., Chen, D.: SimCSE: Simple contrastive learning of sentence embeddings. In: Proceedings of ACL, pp. 6894--6910 (2021)
19. Chen, T., Zhai, X., Ritter, M., Lucic, M., Houlsby, N.: Self-supervised gans via auxiliary rotation loss. In: Proceedings of CVPR, pp. 12154–12163 (2019)
20. Chen, X., Fan, H., Girshick, R., He, K.: Improved Baselines with Momentum Contrastive Learning. arXiv preprint arXiv: 2003.04297 (2020)
21. Devlin, J., Chang, M., Lee, K., Toutanova, K.: BERT: Pre-training of deep bidirectional transformers for language understanding. In: Proceedings of NAACL, pp. 4171–4186 (2019)
22. Zhao, J., Gui, T., Zhang, Q., et al.: A Relation-Oriented Clustering Method for Open Relation Extraction. In Proceedings of ACL, pp. 9707–9718 (2021)
23. Wang, Y., Sun, C., Wu, Y., Zhou, H., Li, L., Yan, J.: ENPAR: enhancing entity and entity pair representations for joint entity relation extraction. In Proceedings of EMNLP, pp. 2877–2887 (2021)
24. Oord, A., Li, Y., Vinyals, O.: Representation learning with contrastive pre-dictive coding. arXiv preprint arXiv: 1807.03748 (2018)
25. Maaten, L., Hinton, G.: Visualizing data using t-sne. J. Mach. Learn. Res. **9**(86), 2579–2605 (2008)
26. Xie, J., Girshick, R., Farhadi, A.: Unsupervised deep embedding for clustering analysis. In: ICML, pp. 478–487 (2016)
27. Han, X., et al.: Fewrel: a large-scale supervised few-shot relation classification dataset with state-of-the-art evaluation. In: Proceedings of EMNLP, pp. 4803–4809 (2018)
28. Elsahar H., et al.: T-rex: a large scale alignment of natural language with knowledge base triples. In: Proceedings of LREC, pp. 3448–3452 (2018)
29. Liu, F., Yan, L., Lin, H., et al.: Element intervention for open relation extraction. In: Proceedings of ACL, pp. 4683–4693 (2021)
30. Marcheggiani, D., Titov, I.: Discretestate variational autoencoders for joint discovery and factorization of relations. In: Proceedings of TACL, pp. 231–244 (2016)
31. Elsahar, H., Demidova, E., Gottschalk, S., Gravier, C., Laforest, F.: Unsupervised Open Relation Extraction. In: Blomqvist, E., Hose, K., Paulheim, H., Ławrynowicz, A., Ciravegna, F., Hartig, O. (eds.) ESWC 2017. LNCS, vol. 10577, pp. 12–16. Springer, Cham (2017). https://doi.org/10.1007/978-3-319-70407-4_3

Security and Privacy

Achieving Provable Byzantine Fault-tolerance in a Semi-honest Federated Learning Setting

Xingxing Tang[1], Hanlin Gu[2], Lixin Fan[2](✉), and Qiang Yang[1,2]

[1] Department of Computer Science and Engineering, HKUST, Hong Kong, China
{xtangav,qyang}@cse.ust.hk
[2] WeBank AI Lab, WeBank, Shenzhen, China
{allengu,lixinfan}@webank.com

Abstract. Federated learning (FL) is a suite of technology that allows multiple distributed participants to collaboratively build a global machine learning model without disclosing private datasets to each other. We consider an FL setting in which there may exist both a) semi-honest participants who aim to eavesdrop on other participants' private datasets; and b) Byzantine participants who aim to degrade the performances of the global model by submitting detrimental model updates. The proposed framework leverages the Expectation-Maximization algorithm first in E-step to estimate unknown participant membership, respectively, of Byzantine and benign participants, and in M-step to optimize the global model performance by excluding malicious model updates uploaded by Byzantine participants. One novel feature of the proposed method, which facilitates reliable detection of Byzantine participants even with HE or MPC protections, is to estimate participant membership based on the performances of a set of randomly generated candidate models evaluated by all participants. The extensive experiments and theoretical analysis demonstrate that our framework guarantees Byzantine Fault-tolerance in various federated learning settings with private-preserving mechanisms.

Keywords: Federated Learning · Byzantine Fault-tolerance · Semi-honest party

1 Introduction

With the increasing popularity of federated learning (FL) in a variety of application scenarios [29], it is of paramount importance to be vigilant against various Byzantine attacks, which aim to degrade FL model performances by submitting malicious model updates [4,11,27]. Effective Byzantine Fault-tolerant methods to thwart such Byzantine attacks have been proposed in literature [5,30,31].

Although numerous Byzantine Fault-tolerant methods have demonstrated effectiveness in defeating attacks under various federated learning settings, however, the majority of existing work is not readily applicable to a critical FL setting in which certain *privacy-preserving mechanisms* e.g., Differential Privacy (DP),

H. Kashima et al. (Eds.): PAKDD 2023, LNAI 13936, pp. 415–427, 2023.
https://doi.org/10.1007/978-3-031-33377-4_32

Table 1. Classification of Byzantine Fault-tolerant methods. -: No Reference.

Protection Mechanism	Byzantine Fault-tolerant Methods					
	Updates-based				Performance-based	
	Robust Statistics	Clustering	Historical	Server-Based	Server Eval.	Client Eval.
No Protection	[5,30]	[23,24]	[2,31]	[8,22]	[26,28]	[16] FedPBF
DP	[19,32]	-	-	-	-	FedPBF
MPC	[14,25]	-	-	-	-	
HE	[20]	-	-	-	-	

Homomorphic Encryption (HE) or Secure Multiparty Computation (MPC)[1] are adopted to protect model updates from disclosing private training data or models. We regard this deficiency as a detrimental shortcoming that renders many Byzantine Fault-tolerant methods useless in practice since federated learning without privacy-preserving mechanisms poses serious privacy leakage risks that defeat the purpose of federated learning in the first place. For instance, it was shown that attackers could exploit unprotected deep neural network model updates to reconstruct training images with pixel-level accuracy [33].

To achieve privacy-preserving and Byzantine Fault-tolerance simultaneously in FL, we propose a ***P****rovable* ***B****yzantine* ***F****ault-tolerant framework in a semi-honest* ***Fed****erated learning setting*, called **FedPBF**, which leverages EM Algorithm [9,10] in E-step to estimate unknown participant membership, respectively, of Byzantine and benign participants, and in M-step to optimize the global model performance by excluding malicious model updates uploaded by Byzantine participants. As compared with *model updates* based methods, the proposed FedPBF is based on model performances of a set of randomly generated candidate models evaluated by all participants via their local dataset, which can be applied to various privacy-preserving mechanisms, e.g., DP, HE and MPC (shown in Table 2). Moreover, the FedPBF uses robust estimation (median) to defence *misreporting* by Byzantine participants (Sect. 2.). Our extensive experiments (Sect. 4) and theoretical analysis (Appendix[2] A) demonstrate that the FedPBF method shows superior model performances in the presence of a large variety of Byzantine attacks and privacy-preserving mechanisms. Table 1 illustrated that the FedPBF can be applied to all FL scenarios with different privacy-preserving mechanisms (e.g., DP, MPC and HE) adopted.

2 Related Work

We classify existing federated learning Byzantine Fault-tolerant methods into two main categories in Table 1: **update-based** and **performance-based** methods.

---

[1] DP adds random noise to data to protect individual privacy while allowing useful data insights [1]. HE is a cryptographic technique that allows computation on encrypted data without the need for decryption, preserving privacy and security [3,18]. MPC is a protocol or technique that enables multiple parties to jointly perform a specific computation task without revealing their private data [6].

[2] Online appendix: https://github.com/TangXing/PAKDD2023-FedPBF.

Updates-based Byzantine Fault-tolerant methods are based on model updates uploaded by the client to detect Byzantine participants. Some methods [5,30] regarded malicious updates as outliers and leveraged **Robust statistics** to filter out malicious updates. Another line of work adopted **clustering** methods [23,24] to distinguish benign and Byzantine participants. Moreover, some **server-based** Byzantine Fault-tolerant methods assume that the server has an additional dataset to evaluate updates uploaded by clients [8,22]. In addition, some methods made use of **historical information** to help correct the statistical bias brought by Byzantine participants during the training, and thus lead to the convergence of optimization of federated learning [2,31]. However, all of the updated-based methods don't consider privacy-preserving mechanisms, such as DP and HE.

Performance-based Byzantine Fault-tolerant methods detect Byzantine participants based on model performance evaluation. Some methods [26,28] assumed the availability of reliable public datasets that can be used by the **server to evaluate** model performances. However, the availability of such server-side public datasets is hardly fulfilled in practice since those server-side datasets are either limited in terms of their data size or their distributions are different from private datasets owned by clients. Moreover, other methods [16] used private datasets on the **client to evaluate** the performance of local updates; however, they didn't consider the existence of the *misreporting* by Byzantine participants.

Byzantine Fault-tolerant & privacy-preserving methods considered Byzantine Fault-tolerant and privacy-preserving at the same time. They combined privacy-preserving mechanisms such as DP [19,32], MPC [14,25] and HE [20] and Byzantine Fault-tolerant methods to address privacy issues and Byzantine attacks simultaneously. However, the methods proposed in [19,32] can only be used to protect the sign of updates sent to the server using DP. Moreover, Ma et al. applied HE to Byzantine problems [20]. Nevertheless, it only allowed for the encryption of the set $\{-1, 0, 1\}$ using HE, which may not be sufficient for the general case of encrypting model updates. Additionally, the approaches presented in [14,25] are designed to be used with MPC protocols for update-based arithmetic operations, which might restrict their applicability to other privacy-preserving mechanisms.

3 The Proposed Method

This section first formally defines the setting and threat model in which a *semi-honest* federated learning setting and an unknown number of *Byzantine* participants (aka, Byzantine clients: this description is used in the following sections.) are assumed. Section 3.2 then delineates the proposed framework, which demonstrates *provable Byzantine Fault-tolerance* in the presence of various privacy-preserving mechanisms (e.g., DP, HE and MPC) adopted by clients to prevent the semi-honest server to eavesdrop private data from clients.

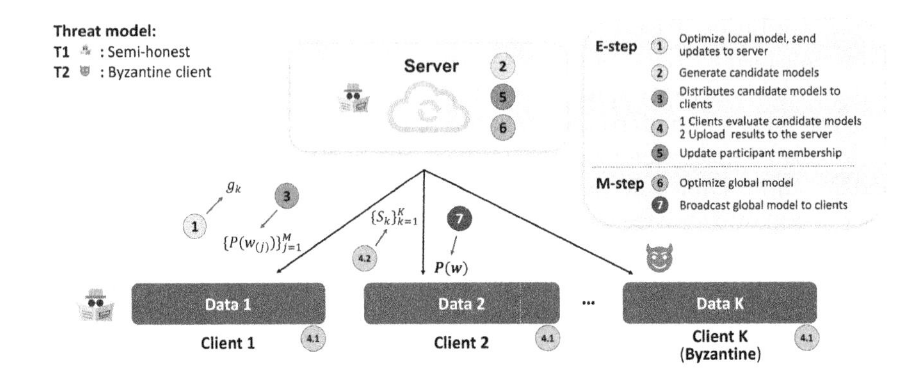

Fig. 1. Overview of threat models and the proposed FedPBF: **E-step** and **M-step**. The details of **E-step** and **M-step** are described in Sect. 3.2.

3.1 The Setting and Threat Model

We consider a *horizontal federated learning* setting (as defined in [29]), in which K clients collaboratively train a global model with weights $\mathbf{w}$ using private datasets residing on each client i.e. $\mathcal{D}_k = \{\mathbf{x}_{k,n}, y_{k,n}\}_{n=1}^{N_k}, k = 1 \cdots K$ and N_k the total number of data points on respective clients:

$$\min_{\mathbf{w}} \sum_{k=1}^{K} \sum_{n=1}^{N_k} F_k(\mathbf{w}, \mathbf{x}_{k,n}, y_{k,n}) = \min_{\mathbf{w}} \sum_{k=1}^{K} F_k(\mathbf{w}, \mathcal{D}_k), \tag{1}$$

where F_k is the loss function w.r.t. weights $\mathbf{w}$ of k_{th} client.

The minimization of Eq. (1) essentially uplifts the global model performance, such that clients are motivated to join the federated learning mission for the benefits of improved model performances. However, one must deal with the following two types of threats that may defeat the purpose of federated learning in the first place (see Fig. 1).

Threat type I – Semi-honest Parties: We assume a *semi-honest* threat model in which either clients or the server may eavesdrop on private data owned by other clients. Therefore, due to the privacy-preserving concern, private data $\mathcal{D}_k$ are never communicated with peer clients or the server. Instead, it is the protected model updates $P(\nabla F_k)$ in Eq. (2) that are sent from each client to the server. $P(\cdot)$ denotes certain privacy-preserving mechanisms, e.g., Differential Privacy or Homomorphic Encryption adopted by clients, to prevent a semi-honest party from inferring information about private data $\mathcal{D}_k$ based on the unprotected local model update ∇F_k[3].

However, as demonstrated throughout this paper, the adoption of such privacy-preserving mechanisms poses severe challenges to the defence of another type of threat, i.e., *Byzantine clients* whose behaviour may significantly degrade global model performances.

[3] Such privacy leakage risks have been demonstrated for particular cases (e.g., see [33]) where attackers can exploit unprotected deep neural network model updates to reconstruct training images with pixel-level accuracy.

Threat type II – Byzantine Clients: We assume that out of K clients, there exist up to f *Byzantine clients*, whose local model updates $\mathbf{g}_k$ may deviate from those of benign clients in an arbitrary manner:

$$\mathbf{g}_k = \begin{cases} P(\nabla F_k(\mathbf{w})) & \text{Benign clients} \\ P(\mathbf{g}_b) & \text{Byzantine clients.} \end{cases} \tag{2}$$

Note that misbehaves of Byzantine clients may be ascribed to different reasons, e.g., network malfunctioning or malicious clients who intentionally aim to degrade the global model performances [4,11,27]. Due to various root causes of Byzantine clients, their identities are often unknown. Moreover, in the case of malicious clients, they may collude and upload detrimental yet disguised model updates that evade many existing Byzantine Fault-tolerant methods (see, e.g., [11,12]). To make things worse, the adoption of certain privacy-preserving mechanisms, e.g., HE or MPC, renders many Byzantine Fault-tolerant methods completely useless, as illustrated in Sect. 4.2.

Our mandate, therefore, is to study a general Byzantine Fault-tolerant federated aggregation scheme that admits *exact Fault-tolerance* as defined below in the presence of privacy-preserving mechanisms[4].

Definition 1 (Exact Fault-tolerance). *Given a set of K model updates $\mathbf{g}_k, k = 1 \cdots K$ with a subset $\mathcal{G}$ consisting of m benign clients, a federated aggregation scheme is said to have* exact Fault-tolerance *if it allows all the benign clients to compute*

$$\mathbf{w}^*_{\mathcal{G}} \in \arg\min_{\mathbf{w}} \sum_{k \in \mathcal{G}} F_k(\mathbf{w}, \mathcal{D}_k). \tag{3}$$

Since the subset $\mathcal{G}$ is a prior unknown, one must estimate its unknown participant membership during the optimization of Eq. (1). Therefore, we regard the detection accuracy of the estimation, as defined below, as a crucial measure of Byzantine Fault-tolerance. We provide the theoretical guarantee in Appendix A that the detection accuracy of our proposed FedPBF converges to 100%.

Definition 2 (Detection Accuracy). *At t_{th} iteration, the Detection Accuracy η_t is defined as the fraction of benign clients over selected clients to aggregate by the server.*

$$\eta_t = \frac{\#(\mathcal{I}^t \cap \mathcal{G})}{K}, \tag{4}$$

where $\mathcal{I}^t$ is the set of clients chosen by the server to aggregate at t_{th} iteration, and K is the number of clients. Moreover, we define the ***averaged detection***

[4] Notions of *exact Fault-tolerance* is previously introduced in [13], in which a comparative elimination (CE) filtered-based scheme was proposed to achieve Byzantine Fault-tolerance under different conditions. We adopt these definitions to prove that the framework proposed in this article does admit these Fault-tolerances in the presence of privacy-preserving mechanisms.

accuracy η *among* T *iterations as:*

$$\eta = \frac{1}{T}\sum_{t=1}^{T}\eta_t \tag{5}$$

3.2 FedPBF

In this section, we illustrate a generic Byzantine Fault-tolerant framework in which an EM algorithm [9,10] (see Fig. 1) is adopted to solve the following problem, where the unknown participant membership $\mathbf{r} \in \{0,1\}^K$ is updated in the E-step (①-⑤), and the global model parameter $\mathbf{w}$ is optimized in the M-step (⑥-⑦) as (see Algorithm 1):

$$\arg\min_{\mathbf{w},\mathbf{r}} \sum_{k=1}^{K} F_k(\mathbf{w},\mathbf{r},\mathcal{D}_k) \tag{6}$$

- ①: Each client optimizes their local model via $\min F_k(\mathbf{w},\mathcal{D}_k)$ and sends the protected model updates $\mathbf{g}_k$ to the server.
- ②: The server first randomly selects M groups Q clients indexed by $\mathcal{I}_j = \{c_{1j},\ldots,c_{Qj}\}, j \in \{1,\ldots,M\}, c_{qj} \in \{1,\ldots,K\}, q \in \{1,\ldots,Q\}$, according to sample probability $\mathbf{p}^t$, which is proportional to the cumulative participant membership summed up from iteration 0 to $t-1$ as follows:

$$\mathbf{p}^t = \sum_{i=0}^{t-1}\mathbf{r}^i / \| \sum_{i=0}^{t-1}\mathbf{r}^i \|_1. \tag{7}$$

  It is noted that we use cumulative participant membership in order to take advantage of the historical evaluation for all clients, i.e., $\{\mathbf{r}^i\}_{i=0}^{t-1}$.
  Then the server generates M protected candidate models as

$$P(\mathbf{w}^t_{(j)}) = P(\mathbf{w}^{t-1}) - \frac{1}{Q}\sum_{k\in\mathcal{I}_j}\mathbf{g}_k, j \in \{1,\cdots,M\}. \tag{8}$$

- ③: The server distributes M protected candidate models $\{P(\mathbf{w}^t_{(j)})\}_{j=1}^{M}$ to all clients.
- ④: Each client executes the following two sub-steps: ④.1 each client decrypts M candidate models to obtain $\mathbf{w}^t_{(j)}$ [5]. They calculate the empirical accuracy $\{S_k \in \mathbb{R}^M\}_{k=1}^{K}$ of M candidate models via their own dataset as:

$$S_k(j) = Acc(\mathbf{w}^t_{(j)},\mathcal{D}_k), j \in \{1,\cdots,M\} \tag{9}$$

  where Acc() represents the accuracy of the candidate model $\mathbf{w}^t_{(j)}$ measured w.r.t. $\mathcal{D}_k$ and $S_k(j)$ represents the j_{th} candidate model measured by the k_{th} client. ④.2 then each clients upload $S_k, k \in \{1,\ldots,K\}$ to the server.

[5] For some protection mechanisms such as DP, this process may cause the loss of model precision.

Note that Byzantine clients may misreport the model accuracy. Our client-side performance-based evaluation method involves each client evaluating all candidate models. The server can use robust filters [5,30] to mitigate the impact of misreporting as long as the ratio of Byzantine clients is below 0.5.

- ⑤: Upon receiving $\{S_k\}_{k=1}^K$, the server picks up the best candidate model:

$$j^* = \arg\max_{j \in \{1,\ldots,M\}} \text{Median}_{k \in \{1,\ldots,K\}} S_k(j) \tag{10}$$

Note that we use the robust filter, i.e. Median, to filter out the Byzantine clients. Other robust filters such as [5,30] could also be applied to our methods. Then the server updates participant membership $\mathbf{r}^t$ as:

$$\mathbf{r}^t(k) = \begin{cases} 1, & \text{if} \quad k \in \mathcal{I}_{j^*}, \text{ where } j^* \text{ is the best candidate model by Eq. 10 ;} \\ 0, & \text{otherwise.} \end{cases} \tag{11}$$

- ⑥: The server optimizes the protected global model $P(\mathbf{w}^t)$ at t_{th} iteration according to participant membership $\mathbf{r}^t$ as:

$$P(\mathbf{w}^t) = P(\mathbf{w}^{t-1}) - \frac{1}{Q}\sum_{k=1}^{K} \mathbf{r}^t(k) P(\mathbf{g}_k^t) \tag{12}$$

- ⑦: The server broadcasts the protected global model $P(\mathbf{w}^t)$ to all clients, which will optimize respective local models in the next iteration.

Algorithm 1. FedPBF

Input: K: the number of clients; $\mathcal{D}_k$: local training datasets of k_{th} client; T: number of global iterations; M: the number of candidate models; Q: the number of aggregated updates for each candidate model; η: the local learning rate; b: the batch size;

Output: Global model $\mathbf{w}$.

1: Initialization: $\mathbf{w}^0 \leftarrow$ random value, participant membership $\mathbf{r}^0 \leftarrow [1, \cdots, 1]$.
2: **for** $t = 1, 2, \cdots, T$ **do**
3: **E Step:**
4: **for** each client k, $k \in [K]$ **do in parallel**
5: Compute the local updates as $\mathbf{g}_k^{t-1} = ModelUpdate(\mathbf{w}_k^t, \mathcal{D}_k, b, \eta)$.
6: Each client sends $\mathbf{g}_k^t$ to the server.
7: **end for**
8: $\mathbf{p}^t = \sum_{i=0}^{t-1} \mathbf{r}^i / \| \sum_{i=0}^{t-1} \mathbf{r}^i \|_1$
9: The server generates the M candidate models $\{P(\mathbf{w}_{(j)}^t)\}_{j=1}^M$ according to $\mathbf{p}^t$ by Eq. (8);
10: The server distributes M candidate models $\{P(\mathbf{w}_{(j)}^t)\}_{j=1}^M$ to all clients.
11: Each clients decrypts and evaluates M candidate models to obtain the accuracies $\{S_k\}_{k=1}^K$ according to Eq. (9) and send them to the server;
12: The server chooses the candidate model with the largest accuracy as Eq. (10). Then the server updates participant membership $\mathbf{r}^t$ by Eq. (11).
13: **M step:**
14: The server updates the global model via Eq. (12);
15: The server broadcast the protected global model $P(\mathbf{w}^t)$ to all clients.
16: **end for**
17: **return** $\mathbf{w}$.

There are three fundamental reasons for the proposed FedPBF to satisfy privacy-preserving and Byzantine Fault-tolerant requirements.

1. FedPBF allows *reliable* detection of Byzantine clients in the presence of privacy-preserving mechanisms (e.g., DP, HE and MPC). This feature is achieved by leveraging client-side datasets to evaluate candidate model performances. Algorithm 1 is thus applicable to all protected model updates, regardless of whatever privacy-preserving mechanisms are adopted.
2. FedPBF allows *efficient* estimation of participant membership $\mathbf{r}$, by sampling multiple candidates group models. This sampling approach is scalable in case the number of clients is large.
3. FedPBF is *robust* to misreported performances reported by Byzantine attackers. This merit is ensured by the robust estimation of the Median filter adopted in Eq. (10) when the ratio of Byzantine clients is less than the breakdown point, i.e. 0.5.

4 Experiments

In this section, we conduct extensive experiments to answer the following three questions: **Question 1:** To what extent does the proposed FedPBF outperform other Byzantine Fault-tolerant methods in federated learning with different privacy-preserving mechanisms (e.g., DP, HE, MPC)? **Question 2:** To what extent the model performance (accuracy) of the proposed FedPBF is affected by the varying Byzantine client proportions, the extent of Non-IID or misreporting conditions by Byzantine clients? **Question 3:** How do hyperparameters, i.e., the number of candidate models M and aggregated updates for each candidate model Q and Byzantine clients ratio, affect the convergence and effectiveness of the proposed FedPBF algorithm in practice (due to the page limit, these results are shown in Appendix B.2)?

4.1 Setup and Evaluation Metrics

Datasets: *MNIST Fashion-MNIST* (FMNIST) and *CIFAR10* are used for image classification tasks. The extent of Non-IID of the dataset is obtained by changing the parameter β from 0.5 to 1 of the Dirichlet distribution $Dir(\beta)$ [17].

Models: *Logistic regression*, *LeNet* and *AlexNet* models are used to train MNIST, FMNIST and CIFAR10 respectively.

Federated Learning Settings: We simulate a horizontal federated learning system with K = 100 clients (MNIST and FMNIST) or K=20 clients (CIFAR10) in a stand-alone machine. The detail of hyper-parameters for training are illustrated in Appendix B.

Federated Learning Privacy-preserving Mechanisms: Privacy-preserving mechanisms: Differential Privacy [1](DP) with the variance of Gaussian noise range between $\sigma^2 = 10^{-4}$ and $\sigma^2 = 10^{-1}$ as in [33], Secure Multiparty Computation (MPC) [7], Homomorphic Encryption (HE) [3] used to protect the privacy of local data.

Byzantine Fault-tolerant Methods: Five existing methods: Krum [5], Median [30], Trimmed Mean [30], Kmeans [24], FLtrust [8], and the proposed method FedPBF are compared in terms of following metrics.

Evaluation Metric: *Model Performance (MP)*, *Averaged Model Performance (AMP)* and *Averaged Detection Accuracy* (Def. 2) of the federated model is used to evaluate model accuracy defending capabilities of different methods.

Byzantine Attacks: We set 10%, 20%, and 35% clients are Byzantine attackers. The following attacking methods are used in experiments: 1) the *same value attack*, where model updates of attackers are replaced by the all ones vector. 2) the *label flipping attack*, where attackers use the wrong label to generate the gradients to upload. 3) the *sign flipping attack*, where local gradients of attackers are shifted by a scaled value -4. 4) the *gaussian attack*, where local gradients at clients are replaced by independent Gaussian random vectors $\mathcal{N}(0, 200)$. 5) the *Lie attack*, which was designed in [4]. 6) the *Fang-v1 attack* [11]. 7) the *Fang-v2 attack* [11]. 8) the *Mimic attack* [15].

4.2 Comparison with Existing Byzantine Fault-tolerant Methods

To answer **Question 1**, we evaluate the Averaged Model Performance (AMP) of five existing Byzantine Fault-tolerant methods and our proposed method under eight attacks. There are three notable observations according to Table 2:

1. When no privacy-preserving mechanisms (No Protection) are applied, the *AMP* of the FedPBF outperforms other methods from 5% to 12% on the MNIST dataset and from 11% and 49% on the FMNIST dataset.
2. When different magnitudes of Gaussian noise are added (DP: the variance of Gaussian noise range between $\sigma^2 = 10^{-4}$ and $\sigma^2 = 10^{-1}$), the FedPBF

Table 2. Averaged model performance (accuracy percentage: %) of different Byzantine Fault-tolerant methods under different privacy-preserving mechanisms (with Non-IID setting $\beta = 0.5$, $Q = 10$, $M = 40$ and 20% Byzantine clients for classification of MNIST and FMNIST). FedAvg W.O. Attack: FedAvg [21] without attack. DP: Differential privacy with Gaussian noise and σ^2 is the variance of Gaussian noise. -: as far as we know, there is no solution existing.

			Krum [5]	Median [30]	Trimmed [30]	Kmeans [24]	FLtrust [8]	**FedPBF (Ours)**	FedAvg [21] W.O. Attack
MNIST	No Protection		78.8±15.4	81.2±15.4	80.3±18.0	82.4±17.5	85.9±5.6	**91.7±0.5**	92.5±0.1
	DP	$\sigma^2 = 10^{-4}$	76.8±25.7	82.5±21.0	83.4±18.4	75.4±29.9	88.1±7.9	**91.5±0.5**	92.4±0.1
		$\sigma^2 = 10^{-3}$	72.1±24.0	79.5±23.7	76.6±26.4	74.5±31.0	86.9±6.9	**90.2±0.1**	92.2±0.1
		$\sigma^2 = 10^{-2}$	63.1±6.5	69.0±33.0	69.1±34.0	73.0±29.1	82.0±9.2	**87.8±0.6**	90.8±0.2
		$\sigma^2 = 10^{-1}$	39.4±9.1	64.3±31.4	64.6±31.5	63.6±28.1	70.9±19.7	**85.8±0.8**	87.2±0.3
	HE		-	-	-	-	-	**91.6±0.4**	92.5±0.1
	MPC		-	-	-	-	-	**91.7±0.3**	92.5±0.1
FMNIST	No Protection		69.9±24.2	39.3±23.5	55.7±31.5	63.3±34.7	76.6±27.2	**88.5±0.6**	90.3±0.6
	DP	$\sigma^2 = 10^{-4}$	65.6±22.9	29.2±26.5	36.1±33.9	67.9±34.3	71.0±24.6	**87.5±1.3**	90.0±0.1
		$\sigma^2 = 10^{-3}$	60.2±20.6	15.8±11.2	33.1±33.8	66.4±34.3	58.2±35.2	**86.3±2.3**	88.2±0.4
		$\sigma^2 = 10^{-2}$	19.7±7.8	10.1±0.6	32.1±31.8	58.4±30.2	43.0±29.5	**81.2±5.8**	83.2±0.3
		$\sigma^2 = 10^{-1}$	10.5±1.1	9.5±1.2	22.5±21.8	31.7±19.9	22.3±16.8	**69.8±2.0**	73.4±0.5
	HE		-	-	-	-	-	**88.8±0.5**	90.2±0.1
	MPC		-	-	-	-	-	**89.1±0.4**	89.8±0.1

significantly outperforms other methods. Especially when the noise increases, the *AMP* of other methods is broken (e.g., *AMP* of FLtrust degrades from 76.6% to 22.3% on FMNIST). However, the *AMP* of our FedPBF doesn't drop seriously under various degrees of noise (e.g., the *AMP* only drops from 91.7% to 85.8% on MNIST).
3. When the HE and MPC are applied, the FedPBF still performs well, i.e., there is a minor loss of the *AMP* compared with the baseline (FedAvg without attack).

4.3 Robustness

In this subsection, we test the robustness of the FedPBF under different Non-IID extents of the local dataset, different Byzantine client percentages and Byzantine clients misreporting types (due to the page limit, these results are shown in Appendix B.1) on CIFAR10 with AlexNet to answer **Question 2**.

1. **Robustness under Different Byzantine Client Percentages:** Figure 2 (left) illustrates the model performance of FedPBF under various attacks for different percentages of Byzantine clients, i.e., 10%, 20%, 35% with IID dataset. It shows that the degradation of model performance of the FedPBF is less than 2% compared with the baseline (FedAvg without attack: blue dotted lines) even the Byzantine client percentage increases to 35%.
2. **Robustness under Heterogeneous Dataset:** In Fig. 2 (right), it is shown that the degradation of model performance of the FedPBF, under various attacks with different clients' datasets Non-IID extents, is less than 1.5% all the time, which indicates the proposed FedPBF is robust under various Non-IID extents.

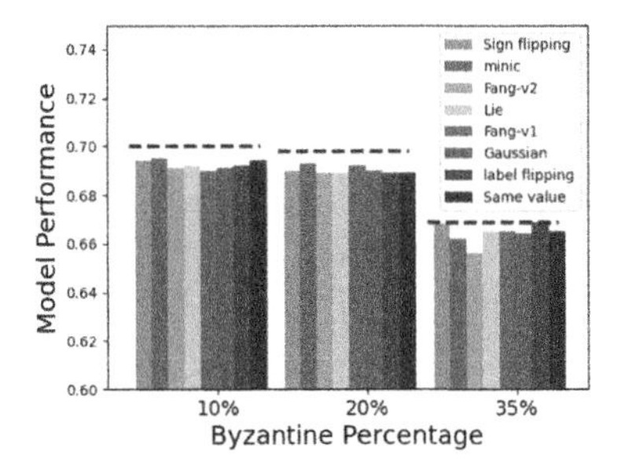

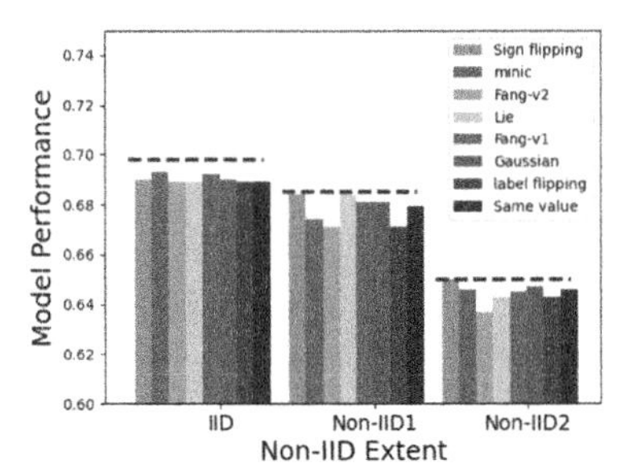

Fig. 2. The model performance of the FedPBF with different Byzantine client percentages (10%, 20% and 35%) and Non-IID extents (IID, Non-IID1 with $\beta = 1$, and Non-IID2 with $\beta = 1$) on CIFAR10 under different attacks. Blue dotted lines represent the baseline meaning FedAvg without attack. (Color figure online)

5 Conclusion

This paper proposed a novel Byzantine Fault-tolerant framework called FedPBF to guarantee Byzantine Fault-tolerance in the presence of protection mechanisms. To our best knowledge, this paper is the first research endeavour that thoroughly investigates the performances of various Byzantine tolerant methods with different protection mechanisms such as DP, HE and MPC being applied. Methodology-wise, we use the Expectation-Maximization algorithm to update the participant membership and optimize the global model performance alternately. The key for the FedPBF applying federated learning with various privacy-preserving mechanisms is that we use *model performance* in E-step to evaluate candidate models at the client side. This novel client-side performance-based evaluation, in tandem with the EM algorithm, constitutes our main contribution to the effective defence of Byzantine attacks in the presence semi-honest FL setting.

Acknowledgments. This work is partly supported by National Key Research and Development Program of China (2020YFB1805501).

References

1. Abadi, M., et al.: Deep learning with differential privacy. In: Proceedings of the 2016 ACM SIGSAC Conference on Computer and Communications Security, pp. 308–318 (2016)
2. Allen-Zhu, Z., Ebrahimianghazani, F., Li, J., Alistarh, D.: Byzantine-resilient non-convex stochastic gradient descent. In: International Conference on Learning Representations (2020)
3. Aono, Y., Hayashi, T., Wang, L., Moriai, S., et al.: Privacy-preserving deep learning via additively homomorphic encryption. IEEE Trans. Inf. Forensics Secur. **13**(5), 1333–1345 (2017)
4. Baruch, G., Baruch, M., Goldberg, Y.: A little is enough: circumventing defenses for distributed learning. Advances in Neural Information Processing Systems 32 (2019)
5. Blanchard, P., El Mhamdi, E.M., Guerraoui, R., Stainer, J.: Machine learning with adversaries: Byzantine tolerant gradient descent. Advances in Neural Information Processing Systems 30 (2017)
6. Bonawitz, K., et al.: Practical secure aggregation for privacy-preserving machine learning. In: proceedings of the 2017 ACM SIGSAC Conference on Computer and Communications Security, pp. 1175–1191 (2017)
7. Boyle, E., Gilboa, N., Ishai, Y.: Function secret sharing. In: Oswald, E., Fischlin, M. (eds.) EUROCRYPT 2015. LNCS, vol. 9057, pp. 337–367. Springer, Heidelberg (2015). https://doi.org/10.1007/978-3-662-46803-6_12
8. Cao, X., Fang, M., Liu, J., Gong, N.Z.: FLTrust: Byzantine-robust federated learning via trust bootstrapping. arXiv preprint arXiv:2012.13995 (2020)
9. Dempster, A.P., Laird, N.M., Rubin, D.B.: Maximum likelihood from incomplete data via the EM algorithm. J. Roy. Stat. Soc.: Ser. B (Methodol.) **39**(1), 1–22 (1977)

10. Dieuleveut, A., Fort, G., Moulines, E., Robin, G.: Federated-EM with heterogeneity mitigation and variance reduction. Adv. Neural. Inf. Process. Syst. **34**, 29553–29566 (2021)
11. Fang, M., Cao, X., Jia, J., Gong, N.: Local model poisoning attacks to {Byzantine-Robust} federated learning. In: 29th USENIX Security Symposium (USENIX Security 20), pp. 1605–1622 (2020)
12. Gu, H., Fan, L., Tang, X., Yang, Q.: FedCut: a spectral analysis framework for reliable detection of byzantine colluders. arXiv preprint arXiv:2211.13389 (2022)
13. Gupta, N., Doan, T.T., Vaidya, N.: Byzantine fault-tolerance in federated local SGD under 2f-redundancy. arXiv preprint arXiv:2108.11769 (2021)
14. He, L., Karimireddy, S.P., Jaggi, M.: Secure byzantine-robust machine learning. arXiv preprint arXiv:2006.04747 (2020)
15. Karimireddy, S.P., He, L., Jaggi, M.: Byzantine-robust learning on heterogeneous datasets via bucketing. arXiv preprint arXiv:2006.09365 (2020)
16. Lai, F., Zhu, X., Madhyastha, H.V., Chowdhury, M.: Oort: efficient federated learning via guided participant selection. In: 15th {USENIX} Symposium on Operating Systems Design and Implementation ({OSDI} 21), pp. 19–35 (2021)
17. Li, Q., Diao, Y., Chen, Q., He, B.: Federated learning on non-IID data silos: an experimental study. arXiv preprint arXiv:2102.02079 (2021)
18. Ma, J., Naas, S.A., Sigg, S., Lyu, X.: Privacy-preserving federated learning based on multi-key homomorphic encryption. Int. J. Intell. Syst. **37**, 5880–5901 (2022)
19. Ma, X., Sun, X., Wu, Y., Liu, Z., Chen, X., Dong, C.: Differentially private byzantine-robust federated learning. IEEE Trans. Parallel Distrib. Syst. **33**(12), 3690–3701 (2022). https://doi.org/10.1109/TPDS.2022.3167434
20. Ma, X., Zhou, Y., Wang, L., Miao, M.: Privacy-preserving byzantine-robust federated learning. Comput. Stand. Interfaces **80**, 103561 (2022)
21. McMahan, B., Moore, E., Ramage, D., Hampson, S., y Arcas, B.A.: Communication-efficient learning of deep networks from decentralized data. In: Artificial intelligence and statistics, pp. 1273–1282. PMLR (2017)
22. Prakash, S., Avestimehr, A.S.: Mitigating byzantine attacks in federated learning. arXiv preprint arXiv:2010.07541 (2020)
23. Sattler, F., Müller, K.R., Wiegand, T., Samek, W.: On the byzantine robustness of clustered federated learning. In: ICASSP 2020–2020 IEEE International Conference on Acoustics, Speech and Signal Processing (ICASSP), pp. 8861–8865. IEEE (2020)
24. Shen, S., Tople, S., Saxena, P.: AUROR: defending against poisoning attacks in collaborative deep learning systems. In: Proceedings of the 32nd Annual Conference on Computer Security Applications, pp. 508–519 (2016)
25. So, J., Güler, B., Avestimehr, A.S.: Byzantine-resilient secure federated learning. IEEE J. Sel. Areas Commun. **39**(7), 2168–2181 (2020)
26. Xie, C., Koyejo, O., Gupta, I.: Zeno: Byzantine-suspicious stochastic gradient descent. arXiv preprint arXiv:1805.10032 24 (2018)
27. Xie, C., Koyejo, O., Gupta, I.: Fall of empires: Breaking byzantine-tolerant SGD by inner product manipulation. In: Uncertainty in Artificial Intelligence, pp. 261–270. PMLR (2020)
28. Xie, C., Koyejo, S., Gupta, I.: Zeno++: robust fully asynchronous SGD. In: International Conference on Machine Learning, pp. 10495–10503. PMLR (2020)
29. Yang, Q., Liu, Y., Chen, T., Tong, Y.: Federated machine learning: concept and applications. ACM Trans. Intell. Syst. Technol. (TIST) **10**(2), 1–19 (2019)
30. Yin, D., Chen, Y., Kannan, R., Bartlett, P.: Byzantine-robust distributed learning: Towards optimal statistical rates. In: International Conference on Machine Learning, pp. 5650–5659. PMLR (2018)

31. Zhang, Z., Cao, X., Jia, J., Gong, N.Z.: FLDetector: defending federated learning against model poisoning attacks via detecting malicious clients. In: Proceedings of the 28th ACM SIGKDD Conference on Knowledge Discovery and Data Mining, pp. 2545–2555 (2022)
32. Zhu, H., Ling, Q.: Bridging differential privacy and byzantine-robustness via model aggregation. arXiv preprint arXiv:2205.00107 (2022)
33. Zhu, L., Liu, Z., Han, S.: Deep leakage from gradients. Advances in Neural Information Processing Systems 32 (2019)

Defending Against Backdoor Attacks by Layer-wise Feature Analysis

Najeeb Moharram Jebreel[1(✉)], Josep Domingo-Ferrer[1], and Yiming Li[2]

[1] Universitat Rovira i Virgili, Av. Països Catalans 26, 43007 Tarragona, Catalonia
{najeeb.jebreel,josep.domingo}@urv.cat
[2] Tsinghua University, Beijing, China
li-ym18@mails.tsinghua.edu.cn

Abstract. Training deep neural networks (DNNs) usually requires massive training data and computational resources. Users who cannot afford this may prefer to outsource training to a third party or resort to publicly available pre-trained models. Unfortunately, doing so facilitates a new training-time attack (*i.e.*, backdoor attack) against DNNs. This attack aims to induce misclassification of input samples containing adversary-specified trigger patterns. In this paper, we first conduct a layer-wise feature analysis of poisoned and benign samples from the target class. We find out that the feature difference between benign and poisoned samples tends to be maximum at a critical layer, which is *not always* the one typically used in existing defenses, namely the layer before fully-connected layers. We also demonstrate how to locate this critical layer based on the behaviors of benign samples. We then propose a simple yet effective method to filter poisoned samples by analyzing the feature differences between suspicious and benign samples at the critical layer. We conduct extensive experiments on two benchmark datasets, which confirm the effectiveness of our defense.

Keywords: Backdoor Detection · Backdoor Defense · Backdoor Learning · AI Security · Deep Learning

1 Introduction

In recent years, deep neural networks (DNNs) have successfully been applied in many tasks, such as computer vision, natural language processing, and speech recognition. However, training DNNs requires massive training data and computational resources, and users who cannot afford it may opt to outsource training to a third-party (*e.g.*, a cloud service) or leverage pre-trained DNNs. Unfortunately, losing control over training facilitates *backdoor attacks* [2,4,9] against DNNs. In these attacks, the adversary poisons a few training samples to cause the DNN to misclassify samples containing pre-defined trigger patterns into an adversary-specified target class. Nevertheless, the attacked models behave normally on benign samples, which makes the attack stealthy. Since DNNs are used

H. Kashima et al. (Eds.): PAKDD 2023, LNAI 13936, pp. 428–440, 2023.
https://doi.org/10.1007/978-3-031-33377-4_33

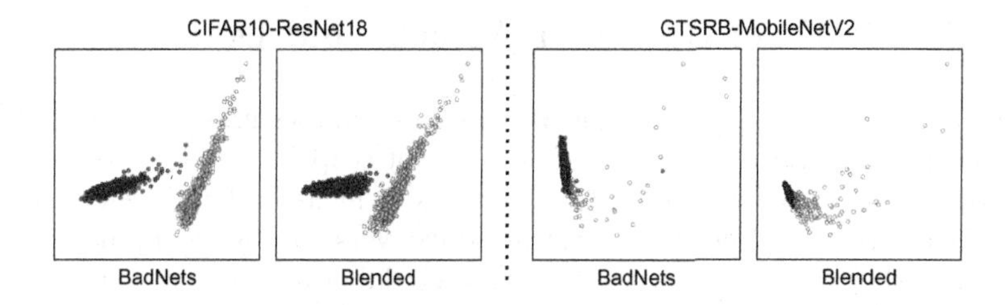

Fig. 1. PCA-based visualization of features of benign (green) and poisoned samples (red) generated by the layer before the fully connected layers of models attacked by BadNets [4] and Blended [2]. As shown in this figure, features of poisoned and benign samples are not well separable on the GTSRB benchmark. (Color figure online)

in many mission-critical tasks (*e.g.*, autonomous driving, or facial recognition), it is urgent to design effective defenses against these attacks.

Among all backdoor defenses in the literature, backdoor detection is one of the most important defense paradigms, where defenders attempt to detect whether a suspicious object (*e.g.*, model or sample) is malicious. Currently, most existing backdoor detectors assume poisoned samples have different feature representations from benign samples, and they tend to focus on the layer before the fully connected layers [1,5,20]. Two intriguing questions arise: **(1)** *Is this layer always the most critical place for backdoor detection?* **(2)** *If not, how to find the critical layer for designing more effective backdoor detection?*

In this paper, we give a negative answer to the first question (see Fig. 1). To answer the second one, we conduct a layer-wise feature analysis of poisoned and benign samples from the target class. We find out that the feature difference between benign and poisoned samples tends to reach the maximum at a critical layer, which can be easily located based on the behaviors of benign samples. Specifically, *the critical layer is the one or near the one that contributes most to assigning benign samples to their true class.* Based on this finding, we propose a simple yet effective method to filter poisoned samples by analyzing the feature differences (measured by cosine similarity) between incoming suspicious samples and a few benign samples at the critical layer. Our method can serve as a 'firewall' for deployed DNNs to identify, block, and trace malicious inputs. In short, our main contributions are four-fold. **(1)** We demonstrate that the features of poisoned and benign samples are not always clearly separable at the layer before fully connected layers, which is the one typically used in existing defenses. **(2)** We conduct a layer-wise feature analysis aimed at locating the critical layer where the separation between poisoned and benign samples is neatest. **(3)** We propose a backdoor detection method to filter poisoned samples by analyzing the feature differences between suspicious and benign samples at the critical layer. **(4)** We conduct extensive experiments on two benchmark datasets to assess the effectiveness of our proposed defense.

2 Related Work: Backdoor Attacks and Defenses

In this paper, we focus on backdoor attacks and defenses in image classification. Other deep learning tasks are out of our current scope.

BadNets [4] was the first backdoor attack, which randomly selected a few benign samples and generated their poisoned versions by stamping a trigger patch onto their images and reassigning their label as the target label. Later [2] noted that the poisoned image should be similar to its benign version for stealthiness; these authors proposed a blended attack by introducing trigger transparency. However, these attacks are with poisoned labels and therefore users can still detect them by examining the image-label relation. To circumvent this, [21] proposed the clean-label attack paradigm, where the target label is consistent with the ground-truth label of poisoned samples. Specifically, in this paradigm, adversarial attacks were exploited to perturb the selected benign samples before conducting the standard trigger injection process. [16] adopted image warping as the backdoor trigger, which modifies the whole image while preserving its main content. Besides, [15] proposed the first sample-specific attack, where the trigger varies across samples. However, such triggers are visible and the adversaries need to control the whole training process. More recently, [12] introduced the first poison-only invisible sample-specific attack to address these problems.

Existing backdoor defenses fall into three main categories: input filtering, input pre-processing, and model repairing. **Input filtering** intends to differentiate benign and poisoned samples based on their distinctive behaviors, like the separability of the feature representations of benign and poisoned samples. For example, [5] introduced a robust covariance estimation of feature representations to amplify the spectral signature of poisoned samples. [23] proposed to filter inputs inspired by the understanding that poisoned images tend to have some high-frequency artifacts. [3] proposed to blend various images on the suspicious one, since the trigger pattern can still mislead the prediction no matter what the background contents are. **Input pre-processing** modifies each input sample before feeding it into the deployed DNN. Its rationale is to perturb potential trigger patterns and thereby prevent backdoor activation. [14] proposed the first defense in this category where they used an encoder-decoder to modify input samples. [17] employed randomized smoothing to generate a set of input neighbors and averaged their predictions. Further, [11] demonstrated that if the location or appearance of the trigger is slightly different from that used for training, the attack effectiveness may degrade sharply. Based on this, they proposed to pre-process images with spatial transformations. **Model repairing** aims at erasing backdoors contained in the attacked DNNs. For example, [8,14,24] showed that users can effectively remove backdoors by fine-tuning the attacked DNNs with a few benign samples. [13] revealed that model pruning can also remove backdoors effectively, because backdoors are mainly encoded in specific neurons. Very recently, [22] proposed to repair compromised models with adversarial model unlearning. In this paper, *we focus on input filtering,* which is very convenient to protect deployed DNNs.

3 Layer-wise Feature Analysis

A deep neural network (DNN) $f(\boldsymbol{x})$ is composed by L layers $f^l, l \in [1, L]$. Each f^l has a weight matrix $\boldsymbol{w}^l$, a bias vector $\boldsymbol{b}^l$, and an activation function σ^l. The output of f^l is $\boldsymbol{a}^l = f^l(\boldsymbol{a}^{l-1}) = \sigma^l(\boldsymbol{w}^l \cdot \boldsymbol{a}^{l-1} + \boldsymbol{b}^l)$, where f^1 takes input $\boldsymbol{x}$ and f^L outputs a vector $\boldsymbol{a}^L$ with C classes. The vector $\boldsymbol{a}^L$ is softmaxed to get probabilities $\boldsymbol{p}$. A DNN has a feature extractor that maps $\boldsymbol{x}$ to latent features, which are input to fully connected layers for classification.

In this paper, we use DNNs as C-class classifiers, where y_i is the ground truth label of $\boldsymbol{x}_i$ and $\hat{y}_i$ is the index of the highest probability in $\boldsymbol{p}_i$. Also, activations of intermediate layers are analyzed for detecting poisoned samples.

We notice that the predictions of attacked DNNs for both benign samples from the target class and poisoned samples are all the target label. The attacked DNNs mainly exploit class-relevant features to predict these benign samples while they use trigger-related features for poisoned samples. We suggest that defenders could exploit this difference to design effective backdoor detection. To explore their main differences, we conduct a layer-wise analysis, as follows.

Definition 1 (Layer-wise centroids of target class features). *Let f' be an attacked DNN with a target class t. Let $X_t = \{\boldsymbol{x}_i\}_{i=1}^{|X_t|}$ be benign samples with true class t, and let $\{\boldsymbol{a}_i^1, \ldots, \boldsymbol{a}_i^L\}_{i=1}^{|X_t|}$ be their intermediate features generated by f'. The centroid of t's benign features at layer l is defined as $\hat{\boldsymbol{a}}_t^l = \frac{1}{|X_t|}\sum_{i=1}^{|X_t|} \boldsymbol{a}_i^l$, and $\{\hat{\boldsymbol{a}}_t^1, \ldots, \hat{\boldsymbol{a}}_t^L\}$ is the set of layer-wise centroids of t's benign features.*

Definition 2 (Layer-wise cosine similarity). *Let $\boldsymbol{a}_j^l$ be the features generated by layer l for an input $\boldsymbol{x}_j$, and let cs_j^l be the cosine similarity between $\boldsymbol{a}_j^l$ and the corresponding t's centroid $\hat{\boldsymbol{a}}_t^l$. The set $\{cs_j^1, \ldots, cs_j^L\}$ is said to be the layer-wise cosine similarities between $\boldsymbol{x}_j$ and t's centroids.*

Settings. We conducted six representative attacks on four classical benchmarks: CIFAR10-ResNet18, CIFAR10-MobileNetV2, GTSRB-ResNet18, and GTSRB-MobileNetV2. The six attacks were BadNets [4], the backdoor attack with blended strategy (Blended) [2], the label-consistent attack (LC) of [21], WaNet [16], ISSBA [12], and IAD [15]. More details on the datasets, DNNs, and attack settings are presented in Sect. 5. Specifically, for each attacked DNN f' with a target class t, we estimated $\{\hat{\boldsymbol{a}}_t^1, \ldots, \hat{\boldsymbol{a}}_t^L\}$ using 10% of the benign test samples labeled as t. Then, for the benign and poisoned test samples classified by f' into t, we calculated the layer-wise cosine similarities between their generated features and the corresponding estimated centroids. Finally, we visualized the layer-wise means of the computed cosine similarities of the benign and poisoned samples to analyze their behaviors.

Results. Figure 2 shows the layer-wise means of cosine similarity for benign and poisoned samples with the CIFAR10-ResNet18 benchmark under the BadNets and ISSBA attacks. As we go deeper into the attacked DNN layers, the gap between the direction of benign and poisoned features gets larger until we reach

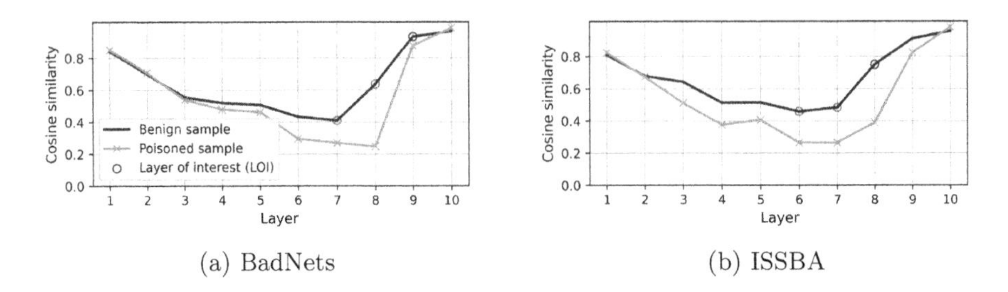

(a) BadNets (b) ISSBA

Fig. 2. Layer-wise behaviors of benign samples from the target class and poisoned samples (generated by BadNets and ISSBA) on CIFAR-10 with ResNet-18

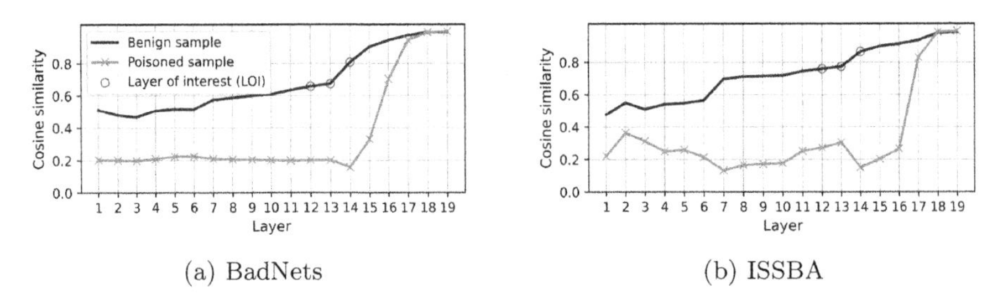

(a) BadNets (b) ISSBA

Fig. 3. Layer-wise behaviors of benign samples from the target class and poisoned samples (generated by BadNets and ISSBA) on GTSRB with MobileNetV2

a specific layer where the backdoor trigger is activated, causing poisoned samples to get closer to the target class. Figure 3 shows the same phenomenon for the GTSRB-MobileNetV2 benchmark. Further, we can see that for BadNets the latent features of benign and poisoned samples are similar in the last layer of the features extractor (*i.e.*, layer 17).

Regardless of the attack or benchmark, when we enter the second half of DNN layers (which usually are class-specific), *benign samples start to get closer to the target class before the poisoned ones, that are still farther from the target class* because the backdoor trigger is not yet activated. This makes the difference in similarity maximum in one of those latter layers, which we call the *critical layer*. In particular, *this layer is not always the one typically used in existing defenses* (*i.e.*, the layer before fully-connected layers). Besides, we show that it is very likely to be either the layer that contributes most to assigning the benign samples to their true target class (which we name the *layer of interest or LOI*, circled in blue) or one of the two layers before the LOI (circled in brown).

Results under other attacks for these benchmarks are presented in the supplementary materials[1]. In those materials, we also provide confirmation that the above distinctive behaviors hold regardless of the datasets or models being used. From the analysis above, we can conclude that focusing on those circled layers can help develop a simple and robust defense against backdoor attacks.

[1] https://www.dropbox.com/s/joyhr978irw344c/supplementary_materilas.pdf?dl=0

Algorithm 1. Identify layer of interest (LOI).

Input: Cosine similarities $\{\hat{cs}_t^{\lfloor L/2 \rfloor}, \ldots, \hat{cs}_t^{L}\}$ for potential target class t

1: $max_{diff} \leftarrow \hat{cs}_t^{\lfloor L/2 \rfloor+1} - \hat{cs}_t^{\lfloor L/2 \rfloor}$; $LOI_t \leftarrow \lfloor L/2 \rfloor + 1$;
2: **for** $l \in \{\lfloor L/2 \rfloor + 2, \ldots, L\}$ **do**
3: $\quad l_{diff} \leftarrow \hat{cs}_t^{l} - \hat{cs}_t^{l-1}$;
4: $\quad$ **if** $l_{diff} > max_{diff}$ **then**
5: $\quad\quad max_{diff} \leftarrow l_{diff}$; $LOI_t \leftarrow l$;
6: **return** LOI_t.

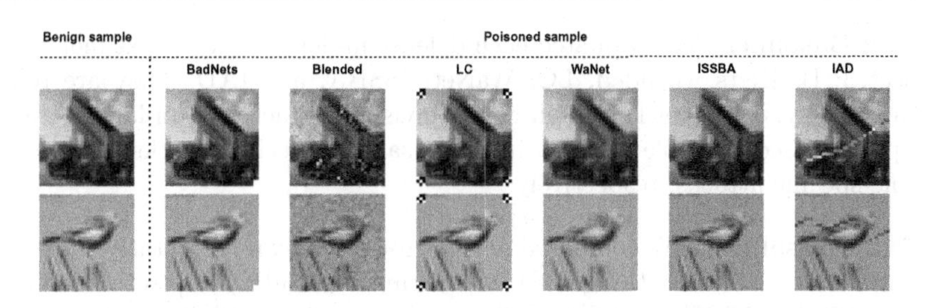

Fig. 4. The example of benign samples and their poisoned versions generated by six representative backdoor attacks.

4 The Proposed Defense

Threat Model. Consider a user that obtains a suspicious trained f_s that might contain hidden backdoors. We assume that the user has limited computational resources or benign samples, and therefore cannot repair f_s. The user wants to defend by detecting at inference time whether a suspicious incoming input $\boldsymbol{x}_s$ is poisoned, given f_s. Similar to existing defenses, we assume that a small set of benign samples X_{val} is available to the user/defender. We denote the available samples that belong to a potential class t as $X_{t_{val}}$. Let $m = |X_{t_{val}}|$ denote the number of available samples labeled as t.

Method Design. Based on the lessons learned in Sect. 3, our method to detect poisoned samples at inference time consists of four steps. **1)** Estimate the layer-wise features' centroids of class t for each of layers $\lfloor L/2 \rfloor$ to L using the class's available benign samples. **2)** Compute the cosine similarities between the extracted features and the estimated centroids, and then compute the layer-wise means of the computed cosine similarities. **3)** Identify the layer of interest (LOI) as per Algorithm 1, sum up the cosine similarities in LOI and the two layers before LOI (sample-wise), and compute the mean and standard deviation of the summed cosine similarities. **4)** For any suspicious incoming input $\boldsymbol{x}_s$ classified as t by f_s, **4.1)** compute its cosine similarities to the estimated centroids in the above-mentioned three layers, and **4.2)** consider it as a potentially poisoned input if its summed similarities fall below the obtained mean by a specific number τ of standard deviations (called threshold in what follows). A detailed pseudocode can be found in the supplementary materials.

5 Experiments

5.1 Main Settings

Datasets and DNNs. In this paper, we use two classic benchmark datasets, namely CIFAR10 [7] and GTSRB [19]. We use the ResNet18 [6] on CIFAR10 and the MobileNetV2 [18] on GTSRB. More details are presented in the supplementary materials. The source code, pre-trained models, and poisoned test sets of our defense are available at https://github.com/NajeebJebreel/DBALFA.

Attack Baselines. We evaluated each defense under the six attacks mentioned in Sect. 3: BadNets, Blended, LC, WaNet, ISSBA, and IAD. They are representative of visible attacks, patch-based invisible attacks, clean-label attacks, non-patch-based invisible attacks, invisible sample-specific attacks, and visible sample-specific attacks, respectively.

Defense Baselines. We compared our defense with six representative defenses, namely randomized smoothing (RS) [17], ShrinkPad (ShPd) [11], activation clustering (AC) [1], STRIP [3], SCAn [20], and fine-pruning (FP) [13]. RS and ShPd are two defenses with input pre-processing; AC, STRIP, and SCAn are three advanced input-filtering-based defenses; FP is based on model repairing.

Attack Setup. For both CIFAR10 and GTSRB, we took the following settings. We used a 2×2 square as the trigger pattern for BadNets (as suggested in [4]). We adopted the random noise pattern, with a 10% blend ratio, for Blended (as suggested in [2]). The trigger pattern adopted for the LC attack was the same used in BadNets. For WaNet, ISSBA, and IAD, we took their default settings. Besides, we set the poisoning rate to 5% for BadNets, Blended, LC, and ISSBA. For WaNet and IAD, we set the poisoning rate to 10%. We implement baseline attacks based on the codes in `BackdoorBox` [10]. More details on settings are given in the supplementary materials. Figure 4 shows an example of poisoned samples generated by different attacks.

Defense Setup. For RS, ShPd and STRIP, we took the settings suggested in [3, 11,17]. For FP, we pruned 95% of the dormant neurons in the last convolution layer and fine-tuned the pruned model using 5% of the training set. We adjusted RS, ShPd, and FP to be used as detectors for poisoned samples by comparing the prediction change before and after applying them to an incoming input. For AC, STRIP, SCAn, and our defense, we randomly selected 10% from each benign test set as the available benign samples. For SCAn, we identified classes with scores larger than e as potential target classes, as suggested in [20]. For our defense, we used a threshold $\tau = 2.5$, which gives a reasonable trade-off between TPR and FPR for both benchmarks.

Evaluation Metrics. We used the main accuracy (MA) and the attack success rate (ASR) to measure attack performance. Specifically, MA is the number of correctly classified benign samples divided by the total number of benign samples, and ASR is the number of poisoned samples classified as the target class

Table 1. Main results (%) on the CIFAR-10 dataset. Boldfaced values are the best results among all defenses. Underlined values are the second-best results.

Attack→	BadNets		Blended		LC		WaNet		ISSBA		IAD		Avg	
Metric→ Defense↓	TPR	FPR	TPR	FPR	TPR	FPR	TPR	FPR	TPR	FPR	TPR	FPR	TPR	FPR
RS	9.84	8.00	7.35	5.76	9.21	7.52	98.48	10.00	8.83	8.72	13.28	6.36	24.50	7.73
ShPd	94.28	13.31	49.72	12.89	69.87	13.18	36.25	17.69	95.22	5.50	42.74	7.56	64.68	11.69
FP	96.10	17.13	96.23	16.16	94.76	17.31	96.01	18.64	98.98	19.53	97.08	22.52	96.53	18.55
AC	**99.52**	31.14	**100.00**	30.69	**100.00**	31.16	**99.18**	32.44	**99.94**	34.22	82.99	31.32	96.94	31.83
STRIP	68.70	11.70	65.20	11.70	66.00	12.80	7.90	12.30	56.20	11.40	2.10	14.00	44.35	12.32
SCAn	96.60	**0.77**	**100.00**	**0.00**	0.02	5.05	98.55	**1.06**	99.89	2.61	84.19	**0.13**	79.88	1.60
Ours	99.38	1.35	**100.00**	1.59	**100.00**	**1.20**	91.04	1.48	98.97	**1.17**	**99.12**	1.26	**98.09**	**1.34**

Table 2. Main results (%) on the GTSRB dataset. Boldfaced values are the best results among all defenses. Underlined values are the second-best results.

Attack→	BadNets		Blended		LC		WaNet		ISSBA		IAD		Avg	
Metric→ Defense↓	TPR	FPR	TPR	FPR	TPR	FPR	TPR	FPR	TPR	FPR	TPR	FPR	TPR	FPR
RS	13.20	22.10	10.12	20.40	9.23	19.15	10.10	17.20	8.61	16.98	17.70	17.60	11.49	18.91
ShPd	94.97	12.16	11.58	10.68	96.16	10.60	66.11	14.81	95.92	8.26	31.07	16.10	65.97	12.10
FP	89.05	18.80	30.56	**3.70**	94.71	50.02	67.12	3.24	94.22	7.05	94.37	5.75	78.34	14.76
AC	0.30	8.84	0.00	5.67	4.83	**5.42**	0.42	25.87	99.06	17.48	43.85	10.73	24.74	12.34
STRIP	32.00	9.00	80.40	10.80	7.40	11.00	34.20	11.40	13.00	13.60	6.60	10.60	28.93	11.07
SCAn	46.05	**2.57**	46.02	4.03	30.45	11.39	54.07	**1.88**	96.85	**0.17**	0.09	19.41	45.59	6.58
Ours	**99.99**	6.23	**100.00**	6.72	**100.00**	5.95	**100.00**	6.49	**100.00**	5.43	**100.00**	**4.67**	**100.00**	**5.92**

divided by the total number of poisoned samples. We adopted TPR and FPR to evaluate the performance of all defenses, where TPR is computed as the number of detected poisoned inputs divided by the total number of poisoned inputs, whereas FPR is the number of benign inputs falsely detected as poisoned divided by the total number of benign inputs.

5.2 Main Results

For each attack, we ran each defense five times for a fair comparison. Due to space limitations, we present the average TPR and FPR in this section. Please refer to our supplementary materials for more detailed results.

As shown in Tables 1 and 2, existing defenses failed to detect attacks with low TPR or high FPR in many cases, especially on the GTSRB dataset. For example, AC failed in most cases on GTSRB, although it had promising performance on CIFAR-10. In contrast, our method had good performance in detecting all attacks on both datasets. There were only a few cases (4 over 28) where our approach was neither optimal nor close to optimal. In these cases, our detection was still on par with state-of-the-art methods, and another indicator (*i.e.*, TPR or FPR) was significantly better than them. For example, when defending against the blended attack on the GTSRB dataset, the TPR of our method was 69.44% larger than that of FP, which had the smallest FPR in this case. These results confirm the effectiveness of our detection.

Table 3. MA% and ASR% under the selected backdoor attacks on the CIFAR10-ResNet18 and the GTSRB-MobileNetV2 benchmarks. Best scores are in bold.

Benchmark↓	Metric↓,Attack→	BadNets	Blended	LC	WaNet	ISSBA	IAD
CIFAR10-ResNet18	MA%	91.45	92.19	91.98	91.13	**94.74**	94.42
	ASR%	97.20	**100.0**	99.96	99.04	**100.0**	99.66
GTSRB-MobileNetV2	MA%	97.00	97.27	97.45	96.09	98.43	**98.81**
	ASR%	95.49	**100.0**	**100.0**	91.82	**100.0**	99.63

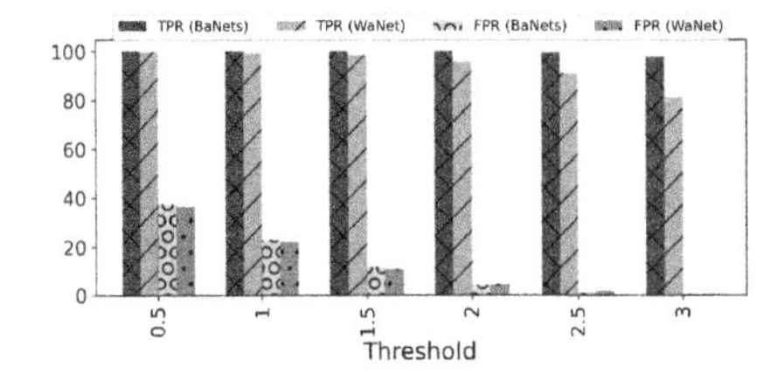

(a) CIFAR10-ResNet18

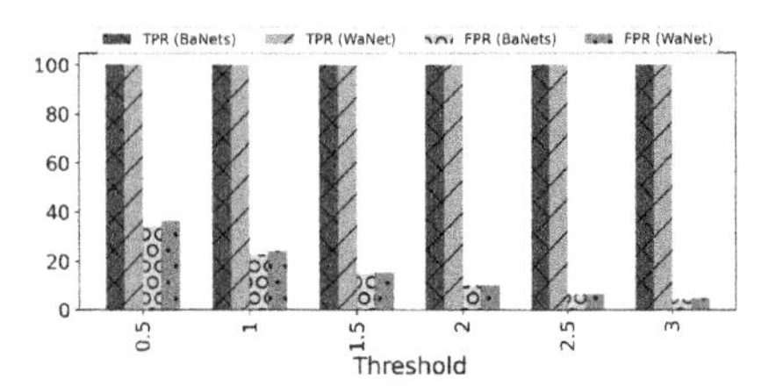

(b) GTSRB-MobileNetV2

Fig. 5. Impact of detection thresholds on TPR (%) and FPR (%)

Table 4. Impact of poisoning rates

Poisoning Rate↓, Metric→	MA (%)	ASR (%)	TPR (%)	FPR (%)
1%	91.52	94.15	99.64	1.25
3%	92.28	96.31	99.32	1.32
5%	91.45	97.20	99.36	1.35
10%	91.45	97.56	99.83	1.62

5.3 Discussions

Performance of Attacks. Table 3 shows the performance of the selected attacks on the CIFAR10-ResNet18 and the GTSRB-MobileNetV2 benchmarks. It can be seen that sample-specific attacks (*e.g.*, ISSBA and IAD) performed better than other attacks in terms of MA and ASR.

Effects of the Detection Threshold. Figure 5 shows the TPRs and FPRs of our defense with threshold $\tau \in \{0.5, 1, 1.5, 2, 2.5, 3\}$ for BadNets and WaNet. It can be seen that a threshold 2.5 is reasonable, as it offers a high TPR while keeping a low FPR. Note that the larger the threshold, the smaller the TPR and FPR. Users should choose the threshold based on their specific needs.

Effects of the Poisoning Rate. We launched BadNets on CIFAR10-ResNet18 using different poisoning rates $\in \{1\%, 3\%, 5\%, 10\%\}$ to study the impact of poisoning rates on our defense. Table 4 shows the attack success rate (ASR) increases

Table 5. Effectiveness of defenses with different features. Latent features denote those generated by the feature extractor that is typically used in existing defenses. Critical features are extracted by our method from the identified layers.

Metric→	TPR (%)		FPR (%)	
Defense↓, Features→	Latent Features	Critical Features	Latent Features	Critical Features
AC	0.3	**96.32**	8.84	**7.67**
SCAn	46.05	**86.19**	2.57	**1.96**
Ours	1.31	**99.99**	**4.93**	6.23

Table 6. Performance of features from individual layers compared to identified layers by our defense. The LOI of WaNet and IAD are 9 and 8, respectively.

Layer		1	2	3	4	5	6	7	8	9	10	Ours
WaNet	TPR (%)	0.00	0.10	0.05	0.00	0.01	0.00	68.82	98.08	59.82	0.00	91.04
	FPR (%)	0.09	0.82	0.24	0.20	0.21	0.04	2.06	1.52	2.06	0.65	1.48
IAD	TPR (%)	19.32	34.03	6.44	30.49	61.09	78.65	88.81	99.65	99.10	2.36	99.12
	FPR (%)	1.65	1.38	1.44	1.60	2.27	1.70	1.29	1.13	1.09	1.24	1.26

with the poisoning rate. However, the poisoning rate has minor effects on our TPR and FPR. These results confirm again the effectiveness of our method.

Effectiveness of Our Layer Selection. We compared the performance of AC, SCAn, and our method at detecting BadNets on the GTSRB-MobileNetV2 benchmark using latent features and critical features. We generated latent features based on the feature extractor (*i.e.*, the layer before fully-connected layers) that is typically adopted in existing defenses. The critical features were extracted by the layer of interest (LOI) used in our method. Table 5 shows that using our features led to significantly better performance in almost all cases. In other words, existing detection methods can also benefit from our LOI selection. Also, we compared the performance of our method on CIFAR10-ResNet18 under WaNet and IAD when using the features of every individual layer, and when using LOI and the two layers before LOI. Table 6 shows that as we approach the critical layer, which was just before LOI with WaNet and at LOI with IAD, the detection performance gets better. Since our method included the critical layer, it also was effective. These results confirm the effectiveness of our layer selection and partly explain our method's good performance.

Effectiveness of Cosine Similarity. We compared the cosine similarity with the Euclidean distance as a metric to differentiate between benign and poisoned samples. In the supplementary materials, we show the cosine similarity gives a better differentiation than the Euclidean distance. This is mostly because the direction of features is more important for detection than their magnitude.

Table 7. Adaptive attack. Top, impact of penalty factor β on MA and ASR. Bottom, impact of penalty factor β on TPR and FPR.

β	0	0.5	0.6	0.7	0.8	0.9	0.91	0.92	0.95
MA (%)	91.45	92.96	92.06	92.65	92.63	90.33	79.97	69.13	10
ASR (%)	97.20	96.72	96.93	96.63	96.29	96.88	96.41	97.36	100

Defense↓	$\beta \rightarrow$ Metric (%)↓	0	0.5	0.6	0.7	0.8	0.9	0.91	0.92	0.95
AC	TPR	99.52	99.20	99.16	45.69	26.26	26.22	23.81	13.38	0.00
	FPR	31.14	29.46	28.85	8.21	7.72	6.21	0.25	7.80	0.00
SCAn	TPR	96.60	96.55	96.60	72.80	56.19	0.00	0.00	0.00	0.00
	FPR	0.77	1.38	4.60	1.14	0.10	0.00	0.00	0.00	0.00
Ours	TPR	99.38	99.41	98.18	97.43	97.52	94.20	24.20	0.00	0.00
	FPR	1.35	1.96	1.44	1.15	0.53	1.40	4.17	0.00	0.00

Resistance to Adaptive Attacks. The adversary may adapt his attack to bypass our defense by optimizing the model's original loss $\mathcal{L}_{org}$ and minimizing the layer-wise angular deviation between the features of the poisoned samples and the features' centroids of the target class's benign samples. We studied the impact of this strategy by introducing the *cosine distance* between the features of poisoned samples and the target class centroids as a secondary loss function $\mathcal{L}_{cd}$ in the training objective function. Also, we introduced a penalty parameter β, which yielded a modified objective function $(1-\beta)\mathcal{L}_{org}+\beta\mathcal{L}_{cd}$. The role of β is to control the trade-off between the angular deviation and the main accuracy loss. We then launched BadNets on CIFAR10-ResNet18 under the modified objective function. Table 7 (top subtable) shows MA and ASR with different penalty factors. We can see that values of $\beta < 0.9$ slightly increased the main accuracy because the second loss acted as a regularizer to the model's parameters, which reduced over-fitting. Also, ASR stayed similar to the non-adaptive ASR (when $\beta = 0$). However, the main accuracy degraded with greater β values, because the original loss function was dominated by the angular deviation loss.

Table 7 (bottom subtable) shows the TPRs and FPRs of AC, SCAn, and our defense with different penalty factors. As β increased (up to $\beta = 0.9$), the TPR of our defense decreased from 99.38% to 94.20% while FPR was almost unaffected. This shows that the adversary gained a small advantage with $\beta = 0.9$. On the other hand, the other defenses achieved limited or poor robustness compared to ours with the same β values. With $\beta \geq 0.91$, AC, SCAn, and our method defense failed to counter the attack. However, looking at Table 7 (top subtable) we can see the main accuracy degraded with these high β values, which made it easy to reject the model due its low performance.

6 Conclusion

In this paper, we conducted a layer-wise feature analysis of the behavior of benign and poisoned samples generated by attacked DNNs. We found that the feature difference between benign and poisoned samples tends to reach the maximum at a critical layer, which can be easily located based on the behaviors of benign samples. Based on this finding, we proposed a simple yet effective backdoor detection to determine whether a given suspicious testing sample is poisoned by analyzing the differences between its features and those of a few local benign samples. Our extensive experiments on benchmark datasets confirmed the effectiveness of our detection. We hope our work can provide a deeper understanding of attack mechanisms, to facilitate the design of more effective and efficient backdoor defenses and more secure DNNs.

Acknowledgment. This research was funded by the European Commission (projects H2020-871042 "SoBigData++" and H2020-101006879 "MobiDataLab"), the Government of Catalonia (ICREA Acadèmia Prize to J.Domingo-Ferrer, grant no. 2021 SGR 00115, and FI_B00760 grant to N. Jebreel), and MCIN/AEI/ 10.13039/501100011033 and "ERDF A way of making Europe" under grant PID2021-123637NB-I00 "CURLING". The authors are with the UNESCO Chair in Data Privacy, but the views in this paper are their own and are not necessarily shared by UNESCO.

References

1. Chen, B., et al.: Detecting backdoor attacks on deep neural networks by activation clustering. In: AAAI Workshop (2019)
2. Chen, X., Liu, C., Li, B., Lu, K., Song, D.: Targeted backdoor attacks on deep learning systems using data poisoning. arXiv preprint arXiv:1712.05526 (2017)
3. Gao, Y., et al.: Design and evaluation of a multi-domain Trojan detection method on deep neural networks. IEEE Trans. Dependable Secure Comput. **19**(4), 2349–2364 (2022)
4. Gu, T., Liu, K., Dolan-Gavitt, B., Garg, S.: BadNets: evaluating backdooring attacks on deep neural networks. IEEE Access **7**, 47230–47244 (2019)
5. Hayase, J., Kong, W.: SPECTRE: defending against backdoor attacks using robust covariance estimation. In: ICML (2021)
6. He, K., Zhang, X., Ren, S., Sun, J.: Deep residual learning for image recognition. In: CVPR (2016)
7. Krizhevsky, A., Hinton, G., et al.: Learning multiple layers of features from tiny images (2009)
8. Li, Y., Lyu, X., Koren, N., Lyu, L., Li, B., Ma, X.: Neural attention distillation: erasing backdoor triggers from deep neural networks. In: ICLR (2021)
9. Li, Y., Jiang, Y., Li, Z., Xia, S.T.: Backdoor learning: a survey. IEEE Transactions on Neural Networks and Learning Systems (2022)
10. Li, Y., Ya, M., Bai, Y., Jiang, Y., Xia, S.T.: BackdoorBox: a python toolbox for backdoor learning. arXiv preprint arXiv:2302.01762 (2023)
11. Li, Y., Zhai, T., Jiang, Y., Li, Z., Xia, S.T.: Backdoor attack in the physical world. In: ICLR Workshop (2021)

12. Li, Y., Li, Y., Wu, B., Li, L., He, R., Lyu, S.: Invisible backdoor attack with sample-specific triggers. In: ICCV (2021)
13. Liu, K., Dolan-Gavitt, B., Garg, S.: Fine-pruning: defending against backdooring attacks on deep neural networks. In: RAID (2018)
14. Liu, Y., Xie, Y., Srivastava, A.: Neural trojans. In: ICCD (2017)
15. Nguyen, T.A., Tran, A.: Input-aware dynamic backdoor attack. In: NeurIPS (2020)
16. Nguyen, T.A., Tran, A.T.: WaNet-imperceptible warping-based backdoor attack. In: International Conference on Learning Representations (2020)
17. Rosenfeld, E., Winston, E., Ravikumar, P., Kolter, Z.: Certified robustness to label-flipping attacks via randomized smoothing. In: ICML (2020)
18. Sandler, M., Howard, A., Zhu, M., Zhmoginov, A., Chen, L.C.: MobileNetV2: inverted residuals and linear bottlenecks. In: CVPR (2018)
19. Stallkamp, J., Schlipsing, M., Salmen, J., Igel, C.: The German traffic sign recognition benchmark: a multi-class classification competition. In: IJCNN (2011)
20. Tang, D., Wang, X., Tang, H., Zhang, K.: Demon in the variant: statistical analysis of *dnns* for robust backdoor contamination detection. In: USENIX Security (2021)
21. Turner, A., Tsipras, D., Madry, A.: Label-consistent backdoor attacks. arXiv preprint arXiv:1912.02771 (2019)
22. Zeng, Y., Chen, S., Park, W., Mao, Z.M., Jin, M., Jia, R.: Adversarial unlearning of backdoors via implicit hypergradient. In: ICLR (2022)
23. Zeng, Y., Park, W., Mao, Z.M., Jia, R.: Rethinking the backdoor attacks' triggers: a frequency perspective. In: ICCV (2021)
24. Zhao, P., Chen, P.Y., Das, P., Ramamurthy, K.N., Lin, X.: Bridging mode connectivity in loss landscapes and adversarial robustness. In: ICLR (2020)

Enhancing Federated Learning Robustness Using Data-Agnostic Model Pruning

Mark Huasong Meng[1,4], Sin G. Teo[1], Guangdong Bai[2(✉)], Kailong Wang[3,4], and Jin Song Dong[4]

[1] Institute for Infocomm Research (I2R), A*STAR, Singapore, Singapore
{menghs,teo_sin_gee}@i2r.a-star.edu.sg
[2] University of Queensland, Brisbane, QLD, Australia
g.bai@uq.edu.au
[3] Huazhong University of Science and Technology, Wuhan, China
wangkl@hust.edu.cn
[4] National University of Singapore, Singapore, Singapore
dcsdjs@nus.edu.sg

Abstract. Federated learning enables multiple data owners with a common objective to participate in a machine learning task without sharing their raw data. At each round, clients train local models with their own data and then upload the model parameters to update the global model. This multi-agent form of machine learning has been shown prone to adversarial manipulation by recent studies. Byzantine attackers impersonated as benign clients can stealthily interrupt or destroy the learning process. In this paper, we propose FLAP, a post-aggregation model pruning technique to enhance the Byzantine robustness of federated learning by effectively disabling the malicious and dormant components in the learned neural network models. Our technique is data-agnostic, without requiring clients to submit their dataset or training output, well aligned with the data locality of federated learning. FLAP is performed by the server right after the aggregation, which renders it compatible with an arbitrary aggregation algorithm and existing defensive techniques. Our empirical study demonstrates the effectiveness of FLAP under various settings. It reduces the error rate by up to 10.2% against the state-of-the-art adversarial models. Moreover, FLAP also manages to increase the average accuracy by up to 22.1% against different adversarial settings, mitigating the adversarial impacts while preserving learning fidelity.

1 Introduction

Federated learning (FL) is a machine learning (ML) technique that collaboratively trains a model from decentralized datasets [14]. Unlike the traditional ML that trains a model using a centralized dataset, FL adopts a distributed paradigm where multiple clients contribute to training a model from their local data. Due to the heterogeneity of the data owned by different parties, FL exhibits the great capacity of mitigating the fairness issue from data bias. On the other hand, FL reverses the stereotype that ML can only be carried out in a computationally

H. Kashima et al. (Eds.): PAKDD 2023, LNAI 13936, pp. 441–453, 2023.
https://doi.org/10.1007/978-3-031-33377-4_34

intensive setting. Through the cluster effect [6], FL enables mobile and edge devices to participate in solving complex real-world problems, such as financial services [13], cybersecurity [25], healthcare [19,28], and knowledge discovery [21].

FL is designed to preserve participants' privacy and locality of their data [10, 14,29]. This process, however, is prone to be manipulated by malicious clients since their data and training processes are not transparent to the server and other participants. In addition, the *Byzantine failure* is a major threat to FL due to its distributed nature. There is no guarantee that every client has faithfully uploaded the trained model to the server. Many attacks exploiting these issues have been discovered by a recent study [5]. For example, *poisoning* is one of the most studied attack methods [1]. Malicious clients can collude with each other and commit a Byzantine attack by intentionally training on adversarial data [4] or directly uploading erroneous model parameters to the server [2,27]. Unfortunately, such distributed poisoning attack is hard to be detected. Notably, an existing study [2] shows that a poisoning attack can be achieved by merely one malicious client launching a one-shot attack in FL.

To tackle the potential attacks, the research community has proposed multiple defense techniques. Most endeavors pursue Byzantine-robust FL through *Byzantine-resilient aggregations.* It is usually achieved by diverse aggregation algorithms, such as multi-Krum [2] and trimmed mean [27], to detect and eliminate the malicious impact on the global model. Unfortunately, as shown by a recent study [5], these Byzantine-resilient aggregations are only effective when the attacker has no extra knowledge about the FL than benign clients. The FL can still be compromised in case the attacker knows information such as the defensive aggregation technique adopted by the server. As a result, some *auxiliary defense* approaches that cooperate with aggregation algorithms are proposed to address their fragility. They are essentially not a part of the conventional FL process but demonstrate promising efforts towards Byzantine robustness.

Recent advances in auxiliary defenses tend to enhance the robustness of the global model prior to the aggregation. Representative studies include taking advantage of a dedicated dataset to exclude certain clients' updates incurring abnormal test accuracy and/or loss values [3,5] and performing a supervised model pruning based on clients' voting [26]. They either require a dataset from a similar distribution of clients' training data, demand the population of attackers among participating clients, or assume participating clients are honest all the time. However, these prerequisites may not be realistic, especially in an adversarial environment. An effective defense technique is needed to strengthen the Byzantine robustness of FL.

In this paper, we propose a post-aggregation defense technique for FL by data-agnostic model pruning named FLAP (FL by data-Agnostic Pruning)[1]. FLAP can be performed by the server independently with no reliance on training data and extra contributions from clients. It is not limited by the estimated population of malicious clients. More importantly, FLAP is designed with a generic FL framework that is compatible with diverse aggregation settings. It can

[1] Our source code is hosted at https://github.com/mark-h-meng/flap.

be deployed either alone or together with existing Byzantine-robust techniques to boost their defenses.

Our design is motivated by an insight that model pruning could disable the insignificant and dormant parameters, which are often introduced by poisoning attacks. FLAP aims to enhance the robustness of the global model in an adversarial environment. Meanwhile, it should preserve learning fidelity to the maximum extent. Therefore, we adopt a conservative pruning strategy rather than the cut-and-re-train approach that has been conventionally applied by centralized learning paradigms. More specifically, our pruning aims to remove the parameters with the most negligible impact on the model's output. To this end, FLAP dynamically measures the effect of deleting a unit to identify pruning candidates. It adopts a scale-based sampling strategy for convolutional (Conv) layers and a cross-layer saliency-based sampling strategy for fully-connected (FC) layers. Our evaluation shows that FLAP is effective in preserving robustness and fidelity against diverse adversarial settings. Meanwhile, it is capable of boosting the state-of-the-art (SOTA) defenses towards a higher degree of Byzantine robustness. Our key contributions are summarized below.

- We propose FLAP, a novel FL pruning technique that does not rely on an estimation of malicious clients' population and makes no request for the cooperation of participating clients. It is implemented as an auxiliary defense for generic FL that can be added to any form of existing FL applications.
- We conduct an empirical study to explore the effectiveness of FLAP in an adversarial environment. We find FLAP can enhance the robustness of FL with different aggregation algorithms while preserving the model fidelity.
- We test FLAP against different advanced adversarial models and compare it with the SOTA defenses. Our empirical study shows that FLAP outperforms the existing defense techniques in all adversarial models and boosts the existing defenses for a higher degree of Byzantine robustness.

2 Related Work

Attacking FL. Poisoning attacks for ML can be categorized into *untargeted attacks* [5] and *targeted attacks* [4,16]. The former is performed to reduce the overall learning accuracy on arbitrary inputs, and the latter aims to precisely misclassify a limited set of classes. An untargeted attack is shown not practical to the extent of FL, as it can be defended at a low cost [20]. Targeted attacks can be further classified into *label-flipping attacks* [9] and *backdoor attacks* [4]. In this paper, we select the targeted label-flipping attacks as the default approach.

Byzantine-Robust FL. Defending FL systems is a widely-studied topic. The dominant baseline defenses include *median* [27], *trimmed mean* [27], *multi-Krum* [2], and *Bulyan* [8]. Among them, median and trimmed mean are *statistical-based aggregations* that assess each client's update independently. multi-Krum and Bulyan are representative *distance-based aggregations*. Those baseline algorithms are later found to be fragile to fine-crafted adversarial models [5].

Recent improvements that address the flaws of baseline defenses mainly focus on proactively detecting malicious clients prior to the aggregation [3,26] and strengthening the existing FL framework to minimize the impact of malicious upload [11,17]. Besides that, there are also auxiliary defenses that co-exist with these baselines and address their shortcomings. The SOTA defenses include the *Error Rate based Rejection* (ERR), the *Loss Function based Rejection* (LFR) and the *Union Rejection* (ERR+LFR) [5]. They are shown effective against various adversarial models but demand a proper estimation of malicious clients' population. Cao et al. [3] proposed another Byzantine-robust FL framework by trust bootstrapping. However, it may require the cooperation of the clients because the server needs to collect a small clean dataset from their training data.

Neural Network Pruning in FL. Pruning is a commonly applied model optimization technique [7]. It is considered useful in FL based on an insight that the defense against targeted attack can be achieved by removing not only the poisoned data but also the activation of adversarial inputs in the model [12,24]. Wu et al. [26] proposed a post-training FL defense by pruning and fine-tuning the global model. However, it relies on a voting process among participating clients, which entails sharing their local models' activation results. That may not be practical because participants may be reluctant to share any knowledge about the models' output based on their own training data, given that disclosing them is prone to a membership attack [18]. In this paper, we study the adoption of pruning that does not rely on the training data, i.e., data-agnostic pruning [15, 22], and therefore can be solely performed by the server without explicitly asking for clients' cooperation.

3 Problem Statement

3.1 Federated Learning

We assume a standard context of FL, in which data is not identically and independently distributed across multiple clients. A client i can only access his/her own data D_i, $i = 1, 2, ..., n$. The server does not have access to clients' data. The learning process is performed in multiple rounds in a synchronous manner.

During an arbitrary round t, clients receive a global model w^{t-1}_{Global} from the server and perform continuous learning with his/her own data D_i, followed by sending the update of the local model's parameters, i.e., $g^t_i = w^t_i - w^{t-1}_{\text{Global}}$, to the server. All the clients' updates would then undergo an aggregation procedure by the server prior to moving to the next round. Let α be the learning rate, the global model can be defined as follows:

$$w^t_{\text{Global}} = w^{t-1}_{\text{Global}} + \alpha \cdot g^t_{\text{Global}} \tag{1}$$

The ultimate goal of FL is to find an aggregated update g^t_{Global} from the clients at a certain round t to result in a minimal loss function $\mathcal{L}\left(D, w^t_{\text{Global}}\right)$ on the joint training dataset $D = \cup^n_{i=1} D_i$.

We assume a standard aggregation named *FedAvg* [14] as the default option to compute the global update unless otherwise specified. FedAvg calculates the average of the clients' updates as the global update, which can be formally defined as $g^t_{\text{Global}} = \sum_{i=1}^{n} \frac{|D_i|}{|D|} g^t_i$, where $|D|$ and $|D_i|$ represent the size of the joint training dataset and the size of the client i's local training set. In addition to FedAvg, we also consider the server to take some defensive measures to thrive for a resilient and safe global model, which we will detail in Sect. 5.

3.2 Threat Model

This paper assumes that the attack of FL adopts a targeted label-flipping attack by model poisoning, which is to maximize the possibility of misclassification of the targeted data samples. The malicious clients are granted no extra privileges than the benign ones, such that they can only access their own training sets and model parameters. However, they can still be manipulated by a single attacker to deploy a Byzantine attack. In other words, malicious clients will collude with each other toward the same attack goal. During an attack round, the malicious clients intentionally learn certain (victim) data $\langle x^v_i, y^v_i \rangle \in D_i$ with a wrong (target) label $y^\mathcal{T}_i$, e.g., learning all digit 1 as digit 7, and learn the remaining data correctly.

Apart from the conventional attack, we also consider that malicious clients can take advantage of the strengthened adversarial models to craft local models [5]. Thus, we anticipate the malicious clients to apply two additional strengthened adversarial models. We brief them below and later assess them in Sect. 5.

Partial knowledge attack. A malicious client can access all other colluding malicious clients' local models and has knowledge about the aggregation rule on the server side. Through analyzing the parameters of other malicious clients, it can craft the local update to influence the direction of the global model update, which subsequently undermines the robustness of the global model.

Full knowledge attack. On the basis of a partial knowledge attack, a malicious client has full access to the training sets and local models of all participating clients, i.e., the entire FL is completely transparent to the attacker. This scenario is helpful for us to assess *the upper bound* of an attacker's capability and estimate the impact of adversarial manipulation.

4 Proposed FLAP

4.1 Approach Overview

Figure 1 shows how FLAP is used in FL. Overall, it is a typical FL process except for the addition of the post-aggregation model pruning at the server side (Phase 4). FL with FLAP executes an iterative process that begins with broadcasting the *global model* (Phase 1). For each client, it performs training over the received global model with its own data set (Phase 2). This training process is supposed to be private so that other clients and the server have no access to it. Upon the completion of training, the client sends the learned *local model* parameters'

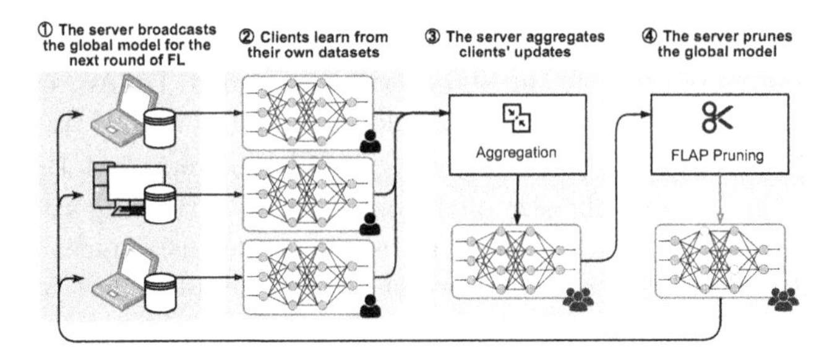

Fig. 1. The workflow of federated learning with FLAP

updates to the server. Once the server has received all participating clients' updates, it aggregates the parameters and produces the global model (Phase 3). Next, FLAP performs model pruning over the newly generated global model (Phase 4) and marks the end of the current round of learning. If the learning is not concluded, the server will broadcast the newly pruned global model to all participating clients and start the next round of learning.

4.2 Model Pruning

Our data-agnostic pruning supports two types of hidden layers of neural network models: the FC layers and the Conv layers. For each time the pruning is launched, FLAP samples the units on the supported hidden layer and nominates a fixed proportion (e.g., 1%) of units to cut. Unlike the conventional data-hungry pruning that adopts an aggressive cut-and-re-train strategy, we design our pruning as a conservative approach that always prefers to remove the units that incur the most insignificant impact to the model output, based on an insight that many attack models stealthily plant their adversarial patterns in those units [12,24]. Although we are given a chance of continuing training in FL, applying a proper conservative pruning technique is helpful in preserving model fidelity. To this end, we propose different sampling strategies for the two layer types to nominate pruning candidates.

Conv Layers Sampling. When handing a Conv layer, FLAP calculates the l_1-norm of all the filters, sorts them, and nominates a few filters with the least norm values as pruning candidates. The norm function is selected as the pruning criteria based on the assumption that the malicious clients tend to exploit the neural network through minor and imperceptible filters, and then manipulate their activation to misclassify. As a result, FLAP removes those filters that have been trained with the least significant values.

FC Layers Sampling. Finding pruning candidates at FC layers is comparably more complicated because the parameters within a FC layer tend to be very similar to each other if we do not take the *cross-layer computation* into account. The least value-based pruning without local retraining usually causes severe performance degradation to the model. FLAP takes advantage of the data-agnostic cross-layer saliency-based pruning that has been studied in [15], where the pruning is conducted in a pair-wise manner. Given a candidate pair of hidden units, we remove one of its units and double the weight of the other unit, which is expected to supersede the role of the pruned unit.

To find the proper units to be pruned, we first iterate all units in a FC layer and form pairs for them. We then assess the *propagated impact* of pruning a pair of hidden units in the middle layer, which is calculated as a range/interval depending on the legitimate value ranges of the model input. The potential *impact of pruning* is jointly measured by the l_1-norm and *entropy* of the propagated impact. When FLAP prunes a FC layer, it sorts all unit pairs by their impact values and removes the pairs with the lowest impact values.

Global Model Pruning. Considering the pruning process is proposed to reinforce the reliability of the FL model, FLAP does not alter the model structure during the pruning. Instead, it zeros out the parameter values of all the pruned units. The pruned model becomes the global model of the next round of FL and will be broadcast to all clients for their local training. Overall, we let p denote the pruning process, and thus, the global model can be defined by a modified form of Eq. 1 as follows:

$$w^t_{\text{Global}} = p\left(w^{t-1}_{\text{Global}} + \alpha \cdot g^t_{\text{Global}}\right) \tag{2}$$

5 Evaluation

5.1 Experiment Setup

We implement FL based on a public repository[2]. The FLAP and the tested adversarial models are implemented based on TensorFlow. All the presented results are the median value observed from at least five repeated executions.

Federated Learning. The simulated FL is composed of 80 participating clients. The distributed training is uniformly configured for each client. Each round of the client's local training consists of two epochs, with the learning rate set to 0.001 and the Adam optimizer applied. FedAvg is selected as the default aggregation algorithm. Our experiments begin with 20 rounds of benign training, which can help the global model maintain stable prediction accuracy. We consider malicious clients to start attacking from the 21st round. By default, 16 out of 80 participating clients (20%) are malicious that collude with each other in conducting model poisoning attacks. We uniformly set the attack goal as misleading the global model to predict the first class as the seventh class.

Pruning. We note that the pruning may not be necessarily carried out for every round of FL. Instead, we stipulate that the server performs pruning from the first round and repeats every five rounds. For each pruning operation, the server zeros out 1% of filters from every Conv layer and 1% of hidden units from every FC layer, by zeroing out their corresponding parameters[3].

[2] https://github.com/pps-lab/fl-analysis.

[3] One unit will be pruned if the layer has less than 100 units.

Models and Datasets. We select FEMNIST and use three different model architectures in our evaluation, including a five-layer MLP, a LeNet-5 ConvNet, and a ResNet-18 model.

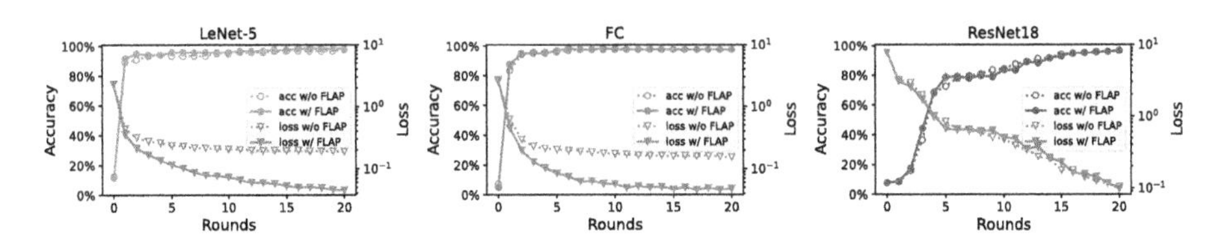

Fig. 2. Test accuracy and loss of FL up to round 20, with and without FLAP

5.2 Benign FL with FLAP

Our first set of experiments aims to investigate if FLAP suits the FL as a post-aggregation optimization. More specifically, we wish to figure out the question "*how does* FLAP *preserve the fidelity of FL in a non-adversarial circumstance?*". To this end, we apply FLAP on all three models under the benign settings and compare the learning process with the FL without it. We evaluate the global model after each round's aggregation and record the test accuracy and loss.

Figure 2 presents the test accuracy and loss value of the first 20 rounds, with and without the equipment of FLAP. We find that the growth of test accuracy of models with FLAP is almost identical with the models without it. We also observe that the adoption of FLAP accelerates the loss descent on LeNet-5 and MLP models. In summary, FLAP shows promising fidelity preservation on all three models. The adoption of FLAP does not impair the learning process.

5.3 FLAP in Adversarial Settings

Next, we explore *whether* FLAP *can boost existing defensive techniques towards Byzantine-robust FL.* We focus on using the ResNet-18 model to compare FLAP with the existing techniques against various adversarial models.

Starting from round 21, we deploy a model poisoning attack for ten rounds to the default setting of FL. We test our approach with two representative Byzantine-resilient aggregation algorithms. We also learn that the malicious clients in FL may strengthen their attack capacity by gaining extra knowledge from the server and benign clients, and therefore we assess our approach with two advanced adversarial models proposed by Fang et al. [5], namely partial knowledge attack and full knowledge attack.

In our experiments, we evaluate the robustness of the global model by calculating the average error rate of consecutive ten rounds of adversarial learning. We also record the average test accuracy of the global model to reflect the overall learning process. The experimental results can be found in Table 1.

Byzantine-Resilient Aggregations. We replace FedAvg with Byzantine resilient aggregations, namely trimmed mean and multi-Krum, and repeat our previous experiments with the default adversarial settings. Both algorithms are configurable with a parameter, which defines the estimated *upper bound* of Byzantine attackers among all participating clients [2]. For each algorithm, we define three modes, named *conservative* (C mode), *perfect* (P mode) and *radical* (R mode). The C mode simulates the server underestimating the existence of malicious clients, the R mode stipulates that the server overestimates the population of malicious clients, and the P mode defines that the server estimates the exact population of malicious clients. These three modes estimate the percentage of malicious clients to be 10%, 30%, and 20%, respectively.

Table 1. Average error rates and test accuracy of FL (ResNet-18) in various adversarial settings, with (shown in **bold** text) and without FLAP (shown in plain text).

Aggregation Rules	Auxiliary Defense	Adversarial Modes		
		Targeted Label Flipping	Partial Knowledge	Full Knowledge
		Error Rate (Lower is Better)*		
FedAvg	Nil	30.8%, **20.0%** (▼)	62.8%, **20.0%** (▼)	40.0%, **20.0%** (▼)
Trimmed Mean†	Nil	(C) 87.0%, **74.7%** (▼)	(C) 72.5%, **62.9%** (▼)	(C) 67.2%, **35.4%** (▼)
		(P) 30.0%, **17.5%** (▼)	(P) 22.9%, **17.5%** (▼)	(P) 38.1%, **33.0%** (▼)
		(R) 11.4%, **9.8%** (▼)	(R) 11.5%, **9.8%** (▼)	(R) 12.7%, **10.6%** (▼)
	ERR+LFR	(C) 59.2%, **20.0%** (▼)	(C) 80.0%, **71.3%** (▼)	(C) 84.6%, **68.3%** (▼)
		(P) 16.8%, **15.6%** (▼)	(P) 16.9%, **15.6%** (▼)	(P) 22.7%, **16.3%** (▼)
		(R) 5.8%, **5.8%** (=)	(R) 5.8%, **4.2%** (▼)	(R) 9.6%, **9.3%** (▼)
Multi- Krum†	Nil	(C) 84.3%, **83.7%** (▼)	(C) 93.3%, **74.8%** (▼)	(C) 83.5%, **73.9%** (▼)
		(P) 22.7%, **20.0%** (▼)	(P) 22.7%, **15.6%** (▼)	(P) 20.6%, **14.2%** (▼)
		(R) 28.8%, **27.3%** (▼)	(R) 28.9%, **19.5%** (▼)	(R) 22.3%, **20.0%** (▼)
	ERR+ LFR	(C) 84.6%, **83.8%** (▼)	(C) 79.2%, **68.3%** (▼)	(C) 83.5%, **75.4%** (▼)
		(P) 22.7%, **20.0%** (▼)	(P) 22.7%, **16.3%** (▼)	(P) 20.6%, **14.2%** (▼)
		(R) 28.7%, **22.3%** (▼)	(R) 27.9%, **21.0%** (▼)	(R) 28.1%, **17.9%** (▼)
		Test Accuracy (Higher is Better)*		
FedAvg	Nil	10.3%, **10.9%** (▲)	10.1%, **10.1%** (=)	9.0%, **9.2%** (▲)
Trimmed Mean†	Nil	(C) 11.5%, **14.6%** (▲)	(C) 13.5%, **17.3%** (▲)	(C) 15.8%, **18.4%** (▲)
		(P) 92.1%, **97.8%** (▲)	(P) 92.3%, **93.2%** (▲)	(P) 15.8%, **18.4%** (▲)
		(R) 94.6%, **95.1%** (▲)	(R) 93.9%, **94.8%** (▲)	(R) 92.5%, **92.1%** (▼)
	ERR+ LFR	(C) 11.2%, **11.0%** (▼)	(C) 11.3%, **16.5%** (▲)	(C) 12.8%, **14.1%** (▲)
		(P) 93.4%, **93.0%** (▼)	(P) 93.4%, **93.9%** (▲)	(P) 93.4%, **93.2%** (▼)
		(R) 95.8%, **96.0%** (▲)	(R) 94.6%, **94.9%** (▲)	(R) 95.6%, **95.6%** (=)
Multi- Krum†	Nil	(C) 34.5%, **56.0%** (▲)	(C) 35.4%, **56.2%** (▲)	(C) 37.6%, **52.2%** (▲)
		(P) 35.6%, **43.5%** (▲)	(P) 36.0%, **44.7%** (▲)	(P) 35.5%, **44.2%** (▲)
		(R) 35.3%, **44.2%** (▲)	(R) 36.0%, **45.9%** (▲)	(R) 35.6%, **46.9%** (▲)
	ERR+ LFR	(C) 34.5%, **56.1%** (▲)	(C) 34.5%, **56.2%** (▲)	(C) 38.8%, **60.9%** (▲)
		(P) 35.5%, **43.4%** (▲)	(P) 36.0%, **44.7%** (▲)	(P) 35.5%, **45.9%** (▲)
		(R) 35.3%, **44.2%** (▲)	(R) 35.3%, **47.7%** (▲)	(R) 35.4%, **46.9%** (▲)

* We use "(▼)", "(=)" and "(▲)" to represent a decrease, no change, and growth, respectively.

† Three parametric settings are adopted in both trimmed mean and multi-Krum aggregation algorithms: (C), (P), and (R) stand for Conservative, Perfect, and Radical modes, respectively.

Our results show that the adoption of Byzantine-resilient aggregations helps reduce the error rate and improve the test accuracy, however, only when the server sufficiently estimates the presence of malicious clients (i.e., the P mode). We learn that any mis-estimiation of the population of malicious clients causes a negative impact on the FL in terms of test accuracy and error rate. This highlights the necessity of defensive techniques that are independent of the server's knowledge regarding the attackers' population.

Our evaluation demonstrates that FLAP can improve the FL in all three modes of the two aggregation algorithms. The adoption of FLAP reduces the error rate by up to 12.5% for the trimmed mean model (P mode) and 2.7% for the multi-Krum model (P mode). On this basis, FLAP also helps FL to better converge as we record a growth of test accuracy at up to 5.7% from the trimmed mean model (P mode) and 21.5% from the multi-Krum model (C mode).

Advanced Adversarial Models. Our next set of experiments aims to investigate *whether* FLAP *makes FL more Byzantine-robust against the SOTA adversarial models.* We simulate the two adversarial models specially designed for FL [5] and evaluate our approach against them. Moreover, we also compare our approach with the prediction-based defenses proposed in the same paper. We take the most radical defense named *ERR+LFR* as the baseline[4].

From Table 1, we observe that the two adversarial models overall stimulate the adversarial effectiveness when a defensive aggregation algorithm is deployed. They incur a higher error rate without significantly impairing the test accuracy and therefore can more stealthily poison the global model. Besides that, the ERR+LFR defense is shown to be effective in most cases, especially when the aggregation algorithms are deployed in P mode and R mode.

We also find that our approach can boost the ERR+LFR as a post-aggregation defense. We record an error rate reduction at up to 10.2% (multi-Krum R mode) and a test accuracy increment at up to 22.1% (multi-Krum C mode) by applying FLAP to the ERR+LFR defense against the full-knowledge attack. Even if we compare the models that are either equipped with ERR+LFR only (i.e., without FLAP) or FLAP only (i.e., absence of ERR+LFR), we find our approach still achieves a lower error rate than the ERR+LFR defense in all scenarios of the multi-Krum setting and two out of six scenarios of the trimmed mean setting. FLAP also manages to outperform the ERR+LFR defense in 14 out of 18 scenarios of two settings with regard to the test accuracy, indicating FLAP better assists the FL towards the learning target.

Summary. FLAP is shown effective towards Byzantine-robust FL in both benign and adversarial environments. It can co-exist with existing defenses including Byzantine-resilient aggregations and auxiliary prediction-based techniques and even outperforms them in most cases. More importantly, FLAP can boost those defenses to achieve a higher degree of Byzantine robustness, especially when the server underestimates the presence of malicious clients.

[4] We assume the perfect estimation that 20% of clients are excluded due to high loss function value and another 20% of clients are excluded due to low accuracy.

6 Discussion

Limitations and Future Work. First, FLAP supports pruning for both FC and Conv layers. For that reason, we may achieve a higher degree of Byzantine robustness if we broaden the FLAP's support for pruning residual blocks. In addition, the server has no access to the training set but owns some data in a similar distribution for testing purposes. That gives us a chance to prune the model in a supervised manner with the testing data. We aim to explore the test set guided pruning in the future.

Broader Impacts. To the best of our knowledge, this is the first work that explores the pruning by the FL server without the reliance on clients' contribution. FLAP does not request any training data or training outputs from clients, therefore it is difficult to be manipulated by malicious participants. It takes place after the aggregation so that it can co-exist with the existing defenses and boost their effectiveness. This paper will help the research community's exploration of more defense techniques to be adopted in FL, and contribute to achieving efficient privacy-preserving machine learning [23, 29].

7 Conclusion

In this paper, we propose FLAP, a post-aggregation pruning technique to boost the Byzantine robustness of FL, based on our insight that pruning can effectively mitigate the unfavorable and malicious parameters learned in adversarial training. We evaluate the proposed FLAP with different models, assess its effectiveness against different adversarial models, and compare it with existing defensive techniques. Our empirical study demonstrates that FLAP can reduce the error rate and preserve the fidelity of FL equipped with different aggregation algorithms under various adversarial settings. FLAP also shows a promising capacity to reinforce the existing defensive techniques against the SOTA adversarial models to achieve a higher degree of Byzantine robustness.

Acknowledgment. This work was supported by The University of Queensland under the NSRSG grant 4018264-617225, Cyber Research Seed Funding, the Global Strategy and Partnerships Seed Funding, and Agency for Science, Technology and Research (A*STAR) Singapore under the ACIS scholarship.

References

1. Bhagoji, A.N., Chakraborty, S., Mittal, P., Calo, S.: Analyzing federated learning through an adversarial lens. In: International Conference on Machine Learning (2019)
2. Blanchard, P., Mhamdi, E.M.E., Guerraoui, R., Stainer, J.: Machine learning with adversaries: Byzantine tolerant gradient descent. In: Guyon, I., von Luxburg, U., Bengio, S., Wallach, H.M., Fergus, R., Vishwanathan, S.V.N., Garnett, R. (eds.) Advances in Neural Information Processing Systems, pp. 119–129 (2017)

3. Cao, X., Fang, M., Liu, J., Gong, N.Z.: Fltrust: Byzantine-robust federated learning via trust bootstrapping. In: Network and Distributed System Security Symposium. The Internet Society (2021)
4. Chen, X., Liu, C., Li, B., Lu, K., Song, D.: Targeted backdoor attacks on deep learning systems using data poisoning. arXiv preprint arXiv:1712.05526 (2017)
5. Fang, M., Cao, X., Jia, J., Gong, N.: Local model poisoning attacks to byzantine-robust federated learning. In: 29th USENIX Security Symposium (2020)
6. Fang, U., Li, J., Akhtar, N., Li, M., Jia, Y.: GOMIC: multi-view image clustering via self-supervised contrastive heterogeneous graph co-learning. World Wide Web, pp. 1–17 (2022)
7. Guan, H., Xiao, Y., Li, J., Liu, Y., Bai, G.: A comprehensive study of real-world bugs in machine learning model optimization. In: Proceedings of the International Conference on Software Engineering (2023)
8. Guerraoui, R., Rouault, S., et al.: The hidden vulnerability of distributed learning in Byzantium. In: International Conference on Machine Learning (2018)
9. Huang, L., Joseph, A.D., Nelson, B., Rubinstein, B.I., Tygar, J.D.: Adversarial machine learning. In: Proceedings of the 4th ACM Workshop on Security and Artificial Intelligence, pp. 43–58 (2011)
10. Jin, C., Wang, J., Teo, S.G., Zhang, L., Chan, C., Hou, Q., Aung, K.M.M.: Towards end-to-end secure and efficient federated learning for XGBoost (2022)
11. Li, T., Hu, S., Beirami, A., Smith, V.: Ditto: Fair and robust federated learning through personalization. In: International Conference on Machine Learning (2021)
12. Liu, K., Dolan-Gavitt, B., Garg, S.: Fine-pruning: defending against backdooring attacks on deep neural networks. In: International Symposium on Research in Attacks, Intrusions, and Defenses, pp. 273–294. Springer (2018)
13. Mahalle, A., Yong, J., Tao, X., Shen, J.: Data privacy and system security for banking and financial services industry based on cloud computing infrastructure. In: IEEE International Conference on Computer Supported Cooperative Work in Design (2018)
14. McMahan, B., Moore, E., Ramage, D., Hampson, S., y Arcas, B.A.: Communication-efficient learning of deep networks from decentralized data. In: International Conference on Artificial Intelligence and Statistics (2017)
15. Meng, M.H., Bai, G., Teo, S.G., Dong, J.S.: Supervised robustness-preserving data-free neural network pruning. In: International Conference on Engineering of Complex Computer Systems (2023)
16. Meng, M.H., Bai, G., Teo, S.G., Hou, Z., Xiao, Y., Lin, Y., Dong, J.S.: Adversarial robustness of deep neural networks: a survey from a formal verification perspective. IEEE Trans. Depend. Secure Comput. (2022)
17. Panda, A., Mahloujifar, S., Bhagoji, A.N., Chakraborty, S., Mittal, P.: Sparsefed: mitigating model poisoning attacks in federated learning with sparsification. In: International Conference on Artificial Intelligence and Statistics (2022)
18. Salem, A., Zhang, Y., Humbert, M., Berrang, P., Fritz, M., Backes, M.: Ml-Leaks: model and data independent membership inference attacks and defenses on machine learning models. In: Network and Distributed System Security Symposium (2019)
19. Shaik, T., et al.: Fedstack: personalized activity monitoring using stacked federated learning. Knowl.-Based Syst. **257**, 109929 (2022)
20. Shejwalkar, V., Houmansadr, A., Kairouz, P., Ramage, D.: Back to the drawing board: a critical evaluation of poisoning attacks on production federated learning. In: IEEE Symposium on Security and Privacy, pp. 1354–1371. IEEE (2022)

21. Song, X., Li, J., Cai, T., Yang, S., Yang, T., Liu, C.: A survey on deep learning based knowledge tracing. Knowl.-Based Syst. **258**, 110036 (2022)
22. Srinivas, S., Babu, R.V.: Data-free parameter pruning for deep neural networks. In: Proceedings of the British Machine Vision Conference (2015)
23. Teo, S.G., Cao, J., Lee, V.C.: DAG: a general model for privacy-preserving data mining. IEEE Trans. Knowl. Data Eng. **32**(1), 40–53 (2018)
24. Wang, B., et al.: Neural cleanse: identifying and mitigating backdoor attacks in neural networks. In: IEEE Symposium on Security and Privacy (SP), pp. 707–723. IEEE (2019)
25. Wang, K., Zhang, J., Bai, G., Ko, R., Dong, J.S.: It's not just the site, it's the contents: intra-domain fingerprinting social media websites through CDN bursts. In: Proceedings of the Web Conference (2021)
26. Wu, C., Yang, X., Zhu, S., Mitra, P.: Mitigating backdoor attacks in federated learning. arXiv preprint arXiv:2011.01767 (2020)
27. Yin, D., Chen, Y., Ramchandran, K., Bartlett, P.L.: Byzantine-robust distributed learning: towards optimal statistical rates. In: International Conference on Machine Learning (2018)
28. Yin, H., Song, X., Yang, S., Li, J.: Sentiment analysis and topic modeling for covid-19 vaccine discussions. World Wide Web **25**(3), 1067–1083 (2022)
29. Zhang, Y., Bai, G., Li, X., Curtis, C., Chen, C., Ko, R.K.: PrivColl: practical privacy-preserving collaborative machine learning. In: European Symposium on Research in Computer Security (2020)

BeamAttack: Generating High-quality Textual Adversarial Examples Through Beam Search and Mixed Semantic Spaces

Hai Zhu[1,3](✉), Qinyang Zhao[2], and Yuren Wu[3]

[1] University of Science and Technology of China, Hefei, China
SA21218029@mail.ustc.edu.cn
[2] Xidian University, Xi'an, China
21151213588@stu.xidian.edu.cn
[3] Ping An Technology (Shenzhen) Co., Ltd., Shenzhen, China
wuyuren134@pingan.com.cn

Abstract. Natural language processing models based on neural networks are vulnerable to adversarial examples. These adversarial examples are imperceptible to human readers but can mislead models to make the wrong predictions. In a black-box setting, attacker can fool the model without knowing model's parameters and architecture. Previous works on word-level attacks widely use single semantic space and greedy search as a search strategy. However, these methods fail to balance the attack success rate, quality of adversarial examples and time consumption. In this paper, we propose **BeamAttack**, a textual attack algorithm that makes use of mixed semantic spaces and improved beam search to craft high-quality adversarial examples. Extensive experiments demonstrate that BeamAttack can improve attack success rate while saving numerous queries and time, e.g., improving at most 7% attack success rate than greedy search when attacking the examples from MR dataset. Compared with heuristic search, BeamAttack can save at most 85% model queries and achieve a competitive attack success rate. The adversarial examples crafted by BeamAttack are highly transferable and can effectively improve model's robustness during adversarial training. Code is available at https://github.com/zhuhai-ustc/beamattack/tree/master

Keywords: Adversarial Examples · Robustness · Natural Language Processing

1 Introduction

In recent years, neural networks have achieved great success in the natural language processing field while being vulnerable to adversarial examples. These adversarial examples are original inputs altered by some tiny perturbations

H. Zhu—This work was done when the author was at Ping An Technology (Shenzhen) Co., Ltd.

H. Kashima et al. (Eds.): PAKDD 2023, LNAI 13936, pp. 454–465, 2023.
https://doi.org/10.1007/978-3-031-33377-4_35

[9,20]. It is worth noting that perturbations are imperceptible to humans but can mislead the model decision. Therefore, it is essential to explore adversarial examples since our goal is to improve the reliability and robustness of the model, especially on some security-critical applications, such as toxic text detection and public opinion analysis [25]. Compared to image and speech attacks [2,22], it is more challenging in crafting textual adversarial examples due to the discrete of natural language. In addition, there are some grammar constraints in the textual adversarial examples:(1)the crafted examples should keep the same meaning as the original texts,(2)generated examples should look natural and grammatical. However, previous works barely conform to all constraints, or satisfy the above constraints at the cost of reducing the attack success rate.

Conventional word-level attack algorithms can be roughly divided into three steps: (1) calculating word importance score according to the changes of class label probabilities after replacing this word, (2) searching synonyms for each origin word, (3) selecting the substitution that reduces the class label probabilities most and replacing origin word with it until model predicts wrong. The problem is that previous works only use a single semantic space to search synonyms, which limits the diversity of substitutions and cut down the search space. In addition, most prior works introduce greedy search to select the best substitution with the maximum change of class label probabilities [9,11]. Greedy search limits the search space and sometimes leads to local optimal solution and word over-substitution. Therefore, some works [20,26] introduce heuristic search to improve attack success rate, at the cost of time-consuming and numerous model queries. In generally, previous works fail to balance the attack success rate, quality of adversarial examples and time consumption.

In this paper, we propose BeamAttack, a textual attack algorithm based on mixed semantic spaces and beam search. Specially, we search substitutions from word embedding space and BERT respectively, and filter out the bad synonyms to improve semantic similarity of adversarial examples, then improve beam search to craft adversarial examples, which greatly expands the search space by controlling beam size. Therefore, it is capable of escaping from local optima within acceptable number of model queries. Furthermore, we evaluate BeamAttack by attacking various neural networks on five datasets. Experiments show that it outperforms other baselines in attack success rate and semantic similarity while saving numerous model queries. Our main contributions are summarized as follows:

- We propose the mixed semantic spaces, making full use of word embedding space and BERT simultaneously to expand the diversity of substitutions and generating high-qualify adversarial examples.
- We propose BeamAttack, a black-box attack algorithm which improves beam search to expand search space and reduce the redundancy word substitution.
- Experiments show that BeamAttack achieves the trade-off results compared with previous works. In addition, adversarial examples crafted by BeamAttack with high semantic similarity, low perturbation, and good transferability.

2 Related Work

We divide the existing textual attack algorithms into char-level, word-level and sentence-level attacks based on granularity. Char-level attacks generate adversarial examples by inserting, swapping or removing characters(such as 'attack' → 'atttack') [6,12], which can be easily rectified by word spelling machine. Sentence-level attacks insert some perturbed sentences into the origin paragraph to confuse models [3]. Nevertheless, these adversarial examples contain many lexical errors.

In order to generate high-quality adversarial examples, word-level attacks have gradually become a prevalent approach. Word-level attacks substitute the origin words with synonyms(such as 'like' → 'love'). Traditional strategies search synonyms from word embedding space. For example, some works [9,14,23] calculate the word saliency and greedily substitute words with synonyms derived from WordNet [16], or utilizing word importance score and replace words with synonyms from counter-fitting word vectors [18]. Recently, researcher [7,11,13] search synonyms from pre-trained language models (e.g. BERT, RoBERTa). The pre-trained language models are trained on massive text data, and predict the masked words. Therefore, it has the ability to predict contextual-aware words.

Above attack algorithms adopt the greedy search, which limits the search space and leads to local optimal solution. Minor work have explored the heuristic search, such as genetic algorithm [20], particle swarm optimization [26]. However, heuristic search is very time-consuming and requires a lot of model queries. Therefore, we propose BeamAttack, searching synonyms from word embedding space and BERT simultaneously, and fine-tuning beam search to expand search space and reduce word-over substitution.

3 Beam Search Adversarial Attack

BeamAttack is divided into three steps. There are word importance calculation, mixed semantic spaces and improved beam search. The overview of BeamAttack is shown in the Fig. 1. Before delving into details, we present the attack settings and problem formulation.

3.1 Black-box Untargeted Attack

The BeamAttack belongs to black-box attacks, it has nothing about model's architecture, parameters and gradients, only class label probabilities are accessible. Given a sentence of n words $\mathcal{X} = [x_1, x_2, \cdots, x_n]$ and label set $\mathcal{Y}$, a well-trained model can classify sentence correctly:

$$\underset{y_i \in \mathcal{Y}}{\operatorname{argmax}} P(y_i|\mathcal{X}) = y_{true} \tag{1}$$

The adversarial example $\mathcal{X}' = [x_1^{'}, x_2^{'}, \cdots, x_n^{'}]$ is crafted to make model predict wrong. In addition, there are some constraints on the word substitution

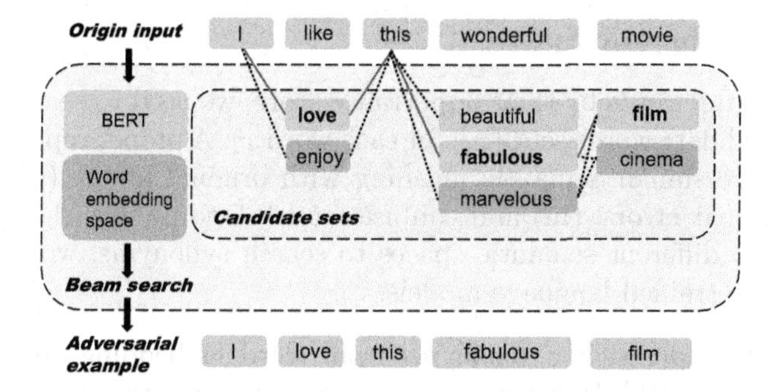

Fig. 1. The overview of BeamAttack. Candidate sets are substitutions generated from BERT and word embedding space. Black lines are beam search paths, wherein red lines are the optimal search path. (Color figure online)

rate(WSR) and semantic similarity(SIM) of the adversarial example. $\mathcal{X}'$ should be close to $\mathcal{X}$ and a human reader hardly differentiate the modifications. The mathematical expression is as follows:

$$\begin{aligned} &\operatorname*{argmax}_{y_i \in \mathcal{Y}} P(y_i|\mathcal{X}') \neq y_{true} \\ &\text{s.t. } \mathrm{SIM}(\mathcal{X}', \mathcal{X}) > \mathcal{L}; \mathrm{WSR}(\mathcal{X}', \mathcal{X}) < \sigma \end{aligned} \tag{2}$$

3.2 Word Importance Calculation

Given a sentence of n words $\mathcal{X} = [x_1, x_2, \cdots, x_n]$, only some important words will affect the prediction results of the model $\mathcal{F}$. In order to measure the importance of x_i, we follow the calculation proposed in TextFooler [9]. We replace x_i with '[oov]'[1] to form $\mathcal{X}/\{x_i\} = [x_1, \cdots, x_{i-1}, [oov], x_{i+1}, \cdots, x_n]$, then word importance of x_i is calculated as follows:

- The predicted label remains the same after replace, *i.e.*, $\mathcal{F}(\mathcal{X}) = \mathcal{F}(\mathcal{X}/\{x_i\}) = y_{true}$,

$$I(x_i) = \mathcal{F}_{y_{true}}(\mathcal{X}) - \mathcal{F}_{y_{true}}(\mathcal{X}/\{x_i\}) \tag{3}$$

- The predicted label is changed after replace, *i.e.*, $\mathcal{F}(\mathcal{X}) = y_{true} \neq y_{other} = \mathcal{F}(\mathcal{X}/\{x_i\})$,

$$\begin{aligned} I(x_i) = &\mathcal{F}_{y_{true}}(\mathcal{X}) - \mathcal{F}_{y_{true}}(\mathcal{X}/\{x_i\}) + \\ &\mathcal{F}_{y_{other}}(\mathcal{X}/\{x_i\}) - \mathcal{F}_{y_{other}}(\mathcal{X}) \end{aligned} \tag{4}$$

where $\mathcal{F}_y(\mathcal{X})$ represents the predicted class label probability of $\mathcal{X}$ by $\mathcal{F}$ on label y. In order to improve the readability and fluency of the adversarial examples, we will filter out stopwords by NLTK[2] after calculating the word importance.

[1] the word out-of-vocabulary.
[2] https://www.nltk.org/.

3.3 Mixed Semantic Spaces

After ranking the words by their importance score, we need to search synonyms, which is a candidate words set $\mathcal{C}(x_i)$ for each word x_i. A proper replacement word should (i) have similar semantic meaning with original input, (ii) avoid some obvious grammar errors, (iii) and confuse model $\mathcal{F}$ to predict the wrong label. There are two different semantic spaces to search synonyms, word embedding spaces and pre-trained language models.

- The former searches for synonyms from word embedding spaces, such as WordNet space [16], HowNet space [5] and Counter-fitting word vectors [18]. Word embedding spaces can quickly generate synonyms with the same meaning as origin word.
- The later searches for synonyms through pre-trained language models(such as BERT). Given a sentence of n words $\mathcal{X} = [x_1, x_2, \cdots, x_n]$, we replace each word x_i with '[MASK]', and get candidate words set $\mathcal{C}(x_i)$ predicted by BERT. Pre-trained language models produce fluent and contextual-aware adversarial examples.

We combine word embedding space and BERT to make full use of the advantage of different semantic spaces. In detail, for each word x_i, we respectively select top N synonyms from word embedding space and BERT to form a candidate words set $\mathcal{C}(x_i)$. To generate high-qualify adversarial examples, we filter out the candidate words set that has different part-of-speech(POS)[3] synonyms with x_i. In addition, for each $c \in \mathcal{C}(x_i)$, we substitute it for x_i to generate adversarial example $\mathcal{X}' = [x_1, \cdots, x_{i-1}, c, x_{i+1}, \cdots, x_n]$, then we measure semantic similarity between $\mathcal{X}$ and adversarial example $\mathcal{X}'$ by universal sentence encoder(USE)[4], which encodes original input $\mathcal{X}$ and adversarial example $\mathcal{X}'$ as dense vectors and use cosine similarity as a approximation of semantic similarity. Only synonyms whose similarity is higher than threshold L will be retained in the candidate words set $\mathcal{C}(x_i)$.

3.4 Improved Beam Search

After filtering out the candidate words set $\mathcal{C}(x_i)$, the construction of adversarial examples is a combinatorial optimization problem as expected in Eq. 2. Previous works use the greedy search since it solely selects the token that maximizes the probability difference, which leads to local optima and word-over substitution.

To tackle this, we improve beam search to give consideration to both attack success rate and algorithm efficiency. Beam search has a hyper-parameter called beam size $\mathcal{K}$. Naive beam search only selects top $\mathcal{K}$ adversarial examples from the current iteration results. In the improved beam search, we merge the output of the last iteration to the current iteration and select top $\mathcal{K}$ adversarial examples as the input of the next iteration jointly. In detail, for each word x_i in the original

[3] https://spacy.io/api/tagger.
[4] https://tfhub.dev/google/universal-sentence-encoder.

text, we replace x_i with the substitution from candidate words set $\mathcal{C}(x_i)$ to generate adversarial examples $\mathcal{X}'$ and calculate the probability differences. The top $\mathcal{K}$ adversarial examples $\mathcal{X}'$ with the maximum probability difference(including the last iteration of top $\mathcal{K}$ adversarial examples) are selected as the input of the next iteration until the attack succeeds or all origin words are iterated. It is worth noting that greedy search is a special case of $\mathcal{K} = 1$. The details of BeamAttack are shown in Algorithm 1.

Algorithm 1. BeamAttack Adversarial Algorithm

Input:Original text $\mathcal{X}$, target model $\mathcal{F}$, semantic similarity threshold $\mathcal{L} = 0.5$ and beam size $\mathcal{K} = 10$, number of words in original text n

Output:Adversarial example $\mathcal{X}_{\text{adv}}$.

1: $\mathcal{X}_{\text{adv}} \leftarrow \mathcal{X}$
2: $set(\mathcal{X}_{\text{adv}}) \leftarrow \mathcal{X}_{\text{adv}}$
3: **for** each word x_i in $\mathcal{X}$ **do**
4: Compute the importance score $I(x_i)$ via Eq.3 and 4.
5: **end for**
6: Sort the words with importance score $I(x_i)$
7: **for** $i = 1$ to n **do**
8: Replace the x_i with [MASK]
9: Generate the candidate set $\mathcal{C}(x_i)$ from BERT and Word Embedding Space
10: $\mathcal{C}(x_i) \leftarrow$ POSFilter($\mathcal{C}(x_i)$) $\cap$ USEFilter($\mathcal{C}(x_i)$)
11: **end for**
12: **for** $\mathcal{X}_{\text{adv}}$ in $set(\mathcal{X}_{\text{adv}})$ **do**
13: **for** c_k in $\mathcal{C}(x_i)$ **do**
14: $\mathcal{X}'_{\text{adv}} \leftarrow$ Replace x_i with c_k in $\mathcal{X}_{\text{adv}}$
15: Add $\mathcal{X}'_{\text{adv}}$ to the $set(\mathcal{X}_{\text{adv}})$
16: **end for**
17: **for** $\mathcal{X}'_{\text{adv}}$ in $set(\mathcal{X}_{\text{adv}})$ **do**
18: **if** $\mathcal{F}(\mathcal{X}'_{\text{adv}}) \neq y_{true}$ **then**
19: **return** $\mathcal{X}'_{\text{adv}}$ with highest semantic similarity
20: **end if**
21: **end for**
22: $set(\mathcal{X}_{\text{adv}}) \leftarrow$ Select top $\mathcal{K}$ adversarial examples in $set(\mathcal{X}_{\text{adv}})$
23: $i \leftarrow i + 1$
24: **if** $i > n$ **then**
25: **break**
26: **end if**
27: **end for**
28: **return** adversarial examples $\mathcal{X}_{\text{adv}}$

4 Experiments

Tasks, Datasets and Models. To evaluate the effectiveness of BeamAttack, we conduct experiments on two NLP tasks, including text classification and text inference. In particular, the experiments cover various datasets, such as MR [19],IMDB [15],SST-2 [21], SNLI [1] and MultiNLI [24]. We train three neural networks as target models including CNN [10], LSTM [8] and BERT [4]. Model parameters are consistent with TextFooler's [9] setting.

Baselines. To quantitatively evaluate BeamAttack, we compare it with other black-box attack algorithms, including TextFooler(TF) [9], PWWS [20], BAE [7], Bert-Attack(BEAT) [13] and PSO [26], wherein TF,PWWS and PSO search synonyms from word embedding spaces, BAE and BEAT search synonyms from BERT. In addition, PSO belongs to heuristics search and other belong to greedy search. These baselines are implemented on the TextAttack framework [17].

Automatic Evaluation Metrics. We evaluate the attack performance by following metrics. Attack Success Rate(ASR) is defined as the proportion of successful adversarial examples to the total number of examples. Word Substitution Rate(WSR) is defined as the proportion of number of replacement words to number of origin words. Semantic Similarity(SIM) is measured by Universal Sentence Encoder(USE). Query Num(Query) is the number of model queries during adversarial attack. The ASR evaluates how successful the attack is. The WSR and semantic similarity together evaluate how semantically similar the original texts and adversarial examples are. Query num can reveal the efficiency of the attack algorithm.

Implementation Details. In our experiments, we carry out all experiments on NVIDIA Tesla P100 16G GPU. We set the beam size $\mathcal{K} = 10$, number of each candidate set $N = 50$, semantic similarity threshold $\mathcal{L} = 0.5$, we take the average value of 1000 examples as the final experimental result.

4.1 Experimental Results

The experiment results are listed in Table 1. It is worth noting that BeamAttack achieves higher ASR than baselines on almost all scenarios. BeamAttack also reduces the WSR on some datasets(MR,IMDB and SST-2). We attribute this superiority to the fine-tuned beam search, as this is the major improvement of our algorithm compared with greedy search. BeamAttack has chance to jump out of the local optimal solution and find out the adversarial examples with lower perturbation by expanding the search space. In terms of model robustness, BERT has better robustness than traditional classifiers(CNN and LSTM), since the attack success rate of attacking BERT is lower than other models.

Semantic Similarity. Except ASR and WSR, fluent and contextual-aware adversarial examples are also essential. Figure 2 plots the semantic similarity of adversarial examples generated by different attack algorithms. Clearly, BeamAttack achieves the highest semantic similarity than other attack algorithms.

Qualitative Examples. To more intuitively contrast the fluency of adversarial examples, we list some adversarial examples generated by different attack algorithms in Table 2. Compared with other methods, Beamattack not only ensures the semantic similarity between replacement words and original words but also successfully misleads the model with the minimum perturbation.

Table 1. The attack success rate and word substitution rate of different attack algorithms on five datasets. The "Origin ACC(%)" denotes the target model's test accuracy on the original inputs.

Datasets	Target Models	Origin ACC	Attack Success Rate(ASR(%))						Word Substitution Rate(WSR(%))					
			TF	PWWS	BAE	BEAT	PSO	BeamAttack	TF	PWWS	BAE	BEAT	PSO	BeamAttack
MR	CNN	80.4	98.81	98.61	98.00	83.31	96.23	**99.87**	17.05	13.22	12.98	15.06	11.53	**8.29**
	LSTM	80.7	98.92	97.92	98.21	84.12	95.32	**99.90**	15.61	13.07	11.71	13.59	10.91	**8.60**
	BERT	90.4	90.54	81.53	90.61	88.36	92.47	**97.88**	20.91	14.67	14.44	15.32	11.93	**9.70**
IMDB	CNN	89.2	**100**	**100**	**100**	99.82	**100**	**100**	2.51	2.23	2.01	3.32	2.43	**2.11**
	LSTM	89.8	99.76	99.47	**100**	99.83	**100**	**100**	3.12	3.11	**2.25**	3.45	2.46	2.43
	BERT	90.9	88.83	86.55	83.96	88.68	89.93	**91.6**	3.81	5.02	7.69	5.66	4.32	**1.65**
SST-2	CNN	82.5	92.37	98.23	95.45	86.44	96.69	**99.88**	17.09	13.10	12.53	15.40	11.47	**8.46**
	LSTM	84.6	93.21	98.48	96.23	86.43	96.42	**100**	17.55	13.53	12.83	15.31	11.45	**8.76**
SNLI	BERT	89.1	96.00	98.42	98.84	98.64	92.51	**99.80**	17.26	13.72	**6.91**	7.80	8.19	13.81
MNLI	BERT	85.1	90.44	94.33	99.23	92.00	83.43	**99.50**	13.93	10.12	**5.45**	5.64	6.65	10.81

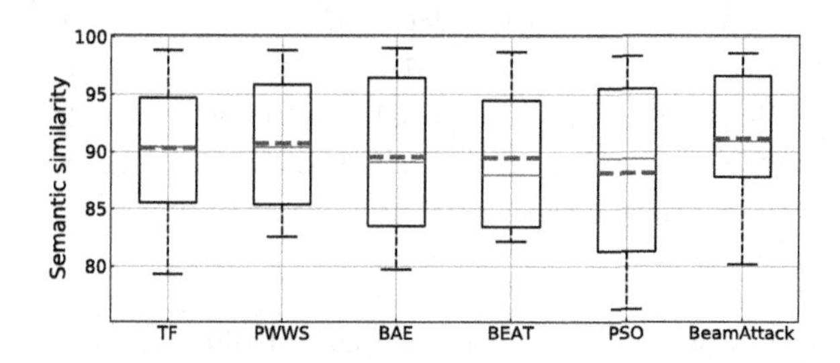

Fig. 2. The semantic similarity between origin inputs and adversarial examples.

Model Query. The number of model queries measures the effectiveness of attack algorithm. Table 3 lists the model queries of various attack algorithms. Results show that although our BeamAttack needs more model queries than greedy search(such as TF), compared with the PSO attack algorithm, which adopts heuristic search, our algorithm obtains competitive results with extremely few model queries.

4.2 Ablation Study

The Effect of Beam Size $\mathcal{K}$. To validate the effectiveness of beam size $\mathcal{K}$, we use BERT as the target model and test on MR dataset with different beam size $\mathcal{K}$. When $\mathcal{K} = 1$, beam search is equal to greedy search. As shown in Table 4, the attack success rate increases gradually with the grow of beam size $\mathcal{K}$.

The Effect of Mixed Semantic Spaces. Another major improvement of our BeamAttack is that substitutions are selected from mixed semantic spaces.

Table 2. The adversarial example crafted by different attack algorithms on MR(BERT) dataset. Replacement words are represented in red.

Origin Text (Positive)	The experience of the roles in the play makes us generates an enormous feeling of empathy for its characters.
BAE (Negative)	The experience of the roles in the play makes us generates an excessive need of empathy for its characters.
PWWS (Negative)	The experience of the roles in the play makes us render an enormous smell of empathy for its eccentric.
TextFooler (Negative)	The experience of the roles in the play makes us leeds an enormous foreboding of empathy for its specs.
BeamAttack (Negative)	The experience of the roles in the play makes us generates an enormous feeling of pity for its characters.

Table 3. The average model queries of different attack algorithms on five datasets. Beam size $\mathcal{K} = 10$

	MR	IMDB	SST-2	SNLI	MNLI
TF	113.8	536.7	146.2	54.1	68.9
PWWS	285.4	3286.5	5054.3	137.7	157.4
BAE	104.2	567.1	171.0	75.5	75.1
BEAT	207.9	585.0	245.6	93.6	119.2
PSO	5124.5	15564.3	3522.8	416.6	1124.8
BeamAttack	650.3	2135.8	584.3	126.0	174.0

As shown in the Table 5, we study the impact of different semantic spaces on different metrics. Compared with single word embedding space or BERT, using both word embedding space and BERT to generate adversarial examples can obtain higher attack success rate, semantic similarity and lower word substitution rate.

4.3 Transferability

The transferability of adversarial examples reflects property that adversarial examples crafted by classifier $\mathcal{F}$ can also fool other unknown classifier $\mathcal{F}'$. We evaluate the transferability on MR dataset across CNN, LSTM and BERT. In detail, we use the adversarial examples crafted for attacking BERT on MR dataset to evaluate the transferability for CNN and LSTM models. As shown in the Fig. 3, the adversarial examples generated by BeamAttack achieve the higher transferability than baselines.

4.4 Adversarial Training

Adversarial training is a prevalent technique to improve the model's robustness by adding some adversarial examples into train data. To validate this, we train the CNN model on the MR dataset and obtains 80.4% test accuracy. Then

Table 4. The effect of beam size $\mathcal{K}$ on MR(BERT) dataset.

Beam Size	ASR(%)	WSR(%)	Similarity(%)	Query
$\mathcal{K} = 1$	89.0	15.5	81.3	**101.3**
$\mathcal{K} = 2$	90.6	15.3	82.0	150.4
$\mathcal{K} = 5$	91.6	**15.1**	82.8	312.6
$\mathcal{K} = 7$	92.3	**15.1**	83.0	411.1
$\mathcal{K} = 10$	**92.6**	**15.1**	**83.1**	516.2

Table 5. The effect of different semantic spaces on MR(BERT) dataset.

semantic space	ASR(%)	WSR(%)	Similarity(%)	Query
Embedding	89.0	15.3	81.2	101.3
BERT	93.6	13.1	82.6	**101.1**
Embedding+BERT	**95.3**	**11.7**	**84.9**	140.3

we randomly generate 1000 MR adversarial examples to its training data and retrain the CNN model. The result is shown in the Table 6, CNN model obtains 83.3% test accuracy, higher than origin test accuracy. Although there is no significant change in ASR, BeamAttack needs to replace more words and more model queries to attack successfully with WSR and model queries increasing. It indicates that adversarial training effectively improves the generalization and robustness of the model.

Table 6. The performance of CNN with(out) adversarial training on the MR dataset.

	Origin ACC(%)	ASR(%)	WSR(%)	SIM(%)	Query
Original	80.4	99.87	8.20	91.08	563.1
Adv.Training	**83.3**	**99.75**	**8.67**	**90.82**	**606.4**

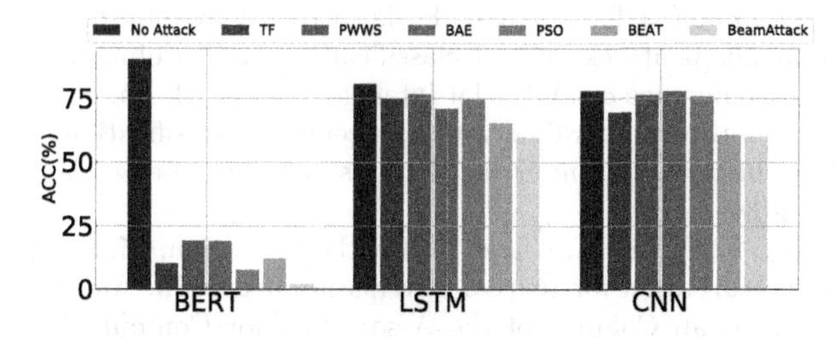

Fig. 3. Transfer attack on MR dataset. Lower accuracy indicates higher transferability (the lower the better).

5 Conclusion

In this paper, we propose an efficient adversarial textual attack algorithm BeamAttack. The BeamAttack makes full use of word embedding space and BERT to generate substitutions and fine-tune beam search to expand search spaces. Extensive experiments demonstrate BeamAttack balances the attack success rate, qualify of adversarial examples and time consumption. In addition, the adversarial examples crafted by BeamAttack are contextual-aware and improve models' robustness during adversarial training.

References

1. Bowman, S.R., Angeli, G., Potts, C., Manning, C.D.: large annotated corpus for learning natural language inference. In: Proceedings of the 2015 Conference on Empirical Methods in Natural Language Processing, pp. 632–642 (2015)
2. Carlini, N., Wagner, D.: Audio adversarial examples: targeted attacks on speech-to-text. In: 2018 IEEE Security and Privacy Workshops, pp. 1–7 (2018)
3. Cheng, M., Yi, J., Chen, P.Y., Zhang, H., Hsieh, C.J.: Seq2sick: evaluating the robustness of sequence-to-sequence models with adversarial examples. In: Proceedings of the AAAI Conference on Artificial Intelligence, vol. 34, pp. 3601–3608 (2020)
4. Devlin, J., Chang, M.W., Lee, K., Toutanova, K.: Bert: Pre-training of deep bidirectional transformers for language understanding. In: Annual Conference of the North American Chapter of the Association for Computational Linguistics: Human Language Technologies, pp. 4171–4186 (2019)
5. Dong, Z., Dong, Q., Hao, C.: Hownet and its computation of meaning. In: Proceedings of the 23rd International Conference on Computational Linguistics: Demonstrations, pp. 53–56 (2010)
6. Gao, J., Lanchantin, J., Soffa, M.L., Qi, Y.: Black-box generation of adversarial text sequences to evade deep learning classifiers. In: 2018 IEEE Security and Privacy Workshops, pp. 50–56 (2018)
7. Garg, S., Ramakrishnan, G.: BAE: BERT-based adversarial examples for text classification. In: Proceedings of the 2020 Conference on Empirical Methods in Natural Language Processing, pp. 6174–6181 (2020)
8. Hochreiter, S., Schmidhuber, J.: Long short-term memory. Neural Comput. **9**, 1735–1780 (1997)
9. Jin, D., Jin, Z., Zhou, J.T., Szolovits, P.: Is BERT really robust? A strong baseline for natural language attack on text classification and entailment. In: Proceedings of the AAAI Conference on Artificial Intelligence, vol. 34, pp. 8018–8025 (2020)
10. Kim, Y.: Convolutional neural networks for sentence classification. In: Proceedings of the 2014 Conference on Empirical Methods in Natural Language Processing, pp. 1746–1751 (2014)
11. Li, D., Zhang, Y., Peng, H., Chen, L., Brockett, C., Sun, M.T., Dolan, B.: Contextualized perturbation for textual adversarial attack. In: Annual Conference of the North American Chapter of the Association for Computational Linguistics: Human Language Technologies, pp. 5053–5069 (2021)
12. Li, J., Ji, S., Du, T., Li, B., Wang, T.: Textbugger: generating adversarial text against real-world applications. In: 26th Annual Network and Distributed System Security Symposium, p. 55 (2019)

13. Li, L., Ma, R., Guo, Q., Xue, X., Qiu, X.: BERT-ATTACK: Adversarial attack against BERT using BERT. In: Proceedings of the 2020 Conference on Empirical Methods in Natural Language Processing, pp. 6193–6202 (2020)
14. Ma, G., Shi, L., Guan, Z.: Adversarial text generation via probability determined word saliency. In: International Conference on Machine Learning for Cyber Security, pp. 562–571 (2020)
15. Maas, A., Daly, R.E., Pham, P.T., Huang, D., Ng, A.Y., Potts, C.: Learning word vectors for sentiment analysis. In: Annual Conference of the North American Chapter of the Association for Computational Linguistics: Human Language Technologies, pp. 142–150 (2011)
16. Miller, George, A.: Wordnet: a lexical database for english. In: Communications of the ACM, vol. 38, pp. 39–41 (1995)
17. Morris, J., Lifland, E., Yoo, J.Y., Grigsby, J., Jin, D., Qi, Y.: Textattack: a framework for adversarial attacks, data augmentation, and adversarial training in NLP. In: Proceedings of the 2020 Conference on Empirical Methods in Natural Language Processing, pp. 119–126 (2020)
18. Mrksic, N., et al.: Counter-fitting word vectors to linguistic constraints. In: Annual Conference of the North American Chapter of the Association for Computational Linguistics: Human Language Technologies, pp. 142–148 (2016)
19. Pang, B., Lee, L.: Seeing stars: exploiting class relationships for sentiment categorization with respect to rating scales. In: Annual Conference of the North American Chapter of the Association for Computational Linguistics: Human Language Technologies, pp. 115–124 (2005)
20. Ren, S., Deng, Y., He, K., Che, W.: Generating natural language adversarial examples through probability weighted word saliency. In: Annual Conference of the North American Chapter of the Association for Computational Linguistics: Human Language Technologies, pp. 1085–1097 (2019)
21. Socher, R., et al.: Recursive deep models for semantic compositionality over a sentiment treebank. In: Proceedings of the 2013 Conference on Empirical Methods in Natural Language Processing, pp. 1631–1642 (2013)
22. Szegedy, C., et al.: Intriguing properties of neural networks. arXiv preprint arXiv:1312.6199 (2013)
23. Wang, Z., Zheng, Y., Zhu, H., Yang, C., Chen, T.: Transferable adversarial examples can efficiently fool topic models. In: Computers & Security, vol. 118, p. 102749 (2022)
24. Williams, A., Nangia, N., Bowman, S.: A broad-coverage challenge corpus for sentence understanding through inference. In: Annual Conference of the North American Chapter of the Association for Computational Linguistics: Human Language Technologies, pp. 1112–1122 (2018)
25. Yang, X., Liu, W., Tao, D., Liu, W.: BESA: Bert-based simulated annealing for adversarial text attacks. In: International Joint Conference on Artificial Intelligence, pp. 3293–3299 (2021)
26. Zang, Y., et al.: Word-level textual adversarial attacking as combinatorial optimization. In: Proceedings of the 58th Annual Meeting of the Association for Computational Linguistics, pp. 6066–6080 (2020)

Semi-supervised and Unsupervised Learning

Reduction from Complementary-Label Learning to Probability Estimates

Wei-I Lin and Hsuan-Tien Lin(✉)

National Taiwan University, Taipei, Taiwan
{r10922076,htlin}@csie.ntu.edu.tw

Abstract. Complementary-Label Learning (CLL) is a weakly-supervised learning problem that aims to learn a multi-class classifier from only complementary labels, which indicate a class to which an instance does not belong. Existing approaches mainly adopt the paradigm of reduction to ordinary classification, which applies specific transformations and surrogate losses to connect CLL back to ordinary classification. Those approaches, however, face several limitations, such as the tendency to overfit. In this paper, we sidestep those limitations with a novel perspective–reduction to probability estimates of complementary classes. We prove that accurate probability estimates of complementary labels lead to good classifiers through a simple decoding step. The proof establishes a reduction framework from CLL to probability estimates. The framework offers explanations of several key CLL approaches as its special cases and allows us to design an improved algorithm that is more robust in noisy environments. The framework also suggests a validation procedure based on the quality of probability estimates, offering a way to validate models with only CLs. The flexible framework opens a wide range of unexplored opportunities in using deep and non-deep models for probability estimates to solve CLL. Empirical experiments further verified the framework's efficacy and robustness in various settings. The full paper can be accessed at https://arxiv.org/abs/2209.09500.

Keywords: complementary-label learning · weakly-supervised learning

1 Introduction

In real-world machine learning applications, high-quality labels may be hard or costly to collect. To conquer the problem, researchers turn to the *weakly-supervised learning* (WSL) framework, which seeks to learn a good classifier with incomplete, inexact, or inaccurate data [14]. This paper focuses on a very weak type of WSL, called *complementary-label learning* (CLL) [3]. For the multi-class classification task, a complementary label (CL) designates a class to which a specific instance does not belong. The CLL problem assumes that the learner receives complementary labels rather than ordinary ones during training, while

Supplementary Information The online version contains supplementary material available at https://doi.org/10.1007/978-3-031-33377-4_36.

H. Kashima et al. (Eds.): PAKDD 2023, LNAI 13936, pp. 469–481, 2023.
https://doi.org/10.1007/978-3-031-33377-4_36

wanting the learner to correctly predict the ordinary labels of the test instances. Complementary labels can be cheaper to obtain. For example, when labeling with many classes, selecting the correct label is time-consuming for data annotators, while selecting a complementary label would be less costly [3]. In this case, fundamental studies on CLL models can potentially upgrade multi-class classification models and make machine learning more realistic. CLL's usefulness also attracts researchers to study its interaction with other tasks, such as generative-discriminative learning [7,10] and domain-adaptation [13].

[3,4] proposed a pioneering model for CLL based on replacing the ordinary classification error with its unbiased risk estimator (URE) computed from only complementary labels assuming that the CLs are generated uniformly. [1] unveiled the overfitting tendency of URE and proposed the surrogate complementary loss (SCL) as an alternative design. [11] studied the situation where the CLs are not generated uniformly, and proposed a loss function that includes a transition matrix for representing the non-uniform generation. [2] argued that the non-uniform generation shall be tackled by being agnostic to the transition matrix instead of including the matrix in the loss function.

The methods mentioned above mainly focused on applying transformation and specific loss functions to the ordinary classifiers. Such a "reduction to ordinary classification" paradigm, however, faces some limitations and is not completely analyzed. For instance, so far most of the methods in the paradigm require differentiable models such as neural networks in their design. It is not clear whether non-deep models could be competitive or even superior to deep ones. It remains critical to correct the overfitting tendency caused by the stochastic relationship between complementary and ordinary labels, as repeatedly observed on URE-related methods [1]. More studies are also needed to understand the potential of and the sensitivity to the transition matrix in the non-uniform setting, rather than only fixing the matrix in the loss function [11] or dropping it [2].

The potential limitations from reduction to ordinary classification motivate us to sidestep them by taking a different perspective—reduction to complementary probability estimates. Our contribution can be summarized as follows.

1. We propose a framework that only relies on the probability estimates of CLs, and prove that a simple decoding method can map those estimates back to correct ordinary labels with theoretical guarantees.
2. The proposed framework offers explanations of several key CLL approaches as its special cases and allows us to design an improved algorithm that is more robust in noisy environments.
3. We propose a validation procedure based on the quality of probability estimates, providing a novel approach to validate models with only CLs along with theoretical justifications.
4. We empirically verify the effectiveness of the proposed framework under broader scenarios than previous works that cover various assumptions on the CL generation (uniform/non-uniform; clean/noisy) and models (deep /non-deep). The proposed framework improves the SOTA methods in those scenarios, demonstrating the effectiveness and robustness of the framework.

2 Problem Setup

In this section, we first introduce the problem of ordinary multi-class classification, then formulate the CLL problem, and introduce some common assumption.

2.1 Ordinary-label Learning

We start by reviewing the problem formulation of ordinary multi-class classification. In this problem, we let K with $K > 2$ denote the number of classes to be classified, and use $\mathcal{Y} = [K] = \{1, 2, \ldots, K\}$ to denote the label set. Let $\mathcal{X} \subset \mathbb{R}^d$ denote the feature space. Let D be an unknown joint distribution over $\mathcal{X} \times \mathcal{Y}$ with density function $p_D(x, y)$. Given N i.i.d. training samples $\{(x_i, y_i)\}_{i=1}^N$ and a hypothesis set $\mathcal{H}$, the goal of the learner is to select a classifier $f: \mathcal{X} \to \mathbb{R}^K$ from the hypothesis set $\mathcal{H}$ that predicts the correct labels on unseen instances. The prediction $\hat{y}$ of an unseen instance x is determined by taking the argmax function on f, i.e. $\hat{y} = \mathrm{argmax}_i\, f_i(x)$, where $f_i(x)$ denote the i-th output of $f(x)$. The goal of the learner is to learn an f from $\mathcal{H}$ that minimizes the following classification risk: $\mathbb{E}_{(x,y)\sim D}\left[\ell(f(x), e_y)\right]$, where $\ell: \mathbb{R}^K \times \mathbb{R}^K \to \mathbb{R}^+$ denotes the loss function, and e_y denote the one-hot vector of label y.

2.2 Complementary-label Learning

In complementary-label learning, the goal for the learner remains to find an f that minimizes the ordinary classification risk. The difference lies in the dataset to learn from. The complementary learner does not have access to the ground-truth labels y_i. Instead, for each instance x_i, the learner is given a complementary label $\bar{y}_i$. A complementary label is a class that x_i does not belong to; that is, $\bar{y}_i \in [K] \backslash \{y_i\}$. In CLL, it is assumed that the complementary dataset is generated according to an unknown distribution $\bar{D}$ over $\mathcal{X} \times \mathcal{Y}$ with density function $\bar{p}_{\bar{D}}(x, y)$. Given access to i.i.d. samples $\{x_i, \bar{y}_i\}_{i=1}^N$ from $\bar{D}$, the complementary-label learner aims to find a hypothesis that classifies the correct ordinary labels on unseen instances.

Next, we introduce the *class-conditional complementary transition assumption*, which is used by many existing work [2–4, 11]. It assumes that the generation of complementary labels only depends on the ordinary labels; that is, $P(\bar{y} \mid y, x) = P(\bar{y} \mid y)$. The transition probability $P(\bar{y} \mid y)$ is often represented by a $K \times K$ matrix, called *transition matrix*, with $T_{ij} = P(\bar{y} = j \mid y = i)$. It is commonly assumed to be all-zeros on the diagonals, i.e., $T_{ii} = 0$ for all $i \in [K]$ in CLL because complementary labels are not ordinary. The transition matrix is further classified into two categories: (a) *Uniform:* In uniform complementary generation, each complementary label is sampled uniformly from all labels except the ordinary one. The transition matrix in this setting is accordingly $T = \frac{1}{K-1}(\mathbf{1}_k - \mathbf{I}_k)$. This is the most widely researched and benchmarked setting in CLL. (b) *Biased:* A biased complementary generation is one that is not uniform. Biased transition matrices could be further classified as invertible ones and noninvertible ones based on its invertibility. The invertibility of a transition

Table 1. Comparison of recent approaches to CLL. $f(x)$ is the probability estimates of x, and ℓ is an arbitrary multi-class loss.

Method	Transformation	Loss Function
URE [3,4]	$\phi = I$	$-(K-1)\ell(f(x), \bar{y}) + \sum_{k=1}^{K} \ell(f(x), k)$
SCL-NL [1]	$\phi = I$	$-\log(1 - f_{\bar{y}}(x))$
Fwd [11]	$\phi(f)(x) = T^{\top} f(x)$	$\ell(\phi(f)(x), \bar{y})$
DM [2]	$\phi(f)(x) = \mathrm{sm}(1 - f(x))$	$\ell(\phi(f)(x), \bar{y})$

matrix comes with less physical meaning in the context of CLL; however, it plays an important role in some theoretical analysis in previous work [1,11].

Following earlier approaches, we assume that the generation of complementary labels follows class-conditional transition in the rest of the paper and that the transition matrix is given to the learning algorithms. What is different is that we do not assume the transition matrix to be uniform nor invertible. This allows us to make comparison in broader scenarios. In real-world scenario, the true transition matrix may be impossible to access. To loosen the assumption that the true transition matrix is given, we will analyze the case that the given matrix is *inaccurate* later. This analysis can potentially help us understand the CLL in a more realistic environment.

3 Proposed Framework

In this section, we propose a framework for CLL based on *complementary probability estimates* (CPE) and *decoding*. We first motivate the proposed CPE framework in Sect. 3.1. Then, we describe the framework and derive its theoretical properties in Sect. 3.2. In Sect. 3.3, we explain how earlier approaches can be viewed as special cases in CPE. We further draw insights for earlier approaches through CPE and propose improved algorithms based on those insights.

3.1 Motivation

To conquer CLL, recent approaches [1–4,11] mainly focus on applying different transformation and surrogate loss functions to the ordinary classifier, as summarized in Table 1. This paradigm of reduction to *ordinary*, however, faces some limitations. For instance, as [1] points out, the URE approach suffers from the large variance in the gradients. Besides, it remains unclear how some of them behave when the transition matrix is biased. Also, those methods only studied using neural networks and linear models as base models. It is unclear how to easily cast other traditional models for CLL. These limitations motivate us to sidestep them with a different perspective—reduction to *complementary* probability estimates.

3.2 Methodology

Overview. The proposed method consists of two steps: In training phase, we aim to find a hypothesis $\bar{f}$ that predicts the distribution of complementary labels well, i.e., an $\bar{f}$ that approximates $P(\bar{y} \mid x)$. This step is motivated by [2,11], which involve modeling the conditional distribution of the complementary labels $P(\bar{y} \mid x)$, and [12], which uses similar idea on noisy-label learning. What is different in our framework is the decoding step during prediction. In inference phase, we propose to predict the label with the closest transition vector to the predicted complementary probability estimates. Specifically, we propose to predict $\hat{y} = \operatorname{argmin}_{k \in [K]} d\left(\bar{f}(x), T_k\right)$ for an unseen instance x, where d denotes a loss function. It is a natural choice to decode with respect to T because the transition vector $T_k = (P(\bar{y} = 1 \mid y = k), \ldots, P(\bar{y} = K \mid y = k))^\top$ is the ground-truth distribution of the complementary labels if the ordinary label is k. In the following paragraph, we provide further details of our framework.

Training Phase: Probability Estimates. In this phase, we aim to find a hypothesis $\bar{f}$ that predicts $P(\bar{y} \mid x)$ well. To do so, given a hypothesis $\bar{f}$ from hypothesis set $\bar{\mathcal{H}}$, we set the following *complementary estimation loss* to optimize:

$$R(\bar{f}; \ell) = \mathbb{E}_{(x,y) \sim \mathcal{D}} \left(\ell(\bar{f}(x), P(\bar{y} \mid x, y))\right) \tag{1}$$

where ℓ can be any loss function defined between discrete probability distributions. By the assumption that complementary labels are generated with respect to the transition matrix T, the ground-truth distribution for $P(\bar{y} \mid x, y)$ is T_y, so we can rewrite Eq. (1) as follows:

$$R(\bar{f}; \ell) = \mathbb{E}_{(x,y) \sim \mathcal{D}} \left(\ell(\bar{f}(x), T_y)\right) \tag{2}$$

The loss function above is still hard to optimize for two reasons: First, the presence of ordinary label y suggests that it cannot be accessed from the complementary dataset. Second, as we only have *one* complementary label per instance, it becomes questionable to directly use the empirical density, i.e., the one-hot vector of the complementary label $e_{\bar{y}}$ to approximate T_y as it may change the objective.

Here we propose to use the Kullback-Leibler divergence for the loss function to solve the two issues mentioned above with the following property:

Proposition 1. *There is a constant C such that*

$$\mathop{\mathbb{E}}_{(x,\bar{y}) \sim \bar{\mathcal{D}}} \ell(\bar{f}(x), e_{\bar{y}}) + C = \mathop{\mathbb{E}}_{(x,y) \sim \mathcal{D}} \ell(\bar{f}(x), T_y) \tag{3}$$

holds for all hypothesis $\bar{f} \in \bar{\mathcal{H}}$ if ℓ is the KL divergence, i.e., $\ell(\hat{y}, y) = \sum_{k=1}^{K} -y_k(\log \hat{y}_k - \log y_k)$.

The result is well-known in the research of proper scoring rules [5,9]. It allows us to replace the T_y by $e_{\bar{y}}$ in Eq. (2) because the objective function only differs by a constant after the replacement. This suggests that minimizing the two objectives is equivalent. Moreover, the replacement makes the objective function accessible through the complementary dataset because it only depends on the complementary label $\bar{y}$ rather than the ordinary one.

Formally speaking, minimizing Eq. (2) becomes equivalent to minimizing the following *surrogate complementary estimation loss (SCEL)*:

$$\bar{R}(\bar{f};\ell) = \mathbb{E}_{(x,\bar{y})\sim\bar{\mathcal{D}}}\left(\ell(\bar{f}(x), e_{\bar{y}})\right) \tag{4}$$

By using KL divergence as the loss function, we have that

$$\bar{R}(\bar{f};\ell) = \mathbb{E}_{(x,\bar{y})\sim\bar{\mathcal{D}}}\left(-\log \bar{f}_{\bar{y}}(x)\right) \tag{5}$$

with $\bar{f}_{\bar{y}}(x)$ being the $\bar{y}$-th output of $\bar{f}(x)$. Next, we can use the following empirical version as the training objective: $\frac{1}{N}\sum_{i=1}^{N} -\log \bar{f}_{\bar{y}_i}(x_i)$. According to the empirical risk minimization (ERM) principle, we can estimate the distribution of complementary labels $P(\bar{y}\,|\,x)$ by minimizing the log loss on the complementary dataset. That is, by choosing $\bar{f}^*$ with $\bar{f}^* = \operatorname{argmin}_{\bar{f}\in\bar{\mathcal{H}}} \frac{1}{N}\sum_{i=1}^{N} -\log \bar{f}_{\bar{y}_i}(x_i)$, we can get an estimate of $P(\bar{y}\,|\,x)$ with $\bar{f}^*$.

In essence, we reduce the task of learning from complementary labels into learning probability estimates for multi-class classification (on the *complementary label space*). As the multi-class probability estimates is a well-researched problem, our framework becomes flexible on the choice of the hypothesis set. For instance, one can use K-Nearest Neighbor or Gradient Boosting with log loss to estimate the distribution of complementary labels. The flexibility becomes superior to the previous methods, who mainly focus on using neural networks to minimize specific surrogate losses. It makes them hard to optimize for non-differentiable models. In contrast, the proposed methods directly enable existing ordinary models to learn from complementary labels.

Inference Phase: Decoding. After finding a complementary probability estimator $\bar{f}^*$ during the training phase, we propose to predict the ordinary label by decoding: Given an unseen example x, we predict the label $\hat{y}$ whose transition vector $T_{\hat{y}}$ is closest to the predicted complementary probability estimates. That is, the label is predicted by

$$\hat{y} = \operatorname*{argmin}_{k\in[K]} d\left(\bar{f}^*(x), T_k\right) \tag{6}$$

where d could be an arbitrary loss function on the probability simplex and T_k is the k-th row vector of T. We use $\operatorname{dec}(\bar{f};d)$ to denote the function that decodes the output from $\bar{f}$ according to the loss function d. The next problem is whether the prediction of the decoder can guarantee a small out-sample classification error $R_{01}(f) = \mathbb{E}_{(x,y)\sim\mathcal{D}}\, I_{f(x)\neq y}$.

We propose to use a simple decoding step by setting L_1 distance as the loss function for decoding:

$$\operatorname{dec}(\bar{f};L_1)(x) = \operatorname*{argmin}_{y\in[K]} \|T_y - \bar{f}(x)\|_1 \tag{7}$$

This choice of L_1 distance makes the decoding step easy to perform and provides the following bound that quantifies the relationship between the error rate and the quality of probability estimator:

Proposition 2. *For any $\bar{f} \in \bar{\mathcal{H}}$, and distance function d defined on the probability simplex Δ^K, it holds that*

$$R_{01}\big(\operatorname{dec}(\bar{f}; d)\big) \leq \frac{2}{\gamma_d} R(\bar{f}; d) \tag{8}$$

where $\gamma_d = \min_{i \neq j} d(T_i, T_j)$ is the minimal distance between any pair of transition vector. Moreover, if d is the L_1 distance and ℓ is the KL divergence, then with $\gamma = \min_{i \neq j} \|T_i - T_j\|_1$, it holds that

$$R_{01}\big(\operatorname{dec}(\bar{f}; L_1)\big) \leq \frac{4\sqrt{2}}{\gamma} \sqrt{R(\bar{f}; \ell)} \tag{9}$$

The proof is in Appendix A.2. In the realizable case, where there is a target function g that satisfies $g(x) = y$ for all instances, the term $R(\bar{f}; \ell_{\mathrm{KL}})$ can be minimized to zero with $\bar{f}^\star : x \mapsto T_{g(x)}$. This indicates that for a sufficiently rich complementary hypothesis set, if the complementary probability estimator is consistent ($\bar{f} \to \bar{f}^\star$) then the L_1 decoded prediction is consistent ($R_{01}\big(\operatorname{dec}(\bar{f}; L_1)\big) \to 0$). The result suggests that the performance of the L_1 decoder can be bounded by the accuracy of the probability estimates of complementary labels measured by the KL divergence. In other words, to obtain an accurate ordinary classifier, it suffices to find an accurate complementary probability estimator followed by the L_1 decoding. Admittedly, in the non-realizable case, $R(\bar{f}; \ell_{\mathrm{KL}})$ contains irreducible error. We leave the analysis of the error bound in this case for the future research.

Another implication of the Proposition 2 is related to the inaccurate transition matrix. Suppose the complementary labels are generated with respect to the transition matrix T', which may be different from T, the one provided to the learning algorithm. In the proposed framework, the only affected component is the decoding step. This allows us to quantify the effect of inaccuracy as follows:

Corollary 1. *For any $\bar{f} \in \bar{\mathcal{H}}$, if d is the L_1 distance and ℓ is the KL divergence, then*

$$R_{01}\big(\operatorname{dec}(f; L_1)\big) \leq \frac{4\sqrt{2}}{\gamma} \sqrt{R(\bar{f}; \ell)} + \frac{2\epsilon}{\gamma}. \tag{10}$$

where $\gamma = \min_{i \neq j} \|T_i - T_j\|_1$ is the minimal L_1 distance between pairs of transition vectors, and $\epsilon = \max_{k \in [K]} \|T'_k - T_k\|_1$ denotes the difference between T' and T.

Validation Phase: Quality of Probability Estimates. The third implication of Proposition 2 is an alternative validation procedure to the unbiased risk estimation (URE) [3]. According to Proposition 2, selecting the best-performing parameter minimizes the right hand side of Eq. (9) among all hyper-parameter choices minimizes the ordinary classification error. This suggests an alternative metric for parameter selection: using the surrogate complementary estimation loss (SCEL) on the validation dataset.

Table 2. A unifying view of earlier approaches and proposed algorithms through the lens of reduction to probability estimates, where U denote the uniform transition matrix. Two versions of Forward Correction are considered: General T denotes the original version in [11], and the Uniform denotes the case when the transition layer is fixed to be uniform. Proof of the equivalence is in Appendix B.

Method	Hypothesis set	Decoder
Fwd (general T) [11]	$\{x \mapsto T^\top f(x;\theta) : \theta \in \Theta\}$	$\text{argmax}_k((T^\top)^{-1}\bar{f}(x))_k$
Fwd (uniform) [11]	$\{x \mapsto U^\top f(x;\theta) : \theta \in \Theta\}$	$\text{argmin}_k \|\bar{f}(x) - U_k\|_1$
SCL [1]	$\{x \mapsto U^\top f(x;\theta) : \theta \in \Theta\}$	$\text{argmin}_k \|\bar{f}(x) - U_k\|_1$
DM [2]	$\{x \mapsto \text{sm}(1 - f(x;\theta)) : \theta \in \Theta\}$	$\text{argmin}_k \|\bar{f}(x) - U_k\|_1$
CPE-I (no transition)	$\{x \mapsto f(x;\theta) : \theta \in \Theta\}$	$\text{argmin}_k \|\bar{f}(x) - T_k\|_1$
CPE-F (fixed transition)	$\{x \mapsto T^\top f(x;\theta) : \theta \in \Theta\}$	$\text{argmin}_k \|\bar{f}(x) - T_k\|_1$
CPE-T (trainable transition)	$\{x \mapsto T(W)^\top f(x;\theta) : \theta \in \Theta, W \in \mathbb{R}^{K \times K}\}$	$\text{argmin}_k \|\bar{f}(x) - T_k\|_1$

Although the proposed validation procedure does not directly estimate the ordinary classification error, it provides benefits in the scenarios where URE does not work well. For instance, when the transition matrix is non-invertible, the behavior of URE is ill-defined due to the presence of T^{-1} in the formula of URE: $\mathbb{E}_{x,\bar{y}}\, e_{\bar{y}} T^{-1} \ell(f(x))$. Indeed, replacing T^{-1} with T's pseudo-inverse can avoid the issue; however, it remains unclear whether the unbiasedness of URE still holds after using pseudo-inverse. In contrast, the quality of complementary probability estimates sidesteps the issue because it does not need to invert the transition matrix. This prevents the proposed procedure from the issue of an ill-conditioned transition matrix.

3.3 Connection to Previous Methods

The proposed framework also explains several earlier approaches as its special cases, including (1) Forward Correction (FWD) [11], (2) Surrogate Complementary Loss (SCL) with log loss [1], and (3) Discriminative Model (DM) [2], which are explained in Table 2 and Appendix B. By viewing those earlier approaches in the proposed framework, we provide additional benefits for them. First, the novel validation process can be applied for parameter selection. This provides an alternative to validate those approaches. Also, we fill the gap on the theoretical explanation to help understand those approaches in the realizable case.

On the other hand, the success of FWD inspires us to reconsider the role of transition layers in the framework. As the base model's output $f(x;\theta)$ is in the probability simplex Δ^K, the model's output $T^\top f(x;\theta)$ lies in the convex hull formed by the row vectors of T. If the transition matrix T provided to the learning algorithm is accurate, then such transformation helps control the model's complexity by restricting its output. The restriction may be wrong, however, when the given transition matrix T is inaccurate. To address this issue, we propose to allow the transition layer to be *trainable*. This technique is also used in label-noise learning, such as [6]. Specifically, we propose three methods in our Complementary Probability Estimates framework: (a) **CPE-I** denotes a model *without* a transition layer (b) **CPE-F** denotes a model with a *fixed* additional layer to T (c) **CPE-T** denotes a model with a *trainable* transition layer. To make the transition layer trainable, we considered a $K \times K$ matrix W. A softmax function was applied to each row of W to transform it into a valid transition matrix $T(W) = \big(\mathrm{sm}(W_1), \mathrm{sm}(W_2), \ldots, \mathrm{sm}(W_K)\big)^\top$. For a base model f, the complementary probability estimates of **CPE-T** for a given instance x would be $T(W)^\top f(x;\theta)$. Note that we use the L_1 decoder for **CPE-I**, **CPE-F**, and **CPE-T**.

4 Experiments

In this section, we benchmark the proposed framework to the state-of-the-art baselines and discuss the following questions: (a) Can the transition layers improve the model's performance? (b) Is the proposed L_1 decoding competitive to MAX? (c) Does the transition matrix provide information to the learning algorithms even if it is inaccurate? We further demonstrate the flexibility of incorporating traditional models in **CPE** in Sect. 4.3 and verify the effectiveness of the proposed validation procedure in the Appendix.

4.1 Experiment Setup

Baseline and Setup. We first evaluate CPE with the following state-of-the-art methods: (a) **URE-GA**: Gradient Ascent applied on the unbiased risk estimator [3,4], (b) **Fwd**: Forward Correction [11], (c) **SCL**: Surrogate Complementary Loss with negative log loss [1], and (d) **DM**: Discriminative Models with Weighted Loss [2]. Following the previous work, we test those methods on MNIST, Fashion-MNIST, and Kuzushiji-MNIST, and use one-layer mlp model (d-500-c) as base models. All models are optimized using Adam with learning rate selected from {1e-3, 5e-4, 1e-4, 5e-5, 1e-5} and a fixed weight decay 1e-4 for 300 epochs. The learning rate for **CPE** is selected with the Surrogate Complementary Estimation Loss (SCEL) on the validation dataset. For the baseline method, it is selected with unbiased risk estimator (URE) of the zero-one loss. It is worth noting that the validation datasets consist of only complementary labels, which is different from some previous works.

Transition Matrices. In the experiment of *clean* transition matrices, three types of transition matrices are benchmarked in the experiment. Besides the uniform transition matrix, following [2,11], we generated two biased ones as follows: For each class y, the complementary classes $\mathcal{Y}\backslash\{y\}$ are first randomly split into three subsets. Within each subset, the probabilities are set to p_1, p_2 and p_3, respectively. We consider two cases for (p_1, p_2, p_3): (a) *Strong*: $(\frac{0.75}{3}, \frac{0.24}{3}, \frac{0.01}{3})$ to model stronger deviation from uniform transition matrices. (b) *Weak*: $(\frac{0.45}{3}, \frac{0.30}{3}, \frac{0.25}{3})$ to model milder deviation from uniform transition matrices. In the experiment of *noisy* transition matrices, we consider the *Strong* deviation transition matrix T_{strong} to be the ground-truth transition matrix, and a uniform noise transition matrix $\frac{1}{K}\mathbf{1}_K$ to model the noisy complementary label generation. We generated complementary labels with the transition matrix

Table 3. Comparison of the testing classification accuracies with different transition matrices (upper part) and different noise levels (lower part).

	MNIST			Fashion-MNIST			Kuzushiji-MNIST		
	Unif.	Weak	Strong	Unif.	Weak	Strong	Unif.	Weak	Strong
URE-GA	90.3 ± 0.2	87.8 ± 0.9	33.8 ± 8.1	79.4 ± 0.7	75.7 ± 2.0	32.3 ± 4.5	65.6 ± 0.8	62.5 ± 1.1	23.3 ± 5.4
SCL	94.3 ± 0.4	**93.8 ± 0.4**	27.5 ± 19.8	82.6 ± 0.4	81.2 ± 0.1	28.5 ± 10.8	**73.7 ± 1.4**	**71.2 ± 2.9**	20.7 ± 4.8
DM	91.9 ± 0.6	90.2 ± 0.3	26.7 ± 4.6	82.5 ± 0.3	80.3 ± 1.1	24.8 ± 5.0	65.6 ± 2.9	64.5 ± 2.7	20.1 ± 3.2
FWD	**94.4 ± 0.2**	91.9 ± 0.3	95.3 ± 0.4	82.6 ± 0.6	**83.0 ± 1.0**	85.5 ± 0.3	73.5 ± 1.6	63.1 ± 2.6	74.1 ± 4.8
CPE-I	90.2 ± 0.2	88.4 ± 0.3	92.7 ± 0.8	81.1 ± 0.3	79.2 ± 0.5	81.9 ± 1.4	66.2 ± 1.0	62.5 ± 0.9	73.7 ± 1.0
CPE-F	**94.4 ± 0.2**	92.0 ± 0.2	**95.5 ± 0.3**	**83.0 ± 0.1**	**83.0 ± 0.3**	**85.8 ± 0.3**	73.5 ± 1.6	64.6 ± 0.5	**75.3 ± 2.6**
CPE-T	92.8 ± 0.6	92.1 ± 0.2	95.2 ± 0.5	**83.0 ± 0.1**	**83.0 ± 0.3**	**85.8 ± 0.3**	63.6 ± 0.4	64.6 ± 0.4	74.2 ± 2.8
	$\lambda = 0.1$	$\lambda = 0.2$	$\lambda = 0.5$	$\lambda = 0.1$	$\lambda = 0.2$	$\lambda = 0.5$	$\lambda = 0.1$	$\lambda = 0.2$	$\lambda = 0.5$
URE-GA	31.8 ± 6.4	27.8 ± 8.2	28.1 ± 4.1	27.3 ± 5.5	28.6 ± 4.1	26.3 ± 2.0	24.5 ± 4.6	21.1 ± 2.2	19.8 ± 2.1
SCL	25.1 ± 11.7	24.7 ± 8.9	23.8 ± 2.7	26.6 ± 9.2	20.6 ± 6.7	23.2 ± 5.7	20.4 ± 4.6	17.3 ± 2.9	16.8 ± 1.6
DM	26.5 ± 9.1	24.6 ± 6.5	22.6 ± 1.3	24.1 ± 5.1	23.6 ± 6.7	22.6 ± 2.9	20.0 ± 3.0	19.2 ± 3.1	18.2 ± 1.6
FWD	88.3 ± 8.7	83.9 ± 10.7	71.6 ± 18.4	**84.8 ± 0.6**	80.2 ± 6.2	62.9 ± 20.1	72.8 ± 5.6	67.6 ± 7.5	54.7 ± 12.4
CPE-I	92.4 ± 0.7	92.0 ± 0.8	87.6 ± 1.4	81.7 ± 1.4	81.3 ± 1.4	78.2 ± 1.5	73.0 ± 0.7	71.6 ± 0.9	62.7 ± 1.6
CPE-F	94.3 ± 0.5	93.6 ± 0.5	89.0 ± 1.4	84.1 ± 0.8	83.0 ± 1.1	78.4 ± 2.5	**76.1 ± 1.3**	73.7 ± 1.5	63.7 ± 1.5
CPE-T	**94.4 ± 0.5**	**93.7 ± 0.5**	**89.6 ± 0.9**	84.1 ± 0.8	**83.2 ± 1.1**	**78.9 ± 2.0**	**76.1 ± 1.3**	**73.9 ± 1.6**	**64.2 ± 1.2**

Table 4. Comparison of testing accuracies of decoders when the baseline models use fixed transition layers. The parameters are selected from the one with smallest SCEL on the validation dataset.

	MNIST			Fashion-MNIST			Kuzushiji-MNIST		
	Unif.	Weak	Strong	Unif.	Weak	Strong	Unif.	Weak	Strong
MAX	94.4 ± 0.2	92.0 ± 0.2	95.5 ± 0.2	83.0 ± 0.1	**83.3 ± 0.2**	**86.1 ± 0.5**	73.5 ± 1.6	**64.8 ± 0.5**	75.3 ± 2.6
L_1	94.4 ± 0.2	92.0 ± 0.2	95.5 ± 0.3	83.0 ± 0.1	83.0 ± 0.3	85.8 ± 0.3	73.5 ± 1.6	64.6 ± 0.5	75.3 ± 2.6
	$\lambda = 0.1$	$\lambda = 0.2$	$\lambda = 0.5$	$\lambda = 0.1$	$\lambda = 0.2$	$\lambda = 0.5$	$\lambda = 0.1$	$\lambda = 0.2$	$\lambda = 0.5$
MAX	**94.4 ± 0.3**	93.5 ± 0.3	84.5 ± 4.1	**85.0 ± 0.3**	**84.0 ± 0.5**	76.5 ± 2.5	**76.4 ± 1.1**	**73.8 ± 1.2**	59.9 ± 3.4
L_1	94.3 ± 0.5	**93.6 ± 0.5**	**89.0 ± 1.4**	84.1 ± 0.8	83.0 ± 1.1	**78.4 ± 2.5**	76.1 ± 1.3	73.7 ± 1.5	**63.7 ± 1.5**

$(1-\lambda)T_{\text{strong}} + \lambda\frac{1}{K}\mathbf{1}_K$, but provided T_{strong} and the generated complementary dataset to the learners. The parameter λ controls the proportion of the uniform noise in the complementary labels. The results are reported in Table 3.

4.2 Discussion

Can Transition Layers Improve Performance? The answer is positive in both clean and noisy experiments. We observe that **CPE-F** and **CPE-T** outperform **CPE-I** in both settings, demonstrating that the transition layer help achieve higher performances, no matter the provided transition matrix is clean or not. Also, we observe that **CPE-T** outperforms **CPE-F** in the noisy setting, especially when the noise factor λ is large. It demonstrates that by making transition layers trainable, the model can potentially fit the distribution of complementary labels better by altering the transition layer. In contrast, **CPE-F** is restricted to a wrong output space, making it underperform **CPE-T**. The difference makes **CPE-T** a better choice for noisy environment.

Is L_1 competitive with MAX? As analyzed in Sect. 3.3, **Fwd** and **CPE-F** only differ in the decoding step, with the former using MAX and the latter using L_1. We provide the testing accuracies of these decoders when the base models are

Table 5. Comparison of testing accuracies of CPE with traditional models. **Boldfaced** ones outperform the baseline methods based on single-layer deep models.

	MNIST			Fashion-MNIST			Kuzushiji-MNIST		
Model	Unif.	Weak	Strong	Unif.	Weak	Strong	Unif.	Weak	Strong
CPE-KNN	93.1± 0.1	92.6± 0.1	94.5± 0.4	79.1± 0.4	77.8± 0.6	79.0± 1.7	**74.9± 0.8**	**73.7± 0.8**	**80.4± 1.3**
CPE-GBDT	86.9± 0.4	86.0± 0.3	90.3± 0.9	79.8± 0.4	78.0± 0.4	81.4± 1.1	60.6± 0.4	56.6± 1.8	68.4± 2.1
	$\lambda = 0.1$	$\lambda = 0.2$	$\lambda = 0.5$	$\lambda = 0.1$	$\lambda = 0.2$	$\lambda = 0.5$	$\lambda = 0.1$	$\lambda = 0.2$	$\lambda = 0.5$
CPE-KNN	93.7± 0.4	93.4± 0.4	**91.9± 1.1**	78.7± 1.9	78.5± 1.9	76.6± 1.9	**77.2± 1.1**	**75.9± 1.6**	**73.2± 1.7**
CPE-GBDT	89.7± 1.0	88.6± 1.2	84.0± 1.7	80.6± 1.7	80.0± 1.6	76.0± 2.2	66.7± 2.4	64.7± 2.4	55.8± 3.1

CPE-F in Table 4. It is displayed that the MAX decoder outperform L_1 in most noiseless settings; however, when the transition matrix is highly inaccurate ($\lambda = 0.5$), we observe that the L_1 decoder outperform the MAX decoder. This suggests that L_1 could be more tolerant to an inaccurate transition matrix. These results reveal that a deeper sensitivity analysis of different decoders, both empirically and theoretically, would be desired. We leave this as future studies.

Discussion of T-agnostic models Among the baseline methods, **URE-GA**, **SCL** and **DM** are ones that does not take T as inputs or assumes T is uniform, which we called T-agnostic models. Those models perform well when the transition matrix is just slightly deviated from the uniform one, but their performances all dropped when the deviation from uniform becomes larger. As we discussed in Sect. 3.3, the result can be interpreted to be caused by their implicit assumption on uniform transition matrices, which brings great performance on uniform transition matrices but worse performance on biased ones. In contrast, we observed that all variations of **CPE** have similar testing accuracies across different transition matrices, demonstrating that **CPE** does exploit the information from the transition matrix that helps the models deliver better performance.

4.3 Learn from CL with Traditional Methods

As discussed in Sect. 3, the proposed framework is not constrained by deep models. We explored the possibility of applying traditional methods to learn from CL, including (a) k-Nearest Neighbor (k-**NN**) and (b) Gradient Boosting Decision Tree (**GBDT**). We benchmarked those models in the same settings and reported the restuls in Table 5. It displays that traditional models, specifically, k-**NN**, outperform all the methods using deep models in Kuzushiji-MNIST, indicating the benefit of the proposed CPE's flexibility in using non-deep models.

5 Conclusion

In this paper, we view the CLL problem from a novel perspective, reduction to complementary probability estimates. Through this perspective, we propose a framework that only requires complementary probability estimates and prove that a simple decoding step can map the estimates to ordinary labels. The framework comes with a theoretically justified validation procedure, provable tolerance in noisy environment, and flexibility of incorporating non-deep models. Empirical experiments further verify the effectiveness and robustness of the proposed framework under broader scenarios, including non-uniform and noisy complementary label generation. We expect the realistic elements of the framework to keep inspiring future research towards making CLL practical.

Acknowlegement. We thank the anonymous reviewers and the members of NTU CLLab for valuable suggestions. The work is partially supported by the National Science and Technology Council via the grants 110-2628-E-002-013 and 111-2628-E-002-018. We also thank the National Center for High-performance Computing (NCHC) of National Applied Research Laboratories (NARLabs) in Taiwan for providing computational resources.

References

1. Chou, Y.T., Niu, G., Lin, H.T., Sugiyama, M.: Unbiased risk estimators can mislead: a case study of learning with complementary labels. In: International Conference on Machine Learning, pp. 1929–1938. PMLR (2020)
2. Gao, Y., Zhang, M.L.: Discriminative complementary-label learning with weighted loss. In: International Conference on Machine Learning, pp. 3587–3597. PMLR (2021)
3. Ishida, T., Niu, G., Hu, W., Sugiyama, M.: Learning from complementary labels. In: Proceedings of the 31st International Conference on Neural Information Processing Systems, pp. 5644–5654 (2017)
4. Ishida, T., Niu, G., Menon, A., Sugiyama, M.: Complementary-label learning for arbitrary losses and models. In: International Conference on Machine Learning, pp. 2971–2980. PMLR (2019)
5. Kull, M., Flach, P.: Novel decompositions of proper scoring rules for classification: score adjustment as precursor to calibration. In: Appice, A., Rodrigues, P.P., Santos Costa, V., Soares, C., Gama, J., Jorge, A. (eds.) ECML PKDD 2015. LNCS (LNAI), vol. 9284, pp. 68–85. Springer, Cham (2015). https://doi.org/10.1007/978-3-319-23528-8_5
6. Li, X., Liu, T., Han, B., Niu, G., Sugiyama, M.: Provably end-to-end label-noise learning without anchor points. In: Meila, M., Zhang, T. (eds.) Proceedings of the 38th International Conference on Machine Learning. Proceedings of Machine Learning Research, vol. 139, pp. 6403–6413. PMLR (18–24 Jul 2021)
7. Liu, J., Hang, H., Wang, B., Li, B., Wang, H., Tian, Y., Shi, Y.: GAN-CL: generative adversarial networks for learning from complementary labels. IEEE Trans. Cybernet. (2021)
8. Wang, D.B., Feng, L., Zhang, M.L.: Learning from complementary labels via partial-output consistency regularization. In: IJCAI, pp. 3075–3081 (2021)
9. Williamson, R.C., Vernet, E., Reid, M.D.: Composite multiclass losses. J. Mach. Learn. Res. **17**(222), 1–52 (2016)
10. Xu, Y., Gong, M., Chen, J., Liu, T., Zhang, K., Batmanghelich, K.: Generative-discriminative complementary learning. In: Proceedings of the AAAI Conference on Artificial Intelligence, vol. 34, pp. 6526–6533 (2020)
11. Yu, X., Liu, T., Gong, M., Tao, D.: Learning with biased complementary labels. In: Proceedings of the European Conference on Computer Vision (ECCV), pp. 68–83 (2018)

12. Zhang, M., Lee, J., Agarwal, S.: Learning from noisy labels with no change to the training process. In: International Conference on Machine Learning, pp. 12468–12478. PMLR (2021)
13. Zhang, Y., Liu, F., Fang, Z., Yuan, B., Zhang, G., Lu, J.: Learning from a complementary-label source domain: theory and algorithms. IEEE Trans. Neural Netw. Learn. Syst. (2021)
14. Zhou, Z.H.: A brief introduction to weakly supervised learning. Nat. Sci. Rev. **5**(1), 44–53 (2018)

Semi-Supervised Text Classification via Self-Paced Semantic-Level Contrast

Yu Xia[1,2(✉)], Kai Zhang[3], Kaijie Zhou[4], Rui Wang[1], and Xiaohui Hu[1]

[1] Institute of Software Chinese Academy of Sciences, Beijing, China
xiayuwilson@163.com
[2] University of Chinese Academy of Sciences, Beijing, China
[3] Realsee, Beijing, China
[4] University of Liverpool, Liverpool, UK

Abstract. Semi-Supervised Text Classification (SSTC) aims to explore discriminative information from unlabeled texts in a self-training manner. These methods pre-train the deep classifier on labeled texts. Recent works further fine-tune the model on the combination of labeled texts and pseudo-labeled texts generated by the pre-trained deep classifier. However, the model's performance largely depends on the quality of pseudo-labels. To tackle such an issue, we propose a novel approach, namely Self-paced Semantic-level Contrastive Learning (S^2CL) for SSTC. S^2CL imposes a self-paced pseudo-label generator to improve the quality of pseudo-labels. We innovatively propose robust supervised learning and semantic-level contrastive learning modules to alleviate the model's over-sensitivity to pseudo-labels' quality. Empirically, S^2CL significantly outperforms the state-of-the-art methods on benchmark datasets with 0.3% - 4.6% improvements on Micro-F1 and 0.3% - 11.1% improvements on Macro-F1. Furthermore, we establish a practical dataset, i.e., *Events39*, to provide a benchmark for evaluating the robustness against domain-shift of SSTC methods. The experimental results demonstrate the effectiveness of S^2CL on Events39.

Keywords: Semi-supervised learning · Contrastive learning · Text classification

1 Introduction

SSTC draws much attention from the community. Conventional approaches solely perform supervised learning on labeled texts, but such approaches do not explore useful information from unlabeled data. With the development of pre-training and fine-tuning frameworks, researchers explore learning the deep representation using labeled and unlabeled texts by following a self-training scheme. Specifically, such an approach first pre-trains the model on labeled texts only, and then the unlabeled texts are assigned corresponding pseudo-labels by the

Supplementary Information The online version contains supplementary material available at https://doi.org/10.1007/978-3-031-33377-4_37.

H. Kashima et al. (Eds.): PAKDD 2023, LNAI 13936, pp. 482–494, 2023.
https://doi.org/10.1007/978-3-031-33377-4_37

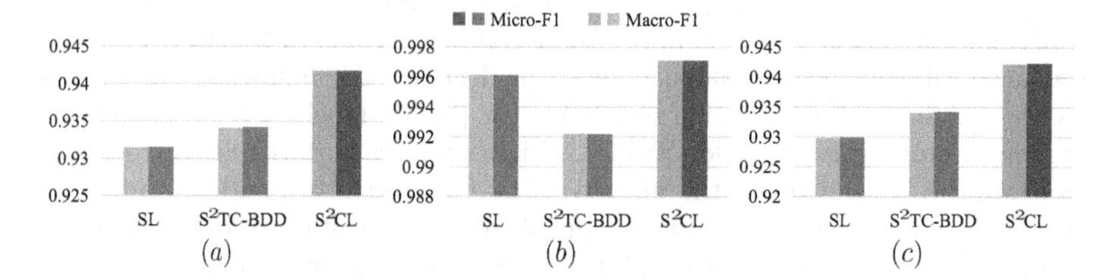

Fig. 1. Performance comparison conducted on the Events39 dataset with a specific train-val-test split, where S^{2}CL is our method, S^{2}TC-BDD [13] denotes a baseline trained on labeled and pseudo-labeled data without meticulous filtering, and SL denotes a baseline only trained on labeled data using the BDD loss [13]. (a) Evaluating the compared methods on the *unseen test* set. (b) Evaluating the methods on the *seen labeled* training set. (c) Evaluating the methods on the *seen unlabeled* training set. S^{2}CL consistently achieves the best performance. S^{2}TC-BDD beats SL in (c) but falls short in (b), which supports that the discriminative information from the labeled training set captured by S^{2}TC-BDD degenerates when training the model on the pseudo-labeled training set without further techniques to improve the quality of pseudo-labels.

pre-trained model. In the regular training phase, the deep classifier trains on the mixture of labeled and pseudo-labeled texts. However, the quality of the pseudo-label is susceptible to the performance of the pre-trained model. Such a learning paradigm may degenerate the discriminability of learned representations for labeled texts. Therefore, the discriminative information from the unlabeled texts is insufficiently explored. We demonstrate a performance comparison in Fig. 1 to support our viewpoint.

Contrastive learning [10,17,20] achieves state-of-the-art in self-supervised representation learning. The fundamental intuition behind the behavior of benchmark contrastive approaches is that pulling an *anchor* and a *positive* examples, i.e., positive pair, together in the hidden space while pushing *negative* samples away from the *anchor*, i.e., negative pair. Self-supervised contrastive methods explore discriminative information from unlabeled data without human annotation. A loss for supervised learning to leverage such a representation capability in supervised learning is proposed [12], which builds on the contrastive self-supervised literature. Unlike self-supervised contrastive learning methods, such an approach explores multiple positive pairs by considering the category information of labeled data. In the field of SSTC, we reckon that *appropriately* performing the supervised contrastive approach to explore *filtered* pseudo-labeled data may promote state-of-the-art performance.

To this end, we propose a novel method, namely **S**elf-paced **S**emantic-level **C**ontrastive **L**earning (S^{2}CL), which *sufficiently* explores discriminative information from unlabeled texts by leveraging a self-paced pseudo-label generator (PLG) to achieve pseudo-labels with high-confidence and jointly performing the robust supervised learning (RSL) and semantic-level contrastive learning (SLCL) modules. Specifically, we propose two variants for PLG: the learnable confidence thresholding filtering and the top-κ confidence filtering, which empower our method to sift out trusted pseudo-labeled texts logically. The issue of the pseudo-label's over-sensitivity to the quality of the pre-trained model can be partially alleviated. For RSL, we apply the BDD loss, proposed by [13], to achieve balanced label angle variances, which can decrease the empirical risk caused by the imbalanced distribution of labeled (or pseudo-labeled) texts. To further explore useful information for SSTC, we innovatively perform SLCL to jointly contrast the labeled texts, the filtered pseudo-labeled texts, and the unlabeled texts, which is achieved by treating multiple data augmentations of the examples from the same class, i.e., data with the same label or pseudo-label, as positives while others as negatives.

To provide a benchmark for evaluating the robustness against domain-shift of SSTC methods, we establish a new dataset, i.e., *Events39*. Unlike the public datasets that mostly contain texts from a single domain, our established dataset includes four significant domains, i.e., Artificial Intelligence, Security, Military Conflicts, and News. Another crucial merit of Events39 is that the self-established dataset has multiple data sources, including news sites, Twitter, etc., which supports that compared with benchmark datasets, Events39 is more various and practical. To compare S^{2}CL with the state-of-the-art SSTC methods, we conduct extensive experiments on benchmark datasets and Events39. The experimental results demonstrate the superiority of S^{2}CL under different amounts of unlabeled data. The **contributions** of our approach are three-fold:

- We propose a self-paced pseudo-label generator with two variants to improve the quality of pseudo-label generation and further alleviate the over-sensitivity of pseudo-labels on the quality of the pre-trained model.
- We innovatively perform robust supervised learning and semantic-level contrastive learning to sufficiently explore the discriminative information in a semi-supervised manner.
- We establish a practical dataset, i.e., Events39, to provide a benchmark for evaluating the robustness against domain-shift of SSTC methods, which contains tweets and news of 39 types representing different types of event collections from four major domains.

2 Related Work

2.1 Semi-Supervised Text Classification

With the development of the pre-trained language model, recent SSTC methods [9,13,18,22] are built on the pre-training and fine-tuning framework, which performs deep representation learning on generic data, followed by supervised learning for downstream tasks. For example, the Variational Auto Encoder (VAE) method [9] first pre-trains a VAE on unlabeled texts and then trains a classifier on the representations of labeled texts computed by the pre-trained VAE. The Virtual Adversarial Training (VAT) method [18] mainly focuses on deep self-training, which can jointly learn deep representation and classifier using both labeled and unlabeled texts in a unified framework. Later, Unsupervised Data Augmentation (UDA) [22] first utilizes data augmentation techniques and applies consistency loss between the predictions of unlabeled texts and corresponding augmented texts. S^2TC-BDD [13] proposes an SSTC method with a balanced depth representation, which applies the angular margin loss, and performs Gaussian linear transformation to achieve balanced label angle variances.

Unlike traditional SSTC methods, we propose a self-paced pseudo-label generator to further improve the quality of pseudo-labels. In addition to using pseudo-labeled data in iterative training for supervised learning, we introduce semantic-level contrastive learning and fully use label information.

2.2 Contrastive Learning

Semi- or self-supervised learning achieves impressive improvement in the fields of image representation learning [3,15,16], graph representation learning [5–7], multi-model learning [14], natural language processing [11], etc. Many recent successes [1,4] are primarily driven by instance-wise contrastive learning, aiming at embedding augmented representations of the same sample close to each other while trying to push away embeddings from different samples. However, the implicit grouping effect of Instance-CL is less stable and more data-dependent since no labels are available. Consequently, the supervised contrastive learning method [12] extends the self-supervised batch contrastive approach to the fully-supervised setting and proposes a loss for supervised contrastive learning. However, rare works conduct in-depth research on SSTC with semi-supervised contrastive learning. Moreover, the utilization of pseudo-labels in current works could be more meticulous.

It should be noted that label information is not required in instance-wise CL, which ignores semantic information in labels and leads to a decrease in the quality of representation learning. To better learn semantic information in labels and pseudo labels, we innovatively propose semantic-level CL in this paper.

3 Methodology

3.1 Overview of S^2CL

As demonstrated in Fig. 2. Given a training dataset $\mathcal{D}$, limited labeled text set $\mathcal{D}_l = \{(\mathbf{x}_i^l, \mathbf{y}_i^l)\}_{i=1}^{N_l}$ and a large unlabeled text set $\mathcal{D}_u = \{\mathbf{x}_j^u\}_{j=1}^{N_u}$ are fed into the encoder $\mathcal{F}$ to obtain the corresponding deep representations f^l and f^u. Specifically, $\mathbf{x}_i^l$ and $\mathbf{x}_j^u$ denote the word sequences of labeled and unlabeled texts. $\mathbf{y}_i^l \in \{0,1\}^K$ denote the corresponding one-hot label vector of $\mathbf{x}_i^l$, where $\mathbf{y}_{ik}^l = 1$ if the text is associated with the $k-th$ label, or vice versa. N_l, N_u, and K denote the numbers of labeled texts, unlabeled texts, and category labels, respectively. We use a one-linear-layer classifier head $\mathcal{C}$ for prediction. Meanwhile, the label features $\{\mathbf{w}_i\}_{i=1}^K$ obtained from $\mathcal{C}$ are sent to PLG with f^u to generate pseudo-labels $\mathbf{y}_i^p$. Specifically, the unlabeled texts with pseudo-labels are further filtered to get trusted pseudo-labeled texts. We then adopt robust supervised learning on representations of pseudo-labeled and labeled data, denoted by f^p and f^l, respectively. Meanwhile, operator R replaces the unlabeled texts with the trusted pseudo-labeled data. We feed all data representations, which now can be

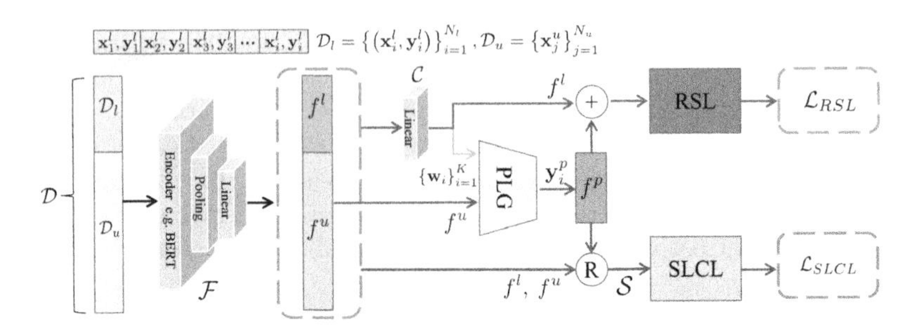

Fig. 2. A visualized pipeline of the proposed S^{2}CL.

represented as $\mathcal{S} = \{f^l, f^u, f^p\}$, into SLCL. In addition, the details of the BERT-based encoder $\mathcal{F}$ and classification head $\mathcal{C}$ are available in the supplementary.

3.2 Self-Paced Pseudo-Label Generator

We introduce the self-paced pseudo-label generator, i.e., PLG, to improve the quality of pseudo-labels. In detail, We propose two variants for PLG:

1) Learnable confidence thresholding filtering. We sift out the trusted pseudo-labels by

$$\tilde{\mathbf{Y}}^p = \mathrm{Cat}^{(+)}\left[\mathrm{Sum}^{(1)}\left[\mathrm{ReLU}\left[\mathbf{Y}^u - \mathrm{Extend}\left(\tau, \mathrm{Shape}\left(\mathbf{Y}^u\right)\right)\right]\right]\right], \quad (1)$$

where $\mathrm{Shape}(\cdot)$ is a matrix shape getter function, $\mathrm{Extend}(\cdot,\cdot)$ is a matrix extending function, $\mathrm{ReLU}[\cdot]$ is an activation function to eliminate the negative elements of the matrix, $\mathrm{Sum}^{(1)}[\cdot]$ is a dimension-wise summation function, $\mathrm{Cat}^{(+)}[\cdot]$ is a concatenation function only performing on the positive vectors, $\mathbf{Y}^u$ denotes the original matrix containing $\{\mathbf{y}_i^p\}_{i=1}^{N_u}$, $\tilde{\mathbf{Y}}^p$ denotes the index of filtered pseudo-labels, and τ is a learnable value. We train τ by back-propagating $\mathcal{L}_{RSL}$. Then, we further derive the filtered pseudo-label matrix by

$$\mathbf{Y}^p = \mathrm{Filter}\left[\mathbf{Y}^u, \mathrm{Index}\left(\tilde{\mathbf{Y}}^p\right)\right], \quad (2)$$

where $\mathbf{Y}^p$ denotes the filtered matrix containing $\{\mathbf{y}_i^p\}_{i=1}^{N_p}$, and $\mathrm{Filter}[\cdot,\cdot]$ is the filtering function that is based on the target index.

2) Top-κ confidence filtering. We propose a simple yet effective filtering approach to derive trusted pseudo-labeled texts by

$$\tilde{\mathbf{Y}}^p = \mathrm{Cat}^{(\kappa)}\left[\mathrm{Sort}\left[\mathrm{Max}^{(1)}\left(\mathbf{Y}^u\right)\right]\right], \quad (3)$$

where $\mathrm{Max}^{(1)}(\cdot)$ is a dimension-wise maximum function, $\mathrm{Sort}[\cdot]$ is a sort function in a descending order, and $\mathrm{Cat}^{(\kappa)}[\cdot]$ is a concatenation function only performing on the top-κ vectors. The filtered pseudo-label matrix is derived by Eq. 2.

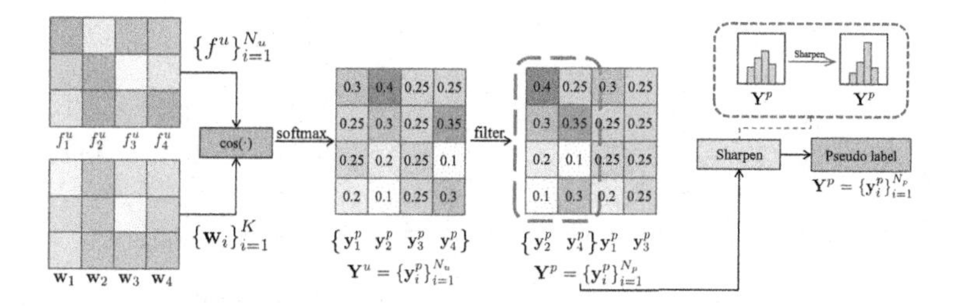

Fig. 3. The structure of top-κ confidence filtering PLG.

Then, to avoid the over-uniformity of the derived pseudo-label distributions $\mathbf{Y}^p$, we perform a sharpen function:

$$\mathbf{y}_i^p = \text{Sharpen}\left(\mathbf{y}_i^p, T\right) = \frac{\left(\mathbf{y}_i^p\right)^{1/T}}{\left\|\left(\mathbf{y}_i^p\right)^{1/T}\right\|_1}, \forall i \in [N_p] \tag{4}$$

where $\|\cdot\|_1$ is the ℓ_1-norm of vectors, and N_p denotes the total number of filtered pseudo-labels. When $T \rightarrow 0$, the pseudo-label distribution tends to be the one-hot vector. After sharpening, we get the pseudo-labels of partial unlabeled data.

As shown in Fig. 3, we first calculate the cosine similarity between f^u and $\mathbf{w}_i$, after normalization by the softmax function, we get $\mathbf{Y}^u = \{\mathbf{y}_i^p\}_{i=1}^{N_u}$, showing probabilities of unlabeled data in different categories. Considering that the quality of the pseudo-label is an important factor affecting the performance of the model performance [13], we further impose the self-paced pseudo-label filtering approach to guarantee the high credibility of pseudo-labels. After filtering, we derive $\mathbf{Y}^p = \{\mathbf{y}_i^p\}_{i=1}^{N_p}$ and then apply a sharpen function with a temperature T to avoid the over-uniformity of the derived pseudo-label distributions of $\mathbf{Y}^p$.

In all the experiments, we adopt top-κ confidence filtering PLG, which outperforms the other variant. Further, we discuss variants of PLG in Sect. 4.4.

3.3 Robust Supervised Learning

In supervised learning, we adopt the BDD loss [13], which alleviates the problem of margin bias problem caused by the large difference between representation distributions of labels in SSTC. BDD loss suppose that the label angles are drawn from each label-specific Gaussian distribution $\left\{\mathcal{N}\left(\mu_k, \sigma_k^2\right)\right\}_{k=1}^K$, and transfer them into the ones with balanced $\left\{\mathcal{N}\left(\mu_k, \widehat{\sigma}^2\right)\right\}_{k=1}^K$, $\widehat{\sigma}^2 = \frac{\sum_{k=1}^K \sigma_k^2}{K}$ by performing the following linear transformations to the angles:

$$\psi_k\left(\theta_{ik}\right) = a_k \theta_{ik} + b_k, \quad \forall k \in [K], \tag{5}$$

where

$$a_k = \frac{\widehat{\sigma}}{\sigma_k}, \quad b_k = \left(1 - a_k\right) \mu_k. \tag{6}$$

With these linear transformations $\{\psi_k(\cdot)\}_{k=1}^K$, all angles become the samples from balanced angular distributions with the same variances, e.g., $\psi_k\left(\theta_{ik}\right) \sim \mathcal{N}\left(\mu_k, \widehat{\sigma}^2\right)$. Accordingly, the BDD loss can be rewritten as:

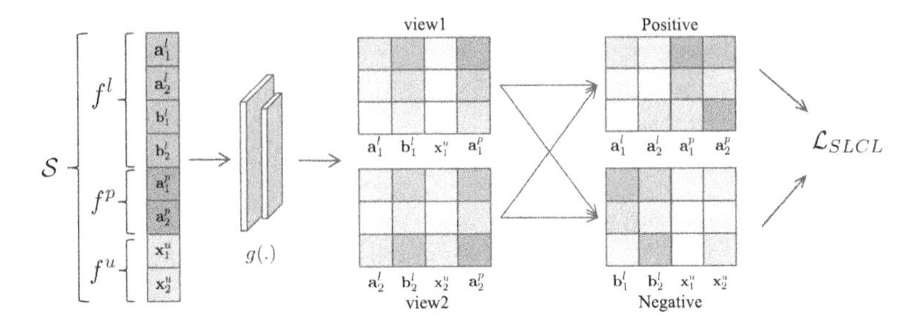

Fig. 4. An example of semantic-level contrastive learning (SLCL).

$$\mathcal{L}_{BDD}\{x_i, y_i; \phi\} = \sum_{k=1}^{K} y_{ik} \log \frac{e^{s(\cos(\psi_k(\theta_{ik}))-y_{ik}m)}}{\sum_{j=1}^{K} e^{s(\cos(\psi_j(\theta_{ij}))-y_{ij}m)}}, \tag{7}$$

where ϕ denotes the model parameters; θ_{ik} is the angle between f_i and $\mathbf{w}_k$; s and m control the rescaled norm and magnitude of cosine margin, respectively, following the value in S^{2}TC-BDD, and

$$\cos(\theta_{ik}) = \frac{f_i^\top \mathbf{w}_k}{\|f_i\|_2 \|\mathbf{w}_k\|_2}, \tag{8}$$

where $\|\cdot\|_2$ denotes the ℓ_2- norm of vectors; f_i and $\mathbf{w}_k$ denote the deep representation of text $\mathbf{x}_i$ and the weight vector of label k, respectively.

3.4 Semantic-Level Contrastive Learning

We apply Semantic-Level Contrastive Learning to the combination of labeled, unlabeled, and trusted pseudo-labeled texts. For data augmentation, two slightly different vector representations are generated after an input text passes the same BERT model twice [8]. We treat these two different representations as data augmentations of the same sample. The given augmentations will first pass a Projection head $g(\cdot)$, which is a two-layer perception (MLP) mapping augmented representations to another latent space where the contrastive loss is calculated. Unlike instance-level contrastive learning, which only treats data augmentations of the same sample as positives, we innovatively propose a semantic-level contrastive learning method. Taking class label information into account results in an embedding space where elements of the same class are more closely aligned than in the self-supervised case [12]. Data with the same label may have similar semantics. Accordingly, we treat many positives per class in addition to many negatives. Positives are drawn from all data augmentations of the same class (data with the same label or pseudo-label), while others are negatives.

As demonstrated in Fig. 4, suppose that all augmented data in certain batch is $\mathcal{S} = \{f^l : \{\mathbf{a}_1^l, \mathbf{a}_2^l, \mathbf{b}_1^l, \mathbf{b}_2^l\}, f^p : \{\mathbf{a}_1^p, \mathbf{a}_2^p\}, f^u : \{\mathbf{x}_1^u, \mathbf{x}_2^u\}\}$. Here, $\mathbf{a}_1^l, \mathbf{a}_2^l$ are features of samples with label $\mathbf{a}$. $\mathbf{a}^p$ are features of those initially unlabeled data but pseudo-labeled as label $\mathbf{a}$ by PLG, which have similar semantic characters with $\mathbf{a}^l$. $\mathbf{x}^u$ denotes the data still unlabeled after PLG. Accordingly, $\{\mathbf{a}_1^l, \mathbf{a}_2^l, \mathbf{a}_1^p, \mathbf{a}_2^p\}$ are treated as positive samples while those unlabeled and different labeled data that possess different semantic characters, i.e., $\{\mathbf{b}_1^l, \mathbf{b}_2^l, \mathbf{x}_1^u, \mathbf{x}_2^u\}$, are considered as negative samples.

Then Semantic-level Contrastive Learning Loss is defined to enforce maximizing the consistency between all positive samples compared with negative samples, which can be rewritten as:

$$\mathcal{L}_{SLCL} = -\sum_{i=1}^{N} \frac{1}{N_{y_i} - 1} \sum_{j=1}^{N} l_{i \neq j} l_{y_i = y_j} \ln \frac{\exp(s_{i,j}/t)}{\exp(s_{i,j}/t) + \sum_{k=1}^{N} l_{y_i \neq y_k} \exp(s_{i,k}/t)}, \tag{9}$$

where N represents a mini-batch size, y_i and y_j represent the label of the anchor sample i and the sample j, respectively. N_{y_i} represents the number of samples whose label is y_i in a mini-batch. $l_{i \neq j} \in \{0, 1\}$, $l_{y_i} = y_j$ and $l_{y_i \neq y_k}$ are similar indicator functions. For instance, $l_{i \neq j} = 1$ if $i \neq j$; otherwise, $l_{i \neq j} = 0$. $s_{i,j}$ is the cosine similarity between the high-level feature vectors of the sample i and the sample j; t is the temperature hyper-parameter.

3.5 Connecting RSL and SLCL

We combine the losses of RSL and SLCL to derive the final loss of S^{2}CL:

$$\mathcal{L}_{S^2CL} = \mathcal{L}_{RSL} + \lambda \cdot \mathcal{L}_{SLCL}, \tag{10}$$

where λ is a hyper-parameter to control the balance between $\mathcal{L}_{RSL}$ and $\mathcal{L}_{SLCL}$.

4 Experiments

4.1 Datasets and Baselines

We conduct experiments on three public datasets: AG News [23], Yelp [23], Yahoo [2], and a self-established dataset, Events39[1].

Self-established Datasets. There are 39 types of data, representing different types of event collections, including Artificial Intelligence, Security, Military Conflicts, and News obtained from Twitter and news sites. The comparison between datasets is shown in Table 1.

Table 1. A brief introduction of the proposed Events39.

Dataset	#Domain	#Class	#Train	#Test
AG News [23]	1	4	120,000	7,600
Yelp [23]	1	5	650,000	50,000
Yahoo [2]	1	10	1,400,000	50,000
Events39	4	39	81,900	35,100

Baselines. We compare our method with the following methods: NB+BE [19]; BERT+AM [21]; VAMPIRE [9]; VAT [18]; UDA [22]; S^{2}TC-BDD [13].

Furthermore, the detailed implementation is available in the supplementary.

[1] https://github.com/XiaYWilson/Events39-Dataset.

Table 2. Experimental results varying the number of labeled texts N_l. The best results are highlighted in **boldface**.

Dataset	AG News				Yelp				Yahoo			
N_u	20,000											
N_l	100		1,000		100		1,000		100		1,000	
Metric-F1	Micro	Macro	Micro	Macro	Micro	Macro	Micro	Macro	Micro	Macro	Micro	Macro
NB+BE	0.834	0.833	0.855	0.855	0.300	0.250	0.355	0.329	0.529	0.489	0.624	0.616
BERT	0.839	0.840	0.878	0.878	0.344	0.324	0.538	0.532	0.564	0.550	0.676	0.671
BERT+AM	0.856	0.856	0.879	0.879	0.399	0.371	0.544	0.535	0.589	0.573	0.679	0.672
VAMPIRE	0.705	0.698	0.833	0.833	0.227	0.144	0.476	0.476	0.389	0.356	0.547	0.545
VAT	0.868	0.867	0.886	0.886	0.224	0.197	0.551	0.548	0.534	0.542	0.685	0.675
UDA	0.855	0.855	0.883	0.883	0.387	0.357	**0.554**	**0.550**	0.576	0.567	0.672	0.666
S^{2}TC-BDD	**0.872**	**0.872**	0.889	0.889	0.417	0.403	0.552	**0.550**	0.618	0.595	0.687	0.680
S^{2}CL	0.866	0.866	**0.892**	**0.892**	**0.438**	**0.423**	**0.554**	0.546	**0.631**	**0.625**	**0.690**	**0.687**

Table 3. Experimental results varying the number of unlabeled texts N_u.

Dataset	AG News						Yelp					
N_l	100											
N_u	200		2,000		20,000		200		2,000		20,000	
Metric-F1	Micro	Macro	Micro	Macro	Micro	Macro	Micro	Macro	Micro	Macro	Micro	Macro
NB+BE	0.696	0.695	0.752	0.751	0.834	0.833	0.307	0.279	0.302	0.286	0.300	0.250
BERT+AM	0.855	0.855	0.856	0.855	0.856	0.856	0.385	0.370	0.393	0.379	0.399	0.371
VAMPIRE	0.329	0.219	0.421	0.341	0.705	0.698	0.238	0.161	0.211	0.124	0.227	0.144
VAT	0.850	0.850	0.870	0.870	0.868	0.867	0.299	0.278	0.294	0.287	0.244	0.197
UDA	0.844	0.843	0.853	0.852	0.855	0.855	0.397	0.344	0.379	0.362	0.387	0.357
S^{2}TC-BDD	**0.857**	**0.857**	0.863	0.864	**0.872**	**0.872**	0.403	0.372	0.417	0.380	0.417	0.403
S^{2}CL	**0.857**	**0.857**	**0.871**	**0.871**	0.866	0.866	**0.407**	**0.394**	**0.433**	**0.422**	**0.437**	**0.423**

4.2 Benchmarking S^{2}CL on Public Datasets

Abundant Unlabeled Data. To evaluate the performance of S^{2}CL under different amounts of data, we conducted experiments in multiple scenarios on benchmark datasets. For all methods, we conduct the experiments with the number of unlabeled texts $N_u = 20,000$ on all datasets and set the number of labeled texts N_l with either 100 or 1,000 to simulate the semi-supervised scenarios of manually annotating a small number of samples. The test set directly uses the original test set. The classification results of both Micro-F1 and Macro-F1 over benchmark datasets are shown in Table 2. Generally speaking, our proposed S^{2}CL outperforms the baselines in most cases.

Rare Labeled Data. As shown in Table 3, S^{2}CL outperforms the existing deep learning methods. The Micro-F1 and Macro-F1 values can be improved by 0.3% to 11.1% compared with the previous best S^{2}TC-BDD method. Compared with self-training methods, VAMPIRE, as a pre-trained method, performs poorly when the unlabeled data is insufficient. In addition, we conduct multiple ablation studies on the Yahoo dataset. The results are shown in Table 5.

4.3 Benchmarking S^{2}CL on Events39

As shown in Table 4, we conduct massive experiments on Events39. Since the number of categories of Events39 is excessively larger than benchmark datasets, we adjust N_l to 1,000. Considering that S^{2}TC-BDD and S^{2}CL outperform other methods on benchmark datasets, we majorly compare them on Events39. S^{2}CL$_{\lambda=0.1}$, S^{2}CL$_{\lambda=0.2}$, and S^{2}CL$_{\lambda=0.5}$ represent the methods that set 0.1, 0.2, and 0.5 as the contrastive learning coefficients, respectively. S^{2}CL can consistently beat S^{2}TC-BDD by wide margins, and S^{2}CL$_{\lambda=0.1}$ outperforms other methods under various settings. A reasonable explanation is that an appropriate coefficient balancing the impacts of RSL and SLCL can improve the performance while noting that S^{2}CL can beat the benchmark method with all settings of λ. Comparative experiments show that the S^{2}CL has an excellent performance in various scale classification problems in news, Q&A websites, social media, and other fields, which supports that S^{2}CL is robust against domain-shift of the SSTC tasks. The results in Fig. 1 and Table 4 further demonstrate the robustness of S^{2}CL.

Table 4. Results on Events39.

N_l	1,000							
N_u	1,000		2,000		10,000		20,000	
Metric	Micro-F1	Macro-F1	Micro-F1	Macro-F1	Micro-F1	Macro-F1	Micro-F1	Macro-F1
S^{2}TC-BDD	0.897	0.897	0.896	0.895	0.873	0.872	0.855	0.854
S^{2}CL$_{\lambda=0.5}$	0.903	0.901	0.898	0.896	0.854	0.853	0.912	0.911
S^{2}CL$_{\lambda=0.2}$	0.910	0.909	0.898	0.897	0.865	0.864	0.897	0.897
S^{2}CL$_{\lambda=0.1}$	**0.920**	**0.918**	**0.915**	**0.914**	**0.883**	**0.882**	**0.917**	**0.916**

4.4 Deepgoing Exploration

Discussion on the Variants of PLG. As defined in Sect. 3.2, we propose two variants for PLG. To evaluate the proposed variants' improvements, we compare them on the Yahoo dataset with $N_l = 100$ and $N_u = 4,000$. As shown in Fig. 5, the two columns in the leftmost indicate learnable confidence thresholding filtering while others indicate top-κ confidence filtering. The learnable confidence thresholding filtering outperforms randomly assigned top-κ confidence filtering when $\kappa = 0.25$, while an elaborately selected κ empowers the top-κ confidence filtering to achieve better performance, e.g., $\kappa = 0.66$. Concretely, we conclude that both variants can improve our method. The learnable confidence thresholding filtering reduces the time consumption of parameter tuning efficiently, but appropriate top-κ confidence filtering further improves S^{2}CL.

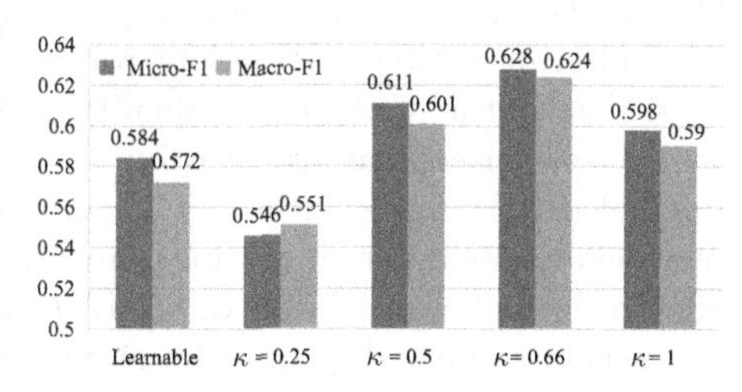

Fig. 5. Results on PLG's variants.

Ablation Study. We conduct ablation research by replacing each module of S^{2}CL. We replace the SLCL with the traditional instance-level contrastive learning, i.e., SICL. Specifically, SICL only treats data augmentations of the same sample randomly chosen from the minibatch as the positive pairs. We then remove contrastive learning with retaining PLG to prove the effectiveness of our label filtering strategy, and the corresponding ablation model is S-P. As shown in Table 5, all variants of S^{2}CL outperform S^{2}TC-BDD, which proves the effectiveness of each part of the proposed method. Especially our method with $N_u = 2,000$ can even beat S^{2}TC-BDD with $N_u = 40,000$, which supports that our method can efficiently model discriminative information from unlabeled data.

Comparing the results of S-P and S^{2}CL, we observe that the complete method keeps consistent improvement over the ablation model, which supports the superiority of SLCL. From the results derived by SICL and S^{2}CL, we observe that the Micro-F1 derived by S^{2}CL increases from 0.617 to 0.631 as N_u changes from 200 to 20,000. At the same time, the improvement trend of SICL is inconsistent, which proves the steady benefit of SLCL trained on increasing unlabeled data.

Table 5. Ablation study on Yahoo.

Metric	N_u	S^{2}TC-BDD	S-P	SICL	S^{2}CL
Micro-F1	200	0.6	0.605	0.606	**0.617**
	400	0.593	0.617	**0.622**	0.617
	2,000	0.624	**0.629**	0.628	0.618
	4,000	0.598	0.611	0.612	**0.628**
	2,0000	0.619	**0.637**	0.636	0.631
	Avg	0.609	0.620	**0.621**	**0.621**
Macro-F1	200	0.583	0.603	0.606	**0.616**
	400	0.586	0.603	0.611	**0.615**
	2,000	0.603	**0.621**	**0.621**	0.614
	4,000	0.59	0.601	0.601	**0.624**
	20,000	0.604	**0.63**	0.628	0.625
	Avg	0.594	0.612	0.613	**0.620**

5 Conclusion

This paper proposes a novel learning paradigm to support SSTC, namely S^{2}CL. Our method improves the quality of the pseudo-labels by utilizing an elaborate label filtering strategy. We adopt robust supervised learning to reduce the oversensitivity of the model to the quality of pseudo-labels. We innovatively leverage semantic-level contrastive learning to explore discriminative information from unlabeled data sufficiently. S^{2}CL empirically outperforms the SSTC baseline methods under different settings, and the ablation study separately proves the effectiveness of each module of S^{2}CL. To provide a benchmark for evaluating the robustness against domain-shift of SSTC methods, we establish Events39 and conduct the comparisons to support the robustness of S^{2}CL.

References

1. Bachman, P., Hjelm, R.D., Buchwalter, W.: Learning representations by maximizing mutual information across views. In: Advances in Neural Information Processing Systems, vol. 32 (2019)
2. Chang, M.W., Ratinov, L.A., Roth, D., Srikumar, V.: Importance of semantic representation. In: Dataless Classification. vol. 2, pp. 830–835. AAAI (2008)

3. Chen, T., Kornblith, S., Norouzi, M., Hinton, G.: A simple framework for contrastive learning of visual representations. In: International Conference on Machine Learning, pp. 1597–1607. PMLR (2020)
4. Chen, X., He, K.: Exploring simple Siamese representation learning. In: Proceedings of the IEEE/CVF Conference on Computer Vision and Pattern Recognition, pp. 15750–15758 (2021)
5. Gao, H., Li, J., Qiang, W., Si, L., Sun, F., Zheng, C.: Bootstrapping informative graph augmentation via a meta learning approach (2022)
6. Gao, H., Li, J., Qiang, W., Si, L., Xu, B., Zheng, C., Sun, F.: Robust causal graph representation learning against confounding effects (2022)
7. Gao, H., Li, J., Qiao, P., Zheng, C.: Weight-aware graph contrastive learning. In: Artificial Neural Networks and Machine Learning - ICANN 2022. Springer Nature Switzerland (2022)
8. Gao, T., Yao, X., Chen, D.: Simcse: simple contrastive learning of sentence embeddings. arXiv preprint arXiv:2104.08821 (2021)
9. Gururangan, S., Dang, T., Card, D., Smith, N.A.: Variational pretraining for semi-supervised text classification. arXiv preprint arXiv:1906.02242 (2019)
10. He, K., Fan, H., Wu, Y., Xie, S., Girshick, R.: Momentum contrast for unsupervised visual representation learning. In: Proceedings of the IEEE/CVF Conference on Computer Vision and Pattern Recognition, pp. 9729–9738 (2020)
11. Jin, Y., Li, J., Lian, Z., Jiao, C., Hu, X.: Supporting medical relation extraction via causality-pruned semantic dependency forest (2022)
12. Khosla, P., et al.: Supervised contrastive learning. Adv. Neural. Inf. Process. Syst. **33**, 18661–18673 (2020)
13. Li, C., Li, X., Ouyang, J.: Semi-supervised text classification with balanced deep representation distributions. In: Proceedings of the 59th Annual Meeting of the Association for Computational Linguistics and the 11th International Joint Conference on Natural Language Processing (Volume 1: Long Papers), pp. 5044–5053 (2021)
14. Li, J., Mo, W., Qiang, W., Su, B., Zheng, C.: Supporting vision-language model inference with causality-pruning knowledge prompt (2022)
15. Li, J., et al.: Information theory-guided heuristic progressive multi-view coding (2021)
16. Li, J., Qiang, W., Zheng, C., Su, B.: RHMC: modeling consistent information from deep multiple views via regularized and hybrid multiview coding. Knowl.-Based Syst. **241**, 108201 (2022)
17. Li, J., Qiang, W., Zheng, C., Su, B., Xiong, H.: MetAug: contrastive learning via meta feature augmentation. In: Proceedings of the 39th International Conference on Machine Learning, pp. 12964–12978 (2022)
18. Miyato, T., Maeda, S.I., Koyama, M., Ishii, S.: Virtual adversarial training: a regularization method for supervised and semi-supervised learning. IEEE Trans. Pattern Anal. Mach. Intell. **41**(8), 1979–1993 (2018)
19. Nigam, K., McCallum, A.K., Thrun, S., Mitchell, T.: Text classification from labeled and unlabeled documents using EM. Mach. Learn. **39**, 103–134 (2000)
20. Qiang, W., Li, J., Zheng, C., Su, B., Xiong, H.: Interventional contrastive learning with meta semantic regularizer. In: Proceedings of the 39th International Conference on Machine Learning (2022)
21. Wang, H., et al.: CosFace: large margin cosine loss for deep face recognition. In: Proceedings of the IEEE Conference on Computer Vision and Pattern Recognition, pp. 5265–5274 (2018)

22. Xie, Q., Dai, Z., Hovy, E., Luong, T., Le, Q.: Unsupervised data augmentation for consistency training. Adv. Neural. Inf. Process. Syst. **33**, 6256–6268 (2020)
23. Zhang, X., Zhao, J., LeCun, Y.: Character-level convolutional networks for text classification. In: Advances in Neural Information Processing Systems, vol. 28 (2015)

Multi-Augmentation Contrastive Learning as Multi-Objective Optimization for Graph Neural Networks

Xu Li(✉) and Yongsheng Chen

Tongji University, Shanghai, China
{lixu,chenyongsheng}@tongji.edu.cn

Abstract. Recently self-supervised learning is gaining popularity for Graph Neural Networks (GNN) by leveraging unlabeled data. Augmentation plays a key role in self-supervision. While there is a common set of image augmentation methods that preserve image labels in general, graph augmentation methods do not guarantee consistent graph semantics and are usually domain dependent. Existing self-supervised GNN models often handpick a small set of augmentation techniques that limit the performance of the model.

In this paper, we propose a common set of graph augmentation methods to a wide range of GNN tasks, and rely on the Pareto optimality to select and balance among these possibly conflicting augmented versions, called **P**areto **G**raph **C**ontrastive **L**earning (**PGCL**) framework. We show that while random selection of the same set of augmentation leads to slow convergence or even divergence, PGCL converges much faster with lower error rate. Extensive experiments on multiple datasets of different domains and scales demonstrate superior or comparable performance of PGCL.

Keywords: graph neural networks · multi-objective Learning · self-supervised learning

1 Introduction

Graph Neural Networks (GNNs) is a powerful model for graph-related data and problems, such as social networks and protein interactions [19,21,22]. It generally includes a message passing component and a combination component to learn graph representations in a supervised way.

Human-annotated graph labels are scarce because annotating complex graphs is tedious, error-prone, and sometimes requires expert knowledge. Recently, self-supervised learning methods show promising results in computer vision [1,7] and natural language processing [4] tasks, achieving comparable or superior results to supervised methods. Inspired by these results, many research studies extend the contrastive learning idea to GNN [16,17,22]. More specifically, the goal is to learn low-dimension representations for graphs by solving predefined contrastive tasks, which typically pull the embeddings of graphs augmented from the same

H. Kashima et al. (Eds.): PAKDD 2023, LNAI 13936, pp. 495–507, 2023.
https://doi.org/10.1007/978-3-031-33377-4_38

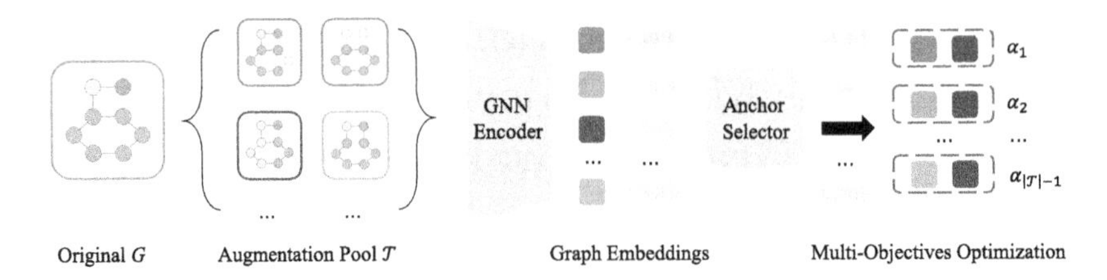

Fig. 1. Overview framework of PGCL. We first generate $|\mathcal{T}|$ different views from an original graph G, and then put them all together into the GNN encoder to obtain their embeddings. The anchor augmentation is chosen by the anchor selector. Finally, we optimize the overall contrastive loss from a multi-objective optimization perspective.

instance (positive pair) closer and push augmented versions of other graphs further away (negative pairs).

Recently, various graph data augmentation methods have been proposed, such as dropping nodes and edges, masking attributes, and adversarial approaches to generate fake graphs [16,26]. However, while most image augmentation techniques preserve semantics in general, e.g., flipping or cropping an image of cat is still a cat, graph augmentation techniques do not necessarily guarantee the same label after transformation [20]. Existing self-supervision GNN methods make various label-consistency assumptions regarding the augmentation techniques they use [22,26].

In addition, unlike image self-supervision where more diverse augmentations tend to help in general, care must be taken if we use multiple graph augmentation methods together. Simply combining multiple augmentation methods may distort the embedding space and lead to conflict [10,25]. We also observe that randomly selecting two augmentations at every epoch can cause oscillation in learning curves or even non-convergence; see Sect. 4.1.

In this paper, we develop a novel Pareto Graph Contrastive Learning (PGCL) framework to incorporate a wide range of graph augmentation techniques for better GNN performance. PGCL consists of two components, a common representation learning backbone to encode various augmented views simultaneously and an anchor selection module to choose one augmented view among the set for loss calculation. Our loss function relaxes the NT-Xent loss [1] commonly used in contrastive learning to take multiple augmented views into account. More specifically, we view different augmentations as separate and possibly conflicting contrastive learning tasks and formulate each learning iteration as a multi-objective optimization problem. Using the Pareto optimality criteria, we try to balance among these learning tasks to achieve good overall performance. Finally, we use the multiple gradient descent algorithm [3] to optimize the model.

We summarize our contribution as follows:

- To our best knowledge, PGCL is the first to study the inconsistency among different augmentations in graph contrastive learning, and use a combina-

tion of augmentation methods instead of handpicked ones to achieve better performance.
- We propose a multi-objective optimization algorithm to incorporate multiple augmentations effectively.
- We propose an augmentation-dependent embedding technique to allow different augmentations to maintain their own feature spaces to relax contrastive loss constraints.
- Extensive experiments show that PGCL outperforms other self-supervision GNN methods on various real-world datasets.

2 Related Work

2.1 Graph Contrastive Learning

Contrastive learning is one of the most popular representation learning algorithms for graphs. The main idea is to enforce embeddings of views augmented from the same graph agree with each other and differ from different instances using a contrastive loss. There are two types of contrastive learning methods based on data augmentation currently: One is studying how to generate positive pairs. For example, ADGCL [16] and LP-Info [27] use a learnable prior to catch crucial graph augmentation information. GraphCL [26] proposes five commonly used simple data augmentation methods to obtain positive samples. Other methods investigate negative sampling strategies. For example, curriculum learning is one way to select negative samples [2,5].

However, none of them consider how to use more data augmentation methods to improve the model. Previous work [6] finds that unlike visual representation learning, contrasting more views without any technique does not improve performance, but it ignores whether there are inconsistencies among various augmentations.

2.2 Multi-objective Optimization

Multi-objective optimization (MOO) addresses the problem of finding a set of Pareto solutions to a composite loss function and finding a gradient descent direction that optimizes all the objectives. It has been successfully applied to a wide range of scenarios, such as reinforcement learning [18], Bayesian optimization [8] and kernel learning [11]. One of the most relevant methods to our work is the multi-gradient descent algorithm, which uses Karush-Kuhn-Tucker (KKT) conditions and provably converges to a point on Pareto set [14].

In our work, we view different graph contrastive losses as multi-objective and apply a gradient-based multi-objective algorithm.

3 Methodology

3.1 Model Overview

The goal of this paper is to explore the problem of inconsistent objectives among different data augmentations. As illustrated in Fig. 1, two key designs distinguish PGCL from the conventional framework. First, PGCL optimizes all augmentation views in one iteration from a multi-objective optimization perspective instead of randomly picking one from an augmentation pool. Second, an augmentation-dependent embedding technique is proposed to relax constraints in the latent space. This helps to reduce the inconsistency among different augmentations in the contrastive loss. Overall, the framework consists of two main components:

Anchor Augmentation Selector. Data augmentation aims at generating more training data via applying certain transformations without semantic changes. It is artificially pre-defined and represented as prior knowledge for semantic invariance. Recent research on visual representation learning shows different views of images help encoders to learn rich representations. Unlike images standard augmentations, e.g., rotating, cropping, etc., applying more augmentations on graphs is not a trivial task. Here, we consider two types: (1) node-space augmentations and change a certain ratio of nodes; (2) structure-space features and operation on initial connectivity by deleting edges and adding edges.

Given a graph $G \in \mathcal{G}$ in the dataset, its augmentation is denoted as $t(G)$, regarded as a distribution defined over $\mathcal{G}$, conditioned on G, and we select the same augmentation pool $\mathcal{T}$ as [25], includeing **NodeDrop** (D), **Subgraph** (S), **EdgePert** (P), **Identical** (I) and **AttrMask** (M). It is worth noting that our framework can handle more data augmentations and has better results with more augmentation without hyperparameter tuning.

For a real-world graph G, it undergoes all graph data augmentations in $\mathcal{T}$ to obtain the augmented views $\{t_i(G)\}_{i=1}^{|\mathcal{T}|}$. Then, we design an anchor selector to choose the anchor augmentation contrasted with other augmented views. Since varying the anchor augmentation may cause significant variance during the training procedure, we fix the anchor once it is selected to ensure stability. To this end, we propose an exploration scheme that aims at choosing the anchor augmentation nearest to everyone and containing the most relevant information with remains. In particular, anchor augmentation is the solution to the following minimizing problem:

$$t_{anchor} = \arg\min_{t_i} - \sum_{i \neq j} I(z_{t_i}; z_{t_j}), \tag{1}$$

where z_{t_i} and z_{t_j} denotes the graph representations respectively. Recall that minimizing the contrastive loss equivalently maximizes the mutual information between latent representations, and we just need to iterate through all the augmentation to determine the anchor. We also notice that too aggressive augmentation, such as NodeDrop in the RDT-B dataset shown in Table 2, being the anchor augmentation leads to the model performance decrease. The $t_{anchor}(G)$ and $t_i(G)$ are regarded as a positive pair, and we use other graphs $t_i(\hat{G}_j)$ in the same batch as negative samples, denoting them as $\{(t_{anchor}(G), t_i(\hat{G}_j))\}_{j=1}^{B}$, where B means the batch size.

After obtaining the positive pairs $(G, t_i(G))$, an L-layer GIN [23] encoder $f(\cdot)$ extracts graph-level representations (h, h_{t_i}) by using a readout layer. Then, a non-linear transformation module called projection head appends, as advocated in [1]. The head projects the representations to a latent space, where contrastive loss is applied. In this paper, we adopt a two-layer perceptron to derive the latent representation.

3.2 Specific Augmentation Contrastive Loss

After the graph-level representation extraction, a minibatch of randomly sampled N graphs generates $N * (|\mathcal{T}| - 1)$ pairs $(z_{t_{anchor}}, z_{t_i})$($|\mathcal{T}|$ refers to the cardinality of the augmentation set). For each particular augmentation. To maximize the mutual information between positive pairs $(z_{t_{anchor}}, z_{t_i})$, we adopt the normalized temperature-scaled cross-entropy loss (NT-Xent). NT-Xent for the n-th graph with augmentation t_i is defined as:

$$L_{t_i} = -\log \frac{\exp(sim(z_n, z_{n,t_i})/\tau)}{\sum_{n'=1,n'\neq n}^{N} \exp(sim(z_n, z_{n',t_i})/\tau)}, \tag{2}$$

where $z_{t_{anchor}}, z_{t_i}$ are re-annotated as z_n, z_{n,t_i} for the n-th graph in the minibatch, τ denotes the temperature parameter. Negative samples are generated from the other N-1 graph within the same minibatch. $sim(z_n, z_{n,t_i})$ denotes the similarity between two views, and we employ the cosine similarity metric as $sim(z_n, z_{n,t_i}) = z_n^T z_{n,t_i}/\|z_n^T\|\|z_{n,t_i}\|$. Minimizing L_{t_i} enforces specific augmented view $t_i(G)$ closer with anchor augmented view $t_{anchor}(G)$ than other negative pairs in the same minibatch. The overall loss is computed across every augmentation method. Here, we assign a contrastive loss for each augmentation with anchor and have $|\mathcal{T}| - 1$ objectives.

3.3 Instantiation of PGCL as Multi-Objective Optimization

Inspired by Pareto optimality [14], we follow the same philosophy to optimize every contrastive loss function without another one being worse off. Specifically, we want all augmented views of the same graph to be similar to the anchor. Hence, we regard each contrastive loss as an objective and optimize together. PGCL framework is instantiated as a multi-objective optimization form:

$$\begin{aligned} \min_{\theta^{enc}, \theta^{proj}} L(\theta^{enc}, \theta^{proj}) = \min L_{t_1}(\theta^{enc}, \theta^{proj}), \ldots, \\ \min L_{t_{|\mathcal{T}|-1}}(\theta^{enc}, \theta^{proj}), \end{aligned} \tag{3}$$

where θ^{enc} is the GNN backbone encoder parameter shared by all the augmentations, and θ^{proj} is the parameter of the projection head. Note that if $|\mathcal{T}|$ equals 2, PGCL degenerates into typical graph contrastive learning, and we refer to this as VanillaGCL. The goal of Eq. 3 is to minimize all the contrastive as much as possible without competition to achieve a Pareto optimal.

Search the Descent Direction. We consider the reduction of $|\mathcal{T}| - 1$ contrastive losses simultaneously, which means we need to identify a direction ω to update vector θ from a given initial point θ^0. Therefore, the directional derivatives of the contrastive loss should all be strictly-positive:

$$\forall i = 1 \ldots |\mathcal{T}| - 1 : (\nabla L_{t_i}(\theta^0), \omega) > 0. \tag{4}$$

Then, ω is one of the descent directions and can be numerically minimized in the convex hall. The parameter θ follows the conventional descent procedure as:

$$\theta^{n+1} = \theta^n - \rho\omega, \tag{5}$$

where ρ determines the step size of descent in each iteration. When ω is small enough or limited iterations run out, the optimization terminates. If ω equals 0, the current point is desired, also called Pareto-stationary. Pareto-stationary is a necessary condition for Pareto optimality and pursues such a point in the convex hull.

In this work, the multiple gradient descent algorithm [3] which leverages the Karush-Kuhn-Tucker (KKT) conditions is used to find a optimization direction. ω is defined as the minimum-norm element in the convex hull $\bar{\mathsf{U}}$ of the contrastive loss gradients, and $\bar{\mathsf{U}}$ is defined as:

$$\bar{\mathsf{U}} = \left\{ u \in \mathbb{R}^N \; |u = \sum_{i=1}^{|\mathcal{T}|-1} \alpha_i \nabla L_i(\theta^0); \alpha_i \geq 0 (\forall i); \sum_{i=1}^{|\mathcal{T}|-1} \alpha_i = 1 \right\}. \tag{6}$$

[3] showed that either ω is equal to 0, which means contrastive loss is Pareto-stationary, or ω is the search direction that can optimize every loss. Identifying a solution in Eq. 6 can be equivalently solving the following quadratic-form constrained minimization problem in $\mathbb{R}^{|\mathcal{T}|-1}$:

$$\min_{\alpha} \; \alpha^T \mathrm{M} \alpha, \quad \text{s.t.} \quad \begin{cases} \|\alpha\|_1 = 1 \\ \alpha \succeq 0 \end{cases}, \tag{7}$$

where $\mathrm{M}_{i,j} = (\nabla_{\theta^{enc}} L_{t_i}(\theta))^T (\nabla_{\theta^{enc}} L_{t_j}(\theta))$ and α is initialized to $(\frac{1}{|\mathcal{T}|-1}, \ldots, \frac{1}{|\mathcal{T}|-1})$

To avoid ω being influenced by the gradients of the small norms in the augmentation family, we normalize the gradients without changing the direction and the problem remains unchanged since the scaling of ω does not affect Eq. 6. Specially, we consider Newton's method-based normalization:

$$u_i = \frac{L_{t_i}(\theta)}{\|\nabla L_{t_i}(\theta)\|_2^2} \nabla L_{t_i}(\theta). \tag{8}$$

Model Optimization. We adopt the Frank-Wolfe algorithm, viewing Eq. 7 as a first-order optimization algorithm, moving towards minimizer of the target function. To be concrete, at k-th iteration, we consider the linear approximation of Eq. 7 and find optimal s_k to minimize $s^T \alpha \mathrm{M}$ under constraints $||s||_1 = 1$ and $s \succeq 0$. Then step size β is determined by solving the problem $\arg\min_\beta(\alpha_k + \beta(s_k - \alpha_k))^T \mathrm{M}(\alpha_k + \beta(s_k - \alpha_k))$. Finally, α is updated by $\alpha_{k+1} = \alpha_k + \beta(s_k - \alpha_k)$, and we apply Eq. 5 to optimize PGCL model. After repeating the above steps, we were able to obtain a descending direction that can optimize all losses at the same time, and avoid competition to a certain degree.

3.4 Augmentation-Dependent Embedding Subspace

Mixing diverse augmentations yields more robust and high-quality embeddings in the latent space. However, this hybridity in latent space may bring up a new problem: aggressive augmentations distort the training distribution more [10,25]. To cope with this problem, we don't enforce the semantic meaning invariant to all kinds of augmentations. Instead, we expect each embedding to maintain its own feature to relax the constraints by contrastive loss.

Due to the complexity of the graph data, it is difficult to measure whether the original graph is semantically consistent with the new graph after transformation. Thus, after an aggressive transformation, the semantic meaning of the new graph may be altered. Instead of projecting them to the invariant space, we allow the embedding to be variant to various augmentations. To this end, we add an augmentation-dependent embedding loss L_{var} to the objective:

$$L_{var} = \sum_{i \neq j} \frac{\lambda}{||z_i - z_j||_2^2} [i \neq anchor][j \neq anchor], \tag{9}$$

where λ is a hyper-parameter and controls to what extent we expect embeddings are different. Hence, the representation will contain both invariant (Eq. 2) and variant (Eq. 9) features which will transfer to downstream tasks regardless of whether the new graph is distinct. We integrate the augmentation-dependent embedding subspace into the PGCL framework, referred to as **PGCL+**.

4 Experiments

4.1 Conflict Among Augmentations

We employ two different field datasets to study the inconsistency among different contrastive losses without any techniques, and the training loss is shown in Figure 2. The same GNN architecture is used as GraphCL [26] in all the settings. In Fig. 2a and Figure 2b, we adopt the IA augmentation pair and observe the loss of unoptimized pairs. Results show that when we force the embeddings of AttrMask view to be similar to the original graph's, it makes other augmented views' far away, such as NodeDrop in PROTEINS dataset and EdgePert on the NCI1 dataset. It implies that different domain attributes vary largely.

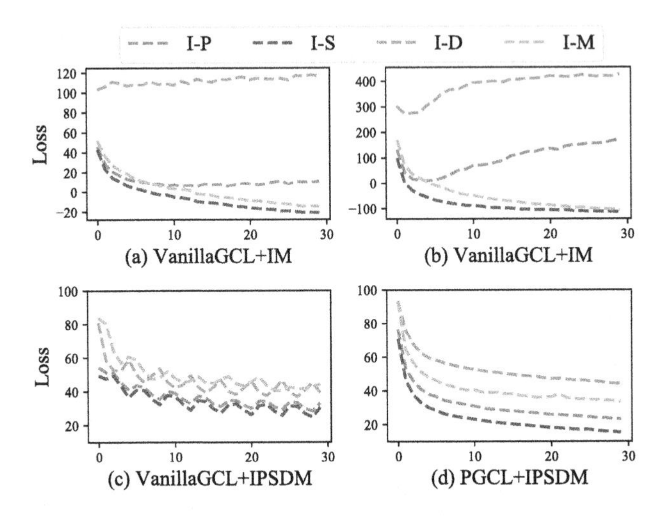

Fig. 2. Training loss versus epoch. (a) Training loss when optimizing IM augmentation pair on PROTEINS dataset., (b) Training loss when optimizing IM augmentation pair on NCI1 dataset, (c) VanillaGCL training loss on NCI1 dataset, (d) PGCL training loss on NCI1 dataset. I: Identity, P: EdgePert, S: Subgraph, M: AttrMask, D: NodeDrop.

In molecules filed, AttrMask and EdgePert are in conflict, but in bioinformatics, they are not. We also find that AttrMask and Subgraph have a high consistency, which means optimizing one could benefit the other.

To further illustrate PGCL's superiority, we also compare it with a model which uses multiple augmentations without any technique on the NCI1 dataset. We iteratively select two augmentations from $\mathcal{T}$ in every epoch, referred to as VanillaGCL. Figure 2c presents the loss trend versus epoch. The loss floats up and down, and when it rises, it means the optimization objectives may be a consistent augmentation. We can see that both IS and IM fall around the sixth epoch, but IP and ID are rising, which is also consistent with Fig. 2a and Fig. 2b. Compared to the VanillaGCL, PGCL has a more stable descent process and a smaller training loss, as shown in Fig. 2d. We also investigate their performance, and we can see from table 1 that PGCL has a significant performance enhancement against VanillaGCL, especially 4% on MUTAG. This also suggests that when applying graph contrastive learning, we need to consider the inconsistency between different objectives and avoid it to get a better model.

4.2 Unsupervised Representation Learning

Experiments Setting. We evaluate our model on seven datasets on TUDataset [12], covering a wide range of tasks. We adopt the same GNN architectures with default hyper-parameters as in GraphCL and λ is set to 0.001. Specifically, our model consists of 3-layer Graph Isomorphism Network (GIN) [23] with 32 hidden dimensions and a sum pooling readout function.

Evaluation Protocol. The unsupervised learning evaluation protocol [26] which we follow consists of two stages: (1) pretraining (using the whole dataset without any labels): only GNN is trained in this stage in a self-supervised way and does not need labels; (2) classification task: the weights of GNN is fixed and we only train a simple SVM using the output of GNN from the stage (1). Here, we use the 10-fold cross validation method to evaluate the performance of SVM.

Compared methods. For comparison, Apart from graph kernel methods WL [15] and DGK [24], we also compare with other state-of-the-art unsupervised methods, containing graph2vec [13], contrastive methods GraphCL [26], JOAO [25], and GraphMAE [9].

Performance Analysis. The results are reported in Table 1. None of the previous work could be the best across all datasets. We find that our model achieves state-of-the-art performance across 5 of 6 datasets and still has a competitive performance on the other datasets.

Compared to JOAO, which uses the same augmentation pool and automatically selects data augmentation pairs, PGCL outperforms JOAO on all seven datasets. Note that JOAO also assigns a weight to each objective function, but it makes these weights tend to be equal. Such prior knowledge may degrade the performance of the model. Experiments show that on MUTAG and RDT-B datasets, PGCL even has a 3.4% and 4.5% improvement, which further illustrates the importance of considering conflict. Besides, PGCL is more computation-efficient since the objective number of PGCL is $|\mathcal{T}| - 1$ while $|\mathcal{T}|^2$ in JOAO.

GraphCL has a pre-fixed augmentation sampling rule, and only two augmentations are used during the training procedure. This exhaustive manual tuning prevents the further application of the model. As can be seen in Table 1, PGCL beats GraphCL in all datasets greatly with a 1.75% improvement, which also illustrates the need to combine more augmentations.

To address the challenge in Sect. 4.1, we then investigate the effect of augmentation-dependent embedding subspace, namely PGCL+. We can find that variant loss improves PGCL's performance, which further echoes our conjecture that excessively enforcing different views to be similar may lead to conflict, and allowing them to be somewhat different can alleviate this problem.

Effect of Anchor Choice. To better understand the effect of anchor augmentation, we conduct experiments on five datasets with four kinds of augmentations, and the results are shown in Table 2. Specifically, we investigate MISP augmentations on MUTAG, DD, and PROTEINS. Since nodes in IMDB-B and RDT-B datasets do not have attributes, we replace AttrMask as NodeDrop, and the bold letter denotes the anchor. We find that different datasets have different preferences for the anchor, none of which could achieve the best through all datasets. EdgePert outperforms in bioinformatics and molecules fields, while Identity prevails in social networks. Besides, results show that if we use an unreasonable augmentation as the anchor, such as NodeDrop on the RDT-B dataset, the performance of the model will be decreased.

Table 1. The performance of different unsupervised representation learning on the TUDataset. Results in bold indicate the best-reported accuracy, and - means the results are not available in the published paper.

Dataset	NCI1	PROTEINS	DD	MUTAG	RDT-B	RDT-M5K
WL	80.01 ± 0.50	72.92 ± 0.56	-	80.72 ± 3.00	68.82 ± 0.41	46.06 ± 0.21
DGK	80.31 ± 0.46	73.30 ± 0.82	-	87.44 ± 2.72	78.04 ± 0.39	41.27 ± 0.18
graph2vec	73.22 ± 1.81	73.30 ± 2.05	-	83.15 ± 9.25	75.78 ± 1.03	47.86 ± 0.26
GraphCL	77.87 ± 0.41	74.39 ± 0.45	78.62 ± 0.40	86.80 ± 1.34	89.53 ± 0.84	55.99 ± 0.28
JOAO	78.36 ± 0.53	74.07 ± 1.10	77.40 ± 1.15	87.67 ± 0.79	86.42 ± 1.45	56.03 ± 0.27
GrpahMAE	**80.40** ± 0.30	75.30 ± 0.39	-	88.19 ± 0.36	88.01 ± 0.19	-
VanillaGCL	78.13 ± 0.81	75.17 ± 0.78	78.56 ± 0.95	87.73 ± 1.50	89.45 ± 0.57	55.77 ± 0.27
PGCL	79.44 ± 0.43	75.82 ± 0.41	79.28 ± 0.60	90.26 ± 0.71	90.56 ± 0.94	56.44 ± 0.25
PGCL+	79.64 ± 0.54	**76.04** ± 0.81	**80.09** ± 1.02	**91.10** ± 0.69	**90.96** ± 0.53	**56.46** ± 0.23

Table 2. Employing different augmentations as the anchor. Bold indicates the anchor augmentation. I: Identity, P: EdgePert, S: Subgraph, M: AttrMask, D: NodeDrop.

	MISP	PISM	SIPM	ISPM
MUTAG	91.25	**91.31**	90.38	89.98
DD	79.79	**80.38**	79.11	79.79
PROTEINS	76.02	**76.45**	75.29	76.19
	DISP	PISD	SIPD	ISPD
IMDB-B	72.50	72.30	72.40	**72.70**
RDT-B	89.30	90.65	90.71	**91.37**

The Number of Objectives. To investigate the effect of the number of objectives, we change the size of $\mathcal{T}$. For the case where the number is 2, PGCL degenerates, and we report the GraphCL accuracy since it elaborately handpicks the best augmentation pair for every dataset. For more augmentations cases, we randomly select augmentations from IPDSM as $\mathcal{T}$ and report the accuracy. Table 3 shows the advantages of combining more objectives. As the number of objectives increases within a certain range, PGCL's performance improves. It is rational that combining more augmentation makes the model more robust, thus generating a better representation.

We notice that on MUTAG and PROTEINS datasets, the best performance is achieved when using four augmentations. This classification accuracy reduction may be caused by the bad augmentations, because sometimes graph augmentation techniques do not guarantee the same label after transformation [20]. Therefore, how to find better augmentations for contrastive learning needs to be studied in future work.

Table 3. Effect of the number of augmentation.

	2 (GraphCL)	3	4	5
MUTAG	86.80	90.41	**91.26**	90.76
NCI1	77.84	78.97	79.47	**79.50**
PROTEINS	74.39	75.65	**76.02**	75.61

5 Conclusion

In this paper, we study the inconsistency when forcing different graph views to be similar when randomly selecting augmentation methods and propose an effective framework for self-supervised learning called PGCL. Extend experiments show that optimizing one augmentation pair may hurt other augmentation pairs, thus distorting the training procedure. We believe this work sheds light on how to use more graph augmentations together in contrastive learning to improve model performance. We conduct comprehensive experiments, which show PGCL improves downstream convergence speed and achieves superior performance than state-of-the-art methods on various datasets.

Acknowledgement. We would like to express our gratitude to the anonymous reviewers for their valuable comments and suggestions that helped improve the quality of this work.

References

1. Chen, T., Kornblith, S., Norouzi, M., Hinton, G.: A simple framework for contrastive learning of visual representations. In: International Conference on Machine Learning, pp. 1597–1607. PMLR (2020)
2. Chu, G., Wang, X., Shi, C., Jiang, X.: CUCO: graph representation with curriculum contrastive learning. In: Proceedings of the Thirtieth International Joint Conference on Artificial Intelligence, IJCAI-21, pp. 2300–2306. International Joint Conferences on Artificial Intelligence Organization (8 2021)
3. Désidéri, J.A.: Multiple-gradient descent algorithm (mgda) for multiobjective optimization. C.R. Math. **350**(5–6), 313–318 (2012)
4. Devlin, J., Chang, M.W., Lee, K., Toutanova, K.: BERT: pre-training of deep bidirectional transformers for language understanding. arXiv preprint arXiv:1810.04805 (2018)
5. Hafidi, H., Ghogho, M., Ciblat, P., Swami, A.: Negative sampling strategies for contrastive self-supervised learning of graph representations. Sig. Process. **190**, 108310 (2022)
6. Hassani, K., Khasahmadi, A.H.: Contrastive multi-view representation learning on graphs. In: International Conference on Machine Learning, pp. 4116–4126. PMLR (2020)
7. He, K., Fan, H., Wu, Y., Xie, S., Girshick, R.: Momentum contrast for unsupervised visual representation learning. In: Proceedings of the IEEE/CVF Conference on Computer Vision and Pattern Recognition, pp. 9729–9738 (2020)

8. Hernández-Lobato, D., Hernandez-Lobato, J., Shah, A., Adams, R.: Predictive entropy search for multi-objective Bayesian optimization. In: International Conference on Machine Learning, pp. 1492–1501. PMLR (2016)
9. Hou, Z., Liu, X., Dong, Y., Wang, C., Tang, J., et al.: GraphMAE: self-supervised masked graph autoencoders. arXiv preprint arXiv:2205.10803 (2022)
10. Jun, H., et al.: Distribution augmentation for generative modeling. In: International Conference on Machine Learning, pp. 5006–5019. PMLR (2020)
11. Li, C., Georgiopoulos, M., Anagnostopoulos, G.C.: Pareto-path multitask multiple kernel learning. IEEE Trans. Neural Netw. Learn. Syst. **26**(1), 51–61 (2014)
12. Morris, C., Kriege, N.M., Bause, F., Kersting, K., Mutzel, P., Neumann, M.: Tudataset: a collection of benchmark datasets for learning with graphs. arXiv preprint arXiv:2007.08663 (2020)
13. Narayanan, A., Chandramohan, M., Venkatesan, R., Chen, L., Liu, Y., Jaiswal, S.: graph2vec: learning distributed representations of graphs. arXiv preprint arXiv:1707.05005 (2017)
14. Sener, O., Koltun, V.: Multi-task learning as multi-objective optimization. arXiv preprint arXiv:1810.04650 (2018)
15. Shervashidze, N., Schweitzer, P., Van Leeuwen, E.J., Mehlhorn, K., Borgwardt, K.M.: Weisfeiler-lehman graph kernels. Journal of Machine Learning Research 12(9) (2011)
16. Suresh, S., Li, P., Hao, C., Neville, J.: Adversarial graph augmentation to improve graph contrastive learning. In: NeurIPS (2021)
17. Tong, Z., Liang, Y., Ding, H., Dai, Y., Li, X., Wang, C.: Directed graph contrastive learning. In: Advances in Neural Information Processing Systems, vol. 34 (2021)
18. Van Moffaert, K., Nowé, A.: Multi-objective reinforcement learning using sets of pareto dominating policies. J. Mach. Learn. Res. **15**(1), 3483–3512 (2014)
19. Velickovic, P., Fedus, W., Hamilton, W.L., Liò, P., Bengio, Y., Hjelm, R.D.: Deep graph infomax. ICLR (Poster) **2**(3), 4 (2019)
20. Wang, Y., Wang, W., Liang, Y., Cai, Y., Liu, J., Hooi, B.: Nodeaug: semi-supervised node classification with data augmentation. In: Proceedings of the 26th ACM SIGKDD International Conference on Knowledge Discovery & Data Mining, pp. 207–217 (2020)
21. Hu, W., et al.: Strategies for pre-training graph neural networks. In: International Conference on Learning Representations (2020). https://openreview.net/forum?id=HJlWWJSFDH
22. Xu, D., Cheng, W., Luo, D., Chen, H., Zhang, X.: InfoGCL: information-aware graph contrastive learning. In: Advances in Neural Information Processing Systems, vol. 34 (2021)
23. Xu, K., Hu, W., Leskovec, J., Jegelka, S.: How powerful are graph neural networks? arXiv preprint arXiv:1810.00826 (2018)
24. Yanardag, P., Vishwanathan, S.: Deep graph kernels. In: Proceedings of the 21th ACM SIGKDD International Conference on Knowledge Discovery and Data Mining, pp. 1365–1374 (2015)
25. You, Y., Chen, T., Shen, Y., Wang, Z.: Graph contrastive learning automated. In: Proceedings of the 38th International Conference on Machine Learning. Proceedings of Machine Learning Research, vol. 139, pp. 12121–12132. PMLR (18–24 Jul 2021). https://proceedings.mlr.press/v139/you21a.html
26. You, Y., Chen, T., Sui, Y., Chen, T., Wang, Z., Shen, Y.: Graph contrastive learning with augmentations. Adv. Neural. Inf. Process. Syst. **33**, 5812–5823 (2020)

27. You, Y., Chen, T., Wang, Z., Shen, Y.: Bringing your own view: graph contrastive learning without prefabricated data augmentations. In: Proceedings of the Fifteenth ACM International Conference on Web Search and Data Mining, pp. 1300–1309 (2022)

Adversarial Active Learning with Guided BERT Feature Encoding

Xiaolin Pang[1(✉)], Kexin Xie[1], Yuxi Zhang[1], Max Fleming[1], Damian Chen Xu[1], and Wei Liu[2]

[1] Salesforce Inc., San Francisco, CA 94105, USA
{xpang,kexin.xie,yuxi.zhang,m.fleming,damian.xu}@salesforce.com
[2] School of Computer Science, University of Technology Sydney, Sydney, NSW 2007, Australia
wei.liu@uts.edu.au

Abstract. Recent advances in BERT-based models has significantly improved the performance of many applications on text data, such as text classification, question answering, e-commerce search and recommendation system, etc. However, the labelling of text data is often complex and time-consuming. While active learning can interactively query and label the data, the effectiveness of existing active learning methods is mostly limited by static text embedding approaches and by the insufficiency of training data. To address this critical problem, in this research we propose a BERT-based adversarial semi-supervised active learning (B-ASAL) model. In our approach, we use generative adversarial modelling and semi-supervised learning to guide the fine-tuning of the BERT and to optimize its corresponding text embeddings and feature encodings. The adversarial generator paired with a semi-supervised classifier guided the BERT model to adjust its feature encoding to best fit the distribution of not only class labels but also the discrimination of labeled and unlabeled data. Moreover, our B-ASAL model selects data points with high uncertainty and high diversity to be labeled using minimax entropy regularization. To our best knowledge, this is the first work that uses adversarial semi-supervised learning joined with active learning to guide and optimize feature encoding. We evaluate our method on various real-world text classification datasets and show that our model outperforms state-of-the-art approaches.

1 Introduction

Advances in deep learning has transformed the field of natural language processing (NLP). BERT-based [4] models are widely used in various NLP tasks: from text classification to question answering to e-commerce search and recommendation etc. Meanwhile, data labelling, a fundamental bottleneck in machine learning, becomes a critical problem due to annotation cost and the need of large amount of labeled data for deep learning NLP tasks. For instance, to build a question answering (QA) model, a human annotator must first read a piece of text and then reason about the answer to the question from context. It is even harder for domain specific labeling task due to the cost of using domain expert.

H. Kashima et al. (Eds.): PAKDD 2023, LNAI 13936, pp. 508–520, 2023.
https://doi.org/10.1007/978-3-031-33377-4_39

In e-commerce, there are very few fine labeled data and professionals are needed to annotate fine labels to map items to fine-grained categories. Therefore, it is necessary to consider how to select more informative samples, so that a better model can be trained with limited labelling capabilities.

Active learning (AL) is one method to collect labeled data cost-efficiently. The goal is to choose the most relevant data points and then query labels from an oracle. Using AL, we can query labels for a small subset of the most relevant documents and immediately train a robust model. For instance, leveraging pre-trained BERT-based language models , task-specific models can be fined-tuned continuously by incorporating newly annotated samples in each iteration to boost the model performance. However, the effectiveness of AL learning methods on NLP tasks is mostly limited by static text embedding, the insufficiency of training data, and the similarity between labeled and unlabeled data distributions.

This research addresses the exact problems above. To address the static text embedding problem, we propose an active learning framework while BERT is fine-tuned in the training progress where the text embedding and feature encoding are both optimized for the training data. To address the problem of the insufficiency of labeled training data, we use adversarial semi-supervised learning to utilize unlabeled data for learning effective representations and for generating new synthetic samples [1] [18]. To discriminate labeled data from unlabeled ones, we incorporate minimax entropy to measure and differentiate the distributions of labeled and unlabeled data. We name our method BERT-based adversarial semi-supervised active learning (B-ASAL). In summary, our contributions in this research are as follows:

- We propose B-ASAL for learning from partially labeled text data. Our B-ASAL model integrates active learning with the fine-tuning of BERT, which guides the BERT to optimize text embedding and feature encoding according to the distribution of the training data.
- We also introduce in the B-ASAL model a generative adversarial network joint with semi-supervised learning, a strategy that can utilize unlabeled data and generalize latent features to select samples for labelling.
- We employ minimax entropy optimization for the unlabeled data to reduce the distribution gap with labeled data while extracting discriminative features for selecting highly representative data samples. Moreover, we also employ conditional entropy maximization in the adversarial network to enhance the robustness and generate uniform-distributed samples.
- We conduct extensive experiments on public datasets and show that our model outperforms state-of-the-art approaches.

2 Related Work

Deep Active Learning (DAL). DAL integrates data labeling and deep model training to improve model performance with minimal amount of labeled data.[1]

[1] In this work, we will only consider the most common pool-based deep active learning.

The scoring function for labeling can be entropy or confidence-score based. Core-set active learning [13] selects a small set of points that approximates the shape of a larger point set using concept of computational geometry. [19] combines clustering with a pre-trained language model (BERT) to select samples. Variational adversarial active learning (VAAL) [14] is proposed as a task-agnostic diversity-based algorithm that samples data points using a discriminator trained adversarially to discern labeled and unlabeled points.

GAN Semi-supervised Learning. Semi-supervised models are able to improve the generalization capability by learning from fewer labeled data points with the help of a large number of unlabeled data points. Semi-Supervised GAN (SS-GAN) [12] extends standard GAN [7] where the labeled data is used to train the discriminator, while the unlabeled data (as well as the ones automatically generated) improve its inner representations. CatGAN [15] proposes categorical GAN for unsupervised and semi-supervised framework by utilizing unlabeled data to learn multi-class classifier. Besides, GAN-BERT [2], a semi-supervised learning model for natural language processing task, enriches the BERT fine-tuning process with a SS-GAN perspective.

Pre-trained BERT. BERT [4] has been used in combination with AL to select representative samples to reduce labelling effort for text classification [5,8]. In [5], it presents a large-scale an empirical study on AL techniques for BERT-based classification, covering a diverse set of AL strategies and datasets on binary text classification. [8] also conducts an empirical study by comparing different uncertainty-based acquisition strategies on two classical NLP multi-class classification datasets.

Entropy regularization. Entropy regularization has been widely used in various deep learning models. In the field of domain adaptation, [11] uses entropy optimization for matching source data to target data distribution. The MAL framework [6] uses the similar idea and proposes a semi-supervised minimax entropy-based active learning algorithm in an adversarial manner for image related tasks. CatGAN [15] and the study of SS-GAN use entropy regularization [3] to improve generation of images conditioned on class assignment.

3 Learning Framework

In this section, we describe our proposed method, the B-ASAL (the BERT-based adversarial semi-supervised active learning) model.

3.1 Problem Formulation

We consider exploiting unlabeled data points and formulate semi-supervised generative adversarial active learning problem as: given an initial labeled data set $\mathcal{S}^l : (\mathcal{X}^l, \mathcal{Y}^l) = \{(x_l, y_l)\}$, where $l \in \{1, ..., m\}$ with size M, and a large unlabeled data pool $\mathcal{S}^u : \mathcal{X}^u = \{(x_u)\}$, where $u \in \{1, ..., n\}$ with size N ($M \ll N$) and $y_l \in \{0, 1\}$ is the class label of x_l for binary classification, or $y_l \in \{1, ..., K\}$ for multi-class classification. We also have a set of generated adversarial data

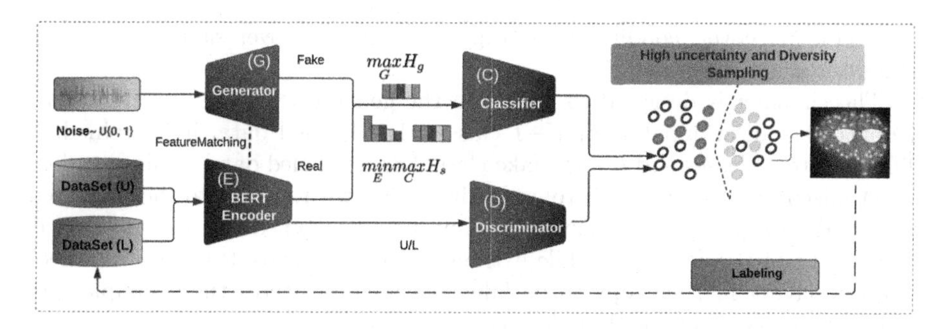

Fig. 1. Workflow of our B-ASAL model. There are Four components in our model: Generator (G), BERT Encoder (E), Classifier (C) and Discriminator (D). Each component and loss function has detailed explanations in Sect. 3.2 and 3.3.

points S^g: $\mathcal{X}^g = \{(x_g)\}$, pairing with true data points to enhance model learning, where x_g is transformed by noise input $\{z_1, ..., z_{m+n}\} \sim U\{0,1\}$ (i.e. p_z) and g $\in \{1, ..., m+n\}$. For all of feature inputs: $\mathcal{X}^l$, $\mathcal{X}^l$ and $\mathcal{X}^g$, we assume they denotes encoded input through encoder. The AL model $\mathcal{M}$ parameterized by $\theta \in \Theta$ is trained on labeled data with their labels, unlabeled data and adversarial data (i.e. $\mathcal{S}^l \cup \mathcal{S}^u \cup \mathcal{S}^g$). This training can be formalized by the optimization problem:

$$\underset{\theta}{\operatorname{argmin}} \mathcal{L}(\theta; y_i | x \in \mathcal{X}^u \cup \mathcal{X}^l \cup \mathcal{X}^g, y \in \mathcal{Y}^l), \tag{1}$$

where $\mathcal{L}$ is the loss function composed of supervised loss trained for labeled data, unsupervised loss trained for unlabeled data and generated fake data. In each AL cycle, trained model $\mathcal{M}$ selects top $k\%$ samples (denoted as $\mathcal{S}^q$ and $\mathcal{S}^q \in \mathcal{S}^u$) constrained by query budget limit and a designed acquisition function $f(x, \mathcal{M})$: $\operatorname{argmax}_{x \in \mathcal{X}^u} f(x, \mathcal{M} | x \in \mathcal{X}^u)$ to obtain their labels from the oracle. $\mathcal{S}^l$ and $\mathcal{S}^u$ are then updated in next cycle, and $\mathcal{M}$ is retrained on $\mathcal{S}^l \cup \mathcal{S}^u \cup \mathcal{S}^g$.

3.2 Proposed Framework: B-ASAL

In this work, we propose a BERT-based adversarial semi-supervised active learning (B-ASAL) framework. We design each possible component to come up with a model learning objective and acquisition strategy. The components are: Generator (G), Classifier (C), Discriminator (D) and BERT Encoder (E) as shown in Fig. 1.

To utilize unlabeled data, we introduce a semi-supervised GAN framework built with BERT fine-tuning across the entire training process. In an adversarial manner, the generator is used to fool the classifier by generating highly realistic data samples. It takes noise input[2] and transforms to map true data distribution. The transformed noise input is treated as $k + 1th$ addition class for the semi-supervised learner. To enhance the robustness and reduce mode collapse, the generator is trained to apply feature matching between generated samples and

[2] Here we generate noise following a uniform distribution (which can be easily replaced by other distributions when needed). We denote noise as: $\{z_1, ..., z_n\} \sim U\{0,1\}$.

real data. Moreover, conditional entropy maximization over samples from the generator is employed as well.

The classifier is designed to pair with the generator and can be treated as a multi-class discriminator for K+1 classes. For labeled data, it is trained to differentiate k classes and $k+1th$ fake class. For unlabeled data, a minimax loss is optimized by performing entropy maximization with respect to the predicted class and entropy minimization with respect to fine-tuned feature encoder. It reduces the distribution gap while extracting discriminative features. We select samples having high entropy to be labeled, which indicates these samples are predicted by the model with high uncertainty.

The discriminator is a binary classifier, we use it to predict whether a sample is labeled or not based on a latent representation from our encoder. We select unlabeled data points with low discriminator scores, which indicates that these samples are sufficiently different from previously labeled ones.

BERT encoder is used as the feature encoder. It is fine-tuned, and the fine-tuning encoded features are through the logit activation layer of the classifier and the discriminator. For labeled data, It is trained to maximize the probability of class assignment from the classifier. It is also trained to differentiate label and unlabeled data from the discriminator. For unlabeled data, it is trained to minimize the entropy to have better discriminative features.

In each AL cycle, samples that have high uncertainty and diversity are selected from unlabeled data for labelling. Detailed steps of our method are shown in Algorithm 1.

3.3 Learning Objective

Now we discuss the overall cost function by incorporating each decomposed component, including generator loss (L_G), discriminator loss (L_D), and classifier loss (L_C). Each type of these losses has supervised loss for labeled data (L_L) and unsupervised loss for unlabeled data (L_U).

Labeled Data Learning BERT Encoder(E) and Classifier (C) are trained to classify labeled data points correctly into $\{1, ..., K\}$ class by both standard cross entropy loss and conditional entropy loss over samples uniformly distributed to K classes from the generator (G) to achieve optimal classification results. The generator (G) generates fake data points belonging to $K + 1th$ class. It tries to minimize the loss between generated fake data points with real data points, including the loss of feature matching and misclassification loss to K classes, while the classifier (C) tries to maximize it. This min-max loss is trained through an adversarial setting and can be denoted as:

$$\mathcal{L}_{\mathrm{L}} = -\underset{G}{min}\underset{C}{max}\mathcal{L}_{\mathrm{C}^l} + \mathcal{L}_{\mathrm{G}^l} \tag{2}$$

The loss function of Classifier (C) ($\mathcal{L}_{\mathrm{C}^l}$):

$$\mathcal{L}_{\mathrm{C}^l} = -\mathbb{E}_{(x,y)\in S^l} log[p(y \leq k|x)] - \mathbb{E}_{z\sim p_z} H_g[p(y \leq k|G(z), C)], \tag{3}$$

where conditional entropy $H_g = -\sum_1^m p(y = k|G(z), C)log[p(y = k|G(z^m), C)]$ and $k \in \{1, ..., K\}$ classes and $m \in M$.

The loss function related to G (L_G) includes feature matching loss to make generated data are very close to the real ones and also considers the error induced by fake data correctly identified by classifier.

$$\mathcal{L}_{\mathrm{G}^l} = ||\mathbb{E}_{x \in S^l} f(x) - \mathbb{E}_{x \in S^g} f(x)||_2^2 - \mathbb{E}_{x \in S^g} log[1 - p(y \leq k|x)], \tag{4}$$

where f is the layer with logits through the classifier and fine-tuning encoder.

Unlabeled Data Learning. When training on the unlabeled data, the unsuprevised loss is $\mathcal{L}_{\mathrm{U}} = \mathcal{L}_{\mathrm{H}}^u + \mathcal{L}_{\mathrm{G}^u}$, where $\mathcal{L}_{\mathrm{H}}^u$ denotes minimax entropy employed on classifier and feature encoder; L_G^u denotes feature matching loss for generated samples paired with unlabeled data, same as first term in L_G^l. They are computed as:

$$\mathcal{L}_{\mathrm{H}}^u = -\underset{E}{min}\underset{C}{max} H_s[p(y \leq k|x)], \tag{5}$$

where the minimax entropy $H_s = -\sum_1^K p(y = k|x)log(p(y = k|x)$ and $k \in \{1, ..., K\}$ classes; we first minimize the entropy in feature encoder to have more discriminative representation and then maximize entropy in classifier to have a more uniform feature representation.

$$\mathcal{L}_{\mathrm{G}}^u = ||\mathbb{E}_{x \in S^u} f(x) - \mathbb{E}_{x \in S^g} f(x)||_2^2, \tag{6}$$

where this part can be combined with first term of Eq. 4 as learning feature matching loss for all of generated samples coming from generator.

Discrimitive Learning for Labeled and Unlabeled Data. The diversity of the data is predicted by a binary classifier (i.e. discriminator denoted as D) that is trained to distinguish between the labeled and unlabeled encoded features. The loss function of D is:

$$\mathcal{L}_{\mathrm{D}} = -\mathbb{E}_{(x,y) \in S^l} log[p(y^l|x^l)] - \mathbb{E}_{(x,y) \in S^u} log[p(y^u|x^u)]. \tag{7}$$

Acquisition Strategy. In our acquisition strategy, we select data points with high diversity and high uncertainty to be labeled . The selection criteria are: (a) *high diversity*: we use the probability associated with the discriminator's (D) predictions as a score to rank samples. The lower the probability, the more confident D is that it comes from the unlabeled pool. (b) *high uncertainty*: the entropy obtained by the classifier on the unlabeled data is used to choose the data points. A higher entropy value is associated with a lower confidence score. The top-k% samples that meet both criteria are selected for labeling.

Algorithm 1. BERT-based Adversarial Semi-Supervised Active Learning (B-ASAL)

Input: labeled data S^l, unlabeled data S^u. Initialize parameters of generator ϕ_G, discriminator γ_D , classifier σ_C and BERT encoder β_E.
Output: Optimized ϕ_G, γ_D , σ_C and β_E
1: **for** i = 1 to epochs **do**
2: Sample batch of size n from S^l labeled and S^u unlabeled data $|S^l| = |S^u| = n$
3: Sample $\{z_1, \ldots, z_{m+n}\} \in S^g$ from the prior P_z
4: Generate encode $E(S^l)$, $E(S^u)$ and $E(S^g)$
5: For labeled data $(x, y \in S^l)$:
6: Compute $\mathcal{L}_C^l$ from Eq. 3
7: Update C by descending:
8: $\sigma_C \leftarrow \sigma_C - \lambda_1 \nabla \mathcal{L}_C^l$
9: For unlabeled data $(x \in S^u)$:
10: Compute $\mathcal{L}_H^u$ from Eq. 5
11: Update E and C by descending/ascending:
12: $\beta_E \leftarrow \beta_E + \lambda_2 \nabla \mathcal{L}_H^u$
13: $\sigma_C \leftarrow \sigma_C - \lambda_3 \nabla \mathcal{L}_H^u$
14: For labeled data $(x, y \in S^l)$ and unlabeled data $(x \in S^u \cup S^g)$:
15: Compute $\mathcal{L}_G$ from Eq. 4 and Eq. 6
16: Update G by descending:
17: $\phi_G \leftarrow \phi_G - \lambda_4 \nabla \mathcal{L}_G$
18: Compute $\mathcal{L}_D$ from Eq. 7
19: Update D by descending:
20: $\gamma_D \leftarrow \gamma_D - \lambda_5 \nabla \mathcal{L}_D$
21: **end for**

4 Experiments

To study the effectiveness of our approach, we evaluate model performance on multiple open public data sets by comparing them with the different sampling strategy.

Datasets: Total of five data sets are used for evaluation: Fine and Coarse Question Classification (QC) [9], Match and Mismatched pair MNLI dataset [16] and Multi-label emotion data [10].

Experiments Settings: We fetch all of the above data from HuggingFace datasets library.We run 3 different seeds and 3 epochs for each experiment. We take the mean of the results. [17]

Performance Evaluation: The model performance is measured by the classification accuracy on balanced datasets and measuring micro-F1, macro-F1 and hamming score on imbalanced datasets/multi-label datasets by varying percentage of labeled data, ranging from $\{1\%, 2\%, ..., 10\%\}$ or $\{0.1\%, 2\%, ..., 5\%\}$ depending on data size.

Acquisition Strategies: To compare with our acquisition function (i.e. B-ASAL), we use three baselines: random sampling (Rdm), diversity sampling (Div) and entropy uncertainty sampling (En).

Table 1. Comparisons of Sampling Methods on Accuracy. The percentages shown in the table (and the same for all tables hereafter) refer to the percentages of training data labeled by active learning.

Method	QC-Coarse						QC-Fine					
	1%	2%	5%	10%	20%	30%	1%	2%	5%	10%	20%	30%
Rdm	21.2	36.3	58.3	84.4	92.8	94.4	8.1	11.6	33.9	56.3	72.0	77.8
En	22.1	36.5	81.1	89.1	93.6	94.5	7.2	13.5	38.7	54.9	67.9	75.9
Div	18.7	35.0	58.7	86.5	92.8	94.6	10.5	11.0	45.0	60.8	71.5	75.2
B-ASAL	**26.2**	**42.6**	**90.4**	**94.5**	**95.5**	**96.3**	**17.3**	**19.2**	**57.4**	**62.4**	**76.9**	**80.8**

Table 2. Comparisons of Sampling Methods on F1

Method	MNLI-mismatch							MNLI-match						
	0.1%	0.2%	0.5%	1%	2%	5%	10%	0.1%	0.2%	0.5%	1%	2%	5%	10%
Rdm	22.3	40.0	46.4	78.3	76.7	85.2	88.3	21.2	21.4	42.6	58.2	80.0	84.6	85.0
cre En	25.0	31.0	40.7	73.7	79.7	86.3	88.1	24.0	29.3	54.0	69.0	82.0	87.3	91.3
Div	25.0	27.0	42.3	73.7	81.3	86.7	87.0	22.7	37.3	43.0	65.7	71.0	88.0	87.7
B-ASAL	**29.0**	**42.3**	**55.7**	**77.0**	**86.7**	**89.7**	**91.7**	**29.3**	**39.7**	**74.3**	**77.7**	**83.3**	**91.0**	94.2

Implementation: For Classifier (C), Discriminator (D), Generator (G), we use the Multi-Layer Perceptron (MLP) neural network with one hidden layer activated by a leaky-relu function followed by a softmax layer for the multi-class prediction and sigmoid layer for multi-label prediction. The dropout is 0.1 after the hidden layer. The input noise vector of G is uniformly distributed. BERT Encoder (E) is loaded from the pre-trained BERT model and fine-tuned through C, D and G.

4.1 Question Answering Classification

Question Classification (QC) dataset [9] has both a six-class (QC-Coarse) and a fifty-class (QC-Fine) version. Both have 5,452 training data and 500 test data. Table 1 shows the experiment output. The accuracy performance of QC-Coarse data can achieve 90%+ when using only 5% labeled data and QC-Fine set achieves around 80% by using 20% labeled data. Our sampling strategy (i.e. B-ASAL) achieves much better performance.

4.2 Multi-Genre Natural Language Inference

The Multi-Genre Natural Language Inference (MultiNLI) corpus is a collection of 433k sentence pairs annotated with textual entailment information [16]. The task is to infer the relationship between the premise and hypothesis in binary classification. We evaluated matched and mismatched data sets. The results are shown in Table 2. The F1-score of mismatch data can achieve 90%+ when using only 10% labeled data, and match set achieves around 80% by using only 2% labeled data. Our sampling strategy (i.e. B-ASAL) consistently shows much better performance over the rest three baselines.

4.3 Multi-Label Emotion Classification

SemEval-2010 Task is for multi-label emotion classification (11 emotions) [10]. Hamming, F1-micro and F1-macro scores are used to evaluate model

Table 3. Comparisons of Sampling Methods on Multi-label Emotion Dataset

Method	Micro-F1					Macro-F1					Hamming				
	1%	2%	5%	10%	20%	1%	2%	5%	10%	20%	1%	2%	5%	10%	20%
Rdm	19.6	25.0	50.7	57.7	60.6	7.51	11.4	27.1	32.4	38.2	11.5	14.8	34.0	40.5	43.5
En	20.6	23.7	41.1	54.8	56.3	7.6	7.96	18.4	26.8	28.2	12.2	13.4	26.5	37.7	38.9
Div	19.6	23.8	41.0	54.3	55.9	7.9	7.98	25.9	26.5	27.2	11.5	13.6	26.4	37.3	38.1
B-ASAL	**23.9**	**33.9**	**54.6**	**58.2**	**60.9**	**8.51**	**13.1**	**27.3**	**32.7**	**38.5**	**13.7**	**20.6**	**37.6**	**40.8**	**43.8**

Table 4. Ablation Studies on Accuracy

Method	QC-Coarse						QC-Fine					
	1%	2%	5%	10%	20%	30%	1%	2%	5%	10%	20%	30%
L-only	20.4	36.0	79.5	91.1	94.2	95.1	8.4	13.4	39.1	59.8	72.9	76.7
No-GAN	25.1	37.3	72.8	79.6	93.2	95.0	13.2	14.1	24.6	43.6	61.7	67.8
BERT	20.1	21.4	50.0	81.6	93.4	95.1	0.8	3.5	17.0	30.5	54.7	64.9
B-ASAL	**26.2**	**42.6**	**90.4**	**94.5**	**95.5**	**96.3**	**17.3**	**19.2**	**57.4**	**62.4**	**76.9**	**80.8**

performance defined by the Task. Our method's performance outperforms the other sampling strategies and with only 20% labeled samples, our method can almost achieve the benchmark performance (as shown in Table 3).

4.4 Further Analysis

To investigate the contribution of each component and understand the benefit of utilizing the unlabeled data points, we designed several types of experiments to show the overall effectiveness of B-ASAL.

Labeled-only vs. (Labeled $\cup$Unlabeled). We study the effectiveness of semi-supervised learning compared to supervised learning by having GAN component. The comparison outputs are shown on the row of L-only vs. the row of B-ASAL in Tables 4 and 5, where L-only denotes only labeled data is used for training with GAN and B-ASAL is our model utilizing semi-supervised learning with GAN. The results show that semi-supervised B-ASAL performs better than supervised GAN where there are only annotated data in training.

With GAN vs. Without GAN. We study the performance of the model with Generator (G) compared to the model without G. The model without G is when B-ASAL only has components E, D and C. The output shows on the row of No-GAN vs. the row of B-ASAL in Table 4 and 5. The results show GAN generates better performance by utilizing unlabeled data and pairing with the classifier.

B-ASAL vs. BERT-only. We compare our model performance with the fully supervised BERT classifier. Results are shown on the row of BERT vs. the row of B-ASAL in Table 4 and 5. Apparently, B-ASAL outperforms supervised classification without utilizing GAN and unlabeled data.

Table 5. Ablation Studies on F1 Score

Method	MNLI-mismatch							MNLI-match						
	0.1%	0.2%	0.5%	1%	2%	5%	10%	0.1%	0.2%	0.5%	1%	2%	5%	10%
L-only	23.3	26.6	44.6	47.6	77.3	85.3	87.6	23.2	26.1	44.3	61.6	78.3	86.0	91.2
No-GAN	22.7	29.3	43.2	65.2	71.5	88.3	89.2	23.4	34.3	39.5	60.4	74.7	88.1	88.5
BERT	22.2	26.4	48.3	58.3	65.5	72.7	73.9	26.3	32.1	48.3	58.2	65.4	70.4	71.9
B-ASAL	**29.0**	**42.3**	**55.7**	**77.0**	**86.7**	**89.7**	**91.7**	**29.3**	**39.7**	**74.3**	**77.7**	**83.3**	**91.0**	**94.2**

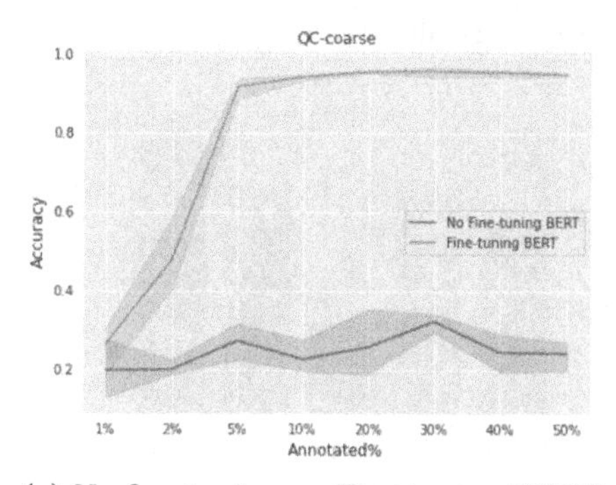

(a) No fine-tuning vs. Fine-tuning BERT

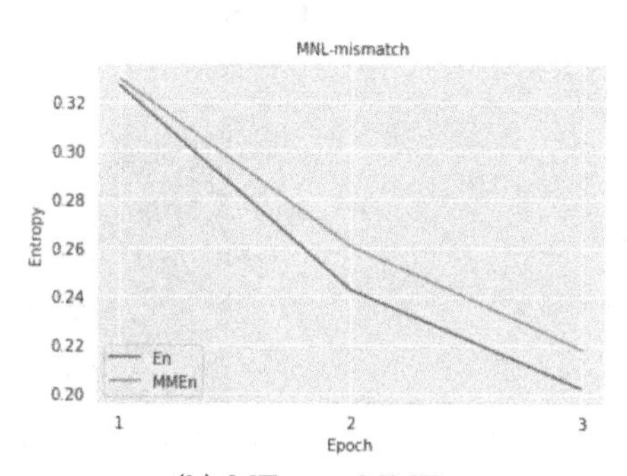

(b) MEn vs. MMEn

Fig. 2. Two of our Ablation Studies

Table 6. Comparisons When Labels are Partially Available at Training

Method	20 out of 50 classes					40 out of 50 classes				
	2%	5%	10%	20%	30%	2%	5%	10%	20%	30%
Random	13.0	29.2	49.4	64.8	75.2	22.7	39.2	57.8	64.8	67.1
En	15.4	27.9	41.5	64.3	64.8	31.1	41.6	59.0	64.4	70.3
Div	15.5	35.0	46.3	65.7	65.9	31.2	48.0	59.6	64.7	71.5
B-ASAL	**16.5**	**38.1**	**50.5**	**66.9**	**75.6**	**36.1**	**40.1**	**60.0**	**66.6**	**75.9**

BERT Encoder Fine-tuning vs. No Fine-tuning. To demonstrate fine-tuning BERT plays an important role throughout the entire B-ASAL training, we study the performance of fine-tuning vs. no fine-tuning (Fig. 2a). It shows the encoder plays an important role not only as a representation encoder but also as a collaborator with component of D and C to achieve the optimal results.

Entropy Optimization. For unlabeled data, we perform entropy minimax optimization. Figure 2(b) plots out the study when minimax optimization (MMEn) is used for our approach vs. when only entropy minimization (MEn) is used for extracting discriminative features. It shows entropy value decreases with the increase of epochs, and the entropy of MMEn is higher than that of MEn, which demonstrates a more effective optimization of the objective function.

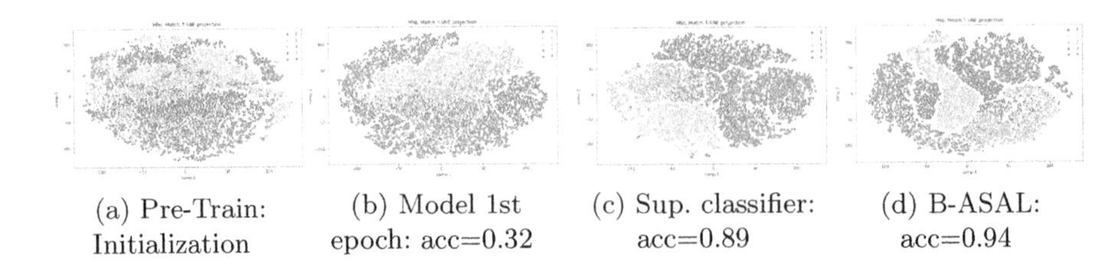

(a) Pre-Train: Initialization (b) Model 1st epoch: acc=0.32 (c) Sup. classifier: acc=0.89 (d) B-ASAL: acc=0.94

Fig. 3. MNL-Match Feature Visualizaton

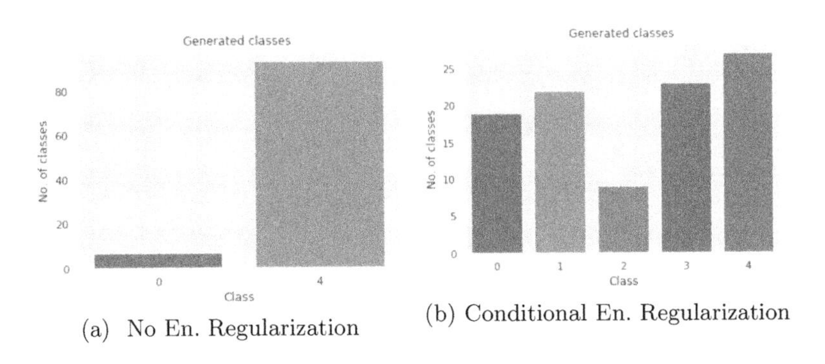

(a) No En. Regularization (b) Conditional En. Regularization

Fig. 4. MNL-mismatch: samples generation from G

Robustness. In our method, Classifier (C) chooses data with high uncertainty for labelling, while Discriminator (D) differentiates labeled data from unlabeled ones. To evaluate the effectiveness, we studied QC-Fine dataset and randomly use 20 classes and 40 classes (out of 50) in the initial training set as a labeled pool. The results are shown in Table 6, which demonstrates our model is less affected when initially labeled data don't well represent the entire data distribution.

Discriminative Feature Visualization. We demonstrate the discriminative features learned from the model. Figure 3 shows the results using 10% labeled MNL-match data for training: (a) feature encoded from Pre-trained Bert before tuning, (b) feature learned at first epoch, (c) feature learned by BERT classifier, and (d) feature learned by B-ASAL. It can be seen that B-ASAL generates more discriminative features.

Sample Generation. We study the Generator capability and distribution coverage of generated samples. We take MNL-mismatch data (5 classes total) as an example and use 5% annotated data to train our model. The sample generation is compared by having conditional entropy regularization on the generator with not having an entropy regularizer. Figure 4(a) shows the histogram of generated classes without conditional entropy regularize, and Fig. 4(b) shows the histogram of generated classes by imposing conditional entropy regularizer. These outputs illustrate that the conditional entropy regularizer helped the generator to generate effective samples to represent true data distribution.

5 Conclusions and Future Work

In this paper, we proposed a BERT encoder-based semi-supervised active learning algorithm, B-ASAL, which guides the fine-tuning of BERT to better fit the training data, creates synthetic data to address data insufficiency problems, and incorporates minimax entropy to differentiate the distribution of labeled data from that of unlabeled data. We introduced a hybrid sampling strategy that selects samples that are most diverse and have high uncertainty from class assignments learned by the multi-class classifier. Our experiments demonstrated significant improvements over the existing state-of-the-art methods. In future, we plan to extend this research to more NLP applications such as question-answering and recommendation systems.

References

1. Chivukula, A.S., Liu, W.: Adversarial deep learning models with multiple adversaries. IEEE Trans. Knowl. Data Eng. **31**(6), 1066–1079 (2018)
2. Croce, D., Castellucci, G., Basili, R.: GAN-BERT: generative adversarial learning for robust text classification with a bunch of labeled examples. In: Proceedings of the 58th Annual Mmeeting of the Association for Computational Linguistics, pp. 2114–2119 (2020)
3. Dai, Z., Yang, Z., Yang, F., Cohen, W.W., Salakhutdinov, R.R.: Good semi-supervised learning that requires a bad GAN. In: Advances in Neural Information Processing Systems, vol. 30 (2017)
4. Devlin, J., Chang, M.W., Lee, K., Toutanova, K.: BERT: pre-training of deep bidirectional transformers for language understanding. arXiv:1810.04805 (2018)
5. Dor, L.E., et al.: Active learning for BERT: an empirical study. In: EMNLP, pp. 7949–7962 (2020)
6. Ebrahimi, S., et al.: Minimax active learning. arXiv preprint arXiv:2012.10467 (2020)
7. Goodfellow, I., et al.: Generative adversarial networks. Commun. ACM **63**(11), 139–144 (2020)
8. Jacobs, P.F., Maillette de Buy Wenniger, G., Wiering, M., Schomaker, L.: Active learning for reducing labeling effort in text classification tasks. In: Benelux Conference on Artificial Intelligence, pp. 3–29. Springer, Cham (2021). https://doi.org/10.1007/978-3-030-93842-0_1
9. Li, X., Roth, D.: Learning question classifiers: the role of semantic information. Nat. Lang. Eng. **12**(3), 229–249 (2006)
10. Mohammad, S.M., Bravo-Marquez, F., Salameh, M., Kiritchenko, S.: Semeval-2018 Task 1: Affect in tweets. In: SemEval-2018. New Orleans, LA, USA (2018)
11. Saito, K., Kim, D., Sclaroff, S., Darrell, T., Saenko, K.: Semi-supervised domain adaptation via minimax entropy. In: Proceedings of the IEEE/CVF International Conference on Computer Vision, pp. 8050–8058 (2019)

12. Salimans, T., Goodfellow, I., Zaremba, W., Cheung, V., Radford, A., Chen, X.: Improved techniques for training GANs. In: Advances in Neural Information Processing Systems, vol. 29 (2016)
13. Sener, O., Savarese, S.: Active learning for convolutional neural networks: a core-set approach. arXiv preprint arXiv:1708.00489 (2017)
14. Sinha, S., Ebrahimi, S., Darrell, T.: Variational adversarial active learning. In: Proceedings of the IEEE/CVF International Conference on Computer Vision, pp. 5972–5981 (2019)
15. Springenberg, J.T.: Unsupervised and semi-supervised learning with categorical generative adversarial networks. arXiv preprint arXiv:1511.06390 (2015)
16. Williams, A., Nangia, N., Bowman, S.R.: A broad-coverage challenge corpus for sentence understanding through inference. arXiv preprint arXiv:1704.05426 (2017)
17. Wolf, T., Debut, L., et al.: Huggingface's transformers: state-of-the-art natural language processing. arXiv:1910.03771 (2019)
18. Yang, P., Liu, W., Yang, J.: Positive unlabeled learning via wrapper-based adaptive sampling. In: Proceedings of the 26th International Joint Conference on Artificial Intelligence, pp. 3273–3279 (2017)
19. Yuan, M., Lin, H.T., Boyd-Graber, J.: Cold-start active learning through self-supervised language modeling. arXiv:2010.09535 (2020)

Theoretical Foundations

Accelerating Stochastic Newton Method via Chebyshev Polynomial Approximation

Fan Sha[1] and Jianyu Pan[2(✉)]

[1] School of Mathematical Sciences, East China Normal University, Shanghai 200241, China
[2] School of Mathematical Sciences, Shanghai Key Laboratory of PMMP, East China Normal University, Shanghai 200241, China
jypan@math.ecnu.edu.cn

Abstract. To solve large scale optimization problems arising from machine learning, stochastic Newton methods have been proposed for reducing the cost of computing Hessian and Hessian inverse, while still maintaining fast convergence. Recently, a second-order method named LiSSA [12] was proposed to approximate the Hessian inverse with Taylor expansion and achieves (almost) linear running time in optimization. The approach is very simple yet effective, but still could be further accelerated. In this paper, we resort to Chebyshev polynomial and its variants to approximate the Hessian inverse. Note that Chebyshev polynomial approximation is broadly acknowledged as the optimal polynomial approximation in the deterministic setting, in this paper we introduce it into the stochastic setting of Newton optimization. We provide a complete convergence analysis and the experiments on multiple benchmarks show that our proposed algorithms outperform LiSSA, which validates our theoretical insights.

Keywords: Stochastic Newton Method · Chebyshev Polynomial · Hessian inverse · Condition Number

1 Introduction

In machine learning, we often model many large scale optimal problems as minimizing an average of m convex functions $f_k : \mathbb{R}^d \to \mathbb{R}$,

$$\min_{x \in \mathbb{R}^d} f(x) \triangleq \frac{1}{m} \sum_{k=1}^{m} f_k(x) \tag{1}$$

Supported by NSFC grant 11771148 and Science and Technology Commission of Shanghai Municipality grant 20511100200, 21JC1402500 and 22DZ2229014.

Supplementary Information The online version contains supplementary material available at https://doi.org/10.1007/978-3-031-33377-4_40.

H. Kashima et al. (Eds.): PAKDD 2023, LNAI 13936, pp. 523–534, 2023.
https://doi.org/10.1007/978-3-031-33377-4_40

where m is the number of samples, d is the dimension of parameter x, and $f_k(x)$ is the loss function with respect to the k-th sample. Such optimization problems are common in machine learning algorithms, such as logistic regressions, smoothed support vector machines, neural networks.

To solve problem (1), many stochastic optimization algorithms have been designed, such as stochastic first-order method and stochastic second-order method, which have the following updating formula:

$$x_{t+1} = x_t - \eta_t H^{-1}(x_t) g(x_t), \quad t = 0, 1, 2, \ldots, \tag{2}$$

where $g(x_t)$ is the gradient of the objective function $f(x)$ at the point x_t, η_t is the step size at the t-th iteration, and $H^{-1}(x_t)$ is set differently in different methods. In the stochastic first-order method, we set $H^{-1}(x_t)$ as an identity matrix. One of the most popular stochastic first-order method is the Stochastic Gradient Descent (SGD) method [15], which has been widely employed to reduce the computational cost per iteration. However, SGD has poor convergence property. In order to accelerate its convergence, many variants have been proposed, such as SVRG [7], SAGA [2], SDCA [16].

Recently, as a typical stochastic second-order method, Stochastic Newton Method (SNM) has received great attention due to its fast convergence rate. For second-order methods, the matrix $H^{-1}(x_t)$ is usually the Hessian inverse or certain constructed Hessian inverse. However, constructing such Hessian matrix and its inverse require a lot of calculations. To conquer this weakness, some sub-sampling techniques which only randomly select a subset of samples to construct a sub-sampled Hessian are proposed to alleviate the computational cost. In [14], the authors proposed to use the Sample Average Approximation (SAA) approach to estimate Hessian-vector multiplications. In [3], a sub-sampled Newton method was proposed where the Hessian approximation is obtained by sub-sampling the true Hessian and then computing a truncated eigenvalue decomposition. It is suggested in [13] to sketch the Hessian using random sub-Gaussian matrices or randomized orthonormal systems. This method requires access to the square root of the true Hessian.

Rather than estimating the Hessian matrix, some stochastic second-order methods are proposed to approximate the Hessian inverse, such as S-BFGS [11] and SB-BFGS [6], which adopt the randomization to the classical L-BFGS algorithm. It is worth mentioning that a second-order method named LiSSA was proposed in [12] to approximate the Hessian inverse in linear running time by polynomial approximation. LiSSA is simple yet effective, which directly approximates the Hessian inverse by combining sub-sampling with Taylor expansion. Moreover, the obtained Hessian inverse is unbiased.

However, we note that the polynomial approximation approach in LiSSA can be further improved. In deterministic case, Chebyshev polynomial has been well known for its superior performance on accelerating the convergence of stationary iterative methods [5]. We expect that introducing Chebyshev polynomial approximation into stochastic Newton method has big potential to accelerate the optimization process. Nevertheless, it is non-trivial to apply Chebyshev polynomial approximation in second-order optimization because it is rarely used in stochastic settings and it is not easy to analyze it theoretically because of the complexity of its iterative formula.

In this paper, we propose to utilize Chebyshev polynomial approximation to accelerate the stochastic Newton method and obtain the **C**hebyshev **P**olynomial accelerated **S**tochastic **N**ewton **M**ethod (CP-SNM). Besides, we also give its two variants which are easier for implementation. Additionally, we give a complete theoretical analysis for the convergence of CP-SNM. We conduct experiments on multiple real datasets and the results show that our new algorithms improve upon the overall running time of LiSSA, which complements our theoretical results.

2 Background

In this section, we describe the background of our proposed methods. We first introduce the polynomial approximation approach proposed in LiSSA, and then give the recursive formula of Chebyshev polynomial approximation which is the essential idea of constructing the Hessian inverse in our proposed algorithms.

2.1 Estimator for the Hessian Inverse in LiSSA

The key idea underlying LiSSA is the following fact: If $A \in \mathbb{R}^{n\times n}$ is non-singular symmetric and $\|A\| < 1$ (in this paper, $\|\cdot\|$ denotes the 2-norm), then the identical equation $A^{-1} = (I-A)A^{-1} + I$ leads to a fixed point iteration scheme

$$X_{j+1} = (I-A)X_j + I, \quad j = 0, 1, 2, \cdots . \tag{3}$$

Based on the above recursive formulation, LiSSA construct an unbiased estimator of Hessian inverse $\nabla^{-2}f$.

Definition 1. *Given j independent and unbiased samples $\{Y_1, \cdots, Y_j\}$ of $\nabla^2 f$, define the estimators $\{\tilde{\nabla}^{-2}f_0, \cdots, \tilde{\nabla}^{-2}f_j\}$ recursively as follows:*

$$\tilde{\nabla}^{-2}f_0 = I \quad \textit{and} \quad \tilde{\nabla}^{-2}f_t = I + (I - Y_t)\tilde{\nabla}^{-2}f_{t-1} \quad \textit{for} \quad t = 1, \cdots, j.$$

Note that $\mathrm{E}[\tilde{\nabla}^{-2}f_j] = \nabla^{-2}f_j = \sum_{i=0}^{j}(I-\nabla^2 f)^i$, we have $\mathrm{E}[\tilde{\nabla}^{-2}f_j] \to \nabla^{-2}f$ as $j \to \infty$, which gives an estimator that is unbiased in the limit.

2.2 Chebyshev Polynomial Approximation

To further improve the approximation of the estimator to A^{-1}, we propose to use the Chebyshev polynomial approach. Let $\sigma(I-A)$ be the collection of all the eigenvalues of $I-A$, and let $[a, b]$ be a interval contains $\sigma(I-A)$, that is, $\sigma(I-A) \subseteq [a, b]$. For a given $\hat{X}_0$, our new estimators $\{\hat{X}_k, k = 1, 2, \ldots\}$ are defined recursively by

$$\hat{X}_{k+1} = \sum_{j=0}^{k+1} a_{k+1,j}\hat{X}_j \quad \text{with} \quad \sum_{j=0}^{k+1} a_{k+1,j} = 1, \quad k = 1, 2, \ldots .$$

Then it holds that

$$\|\hat{X}_{k+1} - A^{-1}\| \leq \max_{\lambda \in \sigma(I-A)} \left| \sum_{j=0}^{k+1} a_{k+1,j} \lambda^j \right| \|X_0 - A^{-1}\| \leq \max_{a \leq t \leq b} |p_{k+1}(t)| \|X_0 - A^{-1}\|, \quad (4)$$

where $p_{k+1}(t) = \sum_{j=0}^{k+1} a_{k+1,j} t^j$ is a $(k+1)$-th degree polynomial satisfying $p_{k+1}(1) = 1$. Our purpose is to find a $(k+1)$-th degree polynomial $q_{k+1}(t)$ satisfying $q_{k+1}(1) = 1$ such that

$$\max_{a \leq t \leq b} |q_{k+1}(t)| = \min_{p(1)=1,\, \deg(p) \leq k+1} \max_{a \leq t \leq b} |p(t)|, \quad (5)$$

where $\deg(p)$ denotes the degree of the polynomial p. It is well known that the unique solution of (5) is given by [4]

$$q_{k+1}(t) = T_{k+1}\left(\frac{2t-b-a}{b-a}\right) \Big/ T_{k+1}\left(\frac{2-b-a}{b-a}\right), \quad (6)$$

where $T_{k+1}(t)$ is the Chebyshev polynomial defined recursively by

$$T_0(t) \equiv 1, \quad T_1(t) = t, \quad T_{k+1}(t) = 2tT_k(t) - T_{k-1}(t), \quad k = 1, 2, \cdots. \quad (7)$$

Taking $p_{k+1}(t) = q_{k+1}(t)$, then we obtain the Chebyshev estimators $\{\hat{X}_k, k = 1, 2, \ldots\}$. By utilizing the three-term recurrence (7), we can compute $\hat{X}_k$ recursively as stated in the following proposition.

Proposition 1. *Let A be a symmetric positive definite matrix, and let α, β be the minimum and maximum eigenvalues of A, respectively. For a given $\hat{X}_0$, we can compute $\hat{X}_k$ recursively by*

$$\begin{cases} \hat{X}_1 = (I - \nu A)\hat{X}_0 + \nu I, \\ \hat{X}_{k+1} = \rho_k \left((I - \nu A)\hat{X}_k + \nu I\right) + (1 - \rho_k)\hat{X}_{k-1}, \quad k = 1, 2, \ldots, \end{cases} \quad (8)$$

where $\nu = \frac{2}{\alpha+\beta}$, $\rho_1 = 2$ and $\rho_k = \left(\left(1 - \frac{1}{4\xi^2}\right)\rho_{k-1}\right)^{-1}$ with $\xi = \frac{\beta+\alpha}{\beta-\alpha}$.

Proof. See proof in Appendix B.2. □

If we set A to be the Hessian matrix, then $\hat{X}_k$ can be used to approximate the Hessian inverse. This is the basic idea of our new algorithm.

3 Our Proposed CP-SNM

Consider the objective function $f(x)$ in problem (1), we can apply the recursion (8) to approximate its Hessian inverse. However, computing its Hessian $\nabla^2 f(x)$ at each Newton iteration step is very expensive for large m. Therefore, at the

k-th Newton iteration step, we sample uniformly $i_{k+1} \in [1, 2, \ldots, m]$ and replace $\nabla^2 f(x)$ with $\nabla^2 f_{i_{k+1}}(x)$, which is an unbiased sample of $\nabla^2 f(x)$.

Denote the minimum and maximum eigenvalues of $\nabla^2 f(x)$ by $\alpha(x)$ and $\beta(x)$, respectively. Then we obtain the stochastic Chebyshev approximation of the Hessian inverse as follows

$$\begin{cases} \tilde{X}_1 = (I - \nu\nabla^2 f_{i_1}(x))\tilde{X}_0 + \nu I, \\ \tilde{X}_{k+1} = \rho_k \left(\left(I - \nu\nabla^2 f_{i_{k+1}}(x)\right) \tilde{X}_k + \nu I \right) + (1-\rho_k)\tilde{X}_{k-1}, \ k = 1, 2, \ldots, \end{cases} \tag{9}$$

where $\nu = \frac{2}{\alpha(x)+\beta(x)}$, $\rho_1 = 2$ and $\rho_k = \left((1 - \frac{1}{4\xi^2})\rho_{k-1}\right)^{-1}$ with $\xi = \frac{\beta(x)+\alpha(x)}{\beta(x)-\alpha(x)}$. We call $\tilde{X}_k$ the k-th stochastic Chebyshev estimator of $\nabla^{-2} f(x)$, which is an unbiased estimator, i.e., $\lim_{k\to\infty} \mathrm{E}[\tilde{X}_k] = \nabla^{-2} f(x)$.

In the following, we introduce two variants of CP-SNM. Let

$$\beta_{\max}(x) \triangleq \max_k \lambda_{\max}\left(\nabla^2 f_k(x)\right) \quad \text{and} \quad \alpha_{\min}(x) \triangleq \min_k \lambda_{\min}\left(\nabla^2 f_k(x)\right),$$

where $\lambda_{\max}(\cdot)$ and $\lambda_{\min}(\cdot)$ denote the maximum and minimum eigenvalues of a given matrix, and define

$$\tilde{\kappa} = \max_x \frac{\beta_{\max}(x)}{\alpha_{\min}(x)}.$$

If we set $\rho_k \equiv \rho = \frac{2}{1+\sqrt{1-\left(\frac{\tilde{\kappa}-1}{\tilde{\kappa}+1}\right)^2}}$, then (9) becomes

$$\tilde{X}_{k+1} = \rho \left(\left(I - \nu\nabla^2 f_{i_{k+1}}(x)\right) \tilde{X}_k + \nu I \right) + (1-\rho)\tilde{X}_{k-1},$$

which is the two-step Richardson iteration, and CP-SNM reduces to the **T**wo-step **R**ichardson accelerated **S**tochastic **N**ewton **M**ethod (TR-SNM).

If we set $\rho_k \equiv 1$, then (9) becomes

$$\tilde{X}_{k+1} = \left(I - \nu\nabla^2 f_{i_{k+1}}(x)\right) \tilde{X}_k + \nu I,$$

which is the one-step Richardson iteration, and CP-SNM reduces to the **O**ne-step **R**ichardson accelerated **S**tochastic **N**ewton **M**ethod (TR-SNM).

The algorithms of CP-SNM, TR-SNM and OR-SNM are described in Algorithm 1. Following LiSSA, we first use some efficient stochastic first-order method (SGD in our algorithms) to produce a guess x_1 so that the function value is shrank to the regime where we can show the linear convergence. Then we carry out the stochastic Newton method where the Hessian inverse is replaced by the stochastic Chebyshev estimator. We control the approximation of the Hessian inverse by two integers: S_1 and S_2, where S_1 controls the number of unbiased estimators (we take the average to get a better estimator) and S_2 controls the number of iteration steps of (9).

Algorithm 1. CP-SNM, TR-SNM and OR-SNM

Input: T (total iteration number of the stochastic Newton method),
T_1 (total iteration number for SGD method to produce x_1),
$f(x) = \frac{1}{m}\sum_{k=1}^{m} f_k(x)$, S_1, S_2, ν, ξ, $\tilde{\kappa}$

Output: x_{T+1}

1: Compute x_1 with SGD method
2: **for** $t = 1$ to T **do**
3: **for** $i = 1$ to S_1 **do**
4: $X[i, 0] = \nabla f(x_t)$
5: $\rho_1 = \begin{cases} 2, & \text{for CP-SNM} \\ \frac{2}{1+\sqrt{1-(\frac{\tilde{\kappa}-1}{\tilde{\kappa}+1})^2}}, & \text{for TR-SNM} \\ 1, & \text{for OR-SNM} \end{cases}$
6: Sample $\tilde{\nabla}^2 f_{[i,1]}(x_t)$ uniformly from $\{\nabla^2 f_k(x_t) \mid 1 \leq k \leq m\}$
7: $X[i, 1] = \nu\nabla f(x_t) + [I - \nu\tilde{\nabla}^2 f_{[i,1]}(x_t)]X[i, 0]$
8: **for** $j = 2$ to S_2 **do**
9: Sample $\tilde{\nabla}^2 f_{[i,j]}(x_t)$ uniformly from $\{\nabla^2 f_k(x_t) \mid 1 \leq k \leq m\}$
10: $\rho_j = \begin{cases} \left(1 - \frac{1}{4\xi^2}\rho_{j-1}\right)^{-1}, & \text{for CP-SNM} \\ \frac{2}{1+\sqrt{1-(\frac{\tilde{\kappa}-1}{\tilde{\kappa}+1})^2}}, & \text{for TR-SNM} \\ 1, & \text{for OR-SNM} \end{cases}$
11: $X[i, j] = \rho_j \left(\nu\nabla f(x_t) + \left(I - \nu\tilde{\nabla}^2 f_{[i,j]}(x_t)\right) X[i, j-1]\right) + (1-\rho_k)X[i, j-2]$
12: **end for**
13: **end for**
14: $X_t = \frac{1}{S_1}\sum_{i=1}^{S_1} X[i, S_2]$, $x_{t+1} = x_t - X_t$
15: **end for**
16: **return** x_{T+1}

4 Theoretical Results

We first introduce some concepts which will be used in the theoretical analysis, and then we present some important lemmas. The most challenging is the derivation of the error bound between our approximated Hessian inverse and its expectation (see Lemma 2), since stochastic form of matrix polynomial lost the overall nature of the deterministic form. Finally we give our main theorem which shows that our algorithm can achieve linear quadratic convergence, and if the condition number is controlled in a certain range, the convergence of our algorithm can be confirmed to outperform LiSSA.

We make the following assumptions for the objective function $f(x)$.

Assumption 1 *(Gradient Lipschitz Continuity and Strong Convexity). The finite sum function $f(x)$ is α-strongly convex and β-smooth, i.e., for all x, y,*

$$\nabla f(x)^\top(y - x) + \frac{\beta}{2}\|y - x\|^2 \geq f(y) - f(x) \geq \nabla f(x)^\top(y - x) + \frac{\alpha}{2}\|y - x\|^2.$$

Assumption 2 *(Lipschitz continuity). For any $i \in \{1, 2, \ldots, m\}$, there exists a constant M, such that*

$$\left\|\nabla^2 f_i(x) - \nabla^2 f_i(y)\right\| \leq M\|x - y\| \quad \textit{holds for all } x, y.$$

Assumption 3 *(Boundness of Hessian) The regularization term has been divided equally and included in $f_k(x)$, and $\nabla^2 f_k(x) \preceq I$ for $k = 1, 2, \ldots, m$.*

By Assumption 1, we can define global condition number for α-strongly convex and β-smooth function $f(x)$ as

$$\kappa \triangleq \frac{\max_x \lambda_{\max}(\nabla^2 f(x))}{\min_x \lambda_{\min}(\nabla^2 f(x))}.$$

Besides $\tilde{\kappa}$ and κ, we also use the following condition numbers to characterize the running time of our algorithms

$$\hat{\kappa}_l = \max_x \frac{\beta_{\max}(x)}{\lambda_{\min}(\nabla^2 f(x))}.$$

The following theorem is a standard concentration of measure result for sums of independent matrices.

Theorem 1 (Matrix Bernstein, [17]). *Consider a finite sequence $\{Z_k\}$ of independent random matrices with dimension $d_1 \times d_2$. Assume that each random matrix satifies*

$$\mathrm{E}[Z_k] = 0 \quad \textit{and} \quad \|Z_k\| \leq R, \quad \textit{almost surely.}$$

Define

$$\sigma^2 \triangleq \max\left\{\left\|\sum_k \mathrm{E}[Z_k Z_k^*]\right\|, \left\|\sum_k \mathrm{E}[Z_k^* Z_k]\right\|\right\}.$$

Then, for all $t \geq 0$, we have

$$Pr\left(\left\|\sum_k Z_k\right\| \geq t\right) \leq (d_1 + d_2)\exp\left(\frac{-t^2/2}{\sigma^2 + Rt/3}\right).$$

As an immediate corollary, we obtain the following result.

Corollary 1. *Consider a finite sequence $\{Z_k\}$ of independent random matrices with dimension $d \times d$. Assume that each random matrix satifies*

$$\mathrm{E}[Z_k] = 0 \quad \textit{and} \quad \|Z_k\| \leq R, \quad \textit{almost surely.}$$

Then, for all $t \geq 0$, we have

$$Pr\left(\left\|\sum_{k=1}^{s} Z_k\right\| \geq t\right) \leq 2d\exp\left(\frac{-t^2/2}{sR^2 + Rt/3}\right).$$

Proof. See proof in Appendix B.1. □

The following lemmas are foundations of our main theorem.

Lemma 1. *Let* $\rho_1 = 2$ *and* $\rho_{j+1} = \left(1 - \frac{1}{4\xi^2}\rho_j\right)^{-1}$. *If* $\beta_{\max}(x) > \alpha_{\min}(x) > 0$, *then we have* $1 < \rho_{j+1} < 2$ *and* ρ_j *monotonously decrease to* $\rho = \frac{2}{1+\sqrt{1-(\frac{\tilde{\kappa}-1}{\tilde{\kappa}+1})^2}}$.

Proof. See proof in Appendix C.1. □

Lemma 2. *Let* $\tilde{\nabla}^{-2} f(x_t)$ *be the average of* S_1 *independent samples* $\left\{\tilde{\nabla}^{-2}_{[i,S_2]} f(x_t), i = 1, 2, \cdots, S_1\right\}$, *which is used for the approximation of* $\nabla^{-2} f(x_t)$ *in Algorithm 1. Then we have that*

$$\left\|\tilde{\nabla}^{-2} f(x_t) - \mathrm{E}[\tilde{\nabla}^{-2} f(x_t)]\right\| \leq \begin{cases} R_1, & \textit{for CP-SNM}, \\ R_2, & \textit{for TR-SNM}, \\ R_3, & \textit{for OR-SNM}, \end{cases} \tag{10}$$

where

$$\begin{aligned} &R_1 \triangleq \begin{cases} 52(1 + \frac{\hat{\kappa}_l}{a-1})a^{S_2}, & \textit{for } a > 1, \\ 52(1 + \hat{\kappa}_l S_2), & \textit{for } a = 1, \\ 52\hat{\kappa}_l(1-a)^{-1}, & \textit{for } a < 1, \end{cases} \quad \textit{with } a \triangleq \left(\frac{\frac{\tilde{\kappa}-1}{\tilde{\kappa}+1}}{\sqrt{3 - (\frac{\tilde{\kappa}-1}{\tilde{\kappa}+1})^2} - 1}\right), \\ &R_2 \triangleq \begin{cases} 20\tilde{a}^{S_2} + 8\hat{\kappa}_l \tilde{a}^{S_2}(\tilde{a}-1)^{-1}, & \textit{for } \tilde{a} > 1, \\ 20 + 8\hat{\kappa}_l S_2, & \textit{for } \tilde{a} = 1, \\ 20\tilde{a}^{S_2} + 8\hat{\kappa}_l(1-\tilde{a})^{-1}, & \textit{for } \tilde{a} < 1, \end{cases} \quad \textit{with } \tilde{a} \triangleq \left(\sqrt{2}+1\right)\frac{\sqrt{\tilde{\kappa}}-1}{\sqrt{\tilde{\kappa}}+1}, \\ &R_3 \triangleq 2S_2 \frac{\tilde{\kappa}}{\tilde{\kappa}+1}\left(\frac{\tilde{\kappa}-1}{\tilde{\kappa}+1}\right)^{S_2-1} + 2\tilde{\kappa}\hat{\kappa}_l. \end{aligned} \tag{11}$$

Proof. See proof in Appendix C.2. □

Lemma 3. *Let* $\tilde{\nabla}^{-2} f(x_t)$ *be the average of* S_1 *independent samples* $\left\{\tilde{\nabla}^{-2}_{[i,S_2]} f(x_t), i = 1, 2, \cdots, S_1\right\}$, *which is used for the approximation of* $\nabla^{-2} f(x_t)$ *in Algorithm 1. For* $\eta_1 > 0$, *if we set*

$$S_2 \geq \begin{cases} \tilde{R}_1 \triangleq \frac{\sqrt{\tilde{\kappa}}+1}{2} \ln \frac{2(\hat{\kappa}_l - 1)}{\eta_1}, & \textit{for CP-SNM}, \\ \tilde{R}_2 \triangleq \left(1 + \sqrt{\tilde{\kappa}}\right) \ln \dfrac{(\hat{\kappa}_l - 1)\left(1 + \frac{4}{e \ln \frac{\sqrt{\tilde{\kappa}}+1}{\sqrt{\tilde{\kappa}}-1}}\right)}{\eta_1}, & \textit{for TR-SNM}, \\ \tilde{R}_3 \triangleq \dfrac{1+\tilde{\kappa}}{2} \ln \dfrac{\hat{\kappa}_l - 1}{\eta_1}, & \textit{for OR-SNM}. \end{cases}$$

then we have that

$$Pr\left(\left\|\tilde{\nabla}^{-2} f(x_t) - \nabla^{-2} f(x_t)\right\| > 3R\sqrt{S_1^{-1} \ln \frac{2d}{\delta}} + \eta_1\right) \leq \delta,$$

where $R = R_1$, R_2 *and* R_3 *for CP-SNM, TR-SNM and OR-SNM, respectively, and* $3\sqrt{S_1^{-1} \ln \frac{2d}{\delta}} < 1$.

Proof. See proof in Appendix C.3. □

Lemma 4. *Let x_t be generated by Algorithm 1. Suppose the conditions of Lemma 3 is fulfilled, then the following inequality holds with probability $1-\delta$:*

$$\|x_{t+1}-x^*\| \leq \gamma\|x_t - x^*\| + M\left\|\nabla^{-2} f(x_t)\right\|\|x_t - x^*\|^2, \tag{12}$$

where $\gamma = 3\hat{\kappa}_l R\sqrt{S_1{}^{-1}\ln\frac{2d}{\delta}} + \hat{\kappa}_l\eta_1$ with $R = R_1$, R_2 and R_3 for CP-SNM, TR-SNM and OR-SNM , respectively.

Proof. See proof in Appendix C.4. □

Theorem 2. *For Algorithm 1, let $FO(M, \hat{\kappa}_l R)$ be the total time required by a first-order algorithm to achieve the least accuracy $\frac{1}{4M\hat{\kappa}_l}$ such that $\|x_1 - x^*\| \leq \frac{1}{4M\hat{\kappa}_l}$, and set the parameters as follows:*

$$\hat{\kappa}_l\eta_1 = \frac{1}{16}, \quad T_1 = FO\left(M, \hat{\kappa}_l R\right), \quad S_1 = O\left(\hat{\kappa}_l^2 R^2 \ln\frac{2d}{\delta}\right), \quad S_2 = O\left(\tilde{R}\right).$$

Then it holds for $t \geq T_1$ with probability $1-\delta$ that

$$\|x_{t+1}-x^*\| \leq \frac{1}{2}\|x_t - x^*\|,$$

and each step of Algorithm 1 runs in time $\tilde{O}(md + \hat{\kappa}_l^2 R^2 \tilde{R} d^2)$.

Moreover, if $f(x)$ is a generalized linear model (GLM) function [12], then each step of Algorithm 1 runs in time $\tilde{O}(md + \hat{\kappa}_l^2 R^2 \tilde{R} d)$, where $R = R_1$, R_2, R_3 and $\tilde{R} = \tilde{R}_1$, $\tilde{R}_2$, $\tilde{R}_3$ for CP-SNM, TR-SNM and OR-SNM, respectively.

Proof. See proof in Appendix A. □

Under the Assumptions 1–3, we can obtain the following results.

Corollary 2. *For a GLM function $f(x)$, Algorithm 1 returns a point x_t with probability at least $1-\delta$ that $f(x_t) \leq \min_x f(x) + \epsilon$ in total time $\tilde{O}(md + \hat{\kappa}_l^2 R^2 \tilde{R} d)\ln(\frac{1}{\epsilon})$, where $R = R_1$, R_2, R_3 and $\tilde{R} = \tilde{R}_1$, $\tilde{R}_2$, $\tilde{R}_3$ for CP-SNM, TR-SNM and OR-SNM, respectively.*

For comparison, we list the runtime of CP-SNM, TR-SNM, OR-SNM and LiSSA in Table 1.

Table 1. Running time comparison

Algorithm	Runtime
CP-SNM	$\tilde{O}(md + \hat{\kappa}_l^2 R_1^2 \tilde{R}_1 d^2)\ln(\frac{1}{\epsilon})$
TR-SNM	$\tilde{O}(md + \hat{\kappa}_l^2 R_2^2 \tilde{R}_2 d^2)\ln(\frac{1}{\epsilon})$
OR-SNM	$\tilde{O}(md + \hat{\kappa}_l^2 R_3^2 \tilde{R}_3 d^2)\ln(\frac{1}{\epsilon})$
LiSSA	$\tilde{O}(md + \tilde{\kappa}^2 \hat{\kappa}_l d^2)\ln(\frac{1}{\epsilon})$

Table 2. Datasets for experiments

Classfication	Data Set	m	d
Logistics	MNIST 4-9	11791	784
Softmax	MUSHROOM	8124	112
	COVERTYPE	100000	54
	RealSIM	72309	20958

Remark 1. Here $\tilde{O}$ hides log factors of κ, d, $\frac{1}{\delta}$. In theory, the bounds on the variances of CP-SNM and TR-SNM are possibly pessimistic, which need to be improved. We leave this for future work. On the other hand, at every Newton step, we can choose a matrix $\tilde{H}$ as a preconditioner for the Hessian H, so that the inner iteration of the Newton method is transformed to solve a system of linear equations $H\tilde{H}^{-1}y = -b$, where the condition number of $H\tilde{H}^{-1}$ can be greatly reduced to a satisfied level as described in Algorithm 3 in [12]. Especially, for GLM functions, the Newton step is in fact to solve a least squares problem. We can take advantage of a low complexity λ-spectral approximation as a preconditioner given in Sect. 5.2 in [12]. These will be our future work.

5 Experiments

In this section, we validate our proposed approach empirically. We use the four datasets tested in [12], namely MNIST 4-9 [8], CoverType [9], Mushroom [1] and RealSIM [10], which are listed in Table 2.

We remark that MNIST 4-9 is a binary classification task, and others are multi-classification tasks, which are optimized with the logistic regression and softmax classification objectives, respectively. All the models are trained with the l_2 regularization, and we plot log(*CurrentValue*-*OptimalValue*) for comparison, where *OptimalValue* is the x_t which first attains the given accuracy baseline. In order to make sure that the norm of the Hessian is bounded, we scale the above data set points to unit norm.

As the testing results in [12] show that LiSSA improves upon the overall time over popular first-order methods, we only compare our proposed methods with LiSSA in this section.

In Fig. 1, we compare the four methods CP-SNM, TR-SNM, OR-SNM and LiSSA in terms of the accuracy achieved versus the number of passes over the data which is named epoch. It shows that, with enough initial SGD steps, all methods are linear convergent. We can find that, all the three new methods outperform LiSSA. Among our proposed three methods, CP-SNM and TR-SNM give similar results, and both of them are better than OR-SNM.

In Fig. 2, we evaluate all the methods for different regularization parameter λ, which shows that our proposed methods consistently outperform LiSSA on

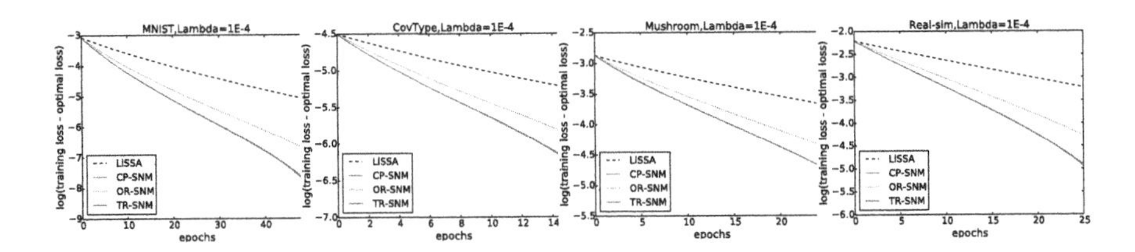

Fig. 1. Performances of CP-SNM, TR-SNM, OR-SNM and LiSSA for different data sets and regularization parameter $\lambda = 10^{-4}$, $S_1 = 1, S_2 = 25$

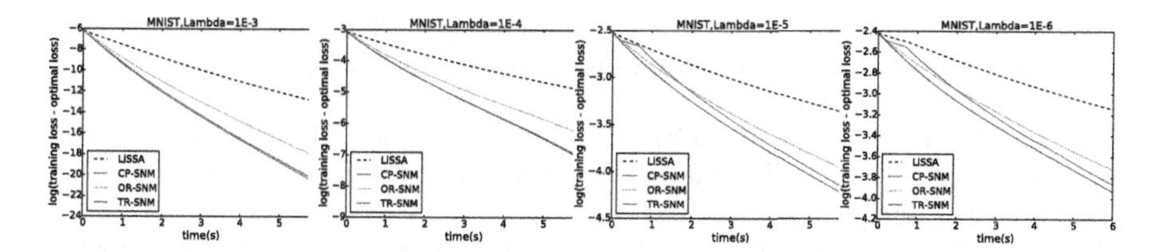

Fig. 2. Performances for different regularization parameter λ and $S_1 = 1, S_2 = 25$.

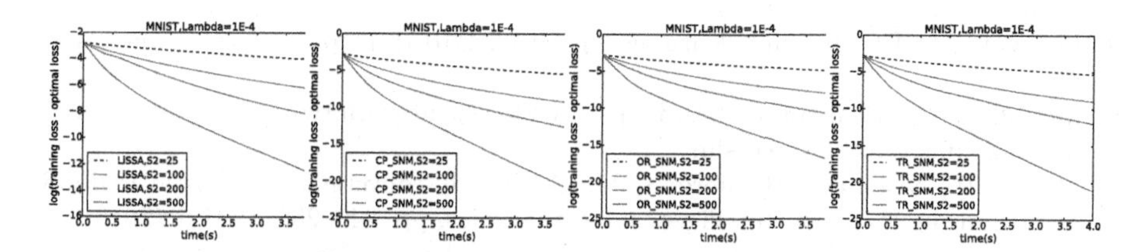

Fig. 3. Convergence rates for different choices of the S_2.

different condition numbers. The results in Fig. 3 show that our proposed methods converge faster with larger S_2. In Figs. 2 and 3, we plot the results for the data set MNIST 4-9 and the results for other data sets are similar.

6 Conclusion

In this paper, we focus on the acceleration of stochastic Newton method by approximating the Hessian inverse with Chebyshev estimators. We propose three methods, CP-SNM, TR-SNM and OR-SNM. A complete convergence analysis is given and experimental results show that our proposed methods consistently outperform LiSSA on the four standard benchmarks.

Acknowledgements. We would like to thank the anonymous reviewers for their constructive comments.

References

1. Blackard, J.A., Dean, D.J.: Comparative accuracies of artificial neural networks and discriminant analysis in predicting forest cover types from cartographic variables. Comput. Electron. Agric. **24**(3), 131–151 (1999)
2. Defazio, A., Bach, F., Lacoste-Julien, S.: SAGA: a fast incremental gradient method with support for non-strongly convex composite objectives. In: Advances in Neural Information Processing Systems, pp. 1646–1654 (2014)
3. Erdogdu, M.A., Montanari, A.: Convergence rates of sub-sampled newton methods, pp. 3034–3042. Advances in Neural Information Processing Systems (2015)
4. Golub, G.H., van Loan, C.F.: Matrix Computations, 2nd edn. The Johns Hopkins University Press, Baltimore and London (1989)

5. Golub, G.H., Varga, R.S.: Chebyshev semi-iterative methods, successive overrelaxation iterative methods, and second order Richardson iterative methods. Numer. Math. **3**, 147–156 (1961)
6. Gower, R., Goldfarb, D., Richtárik, P.: Stochastic block BFGS: squeezing more curvature out of data. In: International Conference on Machine Learning, pp. 1869–1878. PMLR (2016)
7. Johnson, R., Zhang, T.: Accelerating stochastic gradient descent using predictive variance reduction. Adv. Neural. Inf. Process. Syst. **26**, 315–323 (2013)
8. LeCun, Y., Cortes, C.: The MNIST database of handwritten digits (1998). http://yann.lecun.com/exdb/mnist/
9. Lichman, M.: UCI machine learning repository (2013). http://archive.ics.uci.edu/ml
10. McCallum, A.: Real-sim (1997). https://www.csie.ntu.edu.tw/~cjlin/libsvmtools/datasets/binary.html#real-sim
11. Moritz, P., Nishihara, R., Jordan, M.: A linearly-convergent stochastic l-BFGS algorithm. In: Artificial Intelligence and Statistics, pp. 249–258. PMLR (2016)
12. Naman Agarwal, B.B., Hazan, E.: Second-order stochastic optimization for machine learning in linear time. J. Mach. Learn. Res. **18**, 1–40 (2017)
13. Pilanci, M., Wainwright, M.J.: Newton sketch: a linear-time optimization algorithm with linear-quadratic convergence. SIAM J. Optim. **27**(1), 205–245 (2017)
14. Byrd, R.H., Chin, G.M., Neveitt, W., Norcedal, J.: On the use of stochastic Hessian information in optimization methods for machine learning. SIAM J. Optimiz. **21**(3), 977–995 (2011)
15. Robbins, H., Monro, S.: A stochastic approximation method. Ann. Math. Statist. **22**, 400–407 (1951)
16. Shalev-Shwartz, S., Zhang, T.: Accelerated proximal stochastic dual coordinate ascent for regularized loss minimization. In: International Conference on Machine Learning, pp. 64–72. PMLR (2014)
17. Tropp, J.A.: User-friendly tail bounds for sums of random matrices. Found. Comput. Math. **12**(4), 389–434 (2012)
18. Varga, R.S.: Matrix Iterative Analysis, 2nd edn. Englewood Cliff, N.J. Prentice-Hall Inc., (2000). https://doi.org/10.1007/978-3-642-05156-2

Stochastic Submodular Maximization via Polynomial Estimators

Gözde Özcan(✉) and Stratis Ioannidis

Electrical and Computer Engineering Department, Northeastern University, Boston, MA 02115, USA
{gozcan,ioannidis}@ece.neu.edu

Abstract. In this paper, we study stochastic submodular maximization problems with general matroid constraints, which naturally arise in online learning, team formation, facility location, influence maximization, active learning and sensing objective functions. In other words, we focus on maximizing submodular functions that are defined as expectations over a class of submodular functions with an unknown distribution. We show that for monotone functions of this form, the stochastic continuous greedy algorithm [19] attains an approximation ratio (in expectation) arbitrarily close to $(1 - 1/e) \approx 63\%$ using a polynomial estimation of the gradient. We argue that using this polynomial estimator instead of the prior art that uses sampling eliminates a source of randomness and experimentally reduces execution time.

Keywords: submodular maximization · stochastic optimization · greedy algorithm

1 Introduction

Submodular maximization is a true workhorse of data mining, arising in settings as diverse as hyper-parameter optimization [23], feature compression [3], text classification [14], and influence maximization [9,13]. Many of these interesting problems as well as variants can be cast as maximizing a submodular set function $f(S)$, defined over sets $S \subseteq V$ for some ground set V, subject to a matroid constraint. Despite the NP-hardness of these problems, the so-called *continuous-greedy* (CG) algorithm [5], can be used to construct a $1 - 1/e$-approximate solution in polynomial time. Interestingly, the solution is generated by first transferring the problem to the continuous domain, and solving a continuous optimization problem via gradient techniques. The solution to this continuous optimization problem is subsequently rounded (via techniques such as pipage rounding [1] and swap rounding [5]), to produce an integral solution within a $1 - 1/e$ factor from the optimal. The continuous optimization problem solved by the CG algorithm amounts to maximizing so-called *multilinear relaxation* of the original, combinatorial submodular objective. In short, the multilinear relaxation of a submodular function $f(S)$ is its expectation assuming its input

H. Kashima et al. (Eds.): PAKDD 2023, LNAI 13936, pp. 535–548, 2023.
https://doi.org/10.1007/978-3-031-33377-4_41

S is generated via independent Bernoulli trials, and is typically computed via sampling [5,24].

Recently, a series of papers have studied an interesting variant called the *stochastic submodular optimization* setting [2,7,10,11,19,25]. In this setting, the submodular objective function to be optimized is assumed to be of the form of an expectation, i.e., $f(S) = \mathbb{E}_{z \sim P}[f_z(S)]$, where z is a random variable. Moreover, the optimization algorithm does not have access to the a function oracle (i.e., cannot compute the function itself). Instead it can only sample a random instantiation of $f_z(\cdot)$, different each time. This setting is of course of interest when the system or process that f models is inherently stochastic (e.g., involves a system dependent on, e.g., user behavior or random arrivals) and the distribution governing this distribution is not a priori known. It is also of interest when the support of distribution P is very large, so that the expectation cannot be computed efficiently. A classic example of the latter case is influence maximization (c.f. Sect. 3.1), where the expectation $f(S)$ cannot be computed efficiently or even in a closed form, even though samples $z \sim P$ can be drawn.

Interestingly, the fact that the classic continuous greedy algorithm operates in the continuous domain gives rise to a *stochastic continuous greedy* (SCG) method for tackling the stochastic optimization problem [19]. In a manner very similar to stochastic gradient descent, the continuous greedy algorithm can be modified to use *stochastic gradients*, i.e., random variables whose expectations equal the gradient of the multilinear relaxation. In practice, these are computed by sampling *two random variables in tandem*: $z \sim P$, which is needed to generate a random instance f_z, and S, the random input needed to compute the multilinear relaxation. As a result, the complexity of the SCG algorithm depends on the variance due to *both* of these two variables.

We make the following contributions:

- We use polynomial approximators, originally proposed by Özcan et al. [22], to reduce the variance of the stochastic continuous greedy algorithm. In particular, we eliminate one of the two sources of randomness of SCG, namely, sampling S. We do this by replacing the sampling estimator by a deterministic estimator constructed by approximating each $f_z(\cdot)$ with a polynomial function.
- We show that doing so *reduces the variance* of the gradient estimation procedure used by SCG, but introduces a *bias*. We then characterize the performance of SCG in terms of both the (reduced) variance and new bias term.
- We show that for several interesting stochastic submodular maximization problems, including influence maximization, the bias can be well-controlled, decaying exponentially with the degree of our polynomial approximators.
- Finally, we illustrate the advantage of our approach experimentally, over both synthetic and real-life datasets.

2 Related Work

While submodular optimization problems are generally NP-hard, the celebrated greedy algorithm [20] attains a $(1 - 1/e)$ approximation ratio for submodular maximization subject to uniform matroids and a 1/2 approximation ratio for general matroid constraints. As discussed in the introduction, the continuous greedy algorithm [5] restores the $(1 - 1/e)$ approximation ratio by lifting the discrete problem to the continuous domain via the multilinear relaxation.

Stochastic submodular maximization, in which the objective is expressed as an expectation, has gained a lot of interest in the recent years [2,7,25]. Karimi et al. [11] use a concave relaxation method that achieves the $(1-1/e)$ approximation guarantee, but only for the class of submodular coverage functions. Hassani et al. [10] provide projected gradients methods for the general case of stochastic submodular problems that achieve 1/2 approximation guarantee. Mokhtari et al. [19] propose stochastic conditional gradient methods for solving both minimization and maximization stochastic submodular optimization problems. Their method for maximization, Stochastic Continous Greedy (SCG) can be interpreted as a stochastic variant of the continuous greedy algorithm [5,24] and achieves a tight $(1 - 1/e)$ approximation guarantee for monotone and submodular functions.

Our work builds upon and relies on the approach by Özcan et al. [22], who studied ways of accelerating the computation of gradients via a polynomial estimator. Extending on the work of Mahdian et al. [16], Özcan et al. show that submodular functions that can be written as compositions of (a) an analytic function and (b) a multilinear function can be arbitrarily well approximated via Taylor polynomials; in turn, this gives rise to a method for approximating their multilinear relaxation in a closed form, without sampling. We leverage this method in the context of stochastic submodular optimization, showing that it can also be applied in combination with SCG of Mokhtari et al. [19]: this eliminates one of the two sources of randomness, thereby reducing variance at the expense of added bias. From a technical standpoint, this requires controlling the error introduced by the bias of the polynomial estimator, while simultaneously accounting for the variance inherent in SCG, due to sampling instances.

3 Technical Preliminary

Submodularity and Matroids. Given a ground set $V = \{1, \ldots, n\}$ of n elements, a set function $f : 2^V \rightarrow \mathbb{R}_+$ is submodular if and only if $f(B \cup \{e\}) - f(B) \leq f(A \cup \{e\}) - f(A)$, for all $A \subseteq B \subseteq V$ and $e \in V$. Function f is *monotone* if $f(A) \leq f(B)$, for every $A \subseteq B$.

Matroids. Given a ground set V, a matroid is a pair $\mathcal{M} = (V, \mathcal{I})$, where $\mathcal{I} \subseteq 2^V$ is a collection of *independent sets*, for which the following hold: (a) if $B \in \mathcal{I}$ and $A \subset B$, then $A \in \mathcal{I}$, and (b) if $A, B \in \mathcal{I}$ and $|A| < |B|$, there exists $x \in B \setminus A$ s.t. $A \cup \{x\} \in \mathcal{I}$. The *rank* of a matroid $r_{\mathcal{M}}(V)$ is the largest cardinality of its elements, i.e.: $r_{\mathcal{M}}(V) = \max\{|A| : A \in \mathcal{I}\}$. We introduce two examples of matroids:

1. **Uniform Matroids.** The uniform matroid with cardinality k is $\mathcal{I} = \{S \subseteq V, |S| \leq k\}$.
2. **Partition Matroids.** Let $\mathcal{B}_1, \ldots, \mathcal{B}_m \subseteq V$ be a partitioning of V, i.e., $\bigwedge_{\ell=1}^{m} \mathcal{B}_\ell = \emptyset$ and $\bigcup_{\ell=1}^{m} \mathcal{B}_\ell = V$. Let also $k_\ell \in \mathbb{N}, \ell = 1, \ldots, m$, be a set of

cardinalities. A partition matroid is defined as $\mathcal{I} = \{S \subseteq 2^V \mid |S \cap \mathcal{B}_\ell| \leq k_\ell, \text{ for all } \ell = 1, \ldots, m\}$.

3.1 Problem Definition

In this work, we focus on *discrete stochastic submodular maximization* problems. More specifically, we consider set function $f : 2^V \to \mathbb{R}_+$ of the form: $f(S) = \mathbb{E}_{z\sim P}[f_z(S)]$, $S \subseteq V$, where z is the realization of the random variable Z drawn from a distribution P over a probability space (V_z, P). For each realization of $z \sim P$, the set function $f_z : 2^V \to \mathbb{R}_+$ is monotone and submodular. Hence, f itself is monotone and submodular. The objective is to maximize f subject to some constraints (e.g., cardinality or matroid constraints) by only accessing to i.i.d. samples of $f_{z\sim P}$. In other words, we wish to solve:

$$\max_{S\in\mathcal{I}} f(S) = \max_{S\in\mathcal{I}} \mathbb{E}_{z\sim P}[f_z(S)], \tag{1}$$

where $\mathcal{I}$ is a general matroid constraint.

Stochastic submodular maximization problems are of interest in the absence of the oracle that provides the exact value of $f(S)$: one can only access $f_z(S)$, for random instantiations $z \sim P$. A well-known motivational example is contagion propagation in a network (a.k.a., the influence maximization problem [13]). Given a graph with node set V, the reachability of nodes from seeds is determined by sampling sub-graph $G = (V, E)$, via, e.g., the Independent Cascade or the Linear Threshold model [13]. The random edge set, in this case, plays the role of z, and the distribution over graphs the role of P. The function $f_z(S)$ represents the ratio of nodes reachable from the seeds S under the connectivity induced by edges E in this particular realization of z. The goal is to select seeds S that maximize $f(S) = \mathbb{E}_{z\sim P}[f_z(S)]$; both f and f_z are monotone submodular functions; however computing f in a closed form is hard, and $f(\cdot)$ can only be accessed through random instantiations of $f_z(\cdot)$.

3.2 Change of Variables and Multiliear Relaxation

There is a 1-to-1 correspondence between a binary vector $\mathbf{x} \in \{0,1\}^n$ and its support $S = \mathsf{supp}(\mathbf{x})$. Hence, a set function $f : 2^V \to \mathbb{R}_+$ can be interpreted as $f : \{0,1\}^n \to \mathbb{R}_+$ via: $f(\mathbf{x}) \triangleq f(\mathsf{supp}(\mathbf{x}))$ for $\mathbf{x} \in \{0,1\}^n$. We adopt this convention for the remainder of the paper. We also treat matroids as subsets of $\{0,1\}^n$, defined consistently with this change of variables via $\mathcal{M} = \{\mathbf{x} \in \{0,1\}^n : \mathsf{supp}(\mathbf{x}) \in \mathcal{I}\}$. For example, a partition matroid is: $\mathcal{M} = \left\{\mathbf{x} \in \{0,1\}^n \mid \bigcap_{\ell=1}^m \left(\sum_{i\in B_\ell} x_i \leq k_\ell\right)\right\}$. The *matroid polytope* $\mathcal{C} \subseteq [0,1]^n$ is the convex hull of matroid $\mathcal{M}$, i.e., $\mathcal{C} = \mathsf{conv}(\mathcal{M})$.

We define the *multilinear relaxation* of f as:

$$\begin{aligned} G(\mathbf{y}) = \mathbb{E}_{S\sim\mathbf{y}}[f(S)] &= \sum_{S\subseteq V} f(S) \prod_{i\in S} y_i \prod_{j\notin S} (1 - y_j) \\ = \mathbb{E}_{\mathbf{x}\sim\mathbf{y}}[f(\mathbf{x})] &= \sum_{\mathbf{x}\in\{0,1\}^n} f(\mathbf{x}) \prod_{i\in V} y_i^{x_i} (1 - y_i)^{(1-x_i)}, \quad \text{for } \mathbf{y} \in [0,1]^n. \end{aligned} \tag{2}$$

In other words, $G : [0,1]^n \to \mathbb{R}_+$ is the expectation of f, assuming that S is random and generated from independent Bernoulli trials: for every $i \in V$, $P(i \in S) = y_i$. The multilinear relaxation of f satisfies several properties. First, it is indeed a relaxation/extension of f over the (larger) domain $[0,1]^n$: for $\mathbf{x} \in \{0,1\}^n$, $G(\mathbf{x}) = f(\mathbf{x})$, i.e., G agrees with f on integral inputs. Second, it is *multilinear* (c.f. Sect. 3.4), i.e., affine w.r.t. any single coordinate y_i, $i \in V$, when keeping all other coordinates $\mathbf{y}_{-i} = [y_j]_{j \neq i}$ fixed. Finally, in the context of stochastic submodular optimization, it is an expectation that involves *two sources of randomness*: (a) $z \sim P$, i.e., the random instantiation of the objective, *as well as* (b) $\mathbf{x} \sim \mathbf{y}$, i.e., the independent sampling of the Bernoulli variables (i.e., the set S). In particular, we can write:

$$G(\mathbf{y}) = \mathbb{E}_{z\sim P}[G_z(\mathbf{y})], \text{ where } G_z(\mathbf{y}) = \mathbb{E}_{\mathbf{x}\sim\mathbf{y}}[f_z(x)] \text{ is the multilinear relaxation of } f_z(\cdot). \tag{3}$$

3.3 Stochastic Continuous Greedy Algorithm

The stochastic nature of the set function $f(S)$ requires the use the *Stochastic Continuous Greedy (SCG)* algorithm [19]. This is a stochastic variant of the continuous greedy algorithm (method) [24], to solve (1). The SCG algorithm uses a common averaging technique in stochastic optimization and computes the estimated gradient $\mathbf{d}_t$ by the recursion

$$\mathbf{d}_t = (1 - \rho_t)\mathbf{d}_{t-1} + \rho_t \nabla G_{z_t}(\mathbf{y}_t), \tag{4}$$

where ρ_t is a positive step size and the algorithm initially starts with $\mathbf{d}_0 = \mathbf{y}_0 = \mathbf{0}$. Then, it proceeds in iterations, where in the t-th iteration it finds a feasible solution as follows

$$\mathbf{v}_t \in \arg\max_{\mathbf{v}\in\mathcal{C}}\{\mathbf{d}_t^T \mathbf{v}\}, \tag{5}$$

where $\mathcal{C}$ is the matroid polytope (i.e., convex hull) of matroid $\mathcal{M}$. After finding the ascent direction $\mathbf{v}_t$, the current solution $\mathbf{y}_t$ is updated as

$$\mathbf{y}_{t+1} = \mathbf{y}_t + \frac{1}{T}\mathbf{v}_t, \tag{6}$$

where $1/T$ is the step size. The steps of the stochastic continuous greedy algorithm are outlined in Algorithm 1. The (fractional) output of Algorithm 1 is within a $1 - 1/e$ factor from the optimal solution to Problem (1) (see Theorem 2 below). This fractional solution can subsequently be rounded in polynomial time to produce a solution with the same approximation guarantee w.r.t. to Problem (1) using, e.g., either the pipage rounding [1] or the swap rounding [6] methods.

Sample Estimator. The gradient ∇G_{z_t} is needed to perform step (4); computing it directly via Eq. (2). requires exponentially many calculations. Instead, both Calinescu et al. [5] and Mokhtari et al. [19] estimate it via *sampling*.

Algorithm 1. Stochastic Continuous Greedy (SCG)

Require: Step sizes $\rho_t > 0$. Initialize $\mathbf{d}_0 = \mathbf{y}_0 = \mathbf{0}$.
1: **for** $t = 1, 2, \ldots, T$ **do**
2: Compute $\mathbf{d}_t = (1 - \rho_t)\mathbf{d}_{t-1} + \rho_t \nabla G_{z_t}(\mathbf{y}_t)$;
3: Compute $\mathbf{v}_t \in \arg\max_{\mathbf{v} \in \mathcal{C}}\{\mathbf{d}_t^T \mathbf{v}\}$;
4: Update the variable $\mathbf{y}_{t+1} = \mathbf{y}_t + \frac{1}{T}\mathbf{v}_t$;
5: **end for**

In particular, due to multilinearity (i.e., the fact that G_z is affine w.r.t. a coordinate x_i, we have:

$$\frac{\partial G_z(\mathbf{y})}{\partial x_i} = G_z([\mathbf{y}]_{+i}) - G_z([\mathbf{y}]_{-i}), \quad \text{for all } i \in V, \tag{7}$$

where $[\mathbf{y}]_{+i}$ and $[\mathbf{y}]_{-i}$ are equal to the vector $\mathbf{y}$ with the i-th coordinate set to 1 and 0, respectively. The gradient of G can thus be estimated by (a) producing N random samples $\mathbf{x}^{(l)}$, for $l \in \{1, \ldots, N\}$ of the random vector $\mathbf{x}$, and (b) computing the empirical mean of the r.h.s. of (7), yielding

$$\widehat{\frac{\partial G_z(\mathbf{y})}{\partial x_i}} = \frac{1}{N}\sum_{l=1}^{N}\left(f_z([\mathbf{x}^{(l)}]_{+i}) - f_z([\mathbf{x}^{(l)}]_{-i})\right), \quad \text{for all } i \in V. \tag{8}$$

Mokhtari et al. [19] make the following assumptions:

Assumption 1. *Function $f : \{0,1\}^n \to \mathbb{R}_+$ is monotone and submodular.*

Assumption 2. *The Euclidean norm of the elements in the constraint set $\mathcal{C}$ are uniformly bounded, i.e., for all $\mathbf{y} \in \mathcal{C}$, there exists a D s.t. $\|\mathbf{y}\| \leq D$.*

Under these assumptions, SCG combined with the sampling estimator in Eq. (7), yields the following guarantee:

Theorem 1. *[Mokhtari et al. [19]] Consider Stochastic Continuous Greedy (SCG) outlined in Algorithm 1, with $\nabla G_{z_t}(\mathbf{y}_t)$ replaced by $\widehat{\nabla G_{z_t}(\mathbf{y}_t)}$ given by (8). Recall the definition of the multilinear extension function G in (2) and set the averaging parameter as $\rho_t = 4/(t+8)^{2/3}$. If Assumptions 1 & 2 are satisfied, then the iterate $\mathbf{y}_T$ generated by SCG satisfies the inequality*

$$\mathbb{E}\left[G(\mathbf{y}_T)\right] \geq (1 - 1/e)OPT - \frac{15DK}{T^{1/3}} - \frac{f_{\max} r D^2}{2T}, \tag{9}$$

where $OPT = \max_{\mathbf{y} \in \mathcal{C}} G(\mathbf{y})$ and $K = \max\{3\|\nabla G(\mathbf{y}_0) - \mathbf{d}_0\|, 4\sigma + \sqrt{3r} f_{\max} D\}$, where D is the diameter of the convex hull $\mathcal{C}$, $f_{\max}$ is the maximum marginal value of the function f, i.e., $f_{\max} = \max_{i \in \{1,\ldots,n\}} f(\{i\})$, r is the rank of the matroid $\mathcal{I}$, and $\sigma^2 = \sup_{\mathbf{y} \in \mathcal{C}} \mathbb{E}\left[\|\widehat{\nabla G_z(\mathbf{y})} - G(\mathbf{y})\|\right]$, where $\widehat{\nabla G_z}$ is the sample estimator given by Eq. (8).

Thus, by appropriately setting the number of iterations T, we can produce a solution that is arbitrarily close to $1-1/e$ from the optimal (fractional) solution. Again, this can be subsequently rounded (see, e.g., [1,5]) to produce an integer solution with the same approximation guarantee. It is important to note that the number of steps required depends on σ^2, which is a (uniform over $\mathcal{C}$) bound on the variance of the estimator given by Eq. (8). This variance contains *two sources of randomness*, namely $z \sim P$, the random instantiation, and $\mathbf{x} \sim \mathbf{y}$, as multiple such integer vectors/sets are sampled in Eq (8). In general, the variance will depend on the number of samples N in the estimator, and will be bounded (as G is bounded).[1]

3.4 Multilinear Functions and the Multilinear Relaxation of a Polynomial

Recall that a *polynomial* function $p : \mathbb{R}^n \to \mathbb{R}$ can be written as a linear combination of several monomials, i.e.,

$$p(\mathbf{y}) = c_0 + \sum_{\ell \in \mathcal{I}} c_\ell \prod_{i \in \mathcal{J}_\ell} y_i^{k_i^\ell}, \tag{10}$$

where $c_\ell \in \mathbb{R}$ for ℓ in some index set $\mathcal{I}$, subsets $\mathcal{J}_\ell \subseteq V$ determine the terms of each monomial, and , and $\{k_i^\ell\}_{i \in \mathcal{J}_\ell} \subset \mathbb{N}$ are natural exponents. W.l.o.g. we assume that $k_i^\ell \geq 1$ (as variables with zero exponents can be ommited). The degree of the monomial indexed by $\ell \in \mathcal{I}$ is $k^\ell = \sum_{i \in \mathcal{J}_\ell} k_i^\ell$, and the degree of polynomial p is $\max_{\ell \in \mathcal{I}} k^\ell$, i.e., the largest degree across monomials.

A function $f : \mathbb{R}^N \to \mathbb{R}$ is *multilinear* if it is affine w.r.t. each of its coordinates [4]. Alternatively, multilinear functions are polynomial functions in which the degree of each variable in a monomial is at most 1; that is, multilinear functions can be written as:

$$f(\mathbf{y}) = c_0' + \sum_{\ell \in \mathcal{I}} c_\ell' \prod_{i \in \mathcal{J}_\ell} y_i, \tag{11}$$

where $c_\ell \in \mathbb{R}$ for ℓ in some index set $\mathcal{I}$, and subsets $\mathcal{J}_\ell \subseteq V$, again determining monomials of degree *exactly equal to* $|\mathcal{J}_\ell|$. Given a polynomial p defined by the parameters in Eq. (10), let

$$\dot{p}(\mathbf{y}) = c_0 + \sum_{\ell \in \mathcal{I}} c_\ell \prod_{i \in \mathcal{J}_\ell} y_i, \tag{12}$$

be the multilinear function resulting from p, *by replacing all its exponents* $k_i^\ell \geq 1$ *with* 1. We call this function the *multilinearization* of p. The multilinearization of p is inherently linked to its multilinear relaxation:

[1] For example, even for $N = 1$, the submodularity of f_z and Eq. (7) imply that $\sigma^2 \leq 2n\max_{j \in [n]} \mathbb{E}[f_z(\{j\})^2]$ [19], though this bound is loose/a worst-case bound.

Lemma 1 (Özcan et al. [22]). *Let $p : [0,1]^n \to \mathbb{R}$ be an arbitrary polynomial and let $\dot{p} : \mathbb{R}^n \to \mathbb{R}_+$ be its multilinearization, given by Eq. (12). Let $\mathbf{x} \in \{0,1\}^n$ be a random vector of independent Bernoulli coordinates parameterized by $\mathbf{y} \in [0,1]^n$. Then, $\mathbb{E}_{\mathbf{x}\sim\mathbf{y}}[p(\mathbf{x})] = \mathbb{E}_{\mathbf{x}\sim\mathbf{y}}[\dot{p}(\mathbf{x})] = \dot{p}(\mathbf{y})$.*

Proof. Observe that $\dot{p}(\mathbf{x}) = p(\mathbf{x})$, for all $\mathbf{x} \in \{0,1\}^n$. This is precisely because $x^k = x$ for $x \in \{0,1\}$ and all $k \geq 1$. The first equality therefore follows. On the other hand, $\dot{p}(\mathbf{x})$ is the multilinear function given by Eq. (12). Hence $\mathbb{E}_{\mathbf{x}\sim\mathbf{y}}[\dot{p}(\mathbf{x})] = \mathbb{E}_{\mathbf{x}\sim\mathbf{y}}\left[c_0 + \sum_{\ell\in\mathcal{I}} c_\ell \prod_{i\in\mathcal{J}_\ell} x_i\right] = c_0 + \sum_{\ell\in\mathcal{I}} \mathbb{E}_{\mathbf{x}\sim\mathbf{y}}\left[\prod_{i\in\mathcal{J}_\ell} x_i\right] = c_0 + \sum_{\ell\in\mathcal{I}} \prod_{i\in\mathcal{J}_\ell} \mathbb{E}_{\mathbf{x}\sim\mathbf{y}}[x_i] = \dot{p}(\mathbf{y})$, where the second to last equality holds by the independence across x_i, $i \in V$. □

An immediate consequence of this lemma is that the multilinear relaxation of any polynomial function can be computed *without sampling*, by simply computing its multilinearization. This is of particular interest of course for submodular functions that are themselves polynomials (e.g., coverage functions [11]). Özcan et al. extend this to submodular functions that can be written as compositions of a scalar and a polynomial function, by approximating the former via its Taylor expansion. We extend and generalize this to the case of stochastic submodular functions, so long as the latter can be approximated arbitrarily well by polynomials.

4 Main Results

4.1 Polynomial Estimator

To leverage Lemma 1 to the case of stochastic submodular functions, we make the following assumption:

Assumption 3. *For all $z \in V_z$, there exists a sequence of polynomials $\{\hat{f}_z^L\}_{L=1}^\infty$, $\hat{f}_z^L : \mathbb{R}^n \to \mathbb{R}$ such that $\lim_{L\to\infty} |f_z(\mathbf{x}) - \hat{f}_z^L(\mathbf{x})| = 0$, uniformly over $\mathbf{x} \in \{0,1\}^n$, i.e. there exists $\varepsilon_z(L) \geq 0$ such that $\lim_{L\to\infty} \varepsilon_z(L) = 0$ and $|f_z(\mathbf{x}) - \hat{f}_z^L(\mathbf{x})| \leq \varepsilon_z(L)$, for all $\mathbf{x} \in \{0,1\}^n$.*

In other words, we assume that we can asymptotically approximate every function f_z with a polynomial arbitrarily well. Note that there already exists a polynomial function that approximates each f_z *perfectly* (i.e., $\epsilon_z = 0$), namely, its multilinear relaxation G_z. However, the number of terms in this polynomial is exponential in n. In contrast, Assumption 3 requires exact recovery only asymptotically. In many cases, this allows us to construct polynomials with only a handful (i.e., polynomial in n) terms, that can approximate f_z. We will indeed present such polynomials for several applications of interest in Sect. 5. Armed with this assumption, we define an estimator $\widehat{\nabla G_z^L}$ of the gradient of the multilinear relaxation G as follows:

$$\frac{\widehat{\partial G_z^L}}{\partial y_i}\Big|_{\mathbf{y}} \equiv \mathbb{E}_{\mathbf{x}\sim\mathbf{y}}[\hat{f}_z^L([\mathbf{x}]_{+i})] - \mathbb{E}_{\mathbf{x}\sim\mathbf{y}}[\hat{f}_z^L([\mathbf{x}]_{-i})] \overset{\text{Lemma 1}}{=} \dot{\hat{f}}_z^L([\mathbf{y}]_{+i}) - \dot{\hat{f}}_z^L([\mathbf{y}]_{-i}), \text{ for all } i \in V. \tag{13}$$

In other words, our estimator is constructed by replacing the multilinear relaxation G_z in Eq. (7) with the multilinear relaxation of the approximating polynomial $\hat{f}_z$. In turn, by Lemma 1, *the latter can be computed deterministically (without any sampling of the Bernoulli variables* $\mathbf{x} \sim \mathbf{y}$*)*, in closed form: the latter is given by the multilinearization $\dot{\hat{f}}_z^L$ of polynomial $\hat{f}_z^L$.

Nevertheless, our deterministic estimator given by Eq. (13) has a *bias*, precisely because of our approximation of f_z via the polynomial $\hat{f}_z^L$. We characterize this bias via the following lemma:

Lemma 2. *Assume that function* f_z *satisfies Assumption 3. Let* ∇G_z *be the unbiased stochastic gradient for a given* f_z *and let* $\widehat{\nabla G_z^L}$ *be the estimator of the multilinear relaxation given by* (13). *Then,* $\left\|\nabla G_z(\mathbf{y}) - \widehat{\nabla G_z^L}(\mathbf{y})\right\|_2 \leq 2\sqrt{n}\varepsilon_z(L)$, *for all* $\mathbf{y} \in \mathcal{C}$.

The proof can be found in App. A of [21]. Hence, we can approximate ∇G arbitrarily well, uniformly over all $\mathbf{x} \in [0,1]^n$. We can thus use our estimator in the SCG algorithm instead of of the sample estimator of the gradient (Eq. (8)). We prove that this yields the following guarantee:

Theorem 2. *Consider Stochastic Continuous Greedy (SCG) outlined in Algorithm 1. Recall the definition of the multilinear extension function G in* (2). *If Assumption 1 is satisfied and* $\rho_t = 4/(t+8)^{2/3}$, *then the objective function value for the iterates generated by SCG satisfies the inequality*

$$\mathbb{E}[G(\mathbf{y}_T)] \geq (1 - 1/e)OPT - \frac{15DK}{T^{1/3}} - \frac{f_{\max} r D^2}{2T},$$

where $K = \max\{3\|\nabla G(\mathbf{y}_0 - \mathbf{d}_0)\|^2, \sqrt{16\sigma_0^2 + 224\sqrt{n}\varepsilon(L)} + 2\sqrt{r} f_{\max} D\}$, $OPT = \max_{\mathbf{y}\in\mathcal{C}} G(\mathbf{y})$, *r is the rank of the matroid* $\mathcal{I}$, $\varepsilon(L) = \mathbb{E}_{z\sim P}[\varepsilon_z(L)]$, $f_{\max}$ *is the maximum marginal value of the function f, i.e.,* $f_{\max} = \max_{i\in\{1,\ldots,n\}} f(\{i\})$, *and* $\sigma_0^2 = \sup_{\mathbf{y}\in\mathcal{C}} \mathbb{E}_{z\sim P}\left[\|\nabla G(\mathbf{y}) - \nabla G_z(\mathbf{y})\|^2\right]$.

The proof can be found in App. B Our proof follows the main steps of [19] , using however the bias guarantee from Lemma 2; to do so, we need to deal with the fact that our estimator is not unbiased, but also that stochasticity is still present (as variables z are still sampled randomly). This is also reflected in our bound, that contains both a bias term (via $\varepsilon(L)$) and a variance term (via σ_0).

Comparing our guarantee to Theorem 1, we observe two main differences. On one hand, we have replaced the uniform bound of the variance σ^2 with the smaller quantity σ_0^2: the latter is quantifying the gradient variance w.r.t. z, and is thus smaller than σ, that depends on the variance of *both* z *and* $\mathbf{x} \sim \mathbf{y}$. Crucially, σ_0^2 is an "inherent" variance, *independent of the gradient estimation process*: it is the variance due to the randomness z, which is inherent in how we access our stochastic submodular objective and thus cannot be avoided. On the other hand, this variance reduction comes at the expense of introducing a bias term. This, however, can be suppressed via Assumption 3; as we discuss in the next section, for several problems of interest, this can be made arbitrarily small using only a polynomial number of terms in $\hat{f}_z^L$.

5 Problem Examples

In this section, we list several problems that can be tackled through our approach, also summarized in Table 1; these are similar to the problems considered by Özcan et al. [22], but cast into the stochastic submodular optimization setting. All problems correspond to trivially bounded variances σ_0^2 (again, because functions f_z are bounded); we thus focus on determining their bias $\epsilon(L)$. For space reasons, we report Cache Networks (CN) in Table 1, but provide details for it in the [21].

5.1 Data Summarization (SM) [12,15,17]

In data summarization, ground set V is a set of tokens, representing, e.g., words or sentences in a document. A corpus of documents V_z is presented to us sequentially, and the goal is to select a "summary" $S \subseteq V$ that is representative of V_z. The summary should be simultaneously (a) representative of the corpus, and (b) diverse.

To be representative, the summary $S \subset V$ should contain tokens of high value, where the value of a token is document-dependent: for document $z \in V_z$, token $i \in V$ has a value $r_{i,z} \in [0,1]$, where $\sum_i r_{i,z} = 1$. An example of such a value is the term frequency, i.e., the number of times the token appears in the document, divided by the document's length (in tokens). To be diverse, the summary should contain tokens that cover different subjects. To that end, if tokens are partitioned in to subjects, represented by a partition $\{P_j\}_{j=1}^J$ of V, the objective is given by $f(\mathbf{x}) = \mathbb{E}_z(f_z(\mathbf{x}))$ where $f_z(\mathbf{x}) = \sum_{j=1}^J h\left(\sum_{i \in V \cap P_j} r_{i,z} x_i\right)$, and $h(s) = \log(1+s)$ is a non-decreasing concave function. Intuitively, the concavity of h suppresses the selection of similar tokens (corresponding to the same subject), even if they have high value, thereby promoting diversity. Functions f_z (and, thereby, also f) are monotone and submodular, and we can construct polynomial approximators $\hat{f}_z^L$ for them as indicated in Table 1 by replacing h with its Lth-order Taylor approximation around $1/2$, given by:

$$\hat{h}^L(s) = \sum_{\ell=0}^{L} \frac{h^{(\ell)}(1/2)}{\ell!}(s - 1/2)^\ell. \tag{14}$$

Table 1. Summary of problems satisfying Assumption 1& 3.

	Input	$g_z : \{0,1\}^{\lvert V \rvert} \to [0,1]$ $\mathbf{x} \to g_z(\mathbf{x})$	$f_z : \{0,1\}^{\lvert V \rvert} \to \mathbb{R}_+$ $\mathbf{x} \to f_z(\mathbf{x})$	$\hat{f}_z^L : \{0,1\}^{\lvert V \rvert} \to \mathbb{R}_+$ $\mathbf{x} \to \hat{f}_z^L(\mathbf{x})$	Bias $\varepsilon(L)$
SM	Weighted bipartite graph $G = (V \cup P)$ weights $\mathbf{r}_z \in \mathbb{R}_+^n$, and $\sum_{i=1}^n r_{i,z} = 1$	$\sum_{i \in V \cap P_j} r_{i,z} x_i$	$\sum_{j=1}^J h(g_z(\mathbf{x}))$, where $h(s) = \log(1+s)$	$\hat{h}^L(g_z(\mathbf{x}))$, where $\hat{h}^L$ is Eq. (14)	$\frac{1}{(L+1)2^{L+1}}$
IM	Instances $G = (V, E)$ of a directed graph, partitions $P_v^z \subset V$	$\sum_{i \in V} \frac{1}{N}\left(1 - \prod_{u \in P_i^z}(1 - x_u)\right)$	$h(g_z(\mathbf{x}))$ where $h(s) = \log(1+s)$	$\hat{h}^L(g_z(\mathbf{x}))$, where $\hat{h}^L$ is Eq. (14)	$\frac{1}{(L+1)2^{L+1}}$
FL	Complete weighted bipartite graph $G = (V \cup V')$ weights $w_{i_\ell,z} \in [0,1]^{N \times \lvert z \rvert}$	$\sum_{\ell=1}^{N}(w_{i_\ell,z} - w_{i_{\ell+1},z})\left(1 - \prod_{k=1}^{\ell}(1 - x_{i_k})\right)$	$h(g_z(\mathbf{x}))$ where $h(s) = \log(1+s)$	$\hat{h}^L(g_z(\mathbf{x}))$, where $\hat{h}^L$ is Eq. (14)	$\frac{1}{(L+1)2^{L+1}}$
CN	Graph $G = (V, E)$, service rates $\mu \in \mathbb{R}_+^{\lvert z \rvert}$, requests $r \in \mathcal{R}$, P_z path of r, arrival rates $\lambda \in \mathbb{R}_+^{\lvert \mathcal{R} \rvert}$	$\frac{1}{\mu_z}\sum_{r \in \mathcal{R}: z \in p^r} \lambda^r \prod_{k'=1}^{k_{p^r}(v)}(1 - x_{p_k^r, i^r})$	$h(g_z(\mathbf{0})) - h(g_z(\mathbf{x}))$ where $h(s) = s/(1-s)$	$\hat{h}^L(g_z(\mathbf{x}))$, where $\hat{h}^L$ is Eq. (47) in [21]	$\frac{\bar{s}^{L+1}}{1-\bar{s}}$

This is because the composition of polynomial $\hat{f}_z^L$ with polynomial g_z in Table 1 is again a polynomial. We show in [21] that this estimator ensures that f indeed satisfies Assumption 3. Moreover, the estimator bias *decays exponentially* with degree L (see Table 1 and [21]), meaning that polynomial number of terms suffice to reduce the bias to a desired level. A partition matroid can be used with this objective to enforce that no more than k_ℓ sentences come from ℓ-th user, etc.

5.2 Influence Maximization (IM) [8,13]

Given a directed graph $G = (V, E)$, we wish to maximize the expected fraction of nodes reached if we infect a set of nodes $S \subseteq V$ and the infection spreads via, e.g., the Independent Cascade (IC) model [13]. Adding a concave utility to the fraction can enhance the value of nodes reached in early stages. Formally, let z can be a random simulation trace of the IC model, and $P_v^z \subseteq V$ is the set of nodes reachable from v in a random simulation of the IC model. Then, the objective can be written as $f(\mathbf{x}) = \mathbb{E}_{z\sim P}[f_z(\mathbf{x})]$ where $f_z(\mathbf{x}) = h\left(g_z(\mathbf{x})\right), h(s) = \log(1+s)$, and $g_z(\mathbf{x}) = \sum_{v\in V} \frac{1}{N}\big(1 - \prod_{i\in P_v^z}(1 - x_i)\big)$ is the number of infected nodes under seed set $\mathbf{x}$. Since functions $g_z : [0,1]^N \to [0,1]$ are multilinear, monotone submodular and $h : [0,1] \to \mathbb{R}$ is non-decreasing and concave, f satisfies Assumption 1 [22]. Again, we can construct $\hat{f}^L$ by replacing h by $\hat{h}^L$, given by Eq. (14). This again ensures that f indeed satisfies Assumption 3, and the estimator bias again decays exponentially (see Table 1 and [21]). Partition matroid constraints could be used in this setting to bound the number of seeds from some group (e.g., males/females, people in a zip code, etc.).

5.3 Facility Location (FL) [18]

Given a weighted bipartite graph $G = (V \cup V_z)$ and weights $w_{i,z} \in [0,1]$, $i \in V$, $z \in V_z$, we wish to maximize:

$$f(S) = \mathbb{E}_{z\sim P}\left[h(\max_{i\in S} w_{i,z})\right], \tag{15}$$

where $h(s) = \log(1 + s)$. Intuitively, V and V' represent facilities and customers respectively and $w_{v,v'}$ is the utility of facility v for customer v'. The goal is to select a subset of facility locations $S \subset V$ to maximize the total utility, assuming every customer chooses the facility with the highest utility in the selection S; again, adding the concave function h adds diversity, favoring the satisfaction of customers that are not already covered. This too becomes a coverage problem by observing that [11]: $\max_{i\in S} w_{i,z} = \sum_{\ell=1}^{n} (w_{i_\ell,z} - w_{i_{\ell+1},z})\big(1 - \prod_{k=1}^{\ell}(1 - x_{i_k})\big)$, where, for a given $z \in V_z$, weights have been pre-sorted in a descending order as $w_{i_1,z} \geq \ldots \geq w_{i_n,z}$. and $w_{i_{n+1},j} \triangleq 0$. In a manner similar to Sec 5.2, we can show that this function again satisfies Assumption 1 and 3, using again the Lth-order Taylor approximation of h, given by Eq. (14); this will again lead to a bias that decays exponentially (see Table 1 and [21]). We can again optimize such an objective over arbitrary matroids, which can enforce, e.g., that no more than k facilities are selected from a geographic area or some other partition of V.

6 Experiments

We evaluate Alg. 1, with sampling and polynomial estimators over two well-known problem instances (influence maximization and facility location) with real and synthetic datasets. We summarize these setups in Table 2. For a more detailed overview of the datasets and experiment parameters, please refer to [21]. Our code is publicly accessible.[2]

Algorithms. We compare the performance of different estimators. These estimators are: (a) sampling estimator (SAMP) with $N = 1, 10, 20, 100$ and (b) polynomial estimator (POLY) with $L = 1, 2$.

Table 2. Datasets and Experiment Parameters.

instance	dataset	$\|z\|$	$\|S\|$	$\|E\|$	m	k
IM	SBPL	20	400	914	4	1
IM	ZKC	20	34	78	2	3
FL	MovieLens	4000	6041	256	10	2

Metrics. We evaluate the performance of the estimators with their clock running time and via the maximum result ($\max f(\mathbf{y})$) obtained using the best available estimator for a given setting.

Results. The trajectory of the utility obtained at each iteration of the stochastic continuous greey algorithm $f(\mathbf{y})$ is plotted as a function of time in Fig. 1. In Fig. 1(a), we observe that polynomial estimators outperforms sampling estimators in terms of utility. Moreover, POLY1 runs 10 times faster than SAMP20 and runs in comparable time to SAMP1. In Fig. 1(b), POLY2 outperforms all estimators whereas POLY1 underperforms. Finally, in Fig. 1(c) we observe that POLY1 consistently outperforms sampling estimators.

The final outcomes of the objective functions of the estimators are reported as a function of time in Fig. 2. In Fig. 2(a) and 2(b), POLY2 outperforms other estimators in terms of utility. Again in Fig. 2(a), POLY1 outperforms sampling estimators in terms of utility and runs in comparable time to SAMP1 while in Fig. 2(c), POLY1 outperforms sampling estimators both in terms of time and utility. Ideally, we would expect the performance of the estimators to improve as the degree of the polynomial or the number of samples increase. The examples

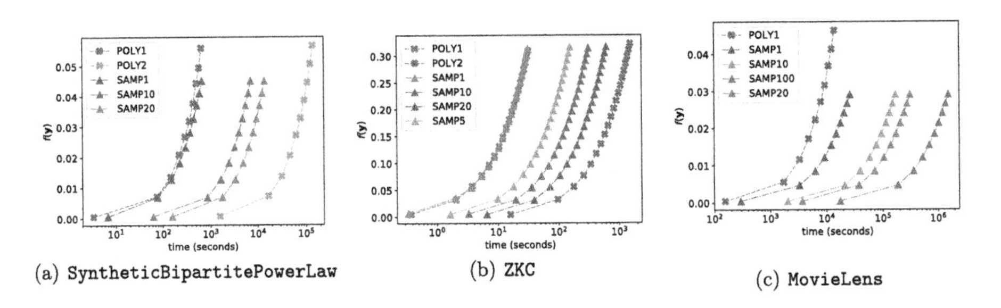

(a) SyntheticBipartitePowerLaw (b) ZKC (c) MovieLens

Fig. 1. Trajectory of the FW algorithm. Utility of the function at the current $\mathbf{y}$ as a function of time is marked for every iteration.

[2] https://github.com/neu-spiral/StochSubMax.

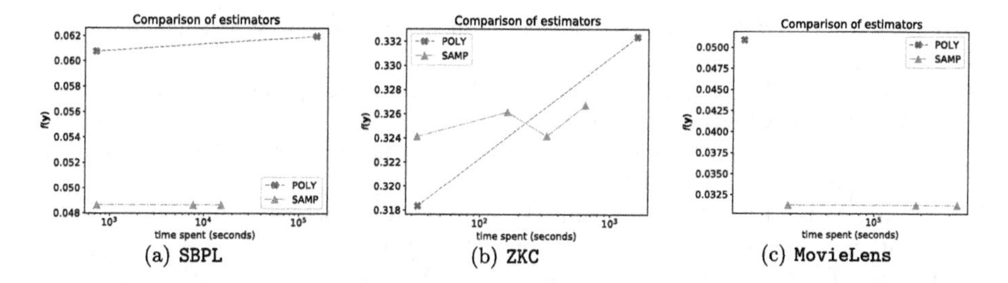

Fig. 2. Comparison of different estimators on different problems. Blue lines represent the performance of the POLY estimators and the marked points correspond to POLY1 and POLY2 respectively. Orange lines represent the performance of the SAMP estimators and the marked points correspond to SAMP1, SAMP10, SAMP20, SAMP100 respectively. (Color figure online)

where this is not always the case can be explained by the stochastic nature of the problem.

7 Conclusions

We show that polynomial estimators can improve existing stochastic submodular maximization methods by eliminating one of the two sources of randomness, particularly the one that stems from sampling. Investigating methodical ways to construct such polynomials can expand the applications of the proposed estimator appearing in this paper. Online versions of stochastic submodular optimization, where performance is characterized in terms of (approximate) regret, are also a possible future research direction.

Acknowledgements. This work is supported by NSF grant CCF-1750539.

References

1. Ageev, A.A., Sviridenko, M.I.: Pipage Rounding: a new method of constructing algorithms with proven performance guarantee. J. Comb. Optim. (2004)
2. Asadpour, A., Nazerzadeh, H., Saberi, A.: Stochastic submodular maximization. In: International Workshop on Internet and Network Economics (2008)
3. Bateni, M., Chen, L., Esfandiari, H., Fu, T., Mirrokni, V., Rostamizadeh, A.: Categorical feature compression via submodular optimization. In: ICML (2019)
4. Broida, J., Williamson, S.: A Comprehensive Introduction to Linear Algebra. Addison-Wesley, Advanced book program (1989)
5. Calinescu, G., Chekuri, C., Pal, M., Vondrák, J.: Maximizing a monotone submodular function subject to a matroid constraint. SICOMP (2011)
6. Chekuri, C., Vondrák, J., Zenklusen, R.: Dependent randomized rounding via exchange properties of combinatorial structures. In: FoCS. IEEE (2010)
7. Chen, L., Hassani, H., Karbasi, A.: Online continuous submodular maximization. In: AISTATS (2018)

8. Chen, W., Wang, Y., Yang, S.: Efficient influence maximization in social networks. In: KDD (2009)
9. Goyal, A., Bonchi, F., Lakshmanan, L.V.: A data-based approach to social influence maximization. VLDB Endowment (2011)
10. Hassani, H., Soltanolkotabi, M., Karbasi, A.: Gradient methods for submodular maximization. NeurIPS (2017)
11. Karimi, M.R., Lucic, M., Hassani, H., Krause, A.: Stochastic submodular maximization: the case of coverage functions. In: NeurIPS17 (2017)
12. Kazemi, E., Mitrovic, M., Zadimoghaddam, M., Lattanzi, S., Karbasi, A.: Submodular streaming in all its glory: tight approximation, minimum memory and low adaptive complexity. In: ICML (2019)
13. Kempe, D., Kleinberg, J., Tardos, É.: Maximizing the spread of influence through a social network. In: KDD (2003)
14. Lei, Q., Wu, L., Chen, P.Y., Dimakis, A., Dhillon, I.S., Witbrock, M.J.: Discrete adversarial attacks and submodular optimization with applications to text classification. MLSys (2019)
15. Lin, H., Bilmes, J.: A class of submodular functions for document summarization. In: ACL (2011)
16. Mahdian, M., Moharrer, A., Ioannidis, S., Yeh, E.: Kelly cache networks. IEEE/ACM Transactions on Networking (2020)
17. Mirzasoleiman, B., Badanidiyuru, A., Karbasi, A.: Fast constrained submodular maximization: personalized data summarization. In: ICML (2016)
18. Mokhtari, A., Hassani, H., Karbasi, A.: Conditional gradient method for stochastic submodular maximization: closing the gap. In: AISTATS (2018)
19. Mokhtari, A., Hassani, H., Karbasi, A.: Stochastic conditional gradient methods: from convex minimization to submodular maximization. In: JMLR (2020)
20. Nemhauser, G.L., Wolsey, L.A., Fisher, M.L.: An analysis of approximations for maximizing submodular set functions-i. Math. Program. (1978)
21. Özcan, G., Ioannidis, S.: Stochastic submodular maximization via polynomial estimators. arXiv preprint arXiv:2303.09960 (2023)
22. Özcan, G., Moharrer, A., Ioannidis, S.: Submodular maximization via Taylor series approximation. In: SDM (2021)
23. Su, F., Zhu, Y., Wu, O., Deng, Y.: Submodular meta data compiling for meta optimization
24. Vondrák, J.: Optimal approximation for the submodular welfare problem in the value oracle model. In: STOC (2008)
25. Zhang, Q., Deng, Z., Chen, Z., Hu, H., Yang, Y.: Stochastic continuous submodular maximization: Boosting via non-oblivious function. In: ICML (2022)

Transfer Learning and Meta Learning

Few-Shot Human Motion Prediction for Heterogeneous Sensors

Rafael Rego Drumond(✉), Lukas Brinkmeyer(✉), and Lars Schmidt-Thieme

University of Hildesheim, Hildesheim, Germany
{radrumond,brinkmeyer,schmidt-thieme}@ismll.uni-hildesheim.de

Abstract. Human motion prediction is a complex task as it involves forecasting variables over time on a graph of connected sensors. This is especially true in the case of few-shot learning, where we strive to forecast motion sequences for previously unseen actions based on only a few examples. Despite this, almost all related approaches for few-shot motion prediction do not incorporate the underlying graph, while it is a common component in classical motion prediction. Furthermore, state-of-the-art methods for few-shot motion prediction are restricted to motion tasks with a fixed output space meaning these tasks are all limited to the same sensor graph. In this work, we propose to extend recent works on few-shot time-series forecasting with heterogeneous attributes with graph neural networks to introduce the first few-shot motion approach that explicitly incorporates the spatial graph while also generalizing across motion tasks with heterogeneous sensors. In our experiments on motion tasks with heterogeneous sensors, we demonstrate significant performance improvements with lifts from 10.4% up to 39.3% compared to best state-of-the-art models. Moreover, we show that our model can perform on par with the best approach so far when evaluating on tasks with a fixed output space while maintaining two magnitudes fewer parameters.

Keywords: Time-series forecasting · Human motion prediction · Few-shot learning

1 Introduction

Time-series forecasting has become a central problem in machine learning research as most collected industrial data is being recorded over time. A specific application for time-series forecasting approaches is human motion prediction (or human pose forecasting), in which a multivariate time-series is given in the form of a human joint skeleton, and the objective is to forecast motion sequences based on previous observations. This area has recently seen various applications ranging from healthcare [21] and smart homes [11] to robotics [14,22]. Deep learning

R. R. Drumond and L. Brinkmeyer—Equal contribution.

H. Kashima et al. (Eds.): PAKDD 2023, LNAI 13936, pp. 551–563, 2023.
https://doi.org/10.1007/978-3-031-33377-4_42

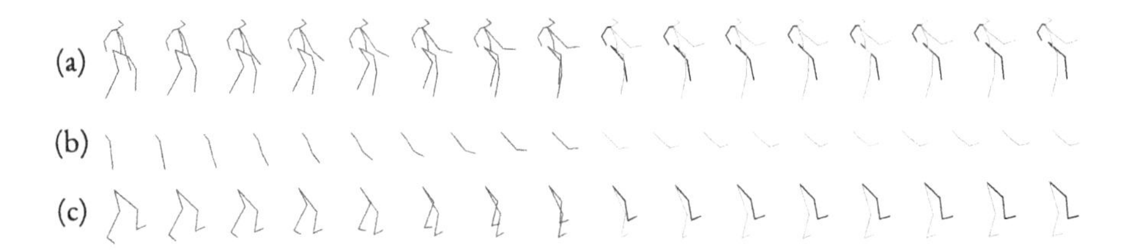

Fig. 1. Examples of three motion prediction tasks with observations (red) and forecasts (teal). (a) being a standard motion prediction on the full graph, while (b) and (c) are forecasts based on only a subgraph of the sensor skeleton learned by the same model. (Color figure online)

methods have shown state-of-the-art performances in the task of human motion prediction in recent years with a focus on popular time-series forecasting models including LSTM's and GRU's [17], temporal autoencoders [3], and more recently transformer-based approaches [15,16]. Moreover, employing graph-based models has shown to be advantageous in cases where the human joint skeleton can be utilized [13].

In few-shot motion prediction, we strive to forecast the motion for previously unseen actions using only a few labeled examples, in contrast to standard human motion prediction, where the training dataset already contains sufficient samples for each action that will be encountered during testing. This can be highly beneficial in practice, as it eliminates the need for such a dataset and allows for a more flexible application. For example, end users can then add new motions by demonstrating an action a few times before the model can accurately classify and forecast future frames. Current approaches for motion prediction are limited to a fixed attribute space such that every observation needs to be recorded across the same set of input sensors. However, an ideal model should be able to cope with only a subset of motion sensors, as not every user should be required to have motion sensors for the full human skeleton. Also, not every action requires information from every possible sensor, e.g., recordings of only the arm for the motion "waving." An example of this is shown in Fig. 1, where a motion prediction for the complete human skeleton, but also partial subgraphs of it, is demonstrated. In few-shot learning, this setup is referred to as learning across tasks with heterogeneous attributes [1,10] and is typically tackled by employing a model which operates on attribute sets (in contrast to vectors) which inherently do not possess any order.

In human motion prediction, the attributes represent sensors distributed on a human skeleton [4,9], meaning they possess order in the form of a graph structure. This information is often used in approaches for classical human motion prediction but not in the current literature for few-shot motion prediction. In a few-shot setting for tasks with heterogeneous sensors, the model would encounter varying graphs in training, similar to classical graph classification approaches [12]. In this chosen scenario, each motion prediction task has a different set of sensors (attributes) that are shared across their subjects, each frame (or pose) corresponds to one time-step, and, finally, the placement of the existing sensors on the subject's body is represented by the task's graph. In this work, we propose the first model for few-shot motion prediction that incorporates the underlying graph information while generalizing across tasks with heterogeneous sensors. We evaluate our approach on different variations of the popular Human3.6M dataset and demonstrate improvements over all related methods. The contributions of this work are the following:

1. We propose the first model for few-shot motion prediction that incorporates the underlying graph structure, while also being the first model for few-shot motion prediction which generalizes to motion tasks with heterogeneous sensors.
2. We conduct the first few-shot human motion experiments on tasks with heterogeneous sensors where we can show significant performance improvements over all related baselines with performance lifts ranging from 10.4% to 39.3%.
3. We demonstrate minor performance improvements over state-of-the-art approaches in the standard experimental setup while maintaining two magnitudes fewer parameters within our model.
4. We also provide code for our method as well as for two of our baselines that have not published a working implementation.

2 Related Work

This work lies in the intersection of few-shot learning (FSL) and human motion prediction. Thus we will discuss the related work of both areas before summarizing the work in the analyzed field. FSL [23] aims to achieve a good generalization on a novel task that contains only a few labeled samples based on a large meta-dataset of related tasks. There are different techniques, including metric-based [20], gradient-based [7], and memory-based approaches [24], that have shown successful results. They typically all involve meta-training across the meta-dataset while performing some adaptation to the test task at hand.

Recently, different works have tried to extend few-shot learning to generalize across tasks that vary in their input [2] or output space [5]. One is to apply permutation-invariant and -equivariant models that operate on sets of elements through the use of deep sets [25]. TimeHetNet [1] extended this approach to perform few-shot time-series forecasting on tasks with a single target variable and a varying amount of covariates. CHAMELEON [2] allows vector data-based tasks to have different shapes and semantics as long as the attributes can be mapped to a common alignment. All these methods, however, did not consider any structural relation between the attributes and operate purely on sets of scalar attributes.

Motion Forecasting (or Pose Forecasting, or Pose Estimation) is the task of predicting the subsequent frames of a sequence of human poses. This data can be collected directly as images, or with accelerometers and gyroscopes [18].

Most approaches naturally rely on standard deep learning methods for time-series forecasting such as Variational Auto Encoders, LSTMs, and recurrent convolution networks [6,11,14,21,22]. These methods are devised for different motion applications that vary in the type of sensors or forecasting length. For example, PEEK [6] and the work of Jalal et al. [11] require only motion data from the arms, while Gui et al. [8] use the rotation of the main joints of the complete human body to predict future time-steps. None of these approaches, however, are designed to handle tasks where the set of motion sensors varies.

There are two recent approaches published for few-shot human prediction that we will focus on in this paper as baselines. PAML [8] consists of the popular meta-learning approach MAML [7] operating on top of the classical motion prediction model RESIDUAL-SUP [17]. It incorporates a simple look-ahead method for the decoder weights based on pre-trained weights on a bigger dataset to fine-tune the model for a new task. MoPredNet [26,27] is a memory-based approach that uses attention and an external memory of pretrained decoder weights to compute the weights for a new task. Although these two methods work with different tasks separated by the human action performed in each pose sequence, they require the same set of sensors for each task.

In this paper, we present GraphHetNet (GHN): a graph-based approach to adapt the TimeHetNet [1] architecture to train across different human motion detection tasks with heterogeneous sensors by integrating information of neighboring sensors through the application of graph convolutional networks [12]. Thus, we can combine both graph and time-series information into our few-shot predictions.

3 Methodology

3.1 Problem Definition

We formulate few-shot motion prediction as a multivariate temporal graph problem. In standard human motion prediction, we are given a graph $\mathcal{G} = (\mathcal{V}, A)$ as predictor data where the vertex set $\mathcal{V}$ consists of C motion sensors $\{1, ..., C\}$ and $A \in \mathbb{R}^{C \times C}$ is a symmetric adjacency matrix representing the edges between sensors with $A_{ij} = 1$ iff sensors i and j are connected by an edge, e.g., an elbow and the shoulder. We also refer to this graph as motion graph, as it contains all the motion sensors. Additionally, we are given a set of node features $X = \{x_{ict}\} \in \mathbb{R}^{I \times T \times C}$ which represent a multivariate time-series with I instances over T time steps for the C motion sensors. We want to forecast the next H time steps given the observed T such that our target is given by $Y \in \mathbb{R}^{I \times H \times C}$.

Extending this formulation to few-shot learning, we are given a set of M tasks $D := \{(D_1^s, D_1^q), ..., (D_M^s, D_M^q)\}$ called meta-dataset where each task consists of support data D^s and query data D^q with $D_m^s := (\mathcal{G}_m, X_m^s, Y_m^s)$ and $D_m^q := (\mathcal{G}_m, X_m^q, Y_m^q)$. The graph is shared across instances of both support and query for a given task. We want to find a model ϕ with minimal expected

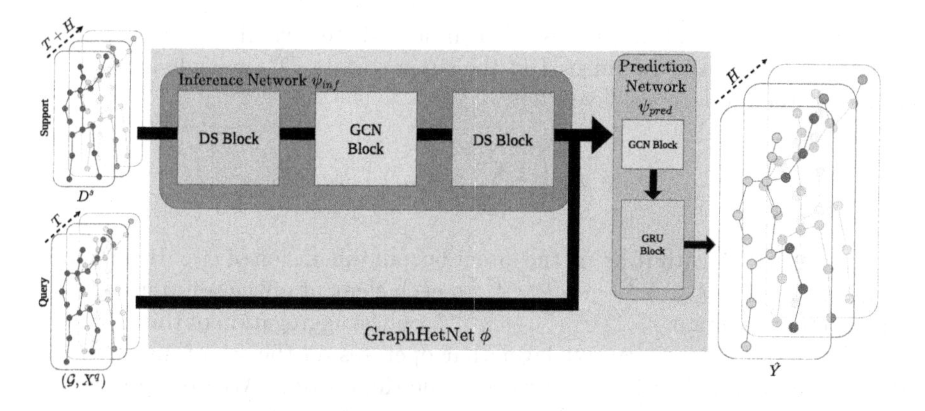

Fig. 2. The pipeline for our proposed approach GraphHetNet. *DS Block* stands for Deep Set Block and *GCN Block* for Graph Convolution Network Block. The network takes the full support data $D^s = (\mathcal{G}, X^s, Y^s)$ and the predictors of the query data $(\mathcal{G}, X^q)$ and outputs a set of outputs $\hat{y}$ represent the next H frames after the T frames of the instances in D^q. Batch dimensions are omitted for simplicity.

forecasting loss over the query data of all tasks when given the labeled support data, and predictor of the query data:

$$\min_{\phi} \frac{1}{M} \sum_{(D^s_m, D^q_m) \in D} \mathcal{L}(Y^q_m, \phi(G_m, X^q_m, D^s)) \tag{1}$$

In the standard setting $\mathcal{G}_m = \mathcal{G}_{m'}$ $\forall m, m' \in M$ $(m \neq m')$, which means that the structure of the graph $\mathcal{G}$ does not vary across the meta-dataset. Thus, each sample of each task contains the same set of motion sensors $\mathcal{V}$ with an identical adjacency matrix A. We want to generalize this problem to tasks with heterogeneous sensors, meaning that the underlying graph structure and the set of vertices vary across tasks ($\mathcal{G}_m \neq \mathcal{G}_{m'}$ $\forall m, m' \in M$ $(m \neq m')$), while it is shared between support and query data of the same task. Thus, the number of motion sensors C is not fixed and depends on the task at hand.

3.2 GraphHetNet

Our model GraphHetNet denoted by ϕ is based on TimeHetNet [1], which uses a set approach for few-shot time-series forecasting with heterogeneous attributes similar to the approach of Iwata et al. [10]. The overall architecture consists of two main components: First, the inference network, which processes the predictor and target data of the support set D^s of a task to generate a latent task representation which should contain useful information to forecast the query instances. Second, the prediction network computes the actual motion forecast for the query set D^q of the task at hand based on its predictors and the task embedding of the support network. In prior approaches [1,10], both components are composed of multiple stacked deep set blocks (*DS Block*), which process the input data as a set of attributes. To compute the embeddings for every single vertex $c \in C$ over the instances I of the support data D^s, a single layer in such a block is then a deep set layer [25]:

$$w_c = g_{\text{DS}} \left(\frac{1}{I} \sum_{i=1}^{I} f_{\text{DS}}(x_{ic}) \right) \ \forall c \in C \tag{2}$$

Here, $w_c \in \mathbb{R}^{T \times K}$ with K being the latent output dimension of g_{DS}. By employing an inner function $f_{\text{DS}} : \mathbb{R}^{T \times 1} \to \mathbb{R}^{T \times K}$ on each element of the set of instances X, and an outer function $g_{\text{DS}} : \mathbb{R}^{T \times K} \to \mathbb{R}^{T \times K}$ on the aggregation of this set, we can model a permutation-invariant layer that operates on the set of instances. The theoretical foundation of this layer lies in the Kolmogorov-Arnold representation theorem, which states that any multivariate continuous function can be written as a finite composition of continuous functions of a single variable and the binary operation of addition [19].

In contrast to previous approaches that operate on heterogeneous attributes, we do not utilize *DS* blocks to aggregate the information across attributes, but only across instances, as our problem's attributes are motion sensors structured in a graph and not in a set. Instead, we include blocks of graph convolutional layers (*GCN Block*) [12] in both the inference and the prediction network. We can then aggregate information across sensors by stacking graph convolutional layers. A single layer in the block is then defined as:

$$u_{ic} = g_{\text{GCN}} \left(\left[x_{ic}, \sum_{j \in N(c)} f_{\text{GCN}}(x_{ij}) \right] \right) \ \forall c \in C \ \forall i \in I \tag{3}$$

where $u_{ic} \in \mathbb{R}^{T \times K}$, $N(c)$ is the set of all vertices that are in the neighborhood of c meaning $A_{cj} = 1$ for every $j \in N(c)$ and $[.]$ is the concatenation along the latent feature axis. The inner function $f_{\text{GCN}} : \mathbb{R}^{T \times 1} \to \mathbb{R}^{T \times K}$ prepares the neighbor embeddings, while the outer function $g_{\text{GCN}} : \mathbb{R}^{T \times 2K} \to \mathbb{R}^{T \times K}$ updates the vertex features of the respective sensor with its aggregated neighbor messages. Note that this layer only captures the information across motion sensors, not instances. The models $g_{\text{GCN}}, f_{\text{GCN}}, g_{\text{DS}}, f_{\text{DS}}$ are Gated recurrent units (GRU) to deal with the temporal information. As shown in Fig. 2, our full model GraphHetNet ϕ consists of the two model components inference and prediction network. The inference network ψ_{inf} processes the full support data D^s to compute the task embeddings across instances and motion sensors. The prediction network ψ_{pred} processes the query data to output the final forecast. Thus, the prediction $\hat{Y}$ of our model for a task m is given by:

$$\hat{Y}_m = \phi(X^q_m, G^q_m, D^s) = \psi_{\text{pred}}(X^q_m, G^q_m, \psi_{\text{inf}}(G^s_m, X^s_m, Y^s_m)) \tag{4}$$

The inference model ψ_{inf} is composed of a *GCN* block in between two *DS* blocks to capture both information across instances and motion sensors. The prediction network ψ_{pred} consists of a *GCN* block, followed by a block of stacked GRU layers (*GRU Block*) which compute the target motion forecast.

4 Results

We conducted multiple experiments on the Human3.6M dataset [4,9], consisting of 17 motion categories recorded for 11 subjects, resulting in 3.6 million frames in total. We want to evaluate our approach for few-shot motion tasks with heterogeneous sensors such that each task contains a subset of the vertices of the full motion graph, with the graph of each task being an induced subgraph of the original one. We also conduct an ablation on the standard few-shot motion prediction setting proposed in prior approaches [8,26,27] that considers homogeneous tasks only, meaning each task contains all sensors in identical order.

4.1 Experimental Setup

In both cases, we have 11 actions in meta-training (directions, greeting, phoning, posing, purchases, sitting, sitting down, taking a photo, waiting, walking a dog, and walking together) and 4 actions in meta-testing (walking, eating, smoking, and discussion). Furthermore, we also utilize the same split across subjects for meta-test and meta-training as proposed by Gui et al. [8]. The task is to forecast the next 10 frames (400ms) given the previous 50 frames (2000ms) across the given set of sensors. A single task consists of five support instances and two query instances which means that the model needs to adapt to a previously unseen action based on five labeled instances only. During meta-training, each meta-batch consists of one task per action totaling 11 tasks. The tasks in the classical setting contain all nonzero angles for each of the 32 joints totaling 54 angles as motion sensors. In our main experiment on heterogeneous sensors, each task has only a subset of the set of all sensors. In particular, we sample an induced subgraph of the original human skeleton graph by selecting a random sensor as the initial root node and then recursively adding a subset of neighboring vertices to the graph, including all edges whose endpoints are both in the current subset. The statistics of the original motion graph of Human3.6M and our sampled induced subgraphs are given in Table 2. The number of unique tasks we sample during our experiments is enormous, as is the number of possible induced subgraphs from a given source graph. We evaluated this empirically by sampling one million tasks from the full graph and found around 842,872 unique tasks, meaning only around 16% of the subgraphs were sampled more than once. This guarantees that many tasks our model encounters during meta-testing are previously unseen. More details on the task sampling procedure are stated in the appendix.

We compare against three non-meta-learning baselines, which are variations of the popular detection network RESIDUAL-SUP [17], which consists of stacked GRU's with residual skip connections: RES-SUP$_{single}$ trains the model on the support data of the test task at hand only while evaluating the query data. RES-SUP$_{all}$ trains the model on the data of all the meta-training actions in standard supervised fashion. In the case of the heterogeneous tasks, the sensor dimension is padded with zeros to 54 since the model is not equipped to deal with heterogeneous sensor sets. The query data of the meta-test tasks is used to evaluate the final performance. RES-SUP$_{trans}$ uses RES-SUP$_{all}$ as a pretrained model to then fine-tune it to the support data of the test task at hand before

Table 1. Results few-shot motion prediction with heterogenous sensors given in Mean Angle Error of different methods on Human3.6M. Best results are in bold, second best are underlined. The percentage improvement is given for our model compared to the respective second-best one.

	walking					smoking				
	80	160	320	400	Avg	80	160	320	400	Avg
RES-SUP$_{single}$ [17]	0.65	1.15	2.06	2.40	1.57	0.76	1.17	1.99	2.02	1.48
RES-SUP$_{all}$ [17]	0.88	1.11	1.17	1.20	1.09	1.47	1.69	1.14	1.4	1.43
RES-SUP$_{trans}$ [17]	0.85	1.18	1.19	1.17	1.09	1.10	1.47	1.73	1.94	1.56
PAML [8]	0.26	0.39	0.56	0.64	0.46	0.58	0.64	0.69	0.83	0.69
TIMEHET [1]	0.23	0.30	0.44	0.53	0.37	0.49	0.52	0.58	0.62	0.55
MOPRED [26,27]	0.26	0.33	0.43	0.52	0.39	0.51	0.52	0.54	0.61	0.54
GHN (ours)	**0.17**	**0.22**	**0.30**	**0.37**	**0.27**	**0.41**	**0.42**	**0.43**	**0.48**	**0.44**
Lift in %	26.1	26.7	30.2	28.8	27.0	16.3	19.2	20.4	21.3	18.5

	discussion					eating				
	80	160	320	400		80	160	320	400	
RES-SUP$_{single}$ [17]	0.97	1.56	1.86	2.67	1.77	0.55	0.93	1.54	1.74	1.19
RES-SUP$_{all}$ [17]	0.96	1.11	1.30	1.44	1.2	0.85	1.03	0.92	1.05	0.96
RES-SUP$_{trans}$ [17]	1.30	1.42	1.68	1.75	1.53	0.68	0.78	0.94	1.03	0.86
PAML [8]	0.35	0.52	0.78	0.91	0.64	0.23	0.28	0.42	0.56	0.37
TIMEHET [1]	0.29	0.42	0.69	0.85	0.59	0.20	0.28	0.41	0.52	0.35
MOPRED [26,27]	0.34	0.42	0.62	0.77	0.54	0.25	0.29	0.39	0.59	0.38
GHN (ours)	**0.22**	**0.30**	**0.55**	**0.69**	**0.44**	**0.14**	**0.17**	**0.25**	**0.34**	**0.22**
Lift in %	24.1	28.6	11.3	10.4	18.5	30.0	39.3	35.9	34.6	37.1

evaluating the query data of it. Furthermore, we compare against the few-shot motion baselines PAML [8], and MOPREDNET [26,27], which both evaluate their approach in the homogeneous setup, as well as TIMEHETNET [1] as it is the first model for time-series forecasting across heterogeneous attributes. Both PAML and MOPREDNET do not have any publicized code (and we could not reach the authors about it), which is why the results for the standard setting are taken from their respective published results. At the same time, we re-implemented both models to evaluate them on the heterogeneous setup. For TIMEHETNET, we utilize the officially published code. We had to adapt it as the original model is built to forecast a single target variable given a set of covariates that span the future time horizon. In contrast, we want to forecast multiple variables in a set without any given future covariates. The adapted version of TIMEHETNET, as well as the reimplementations of PAML and MoPredNet, and the appendix, can also be found in our link: https://github.com/brinkL/graphhetnet. We optimized the hyperparameters of all models via grid search. For our approach, the best found configuration includes two graph convolutional layers per *GCN* block, the *DS* blocks contain three stacked GRUs each, and the number of units per GRU is 64. We optimize our model with Adam and a learning rate of 0.0001.

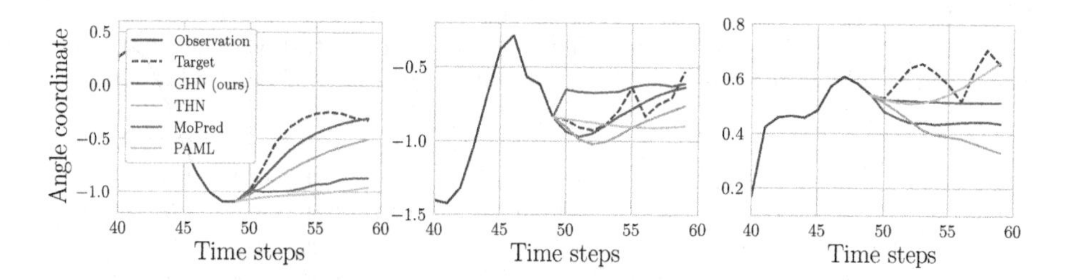

Fig. 3. Examples of three motion predictions in exponential map for GHNand baseline approaches. We sampled two examples where our approach (red) has the lowest error and one where a baseline performs best. Full past horizon is shown in appendix. (Color figure online)

Table 2. Statistics of the full Human3.6M graph and for the subgraphs sampled during training on tasks with heterogeneous attributes.

	Full	Sampled
vertices	54	26.8 ± 12.9
edges per vertex	6.6 ± 3.1	3.9 ± 1.7

Table 3. Number of parameters in the models PAML, MoPred, TIMEHETNET (as THN), and GRAPHHETNET (as GHN, ours) in multiples of 1000.

	PAML	MoPred	THN	GHN
Param	3,373K	40,945K	661K	265K

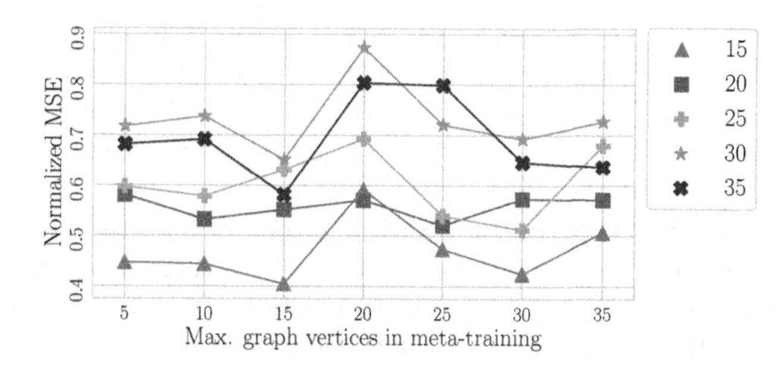

Fig. 4. Each line represents a model evaluated for tasks up to a certain number of sensors in test, while the x-axis shows the maximum number of sensors in meta-training. Results are given in MSE averaged across the normalized results for each action.

4.2 Results

The results for our experiment on few-shot tasks with heterogeneous sensors are shown in Table 1. Our approach outperforms all baselines with significant margins over all actions and time horizons. The performance improvements compared to the respective second best approach range from 10.4 percent for 400ms on the action "*discussion*" to 39.3 percent for the motion prediction at 160ms for the action "*eating.*" The second-best results are shared between TIMEHETNET and MOPREDNET (abbreviated MOPRED in the table). Three examples for

Table 4. Ablation on homogeneous setting: Mean Angle Error of different methods on Human3.6M dataset for standard few-shot motion prediction task with fixed attribute space. The results with * are taken from the published results of Zang et al. [26].

	walking					smoking				
	80	160	320	400	Avg	80	160	320	400	Avg
RES-SUP$_{single}$ [17]*	0.39	0.69	0.97	1.08	0.78	0.27	0.50	0.98	1.00	0.69
RES-SUP$_{all}$ [17]*	0.36	0.61	0.84	0.95	0.69	0.26	0.49	0.98	0.97	0.68
RES-SUP$_{trans}$ [17]*	0.34	0.57	0.78	0.89	0.65	0.26	0.48	0.93	0.91	0.65
PAML [8]*	0.4	0.69	0.97	1.08	0.79	0.34	0.63	1.13	1.12	0.80
TIMEHET [1]	0.32	0.37	0.70	0.94	0.58	0.43	0.46	0.69	0.68	0.57
MOPRED (reimp.)	0.42	0.52	0.77	0.98	0.67	0.48	0.54	0.71	0.94	0.67
MOPRED [26,27]*	0.21	**0.35**	**0.55**	**0.69**	**0.45**	0.26	0.47	0.93	0.9	0.64
GHN (ours)	**0.17**	**0.35**	0.69	0.94	0.54	**0.12**	**0.17**	**0.67**	**0.54**	**0.38**
Lift in %	19.0	0.0	-25.5	-36.2	-20.0	53.8	63.8	28.0	40.0	40.6

	discussion					eating				
Horizon in ms	80	160	320	400		80	160	320	400	
RES-SUP$_{single}$ [17]*	0.32	0.66	0.95	1.09	0.76	0.28	0.50	0.77	0.91	0.62
RES-SUP$_{all}$ [17]*	0.31	0.66	0.94	1.03	0.74	0.26	0.46	0.70	0.82	0.56
RES-SUP$_{trans}$ [17]*	0.30	0.65	0.91	0.99	0.71	0.22	0.35	0.54	**0.69**	0.45
PAML [8]*	0.36	0.72	1.03	1.15	0.82	0.29	0.51	0.8	0.95	0.64
TIMEHET [1]	0.33	0.49	1.00	1.31	0.78	0.28	0.35	0.61	0.91	0.54
MOPRED (reimp.)	0.51	0.67	0.99	1.12	0.82	0.35	0.47	0.62	0.83	0.56
MOPRED [26,27]*	0.29	0.63	**0.89**	**0.98**	**0.70**	0.21	0.34	0.53	**0.69**	0.44
GHN (ours)	**0.19**	**0.42**	0.94	1.25	**0.70**	**0.17**	**0.29**	**0.52**	0.75	**0.43**
Lift in %	34.5	33.3	-5.6	-27.6	0.0	19.0	14.7	1.9	-8.7	2.3

motion forecasts of this experiment are given in the Fig. 3 for two tasks where our approach has the highest performance and one task where a baseline approach performs better. As expected, the motion prediction of GRAPHHETNET is most similar to TIMEHETNET with our method being more accurate. When comparing the model capacity of our approach and the analyzed baselines based on the model parameters illustrated in Table 3, one can see that our model contains significantly fewer parameters, with two magnitudes difference to MOPREDNET [26]. TIMEHETNET is the closest with double the number of parameters. Further experimental results for all actions of the Human3.6M can be found in our appendix.

4.3 Ablations

We also evaluated our model in the standard homogeneous setting where all tasks share a fixed motion graph. This serves the purpose of evaluating whether our model, which is designed for heterogeneous tasks, shows any performance degradation or can perform on par with state-of-the-art approaches in the classical setup. The results are stated in Table 4. We implemented our own version of PAML and MoPredNet, as there is no public implementation in either approach. We received no further information when contacting the original authors. Both the published results and the results for our implementation of MoPredNet are given in the table, as we were not able to replicate the results reported in the publication. For PAML, we only show the reported results as our reimplementation achieves results that match the reported results. Our approach is shown to be on par with our baselines, MoPredNet while showing slight improvements for short-term frames after 80 and 160 ms. At the same time, the model capacity of our model is two magnitudes lower than of MoPredNet and one lower than PAML. Comparing our results to TimeHetNet, we see that convolutional graph layers give significant performance lifts. In a further ablation, we analyzed the influence of the size of sampled subgraphs during meta-training on meta-testing. For this, we repeated our experimental setup but limited the maximum number of nodes in the subgraph from 5 to 35 for meta-training and -testing, respectively. The results in Fig. 4 indicate that our approach is robust to subgraph size in meta-training, with a slight peak when training on tasks up to 20 vertices, demonstrating the model's ability to generalize to larger graphs during testing. This shows how larger tasks correlate to a more difficult motion prediction as the chance to extract useful data from neighbor sensors increases.

5 Conclusion

In this work, we proposed a new approach for few-shot human motion prediction, which generalizes over tasks with heterogeneous motion sensors arranged in a graph, outperforming all related baselines which are not equipped for varying sensor graphs. This is the first approach that allows for the prediction of novel human motion tasks independent of their number of sensors. Moreover, using this model, we can rival state-of-the-art approaches for the standard few-shot motion benchmark on tasks with homogeneous sensors while maintaining a significantly smaller model size which can be crucial for applications of human motion detection as these are often found in mobile and handheld devices. By publicizing all our code, including the baselines reimplementation as well as our benchmark pipeline, we hope to motivate future research in this area.

Acknowledgements. This work was supported by the Federal Ministry for Economic Affairs and Climate Action (BMWK), Germany, within the framework of the IIP-Ecosphere project (project number: 01MK20006D).

References

1. Brinkmeyer, L., Drumond, R.R., Burchert, J., Schmidt-Thieme, L.: Few-shot forecasting of time-series with heterogeneous channels. arXiv preprint arXiv:2204.03456 (Accepted at ECML 2022) (2022)
2. Brinkmeyer, L., Drumond, R.R., Scholz, R., Grabocka, J., Schmidt-Thieme, L.: Chameleon: learning model initializations across tasks with different schemas. arXiv preprint arXiv:1909.13576 (2019)

3. Butepage, J., Black, M.J., Kragic, D., Kjellstrom, H.: Deep representation learning for human motion prediction and classification. In: CVPR, pp. 6158–6166 (2017)
4. Ionescu, C., Fuxin Li, C.S.: Latent structured models for human pose estimation. In: ICCV (2011)
5. Drumond, R.R., Brinkmeyer, L., Grabocka, J., Schmidt-Thieme, L.: HIDRA: head initialization across dynamic targets for robust architectures. In: SIAM SDM, pp. 397–405. SIAM (2020)
6. Drumond, R.R., Marques, B.A., Vasconcelos, C.N., Clua, E.: Peek-an LSTM recurrent network for motion classification from sparse data. In: VISIGRAPP (1: GRAPP), pp. 215–222 (2018)
7. Finn, C., Abbeel, P., Levine, S.: Model-agnostic meta-learning for fast adaptation of deep networks. In: ICML. PMLR (2017)
8. Gui, L.Y., Wang, Y.X., Ramanan, D., Moura, J.M.: Few-shot human motion prediction via meta-learning. In: ECCV (2018)
9. Ionescu, C., Papava, D., Olaru, V., Sminchisescu, C.: Human3.6m: large scale datasets and predictive methods for 3D human sensing in natural environments. IEEE Trans. Pattern Anal. Mach. Intell. **36**(7), 1325–1339 (2014)
10. Iwata, T., Kumagai, A.: Meta-learning from tasks with heterogeneous attribute spaces. Adv. Neural. Inf. Process. Syst. **33**, 6053–6063 (2020)
11. Jalal, A., Quaid, M.A.K., Kim, K.: A wrist worn acceleration based human motion analysis and classification for ambient smart home system. J. Electri. Eng. Technol. **14**(4), 1733–1739 (2019)
12. Kipf, T.N., Welling, M.: Semi-supervised classification with graph convolutional networks. arXiv preprint arXiv:1609.02907 (2016)
13. Li, M., Chen, S., Zhao, Y., Zhang, Y., Wang, Y., Tian, Q.: Dynamic multiscale graph neural networks for 3D skeleton based human motion prediction. In: Proceedings of the IEEE/CVF CVPR (2020)
14. Liu, Z., Liu, Q., Xu, W., Liu, Z., Zhou, Z., Chen, J.: Deep learning-based human motion prediction considering context awareness for human-robot collaboration in manufacturing. Procedia CIRP **83**, 272–278 (2019)
15. Mao, W., Liu, M., Salzmann, M.: History repeats itself: human motion prediction via motion attention. In: Vedaldi, A., Bischof, H., Brox, T., Frahm, J.-M. (eds.) ECCV 2020. LNCS, vol. 12359, pp. 474–489. Springer, Cham (2020). https://doi.org/10.1007/978-3-030-58568-6_28
16. Mao, W., Liu, M., Salzmann, M., Li, H.: Multi-level motion attention for human motion prediction. Int. J. Comput. Vis. **129**(9), 2513–2535 (2021). https://doi.org/10.1007/s11263-021-01483-7
17. Martinez, J., Black, M.J., Romero, J.: On human motion prediction using recurrent neural networks. In: CVPR, pp. 2891–2900 (2017)
18. Parsaeifard, B., Saadatnejad, S., Liu, Y., Mordan, T., Alahi, A.: Learning decoupled representations for human pose forecasting. In: ICCV Workshops (October 2021)
19. Schmidt-Hieber, J.: The kolmogorov-arnold representation theorem revisited. Neural Netw. **137**, 119–126 (2021)
20. Snell, J., Swersky, K., Zemel, R.: Prototypical networks for few-shot learning. In: Advances in Neural Information Processing Systems, vol. 30 (2017)
21. Taylor, W., Shah, S.A., Dashtipour, K., Zahid, A., Abbasi, Q.H., Imran, M.A.: An intelligent non-invasive real-time human activity recognition system for next-generation healthcare. Sensors **20**(9), 2653 (2020)

22. Unhelkar, V.V., et al.: Human-aware robotic assistant for collaborative assembly: integrating human motion prediction with planning in time. IEEE Robot. Autom. Lett. **3**(3), 2394–2401 (2018)
23. Wang, Y., Yao, Q., Kwok, J.T., Ni, L.M.: Generalizing from a few examples: a survey on few-shot learning. ACM Comput. Surv. (CSUR) **53**(3), 1–34 (2020)
24. Yoon, S.W., Seo, J., Moon, J.: TapNet: neural network augmented with task-adaptive projection for few-shot learning. In: ICML, pp. 7115–7123. PMLR (2019)
25. Zaheer, M., Kottur, S., Ravanbakhsh, S., Poczos, B., Salakhutdinov, R.R., Smola, A.J.: Deep sets. In: Advances in Neural Information Processing Systems, vol. 30 (2017)
26. Zang, C., Li, M., Pei, M.: Few-shot human motion prediction using deformable spatio-temporal CNN with parameter generation. Neurocomputing, 513 (2022)
27. Zang, C., Pei, M., Kong, Y.: Few-shot human motion prediction via learning novel motion dynamics. In: IJCAI, pp. 846–852 (2021)

Correction to: Anti-Money Laundering in Cryptocurrency via Multi-Relational Graph Neural Network

Woochang Hyun, Jaehong Lee, and Bongwon Suh

Correction to: Chapter "Anti-Money Laundering in Cryptocurrency via Multi-Relational Graph Neural Network" in: H. Kashima et al. (Eds.): *Advances in Knowledge Discovery and Data Mining*, LNAI 13936, https://doi.org/10.1007/978-3-031-33377-4_10

The originally published version of the chapter 10 contained typesetting errors in Table 2, Table 3, and in Figure 3. These typesetting errors have been corrected

The updated original version of this chapter can be found at https://doi.org/10.1007/978-3-031-33377-4_10

H. Kashima et al. (Eds.): PAKDD 2023, LNAI 13936, p. C1, 2023.
https://doi.org/10.1007/978-3-031-33377-4_43

Author Index

H. Kashima et al. (Eds.): PAKDD 2023, LNAI 13936, pp. 565–567, 2023.
https://doi.org/10.1007/978-3-031-33377-4

Printed in the United States
by Baker & Taylor Publisher Services